E. B. Gareth Jones, Kevin D. Hyde, Ka-Lai Pang (Eds.)
Freshwater Fungi

Also of interest

Biology of Polar Benthic Algae
Christian Wiencke (Ed.), 2010
Series: Marine and Freshwater Botany
ISBN: 978-3-11-022970-7, e-ISBN: 978-3-11-022971-4

Marine Fungi
and Fungal-like Organisms
E. B. Gareth Jones, Ka-Lai Pang (Eds.), 2012
Series: Marine and Freshwater Botany
ISBN: 978-3-11-026398-5, e-ISBN: 978-3-11-026406-7

Advances in Algal Cell Biology
Kirsten Heimann, Christos Katsaros (Eds.), 2012
Series: Marine and Freshwater Botany
ISBN: 978-3-11-022960-8, e-ISBN: 978-3-11-022961-5

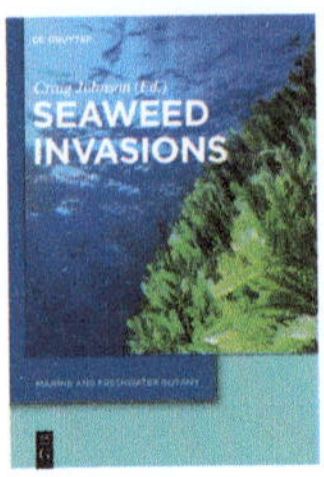

Seaweed Invasions
Craig Johnson (Ed.), 2015
Series: Marine and Freshwater Botany
ISBN: 978-3-11-024065-8, e-ISBN: 978-3-11-024066-5

www.degruyter.com

Freshwater Fungi

and Fungal-like Organisms

Edited by
E. B. Gareth Jones, Kevin D. Hyde and Ka-Lai Pang

DE GRUYTER

Editors
E. B. Gareth Jones
King Saud University
Riyadh 11451
Kingdom of Saudi Arabia
E-Mail: torperadgj@gmail.com

Kevin D. Hyde
Mae Fah Luang University
Chiang Rai
Thailand
E-Mail: kdhyde1@gmail.com

Ka-Lai Pang
Institute of Marine Biology and Center of Excellence for the Oceans
National Taiwan Ocean University
2 Pei-Ning Road
Keelung 20224
Taiwan, R.O.C.
E-Mail: klpang@ntou.edu.tw

ISBN 978-3-11-033345-9
e-ISBN 978-3-11-033348-0

Library of Congress Cataloging-in-Publication data
A CIP catalog record for this book has been applied for at the Library of Congress.

Bibliographic information published by the Deutsche Nationalbibliothek
The Deutsche Nationalbibliothek lists this publication in the Deutsche Nationalbibliografie; detailed bibliographic data are available on the Internet at http://dnb.dnb.de.

Typesetting: Compuscript Limited, Shannon, Ireland
Printing and binding: CPI books GmbH, Leck
Cover image: The cover shows the helicoid conidium of *Helicomyces roseus*, a common freshwater asexual fungus found on submerged cellulosic materials. Photo by Dr Nattawut Boonyuen.

∞ Printed on acid-free paper
Printed in Germany
www.degruyter.com

Preface

This book *Freshwater fungi and Fungal-like Organisms* is the outcome of recent research of the leading world mycologists on selected topics on fungi in streams, rivers, lakes and meltwater. The aim of the book is to bring together the present state of knowledge concerning freshwater fungi, especially topics often neglected in other treatise, to highlight their importance to science and the challenges facing freshwater mycology, as well as to consider future research. The book brings together many subjects not covered in other monographic volumes, such as freshwater yeasts, lichens, freshwater basidiomycetes, fungal-like organisms, their potential industrial application, and their role in the decomposition of complex organic matter in rivers and lakes. We particularly focus on their role as pathogens of commercially important freshwater animals, such as fish, crustaceans and molluscs, the diseases of mammals, including humans, the catastrophic devastation caused by chytrids parasitic on frogs and other reptiles.

Texts in freshwater biology rarely include fungi and fungal-like organisms in their volumes and thus ignore their important contributions and roles in freshwater ecosystems, in particular, their role in the breakdown and sequestration of pollutants in freshwater habitats. Fungi play a major role in the decomposition of complex organic compounds yielding nutrients for other aquatic organisms in the web of life in freshwater ecosystems.

The opening section of the book documents the current knowledge on the phylogeny of obligate freshwater fungi: Ascomycota (Dothideomycetes, Sordariomycetes, and other classes); Basidiomycota; asexual fungi; yeasts; Chytridiomycota and Blastocladiomycota. The second section is devoted to the phylogeny of fungal-like organisms: Microsporidia; Pythiales and Peronosporales. The biodiversity of fungi in selected habitats is reviewed in the third section: economic importance of zoosporic Mesomycetozoean pathogenic in fish; oomycetes and zoosporic organisms of amphibians, fish and freshwater invertebrates; and pythiosis of mammals, including man. The concluding section considers the ecology of lichens, aquatic trichomycetes, fungi found on decaying fronds of palms in a peat swamp, the role of fungi in leaf decomposition and breakdown of wood in tropical streams, yeasts in extreme aquatic environments and the role of fungi in polluted waters. The epilogue considers the importance of fungi and fungal-like organisms, and the direction of future research.

It is expected that this book will be essential reading for mycologists, microbiologists, freshwater biologists and limnologists interested in freshwater fungi and will serve as a useful reference work on their occurrence and role in freshwater ecosystems.

We thank all the authors for agreeing to write for this volume and for delivering their chapters on time. We are much indebted to Frank Gleason for suggestions during the preparation of this volume, especially the section on fungal-like organisms. Our thanks also go to all the staff at De Gruyter for conceiving this volume and for their support in its publication (Simone Witzel, Nicole Karbe, and Hannes Kaden).

Contents

Preface —— v

List of contributing authors —— xvii

E. B. Gareth Jones, Kevin D. Hyde and Ka-Lai Pang
1 Introduction —— 1
1.1 Origin of freshwater fungi and fungal-like organisms —— 4
1.2 Classification of freshwater fungi —— 5
1.3 Estimated number of freshwater fungi —— 6
1.4 World distribution —— 8
1.5 Endophytic fungi —— 8
1.6 Predacious fungi —— 9
1.7 Bioactive compounds —— 10
1.8 Barcoding of freshwater fungi —— 12
1.9 One name one fungus ruling —— 13
1.10 Role of fungi in freshwater habitats —— 14
1.11 Objectives and outline of the volume —— 15
1.12 Phylogeny of true freshwater fungi —— 15
1.13 Phylogeny of fungus-like organisms —— 15
1.14 Biodiversity of freshwater fungi and fungus-like organisms —— 16
1.15 Ecology —— 16
Acknowledgments —— 16
References —— 17

Phylogeny of freshwater fungi —— 23

Carol A. Shearer, Ka-Lai Pang, Satinee Suetrong and Huzefa A. Raja
2 Phylogeny of the Dothideomycetes and other classes of freshwater fissitunicate Ascomycota —— 25
2.1 Introduction —— 25
2.2 Geographical distribution patterns —— 26
2.3 Substrate distribution patterns —— 26
2.4 Morphological adaptations —— 26
2.5 Systematics —— 28
2.5.1 General introduction —— 28
2.5.2 Current phylogenetic placement based on molecular systematics —— 32
2.5.2.1 Dothideomycetes-Pleosporomycetidae-Pleosporales —— 32
2.5.2.2 Pleosporales *incertae sedis* —— 36
2.5.3 Zopfiaceae, Dothideomycetes, family *incertae sedis* —— 38
2.5.4 Dothideomycetes *incertae sedis* —— 38

2.5.4.1 Jahnulales — 38
2.5.4.2 Natipusillales — 39
2.5.4.3 *Minutisphaera* clade — 39
2.5.4.4 Freshwater asexual morphs with affinities to Dothideomycetes — 39
2.6 Conclusions — 40
Acknowledgments — 40
References — 40

Lei Cai, Dian-Ming Hu, Fang Liu, Kevin D. Hyde and E. B. Gareth Jones
3 The molecular phylogeny of freshwater Sordariomycetes and discomycetes — 47
3.1 Introduction — 47
3.2 Materials and methods — 48
3.2.1 Taxon sampling — 48
3.2.2 Phylogenetic analysis — 48
3.3 Discussion — 48
3.3.1 Sordariomycetidae — 56
3.3.1.1 Annulatascaceae — 56
3.3.1.2 Magnaporthales — 60
3.3.1.3 Calosphaeriales — 60
3.3.1.4 Coniochaetales — 61
3.3.1.5 Diaporthales — 61
3.3.1.6 Sordariales — 61
3.3.2 Sordariomycetidae *incertae sedis* — 62
3.3.3 Hypocreomycetidae — 62
3.3.3.1 Savoryellales — 62
3.3.3.2 Microascales — 63
3.3.3.3 Hypocreales — 63
3.3.4 Xylariomycetidae — 64
3.3.4.1 Xylariales — 64
3.3.4.2 Phyllachorales — 64
3.3.4.3 Trichosphaeriales — 64
3.3.5 Discomycetes — 64
3.3.5.1 Helotiales — 64
3.3.5.2 Pezizales — 65
3.3.5.3 Rhytismatales — 66
3.4 Concluding remarks — 66
Acknowledgments — 66
References — 67

E. B. Gareth Jones, Darlene Southworth, Diego Libkind and Ludmila Marvanová
4 Freshwater Basidiomycota — 73
4.1 Group 1 freshwater yeasts — 82

4.1.1 Agaricomycotina — 83
4.1.1.1 Tremellomycetes — 83
4.1.2 Pucciniomycotina — 85
4.1.2.1 Cystobasidiomycetes — 85
4.1.2.2 Microbotryomycetes — 85
4.1.2.3 Microbotryomycetes *Incertae sedis* — 86
4.1.3 Ustilaginomycotina— 87
4.1.3.1 Ustilaginomycetes — 87
4.2 Group 2 filamentous fungi — 87
4.2.1 Agaricomycotina— 87
4.2.1.1 Agaricomycetes — 87
4.2.1.2 Exobasidiomycetes — 92
4.2.1.3 Tremellomycetes — 92
4.2.2 Pucciniomycotina— 92
4.2.2.1 Atractiellomycetes — 92
4.2.2.2 Classiculomycetes — 93
4.2.2.3 Microbotryomycetes — 94
4.2.3 Ustilaginomycotina — 95
4.2.3.1 Ustilaginomycetes — 95
Basidiomycota–*incertae sedis* — 95
4.3 Group 3 endophytes — 99
4.4 Adaptation to freshwater habitats — 99
Acknowledgments — 100
References — 100

Dian-Ming Hu, Lei Cai, E. B. Gareth Jones, Huang Zhang, Nattawut Boonyuen and Kevin D. Hyde
5 Taxonomy of filamentous asexual fungi from freshwater habitats, links to sexual morphs and their phylogeny — 109
5.1 Introduction — 109
5.2 Morphological taxonomy — 110
5.2.1 Hyphomycetes — 110
5.2.2 Coelomycetes — 112
5.2.3 Asexual-sexual connections — 112
5.3 Phylogeny — 113
5.3.1 Dothideomycetes — 114
5.3.1.1 Capnodiales — 114
5.3.1.2 Dothideales — 114
5.3.1.3 Hysteriales — 117
5.3.1.4 Jahnulales — 117
5.3.1.5 Mytilinidiales — 117
5.3.1.6 Pleosporales — 117
5.3.1.7 Tubeufiales — 118

5.3.2 Leotiomycetes — 119
5.3.3 Orbiliomycetes — 120
5.3.3.1 Orbiliales — 120
5.3.4 Sordariomycetes — 120
5.3.4.1 Glomerellales — 123
5.3.4.2 Hypocreales — 123
5.3.4.3 Sordariales — 123
5.3.4.4 Savoryellales — 124
5.4 Discussion — 125
Acknowledgment — 126
References — 126

Martha J. Powell and Peter M. Letcher
6 Phylogeny and characterization of freshwater Chytridiomycota (Chytridiomycetes and Monoblepharidomycetes) — 133
6.1 Introduction — 133
6.2 Chytridiomycetes — 138
6.2.1 Order 1. Chytridiales (Chytridiaceae, Chytriomycetaceae) — 138
6.2.2 Order 2. Spizellomycetales (Spizellomycetaceae, Powellomycetaceae) — 140
6.2.3 Order 3. Rhizophlyctidiales (Rhizophlyctidaceae, Sonoraphlyctidaceae, Arizonaphlyctidaceae, Borealophlyctidaceae) — 141
6.2.4 Order 4. Rhizophydiales (10 families described) — 141
6.2.5 Order 5. Lobulomycetales (Lobulomycetaceae) — 144
6.2.6 Order 6. Cladochytriales (Cladochytriaceae, Nowakowskiellaceae, Septochytriaceae, Endochytriaceae) — 144
6.2.7 Order 7. Polychytriales (no families described) — 146
6.3 *Incertae sedis* — 147
6.4 Monoblepharidomycetes (Harpochytriales, Monoblepharidales, Hyaloraphidiales) — 148
Acknowledgments — 148
References — 148

Phylogeny of fungus-like organisms — 155

Ray Kearney and Frank H. Gleason
7 Microsporidia — 157
7.1 Ecology — 160
7.2 Classification — 162
7.3 Evolutionary origins — 165
7.4 Cell structure and spore significance — 166
7.5 Metabolism — 167
7.6 Genome structure — 168

7.7 Discussion and conclusion — 168
7.8 Further research avenues — 170
References — 171

Agostina V. Marano, Ana L. Jesus, Carmen L. A. Pires-Zottarelli, Timothy Y. James, Frank H. Gleason and Jose I. de Souza

8 Phylogenetic relationships of Pythiales and Peronosporales (Oomycetes, Straminipila) within the "peronosporalean galaxy" — 177

8.1 Introduction — 177
8.2 The monophyly of Chromalveolata and the relationships between heterotrophic straminipile lineages — 178
8.3 Major lineages within the Oomycetes: the "galaxies" — 179
8.4 The "peronosporalean galaxy": a marine origin? — 179
8.5 Ecological and economical significance — 180
8.6 The phylogeny of Pythiales and Peronosporales — 181
8.6.1 Clade 1: Albuginales — 185
8.6.2 Clade 2: Pythiales — 185
8.6.2.1 *Pythiogeton* — 186
8.6.2.2 *Pythium, Lagenidium* and *Phytopythium* — 186
8.6.3 Clade 3: Peronosporales — 187
8.6.3.1 Downy mildews — 188
8.6.3.2 *Phytophthora* and *Peronophythora* — 188
8.6.3.3 *Halophytophthora* and *Salisapilia* — 190
8.7 Conclusions and future perspectives — 192
Acknowledgments — 194
References — 194

Biodiversity of freshwater fungi — 201

Sally L. Glockling, Wyth L. Marshall, Rodolphe E. Gozlan, Agostina V. Marano, Osu Lilje and Frank H. Gleason

9 The ecological and economic importance of zoosporic Mesomycetozoean (Dermocystida) parasites of freshwater fish — 203

9.1 Phylogeny — 203
9.2 Life cycles — 205
9.3 The zoospore — 206
9.4 Symptoms of disease — 207
9.5 Ecological and economic significance — 209
9.6 Discussion and conclusion — 211
Acknowledgment — 212
References — 212

Mohammad N. Sarowar, Marcia Saraiva, Casey N. Jessop, Osu Lilje, Frank H. Gleason and Pieter van West
10 Infection strategies of pathogenic oomycetes in fish — 217
10.1 Introduction — 217
10.2 Taxonomy of oomycetes pathogenic to fish — 221
10.3 Physical adaptation and strategy for infection: macroscopic infection, the face of infection on hosts — 223
10.4 Oomycete zoospores, the first line of attack — 224
10.5 Triggers for zoospore formation, waking up the beast — 225
10.6 Encystment and germination, one step closer to infection — 225
10.7 Repeated zoospore emergence, the back-up plan — 227
10.8 Chemotactic response of zoospores, the specialization — 228
10.9 Proteins and amino acids as substrates for growth — 229
10.10 Sexual reproduction, seeing through the bad times — 231
10.11 Molecular adaptation and strategy in setting infection: microscopic infection — 231
10.12 Host responses to oomycete infections — 233
10.13 The animal trade is responsible for the spread of pathogens into novel and wild ecosystems — 234
10.14 Future perspectives — 235
Acknowledgments — 236
References — 236

Frank H. Gleason, Jodi L. Rowley, Casey N. Jessop and Osu Lilje
11 Zoosporic parasites of amphibians — 245
11.1 Chytridiomycota — 245
11.2 Mesomycetozoea — 247
11.3 Oomycota (oomycetes or water moulds) — 250
11.4 Perkinsozoa — 251
11.5 The Fisher concept of emerging infectious diseases (EIDs) — 252
11.6 Host switching by parasites — 252
11.7 Genetic variation in parasite populations — 254
11.8 Proteases — 255
11.9 International animal trade — 255
11.10 Discussion and conclusion — 256
Acknowledgments — 257
References — 257

Angkana Chaiprasert and Theerapong Krajaejun
12 Pythiosis — 263
12.1 History — 263
12.2 Biology — 263
12.3 Molecular typing — 265

12.4 Epidemiology — 266
12.5 Pathogenesis — 266
12.6 Clinical features — 267
12.6.1 Human pythiosis — 267
12.6.2 Animal pythiosis — 269
12.7 Diagnosis — 269
12.8 Management — 271
12.9 Research direction — 272
Acknowledgment — 273
References — 274

Frank H. Gleason, Sergey A. Karpov, Osu Lilje, Deborah J. Macarthur, Floris F. van Otgen and Telesphore Sime-Ngando
13 Zoosporic parasites of phytoplankton — 279
13.1 The main groups of zoosporic parasites and parasitoids of phytoplankton — 280
13.1.1 Aphelidea — 280
13.1.2 Chytridiomycota — 283
13.1.3 Blastocladiomycota — 293
13.2 Ancient interactions — 294
13.3 Novel food webs — 295
13.3.1 *Vorticella* communities attached to cyanobacterial filaments — 295
13.3.2 Communities involving other protists — 295
13.4 Host parasite dynamics — 296
13.5 Conclusion — 298
Acknowlegments — 299
References — 300

Sally L. Glockling, Agostina V. Marano, Osu Lilje and Frank H. Gleason
14 Zoosporic parasites of freshwater invertebrates — 305
14.1 Parasites in the Blastocladiomycota and Chytridiomycota — 306
14.2 Parasites in the Oomycota — 307
14.3 Parasites in the Mesomycetozoea — 311
14.4 Parasites of crayfish — 312
14.4.1 Crayfish plague — 312
14.4.2 *Psorospermium haekeli* — 313
14.5 Parasites of mosquitoes, blackflies and midges — 313
14.5.1 *Coelomomyces* — 314
14.5.2 *Lagenidium giganteum* — 315
14.5.3 *Pythium* — 316
14.5.4 *Leptolegnia* — 316
14.5.5 *Crypticola* — 316
14.5.6 *Amoebidium* and *Paramoebidium* — 317

14.6 Parasites of *Daphnia* — 317
14.7 Parasites of rotifers and nematodes — 318
14.7.1 *Sommerstorffia spinosa* — 320
14.7.2 *Aquastella* — 321
14.8 Parasites of protozoans — 321
14.9 Discussion — 321
Acknowledgments — 324
References — 324

Ecology — 331

Holger Thüs, André Aptroot and Mark R. D. Seaward
15 Freshwater lichens — 333
15.1 Ecology — 336
15.1.1 Habitats and diversity of freshwater lichens — 336
15.1.2 Collecting and identifying freshwater lichens — 338
15.2 Physiological challenges for freshwater lichens — 339
15.2.1 Water saturation and diffusion resistance — 339
15.3 Freshwater lichens as a food source for other organisms — 343
15.4 Biogeography of freshwater lichens — 344
15.5 Zonation — 345
15.6 Lichen trimlines — 348
15.7 Freshwater lichen communities — 349
15.8 Freshwater lichens as bioindicators — 350
15.9 Water quality — 351
15.10 Conservation — 352
Acknowledgments — 353
References — 353

Robert W. Lichtwardt
16 Aquatic Trichomycetes — 359
16.1 Trichomycetes, an ecological group — 359
16.2 Phylogenetic considerations — 359
16.3 Distribution and success of Trichomycetes — 364
16.4 Variations in symbiotic associations — 365
16.5 Medical implications — 367
Acknowledgments — 368
References — 368

Umpava Pinruan, Aom Pinnoi, Kevin D. Hyde and E. B. Gareth Jones
17 Tropical peat swamp fungi with special reference to palms — 371
17.1 Material and methods — 373
17.1.1 Sample collection — 373
17.2 Results — 373
17.2.1 Abundance of fungi on four palms (*Eleiodoxa conferta*, *Licuala longicalycata*, *Metroxylon sagu* and *Nenga pumila*) — 373

17.2.1.1 *Eleiodoxa conferta* — 379
17.2.1.2 *Licuala longicalycata* — 380
17.2.1.3 *Metroxylon sagu* — 380
17.2.1.4 *Nenga pumila* — 380
17.2.2 Fungal diversity — 381
17.2.3 Percentages overlap in fungal diversity between the four palms — 382
17.3 Conclusion — 383
Acknowledgments — 385
References — 386

Verónica Ferreira, Vladislav Gulis, Cláudia Pascoal and Manuel A. S. Graça
18 Stream pollution and fungi — 389
18.1 The importance of aquatic hyphomycetes in woodland streams — 389
18.2 Effects of nutrient enrichment on stream fungi — 391
18.3 Effects of heavy metals and acidification on stream fungi — 394
18.4 Ecological and toxicological effects of engineered nanoparticles on stream fungi — 395
18.5 Effects of organic xenobiotics on stream fungi — 397
18.6 Effects of thermal pollution on stream fungi — 398
18.7 Effects of the interaction among factors on stream fungi — 403
18.8 Conclusions — 404
Acknowledgments — 404
References — 405

Felix Bärlocher and Kandikere R. Sridhar
19 Association of animals and fungi in leaf decomposition — 413
19.1 History — 413
19.2 Effects of the leaf-fungus complex on invertebrate consumers — 416
19.2.1 Nutritional value of mycelium vs. leaf substrate — 416
19.2.2 Modifications of leaf substrate — 417
19.2.3 Do invertebrates differ in their feeding strategies? — 420
19.2.4 What factors ultimately determine food choice and feeding selectivity? — 421
19.2.5 Stoichiometric considerations — 423
19.2.6 Stimulation of fungi by invertebrate feeding — 424
19.2.7 Anthropogenic changes — 424
19.2.8 Research outside temperate regions — 426
19.3 Effects of invertebrate consumers on the leaf-fungus complex — 428
19.3.1 Invertebrate ingestion of conidia — 429
19.3.2 Invertebrate ingestion of the leaf-fungus complex — 429
19.4 Conclusions — 431
Acknowledgments — 432
References — 432

Diego Libkind, Gabriel Russo and María Rosa van Broock
20 Yeasts from extreme aquatic environments: hyperacidic freshwaters — 443
20.1 Introduction — 443
20.2 The River Agrio-Lake Caviahue acidic aquatic system — 444
20.2.1 Yeast occurrence — 445
20.2.2 Yeast diversity — 446
20.3 Comparative yeast diversity study between RAC and the acidic environments of the Iberian Pyrite Belt (IPB) — 451
20.4 Acidic rock drainage (ARD) yeasts ecoclade — 454
20.5 Physiological aspects of acidophilic yeasts — 456
20.6 Possible ecological roles of yeasts in acidic aquatic environments — 457
20.7 Final remarks — 458
Acknowledgments — 459
References — 460

Nattawut Boonyuen, Somsak Sivichai and E. B. Gareth Jones
21 Decomposition of wood in tropical habitats — 465
21.1 Review of fungal diversity on wood in freshwater streams — 466
21.2 Colonization of 15 timbers exposed at two locations in Thailand — 467
21.2.1 Materials and methods — 467
21.2.2 Results — 468
21.2.3 Rate of decay of selected timbers at two contrasting freshwater ecosystems in Thailand — 472
21.2.4 Discussion — 473
21.2.4.1 Fungal community — 473
21.2.4.2 Decay of wood in freshwater habitats — 474
Acknowledgments — 476
References — 477

Ka-Lai Pang, Kevin D. Hyde and E. B. Gareth Jones
22 Epliogue — 481
22.1 Introduction — 481
22.2 Freshwater fungi — 481
22.3 Freshwater fungus-like organisms — 482
22.4 Knowledge gaps and future work in freshwater mycology — 482
22.5 Conclusions — 486
References — 486

Index — 489

List of contributing authors

André Aptroot
G. van der Veenstraat 107
NL-3762 XK Soest
The Netherlands
E-Mail: andreaptroot@gmail.com

Felix Bärlocher
63B York Street
Dept. Biology
Mt. Allison University
Sackville, NB, E4L 197
Canada
E-Mail: fbaerlocher@mta.ca

Nattawut Boonyuen
National Center for Genetic Engineering and Biotechnology (BIOTEC)
113 Thailand Science Park,
Phahonyothin Road
Khlong Nueng, Khlong Luang
Pathum Thani 12120
Thailand
E-Mail: nattawut@biotec.or.th

María Rosa van Broock
Laboratorio de Microbiología
Aplicada y Biotecnología (MABB)
Universidad Nacional del Comahue
Centro Regional Universitario Bariloche
(CRUB)-CONICET (Consejo Nacional de Investigaciones Científicas y Tecnológicas)
CCT Patagonia Norte. Quintral 1250
(8400) Bariloche
Río Negro
Argentina
E-Mail: queemiegvb@yahoo.com.ar

Lei Cai
State Key Laboratory of Mycology
Institute of Microbiology
Chinese Academy of Sciences
NO.1, Beichen West Road, Chaoyang District
Beijing 100101
China
E-Mail: mrcailei@gmail.com

Angkana Chaiprasert
Department of Microbiology
Faculty of Medicine Siriraj Hospital
Mahidol University
Bangkok 10700
Thailand
E-Mail: angkana.cha@mahidol.ac.th

José I. de Souza
Núcleo de Pesquisa em Micologia
Instituto de Botânica
Av. Miguel Stéfano 3687
04301-012, São Paulo, SP
Brazil
E-Mail: jisouza@yahoo.com.br

Verônica Ferreira
IMAR-CMA
Department of Life Sciences
Faculty of Science and Technology
University of Coimbra
P.O. box 3046
3001-401 Coimbra
Portugal
E-Mail: veronica@ci.uc.pt

Frank H. Gleason
School of Biological Sciences
Level 5, Carslaw (F07)
The University of Sydney
NSW 2006
Australia
E-Mail: frankjanet@ozemail.com.au

Sally Glockling
135 Brodrick Road
Hampden Park
Eastbourne BN22 9RA
UK
E-Mail: sally@glockling.com

Rodolphe E. Gozlan
School of Conservation Sciences
Bournemouth University
Talbot Campus
Fern Barrow
Poole
Dorset BH12 5BB
UK
E-Mail: rgozlan@bournemouth.ac.uk;
rudy.gozlan@ird.fr

Manuel A. S. Graça
IMAR-CMA
Department of Life Sciences
Faculty of Science and Technology
University of Coimbra
P.O. box 3046
3001-401 Coimbra
Portugal
E-Mail: mgraca@ci.uc.pt

Vladislav Gulis
Department of Biology
Coastal Carolina University
P.O. Box 261954
Conway, SC 29528-6054
USA
E-Mail: vgulis@coastal.edu

Kevin D. Hyde
School of Science
Mae Fah Luang University
333 Moo 1, Tambon Tasud
Muang District, Chiang Rai 57100
Thailand
E-Mail: kdhyde1@gmail.com

Dian-Ming Hu
College of Biology and Bioengineering
Jiangxi Agricultural University
Nanchang
Jiangxi Province, 330045
China
E-Mail: hudianming1@gmail.com

Timothy Y. James
Department of Ecology and
Evolutionary Biology
University of Michigan
Kraus Natural Science Bldg., Rm. 1008
830 North University
Ann Arbor, MI 48109-1048
USA
E-Mail: tyjames@umich.edu

Casey N. Jessop
2 School of Biological Sciences
F07, University of Sydeny
Camperdown
NSW 2006
Australia
E-Mail: cjes1587@uni.sydney.edu.au

Ana L. de Jesus
Núcleo de Pesquisa em Micologia
Instituto de Botânica
Av. Miguel Stéfano 3687
04301-012, São Paulo, SP
Brazil
E-Mail: analuciajesus@hotmail.com

E. B. Gareth Jones
Department of Botany and Microbiology
College of Science
King Saud University
Riyadh 11451
Kingdom of Saudi Arabia
E-Mail: torperadgj@gmail.com

Sergey Karpov
St. Petersburg State University
Universitetskaya emb 7/9
St. Petersburg, 199034
Russia
E-Mail: sakarpov4@gmail.com

Ray Kearney
The Department of Infectious Diseases and Immunology
The University of Sydney
NSW 2006
Australia
E-mail: ray.kearney@sydney.edu.au

Theerapong Krajaejun
Department of Pathology
Faculty of Medicine Ramathibodi Hospital
Mahidol University
Bangkok 10400
Thailand
E-Mail: theerapong.kra@mahidol.ac.th

Peter M. Letcher
The University of Alabama
Department of Biological Sciences
1332 SEC, Box 870344
Tuscaloosa, AL 35487
USA
E-Mail: letch006@bama.ua.edu

Diego Libkind
Laboratorio de Microbiología Aplicada y Biotecnología (MABB)
Universidad Nacional del Comahue
Centro Regional Universitario Bariloche (CRUB)-CONICET (Consejo Nacional de Investigaciones Científicas y Tecnológicas)
CCT Patagonia Norte. Quintral 1250
(8400) Bariloche
Río Negro
Argentina
E-Mail: libkindfd@comahue-conicet.gob.ar

Robert Lichwardt
Department of Ecology & Evolutionary Biology
University of Kansas
Lawrence, KS 66049-7534
USA
E-Mail: licht@ku.edu

Osu Lilje
School Biological Sciences
F07, University Sydney
Camperdown
NSW 2006
Australia
E-Mail: osu.lilje@sydney.edu.au

Fang Liu
State Key Laboratory of Mycology
Institute of Microbiology
Chinese Academy of Sciences
NO.1, Beichen West Road, Chaoyang District
Beijing 100101
China
E-Mail: orchid.lf@gmail.com

Deborah J. Macarthur
P.O. Box W29
Watsons Bay
NSW 2030
Australia
E-Mail: deborah.macarthur@sydney.edu.au

Agostina V. Marano
Núcleo de Pesquisa em Micologia
Instituto de Botânica
Av. Miguel Stéfano 3687
04301-012, São Paulo, SP
Brazil
E-Mail: agosvm@hotmail.com

Wyth L. Marshall
BC Centre for Aquatic Health Services
871ª Island Hwy
Campbell River
BC V9W 2C2
Canada
E-Mail: wythmarshall@yahoo.ca

Ludmila Marvanová
Masaryk University, Faculty of Science
Institute of Experimental Biology
Czech Collection of Microorganisms
University Campus Bohunice, Kamenice 753/5
625 00 Brno, Czech Republic.
E-Mail: ludmila@sci.muni.cz

Floris van Ogtrop
Faculty of Agriculture and Environment
C81, University of Sydney,
Sydney, NSW 2006,
Australia
E-mail: floris.vanogtrop@sydney.edu.au

Ka-Lai Pang
Institute of Marine Biology and Center of Excellence for the Oceans
National Taiwan Ocean University
2 Pei-Ning Road, Keelung 20224
Taiwan (R.O.C.)
E-Mail: klpang@ntou.edu.tw

Cláudia Pascoal
CBMA – Centre of Molecular and Environmental Biology
Department of Biology
University of Minho
Campus de Gualtar
4710-057 Braga
Portugal
E-Mail: cpascoal@bio.uminho.pt

Aom Pinnoi
National Center for Genetic Engineering and Biotechnology (BIOTEC)
113 Thailand Science Park,
Phahonyothin Road
Khlong Nueng, Khlong Luang
Pathum Thani 12120
Thailand
E-Mail: aom5736@yahoo.com

Umpava Pinruan
National Center for Genetic Engineering and Biotechnology (BIOTEC)
113 Thailand Science Park,
Phahonyothin Road
Khlong Nueng, Khlong Luang
Pathum Thani 12120
Thailand
E-Mail: umpava328@gmail.com

Carmen L. A. Pires-Zottarelli
Núcleo de Pesquisa em Micologia
Instituto de Botânica
Av. Miguel Stéfano 3687
04301-012, São Paulo, SP
Brazil
E-Mail: zottarelli@uol.com.br

Martha J. Powell
The University of Alabama
Department of Biological Sciences
1332 SEC, Box 870344
Tuscaloosa, AL, 35487
USA
E-Mail: mpowell@biology.as.us.edu

Huzefa Raja
Department of Chemistry and Biochemistry
The University of North Carolina at Greensboro
435 Sullivan Science Building
301 McIver Street
Greensboro, NC 27412
USA
E-Mail: mpowell@biology.as.us.edu

Jodi Rowley
Australian Museum Research Institute
Australian Museum
6 College Street
Sydney NSW 2010
Australia
E-Mail: Jodi.Rowley@austmus.gov.au

Gabriel Russo
Laboratorio de Microbiología Aplicada y Biotecnología (MABB)
Universidad Nacional del Comahue
Centro Regional Universitario Bariloche (CRUB)-CONICET (Consejo Nacional de Investigaciones Científicas y Tecnológicas)
CCT Patagonia Norte. Quintral 1250
(8400) Bariloche
Río Negro
Argentina
E-Mail: gavilano@gmail.com

Telesphore Sime-Ngando
Laboratoire Microorganismes: Génome et Environnement
UMR CNRS 6023
Bât. Biologie A, 24 avenue des Landais, BP 80026
63171 Aubière Cedex
France
E-Mail: Telesphore.SIME-NGANDO@univ-bpclermont.fr

Holger Thüs
Life Sciences Department
Division of Genomic and Microbial Diversity
The Natural History Museum
Cromwell Road
London SW7 5BD
UK
E-Mail: h.thues@nhm.ac.uk

Marcia Saraiva
Aberdeen Oomycete Laboratory
School of Medical Sciences
University of Aberdeen
Foresterhill AB25 2ZD
UK
E-Mail: m.saraiva@abdn.ac.uk

Mohammad N. Sarowar
Aberdeen Oomycete Laboratory
School of Medical Sciences
University of Aberdeen
Foresterhill AB25 2ZD
UK
E-Mail: n.sarowar@bau.edu.bd

Mark R. D. Seaward
Dept. of Achaeological, Environmental & Geographical Sciences (AGES)
University of Bradford
Bradford BD7 1DP
UK
E-Mail: m.r.d.seaward@bradford.ac.uk

Carol A. Shearer
Department of Plant Biology
University of Illinois
Rm. 265 Morrill Hall
505 S. Goodwin Ave.
Urbana, IL 61801
USA
E-Mail: carolshe@illinois.edu

Somsak Sivichai

Darlene Southworth
Department of Biology
Southern Oregon University
1250 Siskiyou Blvd.
Ashland, OR 97520
USA
E-Mail southworth@sou.edu

Kandikere R. Sridhar
Department of Biosciences
Mangalore University
Mangalagangotri 574 199
Mangalore, Karnataka
India
E-Mail: kandikere@gmail.com

Satinee Suetrong
National Center for Genetic Engineering and Biotechnology (BIOTEC)
113 Thailand Science Park,
Phahonyothin Road
Khlong Nueng, Khlong Luang
Pathum Thani 12120
Thailand
E-Mail: satinee.na@gmail.com

Pieter van West
Aberdeen Oomycete Laboratory
School of Medical Sciences
University of Aberdeen
Foresterhill AB25 2ZD
UK
E-Mail: p.vanwest@abdn.ac.uk

Huang Zhang
Faculty of Environmental Science and Engineering
Kunming University of Science and Technology
Kunming, Yunnan Province, 650500
China
E-Mail: zhanghuang2002113@gmail.com

E. B. Gareth Jones, Kevin D. Hyde and Ka-Lai Pang

1 Introduction

Why do we need a book on freshwater fungi? Within the last decade a number of books on freshwater fungi have been published: *Freshwater Mycology* (Tsui and Hyde 2003) and *Genera of Freshwater Fungi* (Cai et al. 2006), or reviews (Wong et al. 1998; Shearer et al. 2007; Wurzbacher et al. 2010, 2012). Most texts on freshwater fungi however, deal with specific or ecological groups, e.g. Ascomycota, chytrids, fungal diseases, ultrastructure of ascospore appendages (Barr 1980; Wong et al. 1999; Shearer and Raja 2010). Thus, none gives the overall picture of the diversity to be found in freshwater habitats. Such vital groups of organisms as *Pythium/Phytophthora* are rarely considered. Many reviews do not include up to date data or ignore certain taxonomic groups (yeasts) or habitats (peat swamp fungi). Freshwater yeast populations have been almost ignored, despite the fact that they are common in aquatic systems (Hagler and Ahearn 1987; Gadanho and Sampaio 2004; Libkind et al. 2010; Fell et al. 2011). There is no overall estimate of the number of freshwater fungi. Jones (2011) undertook such a review of marine fungi which concluded that present day estimates were way out. So what of the estimates of freshwater fungi? This lack of an overall picture of freshwater mycology led Wurzbacher et al. (2010, 2012) to conclude that research on freshwater fungi is much neglected.

Freshwater fungi with unusual branched conidia were reported in the 1890s (Saccardo and Therry 1880; de Wildeman 1884, 1893, 1895; Rostrup 1894) with descriptions of the taxa: *Heliscus lugdunensis*, *Tetracladium marchalianum*, *T. maxilliformis*, *Clavariopsis aquatica* and *Anguillospora longissima* (as *Fusarium longissimum*), but it was not until the early 1940s that it was appreciated that these fungi formed a unique community on senescent leaves in streams and rivers (Ingold 1942, 1944). Since then there has been a steady stream of papers reporting on their occurrence from all over the world (Nawawi 1985; Goh and Hyde 1996). Ingold also drew attention to the existence of abundant ascomycetes in temperate freshwater habitats (Ingold 1951, 1955), while Hyde highlighted their occurrence in tropical locations: Australia (Hyde 1992), the Philippines (Hyde and Wong 2000) and Hong Kong (Tsui et al. 2001). The early 1850s saw the first report of chytrids in freshwater habitats (Braun 1856) which lead to the publication of the book on *Aquatic Phycomycetes* (Sparrow 1960). These early studies culminated in a wealth of data on freshwater fungi as can be seen from the list of taxa in Tab. 1.1, with particular major revisions in their taxonomy in the period 1990–2013 (Shearer 2010; Kurtzman et al. 2011; Powell and Letcher 2012).

Studies of freshwater fungi and fungal-like organisms to date have largely been morphological with sequence data used to help resolve phylogenetic relationships. These studies are largely surveys of different taxonomical or ecological groups with few focusing on an all taxa inventory (Bärlocher 2007; Wurzbacher et al. 2010, 2012).

Tab. 1.1: Classification and estimate of numbers of freshwater fungi and fungal-like organisms.

Phylum	Class	Order	Number of genera/ species
Fungi			
Chytridiomycota	Chytridiomycetes	Chytridiales	18 genera, 385 species
		Rhiziphydiales	18 genera, 244 species
		Spizellomycetales	8 genera, 21 species
		Rhizophlyctidales	4 genera, 40 species
		Cladochytriales	8 genera, 105 species
		Polychytriales	5 genera, 13 species
		Lobulomycetales	4 genera, 5 species
		Synchytrium lineage	4 genera, 33 species
		incertae sedis	28 genera, 100 species
	Monoblepharidomycetes	Monoblepharidales	6 genera, 50 species
Subphylum: Kickxellomycotina	N/A	Asellariales	3 genera, 7 species
		Harpellales	44 genera, 176–212 species
Ascomycota	Arthoniomycetes	Arthroniales	3 genera, 7 species
	Chaetothyriomycetes	Family *incertae sedis*	2 genera, 4 species
	Dothideomycetes	Pleosporales	13 genera, 26 species
		Jahnulales	6 genera, 41 species
		Dothideales	4 genera, 8 species
	Dothideomycetes	Some lichen genera	8 genera, 11 species
	Lecanoromycetes	Baeomycetales	1 genus, 1 species
		Lecanorales	31 genera, 5 species
		Ostropales	3 genera, 18 species
		Peltigerales	4 genera, 20 species
		Pertusariales	2 genera, 5 species
		Teloschistales	3 genera, 9 species
	Lecanoromycetes *incertae sedis*		5 genera, 10 species
	Lecanoromycetes	Families *incertae sedis*	3 genera, 7 species
	Leotiomycetes	Helotiales	16 genera, 25 species
		Rhytismatales	3 genera, 3 species
	Lichinomycetes	Lichinales	16 genera, 43 species
	Sordariomycetes	Amphisphaeriales?	1 genus, 1 species
		Calosphaeriales	1 genus, 1 species
		Choniochaetales	1 genus, 8 species
		Diaporthales	10 genera, 15 species
		Halosphaeriales!	9 genera, 24 species
		Hypocreales	10 genera. 21 species
		Magnaporthales	1 genus, 1 species
		Microascales	2 genera, 2 species

(*continued*)

Tab. 1.1: (*continued*)

Phylum	Class	Order	Number of genera/ species
		Ophiostomatales	1 genus, 1 species
		Phyllachorales	1 genus, 1 species
		Savoryellales	3 genera, 26 species
		Sordariales	8 genera, 24 species
		Trichosphaeriales	1 genus, 1 species
		Xylariales	12 genera, 16 species
	Pezizomycetes	Pezizales	9 genera, 9 species
	Eurotiomycetes	Eurotiales	11 genera 25 species
		Onygenales	1 genus, 1 species
		Pyrenulales	1 genus, 5 species
		Verrucariales	12 genera, 123 species
	Laboulbeniomycetes	Laboulbenials	1 genus, 1 species
Ascomycota *incertae sedis*			3 genera, 7 species
Basidiomycota	Cystobasdiomycetes	Cystobasidiales	2 genera, 6 species
	Classiculomycetes	Classiculales	2 genera 2 species
	Atractiellomycetes	Atractiellales	1 genus, 1 species
	Microbotryomycetes	Sporidiobolales	2 genera, 15 species
		Sporidiobolales *incertae sedis*	2 genera. 5 species
		Leucosporidiales	2 genera, 5 species
		Camptobasidiales	1 genus, 1 species
	Tremellomycetes	Cystofilobasidiales	5 genera, 9 species
		Filobasidiales	1 genus, 1 species
		Holtermanniales	1 genus, 1 species
		Tremellales	4 genera, 27 species
		Trichoporonales	1 genus, 3 species
	Ustilaginomycetes	Ustilaginales	1 genus, 2 species
		Urocystidales	1 genus, 7 species
	Agaricomycetes	Agaricales	4 genera, 4 species
		Atheliales	2 genera, 5 species
		Cantharellales	2 genera, 3 species
		Polyporales	2 genera, 2 species
	Exobasidiomycetes	Doassansiales	3 genera, 3 species
Basidiomycota *incertae sedis*			11 genera, 13 species
Fungal-like organisms			
Mesomycetozoea		Amoebidiales	2 genera, 7–18 species
Kickxellomycotina		Eccrinales	17 genera, 66 species

(*continued*)

Tab. 1.1: (*continued*)

Phylum	Class	Order	Number of genera/ species
		Dermocystida	1 genus, 1 species
		Ichthyosporida	6 genera, 8 species
Oomycota	Saprolegniomycetes	Leptomitales	1 genus, 3 species
		Saprolegniales	4 genera, 23 species
	Peronosporomycetes	Rhipidiales	10 genera
		Peronosporales	3 genera, 40 species
		Pythiales	87 species
		Atkinsiellales	1 genus, 1 species

Complied with data from FWA data base; Martha Powell; Sally Glocking; Robert Lichtwardt; Lei Cai; and data presented in various chapters in this volume.

Neubert et al. (2006) undertook a molecular survey of fungal diversity of the phanerogam *Phragmites australis* and identified 350 distinct operational taxonomic units (OTU). Many were fungal species yet to be identified and similar observations have been made on chytrids in sediments (Luo et al. 2004; Slapeta et al. 2005).

In this introductory chapter, we consider some of the general concepts that apply to all groups of freshwater fungi and fungal-like organisms. They include the composition of major groups and their origin, numbers of species and their distribution. Finally we discuss the organization of this volume and the contents of each major section. Chapters are written by specialists in the field of freshwater mycology and incorporate the latest published data.

1.1 Origin of freshwater fungi and fungal-like organisms

Freshwater fungi may have their origins in the sea while others are migrants from terrestrial habitats (Vijaykrishna et al. 2006; Beakes and Sekimoto 2009; Beakes et al. 2012). Keeling (2009) opined that the stramenopiles form a monophyletic clade with the alveolates as a sister group and evolved some 570 million years ago (mya). All osmotrophic stramenopiles are thought to have originated in the sea and migrated into freshwater as nematodes and insect larvae became terrestrial (Beakes et al. 2012).

Fungi may also have evolved in the sea and migrated into freshwater, e.g. the chytrids *Rozella* (Beakes and Sekimoto 2009; Lara et al. 2009; Jones et al. 2011) and subsequently became terrestrial. Bass et al. (2003) recovered novel lineages of chytrids from environmental DNA from marine ecosystems, supporting the marine affinities of primitive chytrids. Vijaykrishna et al. (2006) were of the opinion that freshwater ascomycetes evolved from terrestrial fungi probably circa 390 mya.

1.2 Classification of freshwater fungi

Jones et al. (2009) published a classification of marine fungi, but nothing similar is available for freshwater fungi with exception of the Fresh Water Ascomycetes (FWA) Database for the Ascomycota (http://fungi.life.illinois.edu/). Hibbett et al. (2007) list six phyla in the Kingdom Fungi, of which four have freshwater representatives: Chytridiomycota, Blastocladiomycota, Ascomycota and Basidiomycota (Tab. 1.2). Estimates of the number of freshwater species are given in Tabs. 1.1 and 1.2, but these are conservative as such lists are not available for many of the groups. For example: there is little data on the number of aquatic Mortierellales, Mucorales or Kickxellales.

The most specious phylum is the Ascomycota with freshwater representatives in 33 orders (622 species and excluding ascomycetous yeasts), while other taxa remain to be referred to orders, e.g. asexual taxa total 531 (Fallah and Shearer 2013) most of these with sexual stages in the Ascomycota (Tab. 1.2). Two orders of the Dothideomycetes support common freshwater fungi: Pleosporales and Jahnulales, the former comprises primarily terrestrial forms, while the Jahnulales is a novel lineage in the phylum with both freshwater and marine species (Suetrong et al. 2011; Zhang et al. 2012). The discomycetes are represented by two orders: Helotiales (16 genera, 25 species) which has many asexual morph connections, and the Rhytismatales which has primarily taxa that are found on leaves of terrestrial plants. Fourteen orders in the Sordariomycetes have freshwater members, with the Microascales (Halosphaeriaceae) and Savoryellales comprising both marine and freshwater taxa (Jones et al. 2009; Boonyuen et al. 2011). The Pezizomycetes and Laboulbeniomycetes are represented by the orders Pezizales and Laboulbeniales, respectively. The order Eurotiales and Verrucariales are members of the phylum Eurotiomycetes, with the lichen order accounting

Tab. 1.2: Estimates of freshwater fungi and fungal-like organisms.

Group	Number of species
Ascomycota	622
Basidiomycota (not yeasts)	41
Chytridiomycota	500–1,289
Asexual taxa	531
Trichomycetes	183
Basidiomycota yeasts	74
Peat swamp fungi	317+
Fungi on *Phragmites* litter	600
Lichens	270
Predacious fungi	40 (200?)
Endophytes	40?
Oomycota	49–87
Total	3,069–4,145

for 13 genera and 123 species, the most speciose genus in the Ascomycota. Cai et al. (See Chapter 3) put the number of freshwater ascomycetes as 622 while in Tab. 1.1 we calculate there are 556, which we regard as a conservative estimate. Many asexual morphs belong in the ascomycetes and are not counted in this table (Jones et al. 2008; Shearer et al. 2009; Hu et al. 2012).

Freshwater Basidiomycota are represented by 19 orders and comprise many yeasts and asexual morphs whose phylogenetic relationships are not fully resolved. Although a few produce fruiting bodies visible to the naked eye, most occur as single cells (yeasts) or as variously branched conidia (See Chapter 4). One hundred and fifteen taxa have been attributed to the Basidiomycota (in 50 genera), the most numerous referred to Tremellales (27 species) and Sporidiobiales (20 species).

The Chytridiomycota may appear well represented by 611 freshwater species, in 103 genera, however, not all these species are to be found in freshwater habitats as many occur as parasites of terrestrial plants and animals, as long as there is a film of water for the development and dispersal of their zoospores. The taxonomical placement of many of these taxa is supported by molecular data. The final phylum is the Blastocladiomycota, a group previously assigned to the Chytridiomycota.

The fungal-like organisms, Straminipila, are represented by the orders Leptomitales, Saprolegniales (Saprolegniomycetes), Rhipidiales and Peronosporales (Peronosporomycetes). With the exception of the Saprolegniales, most taxa referred to these orders are terrestrial and do not occur in fully aquatic habitats (Beakes et al. 2013).

1.3 Estimated number of freshwater fungi

Estimates of freshwater fungi are perfunctory as they do not include all taxa and how the data is calculated not often known, but the figures generally quoted are: 622 species (170 genera) ascomycetes (Cai et al., Chapter 3); more than 531 named hyphomycetes, 226 species (55 genera); trichomycetes (which includes three orders no longer regarded as fungi), while the basidiomycetes and zygomycetes are poorly documented (Misra and Lichwardt 2000; Tsui and Hyde 2003). Shearer et al. (2007) opine there are circa 500 ascomycetes, 405 asexual morphs, aero-aquatic 90, and no basidiomycetes are mentioned, other groups they include are 576 chytrids and 138 members of the Saprolegniales (fungal-like organisms), but no figure for freshwater yeasts (Shearer et al. 2007). Wurzbacher et al. (2011) indicate there are 2,535 freshwater fungi, which include ascomycetes, asexual morphs (which predominate), four basidomycetes (as aero-aquatic fungi), and 200 lichenized fungi. Jones and Pang (2013) estimated the number of freshwater fungi as 548–650 ascomycetes, 660 asexual morphs, 214 trichomycetes. Finally, Gessner and Van Ryckegem (2003) predict there are 20,000 different freshwater/aquatic fungi, based on the assumption that only 5% have been described to date.

These data indicate that: 1) most freshwater records are for temperate areas of the world, 2) basidiomycetes are poorly represented in freshwater habitats, 3) yeasts are poorly documented and require further study, 4) basal lineages and fungal-like organisms (Microsporidia, Rozellida, Blastocladiomycota, Glomeromycota) are poorly known in freshwater environments and knowledge of sediment fungi is only recently attracting attention. Beck-Nielsen and Vindbaek (2001) and Kai and Zhiwei (2006) report significant numbers of Glomeromycota associated with aquatic macrophytes as endo-mycorrhizae or dark septate endophytes, with 90% infection rates (Farmer 1985).

In Tab. 1.1 we present our most recent assessment of the fungi and fungal-like organisms to be found in freshwater habitats. We list 259–622 ascomycete species, excluding yeasts, with the orders Savoryellales (26 species), Eurotiales (25), Pleosporales), Helotiales (25) and Sordariales (24) being the most speciose. Basidiomycetes are more frequent than previously reported with 115 species in 50 genera, many as yeasts often associated with glacial waters (Libkind et al. 2010). The Chytridiomycota may include 611 species (in 103 genera) as many may occur where there is only a film of water. Asexual morphs and trichomycetes may account for a further 500 and 183, respectively.

In Tab. 1.2 ecological groups of fungi are listed with 600 documented from *Phragmites australis* litter (Gessner and Van Ryckegem 2003), 317 reported on senescent palm material in a peat swamp (See Chapter 17), 153 on submerged wood in Hong Kong (Tsui et al. 2001) and 73 species on wood in a tropical stream in Thailand (Sivichai et al. 2002). Not all these fungi are fully identified, thus the figures presented in Tab. 1.2 are underestimates. Freshwater lichens are another group whose numbers are poorly known with Holger et al. (Chapter 15) putting the figure as 244 (55 genera) and 26 (15 genera) for lichen and lichenicolous species, respectively. This brings the total to 3,069–4,145 which we regard as an underestimate as many other habitats and substrates remain to be surveyed.

Despite intensive studies in tropical freshwater habitats our knowledge is still basic. Jones (2011) estimated there were 12,500 marine fungi, of which the total described taxa were 530 species. Equal, or a greater number of freshwater and fungal-like organism await discovery and our studies must be broadened to include a wider range of microhabitats, substrates, greater sampling in selected geographical locations (South Africa, South America), sampling for yeasts and sediments, cryptic species and the use of a wider range of sampling techniques (pyrosequencing). These are topics that will be considered in greater detail in the concluding chapter. Most freshwater fungi are known from senescent shed leaves, submerged wood, endosymbionts of aquatic arthropods, and the culms of macrophytes around lakes and rivers (Gessner and van Ryckegen 2003; Lichwardt 2004; Wurzbacher et al. 2010). Many substrates, such as, flowers, fruits, pollen, algal mats, animal exuviae, remain poorly documented.

1.4 World distribution

Freshwater fungi are an ecological group and are worldwide in their distribution, although some may be restricted to tropical, temperate or cold water habitats, while others are cosmopolitan. However, it may be premature to label them as such, as many are known to occur only from the locality they were collected and described from, e.g. *Jahnula appendiculata* (Pinruan et al. 2002) from a peat swamp in Thailand. Also not all species described from freshwater are resident and may occur on land and washed into aquatic habitats, e.g. *Rogersiomyces okefenokeensis* was described from submerged decaying leaves. This species has subsequently been collected in galleries within the bark of trees attacked by beetles (Dowding 1984; Kirschner and Chen 2003, See Chapter 4). Table 1.3 list examples of species regarded as tropical, temperate and cosmopolitan.

A number of topics on freshwater fungi could not be accommodated as chapters, either due to lack of data or the topic has been recently reviewed. Therefore, they are included here in summary for the sake of completion.

1.5 Endophytic fungi

All aspects of endophyte diversity, systematics, ecology, evolution, metabolite production have been documented with major reviews by Carroll (1988), Petrini (1991),

Tab. 1.3: World distributions of selected freshwater fungal genera.

Fungal genus	Substratum	Reference
Tropical		
Colispora	Herbaceous material	Nawawi and Kuthubutheen (1989)
Obeliospora	Leaf litter	Nawawi and Kuthubutheen (1990)
Stauriella	Wood	Sivichai and Jones (2004)
Thailandiomyces	Senescent palm	Pinruan et al. (2008)
Yinmingella	Submerged wood	Goh and Hyde (1998)
Temperate		
Aquadiscula	*Scirpus atrovirens* litter	Fallah and Shearer (2001)
Ceriospora	Submerged wood	Fallah and Shearer (2001)
Classicula	Leaf litter	Bauer et al. (2003)
Candida aquatica	Foam	Jones and Slooff (1966)
Polycladium	Internodes of *Equisetum fluviatile*	Ingold (1959)
Cosmopolitan		
Annulatascus	Wood	Cai et al. (2006)
Helicomyces	Wood, foam	Kane et al. (2002)
Massarina	Wood	Cai et al. (2006)
Nawawia	Wood, litter	Ho et al. (2002)
Savoryella	Wood	Boonyuen et al. (2011)

Schulz et al. (2006), Arnold (2007) and Suryanarayanan et al. (2009). Few have considered endophytes derived from water plants, riparian tree roots and decaying leaves. Sridhar and Bärlocher (1992) present a broad account of endophytes from aquatic habitats, including freshwater, while Bärlocher (2006) reviews the literature on freshwater endophytes up to 2002. Bärlocher (2006) lists 40 species with *Anguillispora filiformis, A. longissima, Clavariopsis aquaitica, Cylindrocarpon aquaticum, Lunulspora curvula, Heliscus lugdunensis* and *Mycocentrospora* sp. appearing to be the most commonly recorded species. Additional aquatic hyphomycetes were recovered as endophytes by Sati and Belwal (2005), Sokolshi et al. (2006) and Sati et al. (2009) from riparian plants in India: *Acauloage tetaceros, Alatospora acuminta, A. pulchella, Anguillospora crassa, A. longissima, Campylospora purvula, Dwayaangam* sp., *Lemonniera cornuta, L. pseudofloscula, L. terrestris, Pestalotiopsis submersus* and *Tetracladium ninitalense.* Typical terrestrial fungi belonging to genera, such as, *Alternaria, Cladosporium, Penicillium* and *Trichoderma,* have also been isolated as endophytes of riparian plants, such as *Myriophyllum verticillatum, Equisetum arvense,* and *Impatiens chinensis* (Li et al. 2010). Few of these were fully identified and their role in the freshwater environment remains to be evaluated. More recently Sandberg et al. (2014) recovered 226 isolates (from 9,600 plant tissue segments) representing 60 putative species from four aquatic vascular plants in lentic waters: *Elodea bifoliata, Myriophyllum sibiricum, Persicaria amphibian* and *Stuckenia pectinata* in Arizona, USA. They observed a rich phylogenetic diversity with species in 37 genera, 19 families, 13 orders, seven classes and three phyla. The most common taxa belonged to the Eurotiomycetes and Dothideomycetes and the genera *Cladosporium, Penicillium, Alternaria* and *Aspergillus.* Contrary to data presented above, they found no sequence data resembling Ingoldian, aeroaquatic fungi or those from submerged wood, and suggested that endophytes from aquatic plants harbor a previously overlooked group of fungi in freshwater systems.

1.6 Predacious fungi

Predacious fungi are an ecological group comprising fungi: asexual morphs (e.g. *Arthrobotrys obligospora*), Ascomycota (e.g. *Orbilia* cf *luteorubella*), Zygomycota (e.g. *Meristarum asterospermum*), Basidiomycota (*Nematoctonus tylosporus*), Chytridiomycota (e.g. *Catenaria anguillulae*) and Zoopagales (e.g. *Stylopage hadra*), totalling more than 200 species and are worldwide in their distribution (Dayal 2000; Czygier and Bogus 2001; Swe et al. 2011). They are generally soil organisms and occur on a variety of substrata but in aquatic habitats they can easily be missed when compared with aquatic hyphomycetes who are more abundant and morphologically easier to see. They can be found on woody substrata but the presence of nematodes may be necessary before they sporulate on the substratum (Swe et al. 2009). Predacious fungi are often better known for their ability to trap nematodes, but they also trap rotifers, amoebae, tardigrades, and other soil organisms (Cooke 1977). There are many types of predacious fungi based on their mechanisms of attacking their prey: adhesive hyphae characteristic of Zygomycetes (e.g. *Stylopgae hadra*), adhesive branches found in asexual fungi (e.g. *Dactylella cionopaga*), adhesive nets-the most commonly found (e.g. *Arthrobotrys oligospora*), adhesive knobs (e.g. *Dactylaria candida*), constricting and non-constricting rings (e.g. *Dactylaria brochpaga*) and endoparasitic Oomycetes (e.g. *Haptoglossa mirabilis*).

Predacious fungi are generally soil dwelling, but do occur in water, e.g. the new species *Lecophagus antarcticus* was isolated from sediments of a shallow lake on Singy Island, South Orkney, Antarctica (Mcinnes 2003) trapping rotifers and tardigrades. However, they have rarely been studied (Hao et al. 2005). Other predacious fungi isolated from freshwater include: *Dactylella submersa*, *D. dainchiensis*, *D. heptameres*, *Monacrosporium reticulatum*, *M. scaphoides* and *M. tentaculatum* (Peach 1952; Hao et al. 2004). Hao et al. (2005) isolated 35 species from a shallow freshwater lake (Lake Dinachi, China) with *A. oligospora*, *A. musifromis*, *Monacrosporium thaumasium* and *M. longiphorum* the most common. However, most of these have been described originally from terrestrial soil/habitats. Kiziewicz (2004) investigated the diversity and abundance of nematode-trapping fungi in different water reservoirs: springs, lakes, and ponds and the following fungi recovered: *Ancylistes nettrii*, *Arthrobotrys brochopaga*, *A. dactyloides*, *A. oligospora*, *Euryancale sacciospora*, *Sommerstorffia spinosa* (fungus-like organisms: Peronosporomycetes), *Zoopage phanera*, *Zoophagus insidinas* and *Z. tentaculum*.

While most aquatic predacious fungi trap nematodes, others are predatory on rotifers (Saikawa 1986; Kiziewicz and Czeczuga 2003), amoebae (Barko 1975) and fish eggs (Czeczuga and Kiziewicz 1999; Czeczuga and Muszynska 1999). The total number of aquatic predacious fungi/fungus-like organisms is around 40, but other terrestrial species (200) may well occur in freshwater habitats and further studies are required to better understand their biodiversity in such habitats.

1.7 Bioactive compounds

Freshwater fungi have not been intensively studied for their ability to produce bioactive compounds and there are few publications on the subject. One of the earliest records was by Poch et al. (1992) who described a new cytotoxic naphthoquinone dimer kirschsteinin from a freshwater *Kirschsteiniothelia* sp. from submerged wood collected in Chile. Additional compounds included two new chlorinated diphenyl esters 8 and 9 and three known naphthoquinone derivatives. Other compounds documented for freshwater fungi are listed in Tab. 1.4.

A number of studies have screened freshwater fungi for bioactivity, but without characterising the bioactive compound (Cuomo et al. unpublished data; Sati and Arya 2010). For example: Gulis and Stephanovich (1999) screened 28 isolates of aquatic fungi against

Tab. 1.4: Selected examples of bioactive compounds from freshwater hyphomycetes and ascomycetes.

Fungus	Compounds	Activity	Reference
Helicodendron giganteum	Heliconols A–C, polyketides with an unusual reduced furanocyclopentane ring system	Mild antimicrobial	Mudur et al. (2006)
Decaisnella thyridioides	Decaspirones A–E, and the known compound palmarumycin CP	Potent antifungal and antibacterial activity	Jiao et al. (2006a)
Unidentified tubeufiaceous species	Dihydroaltenuenes A, B; dehydroaltenuenes A, B, and five known compounds including isoaltenuene, altenuene, and 5′epialtenuene	Antibiotic activity against Gram-positive bacteria	Jiao et al. (2006b)
Tetrachaetum elegans, Tetracladium marchalianum	Not determined	Antibacterial activity	Arya and Sati (2011)
Helicon richonis	Quinaphthin		Fisher et al. (1988); Adriaens-sens et al. (1994)
Annulatascus triseptatus	Annularins A–H: eight new polyketide metabolites, and two known compounds	Exhibited antibacterial activity	Li et al. (2003)
Pseudohalonectria adversaria	Two new Azaphilones: pseudonectrin A, B	Nematicidal activity	Dong et al. (2006)
Paraniesslia sp.	One new and one known sphingolipids	Nematicidal activity	Dong et al. (2005)
Freshwater derived fungus: YMF 1.01029	Five new preussomerin analogues, and four other known compounds	Nematicidal activity	Dong et al. (2008)
Freshwater derived fungus: YMF 1.01029	Colelomycerones A, B: two new naphthalene containing compounds and three known compounds	Colelomycerones A–B showed antifungal and antibacterial actovotoes	Dong et al. (2009)
Ophioceras dolichostomum	Ophiocerol: a novel macrocylic neolignan and the known compound isoamericanoic acid A	Nematicidal and antifungal activity	Dong et al. (2010)
Anguillospora longissima	Anguillosporal A	Antibacterial and antifungal acrtivity	Harrigan and Aremntrout (1995)

(*continued*)

Tab. 1.4: *(continued)*

Fungus	Compounds	Activity	Reference
Dendrospora tenella	Tenellic acid A–D: new bioactive diphenyl ether derivatives	Antibacterial activity against Gram-positive bacteria	Oh et al. (1999)
Massarina tunicata	Four new Massarigenins A–D, and two massarinins A, B	Antibacterial activity against Gram-positive bacteria	Oh et al. (2002)
Ophioceras venezuelense	Ophiocerin A–D: tetrahydropyran derivatives		Reátegui et al. (2005)
Chaetomium sp. YMF 1.02105	Chaetones A–F: new dibenzo b, e, oxepinone derivatives	Chaetone C showed significant cytotoxicity against five cancer cell lines	Shen et al. (2012)
Camposporium quercicola	Quercilolin		Wang et al. (2008)

16 bacterial isolates. Arya and Sati (2011) demonstrated that the aquatic hyphomycetes *Tetrachaetum elegans* and *Tetracladium marchalianum* showed antibacterial activity but no chemical analysis was undertaken. However, *H. lugdunensis*, *Tetracladium breve* and *T. nainitalense* showed no inhibitory effect against any of the pathogenic bacteria.

Recent topics receiving attention by mycologists are: 1) Barcode of Life Database (BOLD), 2) one fungus one name, and 3) pyrosequencing as an ecological aide to the discovery of taxa and the issue of identification of such sequences. The latter topic will be addressed in the final chapter of this volume.

1.8 Barcoding of freshwater fungi

The Barcode of Life Database was setup with the objective of identifying all known taxa using molecular data. Various genes have been proposed for the barcoding of organisms: in yeasts the D1/D2 domains of LSU rDNA has been used (Kurtzman 1994; Fell et al. 2000). However, for other fungi, partial or the entire ITS region of rDNA has been designated as the most suitable (Rossman 2007; Seifert et al. 2007). It is vital that barcoding of all fungi is undertaken, as it will provide evidence for the origin of freshwater fungi, help in the identification of environmentally derived sequences, especially those from pyrosequencing, and essential for determining the phylogenetic classification of asexual filamentous fungi (Michel et al. 2005; Letournea et al. 2010). Michel et al. (2005) showed that 30 species of freshwater hyphomycetes grouped in four classes of the fungi: Leotiomycetes (22 species), Dothideomycetes (4 species), Sordariomycetes (3 species) and Orbiliomycetes (1 species). Most species grouped in the Helotiales. They confirm the multiple origins of freshwater asexual fungi.

Various contributions have been made in the barcoding of freshwater hyphomycetes (Letournea et al. 2010; Duarte et al. 2012, 2013; Baschien et al. 2013). Letournea et al. (2010) examined intraspecific and interspecific variability in *Tetracladium* species and the suitability of three genes (ITS, COX1, D1/D2) for their barcoding. They concluded that ITS sequences gave the most satisfactory resolution for the *Tetracladium* species. Seena et al. (2010) examined the best sequences for barcoding 19 aquatic hyphomycetes (94 isolates) and proposed the usage of entire ITS1-5.8S-ITS2 rRNA sequences as the barcodes for identifying species of aquatic hyphomycetes (Seena et al. 2010). Subsequently they attempted to elucidate the genetic diversity of *Articulospora tetracladia* of 68 ITS1-5.8S-ITS2 gene sequences obtained from isolates collected from various types of plant litter or foam in streams of the Iberian Peninsula (Seena et al. 2012). Denaturing gradient gel electrophoresis (DGGE) was also used to distinguish *A. tetracladia* genotypes among 50 isolates from the same location. Maximum intraspecific evolutionary divergence between *A. tetracladia* isolates from various regions, streams or substrates was 2.2%, and that ITS barcodes to catalogue the intraspecific diversity with potential application in natural environments was useful. They also concluded that DGGE of the ITS2 gene region proved to be a rapid and less expensive way for assessing intraspecific diversity within the *A. tetracladia* isolates, with results obtained within a span of 5–16 h. Baschien et al. (2013) undertook an extensive sequencing study of aquatic hyphomycetes (36 species) and showed they were dispersed in nine supported clades in the Helotiales (Leotiomycetes). They also demonstrated that the genera *Tricladium, Lemonniera, Articulospora, Anguillospora, Varicosporium, Filosporella* and *Flagellospora* are not monophyletic. They concluded that the adaptation of hyphomycetes to the aquatic environment had evolved independently in multiple phylogenetic lineages within the Leotiomycetes (Baschien et al. 2005).

Duarte et al. (2013) addressed the progress made in DNA sequencing of aquatic hyphomycetes over the past ten years. Major contributions have been made in resolving the asexual/sexual relationships of aquatic fungi but only 72 species have been sequenced out of some 500 species. The majority were shown to belong to the Helotiales (Leotiomycetes) (46 species), with others grouping in the Pleosporales (6 species), Dothideales (1 species), Hypocreales (2 species) and Sordariales (1 species).

1.9 One name one fungus ruling

The desirability of implementing a one name per fungus has been under discussion for over 20 years and endlessly debated at International Mycological Congresses (Cairns 2004, Edinburgh 2008). At the 18th International Botanical Congress in Melbourne in July 2011 the policy of one name, one fungus was adopted, which also applies to fungus-like organisms (Oomycetes, slime moulds). This has been followed up by various meetings to implement this ruling and different groups of fungi were assigned to groups of specialists (Amsterdam, 2012). At present there is no formal document of recommended names for aquatic fungi and it is a task that needs undertaking. As with many fungi, the selection of the recommend name requires consideration. Few freshwater hyphomycetes are linked to their sexual states with Sivichai and Jones (2003) listing 72 connections, while the number currently stands at circa 77 species (See Chapter 5). Selection of the recommend name may therefore be difficult with so few linked to their sexual stage. For example, the sexual state of *Canalisporium* species has only recently been demonstrated as *Ascothailandia* an ascomycete in the Savoryellales (Boonyuen et al. 2011). The genus *Canalisporium* comprises many species with a long history in mycology, and may therefore be the name adopted.

Many aquatic hyphomycetes were described without type material and no living cultures deposited. Hu et al. (Chapter 5) list 77 hyphomycetes with known sexual stages, many only recently identified. Of these, some 20 genera are better known by the asexual state and that name should be accepted, also some are polyphyletic thus confusing the choice of a sexual state name. For example: *Articulospora tetracladia*/*Hymenoscyphus tetracladius*, *Clathrosphaerina zalewski*/*Hyaloscypha zakewskii*, *Dimorphospora foliicola*/*Hymenoscyphus foliicolus*, *Ingoldiella hamata*/*Sistotrema hamatum*, and *Taeniospora gracilis*/*Leptosporomyces galzinii*.

1.10 Role of fungi in freshwater habitats

How crucial are freshwater fungi and what contribution do they make to the microbial food web? Freshwater fungi comprise saprobes, parasites, endophytes and mutualistic taxa, all playing a different role in the breakdown and mineralization of organic matter in freshwater habitats. Freshwater habitats offer a wide range of microhabitats and substrates for fungal colonization: fall of senescent plant leaves (Bärlocher and Kendrick 1974; Bärlocher 1982; Concalves et al. 2013); decaying vascular plants in lakes and streams (Wirsel et al. 2001; Gessner et al. 2007); while yeasts are dependent on soluble polysaccharides in water (Fell 2012). In this volume we do not address the role of freshwater fungi in any detail, but aspects are mentioned in various chapters (See Chapters 19 and 21).

Many freshwater fungal-like organisms are parasites of various organisms and cause serious concern: *Aphanomyces astaci* is the cause crayfish plague, *Saprolegnia* species cause disease of fish and their eggs, often with high mortalities, while *Pythium phragmites* is the causative agent of the decline of the angiosperm *Phragmites* (Reynolds 1988; Nechwatal et al. 2005). Two other *Pythium* species have caused concern: *P. insidiosum* the cause of pythiosis in humans and other mammals and *P. ramorum* an infective pathogen of terrestrial trees (Chaiprasert 2004; Brown and Brasier 2007; See also Chapter 12). *Batrachochytrium dendrobatidis*, a chytrid, is another fungus that caused alarm in the decimation of amphibian population's worldwide (Longcore et al. 1999). These and other diseases caused by fungi and fungal-like organisms are reported in greater detail in Chapters 9 and 10 (fish), 11 (amphibians), 12 (human pathogen) and 14 (small invertebrates).

1.11 Objectives and outline of the volume

The primary objective of this volume is to review the most recent information on the occurrence, distribution, phylogeny, ecology of freshwater fungi. Although there have been a number of review articles recently on the subject, there is no comprehensive analysis of the group as a whole. For example, chytridiomycetes, the ascomycetous and basidiomycetous yeasts, fungal-like organisms are rarely included in books on freshwater fungi. Fungi and fungal-like organisms play a vital role in the ecology of streams, rivers, lakes, and these topics are generally considered in ecological treaties, rather than as a component of freshwater mycology.

The volume is in four sections. The first and second sections deal with the rapid increase in our knowledge of the phylogeny and classification of freshwater fungi and fungal-like organisms, respectively. Biodiversity and ecological studies of freshwater fungi are covered in Sections 3 and 4, respectively. A number of chapters are devoted to pathogens of aquatic animals and their catastrophic effect on aquatic amphibious animals: e.g. *Batrachochytrium dendrobatidis*, the cause of the amphibian disease chytridiomycosis (Longcore et al. 1999). While Fisher et al. (2012) have highlighted the threat of emerging diseases of aquatic animals and plants. The final chapter will highlight gaps in our current knowledge with suggestions for further areas for research.

1.12 Phylogeny of true freshwater fungi

Intensive studies on freshwater fungi at the molecular level have led to a radical change in their classification, a topic dealt with in Section 1. This has led to a better understanding of the relationships of different groups and the realization that many common genera are polyphyletic. This section is devoted to the Dothideomycetes (Chapter 2 by Shearer et al.), Sordariomycetes (Chapter 3 by Cai et al.), Basidiomycota (Chapter 4 by Jones et al.), asexual taxa (Chapter 5 by Hu et al.) and the Chytridiomycota (Chapter 6 by Powell and Letcher). An example of how our knowledge has dramatically changed is the number of freshwater basidiomycetes reported in Chapter 4, where 115 are listed, which can be compared with the few reported by Shearer et al. (2007).

1.13 Phylogeny of fungus-like organisms

Chapters focus on zoosporic organisms, once regarded as true fungi, and include the Cryptomycota, Mesomycetozoea, Oomycota, Hyphochytriomycota, Labyrinthulomycota and Phytomyzea. In this book we include chapters on the Microsporidia (Chapter 7 by Kearney and Gleason) and the Pythiales and Peronosporales (Chapter 8 by Marano et al.). Accounts of marine aquatic fungus-like organisms can be found in Jones and Pang (2012).

1.14 Biodiversity of freshwater fungi and fungus-like organisms

Five chapters are included in this section, concentrating on the ecological and economic importance of these organisms, as many are pathogens of mammals and animals of economic importance. For example, freshwater fungi/fungus-like organisms cause some £25 million losses to fisheries in the USA, annually (Bruno and Wood 1999). Fish parasites (Chapters 9 by Glocking et al., and 10 by Sarowar et al.), amphibian pathogens (Chapter 11 by Gleason et al.), pythiosis of mammals (Chapter 12 by Chaiprasert and Krajaejun), parasites of zooplanktons (Chapter 13 by Gleason et al.) and invertebrate parasites (Chapter 14 by Glocking et al.).

1.15 Ecology

This section deals with ecological groups that occur in freshwater ecosystems and play an important role in the recycling of complex organic matter. Chapter 15 is devoted to freshwater lichens (Thűs et al.) documenting 244 lichen species (in 55 genera) and 26 lichenicolous species (in 26 genera). This greatly increases the 100 estimate previously reported from freshwater habitats (Aptroot and Seaward 2003). Trichomycetes, including both true fungi and fungus-like organisms, are reported in Chapter 16 (Lichtwardt). Peat swamps are unusual habitats for fungi because of their acidic conditions, however, palm fronds are colonised by a wide range of species, many new to science (Chapter 17 by Pinruan et al.). The role of fungi in the decomposition of leaf material and submerged timber test blocks are reported on in Chapters 19 (Bärlocher and Sridhar) and 21 (Boonyuen et al.), respectively. The occurrence of fungi in polluted streams is documented in Chapter 18 (Ferreira et al.), and their tolerance of extreme environments in Chapter 20 (Libkind et al.). The volume concludes with observations on topics for further research and those that have been neglected to date (Chapter 22 by Pang et al.).

Acknowledgments

Gareth Jones is supported by the Distinguished Scientist Fellowship Program (DSFP), King Saud University, Saudi Arabia. We are grateful to: Dr Frank Gleason for many suggestions and in particular his major contribution to the sections on fungus-like organisms, Dr K.R. Sridhar for assistance with literature on aquatic endophytes.

References

Arnold AE. Understanding the diversity of foliar fungal endophytes: progress, challenges, and frontiers. Fungal Biol Rev 2007;21:51–66.
Arya P, Sati SC. Evaluation of endophytic aquatic hyphomycetes for their antagonistic activity against pathogenic bacteria. Inter Res J Microbiol 2011;2:343–347.
Adeiaenssens P, Anson AE, Begley MJ, Fisher PJ, Orrel KG, Webster J, et al. Quinaphthin, a metabolite produced by *Helicoon richonis*. J Chem Soc Perkin Trans 1994;1:2007–2010.
Barr DJS. An outline for the classification of the Chytridiales, and a new order, the Spizellomycetales. Can J Bot 1980;58:2380–2394.
Barko A. Hydromycology – An Overview. Warszawa: Panstwowe Wydawnictwo Naulowe 1975.
Bärlocher F. The contribution of fungal enzymes to the digestion of leaves by *Gammarus fossarum* Koch (Amphipoda). Oecologia 1982;52:1–4.
Bärlocher F. Fungal endophytes in submerged roots. In: Schulz B, Boyle C, Sieber TN, eds. Soil Biology, vol 9, Microbial root endophytes. Springer-Verlag: Berlin 2006:179–190.
Bärlocher F. ed. The Ecology of Aquatic Hyphomycetes. Ecol Studies. Springer-Verlag: Berlin 2007;94:1–225.
Bärlocher F, Kendrick B. Dynamics of the fungal palpation on leaves in a stream. J Ecol 1974;62:761–791.
Baschien C, Tsui CK-M, Culis V, Szewzyk U, Marvaonová L. The molecular phylogeny of aquatic hyphomycetes with affinity to the Leotiomycetes. Fungal Biol 2013;117:660–672.
Bauer R, Begerow D, Oberwinkler F, Marvanová L. *Classicula:* the teleomorph of *Naiadella fluitans*. Mycologia 2003;95:756–764.
Beakes GW, Sekimoto S. The evolutionary phylogeny of oomycetes – insights gained from studies of holocarpic parasites of algae and invertebrates. In: Lamour K, Kamoun S, eds. Oomycete Genetics and Genomics. New York: John Wiley & Sons 2009:1–24.
Beakes GW, Glockling SL, Sekimoto S. The evolutionary phylogeny of the oomycete "fungi". Protoplasma 2012;249:3–19.
Beck-Nielsen D, Vindbaek Madsen T. Occurrence of vesicular-arbuscular mycorrhiza in aquatic macrophytes from lakes and streams. Aquat Bot 2001;71:141–148.
Boonyuen N, Chuassharonnachai C, Suetrong S, Sri-indrasuthdhi V, Sivichai S, Jones EBG, et al. Savoryellales (Hypocreomycetideae, Sordariomycetes): a novel lineage of aquatic ascomycetes inferred from multiple-gene phylogenies of the genera *Ascotaiwania, Ascothailandia* and *Savoryella*. Mycologia 2011;103:1351–1350.
Brown AV, Brasier CM. Colonization of tree xylem by *Phytophthora ramorum, P. kernoviae* and other *Phytophthora* species. Plant Pathol 2007;56:227–41.
Braun A. Über *Chytridium*, eine gattung einzelliger schmarotzergewächse auf algen und infusorium. Gedruckt in der Druckerei der Königl, Berlin 1856.
Cai L, Hyde KD, Tsui CKM. Genera of freshwater fungi. Fungal Diversity Research Series, Hong Kong 2006:1–253.
Carroll GC. Fungal endophytes in stems and leaves: from latent pathogen to mutualistic symbionts. Ecology 1988;69:2–9.
Chairprasert A. Dermatophytes and pythiosis. In: Jones EBG, Tanticharoen M, Hyde KD, eds. Thai Fungal Diversity. Thailand: BIOTEC 2004:213–225.
Cooke R. The Biology of Symbiotic Fungi. London: John Wiley 1977.
Czygier N, Bogus MI. Predacious nematode-destoying fungi. Wiadom Paraztol 2001;47:25–31.
Czeczuga B, Kiziewicz B. Zoosporic fungi growing on the eggs of *Carrasius carrasius* (L.) in oligo- and eutrophic water. Polish J Envir Stud 1999;8:63–66.
Czeczuga B, Muszynska E. Aquatic fungi growing on the eggs of fishes representing 33 Cyprinida taxa (Cypeinidae) in laboratory conditions. Acta Ichth Piscatoria 1999;29:53–72.

Dayal R. Predacious fungi. Published by Ajay Verma, Commonwealth Publ 2000:1–169.
Dong JY, Ru L, et al. Nematicidal sphingolipids from the freshwater fungus *Paraniesslia* sp. YMF1.01400. Europ J Lipid Sci Technol 2005;107:779–785.
Dong J, Zhou1 Y, Li R, Zhou1 W, Li L, Zhu Y, et al. New nematicidal azaphilones from the aquatic fungus *Pseudohalonectria adversaria* YMF1.01019. FEMS Microbiol Lett 2006;264:65–69.
Dong JY, Song HC, et al. Ymf 1029A-E, preussomerin analogues from the fresh-water-derived fungus YMF 1.01029. J Nat Prod 2008;71:952–956.
Dong JY, Song HC, et al. Two unusual naphthalene-containing compounds from a freshwater fungus YMF 1.01029. Chem Biodivers 2009;6:569–577.
Dong JY, Wang L, et al. Ophiocerol, a novel macrocylic neolignan from the aquatic fungus *Ophioceras dolichostomum* YMF1.00988. Nat Prod Res 2010;24:1004–1012.
Dowding P. The evolution of insect-fungus relationships in the primary invasion of forest timber. In: Anderson JM, Raynor ADM, Walton DWH, eds. Invertebrate-microbial Interactions. New York: Cambridge University Press 1984;133–153.
Duarte S, Seena S, Bärlocher F, Cássio F, Pascoal C. Preliminary insights into the phylogeography of six aquatic hyphomycete species. PLOS One 2012;7:1–7.
Duarte S, Seena S, Bärlocher F, Pascoal C, Cássio F. A decade's perspective of DNA sequencing on aquatic hyphomycete research. Fungal Biol Rev 2013;27:1–6.
Fallah PM, Shearer CA. Freshwater ascomycetes: new or noteworthy species from north temperate lakes in Wisconsin. Mycologia 2001;93:566–602.
Farmer A. The occurrence of versicular-arbuscular mycorrhiza in isoetid-type submerged aquatic macrophytes under naturally varying conditions. Aquat Bot 1985;21:245–249.
Fell JW. Yeasts in marine environments. In: Jones EBG, Pang KL. Marine Fungi and Fungal-like Organisms. Berlin: De Gruyter 2012:91–102.
Fell JW, Boekhout T, Fonseca A, Scorzetti G, Statzell-Tallman A. Biodiversity and systematics of basidiomycetous yeast as determined by large-subunit rDNA D1/D2 domain sequence analysis. Int J Syst Evol Microbiol 2000;50:1351–1371.
Fell JW, Statzell-Tallman A, Scorzetti G, Gutiérrez MH. Five new species of yeasts from fresh water and marine habitats in the Florida Everglades. Antonie Leeuwenhoek 2011;99:533–549.
Fisher MC, Henk DA, Briggs CJ, Brownstein JS, Madoff LC, McCraw SL, et al. Emerging fungal threats to animal, plant and ecosystem health. Nat 2012;484:186–194.
Fisher PJ, Anson AE, Webster J. Quinaphthin, a new antibiotic produced by *Helicoon richonis*. Trans Br Mycol Soc 1988;90:499–502.
Gadanho M, Sampaio JP. Occurrence and diversity of yeasts in the Mid-Atlantic Ridge hydrothermal fields near the Azores archipelago. Microbiol Ecol 2005;50:408–417.
Gessner MO, van Ryckegem G. Water fungi as decomposers in freshwater ecosystems. In: Bitton G, ed. Encyclopaedia of Environmental Microbiology. New York: Wiley 2003.
Gessner MO, Gulis V, Kuehn K, Chauvet E, Suberkropp K. Fungal decomposers of plant litter in aquatic ecosystems. In: Kubicak C, Druzhinia I, eds. Environmental and Microbial Relationships: The Mycota. vol 4, Berlin: Springer 2007:301–324.
Goh TK, Hyde KD. *Spadicoides cordanoids* sp. nov., a new dematiaceous hyphomycete from submerged wood in Australia, with a taxonomic review of the genus. Mycologia 1996;88:1022–1031.
Goh TK, Hyde KD. *Yingmingella mitriformis* gen. *et* sp. nov., a new sporodochial hyphomycete from submerged wood in Hong Kong. Can J Bot 1998;76:1693–1697.
Concalves AL, Graca MAS, Canhoto C. The effect of temperature on leaf decomposition and diversity of associated aquatic hyphomycetes depends on the substrate. Fungal Ecol 2013;6:546–553.
Guilis V, Stephanovich AI. Antibiotic effects of some aquatic hyphomycetes. Mycol Res 1999;103:111–115.

Hagler AN, Ahearn DG. Ecology of aquatic yeasts. In: Rose AH, Harrison JS, eds. The Yeasts, Vol 2, Yeasts and the Environment. London: Academic Press 1987:181–205.

Hao YE, Luo J, Zhang KQ. A new aquatic nematode-trapping hyphomycete. Mycotaxon 2004;89:235–239.

Hao Y, Mo M, Su H, Zhang K. Ecology of aquatic nematode-trapping hyphomycetes in southwestern China. Aquatic Microbiol 2005;40:175–181.

Harrigan GG, Aremntrout BL, et al. Anguillosporal, a new antibacterial and antifungal metabolite from the freshwater fungus *Anguillospora longissima*. J Nat Prod 1995;58:1467–1469.

Hibbett DS, Binder M, Bischoff JF, et al. A higher-level phylogenetic classification of the Fungi. Mycol Res 2007;111:509–547.

Ho WH, Yanna, Hyde KD, Hodgkiss IJ. Seasonality ad sequential occurrence of fungi on wood submerged in Tai Po Kau Forest stream, Hong Kong. Fungal Divers 2002;10:5–10.

Hu DM, Cai L, Hyde KD. Three new ascomycetes from freshwater in China. Mycologia 2012;104:1478–1489.

Hyde KD. Tropical Australian freshwater fungi. II. *Annulatascus velatispora* gen. *et* sp. nov, *A. bipolaris* sp. nov. and *Ophioceras olichostomum* (Ascomycetes) Aust Syst Bot 1992;5: 117–124.

Hyde KD, Wong SW. *Annulatacus fusiformis* sp. nov., a new freshwater ascomycete from the Philippines. Mycologia 2000;92:553–557.

Ingold CT. Aquatic hyphomycetes of decaying alder leaves. Tran Br Mycol Soc 1942;25:339–417.

Ingold CT. Some new aquatic hyphomycetes. Tran Br Mycol Soc 1944;28:35–43.

Ingold CT. Aquatic ascomycetes: *Ceriospora caudae-suis* n.sp. and *Ophiobolus typhae*. Tran Br Mycol Soc 1951;34:210–215.

Ingold CT. Aquatic ascomycetes: further species from the English Lake district. Tran Br Mycol Soc 1955;38:157–168.

Ingold CT. *Polycladium equiseti* gen. nov., sp. nov. Trans Br Mycol Soc 1959;42:112–114.

Jiao P, Swenson D ale C, Gloer JB, Campbell J, Shearer CA. Decaspirones A-E, new bioactive Spiroxynaphthalenes from the freshwater aquatic fungus *Decaisnella thyridioides*. J Nat Prod 2006a;69:1667–1671.

Jiao P, Gloer JB, Campbell J, Shearer CA. Altenuene derivatives from an unidentified freshwater fungus in the family Tubeufiaceae. J Nat Prod 2006b;69:612–215.

Jones EBG. Are there more marine fungi to be described? Bot Mar 2011;54:343–354.

Jones EBG, Sloof WC. *Candida aquatica* sp. nov. isolated from water scums. Antonie van Leeuvenhoek 1955;32:223–228.

Jones EBG, Pang KL. eds. Marine Fungi and Fungal-like Organisms. Berlin, Germany: De Gruyter 2012.

Jones EBG, Pang KL. Tropical aquatic fungi. Biodivers Conser 2013;21:2403–2423.

Jones EBG, Sakayaroj J, Suetrong S, Somrithipol S, Pang KL. Classification of marine Ascomycota, anamorphic taxa and Basidiomycota. Fungal Divers 2009;35:1–187.

Jones MDM, Forn I, Gadelha C, Egan M, Bass D, Massana R, et al. Discovery of novel intermediate forms redefines the fungal tree of life. Nature 2011;474:200–203.

Kai W, Zhiwei Z. Occurrence of arbuscular mycorrhizas and dark septate endophytes in hydrophytes from lakes and streams in southwest China. Int Rev Hydriobiol 2006;91:29–37.

Kane DF, Tam WY, Jones EBG. Fungi colonising and sporulating on submerged wood in the River Severn, UK. In: Hyde KD, Jones EBG, eds. Fungal Succession. Fungal Divers 2002;10:45–55.

Keeling PJ. Five things to know about Microsporidia. PLoS Pathog, 2009; 5(9) e1000489. doi:10.1371/journal.ppat.1000489.

Kirschner R, Chen C-J. A new record of *Rogersiomyces okenokeensis* (Basidiomycota) from beetle galleries in pines in Taiwan. Sydowia 2003;55:86–92.

Kiziewicz B. Occurrence of parasitic and predatory fungi and fungus-like organisms in different water reservoirs of Podlasie Province of Poland. Mycol Balcanica 2004;1:159–162.
Kiziewicz B, Czeczuga B. Occurrence and morphology of some predatory fungi, amoebicidal, rotifericial and namatodicdal, in the surface waters of Bialystok region. Wiadom Parazytol 2003;49:281–292.
Kurtzman CP. Molecular taxonomy of the yeasts. Yeast 1994;10:1727–1740.
Kurtzman CP, Fell JW, Boekhout T. The Yeasts, a Taxonomic Study. 5th ed. Vols 1–3. Elsevier: Amsterdam 2011:1–2080.
Lara E, Moriera D, Lopez-Garcia P. The environmental clade LKM11 and *Rozella* from the deepest branching clade of the fungi. Protist 2010;161:116–121.
Letourneau A, Seena S, Marvanová L, Bärlocher F. Potential use of barcoding to identify aquatic hyphomycetes. Fungal Divers 2010;40:51–64.
Li C, Nitka MV, et al. Annularins A–H: New polyketide metabolites from the freshwater aquatic fungus *Annulatascus triseptatus*. J Nat Prod 2003;66:1302–1306.
Li H-Y, Zhao C-A, Liu C-J, Xu X-F. Endophytic fungi diversity of aquatic/riparian plants and their antifungal activity *in vitro*. J Microbiol 2010;48:1–6.
Lichtwardt RW. Trichomycetes: fungi in relationships with Insects and other arthropods. In: Seckbach J, ed. Symbiosis (Vol 4), Netherlands: Springer 2004:575–588.
Libkind D, Sampaio JP, van Broock M. Cystobasdiomycetes yeasts from Patagonia (Argentina): description of *Rhodotorula meli* sp. nov. from glacial meltwater. Inter J Syst Evol Microbiol 2010;60:2251–2256.
Longcore JE, Pessier AP, Nichols DK. *Batrachochytrium dendrobatidis* gen. *et* sp. nov., a chytrid pathogenic to amphibians. Mycol 1999;91:219–227.
McInnes SJ. A predatory fungus (Hyphomycetes: *Lecophagus*) attacking rotifer and tardigrada in maritime Antarctic lakes. Polar Biol 2003;26:79–82.
Michel J, Belliveau R, Bärlocher F. Molecular evidence confirms multiple origins of aquatic hyphomycetes. Mycol Res 2005;109:1407–1417.
Misra JK, Lichtwardt RW. Illustrated genera of Trichomycetes: fungal symbionts of insects and other arthropods. Enfield, NH: Sci Publ Inc.
Mudur SV, Swenson DC, et al. Heliconols A–C: Antimicrobial hemiketals from the freshwater aquatic fungus *Helicodendron giganteum*. Org Lett 2006;8:3191–3194.
Nawawi A. Basidiomycetes with branched water-borne conidia. Bot J Linn Soc 1985;91:51–60.
Nawawi A, Kuthubutheen AJ. A new taxon in *Colispora* (Hyphomycetes) from Malaysia. Mycotaxon 1989;34:497–501.
Nawawi A, Kuthubutheen AJ. *Obeliospora*, a new genus of setose, phialosporous hyphomycetes with appendaged conidia. Mycotaxon 1990;37:395–400.
Neubert K, Mendgen K, Brinkmann H, Wirsel S. Only a few fungal species dominate highly diverse mycofloras associated with the common reed. Appl Enivron Microbiol 2006;7:1118–1128.
Newchwatal K, Wielgloss A, Mendgen K. *Pythium phramitis* sp. nov., a new species close to *P. arrhenomanes* as a pathogen of common reed *Phragmites australis*. Mycol Res 2005; 109:1337–1346.
Oh H, Kwon TO, et al. Tenellic acids A–D: new bioactive diphenyl ether derivatives from the aquatic fungus *Dendrospora tenella*. J Nat Prod 1999;62:580–583.
Oh H, Swenson DC, et al. New bioactive rosigenin analogues and aromatic polyketide metabolites from the freshwater aquatic fungus *Massarina tunicata*. J Nat Prod 2002;66:73–79.
Peach M. Aquatic predacious fungi II. Trans Br Mycol Soc 1952;35:19–23.
Petrini O. Fungal endophytes of tree leaves. In: Andrews J, Hirano D, eds. Microbial Ecology of Leaves. Berlin: Springer 1991:179–197.
Pinruan U, Jones EBG, Hyde KD. Aquatic fungi from peat swamp palms: *Jahnula appendiculata* sp. nov. Sydowia 2002;54:242–247.

Pinruan U, Sakayaroj J, Hyde KD, Jones EBG. *Thailandiomyces bisetulosus* gen. *et* sp. nov. (Diaporthales, Sordariomycetidae) and its anamorphs *Craspedodidymum*, is described based on nuclear SSU and LSU rDNA sequences. Fungal Divers 2008;29:89–98.
Poch GK, Gloer JB, Shearer CA. New bioactive metabolites from a freshwater isolate of the fungus *Kirschsteiniothelia* sp. J Nat Prod 1992;55:1093–1099.
Powell MJ, Letcher PM. From zoospores to molecules: the evolution and systematics of Chytridiomycota. In: Misra JK, Tewari JP, Deshmukh SK, eds. Systematics and Evolution of Fungi. Boca Raton, FL: CRC Press 2012:29–54.
Reátegui RF, Gloer JB, et al. Ophiocerins A-D and ophioceric acid: tetrahydropyran derivatives and an africane sesquiterpenoid from the freshwater aquatic fungus *Ophioceras venezuelense*. J Nat Prod 2005;68:701–705.
Reynolds J. Crayfish extinctions and crayfish plague in central Ireland. Biol Concers 1988;45:279–285.
Rossman A. Report of the planning worshop for all fungi DBA barcoding. Inoculum 2007;58:1–5.
Rostrup E. Mykologiske Meddelelser IV. Bot Tidsskrift 1894;19:36–51.
Saccardo, Therry. Michelia 1880;2(no. 6):132.
Saccardo, Sydow P. Syll fung (Abellini) 1899;14(2):1128.
Saikawa M. Electron microscopy on *Sommerstoffa spinosa*, a water-mold parasitic on rotifers. Mycologia 1986;78:554–561.
Sandberg D, Battista LJ, Arnold AE. Fungal endophytes of aquatic macrophytes: diverse host-generalists characterized by tissue preferences and geographic structure. Microbiol Aquatic Syst 2014; DOI 10.1007/s00248-013-0324-y.
Sati SC, Arya P. Antagonism of some aquatic hyphomycetes against plant pathogenic fungi. Sci World J 2010;10:760–765.
Sati SC, Belwal M. Aquatic hyphomycetes as endophytes of riparian plant roots. Mycologia 2005;97:45–49.
Sati SC, Pargaein N, Belwal M. Diveristy of aquatic hyphomcyetes as root endophytes on pteridophytic plants in Kumaun Himalaya. J Am Sci 2009;5:179–182.
Schulz B, Boyle C, Sieber TN. Microbial Root Endophytes. Berlin: Springer-Verlag.
Seena S, Pascoal C, Marvanová L, Cássio F. DNA barcoding of fungi: a case study using ITS sequences for identifying aquatic hyphomycete species. Fungal Divers 2010;44:77–87.
Seen S, Duarte S, Pascoal C, Cássio F. Intraspecific variation of the aquatic fungus *Articulospora tetracladia*: a ubiquitous perspective. PLoS ONE 2012;7:1–8.
Seifert KA, Samson RA, de Waard JR, Houybraken J, Levesque CA, Moncalvo JM, et al. Prospects for fungus identification using CO1 DNA barcodes, with *Penicillium* as a test case. Proc Natl Acad Sci 2007;104:3901–3906.
Shearer CA. The freshwater ascomycetes. Nova Hedwig 1993;56:1–33.
Shearer CA, Raja HA. Freshwater Ascomycetes Database: http://fungi.life.illinois.edu/.
Shearer CA, Descals E, Kohlmeyer E, Kohlmeyer J, Marvanová L, Padgett DE, et al. Fungal biodiversity in aquatic habitats. Biodivers Conserv 2007;16:49–67.
Shen K-Z, Gao S, et al. Novel dibenzo b,e oxepinones from the freshwater-derived fungus *Chaetomium* sp. YMF 1.02105. Planta Medica 2012;78:1837–1843.
Sivichai S, Jones EBG. Teleomorphic-anamorphic connections of freshwater fungi. In: Tsui CKM, Hyde KD, eds. Hong Kong: Fungal Diversity Research Series 2003;10:259–272.
Sivichai S, Jones EBG. *Stauriella* gen. nov., proposed for a new lignicolous basidiomycetous anamorph from freshwater in Thailand. Sydowia 2004;56:131–136.
Sivichai S, Jones EBG, Hywel-Jones NL. Fungal colonisation of wood in a freshwater stream at Tad Ta Phu, Khao Yai National Par, Thailand. In: Hyde KD, Jones EBG, eds. Fungal Succession, Fungal Divers 2002;10:113–129.

Slapeta J, Moreira D, López-Garcia P. The extent of protist diversity: insights from molecular ecology of freshwater eukaryotes. Proc Roy Soc B Biol Sci 2005;272:2073–2081.
Sokolski S, Piché Y, Chauvet E, Bérybé J. A fungal endophyte of black spruce (*Picea mariana*) needles is also an aquatic hyphomycete. Molc Ecol 2006;15:1955–1962.
Sparrow FK. Aquatic Phycomycetes. 2nd ed. Ann Arbor, MI: University of Michigan Press 1960.
Sridhar KR, Baerlocher F. Aquatic hyphomycetes in spruce roots. Mycologia 1992;84:580–584.
Suetrong S, Boonyuen N, Pang KL, Ueapattanakit J, Klaysuban A, Sri-Indrasutdhi V, et al. A taxonomic revision and phylogenetic reconstruction of the Jahnulales (Dothideomycetes), and the new family Manglicolaceae. Fungal Divers 2011;51:163–188.
Sun, Gao Y-X, Shen K-Z, Xu Y-B, Wanga C-R, Liu H-Y, et al. Antimicrobial metabolites from the aquatic fungus *Delitschia corticola*. Phytochem Lett 2011;4:101–105.
Suryanarayanan TS, Thurumala E, Prakash CP, Rajulu MBG, Thurunavukkarasau N. Fungi from two forests of southern India: a comparative study of endophytes, phellophytes, and leaf litter fungi. Can J Microbiol 2009;55:319–426.
Swe A, Jeewon R, Pointing SB, Hyde KD. Diversity and abundance of nematode-trapping fungi from decaying litter in terrestrial, freshwater and mangrove habitats. Biodivers Conserv 2009;18:1695–1714.
Swe A, Li J, Zhang KQ, Pointing SB, Jeewon R, Hyde KD. Nematode-trapping fungi. Curr Res Environ Appl Mycol 2011;1:1–26.
Vijaykrishna D, Jeewon R, Hyde KD. Molecular taxonomy, origins and evolution of freshwater ascomycetes. Fungal Diver 2006;23:351–390.
Wang L, Dong J, et al. Screening and isolation of antibacterial activities of the fermentative extracts of freshwater fungi from Yunnan Province, China. Ann Microbiol 2008;58:579–584.
de Wildeman E. Notes mycologiques. Fasicle 2. Ann Soc belge Micro 1893;17:35–68.
de Wildeman E. Notes mycologiques. Fasicle 3. Ann Soc belge Micro 1894;18:135–161.
de Wildeman E. Notes mycologiques. Fasicle . Ann Soc belge Micro 1895;19:2011.
Tsui CKM, Hyde KD. Freshwater Mycology. Fungal Diversity Press: Hong Kong 2003:1–350.
Tsui CKM, Hyde KD, Hodgkiss IJ. *Paraniesslia tuberculata* gen. *et* sp. nov., and new records or species of *Clypeosphaeria, Leptosphaeria* and *Astrosphaeriella* in Hong Kong freshwater habitats. Mycologia 2001;893:1002–1009.
Wirsel S, Leibinger W, Ernst M, Merndgen K. Genetic diversity of fungi closely associated with common reed. New Phytol 2001;149:589–598.
Wong SW, Hyde KD, Jones EBG. Ultrastructural studies on the aquatic ascomycetes, *Fluminicola bipolaris* gen. *et* sp. nov. Fungal Divers 1999;2:189–197.
Wong KMK, Goh TK, Hodgkiss IJ, Hyde KD, Ranghoo VM, Tsui CKM, et al. Role of fungi in freshwater ecosystems. Biodivers Conserv 1998;7:1187–1206.
Wurzbacher C, Bärlochedr F, Grossart H-P. Fungi in lake ecosystems. Aquatic Microbial Ecol 2010;59:125–149.
Wurzbacher C, Kerr J, Grossart H-P. Aquatic fungi. In: Grillo O, ed. The Dynamics Processes of Biodiversity—Case Studies of Evolution and Spatial Distribution. Croatia: Tech Open Learning, chapter 10, 2012:227–258.
Zhang Y, Crous P, Schoch CL, Hyde KD. Pleosporales. Fungal Divers 2012;53:1–221.

Phylogeny of freshwater fungi

Carol A. Shearer, Ka-Lai Pang, Satinee Suetrong and
Huzefa A. Raja

2 Phylogeny of the Dothideomycetes and other classes of freshwater fissitunicate Ascomycota

2.1 Introduction

Species comprising the freshwater Ascomycota, exclusive of the yeasts, are filamentous fungi that grow and reproduce on submerged plant debris in freshwater habitats. This chapter deals with one class of the Ascomycota, the Dothideomycetes. The Dothideomycetes are distinguished from other Ascomycota by producing an ascostromatic fruit body and fissitunicate (bitunicate) asci. They reproduce by forming ascospores within an ascus as a result of meiosis and conidia directly on the mycelium, on a conidiophore or within a conidiomata as a result of mitosis. The sexual and asexual states may be separated by time, substrate, habitat, water quality parameters and geography and are considered different species until they are linked by cultural and/or molecular data. When the asexual and sexual states are linked, both states of the fungus are given the same name. As this linking process is ongoing, the nomenclature of the Dothideomycetes is currently in a state of flux. Species of freshwater Dothideomycetes were among the first ascomycetes to be reported from freshwater habitats (Petrak 1925; Webster 1951; Ingold 1955; Eaton and Jones 1970, 1971a, b; Pugh and Mulder 1971; Shearer and Crane 1971; Eaton 1972; Shearer 1972). Since the reviews of freshwater ascomycetes by Dudka (1963) and Shearer (1993), in which the latter study listed 76 Dothideomycetes reported from fresh water, the freshwater Dothideomycetes currently comprise 200 (32%) of 622 species of ascomycetes reported from freshwater habitats (Shearer and Raja 2013). The sexual states of Dothideomycetes occur primarily on submerged woody debris, palm leaves and stems, and emergent aquatic plants. The asexual states, including aquatic, aero-aquatic, dematiaceous hyphomycetes and pycnidial forms, occur on wood, deciduous leaves, emergent macrophytes and palm debris (Goh and Hyde 1996; Wong et al. 1998; Shearer et al. 2007).

Freshwater Dothideomycetes are thought to play an important role, along with other ascomycetes, in decomposing the complex structural materials comprising submerged plant litter (Schoch et al. 2009; Shearer et al. 2009). This activity in turn provides simpler compounds that can be used as a food source, along with the fungal hyphae, by stream invertebrates (See Chapter 21 for details).

2.2 Geographical distribution patterns

The geographical distribution patterns of freshwater Dothideomycetes are based on relatively few comprehensive studies (Shearer and Raja 2013). About 56% occur in temperate latitudes, 35% occur in tropical and subtropical latitudes, while 11% occur in both temperate and tropical latitudes (Shearer and Raja 2013). These geographical distribution patterns suggest that a higher number of freshwater Dothideomycetes occur in temperate latitudes than from the tropics. These data are consistent with the observations of a study on fungal endophytes, which was carried out along a broad latitudinal gradient from Canada to lowland tropical forests of Panama, where the authors found a higher percent of Dothideomycetes from temperate latitudes (Arnold and Lutzoni 2007).

2.3 Substrate distribution patterns

Among the approximately 200 species of freshwater Dothideomycetes that are currently described and/or reported from freshwater habitats, about 50% occur on submerged wood, 33% occur on herbaceous material, and only 17% occur on both substrate types (Shearer and Raja 2013). At this time, it is difficult to predict if the widespread occurrence of freshwater Dothideomycetes on lignicolous substrates is an obvious trend or that this substrate preference is mainly due to a bias in sampling woody substrates over herbaceous material from freshwater habitats. In a broad spatial study on distribution patterns of freshwater ascomycetes, Raja et al. (2009b) note that herbaceous substrates mainly occur in lentic habitats such as lakes, swamps and bogs, whereas woody substrates are commonly found in both lentic and lotic habitats. These factors should obviously be considered before any conclusions are drawn about substrate preferences in freshwater Dothideomycetes.

2.4 Morphological adaptations

Freshwater Dothideomycetes appear to have adapted morphologically to aquatic habitats in a variety of ways (Shearer 1993; Goh and Hyde 1996; Shearer 2001; Tsui and Hyde 2003; Jones 2006; Shearer et al. 2007; Shearer et al. 2009). One type of modification involves the presence on ascospores of viscous, gelatinous sheaths that may enable the ascospores to stick onto substrates in moving water. Species of freshwater Dothideomycetes in *Alascospora, Aliquandostipite, Falciformispora, Isthmosporella, Jahnula, Lindgomyces, Lucidascocarpa, Minutisphaera, Natipusilla,* and *Phaeosphaeria* are all equipped with gelatinous sheaths that are sticky and enlarge in water (Fig. 2.1A–F, H–L, O–P). A number of freshwater Dothideomycetes, such as

Minutisphaera fimbriatispora and *Wicklowia aquatica*, have both a gelatinous sheath as well as appendages (Fig. 2.1 H–I, P). Molecular phylogenetic studies of freshwater Dothideomycetes suggest that these ascospore modifications have evolved multiple times in different lineages and therefore are not reliable taxonomic characters (Shearer et al. 2009; Shearer and Raja 2013).

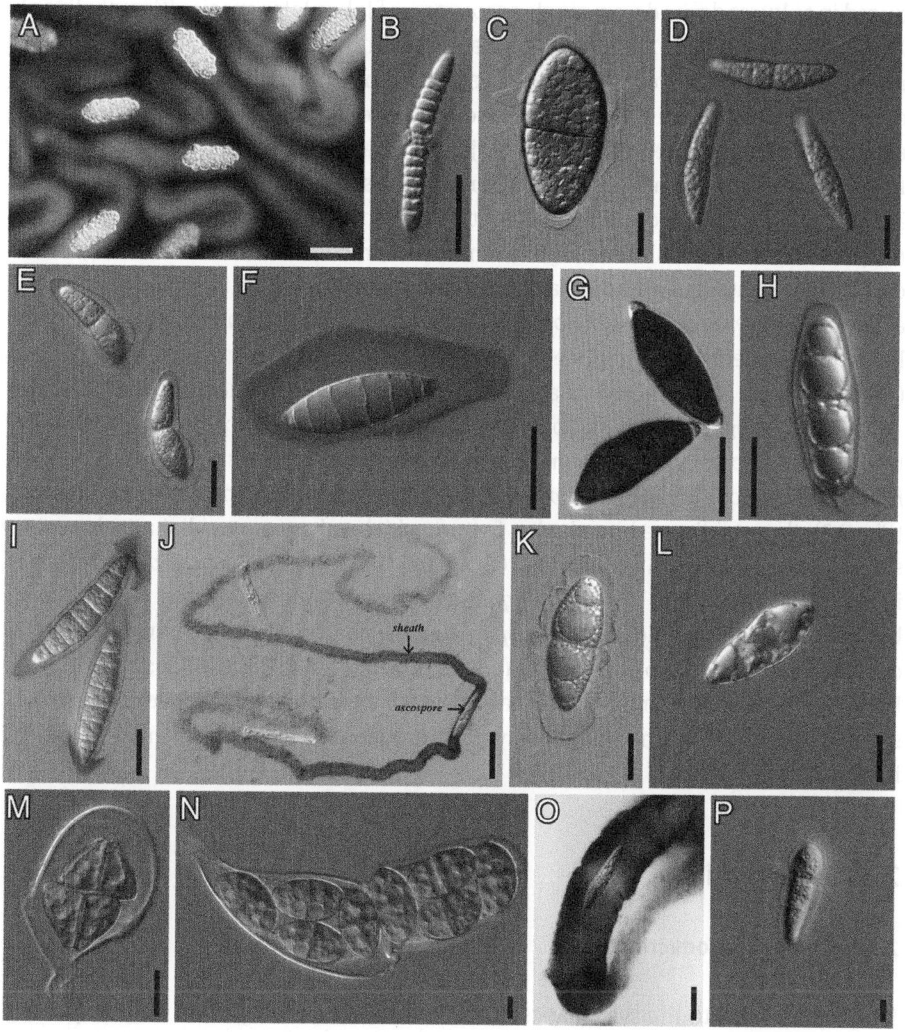

Fig. 2.1: (A–P) A. *Macospora scripicola*. B. *Isthmosporella pulchra*. C. *Jahnula potamophila*. D. *Aliquandostipite crystallinus*. E. *Natipusilla bellaspora*. F. *Lucidascocarpa pulchella*. G. *Jahnula bipileata*. H. *Wicklowia aquatica*. I. *Falciformispora lignatilis*. J. *Phaeosphaeria vilasensis*. K. *Lindgomyces cinctosporus*. L. *Alascospora evergladensis*. M. *Lepidopterella palustris*. N. *Lepidopterella tangerina*. O. *Lindgomyces ingoldianus*. P. *Minutisphaera fimbriatispora*. Scale bars: A–B = 50 µm, C–G, I–M, O = 20 µm, H, N = 10 µm, P = 5 µm.

The cells at the base of the ascomata of all species of Jahnulales produce broad, brown, thick-walled hyphae that spread across the substrate and often connect adjacent ascomata. Initially these broad hyphae were considered to be algal associates (Hyde and Wong 1999), but more recently Pang et al. (2002) considered this unlikely. These thick filaments are connected to and/or develop from peridial cells, and hence, are fungal (Raja and Shearer 2006; Raja et al. 2008). Large brown hyphae also are produced in cultures derived from single ascospores of Jahnulales species (Campbell et al. 2007; Suetrong et al. 2011b). It is hypothesized that these repent, connecting hyphae may play an important role in colonizing and holding fungi onto softened wood in wet or aquatic habitats and thus may represent another adaptive feature for an aquatic existence (Dhanasekaren et al. 2006; Raja and Shearer 2006; Campbell et al. 2007). However, unlike other adaptive characters, such as, those seen is ascospores, which are homoplastic due to convergent evolution, the thick-walled hyphae in members of the Jahnulales is an evolutionarily informative character. All the taxa currently included in the Jahnulales and several mitosporic members within the Jahnulales such as *Brachiosphaera tropicalis*, *Xylomyces chlamydosporus*, and *Speiropsis pedatospora* have thick-walled, broad hyphae on natural substrates as well as in axenic culture (Campbell et al. 2007; Shearer et al. 2009; Sivichai et al. 2011; Suetrong et al. 2011b).

A third category of morphological adaptation to the aquatic environment occurs through the asexual states of some freshwater Dothideomycetes. In the case of *Speiropsis pedatospora* and *Brachiosphaeria tropicalis*, asexual morphs in the Jahnulales (Fig. 2.2), the conidia are branched, a feature which enhances conidial attachment to substrates in flowing water (Ingold 1966; Ingold 1975a, b). *Lemonniera pseudofloscula* in the Pleosporales and *Tumularia aquatica* in the Melanommataceae (Fig. 2.2) also both have branched conidia. *Anguillospora longissima*, a species with long filamentous conidia – also thought to be an adaptation to flowing water (Ingold 1966; Ingold 1975a, b), is aligned with the Amniculicolaceae (Fig. 2.2). It is likely that more aquatic asexual states will be linked to the Dothideomycetes through molecular phylogenetic methods in the future.

2.5 Systematics

2.5.1 General introduction

Early approaches to the systematics of freshwater Dothideomycetes involved placing taxa in existing terrestrial genera with similar morphological characteristics. A classic example of this is *Lindgomyces ingoldianus* (Hirayama et al. 2010). This species was originally placed in the terrestrial genus *Massarina* based on similarities in the ascomata and one-septate ascospores (Shearer and Hyde 1997). With the advent of molecular systematics, freshwater Dothideomycetes were found to be mostly distinct from terrestrial lineages of Dothideomycetes (Schoch et al. 2009; Shearer et al. 2009; Suetrong et al.

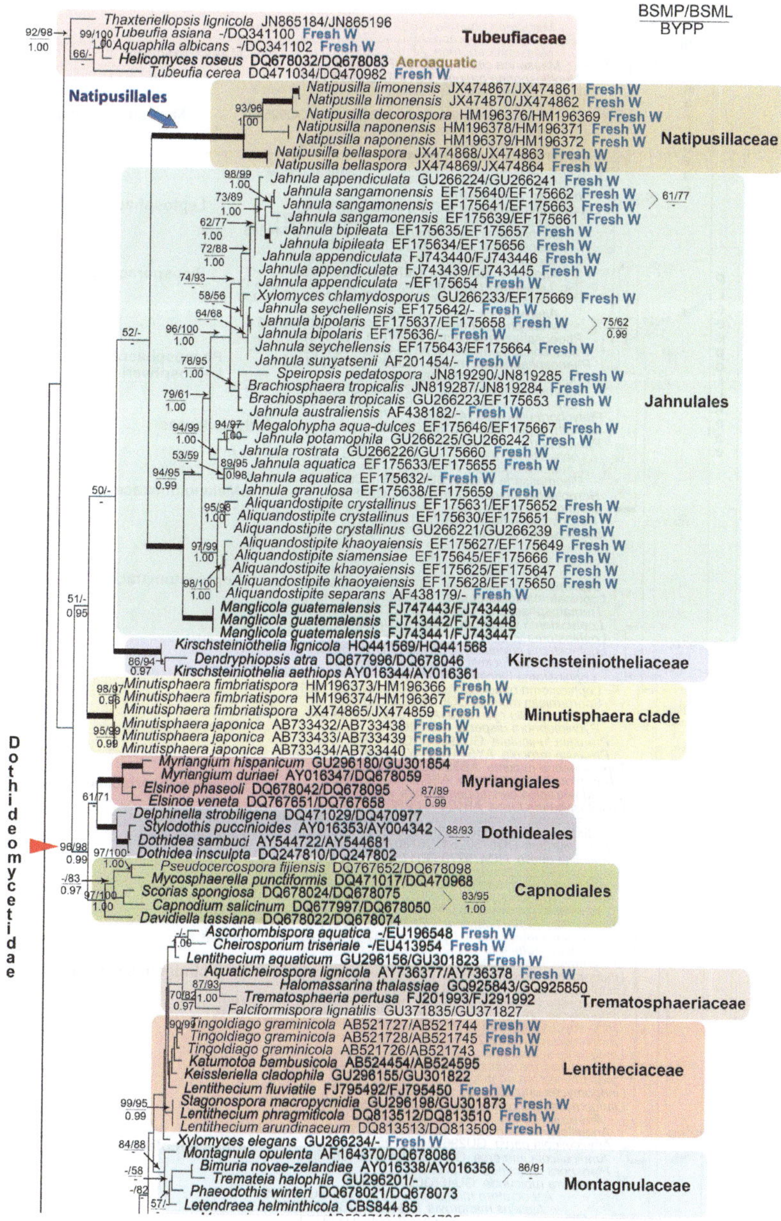

(*Figure continued*)

(*Figure continued*)

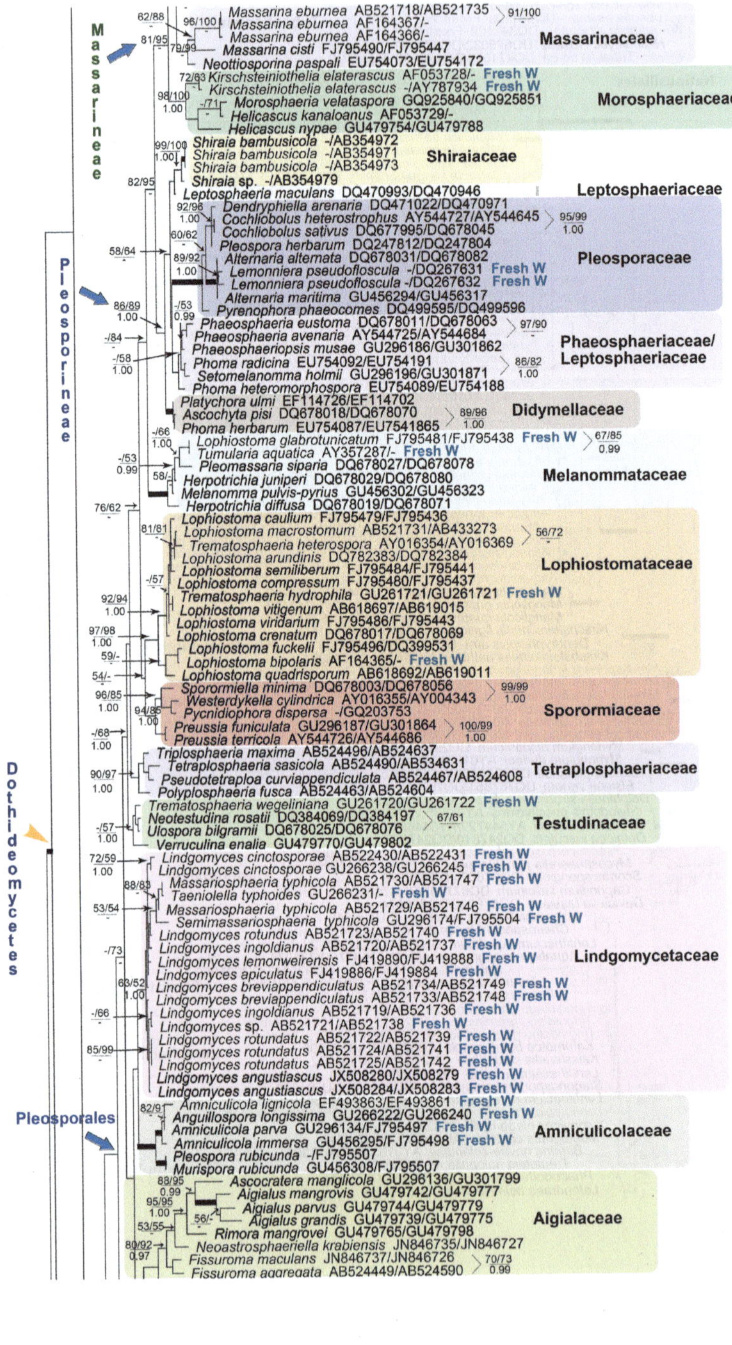

(Figure continued)

Fig. 2.2: One of three MPTs inferred from combined SSU and LSU rDNA sequences of the Dothideomycetes, generated with maximum parsimony (tree length = 5370, C.I. = 0.335, R.I. = 0.803, R.C. = 0.269). Maximum parsimony (BSMP, left) and likelihood (BSML, right) bootstrap values >50% are given above the node. Bayesian posterior probabilities >0.95 are given below each node (BYPP). The internodes that are highly supported by all bootstrap (100%) and posterior probabilities (1.00) are shown as a thicker line.

2011b; Raja et al. 2012; Raja et al. 2013a, b; Hyde et al. 2013); *L. ingoldianus* is now placed in the freshwater family Lindgomycetaceae (Hirayama et al. 2010). Based on molecular and morphological characteristics, the majority of freshwater Dothideomycetes are currently members of the Pleosporales and Jahnulales in the subclass Pleosporomycetidae, while a few taxa are related to the subclass Dothideomycetidae (Mycobank) (Fig. 2.2).

Approximately one third of the freshwater Ascomycota belong to the class Dothideomycetes (~190 species) (Shearer et al. 2009). Based on the relatively small number of taxa of freshwater Dothideomycetes that have been sequenced (18S and/or 28S rRNA genes), they are not a phylogenetically related group. This suggests that they have adapted to freshwater habitats numerous times (Shearer et al. 2009). Four lineages recently have been discovered that consist of freshwater taxa: Jahnulales (Inderbitzin et al. 2001; Pang et al. 2002; Campbell et al. 2007); Lindgomycetaceae (Hirayama et al. 2010; Raja et al. 2011; Raja et al. 2013a); Amniculicolaceae (Zhang et al. 2008; Zhang et al. 2009a); and Natipusillaceae (Raja et al. 2012). More recently a study by Suetrong et al. (2011b) showed that the Jahnulales is comprised of both freshwater species and the marine fungus, *Manglicola guatemalensis* (Suetrong et al. 2010). Taxonomic affiliation for a number of species, however, is still unknown, e.g. *Alascospora evergladensis*, *Ocala scalariformis* and *Wicklowia aquatica*.

In this chapter, we have constructed a new phylogenetic tree showing the major lineages of the freshwater Dothideomycetes based on a combined analysis of the 18S and 28S nrRNA gene sequences retrieved from the NCBI GenBank database. The sequences were aligned with Clustal W V. 1.6 (Thompson et al. 1994). Alignments were further adjusted manually with BioEdit 7.5.0.3 (Hall 2004). The final alignment was again optimized by eye and manually corrected using Se-Al v. 2.0a8 (Rambaut 1996). The phylogenetic analyses of the combined 18S and 28S rDNA dataset were performed using maximum parsimony and Bayesian algorithms as described in Suetrong et al.

(2011b). Maximum parsimony analysis was performed in PAUP* 4.0b10 (Swofford 2002). Maximum likelihood was conducted in RAxML v. 7.2.6 (Stamatakis 2006). A general time reversible (GTR+I+G model) plus invariant sites plus gamma distributed model A tree was obtained by simultaneously running a fast bootstrap search of 1,000 pseudoreplicates followed by a search for the most likely tree under functional setting "a". Representative members of the Geoglossales, namely *Geoglossum difforme* and *Geoglossum simile*, were used as outgroup taxa. Phylogenetic trees were visualized using the program Treeview (Page 1996).

2.5.2 Current phylogenetic placement based on molecular systematics

2.5.2.1 Dothideomycetes-Pleosporomycetidae-Pleosporales

2.5.2.1.1 Amniculicolaceae

Amniculicola was described from submerged wood in France and the 28S nrRNA gene phylogeny referred it to a well-supported group with *Anguillospora longissima*, *Repetophragma ontariense* and *Spirosphaera cupreorufensense* (Zhang et al. 2008). Zhang et al. (2009a) established the Amniculicolaceae to accommodate three genera of freshwater Dothideomycetes: *Amniculicola*, *Murispora* and *Neomassariosphaeria* based on five loci (18S and 28S nrDNA, TEF1, RPB1, RPB2). Although all species grow and produce a purple pigment on wood in water, they differ in morphology. Asci of *Amniculicola* are cylindrical while those of *Murispora* and *Neomassariosphaeria* are clavate. Also ascospores vary in color (hyaline in *Amniculicola* and *N. grandispora*, brown in *N. typhicola* and *Murispora*) and septation (1-septate in *Amniculicola*, multiseptate in *Neomassariosphaeria*, muriform in *Murispora*) (Zhang et al. 2009a).

2.5.2.1.2 Halotthiaceae

Phaeoseptum, typified by *P. aquaticum*, is a recently described freshwater fungus from submerged wood (Zhang et al. 2013). Morphologically, it is similar to the mangrove fungus *Mauritiana rhizophorae* in having a pseudoclypeus, narrowly cellular pseudoparaphyses, eight-spored cylindrical to broadly clavate asci and pale brown ascospores with thickened septa. Phylogenetically, *P. aquaticum* formed a moderately supported group with *M. rhizophorae*, *Halotthia posidoniae* and *Pontoporeia biturbinata*, that was a sister group to the Sporormiaceae but with weak support. A new family, Halotthiaceae, which consists of members from both freshwater and marine habitats was established to accommodate these taxa (Zhang et al. 2013).

2.5.2.1.3 Lentitheciaceae

Zhang et al. (2009c) established the Lentitheciaceae to include the genera *Lentithecium*, *Katumotoa* and *Keissleriella* based on a phylogeny of five loci (18S and 28S nrDNA, TEF1, RPB1, RPB2). Two freshwater *Lentithecium* species, *L. arundinaceum* and *L. aquaticum*, were included. A third freshwater species, *Keissleriella linearis*, grouped with *L. aquaticum* (type species) and subsequently was transferred to *Lentithecium* (Zhang et al. 2009c). *Lentithecium*, however, is not a monophyletic genus. *Tingoldiago* (type species *T. graminicola*) was created to accommodate collections of *Massarina ingoldiana* having ascomata with a flattened base, clavate asci and occurrence on gramineous substrates (Hirayama et al. 2010). Three isolates of *T. graminicola* formed a monophyletic clade and grouped within the Lentitheciaceae (Shearer et al. 2009). *Xylomyces elegans* groups with the Lentitheciaceae but is not supported (Fig. 2.2).

2.5.2.1.4 Lindgomycetaceae

Massarina ingoldiana was found to be a complex of morphologically similar but phylogenetically unrelated species (Hirayama et al. 2010). Consequently in the same study, Hirayama et al. (2010) introduced a new genus, *Lindgomyces*, to accommodate two closely related species, *L. ingoldianus* and *L. rotundatus*, in the new family Lindgomycetaceae. Subsequently several new *Lindgomyces* species have been described: *L. cinctosporae*, *L. apiculatus*, *L. lemonweirensis* and *L. angustiascus*, while *L. breviappendiculatus* was described as a new combination from *Lophiostoma breviappendiculatum* (Raja et al. 2011; Raja et al. 2013a). All seven *Lindgomyces* species constitute a well-supported monophyletic group (Fig. 2.2) (Raja et al. 2013a). Since, the paper by Raja et al. (2013a), a new species, *L. greiosporus*, has been described from submerged wood in France and Spain (Zhang et al. 2014).

2.5.2.1.5 Lophiostomataceae

Lophiostoma is a polyphyletic genus and has been referred to two phylogenetic groups, Lophiostomataceae and Melanommataceae (Shearer et al. 2009; Zhang et al. 2009c) (Fig. 2.2). The freshwater *L. arundinis* and *L. bipolaris* grouped within *Lophiostoma sensu stricto* with the type species of the genus, *L. macrostomum*. *Lophiostoma glabrotunicatum* formed a second *Lophiostoma* group, sister to the Melanommataceae. *Tumularia aquatica*, an aquatic hyphomycete, also formed part of this group along with *Byssothecium circinans* (Zhang et al. 2009c; Shearer et al. 2009). *Byssothecium circinans* was not included in our analyses. Schoch et al. (2009) found that *Quintaria submersa* was phylogentically related to the Lindgomycetaceae rather than the Lophiostomataceae, but this relationship was poorly supported. Liew et al. (2002) found that *Vaginatispora aquatica* formed a well-supported clade with *Massarina armatispora* using the ITS rRNA gene regions. Only the ITS sequence is available for *V. aquatica* and other gene regions are required to confirm the placement of this species in the family. *Massarina australiensis* and *M. corticola* also might be related to this group (Shearer et al. 2009; Zhang et al. 2009c).

2.5.2.1.6 Massarinaceae

The freshwater asexual morph *Aquaticheirospora lignicola* is related to *Massarina eburnea* based on a phylogeny of the nrRNA genes (Kodsueb et al. 2007). Another asexual fungus, *Cheirosporium triseriale,* might also be related to the Massarinaceae but its relationship was not supported (Shearer et al. 2009). Many freshwater *Massarina* species still remain to be sequenced and further taxonomic and systematic work is required to evaluate their placement in this family.

2.5.2.1.7 Melanommataceae

Several isolates of *Byssosphaeria schiedermayriana*, the type species of the genus, grouped well within the Melanommataceae clade, confirming its placement (Mugambi and Huhndorf 2009). *Astrosphaeriella* is unrelated to the Melanommataceae (Tanaka et al. 2009) and polyphyletic, splitting into several clades related to the Aigialaceae (Liu et al. 2011). *Astrosphaeriella stellata*, the only freshwater *Astrosphaeriella* sequenced, grouped with *Delitschia didyma* and *D. winteri* (Liu et al. 2011).

2.5.2.1.8 Montagnulaceae

Paraphaeosphaeria is a polyphyletic genus (Câmara et al. 2001) but the type species, *P. michotii*, which can be found in terrestrial and freshwater habitats, has been confirmed to occur within the Montagnulaceae with robust support (Zhang et al. 2009b). *Paraphaeosphaeria schoenoplecti* is morphologically similar to *P. michotii* in having bi-septate ascospores with a sheath (Wong et al. 2000). At present, no sequence data are available for *P. schoenoplecti*.

2.5.2.1.9 Morosphaeriaceae

This family was created to accommodate several marine and freshwater taxa previously described in the *Massarinaceae*: *Morosphaeria ramunculicola*, *M. velatispora*, *Helicascus kanaloanus*, *H. nypae* and *Kirschsteiniothelia elaterascus* (Suetrong et al. 2009). Due to the distant phylogenetic relationship between *K. elaterascus* and *K. aethiops* (type species), *K. elaterascus* was transferred to *Morosphaeria* (Boonmee et al. 2012).

2.5.2.1.10 Phaeosphaeriaceae

One hundred and ninety-nine names are referred to the genus *Phaeosphaeria* (Index Fungorum). There are 17 freshwater species but some of these also occur in the terrestrial environment (Shearer et al. 2009). *Phaeosphaeria* is morphologically closely related to *Leptosphaeria* and can be separated by ascomal wall structures (Khashnobish and Shearer 1996). Some freshwater *Phaeosphaeria* species cluster well within the family, although scattered in different clades: *P. albopunctata*, *P. alpina*, *P. caricinella*, *P. caricis*,

P. culmorum, *P. eustoma*, *P. halima*, *P. microscopica*, *P. typharum* and *Ophiosphaerella herpotricha* (*sub P. herpotricha*) (Câmara et al. 2002; Schoch et al. 2009; Quaedvlieg et al. 2013). In our tree, three species, *P. eustoma*, *P. avenaria* and *Phaeosphaeriopsis musae* clustered in a single clade (Fig. 2.2).

2.5.2.1.11 Pleosporaceae

Decorospora gaudefroyi is an aquatic species that can be isolated from both freshwater and brackish habitats (Jones et al. 2009; Schoch et al. 2009). Originally described as a *Pleospora* species, *D. gaudefroyi* is phylogenetically unrelated to the type species of *Pleospora*, *P. herbarum*, and it differs in having ascospores enclosed in a sheath with four to five apical extensions (Inderbitzin et al. 2002). *Nymbya scirpicola* (*Macrospora scirpicola*) is morphologically similar to *Pyrenophora* and *Pleospora* and confirmed to be a member of the Pleosporaceae with good support (Woudenberg et al. 2013). Campbell et al. (2006) investigated the phylogeny of several *Lemonniera* species and found that the freshwater species *L. pseudofloscula* grouped within the Dothideomycetes and was later referred to the Pleosporaceae (Shearer et al. 2009). Currently, no sequence data are available for *Clathrospora tirolensis* and several freshwater *Pleospora* species.

2.5.2.1.12 Sporormiaceae

Sporormiella minima and *S. minimoides* belong to the Sporormiaceae with their thick-walled, dark-colored ascospores, which are deeply constricted and may break into part-spores (Barr 2000). Phylogenetically, both species were placed within the *Sporormiella* clade with moderate support based on the 28S and ITS nrRNA, mitochondrial SSU and β-tubulin genes (Kruys and Wedin 2009).

2.5.2.1.13 Trematosphaeriaceae

Falciformispora lignatilis occurs in both freshwater and marine habitats (Raja and Shearer 2008; Jones et al. 2009). Suetrong et al. (2009) sequenced two marine isolates of *F. lignatilis* and they grouped with *Halomassarina thalassiae* and *Trematosphaeria pertusa* in the Trematosphaeriaceae (Suetrong et al. 2011c). *Trematosphaeria* is not a monophyletic genus and the freshwater species, *T. hydrela*, might be related to *T. pertusa* and well placed in this family (Wang et al. 2007). Both *T. biappendiculata* and *T. hydrophila* are phylogenetically unrelated to *T. pertusa* and may belong to the Lophiostomaceae (Shearer et al. 2009; Zhou et al. 2013). Ascospores of *T. biappendiculata* are constricted at all septa, yellowish to olivaceous-brown colored and with appendages, in contrast to those of other *Trematosphaeria* spp. which are constricted at the primary septum only, reddish brown colored and lacking appendages (Tanaka et al. 2005).

2.5.2.1.14 Tubeufiaceae

The Tubeufiaceae is distinctive in the Pleosporales by producing superficial, white, pale to bright ascomata which may darken at maturity (Kodsueb et al. 2006). Based on a sequence analysis of the nrRNA genes (SSU, ITS, LSU), the freshwater species *Acanthostigma scopulum*, *Tubeufia asiana*, *T. cerea*, *T. cylindrothecia* and *T. paludosa* were confirmed to belong in the Tubeufiaceae, but both genera are not monophyletic (Kodsueb et al. 2006; Tsui et al. 2007; Shearer et al. 2009; Promputtha and Miller 2010). Kodsueb et al. (2006) included a sequence of *Boerlagiomyces websteri* in their analysis and it did not group within the Tubeufiaceae. New collections of *B. websteri* may help to verify its phylogenetic placement. This family is now referred to the order Tubeufiales (Boonmee et al. unpublished data).

2.5.2.2 Pleosporales *incertae sedis*

2.5.2.2.1 *Ascominuta* spp.

This genus is characterized by minute ascomata, globose to subglobose, bitunicate asci, one-septate ascospores with a gelatinous sheath. This genus may be phylogenetically related to members of Natipusillaceae. However, there is currently no sequence data available for *Ascominuta* spp. (Ranghoo and Hyde 2000; Hu et al. 2010).

2.5.2.2.2 *Ascorhombispora aquatica* (Cai and Hyde 2007)

Ascorhombispora aquatica was described from submerged bamboo in southern China and is characterized by dark-colored, coriaceous ascomata, saccate asci and dark-colored, rhomboid ascospores with hyaline end cells (Cai and Hyde 2007). In terms of the superficial ascomata and dark-colored ascospores, *A. aquatica* is similar to *Caryospora*, but 18S and 28S nrRNA gene phylogeny suggested that they are not related to *A. aquatica*, falling close to *Dictyosporium* spp. and *Digitodesmium bambusicola* (Cai and Hyde 2007). In another analysis, *A. aquatica* grouped with several species of different familial affiliations: *Montagnula opulenta*, *Phaeodothis winteri* and *Bimuria novae-zelandiae* (Shearer et al. 2009).

2.5.2.2.3 *Alascospora evergladensis* (Raja et al. 2010b)

Alascospora evergladensis is a freshwater ascomycete collected from submerged dead petioles of *Nymphaea odorata* in the Florida Everglades (Raja et al. 2010b). This fungus is unique among genera in the Pleosporales based on a combination of morphological characters that include light brown, translucent, membranous, ostiolate ascomata with dark, amorphous material irregularly deposited on the peridium, especially around the ostiole; globose, fissitunicate, thick-walled asci; septate pseudoparaphyses; and 1-septate ascospores that are hyaline when young and dark upon aging and surrounded by a hyaline gelatinous sheath that is wing-shaped in outline on each side of the ascospore (Fig. 2.1L). Based on morphology, this taxon is similar to species in *Minutisphaera* (Raja et al. 2013b); however, molecular data are necessary to clarify its phylogenetic placement.

2.5.2.2.4 *Boerlagiomyces* spp. (Shearer and Crane 1995; Stanley and Hyde 1997; Crane et al. 1998)

Although, placed in the Tubeufiaceae based on morphology, *Boerlagiomyces* taxa have never been sequenced to determine if they share phylogenetic affinities within the Dothideomycetes. See Tubeufiaceae.

2.5.2.2.5 *Byssothecium flumineum* (Crane et al. 1992)

Based on an unpublished 18S and 28S rDNA phylogeny (C.A. Shearer, unpublished), *B. flumineum* does not group with *B. circinans* (type species), but shows phylogenetic affinities with members of the Lentitheciaceae.

2.5.2.2.6 *Isthmosporella pulchra* (Shearer and Crane 1999)

This species thus far is only known from its type location (from submerged wood in a lake in New York). The fungus does not grow well in culture and has not been sequenced (C.A. Shearer and H. Raja, unpublished).

2.5.2.2.7 *Lepidopterella* spp. (Shearer and Crane 1980; Raja and Shearer 2008)

The type species of *Leptdopterella, L. palustris* is assigned to the family Argynnaceae on the basis of morphological characteristics such as butterfuly shaped ascospores (Shearer and Crane 1980). Based on phylogenetic analysis using rRNA genes (18S and 28S), *L. palustris* is basal to the Mytilinidiales; however, there was no strong nodal support to determine its phylogenetic affinities accurately. The second species, *L. tangerina*, is known only from its type locality thus far and has not been sequenced (Raja and Shearer 2008).

2.5.2.2.8 *Lucidascocarpa pulchella* (Ferrer et al. 2008)

Known only from the type locality and the Peruvian Amazon (C. A. Shearer, unpublished), this species has not yet been successfully sequenced.

2.5.2.2.9 *Ocala scalariformis* (Raja et al. 2009a)

Known only from the type locality, Shearer et al. (2009) showed it was basal to the Janhulales, but without significant nodal support.

2.5.2.2.10 *Wicklowia aquatica* (Raja et al. 2010)

Wicklowia is a recently described genus with unique morphology: dorsiventrally flattened ascomata and an ascospore sheath which is attached at the ascospore base, with a gelatinous curtain extending downward from the base that fragments into filamentous appendages forming a sub-apical fringe (Fig. 2.1H) (Raja et al. 2010a). Morphologically, *Wicklowia* fits well in the Pleosporales and the 28S nrRNA gene phylogeny of its only species, *W. aquatica*, supported this conclusion (Raja et al. 2010a). However, this species did not cluster with any known families in the Pleosporales. In Fig. 2.2, *W. aquatica* groups with the Aigialaceae but without any statistical support.

2.5.3 Zopfiaceae, Dothideomycetes, family *incertae sedis*

Three *Caryospora* species have been described from freshwater environments: *C. callicarpa*, *C. minima* and *C. obclavata* (Shearer et al. 2009). *Caryospora minima* is the only species sequenced in the genus and belongs to the Pleosporales but its relationship with the type species, *C. putaminum*, and other taxa in the order is unknown (Cai and Hyde 2007).

2.5.4 Dothideomycetes *incertae sedis*

2.5.4.1 Jahnulales

Jahnulales is the largest and most distinctive order of freshwater Dothideomycetes. Species of the Jahnulales have the following morphological traits in common: 1) sessile and/or stalked ascomata, 2) peridium of large cells, 3) broad hyphae up to 50 μm wide, 4) ascospores with a gelatinous sheath and/or appendages and 5) dimorphic asci and ascospores. *Jahnula* was tentatively placed in the Dothideales *incertae sedis* despite its unique morphological characteristics (Hawksworth 1984). Inderbitzin et al. (2001) established the family Aliquandostipitaceae to accommodate two new species of *Aliquandostipite*, unaware of the genus *Jahnula*. Pang et al. (2002) established the order Jahnulales to accommodate *Aliquandostipite* and *Jahnula* based on their significant morphological differences from the Pleosporales and sequence phylogeny of the nuclear 18S rRNA gene. Pang et al. (2002) also transferred *Aliquandostipite sunyatsunii* to *Jahnula* and described a new genus, *Patescospora*. Campbell et al. (2007) reexamined the phylogenetic relationships among species of *Jahnula*, *Aliquandostipite*, *Patescospora* and *Megalohypha* using both the 18S and 28S nrRNA genes. They transferred *Jahnula siamensiae* and *Patescospora separans* to *Aliquandostipite* and also referred the asexual morphs *Brachiosphaera tropicalis* and *Xylomyces chlamydosporus* to the order. Currently, there are 42 species in seven genera in the Jahnulales (Suetrong et al. 2011b). *Ocala scalariformis* formed an unsupported branch sister to the rest of the Jahnulales (Shearer et al. 2009), however, it differs from the Jahnulales in having multi-septate ascospores with bipolar gelatinous appendages (Raja et al. 2009a).

2.5.4.2 Natipusillales

The Natipusillaceae includes the freshwater genus *Natipusilla* with four species (Ferrer et al. 2011; Raja et al. 2012). Species of *Natipusilla* have small ascomata, none to few pseudoparaphyses, globose asci and mostly one-septate ascospores. Species in the genus differ in the presence and/or type of sheath or appendage (Ferrer et al. 2011). Phylogenetic analysis of the nrRNA genes of *N. decorospora*, *N. limonensis* and *N. naponensis* suggested that they formed a long branch and may be related to taxa of the Microthyriaceae but this relationship was not supported (Ferrer et al. 2011; Raja et al. 2012). A new order Natipusillales has been established to include members of the freshwater ascomycetes belonging to Natipusillaceae based on 18S and 28S rDNA sequences (Hyde et al. 2013).

2.5.4.3 *Minutisphaera* clade

Minutisphaera was established to accommodate *M. fimbriatispora*, a fungus with dark brown hyphae on the upper part of the ascomata around the ostiole; subglobose, three-cell layered, ostiolate ascomata; septate pseudoparaphyses; fissitunicate clavate to obclavate asci; and one-septate ascospores surrounded by a gelatinous sheath (Ferrer et al. 2011). The description was later emended to include taxa with apothecioid ascomata. *Minutisphaera* appears to be morphologically similar to members of Patellariales; however, ongoing molecular studies suggest that it does not share phylogenetic affinities with *Hysteropatella clavispora*, *H. elliptica* and *Patellaria atrata* (members of Patellariales) (H. Raja et al., unpublished). Therefore, a new family will be established for this unique freshwater ascomycete clade (H. Raja et al., unpublished).

2.5.4.4 Freshwater asexual morphs with affinities to Dothideomycetes

Different ecological groups of freshwater hyphomycetes (asexual morphs) occur in freshwater. These include aquatic hyphomycetes, aeroaquatic hyphomycetes, miscellaneous mitosporic ascomycetes, and coelomycetes (Goh and Hyde 1996; Shearer et al. 2007; See Chapter 5). Many taxa within these ecological groups have phylogenetic affinities with freshwater Dothideomycetes and are nested within newly established freshwater orders and families. These include families such as Aliquandostipitaceae, Aminiculicolaceae, Lindgomycetaceae, Lophiostomataceae, Massarinaceae, Pleosporaceae, and Tubeufiaceae (Shearer et al. 2009). Future systematic studies of freshwater asexual morphs using molecular sequence data will continue to shed light on their phylogenetic relationships within freshwater Dothideomycetes lineages.

2.6 Conclusions

A number of conclusions can be drawn from our molecular sequence analyses of freshwater Dothideomycetes. The first is that freshwater taxa are broadly dispersed on the Dothideomycetes tree and likely invaded fresh water from a variety of origins. Second, as of this writing, numerous clades are strictly aquatic and additional terrestrial and aquatic taxa need to be sequenced in order to determine their patterns of evolution within the Dothideomycetes. Third, there are reliable morphological characters that can be used to define aquatic clades identified by molecular sequence data. Fourth, there is little overlap between the freshwater and marine Dothideomycetes and, overall, fewer marine than freshwater Dothideomycetes. This suggests that sea water has been a greater barrier to the evolution of Dothideomycetes than fresh water. Fifth, given the numerous taxa that are currently *incertae sedis* due to lack of molecular data, there is still much work to be done with respect to isolating and sequencing taxa.

Acknowledgments

Shearer gratefully acknowledges financial support from the National Science Foundation (NSF Grant No. 08-44722) for studies of the freshwater Ascomycetes. Any opinions, findings, and conclusions or recommendations expressed in this publication are those of the author(s) and do not necessarily reflect the views of the National Science Foundation.

References

Abdel-Aziz FA, Abdel-Wahab MA. *Lolia aquatica* gen. *et* sp. nov. (Lindgomycetaceae, Pleosporales), a new coelomycete from freshwater habitats in Egypt. Mycotaxon 2010;114:33–42.

Arnold AE, Lutzoni F. Diversity and host range of foliar fungal endophytes: Are tropical leaves biodiversity hotspots? Ecology 2007;88:541–549.

Barr ME. Notes on coprophilous bitinicate Ascomycetes. Mycotaxon 2000;76:105–112.

Boonmee S, Ko TWK, Chukeatirote E, Hyde KD, Chen H, Cai L, et al. Two new *Kirschsteiniothelia* species with *Dendryphiopsis* anamorphs cluster in Kirschsteiniotheliaceae fam. nov. Mycologia 2012;104:698–714.

Cai L, Hyde KD. *Ascorhombispora aquatica* gen. *et* sp. nov. from a freshwater habitat in China, and its phylogenetic placement based on molecular data. Crypt Mycol 2007;28:291–300.

Câmara MPS, Palm ME, van Berkum P, O'Niell NR. Molecular phylogeny of *Leptosphaeria* and *Phaeosphaeria*. Mycologia 2002;94:630–640.

Câmara MPS, Palm ME, van Berkum P, Stewart EL. Systematics of *Paraphaeosphaeria*, a molecular and morphological approach. Mycol Res 2001;105:41–50.
Campbell J, Ferrer A, Raja HA, Sivichai S, Shearer CA. Phylogenetic relationships among taxa in the Jahnulales inferred from 18S and 28S nuclear ribosomal DNA sequences. Can J Bot 2007; 85:873–882.
Campbell J, Shearer C, Marvanová L. Evolutionary relationships among aquatic anamorphs and teleomorphs: *Lemonniera, Margaritispora*, and *Goniopila*. Mycol Res 2006;110:1025–1033.
Crane JL, Shearer CA, Barr ME. A revision of *Boerlagiomyces* with notes and a key to the saprobic genera of Tubeufiaceae. Can J Bot 1998;76:602–612.
Crane JL, Shearer CA, Huhndorf SM. A new species of *Byssothecium* (Loculoascomycetes) from wood in fresh water. Mycologia 1992;84:235–240.
Dhanasekaren V, Jeewon R, Hyde KD. Molecular taxonomy, origins and evolution of freshwater ascomycetes. Fungal Divers 2006;23:351–390.
Dudka IO. Data on the flora of aquatic fungi of the Ukrainian SSR. II. Aquatic hyphomycetes of Kiev Polessye. Ukr Bot Z 1963;20:86–93.
Eaton RA. Fungi growing on wood in water cooling towers. Int Biod Bull 1972;8:39–48.
Eaton RA, Jones EBG. New fungi on timber from water-cooling towers. Nova Hedwig 1970;19:779–786.
Eaton RA, Jones EBG. The biodeterioration of timber in water cooling towers I. Fungal ecology and decay of wood at Connah's Quay and Ince. Mat Organ 1971a;6:51–80.
Eaton RA, Jones EBG. The biodeterioration of timber in water cooling towers II. Fungi growing on wood in different positions in a water cooling system. Mat Organ 1971b;6:81–92.
Ferrer A, Miller AN, Shearer CA. *Minutisphaera and Natipusilla*: two new genera of freshwater Dothideomycetes. Mycologia 2011;103:411–423.
Ferrer A, Raja HA, Shearer CA. *Lucidascocarpa pulchella*, a new ascomycete genus and species from freshwater habitats in the American tropics. Mycologia 2008;100:642–646.
Goh TK, Hyde KD. Biodiversity of freshwater fungi. J Ind Microbiol Biotech 1996;17:328–345.
Hall T. Bioedit version 6.0.7. Department of Microbiology, North Carolina State University; 2004.
Hawksworth DL. Observations on *Jahnula* Kirschst., a remarkable aquatic pyrenomycete. Sydowia 1984;37:43–46.
Hirayama K, Tanaka K, Raja HA, Miller AN, Shearer CA. A molecular phylogenetic assessment of *Massarina ingoldiana sensu lato*. Mycologia 2010;102:729–746.
Hu DM, Cai L, Chen H, Bahkali AH, Hyde KD. Four new freshwater fungi associated with submerged wood from Southwest Asia. Sydowia 2010;62:191–203.
Hyde KD, Wong SW. Tropical Australian Freshwater Fungi. XV. The ascomycete genus *Jahnula*, with five new species and one new combination. Nova Hedwig 1999;68:489–509.
Hyde KD, Jones, EBG, Liu J-K, Ariyawansa H, Boehm E, Boonmee S, et al. Families of Dothideo-mycetes. Fungal Divers 2013;63:1–313.
Inderbitzin P, Kohlmeyer J, Kohlmeyer B, Berbee ML. *Decorospora*, a new genus for the marine ascomycete *Pleospora gaudefroyi*. Mycologia 2002;94:651–659.
Inderbitzin P, Landvik S, Abdel-Wahab MA, Berbee ML. Aliquandostipitaceae, a new family for two new tropical ascomycetes with unusually wide hyphae and dimorphic ascomata. Am J Bot 2001;88:52–61.
Ingold CT. Aquatic ascomycetes: Further species from the English Lake District. Trans Br Mycol Soc 1955;38:157–168.
Ingold CT. The tetraradiate aquatic fungal spore. Mycologia 1966;58:43–56.
Ingold CT. Convergent Evolution in aquatic fungi - tetraradiate spore. Biol J Linn Soc Lon 1975a;7:1–25.
Ingold CT. An Illustrated guide to the Aquatic and Water-borne Hyphomycetes (Fungi imperfecti) with notes on their Biology. 1975b; Freshwater Biological Association, Scientific Publication No.30, Ferry House, Ambleside, Cumbria, UK.

Jones EBG. Form and function of fungal spore appendages. Mycoscience 2006;47:167–183.

Jones EBG, Sakayaroj J, Suetrong S, Somrithipol S, Pang KL. Classification of marine Ascomycota, anamorphic taxa and Basidiomycota. Fungal Divers 2009;35:1–187.

Khashnobish A, Shearer CA. Phylogentic relationships in some *Leptosphaeria* and *Phaeosphaeria* species. Mycol Res 1996;1000:1355–1363.

Kodsueb R, Jeewon R, Dhanasekaren V, McKenzie EH, Lumyong P, Lumyong S, et al. Systematic revision of Tubeufiaceae based on morphological and molecular data. Fungal Divers 2006;21:105–130.

Kodsueb R, Lumyong S, Ho WH, Hyde KD, Mckenzie EHC, Jeewon R. Morphological and molecular characterization of *Aquaticheirospora* and phylogenetics of Massarinaceae (Pleosporales). Bot J Linn Soc 2007;155:283–296.

Kruys Å, Wedin M. Phylogenetic relationships and an assessment of traditionally used taxonomic characters in the Sporormiaceae (Pleosporales, Dothideomycetes, Ascomycota), utilising multi-gene phylogenies. Syst Biodiv 2009;7:465–478.

Liew ECY, Aptroot A, Hyde KD. An evaluation of the monophyly of *Massarina* based on ribosomal DNA sequences. Mycologia 2002;94:803–813.

Liu J-K, Phookamsak R, Jones EBG, Zhang Y, Ko TWK, Hu H-L, et al. *Astrosphaeriella* is polyphyletic, with species in *Fissuroma* gen. nov., and *Neoastrosphaeriella* gen. nov. Fungal Divers 2011;51:135–154.

Mugambi GK, Huhndorf SM. Molecular phylogenetics of Pleosporales: Melanommataceae and Lophiostomataceae re-circumscribed (Pleosporomycetidae, Dothideomycetes, Ascomycota). Stud Mycol 2009;64:103–121.

Page RDM. Treeview: An application to display phylogenetic trees on personal computers. Comput Appl Biosci 1996;12:357–358.

Pang KL, Abdel-Wahab MA, Sivichai S, El-Sharouney HM, Jones EBG. Jahnulales (Dothideomycetes, Ascomycota): A new order of lignicolous freshwater ascomycetes. Mycol Res 2002;106:1031–1042.

Petrak F. Beiträge zur Pilzflora Sdost-Galiziens und der Zentralkarpaten. Hedwigia 1925;4:64.

Promputtha I, Miller AN. Three new species of *Acanthostigma* (Tubeufiaceae, Dothideomycetes) from Great Smoky Mountains National Park. Mycologia 2010;102:574–587.

Pugh GJF, Mulder JL (1971) Mycoflora associated with *Typha latifolia*. Trans Br Mycol Soc 2010;57:273–282.

Quaedvlieg W, Verkley GJM, Shin HD, Barretto RW, Alfenas AC, Swart WJ, et al. Sizing up *Septoria*. Stud Mycol 2013;75:307–390.

Raja HA, Shearer CA. *Jahnula* species from North and Central America, including three new species. Mycologia 2006;98:319–332.

Raja HA, Carter A, Platt HW, Shearer CA. Freshwater ascomycetes: *Jahnula apiospora* (Jahnulales, Dothideomycetes), a new species from Prince Edward Island, Canada. Mycoscience 2008;49:326–328.

Raja HA, Shearer CA. Freshwater ascomycetes: new and noteworthy species from aquatic habitats in Florida. Mycologia 2008;100:677–682.

Raja HA, Ferrer A, Shearer CA. Freshwater ascomycetes: A new genus, *Ocala scalariformis* gen. *et* sp. nov, and two new species, *Ayria nubispora* sp. nov. and *Rivulicola cygnea* sp. nov. Fungal Divers 2009a;34:79–86.

Raja HA, Schmit JP, Shearer CA. Latitudinal, habitat and substrate distribution patterns of freshwater ascomycetes in the Florida Peninsula. Biodivers Conserv 2009b;18:419–455.

Raja HA, Ferrer A, Shearer CA, Miller AN. Freshwater ascomycetes: *Wicklowia aquatica*, a new genus and species in the Pleosporales from Florida and Costa Rica. Mycoscience 2010a;51:208–214.

Raja HA, Violi HA, Shearer CA. Freshwater ascomycetes: *Alascospora evergladensis*, a new genus and species from the Florida Everglades. Mycologia 2010b;102:33–38.

Raja HA, Tanaka K, Hirayama K, Miller AN, Shearer CA. Freshwater ascomycetes: two new species of *Lindgomyces* (Lindgomycetaceae, Pleosporales, Dothideomycetes) from Japan and USA. Mycologia 2011;103:1421–1432.

Raja HA, Miller AN, Shearer CA. Freshwater ascomycetes: Natipusillaceae, a new family of tropical fungi, including *Natipusilla bellaspora* sp. nov. from the Peruvian Amazon. Mycologia 2012;104:569–573.

Raja HA, Oberlies NH, El-Elimat T, Miller AN, Zelski SE, Shearer CA. *Lindgomyces angustiascus,* (Lindgomycetaceae, Pleosporales, Dothideomycetes), a new lignicolous species from freshwater habitats in the USA. Mycoscience 2013a;54:353–361.

Raja HA, Oberlies NH, Figueroa M, Tanaka K, Hirayama K, Hashimoto A, et al. Freshwater ascomycetes: *Minutisphaera* (Dothideomycetes) revisited, including one new species from Japan. Mycologia 2013b;105:959–976.

Rambaut A. Sequence Alignment Editor. Version 2.0. Department of Zoology, University of Oxford, Oxford 1996.

Ranghoo VM, Hyde KD. *Ascominuta lignicola,* a new loculoascomycete from submerged wood in Hong Kong. Mycoscience 2000;41:1–5.

Schoch CL, Crous PW, Groenewald JZ, Boehm EWA, Burgess TI, de Gruyter J, et al. A class-wide phylogenetic assessment of Dothideomycetes. Stud Mycol 2009;64:1–15.

Shearer CA. Fungi of the Chesapeake Bay and its tributaries. III. The distribution of wood-inhabiting ascomycetes and fungi imperfecti of the Patuxent River. Am J Bot 1972;59:961–969.

Shearer CA. The freshwater ascomycetes. Nova Hedwig 1993;56:1–33.

Shearer CA. The distribution of freshwater filamentous Ascomycetes. In: Mishra JK, Horn BW, eds. Trichomycetes and other fungal groups, vol Robert W. Lichtwardt Commemoration. Plymouth, UK: Science Publishers, Inc. pp 2001;225–292.

Shearer CA, Crane JL. Fungi of the Chesapeake Bay and its tributaries. I. Patuxent River. Mycologia 1971;63:237–260.

Shearer CA, Crane JL. Taxonomy of two cleistothecial ascomycetes with papilionaceous ascospores. Trans Br Mycol Soc 1980;75:193–200.

Shearer CA, Crane JL. *Boerlagiomyces websteri,* a new ascomycete from fresh water. Mycologia 1995;87:876–879.

Shearer CA, Crane JL. Freshwater ascomycetes: *Isthmosporella pulchra* gen. and sp. nov. Mycologia 1999;91:141–144.

Shearer CA, Descals E, Kohlmeyer B, Kohlmeyer J, Marvanová L, Padgett D, et al. Fungal biodiversity in aquatic habitats. Biodivers Conserv. 2007;16:49–67.

Shearer CA, Hyde KD. *Massarina ingoldiana,* a new ascomycete from freshwater habitats. Mycologia 1997;89:114–119.

Shearer CA, Raja HA. Freshwater Ascomycetes and their Anamorphs. http://fungi.life.illinois.edu/. Accessed May, 11 2013.

Shearer CA, Raja HA, Miller AN, Nelson P, Tanaka K, Hirayama K, et al. The molecular phylogeny of freshwater Dothideomycetes. Stud Mycol 2009;64:145–153.

Sivichai S, Sri-Indrasutdhi V, Jones EBG. *Jahnula aquatica* and its anamorph *Xylomyces chlamydosporus* on submerged wood in Thailand. Mycotaxon 2011;116:137–142.

Stamatakis A. RAxML-VI-HPC: Maximum likelihood-based phylogenetic analyses with thousands of taxa and mixed models. Bioinformatics 2006;22:2688–2690.

Stanley SJ, Hyde KD. *Boerlagiomyces grandisporus* sp. nov., a new tropical freshwater ascomycete from the Philippines. Mycol Res 1997;101:635–640.

Suetrong S, Boonyuen N, Pang KL, Ueapattanakit J, Klaysuban A, Sri-indrasutdhi V, et al. A taxonomic revision and phylogenetic reconstruction of the Jahnulales (Dothideomycetes), and the new family Manglicolaceae. Fungal Divers 2011a;51:163–188.

Suetrong S, Boonyuen N, Pang KL, Ueapattanakit J, Klaysuban A, Sri-Indrasutdhi V, et al. A taxonomic revision and phylogenetic reconstruction of the Jahnulales (Dothideomycetes), and the new family Manglicolaceae. Fungal Divers 2011b;51:163–188.

Suetrong S, Hyde KD, Zhang Y, Bahkali AH, Jones EBG. Trematosphaeriaceae fam. nov. (Dothideomycetes, Ascomycota). Crypt Mycol 2011c;32:343–358.

Suetrong S, Sakayaroj J, Phongpaichit S, Jones EBG. Morphological and molecular characteristics of a poorly known marine ascomycete, *Manglicola guatemalensis* (Jahnulales: Pezizomycotina; Dothideomycetes, *incertae sedis*): new lineage of marine ascomycetes. Mycologia 2010;102:83–92.

Suetrong S, Schoch CL, Spatafora JW, Kohlmeyer J, Volkmann-Kohlmeyer B, Sakayaroj J, et al. Molecular systematics of the marine Dothideomycetes. Stud Mycol 2009;64:155–173.

Swofford DL. PAUP*: Phylogenetic Analysis Using Parsimony (* and other methods). Version 4. Sinauer Associates, Sunderland, MA 2002.

Tanaka K, Hatakeyama S, Harada Y. Three new freshwater ascomycetes from rivers in Akkeshi, Hokkaido, northern Japan. Mycoscience 2005;46:287–293.

Tanaka K, Hirayama K, Yonezawa H, Hatakeyama S, Harada Y, Sano T, et al. Molecular taxonomy of bambusicolous fungi: Tetraplosphaeriaceae, a new pleosporalean family with *Tetraploa*-like anamorphs. Stud Mycol 2009;64:175–209.

Thompson JD, Higgin DG, Gibson TJ. CLUSTALW: improving the sensitivity of progressive multiple sequence alignment through sequence weighting, position-specific gap penalties and weight matrix choice. Nucleic Acid Res 1994;22:4673–4680.

Tsui CKM, Hyde KD. (eds) Freshwater Mycology. Fungal Divesity Press, Hong Kong 2003.

Tsui CKM, Sivichai S, Rossman AY, Berbee ML. *Tubeufia asiana*, the teleomorph of *Aquaphila albicans* in the Tubeufiaceae, Pleosporales, based on cultural and molecular data. Mycologia 2007;99:884–894.

Wang HK, Aptroot A, Crous PW, Hyde KD, Jeewon R. The polyphyletic nature of Pleosporales: an example from *Massariosphaeria* based on rDNA and RBP2 gene phylogenies. Mycol Res 2007;111:1268–1276.

Webster J. Graminicolous Pyrenomycetes. I. The conidial stage of *Tubeufia helicomyces*. Trans Br Mycol Soc 1951;34:304–308.

Wong MKM, Goh TK, Hodgkiss IJ, Hyde KD, Ranghoo VM, Tsui CKM, et al. Role of fungi in freshwater ecosystems. Biodivers Conserv 1998;7:1187–1206.

Wong MKM, Goh TK, Hyde KD. *Paraphaeosphaeria schoenoplecti* sp. nov. from senescent culms of *Schoenoplectus littoralis* in Hong Kong. Fungal Divers 2000;4:171–179.

Woudenberg JHC, Groenewald JZ, Binder M, Crous PW. *Alternaria* redefined. Stud Mycol 2013;75:171–212.

Zhang Y, Fournier J, Crous PW, Pointing SB, Hyde KD. Phylogenetic and morphological assessment of two new species of *Amniculicola* and their allies (Pleosporales). Persoonia 2009a;23:48–54.

Zhang Y, Fournier J, Phookamsak R, Bahkali AH, Hyde KD. Halotthiaceae fam. nov. (Pleosporales) accommodates the new genus *Phaeoseptum* and several other aquatic genera. Mycologia 2013;105:603–609.

Zhang Y, Jeewon R, Fournier J, Hyde KD. Multi-gene phylogeny and morphotaxonomy of *Amniculicola lignicola:* a novel freshwater fungus from France and its relationships to the Pleosporales. Mycol Res 2008;112:1186–1194.

Zhang Y, Schoch CL, Fournier J, Crous PW, Gruyter JD, Woudenberg JHC, et al. Multi-locus phylogeny of Pleosporales: a taxonomic, ecological and evolutionary re-evaluation. Stud Mycol 2009b;64:85–102.

Zhang Y, Wang HK, Fournier J, Crous PW, Jeewon R, Pointing SB, et al. Towards a phylogenetic clarification of *Lophiostoma/Massarina* and morphologically similar genera in the Pleosporales. Fungal Divers 2009c;38:225–251.

Zhang Y, Zhang XD, Fournier J, Chen J, Hyde KD. *Lindgomyces greiosporus,* a new aquatic ascomycete from Europe including new records. Mycoscience 2014;55:43–48.

Zhou YG, Gong S, Zhang J, Wang L, Yu X. A new species of the genus *Trematosphaeria* from China. Mycol Prog 2013;13:33–43.

Lei Cai, Dian-Ming Hu, Fang Liu, Kevin D. Hyde and
E. B. Gareth Jones

3 The molecular phylogeny of freshwater Sordariomycetes and discomycetes

3.1 Introduction

Freshwater ascomycetes comprise a diverse taxonomic assemblage of more than 600 species (Shearer and Raja 2013), reviewed by Shearer (1993) and Cai et al. (2003). The numbers of known freshwater ascomycetes have been increasing rapidly, for example, from 423 species in 2001 (Shearer 2001) to 511 in 2003 (Cai et al. 2003) and to 622 in 2010 (Shearer and Raja 2013). Shearer (1993) defined freshwater ascomycetes in a broad ecological sense as "all ascomycetes that occur on submerged or partially submerged substrata in aquatic habitats". Freshwater fungi have also been defined as "fungi that for the whole or part of their life cycle rely on freshwater" (Thomas 1996). The definition of Shearer (1993) has been regarded as the current best working definition (Cai et al. 2003).

Freshwater fungi are a taxonomically diverse group and for this reason are generally difficult to identify (Shearer 1993). Freshwater Sordariomycetes play an important role in ecosystems and many of them have shown potential values in application. This review treats the freshwater Sordariomycetes, which is one of the largest and most important groups of fungi in freshwater habitats. According to Shearer and Raja (2013), freshwater Sordariomycetes account for nearly half of the total known freshwater Ascomycota (307 out of 622), and therefore is the largest group of freshwater ascomycetes. In China, where freshwater fungi have been well-studied, freshwater Sordariomycetes accounted for almost 60% of the total ascomycetes (143 out of 256 species) (Hu et al. 2013). Annulatascaceous species are the most typical and common freshwater Sordariomycetes on submerged wood. The distinctive character of these taxa is the presence of a massive apical ring to the ascus (Wong et al. 1998).

The emergence of molecular technology has significantly improved our understanding on the phylogeny and evolution of freshwater ascomycetes. Most of the new genera and species published in the past 5 years were provided with molecular data, for example: *Ascorhombispora, Achroceratosphaeria, Baipadisphaeria, Conlarium,* and *Cuspidatispora*. The molecular data have enabled us to progress towards a better understanding of the phylogenetic relationships of Sordariomycetes. The modern classification system of the overall Kingdom of Fungi, as well as Sordariomycetes, has been significantly improved with the utilization of molecular phylogenetic data (Hibbett et al. 2007). However, the currently available molecular phylogenetic data are more or less scattered or focused on selected groups.

In this study, we used 28S rDNA sequence data from available freshwater Sordariomycetes to construct a phylogenetic tree, in order to 1) show phylogenetic relationships of typical freshwater Sordariomycetes, such as, members in the morphologically defined family Annulatascaceae, and 2) to understand the phylogenetic placement of other freshwater Sordariomycetes in the major orders and families. Freshwater discomycetes are also briefly reviewed in this chapter.

3.2 Materials and methods

3.2.1 Taxon sampling

Specimens used in the phylogenetic analysis, with strain numbers, GenBank accession numbers and references are listed in Tab. 3.1. In total 131 sequences were used in constructing the phylogenetic tree. The aligned dataset included 856 characters including gaps.

3.2.2 Phylogenetic analysis

Maximum likelihood (ML) analysis method was used to analyze the dataset. RAxML v7.2.6 (Stamatakis 2006) was used for ML analyses. The default setting was used for all parameters in the ML analysis, and statistical support values were obtained using nonparametric bootstrapping with 1,000 replicates.

3.3 Discussion

Our analysis showed that freshwater Sordariomycetes are scattered in three sub-classes, i.e. Sordariomycetidae, Hypocreomycetidae and Xylariomycetidae. Vijaykrishna et al. (2006) suggested that freshwater fungi had evolved from terrestrial fungi through different pathways of adapting to freshwater habitats. Our analysis supports the theory that freshwater fungi are of multiple origins.

Major orders of freshwater Sordariomycetes included in Fig. 3.1 are Amphisphaeriales (1 genus), Calosphaeriales (1 genus), Coniochaetales (1 genus), Diaporthales (10 genera), Hypocreales (9 genera), Magnaporthales (2 genera), Microascales (11 genera), Ophiostomatales (1 genus), Phyllachorales (1 genus), Savoryellales (3 genera), Sordariales (8 genera), Trichosphaeriales (1 genus) and Xylariales (12 genera). Nevertheless, the correct taxonomic placement of a number of freshwater sordariomycetous taxa is still unclear and many have been referred to a class until sequence data are available to confirm their phylogenetic relationship. In this chapter orders for which sequence data is currently available include: Sordariales (Cai et al. 2006b), Microascales (Sakayaroj 2011), Hypocreales (Rossman et al. 2001), Xylariales (Smith et al. 2003), Magnaporthales (Thangkentha et al. 2009) and Savoryellales (Boonyuen et al. 2011). A selection of

Tab 3.1: Taxa and sequences included in the phylogenetic analysis.

Species	Strain/Voucher	28S	Habitats	Citation
Achroceratosphaeria potamia	CBS 125414*	GQ996538	Freshwater	Réblová et al. (2010)
Ambarignomonia petiolorum	CBS 121227*	EU255070	Terrestrial	Sogonov et al. (2008)
Aniptodera chesapeakensis	N/A	U46882	Marine	Spatafora et al. (1998)
Aniptodera lignatilis	A460-1	AY227115	Freshwater	Sakayaroj et al. (2011)
Aniptodera longispora	BBH28304	HQ111008	Freshwater	Sakayaroj et al. (2011)
Annulatascus apiculatus	SS 2424*	JN226107	Freshwater	Boonyuen et al. (2012)
Annulatascus biatriisporus	A 464–3	AY316352	Freshwater	Raja et al. (2003)
Annulatascus hongkongensis	HKUCC 3702*	AF132319	Freshwater	Ranghoo et al. (1999)
Annulatascus nilensis	MF808*	HQ616536	Freshwater	Abdel-Wahab et al. (2011)
Annulatascus velatispora	A 70–18	AY316354	Freshwater	Raja et al. (2003)
Annulatascus velatispora	HKUCC3701*	AF132320	Freshwater	Ranghoo et al. (1999)
Annulatascus velatispora	R 047	AY316355	Freshwater	Raja et al. (2003)
Annulusmagnus triseptatus	A 54–10A	AY590286	Freshwater	Campbell and Shearer (2004)
Annulusmagnus triseptatus	A 413–6	AY590285	Freshwater	Campbell and Shearer (2004)
Aquaticola ellipsoidea	A 411–3	AY316356	Freshwater	Raja et al. (2003)
Aquaticola ellipsoidea	R008	AY590290	Freshwater	Campbell and Shearer (2004)
Aquaticola hongkongensis	HKUCC 3703*	AF132321	Freshwater	Ranghoo et al. (1999)
Aquaticola hyalomura	R 038	AY590291	Freshwater	Campbell and Shearer (2004)
Ascitendus austriacus	CBS 102665*	AF261067	Freshwater	Réblová and Winka (2001)
Ascitendus austriacus	M.R. 2936	GQ996539	Freshwater	Réblová et al. (2010)
Ascitendus austriacus	A 324–1F	AY590294	Freshwater	Campbell and Shearer (2004)
Ascocollumdensa aquatica	HKUCC 3707*	AF132325	Freshwater	Ranghoo et al. (1999)
Ascolacicola aquatica	HKUCC 3704	AF132322	Freshwater	Ranghoo et al. (1999)
Ascosacculus aquaticus	A444-1D	AY227136	Freshwater	Campbell et al. (2003a)
Ascosacculus heteroguttulatus	A108-11B	AY227122	Freshwater	Campbell et al. (2003a)
Ascosacculus heteroguttulatus	A108-7D	AY227121	Freshwater	Campbell et al. (2003a)
Ascotaiwania hughesii	P2-6	AY094189	Freshwater	Campbell and Shearer (2004)
Ascotaiwania mitriformis	HKUCC 3706*	AF132324	Freshwater	Ranghoo et al. (1999)

(*continued*)

Tab. 3.1: (*continued*)

Species	Strain/Voucher	28S	Habitats	Citation
Ascotaiwania sawada	HKUCC 3705	AF132323	Freshwater	Ranghoo et al. (1999)
Ascothailandia grenadoidia	SS 03615*	GQ390267	Freshwater	Sri-Indrasutdhi et al. (2010)
Baipadisphaeria spathulospora	BCC20906	HM134244	Freshwater	Pinruan et al. (2010)
Bertia moriformis	SMH4320	AY695260	Terrestrial	Huhndorf et al. (2004)
Bertia moriformis	SMH3344	AY695261	Terrestrial	Huhndorf et al. (2004)
Bionectria vesiculosa	HMAS183151	HM050302	Terrestrial	Luo and Zhuang (2010)
Bullimyces aurisporus	AF316-1b	JF775590	Freshwater	Ferrer et al. (2012)
Bullimyces communis	AF281-5	JF775587	Freshwater	Ferrer et al. (2012)
Bullimyces costaricensis	AF317-1b	JF775592	Freshwater	Ferrer et al. (2012)
Cataractispora receptaculorum	HKUCC 3710	AF132327	Freshwater	Ranghoo et al. (1999)
Cercophora aquatica	JF 06314	JN673036	Freshwater	Raja et al. (2011)
Cercophora caudata	CBS 606.72	AY999113	Terrestrial	Cai et al. (2005)
Cercophora newfieldiana	SMH 2622	AF064642	Terrestrial	Fernández et al. (2006)
Cercophora terricola	ATCC 200395*	AY780067	Freshwater	Miller and Huhndorf (2005)
Chaetomium globosum	DAOM 221179	JN938908	Terrestrial	Schocha et al. (2012)
Chaetopsinectria chaetopsinae	HMAS 76860	DQ119553	Terrestrial	Luo and Zhuang (2010)
Chaetosphaeria fuegiana	ICMP 15153*	EF063574	Terrestrial	Réblová and Seifert (2007)
Chaetosphaeria innumera	SMH 2748	AY017375	Terrestrial	Huhndorf et al. (2001)
Claviceps purpurea	SAcp11	EF469075	Terrestrial	Sung et al. (2007)
Clohiesia corticola	HKUCC 3712*	AF132329	Freshwater	Ranghoo et al. (1999)
Clypeosphaeria uniseptata	HKUCC 6349	DQ810219	N/A	N/A
Coniochaeta leucoplaca	Jong54	FJ167399	Terrestrial	N/A
Conioscyphascus varius	CBS 113653*	AY484512	Freshwater	Réblová and Seifert (2004)
Conlarium duplumascospora	CGMCC 3.14940	JN936993	Freshwater	Liu et al. (2012)
Conlarium duplumascospora	CGMCC 3.14939	JN936992	Freshwater	Liu et al.(2012)
Corollospora luteola	NBRC31315*	JN941490	Terrestrial	N/A
Cryphonectria japonica	CMW 13749	JN940857	Terrestrial	N/A

(*continued*)

Tab. 3.1: (*continued*)

Species	Strain/Voucher	28S	Habitats	Citation
Cuspidatispora xiphiago	A184-1A*	DQ376251	Freshwater	Miller et al. (2006)
Cyanoannulus petersenii	R044-1a	AY316358	Freshwater	Raja et al. (2003)
Cyanoannulus petersenii	R044-1aB	AY316359	Freshwater	Raja et al. (2003)
Diaporthe eres	AR3538	AF408350	Terrestrial	Castlebury et al. (2002)
Diaporthe passiflorae	CPC 19183	JX069844	Terrestrial	Crous et al. (2004)
Diatrype disciformis	CBS 197.49	DQ470964	Terrestrial	Spatafora et al. (2006)
Dyrithiopsis lakefuxianensis	HKUCC 7303*	AF452047	Freshwater	Jeewon et al. (2003)
Emericellopsis terricola	CBS 120.40*	U57082	N/A	Summerbell et al. (2011)
Esteya vermicola	CBS 115803	EU668903	Terrestrial	Wang et al. (2008)
Fluminicola bipolaris	HKUCC 3717*	AF132332	Freshwater	Ranghoo et al. (1999)
Fusoidispora aquatica	HKU(M) 17484*	AY780365	Freshwater	Vijaykrishna et al. (2005)
Gnomonia borealis	CBS 299.79	EU255169	Terrestrial	Sogonov et al. (2008)
Halosarpheia kandeliae	CY1492	HQ111025	Terrestrial	Sakayaroj et al. (2011)
Halosphaeria appendiculata	CY3485	AY090892	Terrestrial	Pang et al. (2004)
Hydromelitis pulchella	AF284-2	JF775588	Freshwater	Ferrer et al. (2012)
Hypocrea lutea	NBRC104901	JN941457	Terrestrial	Schocha et al. (2012)
Hypocrea rufa	GJS89-127	AY489726	Terrestrial	Castlebury et al. (2004)
Hyponectria buxi	N/A	AY083834	Terrestrial	N/A
Jobellisia guangdongensis	HMAS 251240*	JN936990	Freshwater	Liu et al. (2012)
Jobellisia luteola	SMH2753	AY346286	Terrestrial	Huhndorf et al. (2004)
Lasiosphaeria ovina	CBS958.72	AY587946	Terrestrial	Miller and Huhndorf (2005)
Lasiosphaeria sorbina	CBS 885.85	AY436416	Terrestrial	Miller and Huhndorf (2005)
Lignincola laevis	ATCC 200717	AY225488	Freshwater	Pang et al. (2013)
Lignincola laevis	JK 5180A	U46890	Marine	Spatafora et al. (1998)
Lulworthia medusa	J. K. 5581	AF195637	Terrestrial	Kohlmeyer et al. (2000)
Magnisphaera spartinae	A2211	AY150221	Freshwater	Sakayaroj et al. (2011)
Magnisphaera stevemossago	A409-1B	AY227134	Freshwater	Campbell et al. (2003a)
Magnisphaera stevemossago	A409-4D	AY227135	Freshwater	Campbell et al. (2003a)

(*continued*)

Tab. 3.1: (*continued*)

Species	Strain/Voucher	28S	Habitats	Citation
Melanconis stilbostoma	CBS 109778	AF408374	Terrestrial	Castlebury et al. (2002)
Nais inornata	ATCC 200453	AF539476	Freshwater	Pang et al. (2003)
Natantispora lotica	ATCC 56668	AF396873	Freshwater	Sakayaroj et al. (2011)
Natantispora retorquens	A231-1D	AY227128	Freshwater	Sakayaroj et al. (2011)
Nectria cinnabarina	CBS 125165*	HM484562	Terrestrial	Chaverri et al. (2011)
Neurospora terricola	CBS 298.63*	AY681142	Terrestrial	Cai et al. (2006b)
Nitschkia grevillei	SMH4663	AY346294	Terrestrial	Huhndorf et al. (2004)
Ophioceras aquaticus	IFRDCC 3091*	JQ797433	Freshwater	Hu et al. (2012a)
Ophioceras chiangdaoense	CMU26633*	EU571272	Freshwater	Thongkantha et al. (2009)
Ophioceras commune	M92	JX134688	Freshwater	Luo and Zhang (2013)
Ophioceras dolichostomum	HKUCC10113	DQ341507	Freshwater	Thongkantha et al. (2009)
Ophioceras dolichostomum	CMURp50	DQ341504	Freshwater	Thongkantha et al. (2009)
Ophioceras hongkongense	HKUCC3624	DQ341509	Freshwater	Thongkantha et al. (2009)
Ophioceras leptosporum	CBS168.96	DQ341510	Freshwater	Thongkantha et al. (2009)
Ophioceras leptosporum	CBS 894.70*	JX134690	Freshwater	Luo and Zhang (2013)
Ophioceras tenuisporum	CS652-1	AY346295	Freshwater	Huhndorf et al. (2004)
Ophiostoma karelicum	CMW23099	EU443756	Terrestrial	Linnakoski et al. (2008)
Ophiostoma karelicum	CMW23101	EU443757	Terrestrial	Linnakoski et al. (2008)
Ophiostoma piliferum	AFTOL-ID 910	DQ470955	Terrestrial	Spatafora et al. (2006)
Ophiostoma stenoceras	AFTOL-ID 1038	DQ836904	Terrestrial	Zhang et al. (2006)
Paoayensis lignicola	HKU(M) 17516*	EF622535	Freshwater	Cabanela et al. (2007)
Phaeonectriella lignicola	PP 7008	AY150224	Freshwater	Sakayaroj et al. (2011)
Pleurostoma ootheca	CMU 23858	AY761079	Terrestrial	N/A
Pseudohalonectria lignicola	SMH 2440	AY346299	Freshwater	Huhndorf et al. (2004)
Pseudohalonectria lignicola	YMF1.01288	JX134691	Freshwater	Luo and Zhang (2013)
Pseudohalonectria lutea	CBS 126574	JX066706	Freshwater	Réblová (2013)
Pseudoproboscispora caudae-suis	A336–2D	AY094192	Freshwater	Campbell and Shearer (2004)
Pseudoproboscispora caudae-suis	A40-1A	AY094191	Freshwater	Campbell and Shearer (2004)

(*continued*)

Tab. 3.1: (*continued*)

Species	Strain/Voucher	28S	Habitats	Citation
Riomyces rotundus	AF303-1*	JF775589	Freshwater	Ferrer et al. (2012)
Saccharomyces cerevisiae	NRRL Y-12632	JQ689017	N/A	N/A
Savoryella aquatica	SS00583	HQ446371	Freshwater	Boonyuen et al. (2012)
Savoryella aquatica	SS03801	HQ446372	Freshwater	Boonyuen et al. (2012)
Savoryella lignicola	NF00204	HQ446378	Freshwater	Boonyuen et al. (2012)
Savoryella verrucosa	SS00582	HQ446375	Freshwater	Boonyuen et al. (2012)
Savoryella verrucosa	SS00052	HQ446374	Freshwater	Boonyuen et al. (2012)
Scortechinia acanthostroma	GKM163N	FJ968991	Terrestrial	N/A
Scortechinia acanthostroma	SMH5313	FJ968990	Terrestrial	N/A
Sordaria fimicola	CBS 508.50	AY681160	Terrestrial	Cai et al. (2006b)
Sordaria superba	CBS 784.96	AY681139	Terrestrial	Cai et al. (2006b)
Submersisphaeria aquatica	A354–14	AY094194	Freshwater	Campbell and Shearer (2004)
Submersisphaeria aquatica	A95–1B	AY094193	Freshwater	Campbell and Shearer (2004)
Togninia auqaticoa	IFRDCC 3035*	IFRDCC3035	Freshwater	This study
Togninia fraxinopennsylvanica	ATCC 26664*	AY761083	Terrestrial	N/A
Togninia minima	CBS 6580	AY761082	Terrestrial	N/A
Valsa ambiens	CBS 109491	EU255208	Terrestrial	Sogonov et al. (2008)
Valsella salicis	CBS 109754	EU255210	Terrestrial	Sogonov et al. (2008)
Xylaria acuta	ATCC 56487	AY544676	Terrestrial	N/A
Xylaria hypoxylon	AFTOL-ID 51	AY544648	Terrestrial	N/A

* Indicates type specimens.

species in this class is illustrated in Fig. 3.2. Biodiversity studies of unusual habitats frequently lead to the discovery of new lineages of sordariomycetous taxa, e.g. palm fungi in a peat swamp: *Baipadisphaeria* (Pinruan et al. 2010), *Flammispora* (Pinruan et al. 2004), *Phruensis* (Pinruan et al. 2004) and *Thailandiomyces* (Pinruan et al. 2008).

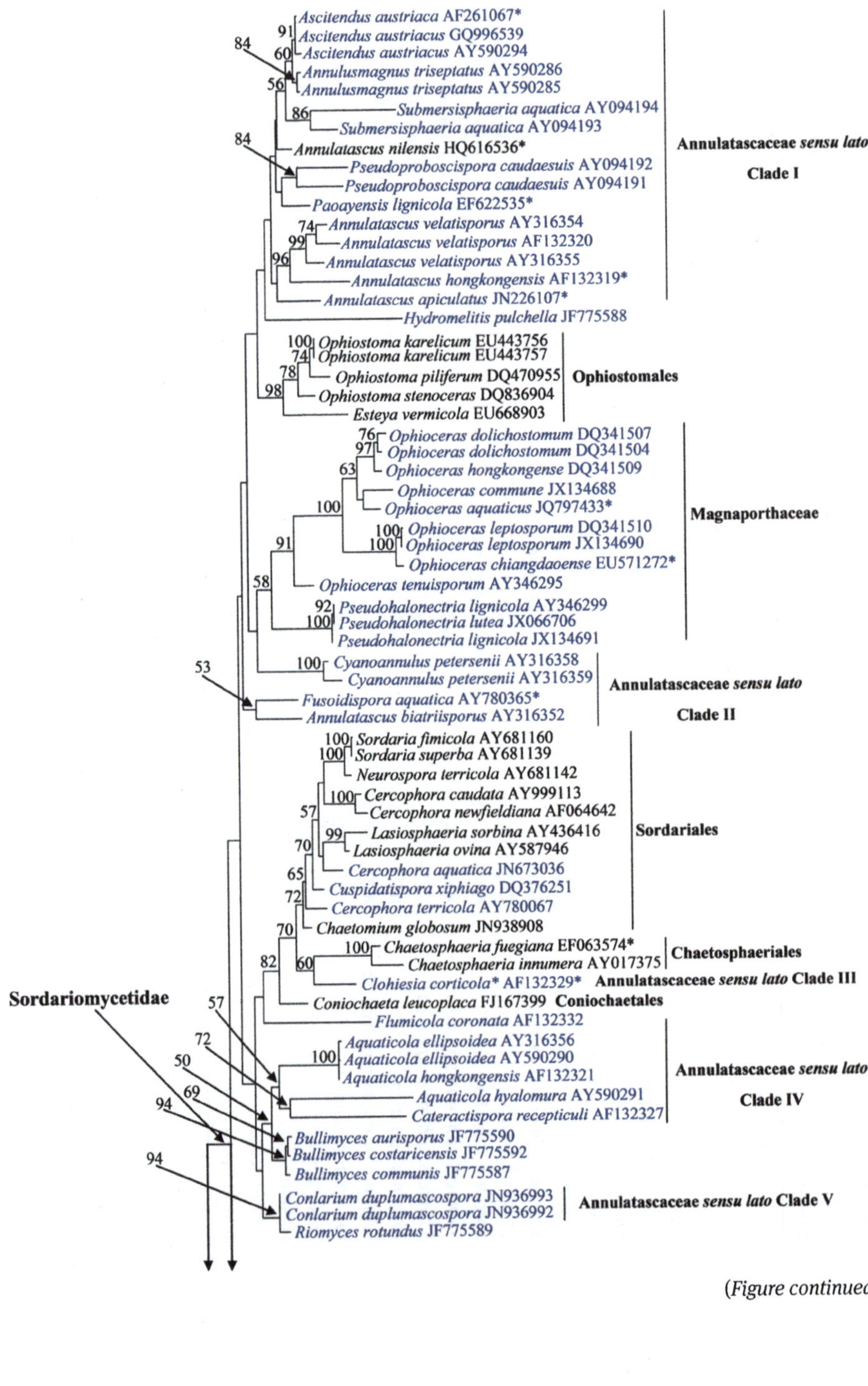

(*Figure continued*)

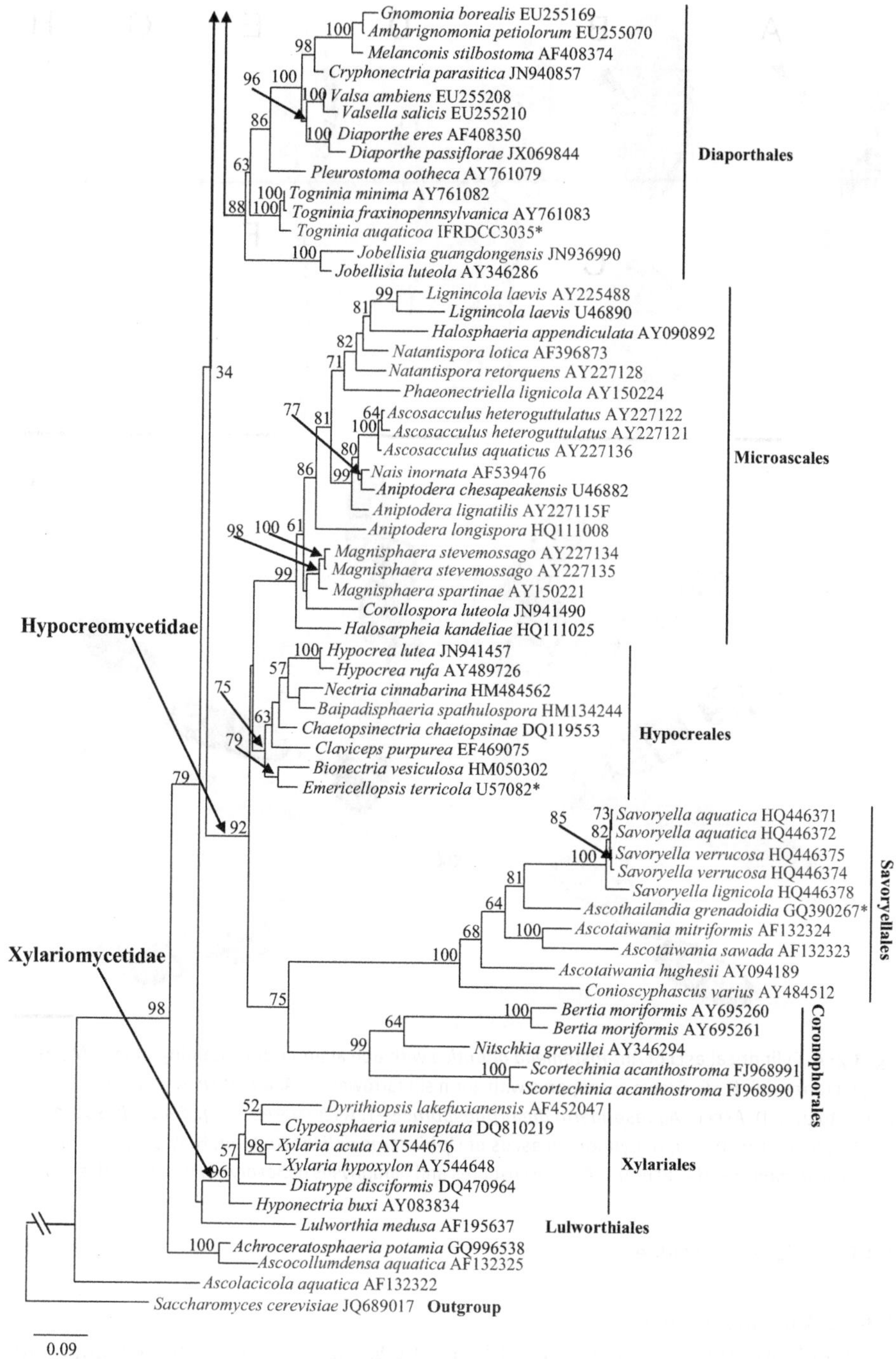

Fig. 3.1: Phylogenetic tree based on 28S rDNA sequences, bootstrap values higher than 50% are indicated on each node. Species known from freshwater are shown in blue. * indicates ex-type strains.

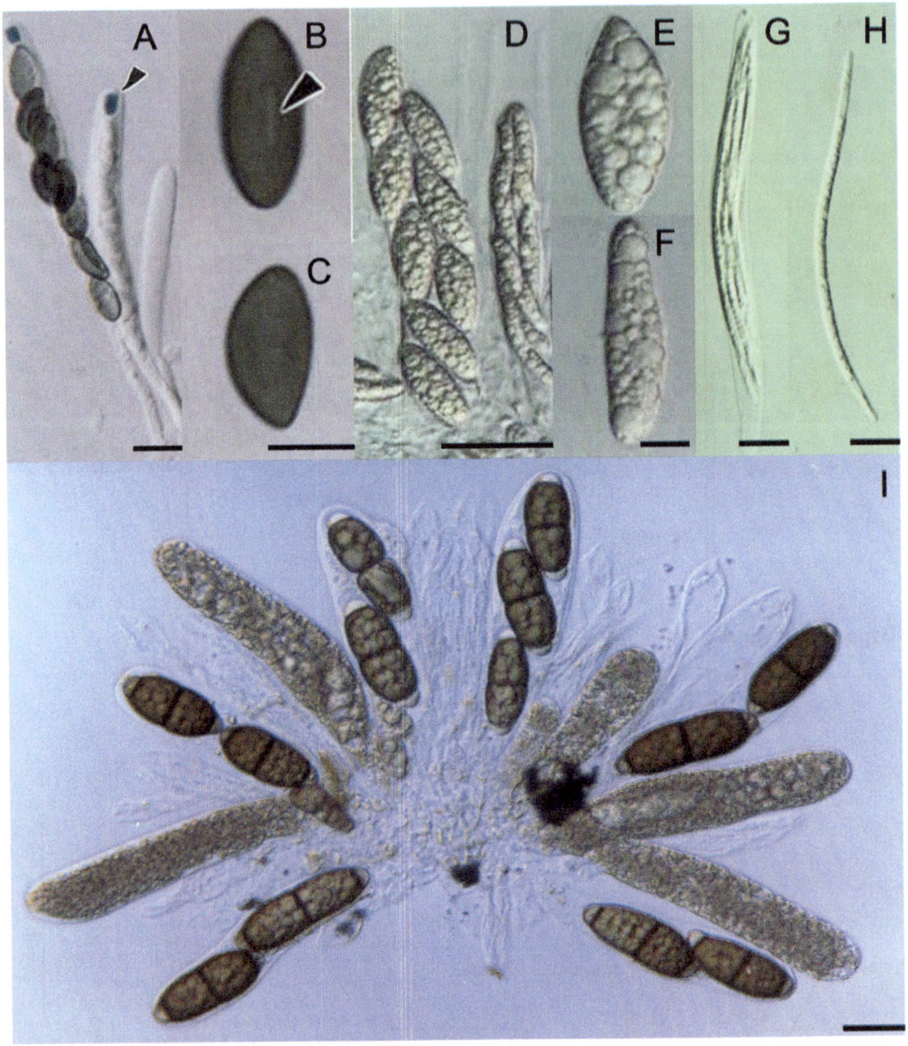

Fig. 3.2: A. Cylindrical asci of *Anthostomella aquatica* with apical apparatus staining blue in Melzer's reagent (arrowed); B—C. Brown ascospores with germ slit (arrowed) of *A. aquatica* (A = 20 µm, B—C = 10 µm); D. Asci of *Aquasphaeria dimorphospora*; E—F. Ascospores of *Aq. dimorphospora* (D = 50 µm, E—F = 10 µm); G. Cylindrical ascus of *Ophioceras dolichostomum*; H. Ascospores of *O. dolichostomum* (G—H = 10 µm); I. Asci, ascospores and paraphyses of *Savoryella verrucosa* (I = 20 µm).

3.3.1 Sordariomycetidae

3.3.1.1 Annulatascaceae

The family Annulaascaceae was established by Wong et al. (1998) to accommodate species of *Annulatascus* (typified by *A. velatispora*) and similar genera. Members of Annulatascaceae are primarily characterized by the presence of a relatively massive, wedge-shaped refractive, bipartite, J-apical ascal ring (Wong et al. 1998). Réblova and Winka (2001) however, demonstrated that species characterized by these characters are phylogenetically polyphyletic. This has been subsequently shown in various studies (Campell et al. 2003; Raja et al. 2003, 2009, 2011; Campbell and Shearer 2004; Vijaykrishna et al. 2005, 2006). The type genus, *Annulatascus*, has also been shown to be polyphyletic, and new genera, such as *Annulusmagnus*, have been established to accommodate the distantly related species *Annulatascus triseptatus*. *Annulusmagnus* is morphologically distinguished from *Annulatascus* as its ascomata have setose hyphae and straw-colored ascospores which are broadly fusiform, and flattened on one side. Ascomatal and ascospore morphology may be more phylogenetically informative, as compared to the ascus and apical ring.

Annulatascaceae is the most commonly encountered group of freshwater Sordariomycetes. However, the taxonomy of Annulatascaceae has long relied on morphology, and the lack of molecular data makes it difficult to confidently assign the family to any existing order (Wong et al. 1998; Ho and Hyde 2000). The family Annulatascaceae was first tentatively placed in Sordariales due to its large refractive apical ascal ring (Wong et al. 1998; Ho and Hyde 2000). Huhndorf et al. (2004) however, used 18S rDNA phylogenetic analysis to show that Annulatascaceae does not belong to Sordariales and could only be placed in Sordariomycetidae *incertae sedis*. Vijaykrishna et al. (2005), based on 28S rDNA data, indicated that Annulatascaceae is not monophyletic but divided into two major clades. Thongkantha et al. (2009) reported the close affinity of Annulatascaceae to Ophiostomatales through analysis based on 18S and 28S rDNA data.

Due to the lack of molecular study it is still difficult to conclude the ordinal and familial placement of *Annulatascus* species. The sequencing and epitypification of the type species of the family, *A. velatispora*, is an essential step towards establishing a taxonomically meaningful framework for this family that is rich in freshwater ascomycetes. The phylogenetic tree generated in the present study showed that as currently defined Annulatascaceae is of multiple origins.

Another hurdle to overcome in understanding the family Annulatascaceae is that many genera established in the group, have single or few morphological characters. For example, *Aquaticola* was established based on a smaller apical ring and biseriate arrangement of ascospores in the asci (Ho et al. 1999); *Brunneosporella* was established generally based on the presence of ascospores with germ pores (Ranghoo et al. 2001); *Cataractispora*, *Diluviocola*, *Fluminicola*, *Proboscispora*, and *Porosphaerellopsis* were established mainly based on ascospore polar appendages, again differentiated by few morphological features (Hyde et al. 1998, 1999; Wong and Hyde 1999; Wong et al. 1999); while *Torrentispora* was established primarily based on the peridium that comprises black, thick-walled cylindrical cells (Hyde et al. 2000). *Vertexicola* was established based on the presence of distoseptate ascospores (Ranghoo et al. 2000). Most genera are united by a relatively large ascal ring, but it has yet to be tested whether molecular data support the inclusion of these genera in one family. A natural classification needs to be established using molecular evidence; however species need to be recollected and sequenced and epitypified if necessary. It is recommended that in future any newly described taxa should be supported with molecular data based on ex-type cultures or type specimens in international culture collections that provide cultures for basic research at no or low cost. However, a meaningful taxonomic assessment of the older genera established before 2005 period, remains to be carried out. Genera established after 2005 are mostly provided wth DNA sequence data. Below we discuss the major groups of Annulatascaceae in the phylogenetic tree (Fig. 3.1).

Clade I of Fig. 3.1 includes six genera, i.e. *Annulatascus*, *Annulusmagnus*, *Ascitendus*, *Paoayensis*, *Pseudoproboscispora* and *Submersisphaeria*. This clade appears to be related to Ophiostomatales, but the statistical support is low. In morphology, these genera (except *Paoayensis*) are characterized by solitary or gregarious, immersed or semi-immersed, brown to black ascomata with long or short necks; paraphyses relatively wide; asci unitunitate, cylindrical, with a massive refractive apical ring; ascospores hyaline to brown, usually septate, smooth or with ornamentation, often with a mucilaginous sheath (Hyde 1992, 1996; Campbell et al. 2003b; Campbell and Shearer 2004; Cabanela et al. 2007). Most of the above characters are in agreement with the original definition of the Annulatascaceae (Wong et al. 1998; Ho and Hyde 2000), and they may represent the Annulatascaceae *sensu stricto*.

Although *Annulusmagnus* has been established to accommodate *Annulatascus triseptatus* that is phylogenetically distant to *A. velatisporus*, the remaining members in *Annulatascus* constitute a polyphyletic group. For example, the *Annulatascus nilensis* type strains clustered distantly to *A. velatispora* and *A. hongkongensis*. *A. apiculatus* also appeared basal to the above clade and is not statistically supported. The genera *Submersisphaeria* and *Pseudoproboscispora* appear in a distinct lineage, this however needs to be further verified by more taxon sampling especially the inclusion of type-derived DNA data. For example, even the two 28S rDNA sequences representing *Submersisphaeria aquatica* (AY094194 and AY094193 from Campbell and Shearer 2004) were found to have even 120 different sites, definitely representing two different species. This is an example showing the importance of using type-derived sequences, as morphologically identified species by even the same scientists might be questionable.

Paoayensis lignicola is interesting as it is characterized by asci with a discoid refractive apical ring, and brown ascospores with a germ slit, more or less similar to members of Sordariales (Cabanela et al. 2007). Since the bootstrap for placement of this species is low, its phylogenetic affinity remains uncertain.

Clade I is not statistically supported, this could either be because of the lack of phylogenetic information in 28S rDNA region or the paraphyletic nature of this group. Further study is needed to resolve the species in this group into a well-supported monophyletic clade. *Annulatascus velatisporus*, the type species of the genus and the family, needs recollecting, sequencing and possibly epitypifying and then ex-type culture and type-derived DNA data could be used to infer its evolutionary relationships, and based on which, a robust taxonomic revision could be made.

Clade II is a grouping comprising *Cyanoannulus*, *Fusoidispora* and *Annulatascus biatriisporus*. Although *Cyanoannulus* is in general agreement with Annulatascaceae in morphology, it appears distantly related, and its taxonomic placement could not be determined. Raja et al. (2003), when establishing *Cyanoannulus*, placed it in Annulatascaceae based on its morphological characters, and in their phylogenetic tree, the bootstrap support uniting *Cyanoannulus* and the type species of *Annulatascus* is relatively low. *Cyanoannulus* is distinctly different from *Annulatascus* in having pale reddish-brown ascomata. Ascomata morphology has been shown to be phylogenetically more informative in the Sordariales (Miller and Huhndorf 2005), and this is likely the case of other orders in the Sordariomycetes.

In morphology, *Fusoidispora* is also in general agreement to the morphological concept of Annulatascaceae but its apical rings are relatively smaller and the ascospores are long cylindrical, different to most of the typical species in Annulatascaceae (Vijaykrishna et al. 2005). In Fig. 3.1, *Annulatascus biatriisporus* is represented by an unverified strain thus its placement awaits further investigation. Morphologically, *A. biatriisporus* is different from the type, *A. velatisporus* in having fusiform spores with swollen ends.

Clade III contains only one strain, the type of *Clohiesia*. When establishing the genus *Clohiesia*, Hyde (1995) suggested the potential placement in families such as, Clypeosphaeriaceae, Lasiosphaeriaceae and Annulatascaceae. Tsui et al. (1998) placed the genus in Annulatascaceae. However, quite recognizable morphological distinctions could be observed between *Clohiesia* species and those members in typical Annulatascaceae. For example, the apical ring of *Clohiesia* species is not as massive as those of typical members of Annulatascaceae, and the ascomata of *Clohiesia* are immersed under a clypeus. Our analysis showed that *Clohiesia* is phylogenetically distant to Annulatascaceae *sensu stricto*. Clade III is very close to Chaetosphaeriales with moderate support. Additional studies bringing in more taxa and multi-locus sequences would help to confirm the relationships of this clade. The phylogenetic placement of *Fluminicola coronata* needs further investigation. The strain used in our analysis is not from a type strain, while it did not cluster with other species provided with polar appendages.

Clade IV included *Aquaticola* and *Cateractispora*, both are morphologically typical Annulatascaceae members (Ho et al. 1999; Ranghoo et al. 1999). *Aquaticola* species in our analysis did not constitute a monophyletic group, and *Aquaticola hyalomura* is clearly distant from the other two species *A. ellipsoidea* and *A. hongkongensis*, with *A. hyalomura* clustering with *Cateractispora recepticuli* (AF132327) with moderate support. We could not find a description of *Cateractispora recepticuli* as no publication source was indicated in MycoBank and Index Fungorum. According to the information given by Ranghoo et al. (1999), this is a specimen with "black perithecia with long necks, cylindrical asci with a relative large refractive apical ring and ascospores with appendages", in good agreement with the genus *Cataractispora* published by Hyde et al. (1999) in the same year. *Cateractispora* was established on its unusual ascospore appendages, and placed in Annulatascaceae based on the massive apical ring (Hyde et al. 1999). Further studies bringing in more taxa will be needed to resolve the phylogenetic relationships of this genus.

Clade V includes *Conlarium* and *Riomyces*, both represented by type-derived sequences. *Conlarium* produces multi-septate ascospores with or without globose or papillary appendages, and an asexual morph with muriform conidia, differing from the other genera in Annulatascaceae *sensu lato*. *Conlarium* appears to be most closely related to *Riomyces*. The latter genus, however, does not have an apical ring or other apparatus in the ascus, further indicating that an apical ring in Annulatascaceae *sensu lato* is a result of convergent evolution from organisms with different origins.

Bullimyces species nested between clade IV and V (Ferrer et al. 2012). The species in this genus are mainly characterized by multi-septate, thick-walled ascospores, with unitunicate asci lacking an apical apparatus. Further molecular data will be necessary to resolve the familial placement of *Bullimyces* (Ferrer et al. 2012), especially those sharing remarkable morphological similarities to such genera as *Luttrellia*.

3.3.1.2 Magnaporthales

In Magnaporthaceae, the freshwater fungi mainly include members from *Ophioceras* and *Pseudohalonectria* with 11 and seven known species, respectively. Both genera are characterized by cylindrical or long cylindrical ascospores and clavate asci (Shearer et al. 1999; Hu et al. 2012a). *Pseudohalonectria* differs in having bright yellow, membranous ascomata, while those of *Ophioceras* species are dark brown to black. Their methods of ascospores discharge are also different. In *Pseudohalonectria* the ascospores (and depending on the conditions) and whole asci are discharged through their beaks and accumulate in masses. In *Ophioceras* species, whole asci are forced up through the neck to the apex. The narrow canal of the beak allows the passage of only one ascus at a time. The ascospores are then violently discharged (Cai et al. 2006a). In Fig. 3.1, *Ophioceras* species and *Pseudohalonectria* species appear to be well separated, indicating the above traditionally used morphological distinctions between the two are phylogenetically informative. *Gaeumannomyces* also belongs to Magnaporthaceae, with *G. graminis* known from freshwater. Magnaporthaceae has been shown to be monophyletic and accommodated in a new order Magnaporthales (Thongkantha et al. 2009). It is most closely related to Diaporthales and Ophiostomatales. In our analysis, and that of Thongkantha et al. (2009), *Ophioceras tenuisporum* was distant from other *Ophioceras* species.

3.3.1.3 Calosphaeriales

This order was introduced by Barr (1983) for some 54 species in 13 genera which have been assigned to two families (Kirk et al. 2008). Shearer and Raja (2013) listed one species assigned to this order, *Erostella minutussima*, found on submerged wood in Hong Kong (Ho et al. 2001) at a low percentage occurrence (0.7%) and further collections are required before it can regarded as a truly freshwater species.

3.3.1.4 Coniochaetales

Shearer and Raja (2013) listed eight *Coniochaeta* species belonging to this order with two described from submerged wood: *C. gigantospora* and *C. renispora* (Crane and Shearer 1995; Raja et al. 2012). This is a well circumscribed genus supported by molecular data (Huhndorf et al. 2004).

3.3.1.5 Diaporthales

Fifteen freshwater species in eleven genera (*Diaporthe, Ditopella, Gnomonia, Gnomoniella, Hyalorostrarum, Jobellisia, Lollipopaia, Phrunensis, Rhamophoria, Thalandiomyces, Tognina*) have been reported for this order (Shearer and Raja 2013), with seven new species originally described from aquatic habitats. These taxa group in four families (Diaporthaceae, Gnomoniaceae, Jobellisiaceae, Togniniaceae) as well as a new lineage within the order. Two new genera, *Phruensis* and *Thalandiomyces*, were described from palm fronds in a peat swamp in Thailand and supported by sequence data (Pinruan et al. 2004, 2008). In a SSU rDNA dataset *Tirisporella beccariana*, a mangrove inhabiting ascomycete (initially classified in the Dothideomycetes), groups with a number of tropical freshwater ascomycetes in a well supported clade: *Thailandiomyces bisetulosus, Lollipopaia minuta* and *Phruensis brunneispora* (Suetrong et al. unpublished data). In an LSU rDNA dataset, *T. beccariana* forms a sister group to *Th. bisetulosus* and two *Jobellisia* species. This lineage groups as a sister clade to the family Jobellisiaceae with moderate support. Jobellisiaceae was introduced by Réblová (2008) for *Jobellisia fraterna* and *J. luteola* and formed a monophyletic clade in a basal position to the Diaporthales, Calosphaeriales and the Togniniaceae. Two *Jobellisia* species have been reported from submerged wood in freshwater habitats: *J. viridifusca* and *J. guangdongensis* (Liu et al. 2012). The taxa *Lollipopaia, Phruensis, Tirisporella* and *Thailandiomyces*, do not group in any family in the Diaporthales, and this clade may constitute a new family in the order (Pinruan et al. 2008; Suetrong et al. unpublished data). Taxa related to this order may well occur in freshwater habitats and further studies are warranted.

3.3.1.6 Sordariales

Some 70 sordariaceous species have been reported from freshwater habitats, primarily on woody substrata (Sheraer and Raja 2013) with most species in the

Lasiosphaeriaceae. Over 17 new species have been described from aquatic habitats; many are tropical and probably can be ascribed to most freshwater mycologists working in the tropics. The most speciose genus is *Chaetomium*, none of which have been described originally from freshwater, and thus may not necessarily be true aquatic species. Not all taxa assigned to this order can be referred to families: e.g. *Cuspidatispora* (Miller et al. 2006). Pang et al. (2013) referred the marine *Saccardoella* species (*S. mangrovei, S. marinospora, S. rhizophorae, S. tiomanensis*) to a new genus and family: *Dyfrolomyces*, Dyfrolomycetaceae, in the Dothideomycetes. Further molecular studies are required to determine the correct assignment of the five freshwater species listed in Shearer and Raja (2013).

3.3.2 Sordariomycetidae *incertae sedis*

As stated above, many taxa assigned to the Sordariomycetidae cannot be refered to either a family or order. For example, Ferrer et al. (2012) described three new genera: *Bullimyces* (3 species), *Riomyces* (2 species) and *Hydromelitis* (1 species) from submerged woody debris in freshwater habitats from Costa Rica, but placement at the family level is uncertain, despite supporting molecular data. The three genera grouped within an unsupported clade consisting of members of Ophiostomatales, Magnaporthales and freshwater Annulatascaceae *sensu lato* and *sensu stricto*. *Flammispora bioteca* is another new genus that cannot be accommodated in any family or order although morphologically it shares a number of features with members of the Halosphaeriaceae (Pinruan et al. 2004). Ascomata are immersed in leaf tissue, with deliquescing asci and 5-septate ascospores with a single polar appendage. This species was collected on decaying leaves of the palm *Licuala longecalycata* in a peat swamp in Thailand, and forms an unsupported sister clade to the Microascales (Pinruan et al. 2004).

3.3.3 Hypocreomycetidae

3.3.3.1 Savoryellales

Savoryellales is a recently established order with most of its members from freshwater and marine environments (Boonyuen et al. 2011). Common species in this order include species accommodated in *Savoryella, Ascotaiwania* and *Ascothailandia*. *Ascotaiwania* and *Savoryella* are the most commonly encountered genera from freshwater habitats, with eight and seven reported species from freshwater, respectively (Cai et al. 2006a; Boonyuen et al. 2011). Freshwater members in Savoryellales are morphologically characterized by having an apical ascus ring and brown ascospores with hyaline end cells or polar pads. While *Ascotaiwania* is different from *Savoryella* in having a relatively massive apical ring; the latter has a small and non-partite ring. *Ascotaiwania* was previously classified in Annulatascaceae, and the fact that it belongs to Savoryellales further shows the multiple origins of the massive apical ring.

Boonyuen et al. (2011) used 28S rDNA, 18S rDNA and *rpb2* regions to analyse the phylogenetic relationships of Savoryellales. The monophyly of *Ascotaiwania* (with exclusion of *A. hughesii*), *Savoryella* and *Ascothailandia* (*Canalisporium*) was generally supported. *Ascothailandia* is a genus recently established for the sexual morph of *Canalisporium*. According to the current one name system ruled by the Melbourne code, these two genera should be combined and *Canalisporium* is likely to take the priority for its much longer history and wider use in literature.

3.3.3.2 Microascales

The Microascales comprises four families: Ceratocystidaceae, Gondwanamycetaceae, Halosphaeriaceae and Microascaceae (Réblová et al. 2011), and of these the Halosphaeriaceae includes a number of freshwater species. In Fig. 3.1, 13 freshwater strains clustered in the family Halosphaeriaceae, including *Lignincola, Natantispora, Phaeonectriella, Ascosacculus, Nais, Aniptodera* and *Magnisphaera*. Other species also found in freshwater habitats include *Fluviatispora, Luttrellia, Oceanitis*, and *Panorbis* (Pang 2012). The Halosphaeriaceae comprises 148 species in 58 genera, with over 97 strains sequenced (Sakayaroj et al. 2011; Pang 2012). These taxa are interesting as many of them have been reported from freshwater, brackish and marine environments. Most of freshwater halosphaeriacious species are found on submerged, trapped wood, or on the culms of macrophytes in rivers, lakes and even water cooling towers (Eaton and Jones 1971). The Microascales forms a strongly supported order in the Hypocreomycetidae in a parsimonious tree generated from 18S, 28S, ITS rDNA and RPB2 sequences (Réblová et al. 2011).

3.3.3.3 Hypocreales

Nineteen species (in nine genera) of freshwater fungi have been reported for the Hypocreales, most known as new species growing on wood (Shearer and Raja 2013). Genera reported include: *Bionectria, Cosmospora, Nectria* (10 strains) and *Nectriella*. *Baipadisphaeria* (on palm fronds in a peat swamp), and *Paraniesslia* (on submerged wood) are new genera and their placement in the Hypocreales is supported by molecular data (Cai and Hyde 2007; Pinruan et al. 2010). *Baipadisphaeria spathulospora*, in common with many hypocreaceous species, has orange ascomata which later become brown (Pinruan et al. 2010). *Nectria lugdunensis* was originally known as the asexual stage (*Heliscus lugdunensis*) on senescent leaves and was linked to its sexual stage by Webster (1959).

3.3.4 Xylariomycetidae

3.3.4.1 Xylariales

Xylariaceous species reported from freshwater habiats number 17, in 13 genera of which three are new and monotypic genera: *Aquadulciospora* (Fallah and Shearer 2001), *Aquasphaeria* (Hyde 1995), and *Mukhakesa* (Udaiyan and Hosagoudar 1992). Genera with more than one species include: *Phomatospora* (3 with *Ph. muskellingensis* a new freshwater species), *Physalospora* (2 with *P. aquatica* a new species) and *Hypoxylon* (2, but these cannot be regarded as truly freshwater species). Hyde and Goh (1998) described *Anthostomella aquatica* from submerged wood in Australia with an asexual morph in the genus *Geniculosporium*, and referred to the Xylariales.

3.3.4.2 Phyllachorales

Only one species in this order has been reported from freshwater, namely *Phyllachora therophila* growing on *Juncus filiformis*, collected in Austria by Magnes and Hafellner (1991).

3.3.4.3 Trichosphaeriales

Minoura and Muroi (1978) collected *Debaryella gracilis* on balsa wood in Japan, the only species recorded for this order.

3.3.5 Discomycetes

3.3.5.1 Helotiales

Freshwater discomycetes include species in the Helotiales, Pezizales and Rhytismatales, orders well supported phylogenetically (Hibbett et al. 2007) and number some 110 species in 34 genera (Shearer and Raja 2013). Some 41 species occur on woody materials (e.g. twigs, catkins) and the remainder on herbaceous plants (leaves, stems) such as, *Carex, Scirpus,* and *Phragmites* spp. Only a few species have been reported from tropical locations (*Cudoniella indica, Hymenoscyphus malawiensis, H. varicosporoides, Orbilia luterobella, Pezoloma rhodocarpa*). The lack of discomycetes in the tropics has been supported by detailed studies of submerged woody material in tropical locations (Ho et al. 2001; Sivichai et al. 2002; See also Chapters 17, 21 this volume). More detailed examination of aquatic herbaceous plants might yield a wider range of species. Not all species reported from aquatic habitats may truly be freshwater taxa, for example, *Hymenoscyphus scutulus* collected on submerged stems of *Scirpus atrovirens* (Fallah and Shearer 2001) is also known from terrestrial habitats. The most speciose freshwater genera are *Belonpsis* (5 species), *Hymenoscyphus* (12), *Lachnum* (4), *Mollisia* (9), *Niptera* (5) and *Vibrissea* (16).

Of the 102 Helotiales species, 22 have been described on substrates collected in freshwater habitats. Most of these taxa have been described based on morphological features. Zhang and Zhuang (2004) examined the phylogenetic affinities of selected *Hymenoscyphus* species and showed they grouped in two clades: clade 1) comprised the lectotype of the genus, *H. fructigenus*, which formed a monophyletic group with *H. caudatus, H. fucatus, H. menthae* and five other species, while clade 2) included *H. ericae, H. rhodolucus* and *Calycina herbarum*. Boonyuen et al. (2006) examined the relationship between *Hymenoscyphus varicosporoides* and its asexual morph *Tricladium indica* (*Cudoniella indica*) confirming its placement in the Helotiales. They also demonstrated that other Ingoldian taxa formed three separate clades in the order: clade B) *Tricladium angulatum, Anguillospora filiformis* and *Varicosporium elodeae*; clade C) *Anguillospora crassa, A. furtiva*, the marine anamorph *Halenospora varia*, and *Tricladium splendens*; clade D) *Tricladium chaetocladium, Varicosporium delicarum*, and *Dimorphospora folicola*, confirming the polyphyletic nature of the various genera.

Baschien et al. (2006), on the basis of 18S rDNA sequences of 36 aquatic hyphomycetes, referred the species *Alatospora acuminata, Anguillospora crassa, A. furtiva, Lemonniera aqutica, L. terrestris, Tetracladium marchalianum, Tricladium angulatum, T. splendens* and *Varisosporium elodeae*, to the Leotiomycetes, Helotiales. In a subsequent paper (Baschien et al. 2013) they confirmed that the aquatic genera *Anguillospora, Articulospora, Filosporella, Flagellospora, Lemonniera, Tricladium* and *Varisosporium* were not monophyletic, with species from the same genus distributed in several different clades. They concluded that adaptation to aquatic habitats had evolved independently in multiple phylogenetic lineages within the Leotiomycetes. In a series of publications the use of the ITS rRNA gene as a barcode for identifying species of aquatic hyphomycetes has been proposed (Letourneau et al. 2010; Seena et al. 2010; Duarte et al. 2012, 2013). See also discussion in Chapter 1 (this volume).

3.3.5.2 Pezizales

The order has eight species in eight genera reported from freshwater environments, all initially described from terrestrial habitats (Shearer and Raja 2013). All occur on woody substrata in temperate locations with two on wood in water cooling towers: *Peziza* sp. (Eaton and Jones 1971) and *Saccobolus beckii* (Udaiyan 1989). More recently Hu et al. (2012b) described the new genus *Aquapeziza* from a small freshwater stream in Yunnan Province, China. The genus is characterized by white epigenous apothecia, ovoid amyloid asci and multi-guttulate single-celled, smooth, globose ascospores.

Phylogenetically, based on 28S rDNA sequences, it groups in a clade comprising *Boudiera dennisii*, *B. tracheia* with three *Pachyella* species in a sister group in the Pezizaceae. *Pachyella babingtonii* and *P. violaceonigra* have previously been reported from freshwater habitats (Ingold 1954).

3.3.5.3 Rhytismatales

Two *Lophodermium* species have been collected from freshwater habitats: *L. alpinum* and *L. scirpinum*, on *Carex nigra* (Ingold 1954) and *Schoenoplectus lacustris* (Magnes and Hafellner 1991), respectively. *Lophodermium* is a large genus of 103 species (MycoBank) occurring especially on conifer needles, or as asymptomatic endophytes of various plants. Ortiz-García et al. (2003) examined the phylogenetic relationships of eleven *Lophodermium* species with those from pine hosts forming a monophyletic sister group to *Lophodermium* species from more distant hosts. No sequence data are available for the two species reported from freshwater habitats.

3.4 Concluding remarks

Members in Annulatascaceae *sensu lato* are the most commonly encountered species of freshwater Sordariomycetes. However, phylogenetic relationships of genera are confused and not well resolved. The massive apical ring used to define Annulatascaceae, has multiple origins and cannot be used to unite all annulatascaceous genera. Annulatascaceae is phylogenetically polyphyletic but a taxonomic revision of this group, like most ascomycete groups, suffers from lack of living cultures and type-derived sequences. The establishment of many genera was based on single character, many of which need to be evaluated for their evolutionary significance. Many freshwater Sordariomycetes are currently not placed in presently known orders or families. This is also a reflection of the lack of investigation of this ecological group that encompasses potentially very high and unknown biodiversity.

To obtain a better understanding of phylogeny in freshwater Sordariomycetes and discomycetes in a natural classification system, further investigation is needed. The new investigation should put high significance in obtaining living cultures. Epitypification is also important for this group, for example the type species of the type genus of Annulatascaceae, i.e. *A. velatispora*, and many other taxa.

Acknowledgments

L.C. and D.M.H. acknowledge NSFC (31322002 & 31093440) for the supporting his study in systematics and biodiversity of freshwater fungi. Gareth Jones is supported by the Distinguished Scientist Fellowship Program (DSFP), King Saud University, Saudi Arabia. Dr. Nattuwut Boonyuen is thanked for preparing Fig. 3.

References

Abdel-Wahab MA, Abdel-Aziz FA, Mohamed SS, Abdel-Aziz AE. *Annulatascus nilensis* sp. nov., a new freshwater ascomycete from the River Nile, Egypt. IMA Fungus 2011;2:1–6.

Barr ME. The ascomycete connection. Mycologia 1983;75:1–13.

Baschien C, Marvanova L, Szewzyk U. Phylogeny of selected aquatic hyphomycetes based on morphological and molecular data. Nova Hedwig 2006;83:311–352.

Baschien C, Tsui CKM, Gulis V, Szewzyk U, Marvanova L. The molecular phylogeny of aquatic hyphomycetes with affinity to the Leotiomycetes. Fungal Biol 2013;117:660–672.

Boonyuen N, Sivichai S, Worapong J, Hywel-Jones N. A molecular phylogenetic study of selected ingoldian species. BRT Research Reports 2006:56–63.

Boonyuen N, Chuaseeharonnachai C, Suetrong S, Sri-Indrasutdhi V, Sivichai S, Jones EBG, et al. Savoryellales (Hypocreomycetidae, Sordariomycetes): a novel lineage of aquatic ascomycetes inferred from multiple-gene phylogenies of the genera *Ascotaiwania*, *Ascothailandia*, and *Savoryella*. Mycologia 2011;103:1351–1371.

Boonyuen N, Sri-Indrasutdhi V, Suetrong S, Sivichai S, Jones EBG. *Annulatascus aquatorba* sp. nov., a lignicolous freshwater ascomycete from Sirindhorn Peat Swamp Forest, Narathiwat, Thailand. Mycologia 2012;104:746–757.

Cabanela MV, Jeewon R, Hyde KD. Morphotaxonomy and phylogeny of *Paoayensis lignicola* gen. et sp. nov.(ascomycetes) from submerged wood in Paoay Lake, Ilocos Norte, the Philippines. Cryptog Mycol 2007;28:301–310.

Cai L, Hyde KD. New species of *Clohiesia* and *Paraniesslia* collected from freshwater habitats in China. Mycoscience 2007;48:182–186.

Cai L, Jeewon R, Hyde KD. Phylogenetic evaluation and taxonomic revision of *Schizothecium* based on ribosomal DNA and protein coding genes. Fungal Divers 2005;19:1–21.

Cai L, Hyde KD, Tsui CK M. Genera of Freshwater Fungi. Thailand: Fungal Diversity Press 2006a:1–261.

Cai L, Jeewon R, Hyde KD. Molecular systematics of *Zopfiella* and allied genera: evidence from multi-gene sequence analyses. Mycol Res 2006b;110:359–368.

Cai L, Zhang KQ, Hyde KD. Freshwater ascomycetes. In: Tsui KM, Hyde KD, eds. Freshwater Mycology. Hong Kong: Fungal Diversity Press 2003:275–326.

Campbell J, Anderson JL, Shearer CA. Systematics of Halosarpheia based on morphological and molecular data. Mycologia 2003a;95:530–552.

Campbell J, Shearer CA. *Annulusmagnus* and *Ascitendus*, two new genera in the Annulatascaceae. Mycologia 2004;96:822–833.

Campbell J, Shearer CA, Crane JL, Fallah PM. A reassessment of two freshwater ascomycetes, *Ceriospora caudae-suis* and *Submersisphaeria aquatica*. Mycologia 2003b;95:41–53.

Castlebury LA, Rossman AY, Jaklitsch WJ, Vasilyeva LN. A preliminary overview of the Diaporthales based on large subunit nuclear ribosomal DNA sequences. Mycologia 2002;94:1017–1031.

Castlebury LA, Rossman AY, Sung GH, Hyten AS, Spatafora JW. Multigene phylogeny reveals new lineage for *Stachybotrys chartarum*, the indoor air fungus. Mycol Res 2004;108:864–872.

Chaverri P, Salgado C, Hirooka Y, Rossman AY, Samuels GJ. Delimitation of *Neonectria* and *Cylindrocarpon* (Nectriaceae, Hypocreales, Ascomycota) and related genera with *Cylindrocarpon*-like anamorphs. Stud Myco 2011;68:57–78.

Crane JL, Shearer CA. A new *Coniochaeta* from fresh water. Mycotaxon 1995;54:107–110.

Crous PW, Groenewald JZ, Risède JM, Simoneau P, Hywel-Jones NL. *Calonectria* species and their *Cylindrocladium* anamorphs: species with sphaeropedunculate vesicles. Stud Mycol 2004; 50:415–430.

Duarte S, Seena S, Bärlocher F, Cássio F, Pascoal C. Preliminary insights into the phylogeography of six aquatic hyphomycete species. PLoS ONE 2012;7:e45289.

Duarte S, Seena S, Bärlocher F, Pascoal C, Cássio F. A decade's perspective on the impact of DNA sequencing on aquatic hyphomycete research. Fungal Biol Rev 2013;27:19–24.
Eaton RA, Jones EBG. The biodeterioration of timber in water cooling towers. I. Fungal ecology and the decay of wood at Connah's Quay and Ince. Material und Organismen 1971;6:51–80.
Fallah PM, Shearer CA. Freshwater ascomycetes: new or noteworthy species from north temperate lakes in Wisconsin. Mycologia 2001;93:566–602.
Fernández FA, Miller AN, Huhndorf SM, Lutzoni FM, Zoller S. Systematics of the genus *Chaetosphaeria* and its allied genera: morphological and phylogenetic diversity in north temperate and neotropical taxa. Mycologia 2006;98:121–130.
Ferrer A, Miller AN, Sarmiento C, Shearer CA. Three new genera representing novel lineages of Sordariomycetidae (Sordariomycetes, Ascomycota) from tropical freshwater habitats in Costa Rica. Mycologia 2012;104:865–879.
Hibbett DS, Binder M, Bischoff JF, Blackwell M, Cannon PF, Eriksson OE, et al. A higher-level phylogenetic classification of the Fungi. Mycol Res 2007;111:509–547.
Ho WH, Hyde KD. A new family of freshwater ascomycetes. Fungal Divers 2000;4:21–36.
Ho WH, Hyde KD, Hodgkiss IJ. Fungal communities on submerged wood from streams in Brunei, Hong Kong, and Malaysia. Mycol Res 2001;105:1492–1501.
Ho WH, Tsui CKM, Hodgkiss IJ, Hyde KD. *Aquaticola*, a new genus of Annulatascaceae from freshwater habitats. Fungal Divers 1999;3:87–97.
Hu DM, Cai L, Hyde KD. Three new ascomycetes from freshwater in China. Mycologia 2012a; 104:1478–1489.
Hu DM, Chen H, Cai L, Bahkali AH, Hyde KD. *Aquapeziza*: a new genus from freshwater and its morphological and phylogenetic relationships to Pezizaceae. Mycologia 2012b;104:540–546.
Hu DM, Liu F, Cai L. Biodiversity of aquatic fungi in China. Mycology 2013;4 :125–168.
Huhndorf SM, Fernández FA, Taylor JE, Hyde KD. Two pantropical ascomycetes: *Chaetosphaeria cylindrospora* sp. nov. and *Rimaconus*, a new genus for *Lasiosphaeria jamaicensis*. Mycologia 2001;93:72–80.
Huhndorf SM, Miller AN, Fernández FA. Molecular systematics of the Sordariales: the order and the family Lasiosphaeriaceae redefined. Mycologia 2004;96:368–387.
Hyde KD. Tropical Australian freshwater fungi. II. *Annulatascus velatispora* gen. *et* sp. nov., *A. bipolaris* sp. nov. and *Nais aquatica* sp. nov. (Ascomycetes). Aust Syst Bot 1992;5:117–124.
Hyde KD. Tropical Australian fresh-water fungi. 7. New genera and species of Ascomycetes. Nova Hedwig 1995;61:119–140.
Hyde KD. Tropical Australian freshwater fungi. X. *Submersisphaeria aquatica* gen et *sp* nov. Nova Hedwig 1996;62:171–175.
Hyde KD, Goh TK. Tropical Australian freshwater fungi. XIII. A new species of *Anthostomella* and its sporodochial *Geniculosporium* anamorph. Nova Hedwig 1998;67:225–233.
Hyde KD, Goh TK. Fungi on submerged wood in the Riviere St Marie-Louis, the Seychelles. S Afr J Bot 1998;64:330–336.
Hyde KD, Ho WH, Jones EB, Tsui CKM, Wong WSW. *Torrentispora fibrosa* gen. sp. nov. (Annulatascaceae) from freshwater habitats. Mycol Re 2000;104:1399–1403.
Hyde KD, Wong SW, Jones EBG. *Cataractispora* gen. nov. with three new freshwater lignicolous species. Mycol Res 1999;103:1019–1031.
Ingold CT. Aquatic ascomycetes: Discomycetes from lakes. Trans Br Mycol Soc 1954;37:1–18.
Jeewon R, Cai L, Liew ECY, Zhang KQ, Hyde KD. *Dyrithiopsis lakefuxianensis* gen. *et* sp. nov. from Fuxian Lake, Yunnan, China, and notes on the taxonomic confusion surrounding *Dyrithium*. Mycologia 2003;95:911–920.
Kirk PM, Cannon PF, Minter DW, Stalpers JA. Dictionary of the Fungi. 10th edition. Wallingford, Oxon, UK:CABI 2008:1–784.

Kohlmeyer J, Spatafora JW, Volkmann-Kohlmeyer B. Lulworthiales, a new order of marine Ascomycota. Mycologia 2000;92:453–458.

Letourneau A, Seena S, Marvanová L, Bärlocher F. Potential use of barcoding to identify aquatic hyphomycetes. Fungal Divers 2010;40:51–64.

Linnakoski R, De Beer ZW, Rousi M, Niemelä P, Pappinen A, Wingfield MJ. Fungi, including *Ophiostoma karelicum* sp. nov., associated with *Scolytus ratzeburgi* infesting birch in Finland and Russia. Mycol Res 2008;112:1475–1488.

Liu F, Hu DM, Cai L. *Conlarium duplumascospora* gen. et sp. nov. and *Jobellisia guangdongensis* sp. nov. from freshwater habitats in China. Mycologia 2012;104:1178–1186.

Luo J, Zhang N. *Magnaporthiopsis*, a new genus in Magnaporthaceae (Ascomycota). Mycologia 2013; 105:1019–1029.

Luo J, Zhuang WY. *Chaetopsinectria* (Nectriaceae, Hypocreales), a new genus with *Chaetopsina* anamorphs. Mycologia 2010;102:976–984.

Magnes M, Hafellner J. Ascomycetes on vascular plants growing on the shores of mountain lakes in the Eastern Alps. Bibliotheca Mycologica 991;139:1–185.

Miller AN, Huhndorf SM. Multi-gene phylogenies indicate ascomal wall morphology is a better predictor of phylogenetic relationships than ascospore morphology in the Sordariales (Ascomycota, Fungi). Molec Phylogen Evol 2005;35:60–75.

Miller AN, Shearer CA, Bartolata M, Huhndorf SM. *Cuspidatispora xiphiago* gen. *et* sp. nov. from an eastern North American creek. Mycoscience 2006;47:218–223.

Minoura K, Muroi T. Some freshwater Ascomycetes from Japan. Trans Mycol Soc Jpn 1978;19:129–134.

Ortiz-García S, Gernandt DS, Stone JK, Johnston PR, Chapela IH, Salas-Lizana R, et al. Phylogenetics of *Lophodermium* from pine. Mycologia 2003;95:846–859.

Pang KL. Phylogeny of the marine Sordariomycetes. In: Marine Fungi and Fungal-like Organisms. Jones EBG, Pang KL, eds. Berlin: De Gruyter 2012.

Pang KL, Jones EG, Vrijmoed LL, Vikineswary S. *Okeanomyces*, a new genus to accommodate *Halosphaeria cucullata* (Halosphaeriales, Ascomycota). Bot J Linn Soc 2004;146:223–229.

Pang KL, Vrijmoed LLP, Jones EBG. Genetic variation within the cosmopolitan aquatic fungus *Lignincola laevis* (Microascales, Ascomycota). Org Divers Evol 2013;13:301–309.

Pang KL, Vrijmoed LLP, Kong RYC, Jones EBG. *Lignincola* and *Nais*, polyphyletic genera of the Halosphaeriales (Ascomycota). Mycol Prog 2003;2:29–36.

Pinruan U, Rungjindamai N, Sakayaroj J, Lumyong S, Hyde KD, Jones EBG. *Baipadisphaeria* gen. nov., a freshwater ascomycete (Hypocreales, Sordariomycetes) from decaying palm leaves in Thailand. Mycosphere 2010;1:53–63.

Pinruan U, Sakayaroj J, Hyde KD, Jones EBG. *Thailandiomyces bisetulosus* gen. *et* sp. nov. (Diaporthales, Sordariomycetidae, Sordariomycetes) and its anamorph *Craspedodidymum*, is described based on nuclear SSU and LSU rDNA sequences. Fungal Divers 2008;29:89–98.

Pinruan U, Sakayaroj J, Jones EBG, Hyde KD. *Flammispora* gen. nov., a new freshwater ascomycete from decaying palm leaves. Stud Mycol 2004;50:381–386.

Réblová M. *Bellojisia*, a new sordariaceous genus for *Jobellisia rhynchostoma* and a description of Jobellisiaceae fam. nov. Mycologia 2008;100:893–901.

Réblová M. Two taxonomic novelties in the Sordariomycetidae: *Ceratolenta caudata* gen. *et* sp. nov. and *Platytrachelon abietis* gen. *et* comb. nov. for *Ceratosphaeria abietis*. Mycologia 2013; 105:462–475.

Réblová M, Fournier J, Hyde KD. *Achroceratosphaeria*, a new ascomycete genus in the Sordariomycetes, and re-evaluation of *Ceratosphaeria incolorata*. Fungal Divers 2010;43:75–84.

Réblová M, Gams W, Seifert KA. *Monilochaetes* and allied genera of the Glomerellales, and a reconsideration of families in the Microascales. Stud Mycol 2011;68:163–191.

Réblová M, Seifert KA. *Conioscyphascus*, a new ascomycetous genus for holomorphs with *Conioscypha* anamorphs. Stud Mycol 2004;50:95–108.
Réblová M, Seifert KA. A new fungal genus, *Teracosphaeria*, with a phialophora-like anamorph (Sordariomycetes, Ascomycota). Mycol Res 2007;111:287–298.
Réblová M, Winka K. Generic concepts and correlations in Ascomycetes based on molecular and morphological data: *Lecythothecium duriligni* gen. *et* sp. nov. with a *Sporidesmium* anamorph, and *Ascolacicola austriaca* sp. nov. Mycologia 2001;93:478–493.
Raja H, Schoch CL, Hustad V, Shearer C, Miller A. Testing the phylogenetic utility of MCM7 in the Ascomycota. MycoKeys 2011;1:63–94.
Raja HA, Campbell J, Shearer CA. Freshwater ascomycetes: *Cyanoannulus petersenii*, a new genus and species from submerged wood. Mycotaxon 2003;88:1–17.
Raja HA, Fournier J, Shearer CA, Miller AN. Freshwater ascomycetes: *Coniochaeta gigantospora* sp. nov. based on morphological and molecular data. Mycoscience 2012;53:373–380.
Raja HA, Schmit JP, Shearer CA. Latitudinal, habitat and substrate distribution patterns of freshwater ascomycetes in the Florida Peninsula. Biodivers Conserv 2009;18:419–455.
Ranghoo VM, Hyde KD, Liew ECY, Spatafora JW. Family placement of *Ascotaiwania* and *Ascolacicola* based on DNA sequences from the large subunit rRNA gene. Fungal Divers 1999;2:159–168.
Ranghoo VM, Hyde KD, Wong SW, Tsui CKM, Jones EBG. *Vertexicola caudatus* gen. *et* sp. nov., and a new species of *Rivulicola* from submerged wood in freshwater habitats. Mycologia 2000; 92:1019–1026.
Ranghoo VM, Tsui CKM, Hyde KD. *Brunneosporella aquatica* gen. et sp. nov., *Aqualignicola hyalina* gen. *et* sp. nov., *Jobellisia viridifusca* sp. nov. and *Porosphaerellopsis bipolaris* sp. nov. (ascomycetes) from submerged wood in freshwater habitats. Mycol Res 2001;105:625–633.
Rossman AY, McKemy JM, Pardo-Schultheiss RA, Schroers HJ. Molecular studies of the Bionectriaceae using large subunit rDNA sequences. Mycologia 2001;93:100–110.
Sakayaroj J, Pang KL, Jones EBG. Multi-gene phylogeny of the Halosphaeriaceae: its ordinal status, relationships between genera and morphological character evolution. Fungal Divers 2011; 46:87–109.
Schoch CL, Seifert KA, Huhndorf S, Robert V, Spouge JL, Levesque CA, et al. Nuclear ribosomal internal transcribed spacer (ITS) region as a universal DNA barcode marker for Fungi. Proc Natl Acad Sci USA 2012;109:6241–6246.
Seena S, Pascoal C, Marvanová L, Cássio F. DNA barcoding of fungi: a case study using ITS sequences for identifying aquatic hyphomycete species. Fungal Divers 2010;44:77–87.
Shearer CA. (1993) The freshwater ascomycetes. Nova Hedwig 2012;56:1–33.
Shearer CA. The distribution of freshwater filamentous Ascomycetes. In: Misra JK, Horn BW, eds. Fungal Groups: Robert W. Lichtwardt Commemoration. Trichomycetes and other. Enfield, New Hampshire, USA: Science Publishers, Inc. 2001:225–292.
Shearer CA, Crane JL, Chen W. Freshwater ascomycetes: *Ophioceras* species. Mycologia 1999; 91:145–156.
Shearer CA, Raja HA.Freshwater Ascomycetes Database: http://fungi.life.illinois.edu/ (Accessed on May 11, 2013).
Sivichai S, Jones EBG, Hywel-Jones N. Fungal colonisation of wood in a freshwater stream at Tad Ta Phu, Khao Yai National Park, Thailand. Fungal Divers 2002;10:113–129.
Smith GJD, Liew ECY, Hyde KD. The Xylariales: a monophyletic order containing 7 families. Fungal Divers 2003;13:175–208.
Sogonov MV, Castlebury LA, Rossman AY, Mejía LC, White JF. Leaf-inhabiting genera of the Gnomoniaceae, Diaporthales. Stud Mycol 2008;62:1–77.
Spatafora JW, Sung GH, Johnson D, Hesse C, O'Rourke B, Serdani M, et al. A five-gene phylogeny of Pezizomycotina. Mycologia 2006;98:1018–1028.

Spatafora JW, Volkmann-Kohlmeyer B, Kohlmeyer J. Independent terrestrial origins of the Halosphaeriales (marine Ascomycota). Am J Bot 1998;85:1569–1580.

Sri-Indrasutdhi V, Boonyuen N, Suetrong S, Chuaseeharonnachai C, Sivichai S, Jones EBG. Wood-inhabiting freshwater fungi from Thailand: *Ascothailandia grenadoidia* gen. *et* sp. nov., *Canalisporium grenadoidia* sp. nov. with a key to *Canalisporium* species (Sordariomycetes, Ascomycota). Mycoscience 2012;51:411–420.

Stamatakis A. RAxML-VI-HPC: maximum likelihood-based phylogenetic analyses with thousands of taxa and mixed models. Bioinformatics 2006;22:2688–2690.

Summerbell RC, Gueidan C, Schroers HJ, De Hoog GS, Starink M, Rosete YA, et al. *Acremonium* phylogenetic overview and revision of *Gliomastix*, *Sarocladium*, and *Trichothecium*. Stud Mycol 2011;68:139–162.

Sung G-H, Hywel-Jones NL, Sung JM, Luangsa-ard JJ, Shrestha B, Spatafora JW. Phylogenetic classification of *Cordyceps* and the clavicipitaceous fungi. Stud Mycol 2007;57:5–59.

Thongkantha S, Jeewon R, Vijaykrishna D, Lumyong S, McKenzie EHC, Hyde KD. Molecular phylogeny of Magnaporthaceae (Sordariomycetes) with a new species, *Ophioceras chiangdaoense* from *Dracaena loureiroi* in Thailand. Fungal Divers 2009;34:157–173.

Thomas K. Australian freshwater fungi. In: Grgurinovic CA, ed. Introductory Volume to the Fungi (Part 2). Fungi of Australia. Canberra, Australia: Australian Biological Resources Study 1996:1–37.

Tsui KM, Hyde KD, Hodgkiss IJ. A new species of *Clohiesia* from Hong Kong. Mycoscience 1998; 39:257–259.

Udaiyan K. Some interesting ascomycetes from water cooling towers. Kavaka 1989;17:11–16

Udaiyan K, Hosagoudar VS. Some interesting fungi from the industrial water cooling towers of Madras II. J Econ Taxon Bot 1989;15:649–666.

Vijaykrishna D, Jeewon R, Hyde KD. *Fusoidispora aquatica*: a new freshwater ascomycete from Hong Kong based on morphology and phylogeny inferred from rDNA gene sequences. Sydowia 2005; 57:267–280.

Vijaykrishna D, Jeewon R, Hyde KD. Molecular taxonomy, origins and evolution of freshwater ascomycetes. Fungal Divers 2006;23:351–390.

Wang CY, Fang ZM, Sun BS, Gu LJ, Zhang KQ, Sung CK. High infectivity of an endoparasitic fungus strain, *Esteya vermicola*, against nematodes. J Microbiol 2008;46:380–389.

Webster. *Nectria lugdunensis* sp. nov., the perfect state of *Heliscus lugdunensis*. Trans Br Mycol Soc 1959;42:322–327.

Wong SW, Hyde KD. *Proboscispora aquatica* gen. et sp. nov., from wood submerged in freshwater. Mycol Res 1999;103:81–87.

Wong SW, Hyde KD, Jones EBG, Moss ST. Ultrastructural studies on the aquatic ascomycetes *Annulatascus velatisporus* and *A. triseptatus* sp. nov. Mycol Res 1999;103:561–571.

Wong SW, Hyde KD, Jones EBG. Annulatascaceae, a new ascomycete family from the tropics. Syst Ascomycetum 1998;16:17–25.

Zhang N, Castlebury LA, Miller AN, Huhndorf SM, Schoch CL, Seifert KA, et al. An overview of the systematics of the Sordariomycetes based on a four-gene phylogeny. Mycologia 2006;98:1076–1087.

Zhang YH, Zhuang WY. Phylogenetic relationships of some members in the genus *Hymenoscyphus* (Ascomycetes, Helotiales). Nova Hedwig 2004;78:3–4.

E. B. Gareth Jones, Darlene Southworth, Diego Libkind and Ludmila Marvanová

4 Freshwater Basidiomycota

Few Basidiomycota have been reported from freshwater habitats with many reported only in their original description (Shearer et al. 2007; Jones and Choeyklin 2008; Fell et al. 2011). As many are known only from a single survey, care must be taken in regarding them as truly freshwater taxa. Many yeasts are undoubtedly transient and further documentation is required. For example, yeasts are common in fresh, marine and polluted runoff water, but the same species may occur on a variety of non-aquatic substrates, e.g. apple fruit (for reviews see Brandão et al. 2010; de Garcia et al. 2010; Starmer and Lachance 2011). Freshwater Basidiomycota are taxonomically diverse including both yeasts and filamentous forms. Most are known primarily by their asexual states with elaborate conidia including forms that are sigmoid (e.g. *Anguillomyces acadiensis*, *Fibulotaeniella canadensis*), tetraradiate (e.g. *Taeniospora descalsii*, *Ingoldiella hamata*), extremely branched (*Dendrosporomyces prolifer*, *D. splendens*) or bizarrely shaped (*Akenomyces costatus*), and identified as Basidiomycota by the presence of clamp-connections or binucleate cells (Nawawi 1985a). Some have been induced to form sexual stages in culture, e.g. *Sistotrema hamatum*/*Ingoldiella hamata* (Nawawi and Webster 1982) and *Naiadella fluitans*/*Classicula fluitans* (Bauer et al. 2003); a few have been classified based on DNA sequence data.

Freshwater Basidiomycota can be found on a variety of substrates, e.g. on culms of *Equisetum* sp. (*Mrakiella aquatica*), on wood (*Mycocalia reticulata*, *Sistotrema hamatum*, *Stauriella aquatica*), in water (*Rhodotorula* spp., *Cryptococcus* spp.), and in foams (*Crucella subtilis*, *Naiadella fluitans*, *Taeniospora gracilis*) (Tab. 4.1; Figs. 4.1–4.2). Freshwater Basidiomycota are few in number similar to their paucity in marine habitats (Hibbett and Binder 2001; Jones and Fell 2012; Jones and Pang 2012). Clearly, with the exception of certain yeasts, Basidiomycota have rarely adapted to aquatic habitats. This may be a consequence of their evolutionary origin as terrestrial organisms and as symbionts of terrestrial plants (Krings et al. 2012). Basidiomycota with ballistospory require air for spore release (Money 1998; Pringle et al. 2005).

There is no definitive estimate of the numbers of species of freshwater Basidiomycota. Jones and Choeyklin (2008) listed 29 freshwater species. In Tab. 4.1 we list 115 species. However, this list is not exhaustive, and more remain to be described, especially freshwater yeasts, which have been poorly documented (Kurtzman et al. 2011). While some freshwater Basidiomycota occur in various places worldwide, others have been isolated rarely or only once. Most have been collected from submerged decaying leaves or submerged woody materials and are able to breakdown plant cell wall polysaccharides, e.g. pectins, cellulose and hemicellulose, suggesting that they are saprotrophic (Chamier 1985; Zemek et al. 1985).

Tab. 4.1: Basidiomycota from freshwater habitats categorized by growth form (yeasts or filamentous fungi). Taxonomic assignment follows MycoBank and Kurtzman et al. (2011).

Classification	Sexual morph	Asexual morph	Substratum	Reference
		Group 1 yeasts		
Agaricomycotina				
Tremellomycetes				
Cystofilobasidiales	*Cystofilobasidium capitatum*		Lake water	Brandão et al. (2011)
Cystofilobasidiales	*Cystofilobasidium infirmominiatum*		Lake water	Brandão et al. (2011)
Cystofilobasidiales	*Cystofilobasidium lacus-mascardii*		Lake water	Libkind et al. (2009)
Cystofilobasidiales	*Cystofilobasidium macerans*		Glacial melt water, lake water	de Garcia et al. (2007); Libkind et al. (2009)
Cystofilobasidiales		*Guehomyces pullulans*	Lake water	Brandão et al. (2011)
Cystofilobasidiales	*Mrakia frigida*		Glacial meltwater	de Garcia et al. (2007)
Cystofilobasidiales		*Mrakiella aquatica*	Foam, *Equisetum* culms	Jones and Slooff (1966)
Cystofilobasidiales		*Udeniomyces pannonicus*	Meltwater rivers	de Garcia et al. (2007)
Cystofilobasidiales		*Udeniomyces pyricola*	Water samples	Fell et al. (2011)
Filobasidiales	*Filobasidium floriforme*		Water samples	Fell et al. (2011)
Holtermanniales		*Holtermanniella festucosa*	Glacial meltwater	de García et al. (2007); Wuczkowski et al. (2011)
Tremellales		*Bullera dendrophila*	Lake water	Brandão et al. (2011)
Tremellales		*Bullera sinensis*	Water samples	Fell et al. (2011)
Tremellales		*Cryptococcus albidus*	Freshwater lake	van Uden and Ahearn (1963)
Tremellales		*Cryptococcus adeliensis*	Glacial meltwater	de García et al. (2007)
		Cryptococcus agrionensis	Glacial meltwater	de Garcia et al. 2010; Russo et al. 2010
Tremellales		*Cryptococcus cistialbidi*	Stagnant water	Inácio et al. (2005)

(*continued*)

Tab. 4.1: *(continued)*

Classification	Sexual morph	Asexual morph	Substratum	Reference
Tremellales		*Cryptococcus cylindricus*	Glacial meltwater	De García et al. (2007)
Tremellales		*Cryptococcus carnescens*	Lake water	Brandão et al. (2011)
Tremellales		*Cryptococcus diffluens*	Lake water; water samples	vanUdenand Ahearn (1963); Brandão et al. (2011); Fell et al. (2011)
Tremellales		*Cryptococcus gastricus*	Freshwater lake	Van Uden and Ahearn (1963)
Tremellales		*Cryptococcus gilvescens*	Glacial meltwater	Turchetti et al.(2008)
Tremellales		*Cryptococcus heveanensis*	Lake water	Brandão et al. (2011)
Tremellales		*Cryptococcus laurentii*	Freshwater lake	Van Uden and Ahearn (1963)
Tremellales		*Cryptococcus magnus*	Lake water	Brandão et al. (2011)
Tremellales		*Cryptococcus saitoi*	Lake water	Brandão et al. (2011)
		Cryptococcus spencermartinsiae	Glacial meltwater	de Garcia et al. 2010; Russo et al. 2010
Tremellales		*Cryptococcus stepposus*	Lake water	Brandão et al. (2011)
Tremellales		*Cryptococcus taeanensis*	Water samples	Fell et al. (2011)
Tremellales		*Cryptococcus tephrensis*	Lake water	Brandão et al. (2011)
Tremellales		*Cryptococcus terreus*	Water samples	Fell et al. (2011)
Tremellales		*Cryptococcus victoriae*	Lake water	Brandão et al. (2011); Vaz et al. (2011)
Tremellales		*Cryptococcus wieringae*	Lake water	Brandão et al. (2011)
Tremellales		*Dioszegia crocea*	Glacial meltwater	De Garcia et al. (2007)
Tremellales		*Dioszegia fristingensis*	Glacial meltwater	De Garcia et al. (2007)
Tremellales		*Dioszegia hungarica*	Glacial meltwater	De Garcia et al. (2007)
Tremellales		*Dioszegia zsoltii*	Water samples	Fell et al. (2011)
Trichoporonales		*Trichosporon aquatile*	Unknown	Sugita (2011)
Trichoporonales		*Trichosporon cutaneum*	Lake water	Van Uden and Ahearn (1963)
Trichoporonales		*Trichosporon moniliiforme*	Antarctic lake	Guého et al. (1992b)

(continued)

Tab. 4.1: *(continued)*

Classification	Sexual morph	Asexual morph	Substratum	Reference
Pucciniomycotina				
Cystobasidiomycetes				
Cystobasidiales		*Rhodotorula minuta*	Water samples	Fell et al. (2011)
Cystobasidiales		*Rhodotorula pinicola*	Lake water	Brandão et al. (2011)
Cystobasidiales		*Rhodotorula slooffiae*	Lake water	Brandão et al. (2011)
Cystobasidiales		*Rhodotorula laryngis*	Glacial meltwater	De Garcia et al. (2007)
Cystobasidiales		*Rhodotorula meli*	Glacial meltwater	Libkind et al. (2010)
Incertae sedis		*Cyrenella elegans*	Periodically inundated sand near lake	Gochenaur (1981); Sampaio GenBank EF450544
Microbotryomycetes				
Leucosporidiales	*Leucosporidium golubevii*		River water	Sampaio et al. (2003)
Leucosporidiales	*Leucosporidium scottii*		River water	Summerbell (1983)
Leucosporidiales		*Leucosporidiella fragaria*	Glacial meltwater	De García et al. (2007)
Leucosporidiales		*Leucosporidiella creatinivora*	Glacial meltwater	De García et al. (2007)
Leucosporidiales		*Leucosporidiella muscorum*	Mountain lake	Libkind et al. (2009)
Sporidiobolales		*Rhodotorula colostri*	Glacial meltwater	De Garcia et al. (2007);
Sporidiobolales		*Rhodotorula mucilaginosa*	Freshwater lake; glacial meltwater	Van Uden and Ahearn (1963); de Garcia et al. (2007)
Sporidiobolales	*Rhodosporidium sphaerocarpum*		Water	Sampaio (2011)
Sporidiobolales	*Rhodosporidium babjevae*		River water	Libkind et al. (2003);
Sporidiobolales	*Rhodosporidium diobovatum*		Water samples	Fell et al. (2011)
Sporidiobolales	*Rhodosporidium toruloides*		Acidic volcanic river	Russo et al. (2008);
Sporidiobolales	*Rhodosporidium kratochvilovae*		Glacial meltwater	Libkind et al. (2003); de García et al. (2007);
Sporidiobolales	*Sporidiobolus ruineniae*		Water samples	Fell et al. (2011)

(continued)

Tab. 4.1: *(continued)*

Classification	Sexual morph	Asexual morph	Substratum	Reference
Sporidiobolales		*Sporobolomyces beijingensis*	Water samples	Fell et al. (2011)
Sporidiobolales	*Sporidiobolus longiusculus*		Lake water	Libkind et al. (2005)
Sporidiobolales		*Sporobolomyces patagonicus*	Lake water	Libkind et al. (2005)
Sporidiobolales	*Sporidiobolus salmonicolor*		Lake water	Libkind et al. (2005)
Sporidiobolales	*Sporidiobolus metaroseus*		Water samples	Libkind et al. (2005); Fell et al. (2011)
Sporidiobolales		*Sporobolomyces ruberrimus*	Water samples, glacial meltwater	Fell et al. (2011); de Garcia et al. (2007)
Incertae sedis		*Rhodotorula cresolica*	Water samples	Fell et al. (2011)
Incertae sedis		*Rhodotorula vanillica*	Water samples	Fell et al. (2011)
Incertae sedis		*Rhodotorula ingeniosa*	Water lake	Sampaio (2011)
Incertae sedis		*Rhodotorula ferulica*	Water	Sampaio and van Uden (1991); Kurtzman and Fell (1998)
Incertae sedis		*Reniforma strues*	Wastewater	Pore and Sorenson (1990); Kurtzman and Fell (2006)
Ustilaginomycotina				
Ustilaginomycetes				
Ustilaginales		*Pseudozyma flocculosa*	Water samples	Fell et al. (2011)
Ustilaginales		*Pseudozyma parantarctica*	Water samples	Fell et al. (2011)
		Group 2 filamentous fungi		
Agaricomycotina				
Agaricomycetes				
Agaricales	*Gloiocephala aquatica*		Submerged culms	Desjardin et al. (1995)
Agaricales	*Mycocalia reticulata*		Submerged wood	Hyde and Goh(1998)
Agaricales	Unnamed *cypheloid*	*Peyronelina glomerulata*	Wood	Yamaguchi et al. (2009)
Agaricales	*Psathyrella aquatica*		Submerged wood or sediment	Frank et al. (2010)

(continued)

Tab. 4.1: *(continued)*

Classification	Sexual morph	Asexual morph	Substratum	Reference
Atheliales	*Leptosporomyces crucelliger*	*Taeniospora descalsii*	Foam	Marvanová and Stalpers (1987); Bernicchia and Gorjón (2010)
Atheliales		*Taeniospora gracilis*	Submerged leaves, foam	Marvanová (1977)
Atheliales		*Taeniospora gracilis* var. *enecta*	Submerged leaves, foam	Marvanová and Stalpers(1987)
Atheliales		*Taeniospora gracilis* var. *gracilis*	Submerged leaves, foam	Marvanová and Stalpers (1987)
Atheliales		*Taeniospora nasifera*	Foam	Marvanová and Bärlocher (1988)
Cantharellales	*Sistotrema hamatum*	*Ingoldiella hamata*	Submerged decaying leaves, foam	Nawawi and Webster (1982)
Cantharellales		*Ingoldiella fibulata*	Foam	Nawawi (1973b, 1985b)
Cantharellales		*Ingoldiella nutans*	*Pinuscontorta* and *Sphagnum* litter, peat bog	Bandoni and Marvanová (1989)
Polyporales	*Limnoperdon incarnatum*		Hardwood twigs, marsh water	Escobar et al. (1976);Hibbett and Binder (2001)
Polyporales	*Bulbillomyces farinosus*	*Aegerita candida*	Decayed wood	EBG Jones, unpublished
Exobasidiomycetes				
Doassansiales	*Burrillia narasimhanii*		Leaves of *Caldesia reniformis*, pond water	Piatek et al. (2011)
Doassansiales	*Pseudodermatosorus alismatis-oligococci*			
Doassansiales	*Rhamphospora nymphaeae*		Leaves of *Caldesia reniformis* pond water	Lotz-Winter et al. (2011)
Tremellomycetes				
Tremellales	*Xenolachne flagellifera*		Logs	Rogers (1947)

(continued)

Tab. 4.1: (*continued*)

Classification	Sexual morph	Asexual morph	Substratum	Reference
Pucciniomycotina				
Atractiellomycetes				
Atractiellales	*Helicogloea angustispora*	*Infundibura adhaerens*	Submerged leaves, foam	Kirschner (2004)
Classiculomycetes				
Classiculales		*Jaculispora submersa*	Submerged leaf litter	Hudson and Ingold (1960); Bauer et al. (2003)
Classiculales	*Classicula fluitans*	*Naiadella fluitans*	*Scirpus microcarpus* leaf litter, foam	Marvanová and Bandoni (1987); Bauer et al. (2003)
Microbotryomycetes				
Camptobasidiales	*Camptobasidium hydrophilum*	*Crucella subtilis*	Foam, submerged leaves	Marvanová and Suberkropp (1990)
Sporidiobolales	*Rogersiomyces okefenokeensis*		Decaying leaves, beetle galleries in bark	Crane and Schoknecht (1978); Kirschner and Chen (2003)
Ustilaginomycotina				
Ustilaginomycetes				
Urocystidales	*Doassansiopsis caldesiae*		Leaves of marsh plant	Piatek et al. (2008)
Urocystidales	*Doassansiopsis tomasii*		Submerged leaves of *Nymphaea nouchali*	Piatek et al. (2008)
Urocystidales	*Doassansiopsis nymphoides*		Submerged leaves of *Nymphaea nouchali*	Piatek et al. (2008)
Urocystidales	*Doassansiopsis ticonis*		Submerged leaves of *Nymphaea nouchali*	Piatek et al. (2008)
Urocystidales	*Doassansiopsis nymphaeae*		Submerged leaves of *Nymphaea nouchali*	Piatek et al. (2008)
Urocystidales	*Doassansiopsis hydrophila*		Submerged leaves of *Potamogeton*	Piatek et al. (2008)
Urocystidales	*Doassansiopsis occulta*		Submerged leaves of *Potamogeton*	Piatek et al. (2008)

(*continued*)

Tab. 4.1: *(continued)*

Classification	Sexual morph	Asexual morph	Substratum	Reference
Basidiomycota				
Incertae sedis		*Akenomyces costatus*	Submerged leaf of *Cladium mariscus*	Voglmayr and Krisai-Greilhuber (1997)
Incertae sedis		*Anguillomyces acadiensis*	Foam	Marvanová and Bärlocher(2000)
Incertae sedis		*Arcispora bisagittaria*	Foam	Marvanová and Bärlocher (1998);
Incertae sedis		*Cruciger lignatillis*	Rotten wood; terrestrial, foam	Kirschner and Oberwinkler (1999)
Incertae sedis		*Dendrosporomyces prolifer*	Submerged leaves. foam	Nawawi et al. (1977b)
Incertae sedis		*Dendrosporomyces splendens*	Decaying leaves, twigs, foam	Nawawi and Webster (1982)
Incertae sedis		*Fibulotaeniella canadensis*	Foam	Marvanová and Bärlocher (1988)
Incertae sedis		*Microstella pluvioriens*	Rain water from tree	Ando and Tubaki (1984)
Incertae sedis		*Nodulospora inconstans*	Foam	Marvanová and Bärlocher (2000)
Incertae sedis		*Stauriella aquatica*	Submerged wood	Sivichai and Jones (2004)
Incertae sedis		*Titaeëlla capnophila*	Rain water from tree, foam	Ando and Tubaki (1985); Descals (1997)
Incertae sedis		*Tricladiomyces geniculatus*	Submerged leaves	Nawawi and Kuthubutheen (1988)
Incertae sedis		*Tricladiomyces malaysianus*	Submerged leaves	Nawawi et al. (1977a); Nawawi (1985b)
		3. Endophytes		
Agaricomycetes				
Cantharellales		*Ceratobasidium* sp.	Submerged plant root, lake	Kohout et al. (2012)
Tremellomycetes				
Tremellales		*Cryptococcus victoriae*	Submerged plant root, lake	Kohout et al. (2012)

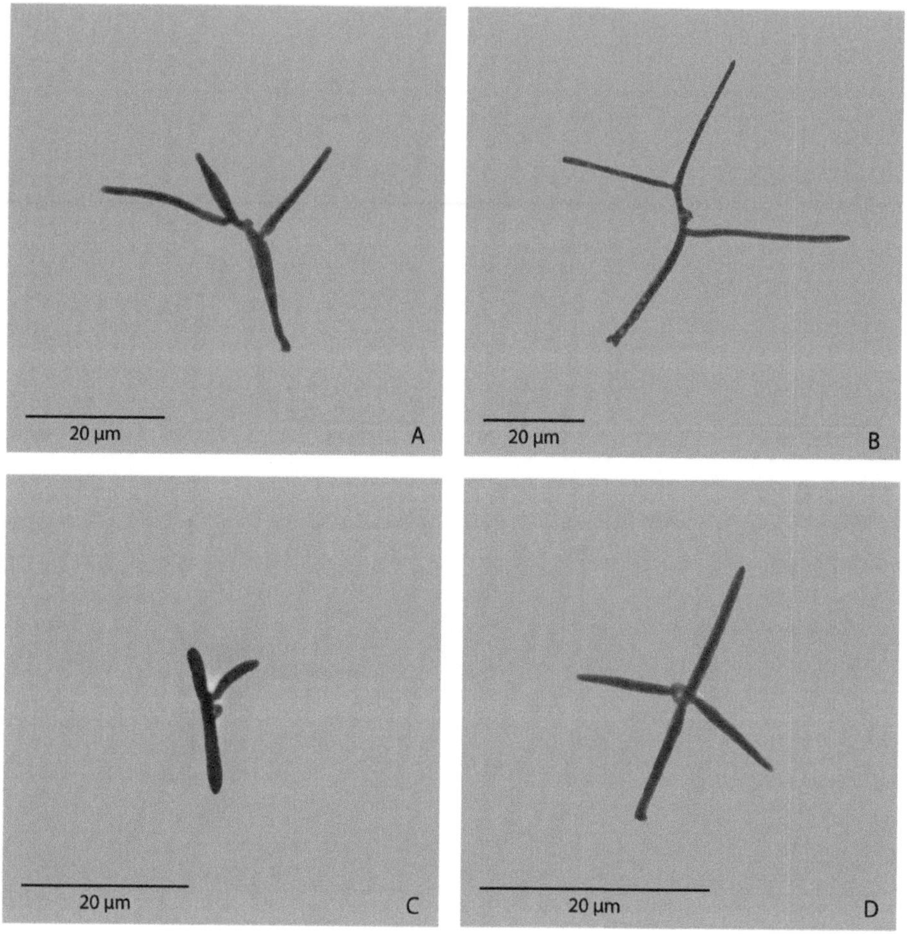

Fig. 4.1: Conidia of freshwater basidiomycetes from pure culture, showing clamps on septa. A. *Taeniospora gracilis* var. *gracilis* from a stream in the UK. Note the single clamp in the middle of the conidial axis between two branches, typical for the genus. The axis is fusoid above and below the clamp; branches are narrower. *B. Taeniospora gracilis* var. *enecta* from New Brunswick, Canada. In contrast to *T. gacilis* var. *gracilis*, the axis and branches are slender and of the same width. C. *Taeniospora nasifera* from a stream in Nova Scotia, Canada. Although it has only a single branch, the clamp in the middle of the axis and colony character suggest congruence with other species of the genus. D. *Taeniospora descalsii* from a river in Devon, UK. A single branch arises from the clamp loop. Photographs by Ludmila Marvanová.

Morphologically freshwater basidiomycetous fungi can be categorised as yeasts (Group 1) or filamentous fungi (Group 2). In addition to the common free-living saprotrophic forms, a few have been isolated from underwater plant parts as endophytes (Group 3) (Tab. 4.1).

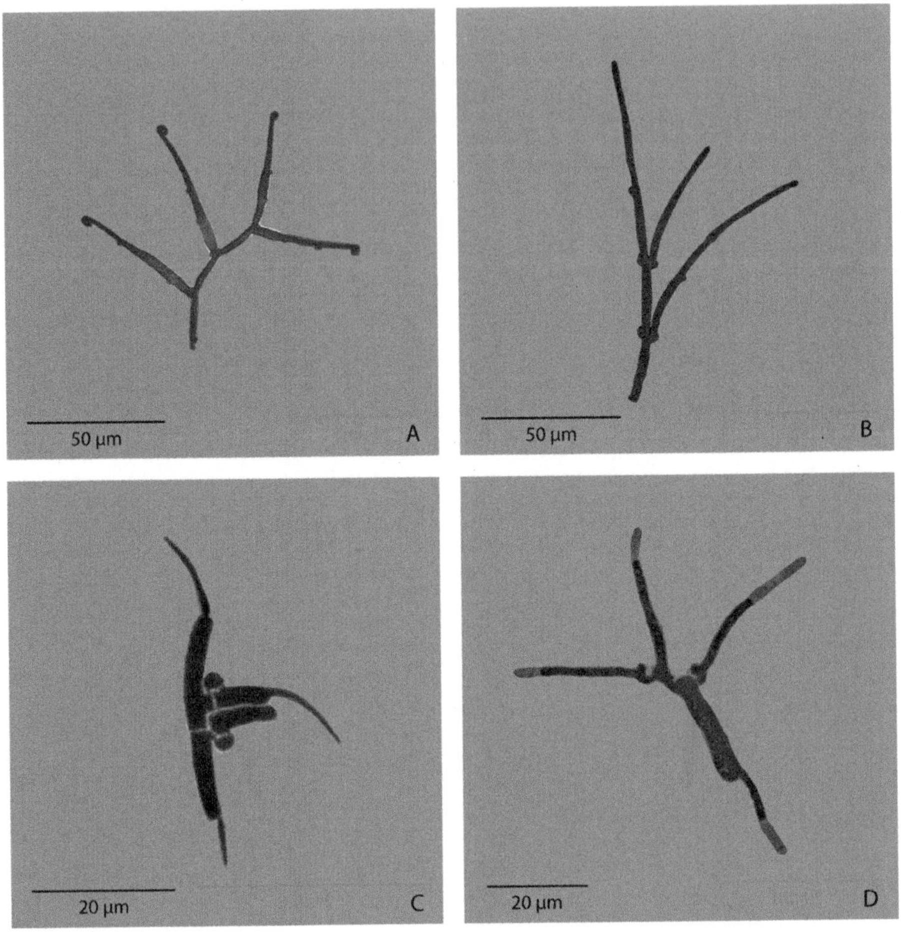

Fig. 4.2: Conidia of some freshwater basidiomycetes from pure culture, showing clamps on septa. A. *Ingoldiella hamata*, a conidium formed on aerial mycelium. The isolate originated from a submerged leathery dicot leaf in a stream on Koh Chang Island, Thailand. Note the branches with hooks at the ends and inconspicuous clamp loops. B. *Ingoldiella nutans*, contrasting in less numerous clamps with distinct loops. The isolate is from the type locality in Canada. C. *Arcispora bisagittaria* from Catamaran Brook, New Brunswick, Canada. The conidium has clamp loops near the base of two tightly parallel branches perpendicular to the axis. D. *Nodulospora inconstans*, irregularly branched conidium obtained in culture isolated from foam in the Cheticamp River (Cape Breton National Park, Nova Scotia, Canada). Note the incomplete clamps at the bases of the conidial branches. Photographs by Ludmila Marvanová.

4.1 Group 1 freshwater yeasts

Yeasts, single-celled and free-floating, occur in freshwater habitats, attached to substrates or within a host (Jones and Slooff 1966; Pore and Sorenson 1990; Libkind et al. 2003; Kurtzman and Fell 2006; Yamaguchi et al. 2009; Fell et al. 2011). Yeasts are poorly documented in most reviews of freshwater fungi (Goh and Hyde 1996; Shearer et al. 2007; Wurzbacker et al. 2010; Wurzbacker et al. 2011). Basidiomycetous yeasts have been reported from freshwater lakes and streams, brackish water, sewage-contaminated water, glacier meltwater, and wastewater (Cooke 1976; Pore and Sorenson 1990; de García et al. 2010; Brandăo et al. 2011), and are often the dominant taxonomic group of yeasts in aquatic ecosystems (de Garcia et al. 2007; Brandão et al. 2010).

Nineteen classes of Basidiomycota have been reported (Bauer et al. 2006; Boekhout et al. 2011), of which four (Exobasidiomycetes, Microbotryomycetes, Tremellomycetes and Ustilaginomycetes) include freshwater yeasts. Polluted waters support a higher yeast population (Sampaio 2004); species richness and population densities are usually highest in freshwater and decrease in marine waters and with increasing depth (Hagler and Ahearn 1987). Many aquatic yeasts remain to be discovered as a study of the yeast populations of the Everglades (Florida, USA) from freshwater to brackish mangrove zones have shown: 74 described species were recovered along with a similar number of undescribed species (Fell et al. 2011). The most commonly isolated freshwater yeasts are *Cryptococcus albidus*, *C. laurentii*, and species of *Rhodotorula*, *Sporobolomyces* and *Trichosporon* (Hagler and Ahearn 1987; Libkind et al. 2003; de Garcia et al. 2007).

Yeasts traditionally have been described based on morphology (basidial form, sexual and asexual morphs, and clamp connections), life cycles, compatible mating type conjugation, ultrastructure (especially septal pore structure), and assimilation or utilization of nutrients (Kirschner 2004; Sampaio 2004; Weiss et al. 2004; Boekhout et al. 2010). Fermentation and assimilation tests are part of a taxonomical description of freshwater basidiomycetous yeasts (Fell et al. 2010). However, these approaches frequently result in confusion due to the dimorphic nature of yeasts. Molecular techniques have greatly aided taxonomic classification (de Garcia et al. 2010; Boekhout et al. 2011). The shift from mostly phenotypic based classifications to mostly molecular based ones has led to modifications in yeast taxonomy that determined the existence of polyphyletic genera such as *Rhodotorula*, *Cryptococcus*, and *Sporobolomyces*. This makes it difficult to draw conclusions about the freshwater distribution and occurrence of yeasts at the genus level.

4.1.1 Agaricomycotina

4.1.1.1 Tremellomycetes

4.1.1.1.1 Cystofilobasidiales

These teleomorphic yeasts produce holobasidia and teliospores, and have dolipore septa lacking parenthesomes. Of the nine genera referred to this order by molecular data, eight have been reported from freshwater: *Mrakia frigida*, *Mrakiella aquatica*, *Guehomyces pullulans*, *Udeniomyces pannonicus*, and four *Cystofilobasidium* species (Kurtzman et al. 1998; Libkind et al 2009; Brandão et al. 2011; Fell et al. 2011). *Udeniomyces pyricola* was isolated from the freshwater segment of the Everglades and *Mrakiella aquatica* from foam collected at the edge of Malham Tarn, UK, and from culms of *Equisetum* sp. (Jones and Slooff 1965; Webster and Davey 1975; Fell et al. 2011). The four *Cystofilobasidium* species were isolated from oligotrophic lake water in Patagonia with low human impact (Libkind et al. 2003). Two of these were recently described: *Cystofilobasidium lacus-mascardii* and *Cyst. macerans*, the sexual morph of *Cryptococcus macerans* (Libkind et al. 2009).

4.1.1.1.2 Filobasidiales

This order forms a well-supported monophyletic group with only one species recorded from freshwater habitats. *Filobasidium floriforme* was recovered in the Everglades, Florida, USA (Fell et al. 2011).

4.1.1.1.3 Holtermanniales

Phylogenetic studies have demonstrated that the genus *Cryptococcus* is polyphyletic with species grouping in the Filobasidiales, Tremellales, Trichosporonales and Cystofilobasidiales, with a further four species related to the genus *Holtermannia* (Wuczkowski et al. 2011). One such species is *Cryptococcus festucosus* which was transferred to *Holtermanniella* (*H. festuscosa*) in a newly introduced order Holtermanniales (Wuczkowski et al. 2011). Using sequence analysis of the D1/D2 region by neighbor-joining method, the order forms a sister clade to the Filobasidiales, Tremellales, Trichosporonales and Cystofilbasidiales in the Tremellomycetes with 100% support. *Holtermanniella festucosa* was reported from glacial melt water by de Garcia et al. (2007) and Wuczkowski et al. (2011).

4.1.1.1.4 Tremellales

In freshwater habitats, the Tremellales are represented by 24 yeast species and the filamentous *Xenolachne flagellifera*. Taxa with dimorphic life cycles show sacculate membranous caps on the dolipore septum, and tremelloid basidia consisting of four cells with longitudinal to oblique walls. The yeast forms include *Bullera* (2 species), *Cryptococcus* (18 species), and *Dioszegia* (4 species) isolated from water samples in the Everglades (Florida, USA) (Fell et al. 2011), lake water samples (van Uden and Ahearn 1963; Brandão et al. 2011), glacial melt waters (Libkind et al. 2003; de Garcia et al. 2007, 2010), and acidic waters (Russo et al. 2010). *Cryptococcus* species accounted for 50% of the total number of strains isolated from glacial melt water rivers in Patagonia (de Garcia et al. 2007), with *Cr. cylindricus* reported from a cold water environment. Two new species, *Cryptococcus spencermartinsiae* and *Cr. agrionensis*, were isolated from glacial waters and acidic environments in Argentina, respectively (de Garcia et al. 2010; Russo et al. 2010). Three *Dioszegia* species were found in glacial melt water at low levels of occurrence and might represent run-off from terrestrial sources (de Garcia et al. 2007; Brandão et al. 2011).

4.1.1.1.5 Trichosporonales

Trichosporon species have been isolated from soil, wood pulp, sludge, clinical samples, and industrial wastewater; *T. cutaneum* was isolated from lake water (Douglas Lake, Michigan, USA) (van Uden and Ahearn 1963). Fell et al. (2000) and Kurtzman and Fell (2006) placed the genus *Trichosporon* in the order Trichosporonales.

4.1.2 Pucciniomycotina

4.1.2.1 Cystobasidiomycetes

4.1.2.1.1 Cystobasidiales

A number of *Rhodotorula* species form a well-supported clade in the Cystobasidiales, including *Rh. minuta,* isolated from two sites in the Everglades (Fell et al. 2011). Three other *Rhodotorula* species have been reported from lake water or glacial melt water (Libkind et al. 2003; de Garcia et al. 2007; Brandão et al. 2011). The taxonomic position of *Cyrenella elegans* is not clear: classified in the Cystobasidiales, it has also been referred to the Naohideales; here we do not assign it to an order (Tab. 4.1). It is dimorphic producing an orange-pigmented yeast stage and conidia with radiating arms, resembling those of aquatic hyphomycetes – features that may adapt them to water dispersal (Bauer et al. 2006). *Cyrenella elegans* produces teliospores on cylindrical transversely septate basidia, but teliospore germination has not been observed (Bauer et al. 2006). This species was isolated from sand scraped off the submerged stipe of *Laccaria trullisata* collected on a sandy shore periodically inundated by lake water (Gochneaur 1981). In pure culture the fungus depends on an external supply of *p*-aminobenzoate and thiamine.

4.1.2.2 Microbotryomycetes

4.1.2.2.1 Leucosporidiales

Five Leucosporidiales species have been recovered from river water or glacial meltwater in Patagonia, the genus *Leucosporidiella* accounting for 20% of total strains at the site. *Leucosporidiella fragaria* was initially isolated by Fell and Statzell-Tallman

(1998) from strawberries and blackcurrants in the UK, thus its freshwater occurrence may be due to run-off.

4.1.2.2.2 Sporidiobolales

In Tab. 4.1, we list 14 species that are referred to the Sporidiobolales including the asexual genera *Rhodotorula* (2 species) and *Sporobolomyces* (3 species), and the sexual genus *Rhodosporidium* (5 species) and *Sporidiobolus* (4 species). Three teleomorphic *Rhodosporidium* species have been reported from freshwater: *Rh. babjevae* and *Rh. kratochvilovae* from glacial meltwater and *Rh. diobovatum* in the Everglades (Libkind et al. 2003; de Garcia et al. 2005; Fell et al. 2011). *Sporidiobolus ruineniae*, *Rhodotorula casssiicola*, *Rh. vanillica* and three *Sporobolomyces* species were retrieved from the Everglades (Fell et al. 2011). Many cold water species were reported from glacial meltwater in Patagonian rivers, including the new species *Rhodotorula meli*, *Sporobolomyces patagonicus* and *Sporidiobolus longiusculus* (Libkind et al. 2005, 2010). The occurrence of *Rhodotorula meli* in meltwater was low suggesting that this may not be the original habitat of the yeast (Libkind et al. 2005). However, it groups with yeasts associated with mainly aquatic habitats.

4.1.2.3 Microbotryomycetes *incertae sedis*

Rhodotorula cresolica, initially described from soil, showed an ability to utilize several benzene compounds and cresols (Middelhoven and Spaaij 1997); it was later reported from water samples (Fell et al. 2011). Further studies are required to resolve its phylogenetic position and taxonomic circumscription.

Rhodotorula vanillica was isolated from stagnant pond water in Portugal (Sampaio 1994) with further strains reported by Fell et al. (2011) from the Everglades. In MycoBank it is included in the Sporidiobolales. *Rhodotorula ingeniosa* has a wide geographical distribution, found in New Zealand, South Korea, Canada, Austria and Portugal. Although the original isolate of this species was from plant material, subsequent collections have been from other substrates, including lake water. Phylogenetically the species is most closely related to *R. vanillica*, based on comparisons of D1/D2 regions of 28S and ITS rDNA sequences (Fell et al. 2000).

Reniforma strues is an asexual species isolated from biofilms in a wastewater treatment plant (Pore and Sorenson 1990); it has a unique morphology with reniform yeast cells. The most comprehensive phylogenetic trees place *R. strues* in an isolated position in the Pucciniomycotina, but an apparent association with species of Cryptomycocolacomycetes and Classiculomycetes might be due to long-branch attraction (Aime et al. 2006).

4.1.3 Ustilaginomycotina

4.1.3.1 Ustilaginomycetes

4.1.3.1.1 Ustilaginales

Pseudozyma was introduced with *P. prolifica* as the type species based on morphology (Bandoni 1985) and molecular data (Begerow et al. 2000). Nine species are referred to the genus (Sampaio 2004). Most are found on plant leaves and flowers, while two species were described from blood samples (Sugita et al. 2003). The genus is the asexual morph of *Ustilago* or *Sporisorium*, with *P. prolifica* the asexual stage of *Ustilago maydis* (Sampaio 2004). Two *Pseudozyma* species (*P. flocculosa, P. parantarctica*) are known from freshwater samples from the Everglades in freshwater wetlands dominated by sawgrass (*Cladium jamaicense*) (Fell et al. 2011).

Pseudozyma prolifica was isolated from *Scirpus microcarpus* litter in a terrestrial habitat. The yeast phase of this species does not consist of typical budding cells, but contains various propagules from short arthroconidia to branched forms resembling conidia of freshwater hyphomycetes (Bandoni 1985). These propagules have not been reported in studies of aquatic hyphomycetes based on detached conidia in water column or in foam, but owing to their similarity to the stauroconidia of freshwater hyphomycetes, they might not have been identified. Species of *Pseudozyma* produce proteinaceous mycotoxins with fungistatic effects (Golubev 2007). *Pseudozyma prolifica* has been found in the gut of cotton bollworms (Noctuidae) (Molnár et al. 2008).

4.2 Group 2 filamentous fungi

4.2.1 Agaricomycotina

4.2.1.1 Agaricomycetes

4.2.1.1.1 Agaricales

Members of the Agaricales have been collected on woody substrates or culms of plants trapped and submerged in streams and rivers. With the exception of *Psathyrella aquatica*, they possess small basidiomes. *Peyronelina glomerulata*, a frequently observed asexual morph linked to the sexual state by molecular data, is an aero-aquatic basidiomycete that sporulates on wood after incubation (Fisher et al. 1976; Yamaguchi et al. 2009). It is an early colonizer of wood, appearing on test blocks after 1 week of incubation in moist chambers (Kane et al. 2002).

Mycocalia reticulata, collected on submerged wood in Lake Barrine, Australia, produces clumps of disc-shaped peridioles under a white cover (Hyde and Goh 1998). It was initially described as *Nidularia* in the Gasteromycetes, a polyphyletic class of unrelated taxa, but a phylogenetic study showed that various gasteroid species belong in the Euagarics clade (Binder and Bresinsky 2002). Two gasteroid lineages were included in this clade: puffballs in the Lycoperdales and birds' nest fungi in the Nidulariales. A second *Mycocalia* species (*M. sphagneti*) has been reported growing on *Juncus effusus* and might be considered semi-aquatic (Cejp and Palmer 1963).

Psathyrella aquatica in the Euagarics clade forms fruiting bodies with gills underwater, unlike other freshwater Agaricales (Fig. 4.3A–C). It grows on submerged wood, attached to stream cobble, or directly from the silt of the river bottom in the Rogue River, Oregon, USA (Frank et al. 2010). In the laboratory, it grew on culture media and fruited on a mixture of grain and alder sawdust in a terrestrial environment (J. L. Frank, unpublished). A phylogenetic study placed *P. aquatica* in the same group as *P. superiorensis* and *P. brooksii* with moderate support. Other species of *Psathyrella* occur on dry or damp terrestrial habitats, but not underwater, indicating a recent adaptation to an underwater habitat.

Fig. 4.3: *Psathyrella aquatica* underwater in the Rogue River (Oregon, USA). A. Sporocarp growing from silt on the river bottom near colonies of the Cyanobacterium *Anabaena* sp. on submerged wood and filaments of the moss *Scleropodium obtusifolium*. Photo courtesy Robert A. Coffan. B. Spores of *Psathyrella aquatica*, attached to basidium showing asymmetric hilar attachment characteristic of ballistospory. Photo courtesy Darlene Southworth. C. Sporocarps with gas bubbles under mature pilei. Photo courtesy Jonathan L. Frank.

Gloiocephala aquatica, a marasmioid fungus found on the bulrush *Scirpus californicus* at La Zeta Lake, Chubut, Argentina, forms small (0.4–0.8 mm) acystidiate basidiomes arising from pseudosclerotial plates that have a lateral stipe, smooth hymenophore and gelatinised tissues (Desjardin et al. 1995). It sporulates under water. *Gloiocephala aquatica* lacks pilocystidia, caulocystidia, and hymenial cystidia. The aquatic habit of *G. aquatica* appears to be a relatively recent adaptation.

Hyphomycetous morphs collected from water and foam are visible only under dissecting or compound microscopes. They grow mostly in submerged tree leaves or woody debris, and produce conidia on conidiophores protruding from the substrate. Known from few species and mostly in pure culture, their basidiomes consist of a few basidia appearing mostly above water. This group comprises fungi with clamped septa on hyphae and on conidia. The ultrastructure of septal pores and phylogenetic affinities through DNA analyses have been studied in few species. In some taxa the relationships of sexual and asexual morphs are known and were determined only in pure culture. The bulk of information on their characters is derived from asexual morphs.

4.2.1.1.2 Atheliales

Two genera with freshwater representatives were originally documented: *Fibulomyces* and *Leptosporomyces* with asexual morphs predominating. However, Bernicchia and Gorjón (2010) reduced *Fibulomyces* to synonymy with *Leptosporomyces* without supporting their decision with molecular data. Basidiomes developed in cultures derived from conidia collected from freshwater. They consist of single, penicillately branched basidiophores with a terminal head of basidia. Conidia are usually branched with a single clamp connection in the middle of the axis.

The first hint of an athelian member of filamentous fungi in freshwater streams was by Ingold (1968) who illustrated conidia assignable to *Taeniospora gracilis* var. *enecta*, collected in the UK from stream foam, without recognizing their basidiomycetous affinity. Three species are known in the asexual genus *Taeniospora*: *T. gracilis* (two varieties: *T. gracilis* var. *gracilis*, Fig. 4.1A and *T. gracilis* var. *enecta*, Fig. 4.1B), *T. descalsii* (Fig. 4.1D) and *T. nasifera* (Fig. 4.1C) (Marvanová 1977; Marvanová and Stalpers 1987). *Taeniospora* species are characterized by a pungent odor emitted from cultures resembling that of the cultivated mushroom *Agaricus bisporus*. In *T. gracilis*, conidia are two-celled, typically with two alternate branches and a single clamp in the middle of the axis. Cells are binucleate, but unicellular conidia without clamps, resembling a minute *Tricladium*, may also occur. The connection between the teleomorph *Leptosporomyces galzinii* and the anamorph morphologically corresponding to *T. gracilis* var. *enecta* (Nawawi et al. 1977a) was doubted by Marvanová and Stalpers (1987) because of differences in colony morphology and the absence of the characteristic odor in *L. galzinii* culture. *Leptosporomyces crucelliger* is the sexual morph of *Taeniospora descalsii* (Marvanová and Stalpers 1987, as *Fibulomyces crucelliger*); the conidia have two opposite branches resembling a cross. *Taeniospora nasifera*, the third species, is holoanamorphic (Marvanová and Bärlocher 1988). Although having single-branched conidia, colony characters justify its accommodation in *Taeniospora*. *Taeniospora* species have been encountered in clean, unpolluted freshwater, in mountains as well as in lowlands. *Taeniospora gracilis* var. *enecta* is recorded frequently from streams in temperate and cold climate worldwide. Its distribution extends to high latitudes in the Northern hemisphere where conidia were collected from a subarctic stream in Sweden and from stream foam in tundra in southern Greenland (Marvanová and Müller-Haeckel 1980; Engblom et al. 1986). At lower latitudes this species survives in streams at high altitudes, e.g. 1248 to 2238 m above sea level in Pakistan (Iqbal and Bhatty 1979). *Taeniospora descalsii* was reported from Europe in pristine streams in the Czech Republic (Marvanová 2001), France (Chauvet 1992), Portugal (Pascoal et al. 2005), UK (Marvanová and Stalpers 1987), Spain (Descals and Rodríguez-Pérez 2002), and Canada (Bärlocher 2000). *Taeniospora nasifera* is known only from Canada (Marvanová and Bärlocher 1988; Bärlocher 2000).

4.2.1.1.3 Cantharellales

Ingoldiella hamata (Fig. 4.2A) is the asexual morph of *Sistotrema hamatum* (Shaw 1972; Nawawi and Webster 1982); the sexual morph was discovered on *Hevea* petioles incubated in a moist chamber and later obtained in pure culture started from conidia. The fungus formed white powdery patches on the substrate, with basidia arising singly from hyphae. Crossing experiments indicated a bipolar mating system (Nawawi and Webster 1982). Two further *Ingoldiella* species, *I. fibulata* and *I. nutans* (Fig. 4.2B), were described with no known sexual morph (Nawawi 1973; Bandoni and Marvanová 1989). Conidia of *Ingoldiella* are larger (axis >110–300 µm) than those of *Taeniospora* (axis < 100 µm). Conidia are multiseptate, with clamps at all septa, and binucleate; lateral branches alternate, and the colonies lack the odor typical of *Taeniospora*. *Ingoldiella hamata* has two conidial forms: underwater conidia are tetraradiate, whereas aerial conidia have two to four alternate branches. Both forms have double hooks at the apices and dolipore septa. *Ingoldiella fibulata* has two alternate branches, resembling conidia of the ascomycetous genus *Tricladium*. It was isolated from stream foam in Malaysia (Nawawi 1973b). *Ingoldiella nutans* is similar to *I. fibulata* in having two lateral branches, but conidia are smaller and stouter. It was described from conifer needles mixed with *Sphagnum* in a peat bog in Canada (Bandoni and Marvanová 1989).

Most *Ingoldiella* species are known from warm to tropical climates. *Ingoldiella hamata* is cosmopolitan; in addition to Australia (Shaw 1972), Malaysia (Nawawi 1985b) and Thailand (Sakayaroj et al. 2005; L. Marvanová unpublished data), it is reported frequently from India (Sridhar and Kaveriappa 1992; Sridhar et al. 1992). Conidia were recorded in Florida, USA (Raja et al. 2009). Conidia of *I. fibulata* are known from Japan and India (Miura 1974; Subramanian and Bhat 1981). *Ingoldiella nutans* is known only from the type locality in British Columbia, Canada.

4.2.1.1.4 Polyporales

Limnoperdon incarnatum was described from submerged hardwood twigs from a freshwater marsh on the shore of Lake Union, Seattle, Washington, USA, and referred to the Gasteromycetes (Escobar et al. 1976). A new family, Limnoperdaceae, was introduced to accommodate the genus because of its unique morphology: unilocular gleba, smooth reddish spores with a beaked pedicel borne on variable basidia with inflated sterigmata, a peridium of hydrophobic dendrophyses, and sporocarps floating on water. The authors queried whether it was truly aquatic or a terrestrial invader of freshwater. Subsequently it has been reported from wood blocks submerged in brackish water (Tubaki 1977) and paddy-field soil (Ito and Yokoyama 1979). McCabe (1979) and Nakagiri and Ito (1991) reported its isolation and development in culture. Basidiomes develop in five stages: 1) young basidiocarps cupulate covered by hyphae, 2) side walls develop to surround the basidiocarp, 3) basidiocarps become completely closed, 4) basidiocarps are depressed and the top flattens, and 5) basidiocarps open by cracks exposing the basidiospores (Nakagiri and Ito 1991). The fruiting bodies of *L. incarnatum* bear a superficial appearance to small terrestrial puffballs (350–1250 × 200–450 μm) that have lost ballistospory. Basidiocarps are produced on the water surface. Nakagiri and Ito (1991) suggest that this is a mechanism for dispersal as in the marine basidiomycete *Nia vibrissa* (Fazzani and Jones 1977).

Molecular data place *L. incarnatum* in the Euagaric clade with high bootstrap support although its phylogenetic relationship is not fully resolved (Hibbett and Binder 2001). Further collections are warranted in order to resolve its phylogenetic relationship within the Euagaric clade. *Limnoperdon incarnatum* resembles *Nia vibrissa,* a marine taxon that grows on wood and has lost ballistospory (Binder et al. 2006). *Nia* differs from *Limnoperdon* in possessing appendaged sporocarps and basidiospores (Jones and Jones 1993; Jones and Choeyklin 2008).

Heavily decayed waterlogged wood, washed up beside streams and rivers, often supports ball-like asexual propagules of *Agerita candida* (sexual morph: *Bulbillomyces farinosus*; E.B.G. Jones unpublished data). They are produced on wet wood above water level but easily detach from the wood with rising water (Eriksson and Ryvarden 1976). These propagules (sclerotia?) are globose to subglobose, up to 0.4 cm in diameter, composed of numerous radiating anastomosing short club-shaped hyphae. Their free-floating propagules resemble aero-aquatic fungi such as *Peyronelina glomerulata* and other helicosporous hyphomycetes.

4.2.1.2 Exobasidiomycetes

4.2.1.2.1 Doassansiales

Rhamphospora nymphaeae is a smut growing embedded in the water plant *Nymphaea alba* initially known from India but also recovered on plants in Frankfurt, Germany and Australia (Lotz-Winter et al. 2011; R. Shivas: http:/www.padil.gov.au). Sori are scattered or gregarious, initially light yellow, later brown spots on the leaves of *N. alba*. Spores are lemon-shaped, papillate at one end, with an appendage at the other end; the asexual morph is unknown.

Two other smuts, assigned to the Doassansiales, are found on the aquatic plant *Caldesia*: *Burrillia narasimhanii* and *Pseudodermatosorus alismatis-oligococci* (Piatek et al. 2008).

4.2.1.3 Tremellomycetes

4.2.1.3.1 Tremellales

It is questionable whether *Xenolachne flagellifera* can be considered to be in a freshwater Basidiomycota as it was described growing on discomycete apothecia "under conditions of high humidity on the lower surface of water-soaked logs in a forest during the wet season" (Rogers 1947). The fungus is characterised by resupinate basidiocarps, with ovoid to ellipsoid basidia on a thin layered subiculum of clamp-bearing hyphae, longitudinally septate hypobasidia with a tubular epibasidium, and long elongate sterigmata that swell at the tips to form oblong basidiospores lacking an apiculus (Rogers 1947). Based on morphological characteristics, i.e. ovate, longitudinally septate basidia that are catenate with sessile basidiospores, the genus is referred to the Sirobasidiaceae, Tremellales, although no molecular data supports its placement.

4.2.2 Pucciniomycotina

4.2.2.1 Atractiellomycetes

4.2.2.1.1 Atractiellales

Helicogloea angustispora along with its asexual morph *Infundibura adhaerens* was classified in the Atractiellales based on ultrastructural features and confirmed by molecular data (Oberwinkler and Bauer 1989; Kirschner 2004). *Infundibura adhaerens*, isolated from decaying leaves in a stream in New Zealand, is a sporodochial fungus with conidia that are hyaline, one-celled, and naviculate, with two asymmetrical mucoid appendages: a foot-like basal one, and a funnel-like, adhesive apical one (Nag Raj and Kendrick 1981). The asexual morph has been reported from Europe, Canada, Brazil, Australia, New Zealand, South Africa, Malaysia, and Borneo while the sexual morph has not been widely reported (Kirschner 2004; L. Marvanová unpublished data). Germinating conidia have been seen in stream foam. Usually two retraction septa are laid down during germination, so such conidia are devoid of cytoplasm and appear tricellular. The retraction septa are good markers indicating affinity to Basidiomycota when clamp connections are absent.

Infundibura adhaerens occurs on plant debris in freshwater and terrestrial habitats in association with fruit bodies of other fungi (Nag Raj and Kendrick 1981; Wu et al. 1997). During phylogenetic analysis of mycobionts in roots of orchids collected in tropical mountain rainforests in the Northern Andes, phylotypes assignable to Atractiellomycetes were found forming a mycorrhizal association (Kottke et al. 2009). They had simple septal pores surrounded by microbodies, characteristic for puccinialean Atractiellales and clustered with *Infundibura adhaerens* and *Helicogloea* sp. based on partial LSU alignments.

4.2.2.2 Classiculomycetes

4.2.2.2.1 Classiculales

Classicula fluitans and *Jaculispora submersa* form a highly supported clade in the Classiculomycetes based on LSU rDNA sequences (Bauer et al. 2006). They are placed as a sister clade of Microbotryomycetes without support (Aime et al. 2006). The unique characters of Classiculomycetes are subapically swollen sterigmata on basidia, tremelloid haustorial cells and simple septal pores surrounded by circular microbodies (Weiss et al. 2006).

In *Classicula fluitans* basidia were obtained only *in vitro* on a piece of agar culture medium floating in standing water (Bauer et al. 2003). Transversely septate basidia appeared above the water in clusters or singly, laterally bearing subapically swollen sterigmata, each producing a single, elongate basidiospore. Whereas the teleomorph is known only from pure culture, the conidial morph, *Naiadella fluitans*, is often reported from freshwater habitats. Morphologically, the conidia are similar to those of *Jaculispora submersa*: both are naviculate with three to four distal setose appendages, but in *J. submersa* the conidial body is apically rostrate, whereas in *N. fluitans* it bears a discrete apical appendage, similar to the lateral ones.

Although no interaction of haustorial cells of *Classicula* or *Jaculispora* with a potential host was observed, the penetration of the haustorial filaments of *Classicula* into hyphae of the same isolate occurred in culture (Bauer et al. 2003). This suggests that these fungi are potentially mycoparasites (cf. Gams et al. 2004).

In *Jaculispora*, no sexual morph has been found. The asexual morph was described from streams in Jamaica without pure culture, thus its basidiomycetous affinity was not recognized (Hudson and Ingold 1960). The first pure culture was isolated from rotten leaves of *Pandanus* in a terrestrial habitat in Okinawa, Japan (Matsushima 1987). Data on distribution are based on detached conidia similar to those of *Naiadella* and hence easily interchangeable. *Jaculispora* was found in tropical and subtropical latitudes, whereas *Classicula* occurs in temperate to warm climates. Collection of *Jaculispora* conidium in a South African river indicates broader ecological amplitude (Webster et al. 1994).

4.2.2.3 Microbotryomycetes

4.2.2.3.1 Camptobasidiales

Basidia of *Camptobasidium hydrophilum* were discovered in pure culture isolated from conidia of *Crucella subtilis*. They are scattered singly on mycelium, usually curved, up to four-celled, with transverse septa, and sessile, drop-shaped or globose basidiospores produced in clusters from single sporogenous loci. *In vitro*, fertile basidia form readily below water where basidiospores discharge tardily. Delayed basidiospore discharge is considered an adaptation against premature detachment by water currents (Marvanová and Suberkropp 1990). Conidia are typically tetraradiate and clampless, with four broadly diverging arms and a swelling in the middle. Germination involves "protoplast migration", leaving behind empty conidia with numerous retraction septa. Chlamydospores formed frequently in culture. Primary septa on hyphae are with clamps, but conidia are clampless. Septal ultrastructure reveals a simple septal pore.

The conidial morph of *C. hydrophilum* is distributed worldwide, from Mediterranean streams through temperate to boreal habitats (Jabiol et al. 2013). Conidia on submerged leaves are documented from North America, where they were frequent in some streams in Alabama, USA, in Canada, and in Australia (Marvanová and Suberkropp 1990; Thomas et al. 1992, illustrated as unknown; Bärlocher and Marvanová 2010). In Europe, conidia were reported from Spain and France (Gessner et al. 1993; Descals and Rodríguez-Pérez 2002).

Although the *Crucella* morph occurs on submerged decaying leaves in streams and can grow in axenic culture on malt agar, it is not capable of growing on sterile leaf material unless this is colonized with other fungi, and even then may show hindered sporulation (Marvanová and Suberkropp 1990; Howe and Suberkropp 1993). In dual culture on weak nutrient agar, hyphae of *Camptobasidium hydrophilum* coiled around hyphae of other aquatic hyphomycetes. Colacosomes, present in some members of Microbotryomycetidae as organelles active in mycoparasitism, are absent in *Camptobasidium* (Weiss et al. 2004).

Phylogenetic affiliation of *Camptobasidium* has been considered in Microbotryomycetidae. Recently, an ordinal classification was proposed with the established order Kriegeriales (Toome et al. 2013). In an ML tree constructed from LSU and ITS regions of rDNA, the Camptobasidiaceae was placed as a sister clade to Kriegeriaceae, with moderate support. Within the Camptobasidiaceae, *C. hydrophilum* forms a highly supported clade with Antarctic yeast. Some inclination to psychrophily of *C. hydrophilum* may be deduced from the study of Turchetti et al. (2011), where it is placed, with low support, on a sister branch of the *Glacyozyma* clade comprising three psychrophilic yeasts.

4.2.2.3.2 Sporidiobolales

Rogersiomyces was introduced for a fungus collected on submerged decayed leaves of angiosperms and *Taxodium ascendens* from the Okefenokee Swamp, Georgia, USA, with *R. okefenokeensis* as the type species (Crane and Schoknecht 1978). The genus is characterised by gymnocarpous basidiomes, fasiculate to synnematous holobasidia, basidia obclavate with prominent truncate sterigmata, basidiospores borne symmetrically and terminally on the basidia, and basidiospores not forcibly discharged. Crane and Schoknecht (1978) referred the species to the Filobasidiaceae, while Cox (1976) suggested the Aphyllophorales; it has also been placed in the Sporodiobolales. Subsequent collections of this fungus, from beetle galleries under tree bark, have enabled isolation of the species to determine septal pore ultrastructure (Dowding 1984; Kirschner and Chen 2003). *Rogersiomyces okefenokeensis* has perforated parenthesomes that place it in the Homobasidiomycetes, but not in the Filobasidiales, which are in a clade with imperforate parenthesomes (Crane and Schoknecht 1978; Hibbett and Thorn 2001).

Kirschner and Chen (2003) drew attention to the similarity of *R. okefenokeensis* to *Hyphobasidiofera malaysiana* described from the bark of a broad-leaved tree (Matsushima and Matsushima 1996). The latter differs from *R. okefenokeensis* by larger basidiospores.

4.2.3 Ustilaginomycotina

4.2.3.1 Ustilaginomycetes

4.2.3.1.1 Urocystidales

Aquatic smut fungi in *Doassansiopsis* are poorly known, especially in the tropics and subtropics (Piatek et al. 2011). Sori are produced within the host tissue of leaves or petioles of various freshwater plants and forming spore balls that may aide infection of the hosts. Thirteen *Doassansiopsis* species are currently accepted with eight reported from freshwater members of the Nymphaeaceae amd Potamogetonaceae (Tab. 4.1). The genus is referred to the family Doassansiopsidaceae (Urocystidales) and supported by ultrastructural studies (species producing haustoria) and molecular data (Begerow et al. 1998, 2007).

Basidiomycota—*incertae sedis*

Diverse species have been collected as asexual conidia from decaying leaves and woody substrata but their sexual states have not been determined. Most have been identified as Basidiomycota by the presence of clamp connections or binucleate cells and dolipore septa. Most information originates from ecological or biodiversity studies often based on free-floating conidia in water or trapped in foam. As there are taxa with clampless conidia (e.g. *Dendrosporomyce*s and *Tricladiomyces*), the basidiomycetous affinity of such taxa may remain unrecognized. Undoubtedly the group of freshwater Basidiomycota will be enriched with new species in the future, especially when molecular techniques (e.g. barcoding, pyrosequencing) become standard methods.

Based on the presence of clamp connections and their location on the thallus, we distinguish three patterns: 1) clamps present on hyphae and conidia (*Arcispora, Fibulotaeniella, Titaeëlla, Nodulospora*), 2) clamps present on hyphae, absent on conidia (*Anguillomyces, Stauriella, Microstella, Cruciger, Akenomyces*), and 3) clamps absent from hyphae and conidia (*Dendrosporomyces, Tricladiomyces*).

Arcispora bisagittaria is very probably related to Atheliales. It has clamps on conidia as well as on mycelial hyphae (Fig. 4.2C). The colony on agar is restricted, growing slowly, beige to brownish, producing a pungent odor resembling that of the cultivated mushroom. After long incubation of a piece of agar culture in standing sterile distilled water, the colony developed an effuse, athelioid basidioma with simple or branched basidiophores bearing doliiform basidia with four sterimata. However, basidiospores were not seen (Marvanová and Bärlocher 1998). The configuration of conidia differs from that in *Taeniospora*; namely, the two branches are on the same side of the axis and the conidium is tri- to tetraseptate. This species was isolated from clean unpolluted streams in Canada and the Czech Republic (Marvanová 2001). Its distribution is probably wider—a conidium was illustrated from a stream in the Pyrenees in Spain (Descals and Moya 1996).

Fibulotaeniella canadensis is known only from stream foam in New Brunswick, Canada (Marvanová and Bärlocher 1988). It has sigmoid or arcuate conidia with a single clamp in the middle. The colonies are slow growing, whitish to pale gray-yellow, with the characteristic odor of cultivated mushrooms. The fungus sporulates profusely in pure culture on agar media, producing conidia in slimy sporodochial aggregations. The pungent odor probably indicates relationships to athelioid fungi.

Stauriella aquatica, found on test blocks of *Dipterocarpus alatus* submerged in a stream at Khao Yai National Park, Thailand, produces colonies on wood that are superficial, sparse, and cream-white (Sivichai and Jones 2004). They bear hyaline conidia comprising four to six cells, each with two to six spines. Because of the diffuse nature of the colony on incubated wood and the small size of the conidia, this fungus can be missed when examining woody substrata. A second species *Stauriella indica* with more complex conidia was collected in plant litter in India (Pratibha et al. 2010). In MycoBank this latter species is classified as Ascomycota along with *S. aquatica*. However, the latter species has distinct clamps on the hyphae (Sivichai and Jones 2004) but Pratibha et al. (2010) do not mention clamp connections in *S. indica*.

Dendrosporomyces and *Tricladiomyces*, with clampless conidia and hyphae, were described from streams in Malaysia (Nawawi 1973a, 1985a; Nawawi et al. 1977b). Basidiomycetous affinity was confirmed by the presence of binucleate cells and dolipore septa. *Dendrosporomyces prolifer, D. splendens* and *Tricladiomyces malaysianus* have septal pores with parenthesomes (Nawawi et al. 1977b; Nawawi 1985a) indicating that they may be related to Agaricomycetes, but their placement within this class awaits confirmation by molecular methods. All these species have conidia with axes at least 300 µm long. *Dendrosporomyces prolifer* and *D. splendens* have multibranched conidia with primary, secondary or rarely tertiary branches. *Tricladiomyces malaysianus* and *T. geniculatus* produce less branched conidia with only primary branches. Sporulation is obtained mostly on leaves collected in streams and then incubated in moist conditions (Nawawi and Kuthubuteen 1998). They seem to be tropical to subtropical and rare; *Dendrosporomyces prolifer* originally isolated from foam is known also from India, where it was recorded from leaves in streams (Sridhar and Kaveriappa 1989) and from foam (Patil et al. 2011).

Dendrosporomyces splendens was found on decaying submerged leaves and twigs and is relatively common in some localities in Malaysia (Nawawi 1973a, as *Varicosporium splendens*; 1985b). It was obtained on moist-incubated leaf litter of the palm *Euterpe edulis* and timber tree *Alchornea triplinervia* in Brazil (Grandi 1998, 1999). *Tricladiomyces malaysianus* was originally described on the basis of cultures derived from conidia collected in stream foam (Nawawi 1974, as *Tricladium malaysianum*), and later reported from decaying submerged leaves in Malaysia (Nawawi 1985b) and from India (Sridhar et al. 1992). Outside tropical or subtropical climates, *T. geniculatus* was isolated from submerged tree leaves collected in a stream in a *Nothofagus* forest in Tierra del Fuego, Argentina, where the water temperature was 3–5°C (Godeas and Arambarri 1992).

Titaeëlla capnophila was originally described as a mycoparasite on mycelium of a sooty mold in a terrestrial habitat (Arnaud 1951). The species seems rare and its natural habitat is not well known. Ando and Tubaki (1985) collected and isolated *T. capnophila* from rainwater dripping off needles of *Pinus densiflora* in Japan. Since then, two collections of conidia in Spain and Portugal were found in stream foam, one in the Pyrenees with conifers and angiosperm trees on the bank (Descals 1997), and one with riparian trees *Pinus pinaster* and *Eucalyptus globulus* (Marvanová et al. 2003). *Titaeëlla capnophila* was reported from naturally colonized leaves (*Eucalyptus* sp.) after exposure in a river in Australia (Bärlocher et al. 2011). The fungus has clamps on mycelia as well as on conidia. Conidia are multiseptate, with secondary clampless septa on the axis and branches; they bear a single primary branch, with one secondary branch originating from the basal cell of the primary, each branch with a clamp at the base. Arnaud (1951) described *Hypochnus capnophilus* as forming simple 'corticiaceous' basidia on sooty mold hyphae but had not seen direct connection with *Titaeëlla capnophila.*

Microstella pluvioriens was obtained from rainwater dripping off needles of *Metasequoia glyptostroboides* in Japan (Ando and Tubaki 1984). This fungus has nonseptate conidia, consisting of a globose central body with short, blunt radiating arms. Conidia are borne on hypha-like conidiophores, not distinguished from mycelium with clamped septa. The fungus is known with certainty only from the original collection. The report of this species in melted ice from water reservoirs in Poland, together with conidia of several tropical aquatic hyphomycetes (Czeczuga and Orlowska 1999) seems questionable, likewise its abundant occurrence on free-floating pollen grains and fern spores in water (Czeczuga and Orlowska 2001). A synanamorph assignable to *Microstella pluvioriens* was seen in a culture of *Resupinatus applicatus* (Tricholomataceae) supplemented with nematodes (Thorn and Barron 1986).

Anguillomyces acadiensis and *Nodulospora inconstans*, species of monotypic genera, were described from lotic waters in two Canadian Maritime Provinces, Nova Scotia and New Brunswick. Although the conidia and conidiogenous structures differ, colonies are similar with relatively robust hyphae up to 5 µm wide, producing soft white sclerotia in culture. Germination of conidia involves retraction septa and emptying of cytoplasm into germ tubes like some members of the 'heterobasidiomycetes'. Another common character of both species is the probable dependence on metabolites of other fungi during the germination of conidia and early stages of growth. These phases are more vigorous in the vicinity of other species of aquatic hyphomycetes (L. Marvanová, unpublished data). Both species were isolated from foam. *Anguillomyces acadiensis* has relatively large scolecoid conidia without septa. A clamped septum in the middle of the conidium appears just before germination. *Nodulospora inconstans* differs from *A. acadiensis* through branched conidia with irregularly located clamps on the conidial body, often near the hilum or at the base of branches (Fig. 4.2D).

Cruciger lignatilis has clamps on the mycelium, and its conidia, resembling a cross with two opposite branches on an erect axis, are aseptate and clampless (Kirchner and Oberwinkler 1999). The fungus was found on a rotting branch on the ground. It formed white to pale violet sporodochia with long branched conidiophores producing conidia successively on proliferating clamp loops on conidiophore tips. The ultrastructure of septal pore showed a dolipore septum with perforated parenthesome, typical for Agaricomycetes. A similar fungus was collected in Hungary on a rotting twig in a hardwater stream (Gönczöl and Révay 1992, as unknown). It formed slimy colonies with masses of cruciform conidia similar in size and shape to those of *C. lignatilis*, but stouter and occasionally bearing a clamp vestige at the base of the axis. A typical conidium was reported from foam from the Ysper stream in Austria (Marvanová and Gulis 2000, as unknown).

Akenomyces costatus is a rare member of potentially aeroaquatic Basidiomycota collected on submerged leaves of *Cladium mariscus* in Austria (Voglmayr and Krisai-Greilhuber 1997). The fungus was described from *Carex riparia* in France (Arnaud 1954) indicating a moist habitat. There have been a few isolations from wheat fields, but pathogenicity tests performed on wheat and barley gave negative results (Hornby 1984). The fungus forms only sclerotia; no other propagules were observed. Clamps occur on mycelium; sclerotia are dark brown, obovoid, up to 400 µm long and 150 µm wide. In the central part they consist of tightly interwoven hyphae, capable of germination, surrounded by vertical thick-walled pallisade hyphae. The sclerotium is covered by colourless encrusted hyphae and is hydrophobic (Voglmayr and Krisai-Greilhuber 1997).

4.3 Group 3 endophytes

Leaf and root endophytes are well known in terrestrial habitats (Rodriguez et al. 2009). These are predominantly in the Ascomycota with a few taxa in the Basidiomycota in the subphyla Agaricomycotina, Pucciniomycotina and Ustilaginomycotina. The relationships between terrestrial leaf endophytes and the endophytes associated with leaves that have fallen into freshwater is unknown.

Many plant roots, especially of trees growing along rivers, e.g. *Alnus glutinosa*, and *Betula papyrifera*, have been surveyed for freshwater endophytes (Fisher et al. 1991; Sridhar and Bärlocher 1992; Raviraja et al. 1996; Marvanová et al. 1997). These also are rarely basidiomycetous (Sati and Belwal 2005; Bärlocher 2006; Sokolski et al. 2006; Sati et al. 2009; Li et al. 2010).

The rarity of basidiomycetous endophytes may result from poor detection of sterile isolates from host plants (Evans et al. 2003; Rungjindamai et al. 2008; Pinruan et al. 2010). Hyphae in sterile cultures need to be examined closely for clamp connections or minute basidiomes on the sides of Petri dishes (Rungjindamai et al. 2008). Also isolates need selective media to encourage sporulation. However, molecular methods also identified few Basidiomycota among freshwater root endophytes. Kohout et al. (2012) screened roots of five submerged aquatic plants (*Isoëtes echinispora, I. lacustris, Littorella uniflora, Lobelia dortmanna, Subularia aquatica*) for endophytic fungi using ITS sequences. In addition to sequences corresponding to ascomycetous aquatic hyphomycetes [e.g. *Nectria lugdunensis* (dominant), *Tetracladium furcatum, Varicosporium elodeae*], they recovered sequences of two Basidiomycota: *Ceratobasidium* sp. 2 (on *Isoëtes echinospora*), and *Cryptococcus victoriae* (on *Isoëtes lacustris*).

4.4 Adaptation to freshwater habitats

Aquatic Basidiomycota may have evolved from terrestrial habitats to periodically immersed and then to fully submerged substrata, with the loss of basidiospory, the evolution of appendaged spores and enclosed fruit bodies (Hibbett and Binder 2001). While sexual morphs are more prevalent in marine species, freshwater Basidiomycota tend to produce asexual spores with branched or sigmoid spores (Jones and Choeyklin 2008; Jones and Fell 2012). Most freshwater Basidiomycota have asexual spores that aid dispersal in water (Hyde and Goh 2003). Spores of freshwater hyphomycetes have evolved mechanisms of attachment that include conidia covered in mucilage, release of mucilage when they contact the substratum, and development of appressoria for secondary attachment (Au et al. 1996a, b).

Freshwater habitats, particularly continuous streams and lakes, extend the availability of moist or wet habitats during dry seasons. For this reason, the freshwater habitat seems conducive to fungal growth, with a longer wet season, abundant decaying organic matter, and higher oxygen levels, at least in shallow lakes and aerated streams. However, the aquatic habitat creates a challenge for ballistosporic spore release. As long as fruiting bodies are above the water surface, they can discharge spores, but if gilled caps or hymenophores remain underwater, the necessary drying of a water film would be inhibited (Pringle et al. 2005). Yet basidiospores of *Psathyrella aquatica* show the asymmetrically positioned hilar appendix characteristic of ballistospores (Fig. 4.3B). Gas pockets were observed under caps of *P. aquatica* (Fig. 4.3C) as well as under hymenophores of *Gloiocephala menierii*, attached to *Carex* stems near the mud-air interface (Redhead 1981; Frank et al. 2010). These gas pockets were present underwater and remained briefly when the fruiting body of *P. aquatica* was lifted out of the water into the air, where-upon the bubble burst leaving a wedge-shaped pattern of spores on fingers or paper (Frank et al. 2010). Trapped bubbles may create the gaseous atmosphere needed for ballistosporic discharge underwater.

With the exception of *Coprinus* sp. and *Psathyrella aquatica*, most freshwater basidiomycetes have reduced fruit bodies that are hydrophobic, and the hymenium is not exposed to water (Nakagiri and Ito 1991). In many cases, fruiting bodies form when the substratum is terrestrial and not submerged in water, in a similar fashion to discomycetes in aquatic habitats. Freshwater fungi have evolved diverse adaptive mechanisms for survival and dispersal of their spores.

Acknowledgments

E.B. Gareth Jones is supported by the Distinguished Scientist Fellowship Program (DSFP), King Saud University, Saudi Arabia. Darlene Southworth thanks Jonathan L. Frank and Robert A. Coffan for collaboration on *Psathyrella aquatica*. Ludmila Marvanová acknowledges financial support from the Czech Collection of Microorganisms and thanks Jarmila Marvanová for help with photomicrographs. We are grateful to Dr. Roland Kirschner for drawing attention to the studies of Piatek et al. on *Doassansiopsis*.

References

Aime MC, Matheny PB, Henk DA, Frieders EM, Nilsson RH, Piepenbrink M, et al. An overview of the higher level classification of Pucciniomycotina based on combined analyses of nuclear large and small subunit rDNA sequences. Mycologia 2006;98:896–905.

Ando K, Tubaki K. Some undescribed hyphomycetes in the rain drops from intact leaf-surface. Trans Mycol Soc Japan 1984;25:21–37.

Ando K, Tubaki K. Three new hyphomycetes from Japan: *Anthopsis microspora*, *Scutisporus brunneus*,and *Titaeëlla capnophila*. Trans Mycol Soc Japan 1985;26:151–160.

Arnaud G. Les boucles mycéliennes des eumycètes et al philogenie des urédinées. Bull Soc Mycol France 1951;67:173–198.

Au DWT, Jones EBG, Moss ST. Spore attachment and extracellular mucilage of aquatic hyphomycetes. In: Callow M, ed. Biofouling 10, Singapore: Hardwood Academic Publishers 1996a;123–140.
Au DWT, Jones EBG, Moss ST, Hodgkiss IJ. The role of mucilage in the attachment of conidia, germ tubes and appressoria in the saprobic aquatic hyphomycetes *Lemonniera aquatica* and *Mycocentrospora filiformis*. Can J Bot 1996b;74:1789–1800.
Bandoni RJ. On an undescribed, pleomorphic hyphomycete from litter. Bot J Linn Soc 1985;91:37–43.
Bandoni RJ, Marvanová L. On a new species of *Ingoldiella*. Mycologia, 1989;81:42–46.
Bärlocher F. Fungal endophytes in submerged roots. In: Schulz B, Boyle C, Sieber TN, eds. Soil Biology. Springer Verlag: Berlin 2006;9:179–190.
Bärlocher F. Water-borne conidia of aquatic hyphomycetes: seasonal and yearly patterns in Catamaran Brook, New Brunswick, Canada. Can J Bot 2000;78:157–167.
Bärlocher F, Marvanová L. Aquatic Hyphomycetes (Deuteromycotina) of the Atlantic maritime ecozone. In: McAlpine DF, Smith IM, eds. Assessment of Species Diversity in the Atlantic Maritime Ecozone. Ottawa: NRC Research Press 2010:71–106.
Bärlocher F, Stewart M, Ryder DS. Analyzing aquatic fungal communities in Australia: impacts of sample incubation and geographic distance of streams. Czech Mycol 2011;63:113–132.
Bauer R, Begerow D, Oberwinkler F, Marvanová L. *Classicula*: the teleomorph of *Naiadella fluitans*. Mycologia 2003;95:756–764.
Bauer R, Begerow D, Sampaio JP, Weiss M, Oberwinkler F. The simple-septate basidiomycetes: a synopsis. Mycol Prog 2006;5:41–66.
Begerow D, Bauer R, Boekhout T. Phylogenetic placements of ustilaginomycetous anamorphs as deduced from nuclear LSU rDNA sequences. Mycol Res 2000;104:53–60.
Begerow D, Bauer R, Oberwinkler F. Phylogenetic studies on nuclear large subunit ribosomal DNA sequences of smut fungi and related taxa. Can J Bot 1998;75:2045–2056.
Begerow D, Stoll M, Bauer R. A phylogenetic hypothesis of Ustilaginomycotina based on multiple gene analyses and morphological data. Mycologia 2007;98:906–916.
Bernicchia A, Gorjón SP. Fungi Europaei. Corticiaceae s.l. Edizioni Candusso, Alassio Italia, 2010;12:1–1008.
Binder M, Bresinsky A.Derivation of a polymorphic lineage of Gasteromycetes from boletoid ancestors. Mycologia 2002;94:85–98.
Binder M, Hibbett DS, Wang Z, Farnham WF. Evolutionary relationships of *Mycaureola dilsea* (Agaricales), a basidiomycete pathogen of a subtidal rhodophyte. Am J Bot 2006;93:547–556.
Boekhout T, Fonseca A, Sampaio JP, Bandoni RJ, Kwon-Chung KJ. Discussion of teleomorphic and anamorphic Basidiomycetous Yeasts. In: Kurtzman CP, Fell JW, Boekhout T, eds. The Yeasts: A Taxonomic Study. Amsterdam: Elsevier 2010:1339–1372.
Brandão LR, Rosa CA, Medeiros AO. Yeast diversity in freshwater ecosystems. In: Daniels JA, ed. Advances in Environmental Research. Hauppauge, NY: Science Publisher Inc. 2011;5: 207–222.
Brandão LR, Libkind D, Vaz ABM, Santo LCE, Moliné M, de Garcia V, et al. Yeasts from an oligotrophic lake in Patagonia (Argentina): diversity, distribution and synthesis of photoprotective compounds and extracellular enzymes. FEMS Microbiol Ecol 2011;76:1–13.
Cejp K, Palmer JT. The genera *Nidularia* Fr. and *Mycocalia* J. T. Palmer in Czechoslovakia and *Mycocalia sphagneti* J. T. Palmer sp. nov. from England. Č Mykol 1963;17:113–126.
Chamier AC. Cell-wall degrading enzymes of aquatic hyphomycetes – a review. Bot J Linn Soc 1985;91:67–81.
Chandrashekar KR, Kaveriappa KM. Production of extracellular enzymes by aquatic hyphomycetes. Folia Microbiol 1988;33:53–58.
Chandrashekar KR, Sridhar KR, Kaveriappa KM. Palatability of rubber leaves colonized by aquatic hyphomycetes. Arch Hydrobiol 1989;115:361–369.

Chauvet E. Dynamique saisonnière des spores d'hyphomycètes aquatiques de quatre rivières. Nova Hedwig 1992;54:379–395.

Cooke WEB. Fungi in sewages. In: Jones EBG, ed. Recent Advances in Aquatic Mycology. New York: Wiley 1976:389–434.

Cox DE. A new Basidiomycete with anomalous basidia. Mycologia 1976;68:481–510.

Crane JL, Schoknecht JD. *Rogersiomyces*, a new genus in the Filobasidiaceae (Homobasidiomy cetes) from an aquatic habitat. Am J Bot 1978;65:902–906.

Czeczuga B, Orlowska M. Hyphomycetes in the ice of water reservoirs. Rocz Akad Med Bialymst 1999;44:64–75.

Czeczuga B, Orlowska M. Hyphomycetes species on floating plant spores and pollen. Acta hydrochim hydrobiol 2001;29:100–110.

de Garcia V, Brizzio S, Libkind D, Buzzini P, van Broock M. Biodiversity of cold-adapted yeasts from glacial meltwater rivers in Patagonia, Argentina. FEMS Microbiol Ecol 2007;59:331–341.

de Garcia V, Brizzio S, Russo G, Rosa CA, Boekhout T, Theelen B, Libkind D, van Broock M. *Cryptococcus spencermartinsiae* sp. nov., a basidiomycetous yeast isolated from glacial waters and apple fruits. Inter J Syst Evol Microbiol 2010;60:707–711.

Descals E. Ingoldian fungi from the Catalan Pyrenees: pure culture studies. Mycotaxon 1997;63:431–466.

Descals E, Moya O. Fungal spora in streams of the Catalan Pyrenees. Bol Soc Micol Madrid 1996;21:145–174.

Descals E, Rodríguez-Pérez J. Bases Corológicas De Flora Micológica Ibérica. Números 1933–2069. In: Pando F, ed. Cuadernos de trabajo de flora micológica Ibérica. Madrid: CSIC 2002;18:5–195.

Desjardin DE, Martinez-Peck L, Raichenberg M. An unusual psychrophilic aquatic agaric from Argentina. Mycologia 1995;87:547–550.

Dowding P. The evolution of insect-fungus relationships in the primary invasion of forest timber. In: Invertebrate-Microbial Interactions. Anderson JM, Raynor ADM, Walton DWH, eds. New York: Cambridge University Press 1984:133–153.

Engblom L, Lingdell PE, Marvanová L, Müller-Haeckel A. Foam spora in running waters of southern Greenland. Polar Res 1986;4:47–51.

Eriksson J, Ryvarden L. The Corticiaceae of north Europe. Vol 4. Oslo, Norway: Fungiflora 1976: 1–339.

Escobar GA, McCabe DE, Harpel CW. *Limnoperdon*, a floating gasteromycete isolated from marshes. Mycologia 1976;87:874–880.

Evans HC, Holmes KA, Thomas SE. Endophytes and myco-parasites associated with an indigenous forest tree, *Theobroma gileri*, in Ecuador and preliminary assessment of their potential as biocontrol agents of cocoa diseases. Mycol Progr 2003;2:149–160.

Fazzani K, Jones EBG. Spore release and dispersal in marine and brackish water fungi. Mat Organ 1977;12:235–248.

Fell JW, Boekhout T, Fonesca A, Scorzetti G, Statzell-Tallman A. Biodiversity and systematics of basidiomycetous yeasts as determined by large-subunit rDNA D1/D2 domain sequence analysis. Int J Syst Evol Microbiol 2000;50:1351–1371.

Fell JW, Statzell-Tallman A. *Cryptococcus* Vuillemin. In: Kurtzman CP, Fell JW, eds. The Yeasts, a Taxonomic Study. Amsterdam: Elsevier 1998:472–767.

Fell JW, Statzell-Tallman A, Scorzetti G, Gutiérrez MH. Five new species of yeasts from fresh water and marine habitats in the Florida Everglades. Antonie Leeuwenhoek 2011;99:533–549.

Fisher PJ, Webster J, Kane DF. *Peyronelina glomerulata* from submerged substrata in Britain. Trans Br Mycol Soc 1976;67:351–354.

Fisher PJ, Petrini O, Webster J. Hyphomycetes and other fungi in living aquatic and terrestrial roots of *Alnus glutinosa*. Mycol Res 1991;95:543–547.

Frank JL, Coffan RA, Southworth D. Aquatic gilled mushrooms: *Psathyrella* fruiting in the Rogue River in southern Oregon. Mycologia 2010;102:93–107.

Gams W, Diederich P, Põldmaa K. Fungicolous fungi. In: Mueller GM, Bills GF, Foster MS, eds. Biodiversity of Fungi: Inventory and Monitoring Methods. San Diego, CA: Elsevier 2004:343–392.

Gessner MO, Thomas M, Jean-Louis AM. Chauvet E. Stable successional patterns of aquatic hyphomycetes on leaves decaying in a summer cool stream. Mycol Res 1993; 97: 163–172.

Gochenaur SE. *Cyrenella elegans* gen. *et* sp. nov., a dikaryotic anamorph. Mycotaxon 1981;13:267–277.

Godeas AM, Arambarri AM. Additions to the aquatic hyphomycetes of Tierra del Fuego Argentina. Mycotaxon 1992;43:157–161.

Goh TK, Hyde KD. Biodiversity of freshwater fungi. J Indust Microbiol 1996;17:328–345.

Golubev WI. Mycocinogeny in smut yeast-like fungi of the genus *Pseudozyma*. Microbiology 2007;76:719–722

Gönczöl J, Révay Á. Aquatic Hyphomycetes in softwater and hardwater streams of the Aggtelek National Park, NE Hungary. Ann hist-nat mus natl Hung 1992;84:17–31.

Grandi RAP. Hyphomycetes decompositores do folhedo de *Alchornea triplinervia* (Spreng) Müll Arg Hoehnea 1998;25:133–148.

Grandi RAP. Hyphomycetes decompositores do folhedo de *Euterpe edulis* Mart. Hoehnea 1999;26:87–101.

Guého E, Smith MT, de Hoog GS, Billon-Grand G, Christen R, Battenberg-van der Vegte WH. Contributions to a revision of the genus *Trichosporon*. Antonie Leeuwenhoek 1992;621:289–316.

Hagler AN, Ahearn DG. Ecology of aquatic yeasts. In: Rose AH, Harrison JS, eds. The Yeasts, 2nd edn, vol. 1. London: Academic Press 1987:181–205.

Hibbett DS, Binder M. Evolution of marine mushrooms. Biol Bull 2001;201:319–322.

Hibbett DS, Thorn RG. Basidiomycota: Homobasidiomycetes. In: McLaughlin DJ, McLaughlin EG, Lemke PA, eds. The Mycota, Vol. VIIB, Systematics and Evolution. New York: Springer 2001:121–168.

Hornby D. *Akenomyces costatus* sp. nov. and the validation of *Akenomyces* Arnaud. Trans Br Mycol Soc 1984;82:653–664.

Howe MJ, Suberkropp K. Effects of mycoparasitism on an aquatic hyphomycete growing on leaf litter. Mycologia 1993;85:898–901.

Hudson HJ, Ingold CT. Aquatic hyphomycetes from Jamaica. Trans Br Mycol Soc 1960;43:469–478.

Hyde KD, Goh WH. Fungi on submerged wood in Lake Barrine, North Queensland, Australia. Mycol Res 1998;102:739–749.

Hyde KD, Goh WH. Adaptations for dispersal in filamentous freshwater fungi. Fungal Divers Res Series 2003;10:21–258.

Inácio, J, Portugal L, Spencer-Martins I, Fonseca A. Phylloplane yeasts from Portugal: seven novel anamorphic species in the Tremellales lineage of the Hymenomycetes (Basidiomycota) producing orange-coloured colonies. FEMS Yeast Res 2005;5:1167–1183.

Ingold CT. More spores from rivers and streams. Trans Br Mycol Soc 1968;51:137–143.

Ito T, Yokoyama T. Distribution of *Limnoperdon incarnatum* Escobar in rice paddy field soils. In: Proceedings of the 23rd Annual Meeting of the Mycological Society Japan, Otsu May 25–26 1979, p. 75 (Abstr).

Iqbal SH, Bhatty SF. Conidia from stream foam. Trans mycol Soc Japan 1979;20:83–91.

Jabiol J, Bruder A, Gessner MO, Makkonen M, Mckie BG, Peeters ETHM, Voss VCA, Chauvet E. Diversity patterns of leaf-associated aquatic hyphomycetes along a broad latitudinal gradient. Fungal Ecol 2013;6:439–448.

Jones AM, Jones EBG. Observations on the marine gasteromycete *Nia vibrissa*. Mycol Res 1993;97:1–6.

Jones EBG, Fell JW. Basidiomycota. In: Jones EBG, Pang K, eds. Marine and Fungal-like Organisms. Germany: Walter De Gruyter 2012:49–63.

Jones EBG, Choeyklin R. Ecology of marine and freshwater basidiomycetes. In: Boddy L, Frankland JC, van West P, eds. Ecology of Saprotrophic Basidiomycetes. London: Elsevier 2008:301–324.

Jones EGB, Pang KL, eds. Marine Fungi. Berlin: Walter de Gruyter 2012.

Jones EBG, Slooff WC. *Candida aquatica* sp. nov. isolated from water scums. Antonie Leeuvenhoek 1966;32:223–228.

Kane DF, Tam WY, Jones EBG. Fungi colonising and sporulating on submerged wood in the River Severn, UK. Fungal Divers 2002;10:45–55.

Kirschner R. Sporodochial anamorphs of species of *Helicogloea*. In: Agerer R, Piepenbring M, Blanz P, eds. Frontiers in Basidiomycete Mycology. Eching, Germany: IHW Verlag 2004:165–178.

Kirschner R, Chen C-J. A new record of *Rogersiomyces okenokeensis* (Basidiomycota) from beetle galleries in pines in Taiwan. Sydowia 2003;55:86–92.

Kirschner R, Oberwinkler F. A new basidiomycetous anamorph genus with cruciform conidia. Mycoscience 1999;40:345–348.

Kohout P, Sýkorová Z, Čtvrtlíková M, Rydlová J, Suda J, Vohník M, et al. Surprising spectra of root-associated fungi in submerged aquatic plants. FEMS Microbiol Ecol 2012;80:216–235.

Kottke I, Suárez JP, Herrera P, Cruz D, Bauer R, Haug I, et al. Atractiellomycetes belonging to the ŕust΄lineage (Pucciniomycotina) form mycorrhizae with terrestrial and epiphytic neotropical orchids. Proc R Soc B 2009;doi: 10.1098/rspb.2009.1884.

Krings M, Taylor TN, Dotzler N. Fungal endophytes as a driving force in land plant evolution: evidence from the fossil record. In: Southworth D, ed. Plant-Fungal Interactions. Chichester, UK: Wiley-Blackwell 2012:5–28.

Kurtzman CP, Fell JW, Boekhout T. The Yeasts, A Taxonomic Study, vol 1. London: Elsevier 1998.

Kurtzman CP, Fell JW. Yeast systematics and phylogeny-implications of molecular identification methods for studies in ecology. In: Carlos R, Gabor P, eds. Biodiversity and Ecophysiology of Yeasts. Germany: Springer 2006:11–30.

Kurtzman CP, Fell JW, Boekout T. The Yeasts, 5th edn. Vols 1–3. Amsterdam: Elsevier 2011:1–2080.

Li HY, Zhao CA, Liu CJ, Xu XF. Endophytic fungi diversity of aquatic/riparian plants and their antifungal activity in vitro. J Microbiol 2010;48:1–6.

Libkind D, Brizzio S, Ruffini A, Gadanho M, van Broock MR, Sampaio JP. Molecular characterization of carotenogenic yeasts from aquatic environments in Patagonia, Argentina. Antonie Leeuwenhoek 2003;84:313–322.

Libkind D, Gadanho M, van Broock M, Sampaio JP. *Sporidiobolus longiusculus* sp. nov. and *Sporobolomyces patagonicus* sp. nov., novel yeasts of the Sporidiobolales isolated from aquatic environments in Patagonia, Argentina. Inter J Syst Evol Microbiol 2005;55:503–509.

Libkind D, Gadanho M, van Broock M, Sampaio JP. *Cystofilobasidium lacus-mascardii* sp. nov., a basidiomycetous yeast species isolated from aquatic environments of the Patagonian Andes, and *Cystofilobasidium macerans* sp. nov., the sexual stage of *Cryptococcus macerans*. Inter J Syst Evol Microbiol 2009;59:622–630.

Libkind D, Sampaio JP, van Broock M. Cystobasidiomycetes yeasts from Patagonia (Argentina): description of *Rhodotorula meli* sp. nov. from glacial meltwater. Inter J Syst Evol Microbiol 2010;60:2251–2256.

Lotz-Winter H, Hofmann T, Kirschner R, Kursawe M, Trampe T, Piepenbring M. Pilze im Botanischen Garten der Universität Frankfurt am Main. Z Mykol 2011;77:89–122.

Marvanová L. *Taeniospora gracilis* gen. *et* sp. nov. Trans Br Mycol Soc 1977;69:146–148.

Marvanová L. Streamborne fungal spora in running waters of the Bohemian forest. Silva Gabreta 2001;7:147–164.

Marvanová L, Bandoni RJ. *Naiadella fluitans* gen. *et* sp. nov.: a conidial basidiomycete. Mycologia 1987;79:578–586.

Marvanová L, Bärlocher F. Hyphomycetes from Canadian streams. 1. Basidiomycetous anamorphs. Mycotaxon 1988;32:339–351.
Marvanová L, Bärlocher F. Hyphomycetes from Canadian streams. III. *Arcispora bisagittaria* gen. et sp. nov. Mycologia 1998;9:531–536.
Marvanová L, Bärlocher F. Hyphomycetes from Canadian streams. V. Two new conidial basidiomycetes. Mycotaxon 2000;75:409–423.
Marvanová L, Gulis VI. Notes on aquatic hyphomycetes and streamborne spora from Austria. Österr Z Pilzk.2000;9:125–140.
Marvanová L, Müller-Haeckel A. Water-borne spores in foam in a subarctic stream system in Sweden. Sydowia Ann Mycol 1980;33:210–220.
Marvanová L, Pascoal C, Cássio F. New and rare hyphomycetes from streams of northwest Portugal. Part I. Cryptogam Mycol 2003;24:339–358.
Marvanová L, Stalpers JA. The genus *Taeniospora* and its teleomorphs. Trans Br Mycol Soc 1987;89:489–498.
Marvanová L, Suberkropp K. *Camptobasidium hydrophilum* and its anamorph *Crucella subtilis*: a new heterobasidiomycete from streams. Mycologia 1990;82:208–217.
Marvanová L, Fisher PJ, Descals E, Bärlocher F. *Fontanospora fusiramosa* sp. nov., a hyphomycete from live tree roots and from stream foam. Czech Mycol 1997;50:3–11.
Matsushima T. Mycological Memoirs no. 5. Matsushima Fungal Collection. Japan: Kobe 1987:1–100.
Matsushima K., Matsushima T. Fragmenta Mycologica II. In: Matsushima T. Matsushima Mycol Mem no 9. Japan: Kobe 1996:31–40.
McCabe DE. Synchronous production and developmental history of sporocarps of *Limnoperdon incarnatum*. Mycologia 1979;71:899–907.
Middelhoven WJ, Spaaij F. *Rhodotorula cresolica* sp. nov., a cresol-assimilating yeast species isolated from soil. Inter J Syst Evol Microbiol 1994;47:324–327.
Miura K. Stream spora in Japan. Trans Mycol Soc Japan 1974;15:289–308.
Molnár O, Wuczkowski M, Prillinger H. Yeast biodiversity in the guts of several pests on maize; comparison of three methods: classical isolation, cloning and DGGE. Mycol Prog 2008;7: 111–123.
Money NP. More g's than the Space Shuttle: ballistospore discharge. Mycologia 1998;90:547–558.
Nag Raj TR, Kendrick B. *Infundibura*, a new hyphomycete with unique appendages. Can J Bot 1981;59:542–546.
Nakagiri A, Ito, T. Basidiocarp development of the cyphelloid gasteroid aquatic basidiomycetes *Halocyphina villosa* and *Limnoperdon incarnatum*. Can J Bot 1991;69:2320–2327.
Nawawi A. A new species of *Varicosporium* from Malaysia. Nova Hedwig 1973a;24:39–43.
Nawawi A. Two clamp-bearing aquatic fungi from Malaysia. Trans Br Mycol Soc 1973b;61:521–528.
Nawawi A. Two new *Tricladium* species.Trans Br Mycol Soc 1974;63:267–272.
Nawawi A. Basidiomycetes with branched, water-borne conidia. Bot J Linn Soc 1985a;91:51–60.
Nawawi A. Aquatic hyphomycetes and other water borne fungi from Malaysia. Malay Nat J 1985b;39:75–134.
Nawawi A, Kuthubutheen AJ. *Tricladiomyces geniculatus* sp. nov., a conidial basidiomycete. Trans Br Mycol Soc 1988;90:670–673.
Nawawi A, Webster J. *Sistotrema hamatum* sp. nov., the teleomorph of *Ingoldiella hamata*. Trans Br Mycol Soc 1982;78:287–291.
Nawawi A, Descals E, Webster J. *Leptosporomyces galzinii,* the basidial state of a clamped branched conidium from freshwater. Trans Br Mycol Soc 1977a;68:31–36.
Nawawi A, Webster J, Davey RA. *Dendrosporomyces prolifer* gen. *et* sp. nov., a basidiomycete with branched conidia. Trans Br Mycol Soc 1977b;68:59–63.
Oberwinkler F, Bauer R. The systematics of gasteroid, auricularioid heterobasidiomycetes. Sydowia 1989;41:224–256.

Pascoal C, Marvanová L, Cássio F. Aquatic hyphomycete diversity in streams of northwest Portugal. Fungal Divers 2005;19:109–128.
Patil VR, Patil SY, Nemade LC, Borse BD. Aquatic fungi from Buldhana district. Curr Bot 2011;2: 56–58.
Piatek M, Vánky K, Mossebo DC, Piatek J. *Doassansiopsis caldesiae* sp. nov. and *Doassansiopsis tomasii*: two remarkable smut fungi from Cameroon. Mycologia 2008;100:662–672.
Pinruan U, Rungjindamai N, Choeyklin R, Lumyong S, Hyde KD, Jones EBG. Occurrence and diversity of basidiomycetous endophytes from the oil palm, *Elaeis guineensis* in Thailand. Fungal Divers 2010;41:71–88.
Pore RS, Sorenson WG. *Reniforma strues*, a new yeast from wastewater. Mycologia 1990;82:549–553.
Pratibha J, Raghukumar S, Bhat DJ. New species of *Dendryphiopsis* and *Stauriella* from Goa, India. Mycotaxon 2010;113:297–303.
Pringle A, Patek SN, Fischer M, Stolze J, Money NP. The captured launch of a ballistospore. Mycologia 2005;97:866–871.
Raja HA, Schmit JP, Shearer CA. Latitudinal, habitat and substrate distribution patterns of freshwater ascomycetes in the Florida penninsula. Biodivers Conserv 2009;18:419–455.
Raviraja NS, Sridhar KR, Bärlocher F. Endophytic aquatic hyphomycetes of roots of plantation crops and ferns from India. Sydowia 1996;48:152–160.
Redhead SA. Agaricales on wetland Monocotyledoneae in Canada. Can J Bot 1981;59:574–589.
Rodriguez RJ, White JF Jr, Arnold AE, Redman RS. Fungal endophytes: diversity and functional roles. N Phytol 2009;182:314–330.
Rogers DP.A new gymnocarpous heterobasidiomycete with gasteromycetous basidia. Mycologia 1947;39:562–564.
Rungjindamai N, Pinruan U, Choeyklin R, Hattori T, Jones EBG. Molecular characterization of basidiomycetous endophytes isolated from leaves, rachis and petioles of the oil palm, *Elaeis guineensis*, in Thailand. Fungal Divers 2008;33:139–161.
Russo G, Libkind D, Sampaio JP, van Broock M. Yeast diversity in the acidic Rio Agrio-Lake Caviahue volcanic environment (Patagonia, Argentina). FEMS Microbiol Ecol 2008;65:415–424.
Russo G, Libkind D, Ulloa RJ, de Garcia V, Sampaio JP, van Broock M. *Cryptococcus agrionensis* sp. nov., a basidiomycetous yeast of the acidic rock drainage ecoclade, isolated from an acidic aquatic environment of volcanic origin. Inter J Syst Evol Microbiol 2010;60:996–1000.
Sakayaroj J, Phongpaichit S, Jones EBG. Viability and biodiversity of aquatic hyphomycetes in foam at Ton Nga Chang Wildlife -Sanctuary, Songkhla, Southern Thailand. Fungal Divers 2005;18:135–145.
Sampaio JP. Utilization of low molecular weight lignin-related aromatic compounds for the selective isolation of yeasts: *Rhodotorula vanillica*, a new basidiomycetous yeast species. Syst Appl Micobiol 1994;17:613–619.
Sampaio JP. Diversity, phylogeny and classification of basidiomycetous yeasts. In: Agerer R, Piepenbring M, Blanz P, eds. Frontiers in Basidiomycete Mycology. Eching, Germany: IHW Verlag 2004:49–80.
Sampaio JP. *Rhodotorula*. In: Kurtzman CP, Fell JW, Boekhout T, eds. The Yeasts, 5th edn. Amsterdam: Elsevier 2011:1873–1928.
Sampaio JP, Gadanho M, Bauer R, Weisst M. Taxonomic studies in the Microbotryomycetidae: *Leucosporidium golubevii* sp. nov., *Leucosporidiella* gen. nov. and the new orders Leucosporidiales and Sporidiobolales. Mycol Prog 2003;2:53–68.
Sampaio JP, van Uden N. *Rhodotorula ferulica* sp. nov., a yeast that degrates ferulic acid and other phenolic compounds. Syst Appl Microbiol 1991;14:146–149.
Sati SC, Belwal M. Aquatic hyphomycetes as endophytes of riparian plant roots. Mycologia 2005;97:45–49.

Sati SC, Arya P, Belwal M. *Tetracladium nainitalense* sp. nov., a root endophytes from Kumaun Himalaya, India. Mycologia 2009;101:692–695.

Shaw DE. *Ingoldiella hamata* gen. *et* sp. nov., a new fungus with clamp connections from a stream in North Queensland. Trans Br Mycol Soc 1972;59:255–259.

Shearer CA, Descals E, Kohlmeyer B, Kohlmeyer J, Marvanová L, Padgett D, et al. Fungal biodiversity in aquatic habitats. Biodivers Conserv 2007;16:49–67.

Sivichai S, Jones EBG. *Stauriella* gen. nov., proposed for a new lignicolous basidiomycetous anamorph from freshwater in Thailand. Sydowia 2004;56:131–136.

Sokolski S, Piche Y, Chauvet E, Bérubé JA. A fungal endophyte of black spruce (*Picea mariana*) needles is also an aquatic hyphomycete. Mol Ecol 2006;15:1955–1962.

Sridhar KR, Barlocher F. Aquatic hyphomycetes in spruce roots. Mycologia 1992;84: 580–584.

Sridhar KR, Kaveriappa KM. Water-borne hyphomycete flora of two freshwater streams. Environ Ecol 1989;7:771–772.

Sridhar KR, Kaveriappa KM. Aquatic Hyphomycetes of Western Ghat streams, India. Sydowia 1992;44:66–77.

Sridhar KR, Chandrashekar KR, Kaveriappa KM. Research on the Indian subcontinent. In: Bärlocher F, ed. The Ecology of Aquatic Hyphomycetes. Germany: Springer 1992:182–211.

Starmer WT, Lachance M-A. Yeast ecology. In: The yeasts. 5th edn. Amsterdam: Elsevier 2011:65–86.

Subramanian CV, Bhat DJ Conidia from freshwater foam samples from the Western Ghats, Southern India. Kavaka 1981;9:45–62.

Sugita T. *Trichosporon*. In: Kurtzman CP, Fell JW, Boekhout T, eds. The Yeasts, 5th edn. Amsterdam: Elsevier 2011:2015–2062.

Sugita T, Takashima M, Poonwan N, Mekha N, Malauthao K, Thungmuthasawat B, et al. The first isolation of ustilaginomycetous anamorphic yeasts, *Pseudozyma* species, from patient's blood and a description of two new species: *P. parantarctica* and *P. thailandica*. Microbiol Immun 2003;47:183–190.

Summerbell RC. The heterobasidiomycetous yeast genus *Leucosporidium* in an area of temperate climate. Can J Bot 1983;61: 1402–1410.

Thomas K, Chilvers GA, Norris RH. Aquatic hyphomycetes from different substrates: substrate preference and seasonal occurrence. Aust J Mar Fresh Res 1992;43:491–509.

Thorn RG, Barron GL. *Nematoctonus* and the tribe Resupinatae in Ontario, Canada. Mycotaxon 1986;25:321–453.

Toome M, Roberson RW, Aime MC. *Meredithblackwellia eburnea* gen et sp. nov., Kriegeriaceae fam. nov. and Kriegeriales ord. nov. – toward resolving higher-level classification in Microbotryomycetes. Mycologia 2013;105:486–495.

Tubaki K. Brackish water fungi and their relationships to marine fungi. In: Proceedings of the 2nd Mycological Congress, Tampa, Forida, August 27–September 3 1977, p 681.

Turchetti B, Buzzini P, Goretti M, Branda E, Diolaiuti G, D'Agata C, et al. Psychrophilic yeasts in glacial environments of Alpine glaciers. FEMS Microbiol Ecol 2008;63:73–83.

Turchetti B, Skye R, Hall T, Connell LB, Branda E, Buzzini P, et al. Psychrophilic yeasts from Antarctica and European glaciers: description of *Glaciozyma* gen nov., *Glaciozyma martinii* sp. nov and *Glaciozyma watsonii* sp.nov. Extremophiles 2011;15:573–586.

Van Uden N, Ahearn DC. Occurrence and population densities of yeast species in a fresh-water lake. Antonie van Leeuwenhoek 1963;29:308–312.

Vaz ABM, Rosa LH, Vieira MLA, de Garcia V, Brandão LR, Teixeira LCRS, Moliné M, Libkind D, van Broock M, Rosa CA. The diversity, extracellular enzymatic activities and photoprotective compounds of yeasts isolated in Antarctica. Brazilian J Microbiol 2011;42:937–947.

Voglmayr H, Krisai-Greilhuber I. *Akenomyces costatus*, an interesting basidiomycetous anamorph with unknown affinities. Österr Z Pilzk 1997;6:61–66.

Webster J, Davey RA. Sedimentation rates and trapping efficiency of cells of *Candida aquatica*. Trans Br Mycol Soc 1975;64:437–440.
Webster J, Marvanová L, Eicker A. Spores from foam from South African rivers. Nova Hedwig 1994;57:379–398.
Weiss M, Bauer R, Begerow D. Spotlights on Heterobasidiomycetes. In: Agerer R, Piepenbring M, Blanz P, eds. Frontiers in Basidiomycete mycology. IHW Verlag, Eching, Germany, 2004:7–48.
Wu W, Sutton BC, Gange AC. Notes on three fungicolous fungi: *Anastomyces microsporus* gen et sp. nov., *Idriella rhododendri* sp. nov. and *Infundibura adhaerens*. Mycol Res 1997;101:1318–1322.
Wuczkowski M, Passoth V, Turchetti B, Andersson AC, Olstorpe M, Laitila A, et al. Description of *Holtermanniella* gen. nov., including *Holtermanniella takashimae* sp. nov. and four new combinations, and proposal of the order Holtermanniales to accommodate tremellomycetous yeasts of the *Holtermannia* clade. Int J Syst Evol Microbiol 2011;61:680–9.
Wurzbacher C, Bärlocher F, Grossart H-P. Fungi in lake ecosystems. Aquatic Microbial Ecol 2010;59:125–149.
Wurzbacher C, Kerr J, Grossart H-P. Aquatic Fungi. In: Grillo O, Venora G, eds. The Dynamical Processes of Biodiversity – Case Studies of Evolution and Spatial Distribution. In Tech open access. Available at: http://www.intechopen.com/articles/show/title/aquatic-fungi. 2011:227–258.
Yamaguchi K, Degawa Y, Nakagiri A. An aero-aquatic fungus, *Peyronelina glomerulata*, is shown to have teleomorphic affinities with cyphelloid basidiomycetes. Mycoscience 2009;50:156–164.
Zemek J, Kuniak L, Marvanová L, Kadlečíková B. Hydrolytic enzymes in aquatic hyphomycetes. Folia Microbiol 1985;30:363–372.

Dian-Ming Hu, Lei Cai, E. B. Gareth Jones, Huang Zhang, Nattawut Boonyuen and Kevin D. Hyde

5 Taxonomy of filamentous asexual fungi from freshwater habitats, links to sexual morphs and their phylogeny

5.1 Introduction

Asexual fungi (also referred to as anamorphic fungi or mitosporic fungi) are generally divided into three major groups according to their fruiting structures and other morphology: 1) asexual fungi produce conidia directly on hyphae (hyphomycetes) lacking pycnidial or acervular fruiting-bodies, mostly ascomycetous with some basidiomycetes; 2) asexually reproducing yeasts, including both ascomycetes and basidiomycetes; 3) pycnidial and acervular fungi (coelomycetes), mostly ascomycetous, a few basidiomycetous (Seifert et al. 2011; See also Chapters 2, 4). In freshwater habitats, all three groups have been reported from various locations (Hu et al. 2013).

Many freshwater filamentous asexual fungi have been reported from freshwater habitats (Hudson and Ingold 1960; Cai et al. 2002a, b; Hu et al. 2007; Cai et al. 2008, 2010; Czeczuga et al. 2010; Hu et al. 2010a, b; Patil 2013). These fungi are an ecological group and taxonomically diverse. According to Goh and Hyde (1996), freshwater hyphomycetes fall into four subgroups:

1. Ingoldian fungi: abound in fast flowing tree-lined streams, babbling brooks, and well aerated lakes, growing on submerged leaves and twigs, but are relatively sparse on woody substrates,
2. Aero-aquatic hyphomycetes: more usually found in stagnant ponds, ditches, or slow-running streams and are capable of vegetative growth on submerged leaves or woody substrates under semi-anaerobic conditions,
3. Terrestrial-aquatic hyphomycetes: a number of conidial fungi isolated from rain drops associated with intact terrestrial plant parts, such as the leaf-surfaces or rainwater draining from intact tree trunks, and,
4. Submerged-aquatic hyphomycetes: a heterogeneous assemblage of fungi growing on submerged decaying plant materials.

These "biological groups" are not natural groupings, but help categorize the fungi. Both freshwater hyphomycetes and some coelomycetes have been reported from freshwater habitats (Zhang et al. 2012).

5.2 Morphological taxonomy

5.2.1 Hyphomycetes

Aquatic hyphomycetes are broadly defined as asexual morphs typically with relatively large branched (often stauro-form) and scolecoform conidia occurring mainly in lotic waters (Shearer et al. 2007). Their common substrate is submerged decaying tree leaves and woody debris. To date, 531 freshwater hyphomycetes have been reported from freshwater habitats (Shearer et al. 2013).

Morphologically freshwater filamentous asexual fungi can be basically divided into two groups: 1) conidiophores absent or reduced, e.g. Ingoldian fungi and some terrestrial-aquatic hyphomycetes; 2) conspicuous conidiophores present, e.g. submerged-aquatic hyphomycetes. The conidiophores of most Ingoldian fungi are reduced, while their conidia are commonly branched or sigmoid (Fig. 5.1). Many of the branched conidia are tetra-radiate, e.g. *Articulospora*, *Tetracladium* and *Tricladium*; some of the branched conidia have other shapes, e.g. *Dendrospora*, *Polycladium* and *Varicosporium*. Sigmoid conidia are also commonly encountered in freshwater habitats (Ingold 1958; Liu et al. 1992; Chan et al. 2000), e.g. *Anguillospora*, *Flagellospora*, *Lunulospora* and *Mycocentrospora*. Many aero-aquatic hyphomycetes also lack conspicuous conidiophores, e.g. *Alatosessilispora*, *Arborispora*, *Curucispora* and *Tricladiella* (Goh and Hyde 1996).

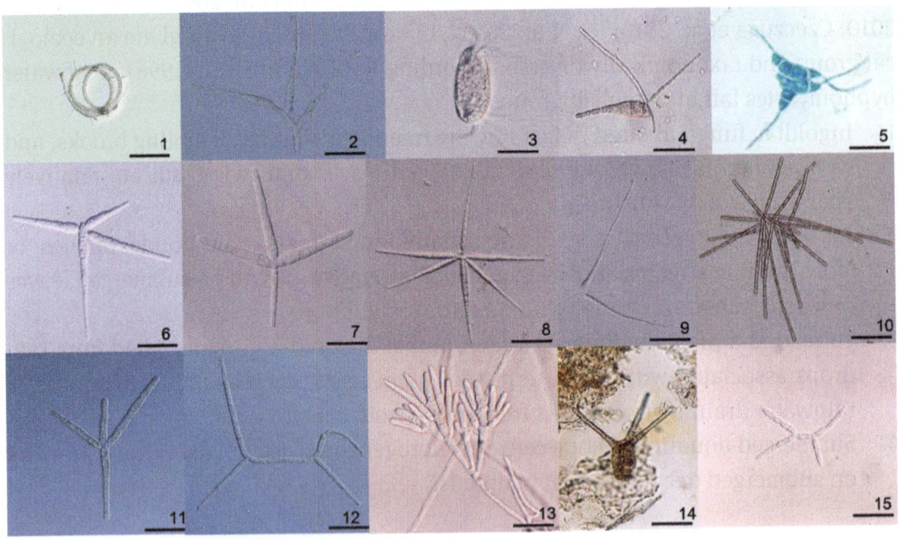

Fig. 5.1: Conidia of some Ingoldian fungi: 1. *Anguillospora* sp.; 2. *Ingoldiella hamata*; 3. Unidentified Ingoldian fungus; 4. *Culicidospora* sp.; 5. *Campylospora chaetocladia*; 6. *Triscelophorus monosporus*; 7. *Flabellospora* sp.; 8. *Flabellospora verticillata*; 9. *Lunulospora curvula*; 10. *Dendrospora* cf. *erecta*; 11. *Articulospora*-like; 12. *Tricladium* sp.; 13. *Tetracladium* sp.; 14. *Tetraploa aristata*; 15. *Phalangispora*-like. Scale bars: 1–7 = 10 µm; 8 = 15 µm; 9 = 20 µm; 10–13 = 10 µm; 14 = 20 µm; 15 = 10 µm.

Most submerged-aquatic hyphomycetes possess conspicuous conidiophores (Fig. 5.2), and colonize submerged woody substrates. Traditionally, these fungi were classified mainly based on their characters of conidiophores, conidiogenous cells and conidia. The morphology of conidiophores of submerged-aquatic hyphomycetes varies widely. Many of them are mononematous (e.g. *Acrogenospora*, *Dactylaria*, *Dictyochaeta*, *Ellisembia*, *Mariannaea*, *Monotosporella* and *Sporoschisma*), some are sporodochial (e.g. *Canalisporium*, *Cheiromyces* and *Dictyosporium*), and some are synnematous (e.g. *Menisporopsis* and *Phaeoisaria*). Pigmentation of conidiophores is also taxonomically important. The conidiophores of submerged-aquatic hyphomycetes are mainly brown (e.g. *Acrogenospora*, *Ellisembia*, *Monotosporella* and *Sporoschisma*) or hyaline (e.g. *Aquaphila* and some species of *Dactylaria*).

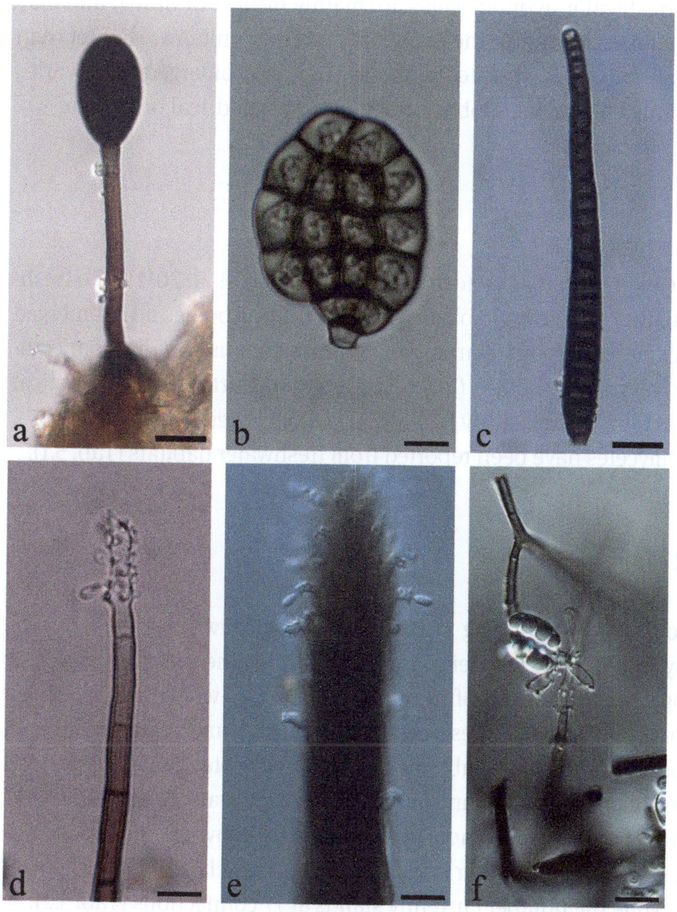

Fig. 5.2: Some species of freshwater hyphomycetes from submerged wood. a. *Acrogenospora ellipsoidea*; b. *Canalisporium jinghongensis*; c. *Ellisembia adscendens*; d. *Dactylaria biguttulata*; e. *Phaeoisaria clematidis*; f. *Neta patuxentica*. Scale bars: a = 30 μm, b = 10 μm, c = 30 μm, d–f = 10 μm.

Conidiogenous cell structures are important in the identification at the generic level; some of them produce only one conidium on one cell (e.g. *Acrogenospora, Ellisembia, Monotosporella, Aquaphila* and *Sporoschisma*), while others produce more than one conidium (e.g. *Phaeoisaria, Dactylaria* and *Neta*). The shape of conidiogenous cells varies: dentical (e.g. *Dactylaria* and *Neta*), blastic (e.g. *Aquaphila, Bactrodesmium, Brachydesmiella* and *Berklesmium*), and phialidic (e.g. *Chloridium* and *Cryptophiale*). Other conidiogenous cell features, such as pigmentation, branching, and spore wall ornamentation are also commonly used in the identification of freshwater hyphomycetes.

Conidial morphology helps in the delineation of genera: conidial dimension, pigmentation, shape, septation, conidial wall ornamentation, and the presence of appendages. Conidial dimensions are important in the identification of hyphomycetes species, for example, many *Dactylaria* species are morphologically similar; but the dimensions of the conidia are always key characters in distinguishing species (Goh and Hyde 1997). Conidia of freshwater hyphomycetes are mainly hyaline or brown and may vary at different growth stages. Normally, the pigments of young conidia are lighter than old conidia. The conidial shape of freshwater hyphomycetes on submerged wood varies widely, but they are mainly ellipsoidal, globose, subglobose, cylindrical or clavate.

5.2.2 Coelomycetes

Coelomycetes are also called pycnidial or acervular fungi (Seifert et al. 2011). The freshwater coelomycetes usually form brown to black pycnidial fruiting bodies on submerged woody debris and stems of herbaceous plants, and produce various conidia from conidiogenous cells inside the fruiting bodies (Fig. 5.3). Due to a dearth of morphological characters available, it is difficult to identify freshwater coelomycetes (Zhang et al. 2012). Currently, only 16 coelomycetes have been reported from freshwater habitats (Tab. 5.1).

5.2.3 Asexual-sexual connections

The asexually, or mitotically, reproducing structures of fungi have been called anamorphs; while sexually reproducing, or meiotic, counterparts of the same life cycle are called teleomorphs. Together, these forms of sporulation make up a whole fungus, called holomorph (Seifert et al. 2011). Some species possess both asexual and sexual states and may occur simultaneously on the same substrate. Most ascomycete species reproduce more frequently via asexual means with many only known in that state. Hyde et al. (2011) list 2,873 asexual genera, of which 1,728 genera (60.2%) have no known sexual state. Few freshwater hyphomycetes are linked to their sexual states with Sivichai and Jones (2003) listing 72 connections, while the number currently stands at 77 connections (Tab. 5.2).

Various techniques for obtaining asexual-sexual connections have been reported (Webster and Descals 1979; Reynolds 1987; Sivichai and Jones 2003). They include 1) isolation of the asexual stage and sporulation of the sexual state in culture,

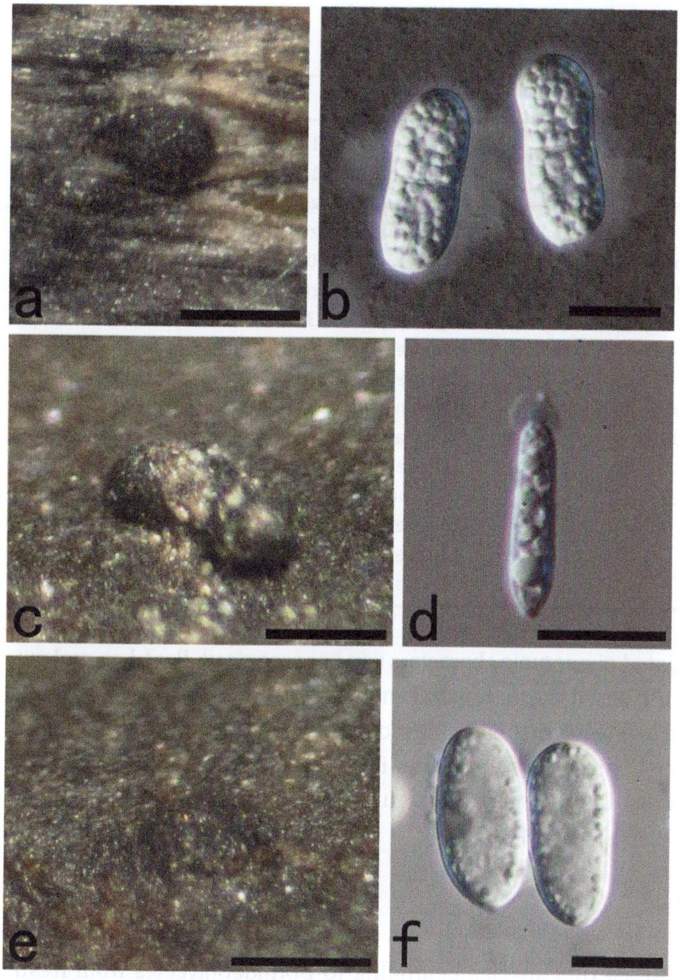

Fig. 5.3: Species of freshwater coelomycetes. a–b. *Clohesyomyces aquaticus*. a. Pycnidium. b. Conidia. c–d. *Acrocalymma aquatica*. c. Pycnidia. d. Conidium. e–f. *Aquasubmersa mircensis*. e. Pycnidium. f. Conidia. Scale bars: a, c = 300 µm; c = 200 µm; b, d, f = 10 µm.

2) growth of the two morphs on natural substrates and 3) sequence data placing asexual stages in known families of ascomycetes or basidiomycetes, however this rarely links the asexual stage to the sexual species.

5.3 Phylogeny

Freshwater filamentous asexual fungi have been traditionally classified based on morphological characters. However, the taxonomy of asexual fungi has been in a state of transition. Modern classification emphasizes phylogenetic affinities, which is a sensible trend for future schemes of asexual fungi, providing that phylogenetically defined

Tab. 5.1: List of selected freshwater coelomycetes.

Species	References
Acrocalymma aquatica	Zhang et al. (2012)
Aquasubmersa mircensis	Zhang et al. (2012)
Asterosporium asterospermum	Révāy and Gonczol (1990)
Chaetomella raphigera	Révāy and Gonczol (1990)
Chaetospermum chaetosporium	Shearer et al. (2013)
Coeloanguillospora appalachiensis	Czeczuga and Orlowska (1996)
Clohesyomyces aquaticus	Hyde (Hyde 1993)
Diplolaeviopsis ranula	Shearer et al. (2013)
Lolia aquatica	Shearer et al. (2013)
Monichaetiopsis lakefuxianensis	Jeewon et al. (2003)
Neoheteroceras macrosporum	Yonezawa and Tanaka (2008)
Pestalotiopsis submersus	Shearer et al. (2013)
Pseudobasidiospora caroliniana	Dyko and Sutton (1978)
Pseudorobillarda sojae	Shearer et al. (2013)
Tiarosporella paludosa	Shearer et al. (2013)
Trematophoma lignicola	Shearer et al. (2013)

groups exist and associated phenotypic characters can be recognized. Based on molecular phylogeny and culture based asexual-sexual connections, some freshwater filamentous asexual fungi have been integrated into the modern classification, with most belonging to Ascomycota (See Chapter 4 for freshwater Basidiomycota). We will only discuss the phylogeny of freshwater filamentous asexual fungi in Ascomycota in this chapter.

5.3.1 Dothideomycetes

Table 5.3 lists some 120 asexual species belonging to Dothideomycetes from freshwater habitats which group in six orders, i.e. Capnodiales, Dothideales, Hysteriales, Jahnulales, Mytilinidiales and Pleosporales.

5.3.1.1 Capnodiales

Four freshwater hyphomycetes have been assigned to the order Capnodiales, i.e. *Cladosporium cladosporioides*, *C. oxysporum*, *Pseudocercosporella indica*, and *Tripospermum porosporiferum* (Crous et al. 2009; Bensch et al. 2010). Whether these are true freshwater species remains to be determined.

5.3.1.2 Dothideales

One freshwater species groups in the Dothideales, *Aureobasidium pullulans* (Smith et al. 1989; Tokumasu et al. 1994; Lygis et al. 2005). Based on an analysis of ITS

Tab. 5.2: Asexual-sexual connections based on cultures, updated from Sivichai and Jones (2003).

Asexual state	Sexual state	References
Actinospora megalospora	*Miladina lechithina*	Descals and Webster (1978)
Alternaria scirpicola	*Pleospora scirpicola*	Sivanesan (1984)
Anavirga dendromorpha	*Apostemidium torrenticola*	Hamad and Webster (1988)
Anavirga dendromorpha	*Vibrissea flavovirens*	Hamad and Webster (1988)
Anguillospora crassa	*Mollisia uda*	Webster (1961)
Anguillospora furtiva	*Pezoloma* sp.	Webster and Descals (1979)
Anguillospora fustiformis	*Rutstroemia* sp.	Abdullah et al. (1981)
Anguillospora longissima	*Massarina* sp.	Willoughby and Archer (1973)
Anguillospora rosea	*Orbilia* sp.	Webster and Descals (1979)
Anguillospora sp.	*Hymenoscyphus imberbis*	Sivichai and Jones (2003)
Anguillospora sp.	*Loramyces juncicola*	Digby and Goos (1987)
Aquaphila albicans	*Tubeufia* sp.1	Sivichai and Jones (2003)
Articulospora tetracladia	*Hymenoscyphus tetracladius*	Abdullah et al. (1981)
Chaetopsina fulva	*Cosmospora chaetopsinae*	Sivichai and Jones (2003)
Chaetopsina polyblastiae	*Cosmospora chaetopsinae-polyblastiae*	Sivichai and Jones (2003)
Chaetopsina sp.	*Cosmospora* sp.1	Sivichai and Jones (2003)
Clathrosphaerina zalewski	*Hyaloscypha zakewskii*	Descals and Webster (1976)
Clavariospsis aquatica	*Massarina* sp.	Webster and Descals (1979)
Coniothyrium scirpi	*Paraphaeosphaeria michotii*	Webster (1955a)
Crucella subtilis	*Camptobasidium hydrophilum*	Marvanová and Stalpers (1987)
Cylindrocarpon ianthothele	*Nectria discophora*	Booth (1966)
Cylindrocarpon lucidum	*Nectria lucidum*	Booth (1966)
Cylindrocarpon sp.1	*Nectria* sp.1	Sivichai and Jones (2003)
Cylindrotrichum clavatum	*Reticulascus clavatus*	Réblová et al. (2011)
Dacylaria sp.2	Unidentified ascomycete	Sivichai and Jones (2003)
Denticulate anamorph	*Pyxidiophora spinulo-rostrata*	Webster and Hawksworth (1986)
Dimorphospora foliicola	*Hymenoscyphus foliicolus*	Abdullah et al. (1981)
Ellisembia brachypus	Unidentified ascomycete	Sivichai and Jones (2003)
Filosporella sp.	*Mollisia* sp.	Webster and Descals (1979)
Flagellospora curta	*Nectria curta*	Webster (1993)
Flagellospora penicillioides	*Nectria penicillioides*	Ranzoni (1956)
Fusarium solani	*Haematonectria haematococca*	Booth (Booth 1960; 1971)
Geniculospora grandis	*Hymenoscyphus africanus*	Descals et al. (1984)
Geniculosporium sporodochiale	*Anthostomella aquatica*	Hyde and Goh (1998)
Helicodendron giganteum	*Mollisia gigantea*	Fisher and Webster (1983)
Helicodendron paradoxum	*Hymenoscyphus paradoxus*	Fisher and Webster (1983)
Helicodendron tubulosum	*Labertella tubulosa*	Abdullah and Webster (1981)
Helicomyces roseus	*Tubeufia cylindrothecia*	Sivichai and Jones (2003)
Helicosporium phragmitis	*Tubeufia paludosa*	Webster (1951)
Heliscus lugdunensis	*Nectria lugdunensis*	Webster (1959)
Ingoldiella hamata	*Sistotrema hamatum*	Nawawi and Webster (1982)
Monodictys-like hyphomycete	*Conlarium duplumascospora*	Liu et al. (2012)

(*continued*)

Tab. 5.2: (*continued*)

Asexual state	Sexual state	References
Monotosporella state of *A. mitriformis*	*Ascotaiwania mitriformis*	Ranghoo and Hyde (1998)
Monotosporella state of *A. sawada*	*Ascotaiwania sawada*	Sivichai and Jones (2003)
Penicillium dangeardii	*Talaromyces flavus* var. *flavus*	Pitt (1979)
Phaeoisaria clematidis	Unidentified ascomycete	Sivichai and Jones (2003)
Phaeoseptoria airae	*Phaeosphaeria microscopica*	Webster (1955a)
Phialophola cf. *cyclamen*	*Oxydothis* sp.1-like	Sivichai and Jones (2003)
Phoma piskorzii	*Leptosphaeria acuta*	Sivanesan (1984)
Phomopsis sp. 1-like	*Delitschia* sp.1	Sivichai and Jones (2003)
Phomopsis sp.2-like	*Diaporthe* sp.1	Sivichai and Jones (2003)
Phomopsis sp.3-like	*Hysterographium* sp.1	Sivichai and Jones (2003)
Pseudaegerita sp.	*Hyaloscypha lignicola*	Abdullah and Webster (1983)
Pyricularia aquatica	*Massarina aquatica*	Webster (1965)
Scolecosporiella typhae	*Phaeosphaeria typharum*	Webster (1955b)
Scopulariopsis lunaspora	*Microascus lunasporus*	Udaiyan (1989)
Sepedonium-like sp.	*Corynascus sepedonium*	Udaiyan (1989)
Sesquicillium sp.	*Gnomonia papuana*	Sivanesan and Shaw (1977)
Sporodochial sp.1	Unidentified discomycete sp.9	Sivichai and Jones (2003)
Sporoschisma saccardoi	*Melanochaeta hemipsila*	Sivichai et al. (2000)
Sporoschisma uniseptatum	*Melanochaeta garethyonesii*	Sivichai et al. (2000)
Taeniolella-like	*Chaetorostrum quincemilensis*	Zelski et al. (2011)
Taeniospora descalsii	*Fibulomyces crucelliger*	Marvanová and Stalpers (1987)
Taeniospora gracilis	*Leptosporomyces galzinii*	Nawawi et al. (1977)
Taeniospora gracilis var. *enecta*	*Fibulomyces* sp.	Marvanová and Stalpers (1987)
Tetraploa aristata	*Massarina tetraploa*	Scheuer (1991)
Trichocladium uniseptatum	*Ascolacicola aquatica*	Ranghoo and Hyde (1998)
Tricladium chaetocladium	*Hydrocina chaetocladia*	Webster et al. (1991)
Tricladium indicum	*Cudoniella indica*	Webster et al. (1995)
Tricladium splendens	*Hymenoscyphus splendens*	Abdullah et al. (1981)
Unidentified hyphomycete	*Hymenoscyphus malawienis*	Fisher and Spooner (1987)
Unidentified hyphomycete	*Microascus* sp.1-like	Sivichai and Jones (2003)
Unidentified hyphomycete	*Pseudohalonectria phialidica*	Shearer (1989)
Unidentified hyphomycete	Unidentified discomycete sp.7	Sivichai and Jones (2003)
Varicosporium sp.	*Hymenoscyphus varicosporoides*	Sivichai and Jones (2003)
Xylomyces chlamydosporu	*Jahnula aquaticac*	Sivichai et al. (2011)
Xylomyces sp.	*Jahnula siamensiae*	Pang et al. (2002)

Tab. 5.3: Number of freshwater hyphomycetous species in each order of Dothideomycetes.

Capnodiales	Dothideales	Hysteriales	Jahnulales	Mytilinidiales	Pleosporales	*Incertae sedis*
4	1	6	6	7	73	23

sequences, Hambleton et al. (2003) showed that three strains of *A. pullulans* nest together on a branch distinct from *Dothidea*. However, this species is common in other habitats and is not truly an aquatic species.

5.3.1.3 Hysteriales

Acrogenospora is a common genus reported on submerged wood with six species reported from freshwater habitats, i.e. *A. ellipsoidea, A. gigantospora, A. ovalia, A. sphaerocephala, A. subprolata*, and *A. verrucospora* (Goh et al. 1998; Tsui et al. 2003; Zhu et al. 2005; Hu et al. 2010; Kurniawati et al. 2010; Sridhar et al. 2010). The related sexual genus of *Acrogenospora* is *Farlowiella* (Goh et al. 1998) which is generally assigned to Hysteriales (Boehm et al. 2009). However, recent molecular data (Schoch et al. 2006b; Boehm et al. 2009) support the transfer of the genus *Farlowiella* from the Hysteriales, and its current placement as Pleosporomycetidae gen. *incertae sedis.*

5.3.1.4 Jahnulales

The Jahnulales was established by Pang et al. (2002) mainly composed of lignicolous freshwater ascomycetes. Studies by Campbell et al. (2007), Shearer et al. (2009) and Sivichai et al. (2011) showed that the asexual state *Xylomyces* groups with *Jahnula* species. Three mitosporic genera: *Brachiosphaera* (2 species), *Speiropsis* (1 species) and *Xylomyces* (6 species) group in family Aliquandostipitaceae (Jahnulales) with high statistical support (Suetrong et al. 2011). Furthermore Sivichai et al. (2011) confirmed that *Jahnula aquatica* is the sexual state of *Xylomyces chlamydosporus*, using a culture approach.

5.3.1.5 Mytilinidiales

Seven freshwater *Taeniolella* species group in the Mytilinidiales with *Glyphium* as the sexual state (Kirk et al. 2008). However, recent molecular data showed that *Taeniolella typhoides* occurred in a well-supported group with members of Lindgomycetaceae in the Pleosporales (Shearer et al. 2009).

5.3.1.6 Pleosporales

Seventy-three freshwater hyphomycetes are referred to the Pleosporales, with *Dictyosporium* (22 species) and *Sporidesmium* (17 species) the most speciose genera in the order (Cai et al. 2003; Cai and Hyde 2007; Cai et al. 2008; Wongsawas et al. 2009a; Hu et al. 2010). The genera *Dictyosporium* and *Sporidesmium*, together with *Cheirosporium, Cheiromyces, Cheiromycina, Digitomyces, Digitodesmium* and *Pseudodictyosporium,* producing pigmented cheiroid or digitate conidia, are called "cheirosporous hyphomycetes" (Cai et al. 2008; Kirschner et al. 2013). Cheirosporous hyphomycetes are commonly reported on submerged wood in various freshwater habitats (Cai et al. 2003, 2008; Cai and Hyde 2007; Wongsawas et al. 2009a; Hu et al. 2010). Molecular analyses indicate that this peculiar conidial morphology might be of some phylogenetic significance, hitherto confined mainly to a clade within the Pleosporales (Cai et al. 2008; Kirschner et al. 2013). *Aquaticheirospora* is a newly established (Kodsueb et al. 2007) "cheirosporous hyphomycetes" genus from a freshwater habitat, and it is closely related to the ascomycete family Massarinaceae (Pleosporales) in a phylogenetic tree based on 18S rDNA, 28S rDNA and ITS data.

Some other freshwater hyphomycetes are also related to Pleosporales as revealed by molecular phylogeny. Phylogenies revealed that *Berkleasmium micronesicum* and *B. nigroapicale* were related to *Westerdykella cylindrica* and *Sporormia australis*, which are members of the family Sporormiaceae (Pleosporales) (Pinnoi et al. 2007). *Berkleasmium* species are commonly reported from freshwater habitats, e.g. *B. concinnum* in USA (Shearer and Crane 1986) and China (Tsui et al. 2000), *B. corticola* in China (Tsui et al. 2000; Cai et al. 2002), and *B. zhejiangense* in China (Wongsawas et al. 2009b). The genus *Curvularia* was placed in Pleosporaceae based on molecular phylogeny (Zhang et al. 2011; Manamgoda et al. 2012), and some species of this genus have been reported from freshwater habitats, e.g. *C. inaequalis* and *C. protuberata* in USA (Shearer 1972).

Recently, three newly described freshwater coelomycetes (*Acrocalymma aquatica*, *Aquasubmersa mircensis* and *Clohesyomyces aquaticus*) were placed in the Pleosporales based on phylogenetic studies using SSU and LSU sequences (Zhang et al. 2012).

Some Ingoldian fungi were also assigned to this order. For example, molecular phylogeny indicated that *Anguillospora longissima* was a member of the Pleosporales (Baschien et al. 2006). *Goniopila monticola* and *Lemonniera pseudofloscula* also grouped in the Pleosporales based on molecular phylogeny using SSU rDNA sequences (Campbell et al. 2006; Shearer et al. 2009).

5.3.1.7 Tubeufiales

The "helicosporous hyphomycete" genera *Helicoma*, *Helicomyces* and *Helicosporium* are commonly encountered in freshwater habitats (Lane and Shearer 1984; Goh and Hyde 1996; Kane et al. 2002; Sivichai et al. 2002; Tsui et al. 2003). They are asexual states of genera in *Tubeufiaceae* (Goos 1987). Tsui et al. (2006) assessed their phylogenetic relationships from ribosomal sequences of ITS, 5.8S and partial LSU regions, and found that 45 isolates from the three genera were closely related and were within the sexual genus *Tubeufia* (Tubeufiaceae). The freshwater hyphomycete *Aquaphila albicans*, reported on submerged wood, was also assigned to Tubeufiaceae (Tsui et al. 2007; Shearer et al. 2009). More recently Boonmee et al. (2011) have undertaken a revision of the sexual stages assigned to the Tubeufiaceae and included a number of asexual species as well. They demonstrated that many asexual genera were polyphyletic, e.g. *Helicosporium* grouping in *Tubeufia paludosa* clade (*H. linderi*), and *Acanthostigma minutum* clade (*H. guianense*). Hyde et al. (2011) list the following asexual helicosporus, staurosporous or dictyosporous species in the Tubeufiaceae: *Annelloporella, Aquaphila, Arneomyces, Guelichia, Helicoma, Helicoon, Helicomyuces, Helicosporium, Kamalomyces, Pendulispora, Peziotrichum, Tetracrium, Titaea* and *Xenosporium*, many reported from freshwater habitats.

5.3.2 Leotiomycetes

Baschien et al. (2006) studied the phylogenetic relationships of 11 freshwater hyphomycetes, based on SSU, LSU and ITS sequences, and placed nine in the class Leotiomycetes. Duarte et al. (2013) reviewed the molecular phylogenetic studies over the past decade, and concluded that 46 aquatic hyphomycetes belong in the order Helotiales (Leotiomycetes). Baschien et al. (2013) examined the phylogenetic relationships among various genera of aquatic hyphomycetes belonging to the Leotiomycetes using sequences of internal transcribed spacer (ITS) and large subunit (LSU) regions of rDNA generated from 42 pure cultures including 19 ex-types, and 75 species of aquatic hyphomycetes and their teleomorphs are associated with the Helotiales (Leotiomycetes). To date, 77 aquatic hyphomycetes can be accommodated in the Helotiales (Tab. 5.4).

Tab. 5.4: Aquatic hyphomycetes species belonging to Helotiales based on sequence data [after Baschien et al. (2013)].

Species	Species
Alatospora acuminata	*Loramyces* sp.
A. constricta	*L. terrestris*
A. crassipes	*L. macrosporus*
A. pulchella	*Margaritispora aquatica*
Anguillospora crassa	*Miniancora allisoniensis*
A. filiformis	*Mycoarthris corallina*
A. furtive	*Mycofalcella calcarata*
Arbusculina fragmentans	*Tetrachaetum elegans*
Articulospora atra	*T. apiense*
A. tetracladia	*T. breve*
Cladochasiella divergens	*Tetracladium furcatum*
Cudoniella indica	*T. marchalianum*
Dimorphospora foliicola	*T. maxilliforme*
Dwayaangam colodena	*T. palmatum*
Filosporella cf. *annelidica*	*T. setigerum*
F. exilis	*Tricladium alaskense*
F. fistucella	*T. angulatum*
F. versimorpha	*T. attenuatum*
Flagellospora curvula	*T. biappendiculatum*
F. fusarioides	*T. castaneicola*
F. leucorhynchos	*T. caudatum*

(*continued*)

Tab. 5.4: (*Continued*)

Species	Species
F. saccata	*T. chaetocladium*
Flagellospora sp.1	*T. curvisporum*
Flagellospora sp. 2	*T. indicum*
Fontanospora eccentrica	*T. kelleri*
F. fusiramosa	*T. minutum*
Geniculospora grandis	*T. obesum*
Gorgomyces honrubiae	*T. patulum*
G. hungaricus	*T. procerum*
Gyoerffyella cf. *craginiformis*	*T. splendens*
G. entomobryoides	*T. terrestre*
G. gemellipara	*Varicosporium delicatum*
G. rotula	*V. elodeae*
G. tricapillata	*V. giganteum*
Hydrocina chaetocladia	*V. scoparium*
Lemonniera aquatica	*V. trimosum*
L. centrosphaera	*Variocladium giganteum*
L. cornuta	*Ypsilina graminea*
L. filiformis	

5.3.3 Orbiliomycetes

5.3.3.1 Orbiliales

Orbiliales were previously classified in Helotiales based on morophlogical characters. However, molecular phylogenetic studies of rDNA did not support this taxon as a monophyletic group, and a new order and class were erected to accommodate this taxon (Spatafora et al. 2006). *Arthrobotrys* species are asexual fungi linked to *Orbilia* (Orbiliales) (Spatafora et al. 2006), and are commonly reported in freshwater habitats, e.g. *A. musiformis* in the USA (Shearer 1972), *A. oligospora* in the UK (Kane et al. 2002), and *A. superba* in the USA (Shearer and Crane 1986).

The genus *Anguillospora* has been shown to be polyphyletic with species distributed among the Leotiomycetes, Dothideomycetes and Orbiliomycetes (Michel et al. 2005). Belliveau and Bärlocher (2005) confirmed the placement of *A. rosea* in *Orbilia* (Orbiliales), based on SSU rDNA sequences. *Anguillospora rosea* appears to be a common species in aquatic habitats and was reported from Ireland (Harrington 1997), Hungary (Gönczöl et al. 2004) and Canada (Sokolski et al. 2006).

5.3.4 Sordariomycetes

A phylogenetic tree has been constructed, based on 28S sequences downloaded from GenBank (Tab. 5.5) to determine the phylogeny of selected freshwater filamentous asexual fungi. One representative taxon of the order Dothideales (*Dothidea insculpta*)

Tab. 5.5: Taxa and sequences included in the phylogenetic analysis.

Taxa	Sources	28S	Habitats	References
Ambarignomonia petiolorum	CBS 121227	EU255070	Terrestrial	Sogonov et al. (2008)
Bertia moriformis	SMH4320	AY695260	Terrestrial	Huhndorf et al. (2004)
Bertia moriformis	SMH3344	AY695261	Terrestrial	Huhndorf et al. (2004)
Chaetosphaeria capitata	N/A	AF466061	Terrestrial	Fernández et al. (2006)
Chaetosphaeria fuegiana	ICMP 15153	EF063574	Terrestrial	Réblová and Seifert (2007)
Coniochaeta polymorpha	CBS 132722	HE863327	Terrestrial	Khan et al. (2013)
Conioscypha lignicola	CBS 335.93	AY484513	Terrestrial	Réblová and Seifert (2004)
Conioscypha varia	CBS 113653	AY484512	Terrestrial	Réblová and Seifert (2004)
Cylindrotrichum clavatum	CBS 125296	GU180643	Freshwater	Réblová et al. (2011)
Cylindrotrichum clavatum	CBS 125239	GU180649	Freshwater	Réblová et al. (2011)
Cylindrotrichum clavatum	CBS 125297	GU180650	Freshwater	Réblová et al. (2011)
Diaporthe eres	AR3538	AF408350	Terrestrial	Castlebury et al. (2002)
Diaporthe passiflorae	CPC 19183	JX069844	Terrestrial	Crous et al. (2012)
Dictyochaeta cylindrospora	MUCL 39171	EF063575	Freshwater	Réblová and Seifert (2007)
Dothidea insculpta	CBS 189.58	NG_027643	Terrestrial	Schoch et al. (2006a)
Glomerella acutata	N/A	FJ588235	Terrestrial	Réblová et al. (2011)
Glomerella cingulata	N/A	DQ286199	Terrestrial	Réblová et al. (2011)
Helicoön farinosum	DAOM 241947	JQ429230	Freshwater	Réblová et al. (2012)
Hypocrea lutea	NBRC104901	JN941457	Terrestrial	Schocha et al. (2012)
Hypocrea rufa	DAOM JBT1003	JN938865	Terrestrial	Schocha et al. (2012)
Kylindria peruamazonensis	CBS 421.95	HM237325	Terrestrial	Réblová et al. (2011)
Lulworthia medusa	J. K. 5581	AF195637	Terrestrial	Kohlmeyer et al. (2000)
Magnisphaera spartinae	A2211	AY150221	Freshwater	Sakayaroj et al. (2011)
Magnisphaera stevemossago	A4091B	AY227135	Freshwater	Sakayaroj et al. (2011)
Meliola centellae	VIC31244	JQ734545	Terrestrial	Pinho et al. (2013)
Monilochaetes infuscans	CBS 870.96	GU180644	Terrestrial	Réblová et al. (2011)

(*continued*)

Tab. 5.5 *(continued)*

Taxa	Sources	28S	Habitats	References
Nectria cinnabarina	CBS 125165	HM484562	Terrestrial	N/A
Ophiostoma piliferum	AFTOL-ID 910	DQ470955	Terrestrial	Spatafora et al. (2006)
Pleurotheciella centanaria	DAOM 229631	JQ429234	Freshwater	Réblová et al. (2012)
Pleurotheciella rivularia	CBS 125238	JQ429232	Freshwater	Réblová et al. (2012)
Pleurotheciella rivularia	CBS 125237	JQ429233	Freshwater	Réblová et al. (2012)
Pleurothecium recurvatum	DAOM 230069	JQ429238	Freshwater	Réblová et al. (2012)
Pleurothecium semifecundum	CBS 131482	JQ429239	Freshwater	Réblová et al. (2012)
Podospora fibrinocaudata	CBS 315.91	AY780074	Terrestrial	Miller and Huhndorf (2005)
Sordaria fimicola	CBS 508.50	AY681160	Terrestrial	Cai et al. (2006)
Sterigmatobotrys macrocarpa	CBS 113468	GU017316	Freshwater	Réblová et al. (2012)
Sterigmatobotrys macrocarpa	PRM 915682	GU017317	Terrestrial	Réblová et al. (2012)
Trichosphaeria pilosa	089319b	AY590297	Terrestrial	Campbell and Shearer (2004)
Xylaria acuta	ATCC 56487	AY544676	Terrestrial	N/A
Xylaria hypoxylon	ATCC 42768	U47841	Terrestrial	N/A

was used as the out-group taxon. In total 42 sequences, including 13 sequences from freshwater filamentous asexual fungi were used in the construction of the phylogenetic tree. The aligned dataset included 749 characters including gaps. Maximum likelihood (ML) analysis method was used to analyze the dataset. RAxML v7.2.6 (Stamatakis 2006) was used for ML analyses. Default setting was used for all parameters in the ML analysis, and statistical support values were obtained using nonparametric bootstrapping with 1,000 replicates. The results are shown in Fig. 5.4.

In the phylogenetic tree (Fig. 5.4), the strains of freshwater hyphomycetes nested in three clades; Clade 1 comprised *Conioscypha lignicola, Helicoon farinosum, Pleurotheciella rivularia, Pleurothecium recurvatum* and *Sterigmatobotrys macrocarpa* in a well-supported clade, closely related to the order Hypocreales. Clade 2 comprised three strains of *Cylindrotrichum clavatum* which nested in the Glomerellales, and Clade 3 included *Dictyochaeta cylindrospora* in a unique clade unrelated to any order in the Sordariomycetes (Fig. 5.4).

Many freshwater filamentous asexual fungi have been referred to Sordariomycetes based on molecular studies. These species were mainly assigned to the orders: Chaetosphaeriales, Coniochaetales, Glomerellales, Hypocreales, Lulworthiales, Microascales, Savoryellales, Sordariales, Trichosphaeriales, and Xylariales, while many remain unassigned to any known order.

5.3.4.1 Glomerellales

Réblová et al. (2011) introduced this order and included three families: Glomerellaceae, Australiascaceae and Reticulascaceae. *Cylindrotrichum clavatum* nested in the family Reticulascaceae, a species reported on submerged wood (Réblová et al. 2011) (Fig. 5.4).

5.3.4.2 Hypocreales

A number of filamentous asexual fungi from freshwater habitats are referred to this order. Cai et al. (2010) described a new freshwater hyphomycete *Mariannaea aquaticola*, and phylogenetic studies based on 28S and ITS sequences indicated its affinity to Nectriaceae (Hypocreales). Duarte et al. (2013) placed two aquatic hyphomycetes species (*Flagellospora curta* and *Heliscus lugdunensis*) in Hypocreales based on molecular studies using SSU and LSU sequences.

5.3.4.3 Sordariales

Belliveau and Bärlocher (2005), based on SSU sequences, referred the common aquatic hyphomycete *Lunulospora curvula* in this order, a species widely reported in both temperate and tropical locations growing on senescent shed leaves (Ingold 1942; Belwal et al. 2008; Sridhar et al. 2010).

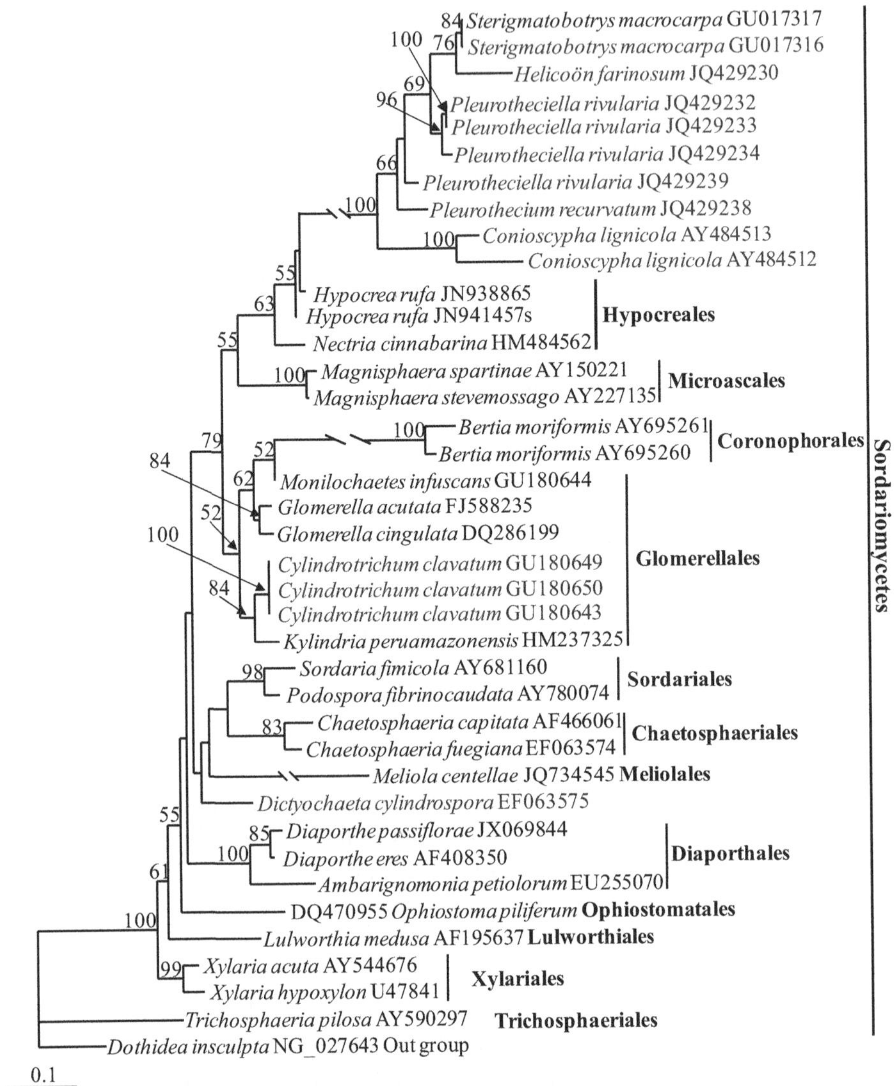

Fig. 5.4: Phylogenetic tree (ML) based on 28S rDNA sequences downloaded from NCBI, bootstrap value higher than 50% is indicated on each node. Filamentous asexual fungi known from freshwater are shown in blue.

5.3.4.4 Savoryellales

Sri-Indrasutdhi et al. (2010) introduced the genus *Ascothailandia* (*A. grenadoidia*) to accommodate a perithecial ascomycete growing on submerged wood with the ascospores giving rise to an asexual morph in the genus *Canalisporium*. This, and five other *Canalisporium* species, grouped in a sister clade to the Hypocreales and

Microascales. Subsequently, Boonyuen et al. (2011) introduced the order Savoryellales to accommodate *Ascothailandia* and two other wood inhabiting ascomycetes *Ascotaiwania* and *Savoryella*, genera that include both freshwater and marine taxa. Two *Monotosporella* species, also reported from submerged wood, also grouped in the Savoryellales. Under the one fungus one name rule, the asexual genus *Canalisporium* may have priority over the more recently described sexual genus *Ascothailandia*. At the time of writing however, there is no authoritative account of accepted names for freshwater fungi.

5.4 Discussion

Traditional taxonomy of freshwater filamentous asexual fungi has been based on the morphology and development of conidia and the morphology of conidiophores. Since many morphological characters of freshwater filamentous asexual fungi are formed due to convergent evolution, such as tetraradiate conidia (Belliveau and Bärlocher 2005), they are often phylogenetically non-informative. Recent studies based on molecular data revealed that many taxa of freshwater filamentous asexual fungi are polyphyletic (Nikolcheva and Barlocher 2002; Baschien et al. 2006; Tsui and Berbee 2006; Baschien et al. 2013).

Molecular based studies have resolved aspects of the phylogeny and evolution of aquatic hyphomycetes that were inaccessible by conventional methods (Duarte et al. 2013). Vijaykrishna et al. (2006) studied the molecular taxonomy, origins and evolution of freshwater ascomycetes, and the results indicated that freshwater ascomycetes have evolved from terrestrial fungi and mainly occur in three classes. The adaptation to populate freshwater substrates has occurred in several lineages, and the earliest possible date, when fungi became adapted to freshwater habitation is estimated at 390 million years ago. The study based on molecular data showed that the adaptation of aquatic hyphomycetes to colonize the aquatic environment has evolved independently in multiple phylogenetic lineages within the Leotiomycetes (Baschien et al. 2013). DNA sequence data will continue to expand the knowledge on the evolution and ecology of aquatic hyphomycetes (Duarte et al. 2013).

However, only a few taxa of freshwater filamentous asexual fungi have been reassessed based on molecular data, and most taxa are still classified according to their morphological characters. Currently there is an urgent need to complete the following work:

1. Recollect and isolate freshwater filamentous asexual fungi into axenic cultures. It is relatively easy to extract DNA and generate gene sequences from cultures, for phylogenetic studies. Thus species should be recollected and studied.
2. Genomic DNA should be extracted from fresh materials on natural substrata when pure cultures are not available.

Acknowledgment

E. B. Gareth Jones is supported by the Distinguished Scientist Fellowship Program (DSFP), King Saud University, Saudi Arabia.

References

Abdullah SK, Descals E, Webster J. Teleomorphs of three aquatic Hyphomycetes. Trans Br Mycol Soc 1981;77:475–483.

Abdullah SK, Webster J. The aero-aquatic genus *Pseudaegerita*. Trans Br Mycol Soc 1983;80:247–254.

Abdullah SK, Webster J. *Lambertella tubulosa* sp. nov., teleomorph of *Helicodendron tubulosum*. Trans Br Mycol Soc 1981;76:261–263.

Baschien C, Marvanová L, Szewzyk U. Phylogeny of selected aquatic hyphomycetes based on morphological and molecular data. Nova Hedwig 2006;83:311–352.

Baschien C, Tsui CKM, Gulis V, Szewzyk U, Marvanová L. The molecular phylogeny of aquatic hyphomycetes with affinity to the Leotiomycetes. Fungal Biol 2013;117:660–672.

Belliveau MJR, Bärlocher F. Molecular evidence confirms multiple origins of aquatic hyphomycetes. Mycol Res 2005;109:1407–1417.

Belwal M, Sati SC, Arya P. Temperature tolerance of water borne conidial fungi in freshwater streams of Central Himalaya. Natl Acad Sci Lett 2008;31:175–179.

Boehm EWA, Schoch CL, Spatafora JW. On the evolution of the Hysteriaceae and Mytilinidiaceae (Pleosporomycetidae, Dothideomycetes, Ascomycota) using four nuclear genes. Mycol Res 2009;113:461–479.

Boonmee S, Zhang Y, Chomnunti P, Chukeatirote E, Tsui CKM, Bahkali AH, et al. Revision of lignicolous Tubeufiaceae based on morphological reexamination and phylogenetic analysis. Fungal Divers 2011;51:63–102.

Boonyuen N, Chuaseeharonnachai C, Suetrong S, Sri-Indrasutdhi V, Sivichai S, Jones EBG, et al. Savoryellales (Hypocreomycetidae, Sordariomycetes): a novel lineage of aquatic ascomycetes inferred from multiple-gene phylogenies of the genera *Ascotaiwania*, *Ascothailandia*, and *Savoryella*. Mycologia 2011;103:1351–1371.

Booth C. The genus *Cylindrocarpon*. Mycol Papers 1966;104:1–56.

Booth C. The Genus *Fusarium*. Kew: Commonwealth Mycological Institute 1971.

Booth C. Studies of Pyrenomycetes: V. Nomenclature of some *Fusaria* in relation to their nectroid perithecial states. Mycol Papers 1960;74:1–16.

Cai L, Guo XY, Hyde KD. Morphological and molecular characterisation of a new anamorphic genus *Cheirosporium*, from freshwater in China. Persoonia 2008;20:53–58.

Cai L, Hyde KD. Anamorphic fungi from freshwater habitats in China: *Dictyosporium tetrasporum* and *Exserticlava yunnanensis* spp. nov., and two new records for *Pseudofuscophialis lignicola* and *Pseudobotrytis terrestris*. Mycoscience 2007;48:290–296.

Cai L, Jeewon R, Hyde KD. Phylogenetic investigations of Sordariaceae based on multiple gene sequences and morphology. Mycol Res 2006;110:137–150.

Cai L, Kurniawati E, Hyde KD. Morphological and molecular characterization of *Mariannaea aquaticola* sp. nov. collected from freshwater habitats. Mycol Prog 2010;9:337–343.

Cai L, Tsui C, Zhang K, Hyde KD. Aquatic fungi from Lake Fuxian, Yunnan, China. Fungal Divers 2002;9:57–70.

Cai L, Zhang KQ, McKenzie EHC, Lumyong S, Hyde KD. New species of *Canalisporium* and *Dictyosporium* from China and a note on the differences between these genera. Cryptogamie Mycol 2003;24:3–11.

Campbell J, Shearer C, Marvanová L. Evolutionary relationships among aquatic anamorphs and teleomorphs: *Lemonniera*, *Margaritispora*, and *Goniopila*. Mycol Res 2006;110:1025–1033.

Campbell J, Shearer CA. *Annulusmagnus* and *Ascitendus*, two new genera in the Annulatascaceae. Mycologia 2004;96:822–833.

Campbell J, Shearer CA, Ferrer A, Raja HA, Sivichai S. Phylogenetic relationships among taxa in the Jahnulales inferred from 18S and 28S nuclear ribosomal DNA sequences. Can J Bot 2007;85:873–882.

Castlebury LA, Rossman AY, Jaklitsch WJ, Vasilyeva LN. A preliminary overview of the Diaporthales based on large subunit nuclear ribosomal DNA sequences. Mycologia 2002;94:10–17.

Crous PW, Summerell BA, Shivas RG, Burgess TI, Decock CA, Dreyer LL, et al. Fungal planet description sheets: 107–127. Persoonia 2012;28:138–182.

Czeczuga B, Orlowska M. Hyphomycetes in twenty springs of the Knyszyn-Bialystok Forest in various seasons. Internationale Revue der gesamten Hydrobiologie und Hydrographie 1996;81:417–433.

Descals C, Webster J. *Hyaloscypha*: perfect state of *Clathrosphaerina zalewsk*ii. Trans Br Mycol Soc 1976;67:525–528.

Descals E. Diagnostic characters of propagules of Ingoldian fungi. Mycol Res 2005;109:545–555.

Descals E, Fisher PJ, Webster J. The *Hymenoscyphus* teleomorph of *Geniculospora grandis*. Trans Br Mycol Soc 1984;83:541–546.

Descals E, Webster J. *Miladina lecithina* (Pezizales), the ascigerous state of *Actinospora megalospora*. Trans Br Mycol Soc 1978;70:466–472.

Digby S, Goos RD. Morphology, development and taxonomy of *Loramyces*. Mycologia 1987;79:821–831.

Duarte S, Seena S, Bärlocher F, Pascoal C, Cássio F. A decade's perspective on the impact of DNA sequencing on aquatic hyphomycete research. Fungal Biol Rev 2013;27:19–24.

Dyko BJ, Sutton BC. Two new genera of water-borne coelomycetes from submerged leaf litter. Nova Hedwig 1978;29:167–178.

Fernández FA, Miller AN, Huhndorf SM, Lutzoni FM, Zoller S. Systematics of the genus *Chaetosphaeria* and its allied genera: morphological and phylogenetic diversity in north temperate and neotropical taxa. Mycologia 2006;98:121–130.

Fisher PJ, Spooner B. Two new ascomycetes from Malawi. Trans Br Mycol Soc 1987;88:47–54.

Fisher PJ, Webster J. The teleomorphs of *Helicodendron giganteum* and *H. paradoxum*. Trans Br Mycol Soc 1983;81:656–659.

Gönczöl J, Révay Á, Papp B. Aquatic hyphomycetes in two streams differing in discharge and distribution of leaf litter. Studia Bot Hung 2004;35:45–58.

Goh TK, Hyde KD. Biodiversity of freshwater fungi. J Ind Microbiol Biot 1996;17:328–345.

Goh TK, Hyde KD. A revision of *Dactylaria*, with description of *D. tunicata* sp. nov. from submerged wood in Australia. Mycol Res 1997;101:1265–1272.

Goh TK, Hyde KD, Tsui KM. The hyphomycete genus *Acrogenospora*, with two new species and two new combinations. Mycol Res 1998;102:1309–1315.

Goos RD. Fungi with a twist: the helicosporous hyphomycetes. Mycologia 1987;79:1–22.

Hamad SR, Webster J. *Anavirga dendromorpha*, anamorph of *Apostemidium torrenticola*. Sydowia 1988;40:60–64.

Hambleton S, Tsuneda A, Currah RS. Comparative morphology and phylogenetic placement of two microsclerotial black fungi from *Sphagnum*. Mycologia 2003;95:959–975.

Harrington TJ. Aquatic hyphomycetes of 21 rivers in southern Ireland. Biol Environ: Proc R Irish Acad 1997;97B:139–148.
Hu D-M, Liu F, Cai L. Biodiversity of aquatic fungi in China. Mycology 2013;4:125–168.
Hu DM, Cai L, Chen H, Bahkali AH, Hyde KD. Four new freshwater fungi associated with submerged wood from Southwest Asia. Sydowia 2010;62:191–203.
Huhndorf SM, Miller AN, Fernández FA. Molecular systematics of the Coronophorales and new species of *Bertia*, *Lasiobertia* and *Nitschkia*. Mycol Res 2004;108:1384–1398.
Hyde KD. Tropical Australian freshwater fungi. VI. *Tiarosporella paludosa* and *Clohesyomyces aquaticus* gen. et sp. nov. Coelomycetes). Aust Syst Bot 1993;6:169–173.
Hyde KD, Goh TK. Tropical Australian freshwater fungi XIII. a new species of *Anthostomella* and its sporodochial *Geniculosporium* anamorph. Nova Hedwig 1998;67:225–233.
Hyde KD, McKenzie EHC, KoKo TW. Towards incorporating anamorphic fungi in a natural classification–checklist and notes for 2010. Mycosphere 2011;2:1–88
Ingold C. Aquatic hyphomycetes of decaying alder leaves. Trans Br Mycol Soc 1942;25:339–417.
Jeewon R, Cai L, Liew ECY, Zhang KQ, Hyde KD. *Dyrithiopsis lakefuxianensis* gen. et sp. nov. from Fuxian Lake, Yunnan, China, and notes on the taxonomic confusion surrounding *Dyrithium*. Mycologia 2003;95:911–920.
Kane DF, Tam WY, Jones EBG. Fungi colonising and sporulating on submerged wood in the River Severn, UK. Fungal Divers 2002;10:45–55.
Khan Z, Gené J, Ahmad S, Cano J, Al-Sweih N, Joseph L, et al. *Coniochaeta polymorpha*, a new species from endotracheal aspirate of a preterm neonate, and transfer of *Lecythophora* species to *Coniochaeta*. Antonie van Leeuwenhoek 2013;104:243–252.
Kirk PM, Cannon PF, Minter DW, Stalpers JA. Dictionary of the Fungi (10th edition). Trowbridge: Cromwell Press 2008;771pp.
Kirschner R, Pang K-L, Jones EBG. Two cheirosporous hyphomycetes reassessed based on morphological and molecular examination. Mycol Prog 2013;12:29–36.
Kodsueb R, Lumyong S, Ho WH, Hyde KD, Mckenzie EH, Jeewon R. Morphological and molecular characterization of *Aquaticheirospora* and phylogenetics of Massarinaceae (Pleosporales). Bot J Linn Soc 2007;155:283–296.
Kohlmeyer J, Spatafora JW, Volkmann-Kohlmeyer B. Lulworthiales, a new order of marine Ascomycota. Mycologia 2000;92:453–458.
Kurniawati E, Zhang HA, Chukeatirote E, Sulistyowati L, Moslem MA, Hyde KD. Diversity of freshwater ascomycetes in freshwater bodies at Amphoe Mae Chan, Chiang Rai. Cryptogamie Mycol 2010;31:323–331.
Lane LC, Shearer CA. *Helicomyces torquatus*, a new hyphomycete from Panama. Mycotaxon 1984;19:291–297.
Liu F, Hu DM, Cai L. *Conlarium duplumascospora* gen. *et*. sp. nov. and *Jobellisia gregariusca* sp. nov. from freshwater habitats in China. Mycologia 2012;104:1178–1186.
Lygis V, Vasiliauskas R, Larsson K-H, Stenlid J. Wood-inhabiting fungi in stems of *Fraxinus excelsior* in declining ash stands of northern Lithuania, with particular reference to *Armillaria cepistipes*. Scand J Forest Res 2005;20:337–346.
Manamgoda DS, Cai L, McKenzie EHC, Crous PW, Madrid H, Chukeatirote E, et al. A phylogenetic and taxonomic re-evaluation of the *Bipolaris-Cochliobolus-Curvularia* complex. Fungal Divers 2012;56:131–144.
Marvanová L, Stalpers JA. The genus *Taeniospora* and its teleomorphs. Trans Br Mycol Soc 1987;89:489–498.
Miller AN, Huhndorf SM. Multi-gene phylogenies indicate ascomal wall morphology is a better predictor of phylogenetic relationships than ascospore morphology in the Sordariales (Ascomycota, Fungi). Mol Phylogenet Evol 2005;35:60–75.

Nawawi A, Descals E, Webster C, Webster J. *Leptosporomyces galzinii*, the basidial state of a clamped branched conidium from fresh water. Trans Br Mycol Soc 1977;68:31–36.

Nawawi A, Webster J. *Sistotrema hamatum* sp. nov., the teleomorph of *Ingoldiella hamata*. Trans Br Mycol Soc 1982;78:287–291.

Pang KL, Abdel-Wahab MA, Sivichai S, El-Sharouney HM, Jones EBG. Jahnulales (Dothideomycetes, Ascomycota): a new order of lignicolous freshwater ascomycetes. Mycol Res 2002;106:1031–1042.

Pinho DB, Firmino AL, Ferreira-Junior WG, Pereira OL. An efficient protocol for DNA extraction from Meliolales and the description of *Meliola centellae* sp. nov. Mycotaxon 2013;122:333–345.

Pinnoi A, Jeewon R, Sakayaroj J, Hyde KD, Jones EGB. *Berkleasmium crunisia* sp. nov. and its phylogenetic affinities to the Pleosporales based on 18S and 28S rDNA sequence analyses. Mycologia 2007;99:378–384.

Pitt JI. The Genus *Penicillium* and its Teleomorphic States *Eupenicillium* and *Talaromyces*. London: Academic Press 1979;634pp.

Réblová M, Gams W, Seifert KA. *Monilochaetes* and allied genera of the Glomerellales, and a reconsideration of families in the Microascales. Stud Mycol 2011;68:163–191.

Réblová M, Seifert KA. *Conioscyphascus*, a new ascomycetous genus for holomorphs with *Conioscypha* anamorphs. Stud Mycol 2004;50:95–108

Réblová M, Seifert KA. A new fungal genus, *Teracosphaeria*, with a *Phialophora*-like anamorph (Sordariomycetes, Ascomycota). Mycol Res 2007;111:287–298.

Réblová M, Seifert KA, Fournier J, Štěpánek V. Phylogenetic classification of *Pleurothecium* and *Pleurotheciella* gen. nov. and its *Dactylaria*-like anamorph (Sordariomycetes) based on nuclear ribosomal and protein-coding genes. Mycologia 2012;104:1299–1314.

Révãy A, Gonczol J. Longitudinal distribution and colonization patterns of wood-inhabiting fungiin a mountain stream in Hungary. Nova Hedwig 1990;51:505–520.

Ranghoo VM, Hyde KD. Ascomycetes from freshwater habitats: *Ascolacicola aquatica* gen *et* sp nov and a new species of *Ascotaiwania* from wood submerged in a reservoir in Hong Kong. Mycologia 1998;90:1055–1062.

Ranzoni FV. The perfect stage of *Flagellospora penicillioides*. Am J Bot 1956;43:13–17.

Sakayaroj J, Pang K-L, Jones EBG. Multi-gene phylogeny of the Halosphaeriaceae: its ordinal status, relationships between genera and morphological character evolution. Fungal Divers 2011;46:87–109.

Scheuer CH. *Massarina tetraploa* sp. nov., the teleomorph of *Tetraploa aristata*. Mycol Res 1991;95:126–128.

Schoch CL, Kohlmeyer J, Volkmann-Kohlmeyer B, Tsui CKM, Spatafora JW. The halotolerant fungus *Glomerobolus gelineus* is a member of the Ostropales. Mycol Res 2006a;110:257–263.

Schoch CL, Shoemaker RA, Seifert KA, Hambleton S, Spatafora JW, Crous PW. A multigene phylogeny of the Dothideomycetes using four nuclear loci. Mycologia 2006b;98:1041–1052.

Schoch CL, Seifertb KA, Huhndorfc S, Robertd V, Spougea JL, Levesque CA. Nuclear ribosomal internal transcribed spacer (ITS) region as a universal DNA barcode marker for Fungi. PNAS 2012;109:6241–6246.

Seifert KA, Morgan-Jones G, Gams W, Kendrick B. The Genera of Hyphomycetes. Utrecht: CBS-KNAW Fungal Biodiversity Centre 2011.

Shearer C. *Pseudohalonectria* (Lasiosphaeriaceae), an antagonistic genus from wood in freshwater. Can J Bot 1989;67:1944–1955.

Shearer CA. Fungi of the Chesapeake Bay and its tributaries. III. The distribution of wood-inhabiting Ascomycetes and fungi Imperfecti in the Patuxent River. Am J Bot 1972;59:961–969.

Shearer CA, Crane JL. Illinois fungi XII. Fungi and Myxomycetes from wood and leaves submerged in southern Illinois swamps. Mycotaxon 1986;25:527–538.

Shearer CA, Descals E, Kohlmeyer B, Kohlmeyer J, Marvanova L, Padgett D, et al. Fungal biodiversity in aquatic habitats. Biodivers Conserv 2007;16:49–67.
Shearer CA, Fallah PM, Ferrer A, Raja HA, Schmit JP. Freshwater ascomycetes and their anamorphs. URL: www.fungi.life.illinois.edu (accessed December 2013). 2013.
Shearer CA, Raja HA, Miller AN, Nelson P, Tanaka K, Hirayama K, et al. The molecular phylogeny of freshwater Dothideomycetes. Stud Mycol 2009:145–153.
Sivanesan A. *Acantharia, Gibbera* and their anamorphs. Trans Br Mycol Soc 1984;82:507–529.
Sivanesan A, Shaw DE. *Gnomonia papuana* sp. nov. with a *Sesquicillium* conidial state. Trans Br Mycol Soc 1977;68:85–90.
Sivichai S, Hywel-Jones NL, Somrithipol S. Lignicolous freshwater Ascomycota from Thailand: *Melanochaeta* and *Sporoschisma* anamorphs. Mycol Res 2000;104:478–485.
Sivichai S, Jones EBB. Teleomorphic-anamorphic connections in freshwater fungi. Fungal Divers Research Series 2003;10:259–272.
Sivichai S, Jones EBG, Hywel-Jones N. Fungal colonisation of wood in a freshwater stream at Tad Ta Phu, Khao Yai National Park, Thailand. Fungal Divers 2002;10:113–129.
Sivichai S, Sri-Indrasutdhi V, Jones EGB. *Jahnula aquatica* and its anamorph *Xylomyces chlamydosporus* on submerged wood in Thailand. Mycotaxon 2011;116:137–142.
Smith CS, Chand T, Harris RF, Andrews JH. Colonization of a submersed aquatic plant, Eurasian water milfoil (*Myriophyllum spicatum*), by fungi under controlled conditions. Appl Environ Microb 1989;55:2326–2332.
Sogonov MV, Castlebury LA, Rossman AY, Mejía LC, White JF. Leaf-inhabiting genera of the Gnomoniaceae, Diaporthales. Stud Mycol 2008;62:1–77.
Sokolski S, Piche Y, Chauvet É, BÉRubÉ JA. A fungal endophyte of black spruce (*Picea mariana*) needles is also an aquatic hyphomycete. Molec Ecol 2006;15:1955–1962.
Spatafora JW, Sung G-H, Johnson D, Hesse C, O'Rourke B, Serdani M, et al. A five-gene phylogeny of Pezizomycotina. Mycologia 2006;98:1018–1028.
Sri-Indrasutdhi V, Boonyuen N, Suetrong S, Chuaseeharonnachai C, Sivichai S, Jones EBG. Wood-inhabiting freshwater fungi from Thailand: *Ascothailandia grenadoidia* gen. et sp. nov., *Canalisporium grenadoidia* sp. nov. with a key to *Canalisporium* species (Sordariomycetes, Ascomycota). Mycoscience 2010;51:411–420.
Sridhar KR, Karamchand KS, Hyde KD. Wood-inhabiting filamentous fungi in 12 high-altitude streams of the Western Ghats by damp incubation and bubble chamber incubation. Mycoscience 2010;51:104–115.
Stamatakis A. RAxML-VI-HPC: maximum likelihood-based phylogenetic analyses with thousands of taxa and mixed models. Bioinformatics 2006;22:2688–2690.
Suetrong S, Boonyuen N, Pang KL, Ueapattanakit J, Klaysuban A, Sri-Indrasutdhi V, et al. A taxonomic revision and phylogenetic reconstruction of the Jahnulales (Dothideomycetes), and the new family Manglicolaceae. Fungal Divers 2011;51:163–188.
Tokumasu S, Aoki T, Oberwinkler F. Fungal succession on pine needles in Germany. Mycoscience 994;35:29–37.
Tsui CKM, Hyde KD, Fukushima K. Fungi on submerged wood in the Koito River, Japan. Mycoscience 2003;44:55–59.
Tsui CKM, Hyde KD, Hodgkiss IJ. Biodiversity of fungi on submerged wood in Hong Kong streams. Aquat Microb Ecol 2000;21:289–298.
Tsui CKM, Sivichai S, Berbee ML. Molecular systematics of *Helicoma, Helicomyces* and *Helicosporium* and their teleomorphs inferred from rDNA sequences. Mycologia 2006;98:94–104.
Tsui CKM, Sivichai S, Rossman AY, Berbee ML. *Tubeufia asiana*, the teleomorph of *Aquaphila albicans* in the Tubeufiaceae, Pleosporales, based on cultural and molecular data. Mycologia 2007;99:884–894.

Udaiyan K. Some interesting ascomycetes from water-cooling towers. Kavaka 1989;17:11–16.
Vijaykrishna D, Jeewon R, Hyde KD. Molecular taxonomy, origins and evolution of freshwater ascomycetes. Fungal Divers 2006;23:351–390.
Webster J. Graminicolous pyrenomycetes. V. Conidial states of *Leptosphaeria michotii, L. microscopica, Pleospora vagans* and the perfect state of *Dinemasporium graminum*. Trans Br Mycol Soc 1955a;38:347–365.
Webster J. Graminicolous pyrenomycetes: I. The conidial stage of *Tubeufia helicomyces*. Trans Br Mycol Soc 1951a;34:304–308.
Webster J. *Hendersonia typhae* the conidial state of *Leptosphaeria typharum*. Trans Br Mycol Soc 1955b;38:405–408.
Webster J. The *Mollisia* perfect state of *Anguillospora crassa*. Trans Br Mycol Soc 1961;44:559–564.
Webster J. *Nectria curta* sp. nov., (Ascomycetes, Hypocreales) an aquatic fungus and its *Flagellospora anamorph*. Nova Hedwig 1993;56:455–464.
Webster J. *Nectria lugdunensis* sp. nov., the perfect state of *Heliscus lugdunensis*. Trans Br Mycol Soc 1959;42:322–327.
Webster J. The perfect state of *Pyricularia aquatica*. Trans Br Mycol Soc 1965;48:449–452.
Webster J, Descals E. The teleomorphs of water-borne Hyphomycetes from freshwater. In: Kendrick WB, editor. The Whole Fungus. Ottawa: National Museum of Natural Sciences 1979;419–451.
Webster J, Eicker A, Spooner BM. *Cudoniella indica* sp. nov. (Ascomycetes, Leotiales), the teleomorph of *Tricladium indicum*, an aquatic fungus isolated from a South African river. Nova Hedwig 1995;60:493–498.
Webster J, Hawksworth DL. *Pyxidiophora spinulorostrata*, a new species with denticulate conidiophores from submerged twigs in Southwest England. Trans Br Mycol Soc 1986;87:77–79.
Webster J, Scheuer CH, Om-Kalthoum Khattab S. *Hydrocina chaetocladia* gen. *et* sp. nov., the teleomorph of *Tricladium chaetocladium*. Nova Hedwig 1991;52:65–72.
Willoughby LG, Archer JF. The fungal spora of a freshwater stream and its colonization pattern on wood. Freshwater Biol 1973;3:219–239.
Wongsawas M, Wang HK, Hyde KD, Lin FC. *Dictyosporium zhejiangense* sp. nov., a new freshwater anamorphic fungus from China. Cryptogamie Mycol 2009a;30:355–362.
Wongsawas M, Wang HK, Hyde KD, Lin FC. Two new hyphomycetes from submerged wood collected in China. Sydowia 2009b;61:345–351.
Yonezawa H, Tanaka K. A second species of *Neoheteroceras* and additional characters of the genus. Mycoscience 2008;49:152–154.
Zelski SE, Raja HA, Miller AN, Shearer CA. *Chaetorostrum quincemilensis*, gen. *et* sp. nov., a new freshwater ascomycete and its *Taeniolella*-like anamorph from Peru. Mycosphere 2011;2:593–600.
Zhang H, Hyde KD, Mckenzie EH, Bahkali AH, Zhou D. Sequence data reveals phylogenetic affinities of *Acrocalymma aquatica* sp. nov., *Aquasubmersa mircensis* gen. *et* sp. nov. and *Clohesyomyces aquaticus* (freshwater Coelomycetes). Cryptogamie Mycol 2012;33:333–346.
Zhang Y, Crous PW, Schoch CL, Hyde KD. Pleosporales. Fungal Divers 2012;53:1–221.
Zhu H, Cai L, Hyde KD, Zhang KQ. A new species of *Acrogenospora* from submerged Bamboo in Yunnan, China. Mycotaxon 2005;92:383–386.

Martha J. Powell and Peter M. Letcher

6 Phylogeny and characterization of freshwater Chytridiomycota (Chytridiomycetes and Monoblepharidomycetes)

6.1 Introduction

Modern phylogenetic analyses of gene molecular sequences and zoospore ultrastructural characters have revolutionized our understanding of relationships among Chytridiomycota and revealed unexpected diversity among these fungi. Seven monophyletic orders (Fig. 6.1) have now been segregated from what Sparrow (1960) considered as the single order Chytridiales, and additional orders are still to be characterized (James et al. 2006; Powell and Letcher 2012). In Chytridiomycota we include two classes, Chytridiomycetes (chytrids) and Monoblepharidomycetes (monoblephs) (Hibbett et al. 2007). Doweld (2001) raised monoblephs to a phylum, but in this chapter we retain it as a class because it is a sister of chytrids in most phylogenetic analyses (James et al. 2006) and, in contrast to other posteriorly uniflagellate zoosporic fungi (Neocallimastigomycota, Blastocladiomycota), both chytrids and monoblephs have centric mitosis with open polar fenestrae (Powell 1980; Dolan and Fuller 1985; Powell and Letcher 2012). Phylogenetic analyses of molecular sequence data and zoospore ultrastructure characters are mutually compatible (James et al. 2006). Consequently, the current classification scheme of Chytridiomycota and the circumscription of orders are based upon molecular monophyletic relations and constellation of zoospore ultrastructural characters (Fig. 6.2).

Alexander Braun first recognized the fungal nature of Chytridiomycota when he described *Chytridium olla*, an obligate parasite of oogonia of the green alga, *Oedogonium* (Braun 1851). Since that early beginning, the concept of Chytridiomycota has expanded to include additional parasites of plants and algae (Figs. 6.7, 6.9, 6.12) as well as microinvertebrates, oomycetes, cyanobacteria (Fig. 6.10) and other fungi. These fungi also function as saprotrophs of refractory materials such as pollen (Fig. 6.5), cellulose, chitin (Fig. 6.13), and keratin (Sparrow 1960; Karling 1977). The first reported mutualistic partnership between a chytrid and alga involves cross-feeding responses between *Rhizidium phycophilum* and a coccoid green alga (Picard et al. 2013). The relatively recent awareness of the vertebrate pathogen, *Batrachochytrium dendrobatidis*, and its devastating impact on amphibian populations and onslaught of frog species mass extinctions (Longcore et al. 1999; Rosenblum et al. 2008; Garmyn et al. 2012; Greenspan et al. 2012; Schloegel et al. 2012) hints that there is more to be discovered about the biodiversity of Chytridiomycota.

Chytridiomycota is characterized by the production of posteriorly uniflagellate zoospores (Figs. 6.3, 6.4) which function in the dispersal of the fungus onto new substrates or hosts (Figs. 6.5, 6.8). The zoospore is a dispersal stage only, relying totally upon endogenous reserves, primarily lipids, for energy (Powell 1976, 1978). Because of the zoospore's dependency on water for dispersal, in a sense all Chytridiomycota could be considered aquatic fungi. Many are found exclusively in freshwater and are clearly freshwater fungi, especially those that are parasites on aquatic algae, microinvertebrates, and oomycetes. Phylogenetic analyses of molecular sequences of phylotypes in environmental samples collected from lakes suggest that Chytridiomycota are major fungi within the photic layer of the pelagic zone (Lefèvre et al. 2008, 2012; Wurzbacher et al. 2010; Monchy et al. 2011; Kagami et al. 2012). The seasonal impacts of chytrid parasites on phytoplankton populations and biological productivity in aquatic habitats are broadly documented (Waterhouse 1942; Canter and Lund 1948, 1951; Canter 1954; Canter and Jaworski 1986; Powell 1993; Holfeld 2000; Dileo et al. 2010; Lefèvre et al. 2012; Gerphagnon et al. 2013), but establishing chytrids as important players in food webs and as transformers of inedible biomass into palatable forms are emerging research areas (Kagama et al. 2007, 2011, 2012; Gleason et al. 2008; Miki et al. 2011; Sime-Ngando et al. 2011). Chytridiomycota can also be readily

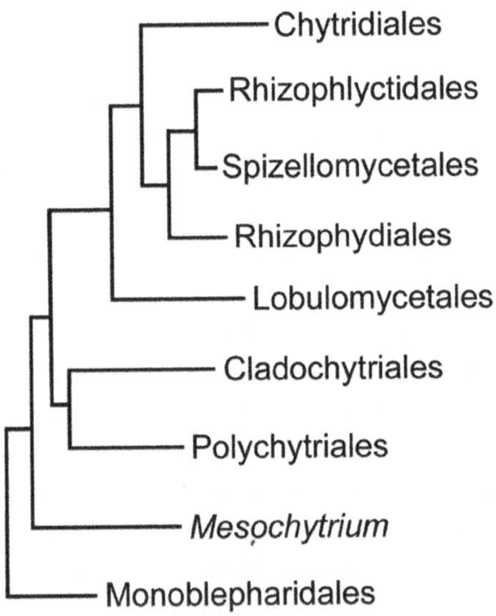

Fig. 6.1: Phylogeny of Chytridiomycota based on analyses of rDNA sequences as in James et al. (2006). Monoblepharidomycetes (Monoblepharidales) is sister of Chytridiomycetes for which seven orders have been circumscribed based on molecular monophyly and constellation of zoospore ultrastructural characters. The green algal parasite, *Mesochytrium*, is in the phylogenetic tree as one of several taxa for which relationships are not yet well supported and placement into an order has not been determined.

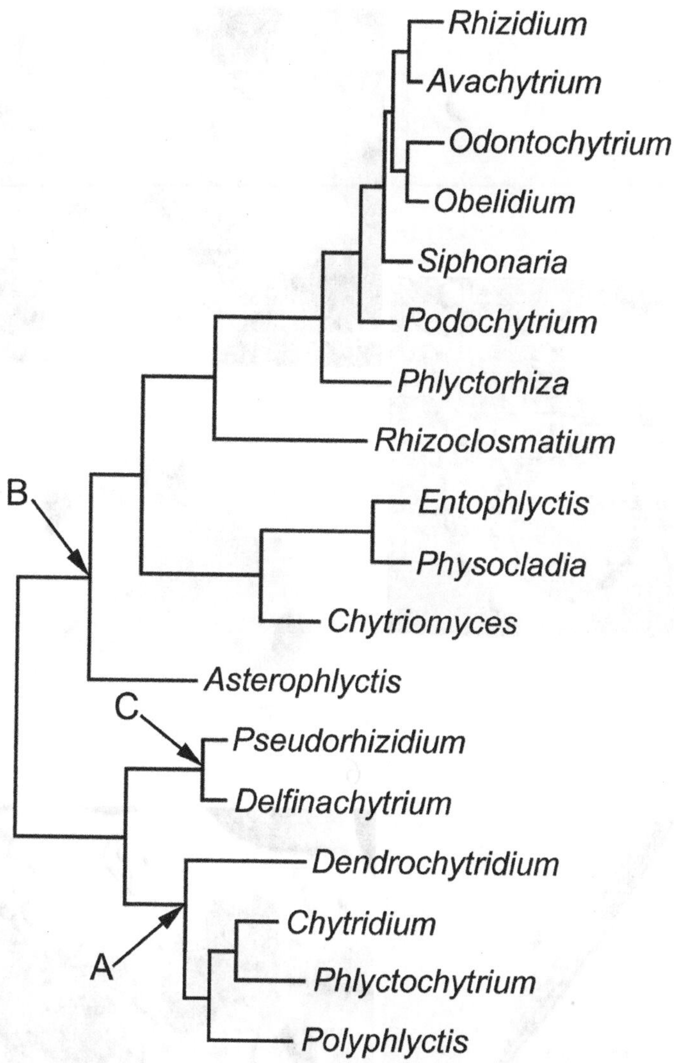

Fig. 6.2: Phylogeny of members of Order Chytridiales based on rDNA sequences as in Vélez et al. (2013). Three sub-clades (A, B, C) are resolved and each has zoospores with unique suites of ultrastructural characters: A includes four genera in Chytridiaceae; B includes 12 genera in Chytriomycetaceae; C includes two genera as *incertae sedis* and awaits greater sampling to populate the sub-clade before a family should be described.

isolated from submerged and moist soil at the water's edge, a zone which is nutrient rich. We now also recognize chytrids as soil fungi, and some species have been isolated only from soil and dung (Wakefield et al. 2010; Simmons et al. 2012). Others have been isolated from detritus that accumulates in tree canopies (Longcore 2005). Thus, in this chapter we will discuss the phylogeny of Chytridiomycota in general, and then focus on representatives that are typically located in and near bodies of freshwater.

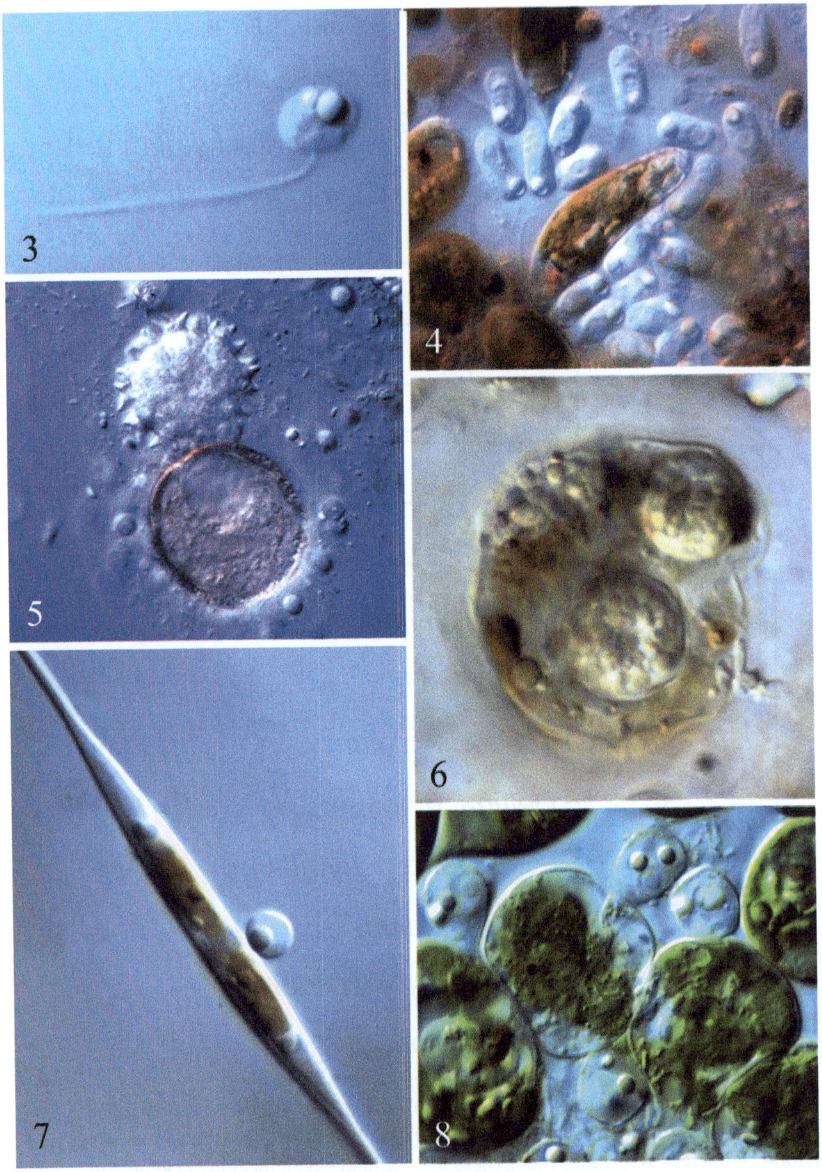

Figs. 6.3–6.8: Fig. 6.3. Chytrid zoospore with a single, posteriorly directed flagellum. A highly refractive lipid globule is visible at the anterior end of the spore. **Fig. 6.4** Swimming among host *Euglena* cells which are green, zoospores of *Polyphagus euglenae* are elongate. A conspicuous, highly refractive lipid globule is visible at the posterior end of each zoospore. **Fig. 6.5** Attached to a spherical sweet gum pollen grain, a sporangium of *Phlyctochytrium aureliae* is ornamented with bipartite teeth. Just below the sporangium and within the pollen grain, a prominent subsporangial swelling (apophysis) is visible. **Fig. 6.6** Two rounded-endobiotic-holocarpic thalli of *Olpidium* within the degenerating cell of an alga. **Fig. 6.7** Zoospore of a chytrid recently encysted at the surface of a host algal cell. **Fig. 6.8** Young thallus of *Polyphagus euglenae* penetrating the surface of a rounded *Euglena* cell with a germ tube extending from the enlarging zoospore cyst, which will become a prosporangium.

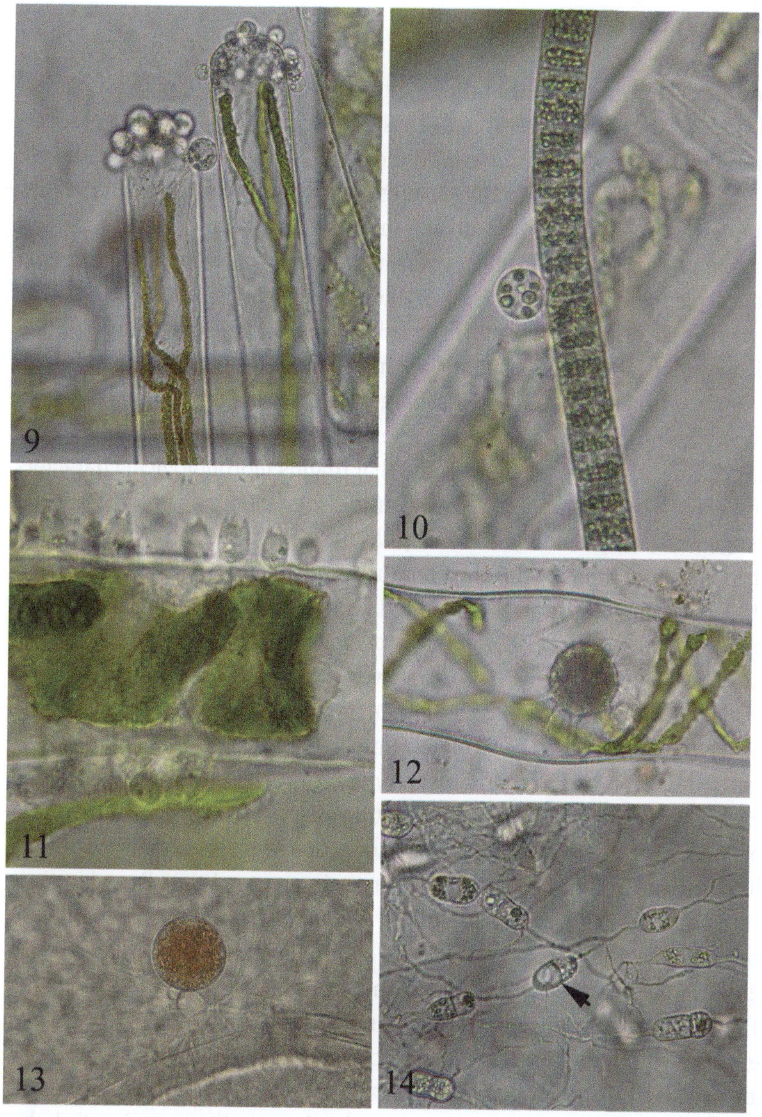

Figs 6.9–6.14: Fig. 6.9 Parasitic chytrid thalli cluster at the tip of a green algal filament. Notice the thin chytrid rhizoids that extend through host cytoplasm toward chloroplasts. **Fig. 6.10** Chytrid thallus infecting a filamentous cyanobacterium. **Fig. 6.11** A row of *Phlyctochytrium planicorne* thalli is visible infecting the filament of a green alga. Distinctive apical teeth-like appendages extend from the sporangium of the chytrid. **Fig. 6.12.** Endobiotic, spiney prosorus of *Micromyces* sp. parasitic in a *Spirogyra* filament. **Fig. 6.13.** Thallus of *Rhizoclosmatium aurantiacum* growing on shrimp chitin. The sporangium is orange and spherical, and the rhizoidal system consists of a fusiform apophysis from which progressively finer branches extend. **Fig. 6.14.** Rhizomycelium of *Cladochytrium replicatum* growing in agar nutrient media and producing branching tubular filaments with characteristic swellings (turbinate cells) that are crossed with septa (arrow). Growth is indeterminate and can be extensive.

6.2 Chytridiomycetes

Thalli of chytrids are relatively simple, but their pattern of thallus development is varied (Blackwell et al. 2006). The thallus may consist of a single sporangium, either with (eucarpic, monocentric, Fig. 6.5) or without rhizoids (holocarpic, monocentric, Fig. 6.6). Producing a more extensive thallus, especially on substrates such as cellulose and chitin, there may be many sporangia along a branched and tubular rhizomycelium (eucarpic, polycentric, Fig. 6.14). Holocarpic thalli are always endobiotic within the substrate or host (Fig. 6.6), and eucarpic thalli may be epibiotic or endobiotic. Septa are produced only to delimit sporangia (which are multinucleate) from rhizoids (which are anucleate) or along polycentric rhizomycelia, and septa contain plasmodesmata (Powell 1974). Although morphological characters, such as mode of discharge, thallus complexity, and pattern of development, were the bases for the classical systematics of Chytridiomycota (Sparrow 1960), molecular phylogenetics has demonstrated that thallus morphological characters are often convergent. The genus *Entophlyctis* illustrates the taxonomic challenges that still await investigators of Chytridiomycota. *Entophlyctis* is characterized by an endobiotic, monocentric, eucarpic thallus with the sporangium derived from the enlargement of the germ tube which emerges from the encysted zoospore body (Karling 1977). Most species of *Entophlyctis* are aquatic, either growing parasitically on algae or saprotrophically on aquatic higher plants (Sparrow 1960). However, based on zoospore ultrastructure of the soil-inhabiting *Entophlyctis* species, Barr (1980) transferred the genus from the Chytridiales into the newly erected Spizellomycetales. Researchers later realized that the type of the genus was aquatic and had a chytridialean-type zoospore. Consequently, the new genus *Powellomyces* was erected for the soil-inhabiting species, with aquatic-inhabiting *Entophlyctis* species returning to the Chytridiales (Longcore 1995). Recent investigations of zoospore ultrastructure and molecular sequence analyses reveal that *Entophlyctis* is still polyphyletic. In molecular phylogenetic studies, *Entophlyctis helioformis* is in the Rhizophydiales (James et al. 2006; Letcher et al. 2006) and *E. luteolus* is in the Chytridiales (Letcher et al. 2005; James et al. 2006). The type of the genus, *E. cienkowskiana* (Clements and Shear 1931), has not been studied molecularly or ultrastructurally.

Thus, one of the difficulties in the modern classification of chytrids is harmonizing the numerous described species that are rare and have not been observed in recent times, some of which are types of genera, with what we discover about phylogenetic relationships of chytrids that have been collected and characterized molecularly and ultrastructurally. Consequently when we categorize a genus as a member of an order, this placement may change later if not based on the type species.

6.2.1 Order 1. Chytridiales (Chytridiaceae, Chytriomycetaceae)

With the rediscovery and culture of *Chytridium olla*, the type species for Chytridiomycota, it was possible to establish the boundaries of a monophyletic Chytridiales

(Vélez et al. 2011). As currently delineated, the Chytridiales includes 18 genera (Fig. 6.2), including monocentric and polycentric taxa (Letcher et al. 2005; Vélez et al. 2011, 2013). Molecular phylogenies resolve three well-supported sub-clades and each has a zoospore with characteristic suite of characters. Two of the sub-clades are well studied, and zoospore ultrastructure defines each as a family, Chytridiaceae and Chytriomycetaceae (Barr and Hartmann 1976; Longcore 1992b, 1995; Letcher and Powell 2005b; Letcher et al. 2005, 2012a, 2014; Vélez et al. 2011, 2013). The third sub-clade represents a third family, but greater taxon representation is needed before formal description should be made for a family (Powell et al. 2013; Vélez et al. 2013).

Members of Chytridiales are predominantly freshwater chytrids, some such as *Asterophlyctis* and *Podochytrium* being found only in freshwater, and others such as *Chytriomyces hyalinus* and *Phlyctochytrium aureliae* capable of growth in freshwater or soils. A few members (Fig. 6.2) have been reported only in soil, such as *Rhizidium phycophilum* (Picard et al. 2009) and *Dendrochytridium crassum* (Letcher et al. 2014). Some genera in this order are nutritional generalist as is *Chytriomyces hyalinus* which has been reported isolated on chitin, pollen, and keratin (Miller 1965; Davis et al. 2013). In contrast, members of certain subclades are substrate specialists. Placing together in a sub-clade (Fig. 6.2), *Phlyctorhiza, Podochytrium, Siphonaria* and *Obelidium*, are collected directly from lakes, streams, and submerged muds on insect exuviae (empty chitinous integuments of aquatic insects) and some may substantially utilize the fluids released during metamorphosis as well as chitin (Sparrow 1937; Willoughby 1961).

Four genera are in Chytridiaceae and all of these are monocentric (Fig. 6.2, subclade A). *Chytridium* and *Phlyctochytrium,* which typically produce a subsporangial swelling on the rhizoid (apophysis), group together and are distinguished by differences in discharge apparatus, operculate versus inoperculate. The recent observation of an endobiotic resting spore in *P. aureliae* (Letcher et al. 2012a) brings into question the distinction of *Chytridium* which characteristically produces endobiotic resting spores (Sparrow 1960). As currently represented this family includes freshwater obligate algal parasites (*Chytridium olla*), facultative algal parasites (*Chytridium lagenaria, Phlyctochytrium bullatum, P. planicorne*) and saprotrophs (*Phlyctochytrium aureliae*). It is common to find entire surfaces of algal filaments infected with chytrids, such as *P. planicorne* (Fig. 6.11). Expanding the diversity of habitats in which these chytrids are found, *Dendrochytridium crassum* was isolated from tree-top canopy debris (Longcore 2005; Letcher et al. 2014).

Twelve genera are placed into the Chytriomycetaceae (Fig. 6.2, sub-clade B). The majority of the genera are monocentric with operculate or inoperculate zoospore discharge. *Physocladia* is polycentric with a thallus that resembles those of the Cladochytriaceae, demonstrating that thallus structure may be convergent. Significantly *Physocladia* groups with *Entophlyctis luteolus,* which although monocentric, thallus development also involves migration of the nucleus out of the zoospore cyst (Longcore 1995; James et al. 2006). *Phlyctorhiza* begins monocentric development, but may secondarily become polycentric. *Chytriomyces hyalinus* and species of *Rhizoclosmatium* (Fig. 6.13) are typically the most commonly isolated chytrids on chitin from a range of aquatic habitats (Miller 1965; Letcher and Powell 2002; Davis et al. 2013).

A third sub-clade (Fig. 6.2, sub-clade C) includes two recently described genera and circumscription of a family requires greater sampling of diversity within this group. Although both are isolated on pollen bait, *Pseudorhizidium* is characteristically isolated from soil (Powell et al. 2013) and *Delfinachytrium* (Vélez et al. 2013) has only been reported from aquatic habitats.

Genetically diverse phylotypes phylogenetically placing in the Chytridiales are widely detected in environmental samples from lakes (Lepère et al. 2008; Kagami et al. 2012; Lefèvre et al. 2012). Thus, Chytridiales is a major representative of chytrids in freshwater as saprotrophs and as parasites of algae.

6.2.2 Order 2. Spizellomycetales (Spizellomycetaceae, Powellomycetaceae)

The Spizellomycetales was the first order segregated from the Chytridiales based on zoospore ultrastructure and was done prior to molecular sequencing of chytrid genes (Barr 1980). Later gene sequence analyses supported the monophyly of most members Barr (1980, 2001) classified in this order. Among the exceptions are *Olpidium* and *Rozella* (James et al. 2006; Wakefield et al. 2010). Molecular-based phylogenetic analyses unexpectedly placed *Olpidium* species outside the Chytridiomycota and *Rozella* species basal to all other fungi (James et al. 2006); but because the type species for these genera have not been studied phylogenetically, final placement of these genera is uncertain. For example, *Olpidium* species studied molecularly are pathogens of plant roots (James et al. 2006; Sekimoto et al. 2011), whereas the type species of the genus is an endoparasite of algae (Sparrow 1960). Additionally *Caulochytrium* (Powell 1981a, 1981c) does not group within the Spizellomycetales in molecular phylogenetic analyses (Karpov et al. 2010). As currently circumscribed, Spizellomycetales consists of species that are most commonly isolated from soils, including dung. The thalli are eucarpic and monocentric. Members of the Spizellomycetaceae produce sporangia directly from the encysted zoospore and typically generate several to many inoperculate discharge pores or tubes (Wakefield et al. 2010).

In members of the Powellomycetaceae the sporangium arises from enlargement of the germ tube, with the zoospore cyst often remaining attached to the sporangium as an appendage (Simmons 2011; Simmons and Longcore 2012). Despite the prevalence of this order as soil chytrids, molecular sequence analyses of environmental samples from lakes (Lefranc et al. 2005) detected phylotypes of this order (related to *Spizellomyces acuminatus*). It is possible that phylotypes occur in aquatic sites as terrestrial debris washed or blown into lakes. Thus, as an order the Spizellomycetales are not known to have a large role in freshwater habitats.

6.2.3 Order 3. Rhizophlyctidales (Rhizophlyctidaceae, Sonoraphlyctidaceae, Arizonaphlyctidaceae, Borealophlyctidaceae)

The Rhizophlyctidales was separated from the Spizellomycetales (Barr 1980), and three additional genera were described based on uniqueness of zoospore ultrastructure and molecular-based phylogenetic analyses (Letcher et al. 2008a; Powell and Letcher 2012). In analyses of ribosomal gene sequences, Rhizophlyctidiales is a sister clade of the Spizellomycetales; but their thalli, with multiple rhizoids extending from sporangia, distinguish them from the Spizellomycetales, which typically have rhizoids originating from the base of the sporangium, often with a subsporangial swelling (Wakefield et al. 2010). Four sub-clades corresponded to four variations in zoospore ultrastructural motifs and are the bases for the description of four families within the order (Letcher et al. 2008a; Powell and Letcher 2012). Most species produce monocentric thalli with several inoperculate discharge openings or tubes. Under certain growth conditions, *Rhizophlyctis rosea* may develop polycentric thalli, but these do not become extensive. A polycentric isolate identified as *Catenomyces persicinus* phylogenetically clustered into this clade (James et al. 2006), but the ultrastructure of its zoospore has not been characterized. Thus, its taxonomic placement is still uncertain (Powell and Letcher 2012).

Members of this clade are commonly collected from soil on cellulose baits, are often isolated from grass compost piles (Willoughby 2001), and are considered soil fungi (Letcher et al. 2008a). Laboratory simulations demonstrate that they are resilient to soil desiccation (Willoughby 2001; Gleason et al. 2004). *Rhizophlyctis rosea* is considered one of the most ubiquitous soil chytrids (Willoughby 2001). *Sonoraphlyctis ranzonii* has been isolated from arid soils from two continents (Letcher et al. 2008a). *Arizonaphlyctis lemmonensis* was isolated from soil from a dried stream bed. *Borealophlyctis paxensis* was isolated on pollen from arid forest soil at 700 m in British Columbia. If *Catenomyces persicinus* proves to be a member of this order, it will be the only member with polycentric thalli and regularly located in aquatic habitats. Detection of phylotypes similar to *R. rosea* in lake pelagic zones (Kagami et al. 2012) is curious because this chytrid has not been cultured from this location. Sources of these phylotypes can only be hypothesized as coming in from terrestrial sources as air-born sporangia or resting spores. Thus, at this time this order is considered important in cellulose degradation in soils and not significant as aquatic chytrids.

6.2.4 Order 4. Rhizophydiales (10 families described)

The Rhizophydiales was delimited based on a distinct zoospore ultrastructure and monophyly of the lineage (James et al. 2006; Letcher et al. 2006). It was the first order (Letcher et al. 2006) to be segregated from the polyphyletic Chytridiales *sensu* Barr (2001) based on molecular phylogenetics. At that time, three families (Kappamycetaceae, Terramycetaceae, and Rhizophydiaceae) of primarily soil-inhabiting isolates were delineated (Letcher et al. 2006). A subsequent study of aquatic isolates from North America and Argentina (Letcher et al. 2008c) resulted in delineation of seven additional families (Alphamycetaceae, Angulomycetaceae, Aquamycetaceae, Globomycetaceae, Gorgonomycetaceae, Pateramycetaceae, and Protrudomycetaceae); additional lineages that may be under-sampled or exist as a singular isolate (e.g. *Batrachochytrium dendrobatidis*) remain family *incertae sedis*. The nearest neighbor to *B. dendrobatidis* is *Homolaphlyctis polyrhiza*, which was isolated on cellulose from lake water containing plant debris, and is similarly not placed formally into a family yet (Longcore et al. 2011). The ability of *B. dendrobatidis* to parasitize vertebrates is believed to be derived from horizontal gene transfers from bacteria and oomycetes, and not directly evolving within lineages in the Rhizophydiales (Sun et al. 2011). Thus, the order currently consists of 10 families and 18 genera (14 in families and 4 *incertae sedis*) (Longcore et al. 1999; Longcore 2004; Letcher et al. 2004, 2006, 2008b, 2008c, 2012b; Letcher and Powell 2005a; Powell et al. 2011).

Members of this order are monocentric and eucarpic. In the majority of taxa, the nucleus remains in the encysted zoospore as it enlarges to form the sporangium (endogenous development); however, *Entophlyctis helioformis* also belonged in this clade in molecular phylogenetic analyses and the germ tube enlarges to produce the sporangium (exogenous development). Rhizophydiales is based on *Rhizophydium globosum*, an early-described chytrid (Braun 1855) in which: the sporangium is spherical, epibiotic, and inoperculate; the rhizoidal system is branched and extensive; and the resting spore is epibiotic. For more than 150 years *Rhizophydium* was based on thallus morphological features, and over 220 species have been described with a range of thallus morphologies, substrates, hosts, and habitats (Letcher and Powell 2012). A majority of species has a sporangium that is spherical or sub-spherical throughout development, but numerous taxa are more morphologically diverse with sporangia that may be urceolate, angular, pyriform, obpyriform, cylindrical, or fusiform. Although at first considered to comprise members that were inoperculate, two operculate genera have now been described in this order, *Pateramyces* and *Operculomyces* (Letcher et al. 2008c; Powell et al. 2011). Whether the cellular mechanism for operculum formation is similar in each lineage in which operculate discharge occurs is not known, but ultrastructural studies suggest there may be differences (Beakes et al. 1992). Within this lineage, however, structure of the discharge opening is valuable at the generic level (Powell et al. 2011). Rhizoids may be sparingly to extensively branched, or unbranched and peg-like, and they do not produce a prominent apophysis (subsporangial swelling of the rhizoid). Resting spores have been observed for many taxa. Sexual recombination, although infrequently reported, typically involves transfer of the contents of one gametangium to an adjacent gametangium that develops into a resting spore functioning as a zygote (Sparrow 1960). Recently the genus has been summarized and revised (Letcher and Powell 2012).

Studies on the zoospore ultrastructure of several species of *Rhizophydium* (Barr and Hadland-Hartmann 1978) resulted in a representative, composite zoospore for the genus (Barr 1980). More recently, zoospore ultrastructural examinations have revealed approximately 20 zoospore sub-types based on unique suites of ultrastructural character states. Zoospore sub-types correlate with monophyletic lineages in molecular phylogenies, which serve as the basis for generic delineation (Powell and Letcher 2012). Characters considered for distinction of zoospore sub-types include the microbody lipid-globule complex (MLC), cisterna (Powell 1976, 1978), microtubular root, location of the nucleus relative to the ribosomal aggregation, number of mitochondria, morphology of the microbody, morphology of the kinetosome, the kinetosome-associated structure, the fibrillar bridge between the kinetosome and non-flagellated centriole, and the zone of convergence in the fibrillar bridge (Powell and Letcher 2012).

Members are parasitic on various algae and saprotrophic on pollen. As an extensive and diverse assortment of organisms, some members of Rhizophydiales are typically found in terrestrial habitats as saprotrophs on pollen grains (e.g. Terramycetaceae and Kappamycetaceae); many might be considered amphibious (e.g. Angulomycetaceae and some lineages in Rhizophydiaceae); and some are confined to aquatic habitats (e.g. Gorgonomycetaceae, Protrudomycetaceae). Those which are considered aquatic are often parasites of algae, desmids, chrysophytes, dinoflagellates, diatoms, protists, other chytrids, platyhelminthes eggs, and rotifers; or saprotrophs on cellulose, chitin, keratin, fern microspores, and gymnosperm and angiosperm pollen grains (Sparrow 1960).

Numerous members of the order have been reported from aquatic habitats (lakes, rivers, streams) as obligate parasites and are frequently detected in molecular phylogenetic analyses of lakes (Lefèvre 2007, 2012). As examples: *Rhizophydium rostellatum* occurs on *Spirogyra* spp., *Rhizophydium planktonicum* and *R. tetragenum* occur on the diatom *Asterionella formosa*, *R. vampyrellae* parasitizes the protozoan *Vampyrella* sp., and *Rhizophydium chytriomycetis* occurs as a parasite on the chytrids *Chytriomyces hyalinus* and *C. aureus*. *Batrachochytrium dendrobatidis* is a virulent pathogen from aquatic habitats that has devastated amphibian populations around the world, and is the causative agent of multiple extinctions. *Gorgonomyces haynaldii* is typical of members of Rhizophydiales found in aquatic habitats as saprotrophic on pollen grains. It is extensively distributed, having been observed or isolated from lakes, rivers, and marshlands. Morphologically it is distinctive, having a spherical sporangium with numerous elongate discharge tubes on artificial media (Letcher et al. 2008c). *Coralloidiomyces digitatus* (Letcher et al. 2008b), found in an oligotrophic lake in Patagonia, is a saprotroph on pollen grains. It has a distinctive lobose, branched sporangium and sparsely branched rhizoids. *Entophlyctis helioformis* is placed in this order, and a related phylotype has been identified in lake samples from the United States and Japan (Kagami et al. 2012; Lefèvre et al. 2012). Molecular sequence analyses of lake environmental samples identify unique clades that fall within this order and indicate uncharacterized diversity (Lefèvre et al. 2007, 2012; Lepère et al. 2008; Monchy et al. 2011). Thus, Rhizophydiales is an important order in freshwater habitats.

6.2.5 Order 5. Lobulomycetales (Lobulomycetaceae)

This order was established for the proper phylogenetic placement of *Chytriomyces angularis* which was not monophyletic with the type species, *Chytriomyces hyalinus* (Letcher and Powell 2002), in analyses of gene sequences (Letcher et al. 2005; James et al. 2006). All members of this monophyletic order are monocentric and some occur in aquatic habitats, but most are found in soils, including soils from extreme environments and herbivore dung (Simmons et al. 2009, 2012). None have been isolated from the pelagic zone of lakes. The only parasitic member of this order based on sequence analysis is a marine parasite of brown algae (Simmons et al. 2009), but the zoospore of *Chytridium polysiphoniae*, a potential member of this order, has not been investigated to verify its proper placement within an order. The order as circumscribed includes four genera (*Alogomyces, Lobulomyces, Clydaea* and *Maunachytrium*), and two of these genera have been reported from freshwater habitats. *Lobulomyces* produces a distinctive monocentric, operculate, lobate sporangium with a single narrow tubular rhizoid, and two species have been described. *Lobulomyces angularis* has been isolated from littoral areas of lakes in Maine on *Oedogonium* filaments (possible as a weak parasite) and pollen incubated *in situ* in bait bags, and a related phylotype has been detected in lakes in Japan (Kagami et al. 2012). This chytrid can also be isolated from *Sphagnum* in lakes (Longcore 1992a). *Lobulomyces poculatus* sporangium is obpyriform and ornamented with overlapping cupules (Willoughby and Townley 1961) and has been collected on keratin and pollen from peaty acidic soils. *Clydaea vesicular* was isolated on pollen from soil and creek bank mud (Simmons et al. 2009). Thus, Lobulomycetales is most properly viewed as an amphibious and soil-inhabiting order of saprotrophs.

6.2.6 Order 6. Cladochytriales (Cladochytriaceae, Nowakowskiellaceae, Septochytriaceae, Endochytriaceae)

The Cladochytriales as circumscribed is molecularly monophyletic and comprises eight genera, all of which, except *Allochytridium,* have species reported from freshwater habitats (Mozley-Standridge et al. 2009; Steiger et al. 2011). *Allochytridium* has only been reported from soil, and the other genera also include some species that are soil inhabiting. Molecular phylogenetic analyses of ribosomal gene sequences resolve four sub-clades on which the four families are based (Mozley-Standridge et al. 2009). In these analyses, isolates of *Cladochytrium* and *Nowakowskiella* were each monophyletic. *Cylindrochytridium* is a monotypic genus (Steiger et al. 2011). The other five genera as currently recognized were not monophyletic in the study by Mozley-Standridge et al. (2009).

Morphologically this clade is diverse and includes monocentric and polycentric members. All release zoospores through operculate openings except for *Cladochytrium* where discharge is inoperculate. Several morphological characteristics of the vegetative thallus unite this order: 1) nuclear migration is involved in thallus development for both polycentric and monocentric forms; and 2) the rhizoidal system is irregular in contour and may be catenulate or have spindle-shaped (turbinate) swellings (Fig. 6.14). Zoospore ultrastructure is similar to that found in the Chytridiales, but is distinguished by fibril linkers connecting individual microtubules of the root extending from the side of the kinetosome toward the fenestrated cisterna of the MLC (Lucarotti 1981; Barr 1986; Barr and Désaulniers 1987). Members of the Cladochytriales are saprotrophs and readily isolated on cellulosic baits (Couch 1939; Sparrow and Barr 1955) and are found on decaying plant material in aquatic habitats as well as in soil. In culture-based isolation studies, they are typically found at the water's edge in submerged muds or on moribund plants and green algal filaments. There are reports that some, such as *Endochytrium*, are weakly parasitic on algae (Sparrow 1960), and Sparrow (1931) described *Cladochytrium nowakowskii* as an "extremely virulent parasite" of filamentous green algae.

Cladochytrium with an inoperculate discharge opening rather than an operculate opening is also distinguished from the other polycentric members of the order such as *Nowakowskiella* because of the septa produced across turbinate swellings along the rhizomycelium (Fig. 6.14). Some other members, although they are primarily polycentric, are capable of monocentric development as in *Septochytrium*. *Septochytrium* is recognizable because septa are scattered along its irregularly constricted rhizoidal system. One aquatic species in this genus, *S. plurilobulum*, is found in decaying maple leaves at the lakes' edge. Production of a rhizomycelium allows these chytrids to colonize large areas of decaying plant and algal tissue. *Catenochytridium*, *Cylindrochytridium*, *Endochytrium*, and *Nephrochytrium* are predominantly monocentric and development involves migration of the nucleus out of the zoospore cyst at some point during thallus development. *Nephrochytrium* and *Endochytrium* produce germ tubes from the encysted zoospores into which the nucleus migrates: in *Nephrochytrium* the tip of the germ tube enlarges into an apophysis from which the sporangium buds; in *Endochytrium* the tip of the germ tube enlarges into a sporangium from which rhizoids emerge. In *Catenochytrium* the sporangium buds from the encysted zoospore which has already established the rhizoidal system and apophysis (Sparrow 1960). Thallus development of *Cylindrochytridium* is a variation of this pattern; the nucleus remains in the zoospore cyst at first and then produces an elongate sporangium into which the nucleus migrates and divides (Steiger et al. 2011).

As an order Cladochytriales is most obvious in aquatic systems as decomposers of cellulose, especially moribund plant material that accumulates in the littoral region of lakes. In fact molecular analyses of environmental samples of lake water detected a clade inclusive of Cladochytriales which was genetically diverse (Lepère et al. 2008).

6.2.7 Order 7. Polychytriales (no families described)

The Polychytriales is the most recent of the monophyletic orders newly delineated from a polyphyletic Chytridiales *sensu* Barr (2001) and is composed of three sub-clades (Longcore and Simmons 2012). All described species grow on chitinous substrates and produce monocentric or polycentric thalli characterized by rhizoids radiating from multiple sites on sporangia (rhizophlyctoid alliance of Dogma 1973). All genera release zoospores from sporangia through inoperculate openings except for *Karlingiomyces asterocystis,* which produces an operculate discharge pore. Polycentric members of Polychytriales and Cladochytriales can be distinguished because the Polychytriales lack swellings and turbinate cells along their rhizomycelia. At the ultrastructural level, zoospores are characterized by a nonflagellated centriole as long or longer that its width (whereas in other orders it is wider than long). By using chitinous baits, five genera in the Polychytriales have been isolated from soil and water samples containing organic debris or collected directly from baits incubated *in situ* in bait bags (Longcore 1993; Longcore and Simmons 2012). No parasitic members have been identified in this order.

Two polycentric and one monocentric species cluster together in a sub-clade (Longcore and Simmons 2012). A monotypic genus, *Polychytrium aggregatum* is polycentric and saprotrophic on decaying chitinous and sometimes cellulosic substrates in bogs and streams but may also be found in soil. Both smooth and tuberculate sporangia are formed laterally along a coarse and extensive rhizomycelium. In streams *P. aggregatum* may have a role in the successional decomposition of submerged leaves (Schoenlein-Crusius et al. 1998). *Lacustromyces hiemalis* is also polycentric but its sporangia are intercalary along the robust rhizomycelium. *Lacustromyces hiemalis* has only been isolated from lakes, ponds and rivers in Maine (USA), usually when water temperatures were less than 11 C (Longcore 1993; Longcore and Simmons 2012), suggesting that some species of chytrids may be geographically limited in distribution. *Neokarlingia chitinophila* monocentric thalli produce numerous discharge tubes and are reported from moist soil and water (Sparrow 1960; Longcore and Simmons 2012). This species is an example of the amphibious habit of some chytrids.

Arkaya currently includes two described species in another well-supported clade, and their zoospores differ ultrastructurally from other members of the order because a fenestrated cisterna is part of the microbody lipid globule complex (Powell 1978; Longcore and Simmons 2012). *Arkaya* superficially resembles species of *Rhizophlyctis*, a genus phylogenetically placed in another order. Thalli bear multiple rhizoids emerging from the sporangium and one to several inoperculate discharge papillae, which each contains a hyaline plug prior to zoospore discharge. Molecular phylogenetic analyses suggest that this newly described monocentric genus is genetically diverse. *Arkaya lepida* was isolated from a bait-bag of chitin incubated in a river (Longcore and Simmons 2012). The second species, *A. serpentina,* is not aquatic and was isolated from sandy soil and leaf litter. It is distinctive because of the numerous hair-like setae radiating from its sporangia.

Karlingiomyces asterocystis is sister of the other two sub-clades, is the type species for the genus, and the only operculate member of the order. The species name is based on the peg-like spines at the surface of resting spores. This monocentric species, most commonly isolated on chitin, has limited growth on keratin and no growth on cellulose. It also appears amphibious because it is isolated from moist soil and water.

Genera of this order have not been detected in pelagic regions and in general can be considered as inhabitants of littoral areas of lakes and sites along the water's edge. Thus as an order, Polychytriales is saprotrophic and important in the decomposition of chitin which accumulates in the littoral region.

6.3 *Incertae sedis*

There are still lineages in the Chytridiomycota to characterize but should await greater representation. Among these are aquatic chytrids, including the obligate parasite of a coccoid green alga, *Mesochytrium penetrans*, the sole member of a lineage (Karpov et al. 2010). Aquatic members of the family Synchytriaceae (Sparrow and Barr 1955; Sparrow 1960) have not been studied molecularly or ultrastructurally, although there are numerous algal parasites, such as *Micromyces* (Fig. 6.12). Only plant pathogenic members of Synchytriaceae have been included in molecular phylogenetic studies (James et al. 2006).

On the other hand some aquatic chytrid isolates have been characterized either ultrastructurally or molecularly, but not the same isolate with both approaches. The 18S ribosomal RNA gene sequence of the widely researched freshwater planktonic diatom parasite, *Zygorhizidium planktonicum* strain FMS 34600 (Lefèvre et al. 2010, FJ799984), would place this isolate within the Chytridiales; but earlier studies of zoospore ultrastructure attributed to this same species would place it outside of the Chytridiales (FBA clone C9 in Beakes et al. 1988; Doggett and Porter 1995). *Polyphagus euglenae* is a parasite of *Euglena* (Figs. 6.4, 6.8) and has a distinctive zoospore (Powell 1981b), but its molecular phylogentic placement remains unresolved. Although there is a sequence (James et al. 2006) for the aquatic chytrid *Blyttiomyces* commonly found in bogs (Sparrow and Barr 1955; Blackwell et al. 2011), it was not from a pure culture; and its zoospore has not been studied ultrastructurally. Thus, we have only started to reveal the diversity of chytrids in freshwater and greater sampling in both the littoral and pelagic regions is needed.

6.4 Monoblepharidomycetes (Harpochytriales, Monoblepharidales, Hyaloraphidiales)

The monoblephs are typically sister of Chytridiomycetes in molecular phylogenetic analyses (James et al. 2006) and both have centric mitosis with the nuclear poles open (Powell 1975, 1980; Dolan and Fuller 1985; Powell and Letcher 2012). One ultrastructural feature of monobleph zoospores distinctive from that of chytrids is that the MLC cisterna (=rumposome) backs the microbody rather than the lipid globule (Dorward and Powell 1980; Powell and Letcher 2012). Only limited phylogenetically based revisionary taxonomy has been accomplished with this group because there has been no widely sampled molecular phylogenetic analysis. In general the classical classification has been supported in the few molecular studies reported (Paquin et al. 1997; Chambers 2003; James et al. 2006). The placements of *Monoblepharis* (the type for the class) and *Monoblepharella* in the same family, Monoblepharidaceae, as well as the separation of *Gonapodya* into another family, Gonapodyaceae, were supported (Chambers 2003). *Gonapodya*, however may be polyphyletic. One of the most notable discoveries in the group is that *Hyaloraphidium curvatum*, earlier considered a colorless green alga, was actually a basal member of this class (Ustinova et al. 2000; Forget et al. 2002). Uniquely among members of this class, *H. curvatum* produces autospores rather than zoospores. Thalli in this group range from simple short filaments as with *Harpochytrium* and *Hyaloraphidium* to mycelial as with *Monoblepharella* (Powell and Letcher 2012). Most members are found in freshwater, especially on submerged fruits and twigs in cool waters with pHs in slightly acidic range (Perrott 1955, 1958, 1960; Sparrow and Barr 1955; Sparrow 1960). In contrast some members are collected in tropical soils on hempseeds. None are parasitic or found in marine habitats. One feature that distinguishes the mycelium of monoblephs from other filamentous heterotrophs is the foamy appearance of its cytoplasm due to the highly vacuolated nature of its cytoplasm. The monobleps, thus, serve as biodegraders of plant material in shallow freshwater habitats.

Acknowledgments

Appreciation is expressed to Dr. Will H. Blackwell for assistance in collecting material photographed in Figs. 6.9 and 6.12. This study was supported in part by a grant from the National Science Foundation REVSYS DEB-00949305.

References

Barr DJS. An outline for the reclassification of the Chytridiales, and for a new order, the Spizellomycetales. Can J Bot 1980;58:2380–2394.

Barr DJS. *Allochytridium expandens* rediscovered: morphology, physiology and zoospore ultrastructure. Mycologia 1986;78:439–448.

Barr DJS. Chytridiomycota. In: McLaughlin DJ, McLaughlin EG, Lemke PA, eds. The Mycota, Systematics and Evolution. Vol VIIA. New York: Springer-Verlag 2001:93–112.

Barr DJS, Désaulniers NL. *Allochytridium luteum* n. sp.: morphology, physiology and zoospore ultrastructure. Mycologia 1987;79:193–199.

Barr DJS, Hadland-Hartmann VE. Zoospore ultrastructure in the genus *Rhizophydium* (Chytridiales). Can J Bot 1978;56:2381–2404.

Barr DJS, Hartmann VE. Zoospore ultrastructure of three *Chytridium* species and *Rhizoclosmatium globosum*. Can J Bot 1976;54:2000–2013.

Beakes GW, Canter HM, Jaworski GHM. Zoospore ultrastructure of *Zygorhizidium affluens* and *Z. planktonicum*, two chytrids parasitizing the diatom *Asterionella formosa*. Can J Bot 1988;66:1054–1067.

Beakes GW, Canter HM, Jaworski GHM. Ultrastructural study of operculation (discharge apparatus) and zoospore discharge in zoosporangia of *Zygorhizidium affluens* and *Z. planktonicum*, chytrid parasites of *Asterionella formosa*. Mycol Res 1992;96:1060–1067.

Blackwell WH, Letcher PM, Powell MJ. Thallus development and the systematics of Chytridiomycota: an additional developmental pattern represented by *Podochytrium*. Mycotaxon 2006;97:91–109.

Blackwell WH, Letcher PM, Powell MJ, Vélez CG. The occurrence of *Blyttiomyces spinulosus* in Alabama and Argentina, and comments on the genus *Blyttiomyces* (Chytridiomycota). Phytologia 2011;93:304–315.

Braun A. Betrachtungen über die Erscheinung der Verjüngung in der Natur, insbesondere in der Lebens-und Bildungsgeschichte der Pflanze. Leipzig 1851.

Braun A. Über *Chytridium*, eine Gattung einzelner Schmarotzergewächse auf Algen und Infusorien. Monatsber Berlin Akad 1855;378–384

Canter HM. Fungal parasites of the phytoplankton. III. Studies on British chytrids. Trans Br Mycol Soc 1954;37:111–133.

Canter HM, Jaworski GHM. A study on the chytrid *Zygorhizidium planktonicum* Canter a parasite of the diatoms *Asterionella* and *Synedra*. Nova Hedwig 1986;43:269–298.

Canter HM, Lund JWG. Studies on plankton parasites. I. Fluctuations in the numbers of *Asterionella formosa* Haas. In relation to fungal epidemics. New Phytol 1948;47:238–261.

Canter HM, Lund JWG. Studies on plankton parasites. III. Examples of the interaction between parasitism and other factors determining the growth of diatoms. Ann Bot (Oxford) 1951;15:359–371.

Chambers JG. Ribosomal DNA, secondary structure, and phylogenetic relationships among the Chytridiomycota. Ph.D. dissertation. University of Alabama, Tuscaloosa, AL 2003.

Clements FE, Shear CL. The Genera of Fungi. New York: H.W. Wilson Co., 1931.

Couch JN. Technic for collection, isolation, and culture of chytrids. J Elisha Mitchell Sci Soc 1939; 55:208–214.

Davis WJ, Letcher PM, Powell MJ. Chytrid diversity of Tuscaloosa County, Alabama. Southeastern Naturalist 2013;12:666–683.

Dileo K, Donat K, Min-Venditti A, Dighton J. A correlation between chytrid abundance and ecological integrity in New Jersey pine barrens waters. Fungal Ecol 2010;3:295–301.

Doggett MS, Porter D. Further evidence for host-specific variants in *Zygorhizidium planktonicum*. Mycologia 1995;87:161–171.

Dogma IJ. Developmental and taxonomic studies on rhizophlyctoid fungi, Chytridiales I. Dehiscence mechanisms and generic concepts. Nova Hedwig 1973;24:393–411.

Dolan TE, Fuller MS. The ultrastructure of nuclear division in *Monoblepharella* sp. Mycologia 1985;77:791–809.

Dorward DW, Powell MJ. Microbodies in *Monoblepharella* sp. Mycologia 1980;72:549–557.

Doweld A. *Prosyllabus tracheophytorum*: Tentamen systematis plantarum vascularium (Tracheophyta). Moscow: A. Geos 2001: 111 pp. ISBN: 5:-89118-283-1.

Forget L, Ustinova J, Wang Z, Huss VAR, Lang BF. *Hyaloraphidium curvatum*: a linear mitochondrial genome, tRNA editing, and an evolutionary link to lower fungi. Molec Biol Evol 2002;19:310–319.

Garmyn A, Van Rooij P, Pasmans F, Hellebuyck T, Van Den Broeck W, Haesebrouck F, et al. Waterfowl: potential environmental reservoirs of the chytrid fungus *Batrachochytrium dendrobatidis*. PLoS ONE 2012;7(4):e35038, doi:10.1371/journal.pone.0035038

Gerphagnon M, Latour D, Colombet J, Sime-Ngando T. Fungal parasitism: Life cycle, dynamics and impact on Cyanobacterial blooms. PLoS ONE 2013;8(4):e60894. doi:10.1371/journal.pone.0060894.

Gleason FH, Letcher PM, McGee PA. Some Chytridiomycota in soil recover from drying and high temperatures. Mycol Res 2004;108:583–589.

Gleason FH, Kagami M, Lefèvre E, Sime-Ngando T. The ecology of chytrids in aquatic ecosystems: roles in food web dynamics. Fungal Biol Rev 2008;22:17–25.

Greenspan SE, Longcore JE, Calhoun AJK. Host invasion by *Batrachochytrium dendrobatidis*: fungal and epidermal ultrastructure in model anurans. Dis Aquat Org 2012;100:201–210.

Hibbett DS, Binder M, Bischoff J, Blackwell M, Cannon P, Eriksson O, et al. A higher level phylogenetic classification of the fungi. Mycol Res 2007;111:509–547.

Holfeld H. Relative abundance, rate of increase, and fungal infections of freshwater phytoplankton. J Plankton Res 2000;22:987–995.

James TY, Letcher PM, Longcore JE, Mozley-Standridge SE, Porter D, Powell MJ, et al. A molecular phylogeny of the flagellated Fungi (Chytridiomycota) and description of a new phylum (Blastocladiomycota). Mycologia 2006;98:860–871.

Kagami M, deBruin A, Ibelings BW, van Donk E. Parasitic chytrids: their effects on phytoplankton communities and food-web dynamics. Hydrobiol 2007;578:113–129.

Kagami M, Helmsing NR, van Donk E. Parasitic chytrids could promote copepod survival by mediating material transfer from inedible diatoms. Hydrobiol 2011;569:49–54.

Kagami M, Amano Y, Ishii N. Community structure of planktonic fungi and the impact of parasitic chytrids on phytoplankton in Lake Inba, Japan. Microbial Ecol 2012;63:358–368.

Karling JS. Chytridiomycetarum Iconographia. Monticello, NY: Lubrecht and Cramer 1977.

Karpov SA, Letcher PM, Mamkaeva MA, Mamkaeva KA. Phylogenetic position of the genus *Mesochytrium* (Chytridiomycota) based on zoospore ultrastructure and sequences from the 18S and 28S rRNA gene. Nova Hedwig 2010;90:81–94.

Lefèvre E, Bardot C, Noel C, Carrias J, Viscogliosi E, Amblard C, et al. Unveiling fungal zooflagellates as members of freshwater picoeukaryotes: evidence from a molecular diversity study in a deep meromictic lake. Environ Microbiol 2007;9:61–71.

Lefèvre E, Roussel B, Amblard C, Sime-Ngando T. The molecular diversity of freshwater picoeukaryotes reveals high occurrence of putative parasitoids in the plankton. PLoS ONE 2008;3(6):e2324, doi:10.1371/journal.pone.0002324

Lefèvre E, Jobard M, Venisse J-S, Bec A, Kagami M, Amblard C, et al. Development of real-time PCR assay for quantitative assessment of uncultured freshwater zoosporic fungi. J Microbiol Methods 2010:81:69–76.

Lefèvre E, Letcher PM, Powell MJ. Temporal variation of the small eukaryotic community in two freshwater lakes: emphasis on the zoosporic fungal community. Aquat Microb Ecol 2012;67:91–105.

Lefranc M, Thénot A, Lepère C, Debroas D. Genetic diversity of small eukaryotes in lakes differing by their trophic status. Appl Environ Microbiol 2005;71:5935–5942.

Lepère C, Domaizon I, Debroas D. Unexpected importance of potential parasites in the composition of the freshwater small-eukaryote community. Appl Environ Microbiol 2008;74:2940–2949.

Letcher PM, Powell MJ. A taxonomic summary of *Chytriomyces* (Chytridiomycota). Mycotaxon 2002;84:447–487.

Letcher PM, Powell MJ. *Kappamyces*, a new genus in the Chytridiales (Chytridiomycota). Nova Hedwig 2005a;80:115–133.

Letcher PM, Powell MJ. Phylogenetic position of *Phlyctochytrium planicorne* (Chytridiales, Chytridiomycota) based on zoospore ultrastructure and partial nuclear LSU rRNA gene sequence analysis. Nova Hedwig 2005b;80:134–146.

Letcher PM, Powell MJ. A Taxonomic Summary and Revision of *Rhizophydium* (Rhizophydiales, Chytridiomycota). Tuscaloosa, AL: University Printing 2012.

Letcher PM, Powell MJ, Chambers JG, Holznagel WE. Phylogenetic relationships among *Rhizophydium* isolates from North America and Australia. Mycologia 2004;96:1339–1351.

Letcher PM, Powell MJ, Chambers JG, Longcore JE, Churchill PF, Harris PM. Ultrastructural and molecular delineation of the Chytridiaceae (Chytridiales). Can J Bot 2005;83:1561–1573.

Letcher PM, Powell MJ, Churchill PF, Chambers JG. Ultrastructural and molecular phylogenetic delineation of a new order, the Rhizophydiales (Chytridiomycota). Mycol Res 2006;110:898–915.

Letcher PM, Powell MJ, Barr DJS, Churchill PF, Wakefield WS, Picard KT. Rhizophlyctidales – a new order in Chytridiomycota. Mycol Res 2008a;112:1031–1048.

Letcher PM, Powell MJ, Viusent MC. Rediscovery of an unusual chytridiaceous fungus new to the order Rhizophydiales. Mycologia 2008b;100:325–334.

Letcher PM, Vélez CG, Barrantes ME, Powell MJ, Churchill PF, Wakefield WS. Ultrastructural and molecular analyses of Rhizophydiales (Chytridiomycota) isolates from North America and Argentina. Mycol Res 2008c;112:759–782.

Letcher PM, Powell MJ, Picard KT. Zoospore ultrastructure and phylogenetic position of *Phlyctochytrium aureliae* Ajello is revealed (Chytridiaceae, Chytridiales, Chytridiomycota). Mycologia 2012a;104:410–418.

Letcher PM, Vélez CG, Schultz S, Powell MJ. New taxa are delineated in Alphamycetaceae (Rhizophydiales, Chytridiomycota). Nova Hedwig 2012b;94:9–29.

Letcher PM, Longcore JE, Powell MJ. *Dendrochytridium crassum* gen. et sp. nov., a new taxon in Chytridiales with unique zoospore ultrastructure. Mycologia 2014;106:145–153.

Longcore JE. Morphology and zoospore ultrastructure of *Chytriomyces angularis* sp. nov. (Chytridiales). Mycologia 1992a;84:442–451.

Longcore JE. Morphology and zoospore ultrastructure of *Podochytrium dentatus* sp. nov. (Chytridiales). Mycologia 1992b;84:183–192.

Longcore JE. Morphology and zoospore ultrastructure of *Lacustromyces hiemalis* gen. et sp. nov. (Chytridiales). Can J Bot 1993;71:414–425.

Longcore JE. Morphology and zoospore ultrastructure of *Entophlyctis luteolus* sp. nov. (Chytridiales): Implications for chytrid taxonomy. Mycologia 1995;87:25–33.

Longcore JE. *Rhizophydium brooksianum* sp. nov., a multipored chytrid from soil. Mycologia 2004;96:162–171.

Longcore JE. Zoosporic fungi from Australian and New Zealand tree-canopy detritus. Aust J Bot 2005;53:259–272.

Longcore JE, Simmons DR. The Polychytriales ord. nov. contains chitinophilic members of the rhizophlyctoid alliance. Mycologia 2012;104:276–294.

Longcore JE, Barr DJS, Désaulniers N. *Powellomyces*, a new genus in the Spizellomycetales. Can J Bot 1995;73:1385–1390.

Longcore JE, Pessier AP, Nichols D. *Batrachochytrium dendrobatidis* gen. et sp. nov., a chytrid pathogenic to amphibians. Mycologia 1999;91:219–227.

Longcore JE, Letcher PM, James TY. *Homolaphlyctis polyrhiza* gen. *et* sp. nov., a species in the Rhizophydiales (Chytridiomycetes) with multiple rhizoidal axes. Mycotaxon 2011;118:433–440.

Lucarotti CJ. Zoospore ultrastructure of *Nowakowskiella elegans* and *Cladochytrium replicatum*. Can J Bot 1981;59:137–148.

Miki T, Takimoto G, Kagami M. Role of parasitic fungi in aquatic food webs: A theoretical approach. Freshwater Biol 2011;56:173–1183.
Miller CE. Annotated list of aquatic phycomycetes from Mountain Lake Biological Station, Virginia. Va J Sci 1965:14:219–228.
Monchy S, Sanciu G, Jobard M, Rasconi S, Gerphagnon M, Chabé M, et al. Exploring and quantifying fungal diversity in freshwater lake ecosystems using rDNA cloning/sequencing and SSU tag pyrosequencing. Environ Microbiol 2011;13:1433–1453.
Mozley-Standridge SE, Letcher PM, Longcore JE, Porter D, Simmons DR. Cladochytriales – a new order in Chytridiomycota. Mycol Res 2009;113:498–507.
Paquin B, Laforest MJ, Forget L, Roewer I, Wang Z, Longcore J, et al. The fungal mitochondrial genome project: evolution of fungal mitochondrial genomes and their gene expression. Curr Genet 1997;31:380–395.
Perrott PE. The genus *Monoblepharis*. Trans Br Mycol Soc 1955;38:247–282.
Perrott PE. Isolation and pure culture of *Monoblepharis*. Nature 1958;182:1322–1324.
Perrott PE. The ecology of some aquatic phycomycetes. Trans Br Mycol Soc 1960:43:19–30.
Picard KT, Letcher PM, Powell MJ. *Rhizidium phycophilum*, a new species in the Chytridiales. Mycologia 2009;101:696–706.
Picard KT, Letcher PM, Powell MJ. Evidence for a facultative mutualist nutritional relationship between the green coccoid alga *Bracteacoccus* sp. (Chlorophyceae) and the zoosporic fungus *Rhizidium phycophilum* (Chytridiomycota). Fungal Biol 2013;117:319–328.
Powell MJ. Fine structure of plasmodesmata in a chytrid. Mycologia 1974;66:606–614.
Powell MJ. Ultrastructural changes in nuclear membranes and organelle associations during mitosis of the aquatic fungus *Entophlyctis* sp. Can J Bot 1975;53:627–646.
Powell MJ. Ultrastructure and isolation of glyoxysomes (microbodies) in zoospores of the fungus *Entophlyctis* sp. Protoplasma 1976;89:1–27.
Powell MJ. Phylogenetic implications of the microbody-lipid globule complex. BioSystems 1978;10:167–180.
Powell MJ. Mitosis in the aquatic fungus *Rhizophydium sphaerotheca* (Chytridiales). Am J Bot 1980;67:839–853.
Powell MJ. Structure of the interface between the haustorium of *Caulochytrium protostelioides* and the hyphal cytoplasm of *Cladosporium cladosporioides*. J Elish Mitch Sci Soc 1981a;97:171–182.
Powell MJ. Ultrastructure of *Polyphagus euglenae* zoospores. Can J Bot 1981b;59:2049–2061.
Powell MJ. Zoospore structure of the mycoparasitic chytrid *Caulochytrium protostelioides* Olive. Am J Bot 1981c;68:1074–1089.
Powell MJ. Looking at mycology with a Janus face. A glimpse at Chytridiomycetes active in the environment. Mycologia 1993;85:1–20.
Powell MJ, Letcher PM. From zoospores to molecules: the evolution and systematics of Chytridiomycota. In: Misra JK, Tewari JP, Deshmukh SK, eds. Systematics and Evolution of Fungi. Boca Raton, FL: CRC Press 2012:29–54.
Powell MJ, Letcher PM, Longcore JE. *Operculomyces* is a new genus in the order Rhizophydiales. Mycologia 2011;103:854–862.
Powell MJ, Letcher PM, Longcore JE. *Pseudorhizidium* is a new genus with distinct zoospore ultrastructure in the order Chytridiales. Mycologia 2013;105:496–507.
Rosenblum EB, Stajich JE, Maddox N, Eisen MB. Global gene-expression profiles for life stages of the deadly amphibian pathogen *Batrachochytrium dendrobatidis*. Proc Natl Acad Sci USA 2008;105:17034–17039.
Schloegel LM, Toledo LF, Longcore JE, Greenspan SE, Vieira CA, Lee M, et al. Novel, panzootic and hybrid genotypes of amphibian chytridiomycosis associated with the bullfrog trade. Mol Ecol 2012;21:5162–5177, doi:10.1111/j.1365-294X.2012.05710.x

Schoenlein-Crusius IH, Milanez AI. Fungal succession on leaves of *Alchornea triplinervia* (Spreng.) Muell. Arg. submerged in a stream of an Atlantic Rainforest in the state of São Paulo, Brazil. Revta Brasil Bot. 1998;21(3):ISSN 1806-9959. http://dx.doi.org/10.1590/S0100-84041998000300003.

Sekimoto S, Rochon D, Long JE, Dee JM, Berbee ML. A multigene phylogeny of *Olpidium* and its implications for early fungal evolution. BMC Evol Biol. 2011;11:331, doi:10.1186/1471-2148-11-331

Sime-Ngando T, Lefèvre E, Gleason FH. Hidden diversity among aquatic heterotrophic flagellates: ecological potential of zoosporic fungi. Hydrobiol 2011;659:5–22.

Simmons DR. Phylogeny of Powellomycetaceae fam. nov. and description of *Geranomyces variabilis* gen et comb nov. Mycologia 2011;103:1411–1420.

Simmons DR, Longcore JE. *Thoreauomyces* gen. nov., *Fimicolochytrium* gen. nov. and additional species in *Geranomyces*. Mycologia 2012;104:1229–1243, doi:10.3852/12-015

Simmons DR, James TY, Meyer AF, Longcore JE. Lobulomycetales, a new order in the Chytridiomycota. Mycol Res 2009;113:450–460.

Simmons DR, Letcher PM, Powell MJ, Longcore JE. *Alogomyces tanneri* gen. *et* sp. nov., a chytrid in Lobulomycetales from horse manure. Mycologia 2012;104:157–163.

Sparrow FK. Two new chytridiaceous fungi from Cold Spring Harbor. Am J Bot 1931;18:615–623.

Sparrow FK. Some chytridiaceous inhabitants of submerged insect exuviae. Proc Am Phil Soc 1937;78:23–53.

Sparrow FK. Aquatic Phycomycetes, 2nd ed. Ann Arbor, MI: University of Michigan Press, 1960.

Sparrow FK, Barr ME. Additions to the phycomycete flora of the Douglas Lake region I. New taxa and records. Mycologia 1955;47:546–556.

Steiger RA, Simmons DR, Longcore JE. *Cylindrochytridium johnstonii* is a member of the Cladochytriales. Mycotaxon 2011;118:293–302.

Sun G, Yang Z, Kosch T, Summers K, Huang J. Evidence for acquisition of virulence effectors in pathogenic chytrids. BMC Evol Biol 2011;11:195, doi:10.1186/1471–2148/11/195

Ustinova I, Krienitz L, Huss VAR. *Hyaloraphidium curvatum* is not a green alga, but a lower fungus; *Amoebidium parasiticum* is not a fungus, but a member of the DRIPs. Protist 2000;151:253–262.

Vélez CG, Letcher PM, Schultz S, Powell MJ, Churchill PF. Molecular phylogenetic and zoospore ultrastructural analyses of *Chytridum olla* establish the limits of a monophyletic Chytridiales. Mycologia 2011;103:118–130.

Vélez CG, Letcher PM, Schultz S, Mataloni G, Lefèvre E, Powell MJ. 2013. Three new genera in Chytridiales from aquatic habitats in Argentina. Mycologia 2013;105:1251–1265.

Wakefield WS, Powell MJ, Letcher PM, Barr DJS, Churchill PF, Longcore JE, et al. A molecular phylogenetic evaluation of the Spizellomycetales. Mycologia 2010;102:596–604.

Waterhouse GM. Some water moulds of the Hogsmill River collected from 1937 to 1939. Trans Br Mycol Soc 1942;25:315–325.

Willoughby LG. Chitinophilic chytrids from lake muds. Trans Br Mycol Soc 1961;44:586–592.

Willoughby LG. The activity of *Rhizophlyctis rosea* in soil: some deductions from laboratory observations. Mycologist 2001;15:113–117.

Willoughby LG, Townley PJ. Two new saprophytic chytrids from the Lake District. Trans Br Mycol Soc 1961;44:177–184.

Wurzbacher CM, Barlocher F, Grossart H-P. Fungi in lake ecosystems. Aquat Microb Ecol 2010;59:125–149.

Phylogeny of fungus-like organisms

Ray Kearney and Frank H. Gleason

7 Microsporidia

The phylum *Microsporidia* comprises more than 1,500 formally described unicellular species in over 187 genera (Keeling and Fast 2002; Corradi and Keeling 2009; Li et al. 2012; Vávra and Lukeš 2013). Whilst almost half of the microsporidian genera infect aquatic hosts (Stentiford et al. 2013b), the focus until recently has been on terrestrial hosts, including humans.

Microsporidia are emerging diverse pathogens in many aquatic systems. Their extreme morphological plasticity and a capacity to infect virtually all organs provide opportunities to bring change not only at the individual level but also within communities of aquatic hosts (Stentiford et al. 2013a). In the context of global change, such extraordinary potential to adapt highlights the need for stringent research and surveillance of their potential to cause emergent diseases not only in aquatic hosts but in humans as well.

In humans, *Enterocytozoon bieneusi* is a common pathogen in immuno-compromised patients. It is noteworthy that its closest relatives are the marine crab pathogen *Enterospora canceri*, a marine shrimp pathogen (*Enterocytozoon hepatopenaeii*) and certain fish pathogens (Stentiford et al. 2013b). Evidence confirms that some microsporidia can cycle between crustaceans and vertebrates (Nylund et al. 2011). Such close relationships raise questions of the possibility of aquatic invertebrates being an important source of certain zoonotic infections of humans.

Excluding genera which infect aquatic life stages of insects, at least 21 genera infect aquatic non-arthropod invertebrates, protists as well as hyperparasites of aquatic hosts and 50 genera infect aquatic arthropods. In addition, approximately 20 genera infect bony fish (class Osteichthyes) but none has been reported from cartilaginous fish (see Tabs 7.1, 7.2 and 7.3). These aquatic hosts may inhabit freshwater, brackish as well as marine environments. With such a capacity to infect a wide taxonomic range of hosts, most of which are poorly studied, there is likely to be many thousands of microsporidian taxa yet undescribed from aquatic hosts.

Microsporidia are spore-forming obligate intracellular parasites of eukaryotic hosts. As biotrophs each parasite derives nutrients from the living tissues of another organism (its host). Some are hyperparasites, i.e. a parasite whose host is a parasite (Stentiford et al. 2013b). In earlier studies, the phylum *Microsporidia* has been either ignored in eukaryote systematics or treated as a group of protozoa (e.g. Levine et al. 1980; Cavalier-Smith 1993; Wittner and Weiss 1999). They were once thought to be protists but are now known to be related to fungi (Cuomo et al. 2012; James et al. 2013). Whilst Microsporidia appear to be closely related to fungi, the question of whether they should be included within the Kingdom Fungi remains contentious (James et al. 2006).

Tab. 7.1: Genera of Microsporidia infecting bony fish hosts (host class – Osteichthyes).

Genus	Genus
1. *Amazonspora*	11. *Myosporidium*
2. *Dasyatispora*	12. *Neonosemoides*
3. *Glugea*	13. *Nucleospora*
4. *Heterosporis*	14. *Ovipleistophora*
5. *Ichthyosporidium*	15. *Paranucleospora*
6. *Kabatana*	16. *Pleistophora*
7. *Loma*	17. *Potaspora*
8. *Microfilium*	18. *Pseudoloma*
9. *Microgemma*	19. *Spraguea*
10. *Microsporidium*	20. *Tetramicra*

Adapted from Stentiford et al. (2013b).

Tab. 7.2: Genera of Microsporidia infecting aquatic arthropod hosts.

Genus	Host class	Genus	Host class
1. *Abelspora*	*Malacostraca*	26. *Holobispora*	*Maxillopoda*
2. *Agglomerata*	*Branchiopoda*	27. *Inodosporus*	*Malacostraca*
3. *Agmasoma*	*Malacostraca*	28. *Lanatospora*	*Maxillopoda*
4. *Alfvenia*	*Maxillopoda*	29. *Larssonia*	*Branchiopoda*
5. *Ambylospora*	*Maxillopoda*	30. *Marssoniella*	*Maxillopoda*
6. *Ameson*	*Malacostraca*	31. *Mrazekia*	*Malacostraca*
7. *Baculea*	*Branchiopoda*	32. *Myospora*	*Malacostraca*
8. *Berwaldia*	*Branchiopoda*	33. *Nadelspora*	*Malacostraca*
9. *Binucleata*	*Branchiopoda*	34. *Nelliemelba*	*Maxillopoda*
10. *Binucleospora*	*Ostracoda*	35. *Norlevinea*	*Branchiopoda*
11. *Cougourdella*	*Maxillopoda*	36. *Nosema*	*Branchiopoda*
12. *Cucumispora*	*Malacostraca*	37. *Octosporea*	*Branchiopoda*
13. *Desomozoon*	*Maxillopoda*	38. *Ordospora*	*Branchiopoda*
14. *Dictyocoela*	*Malacostraca*	39. *Ormieresia*	*Malacostraca*
15. *Encephalitozoon*	*Maxillopoda*	40. *Parathelohania*	*Maxillopoda*
16. *Enterocytozoon*	*Malacostraca*	41. *Paranucleospora*	*Maxillopoda*
17. *Enterospora*	*Malacostraca*	42. *Perezia*	*Malacostraca*
18. *Facilispora*	*Maxillopoda*	43. *Pleistophora*	*Malacostraca*
19. *Fibrillanosema*	*Malacostraca*	44. *Pyrotheca*	*Maxillopoda*
20. *Flabelliforma*	*Ostracoda*	45. *Thelohania*	*Malacostraca*
21. *Glugoides*	*Branchiopoda*	46. *Trichotuzetia*	*Maxillopoda*
22. *Gurleya*	*Branchiopoda*	47. *Triwangia*	*Malacostraca*
23. *Gurleyides*	*Branchiopoda*	48. *Tuzetia*	*Maxillopoda*
24. *Hamiltosporidium*	*Branchiopoda*	49. *Vairimorpha*	*Malacostraca*
25. *Hepatospora*	*Malacostraca*	50. *Vavraia*	*Malacostraca*

Microsporidian genera infecting aquatic life stages of insects are excluded from list.
Adapted from Stentiford et al. (2013b).

Tab 7.3: Genera of Microsporidia infecting invertebrate hosts other than arthropod, and including free-living protists and hyperparasites of hosts from aquatic habitats.

Genus	Group	Genus	Group
1. *Microsporidium*	Acanthocephala	16. *Microsporidium*	Rotifera
2. *Napamichum*	Arachnida	17. *Pseudonosema*	Phylactolaemata
3. *Hrabyeia*	Oligochaeta	18. *Trichonosema*	Phylactolaemata
4. *Rectispora*	Oligochaeta	19. *Bryonosema*	Phylactolaemata
5. *Kinorhynchospora*	Kinorhyncha	20. *Schoedera*	Phylactolaemata
6. Undescribed taxon	Monogenea	21. *Nosema*	Digenea
7. *Bacillidium*	Oligochaeta	22. *Bacillidium*	Oligochaeta
8. *Ciliatosporidium*	Ciliophora	23. *Nosema*	Digenea
9. *Unikaryon*	Cestoda	24. *Nosemoides*	Gregarinasina
10. *Wittmannia*	Dicyemida	25. *Amphiamblys*	Gregarinasina
11. *Steinhausia*	Mollusca	26. *Nosema*	Cestoda
13. *Microsporidium*	Clitellata	27. Undescribed taxon	Amoebozoa
14. *Amphiacantha*	Gregarinasina	28. *Euplotespora*	Ciliophora
15. Undescribed taxon	Mollusca	29. *Sporanauta*	Nematoda
16. Undescribed taxon	Hydrozoa	30. *Metchnikovella*	Gregarinasina

Microsporidian genera infecting aquatic life stages of insects are excluded from list.
Adapted from Stentiford et al. (2013b).

The strategies used by these obligate intracellular pathogens to thrive within host cells remain puzzling, despite being responsible for significant medical and agricultural problems. Microsporidia can infect and kill animals from virtually every phylum, with few treatments available (Keeling and Fast 2002; Williams 2009). Microsporidian genomes encode a greatly reduced metabolic potential compared with other eukaryotes (Keeling and Corradi 2011). Despite such reduction in genetic information, these microbes can still undergo dramatic proliferation within host cells in ways that continue to be unraveled.

For a long time Microsporidia were considered to be amitochondriate, until a tiny mitochondrion-derived organelle called the mitosome was detected (Williams et al. 2002). The molecular function of this organelle remains poorly understood. The mitosome has no genome, so it must import all its proteins from nuclear genes via the cytosol (Burri et al. 2006).

The highly divergent sequences of Microsporidia are a major challenge in their phylogenetic assignment (Stiller and Hall 1999; Katinka et al. 2001). A comparison of many shared genes among Microsporidia and Fungi could help to resolve some of these controversies regarding their phylogeny (Cuomo et al. 2012; James et al. 2013). The details of their life cycle can be complex as they vary from species to species. However, they share general features in their life cycle such as infection using a polar tube, intracellular replication as meronts and differentiation into spores. Whether they are haploid or diploid and whether they undergo a mating and meiosis cycle are questions not yet clearly resolved.

Here we briefly discuss various genomic studies and analyse results to better understand the evolutionary strategies that led to the emergence of this large phylum of important obligate intracellular pathogens. This account will also include a brief overview of their ecology, unique structure and classification to provide insights into events that may have facilitated their emergence by adopting strategies for invasive growth shared elsewhere in the animal kingdom.

7.1 Ecology

Microsporidia infect a wide range of invertebrate and vertebrate species, including humans (Lobo et al. 2012) and are common parasites of both fish (Lom and Nilsen 2003) and arthropods. To date there are 14 microsporidian species which are known to infect humans, including malnourished children and the elderly but primarily affect those with immuno-compromised disorders, AIDS as well as organ transplants (Cali et al. 1993; Didier et al. 2005; Didier and Weiss 2011). Microsporidia were only rarely recognized as causes of disease in humans until the AIDS pandemic (Didier and Weiss 2011).

Genotype distribution of *Enterocytozoon bieneusi* differs by geography and by the type of host immunity (Matos et al. 2012; Mori et al. 2013). In Australia, only genotype B has been reported as a causative genotype (Stark et al. 2009). In European countries, genotype B has been most frequently detected followed by genotypes A and C (Breitenmoser et al. 1999; Dengjel et al. 2001). In Thailand, three genotypes were identified: genotype D predominated, followed by EbpC, and then novel genotype ETMK1 (Mori et al. 2013). The first two genotypes have zoonotic potential.

Human infections are predominantly associated with symptoms of wasting and diarrhoea. Transmission is believed to occur primarily through fecal-oral routes with sources of infection including other infected humans and animals, as well as contaminated water and food. The routes of infection are principally ingestion, inhalation and probably direct inoculation of infectious spores from the environment (Didier and Weiss 2006, 2011). *Enterocytozoon bieneusi* infections in immuno-competent hosts, in contrast to those in immuno-deficient persons, are usually self-limiting (Mori et al. 2013).

Microsporidia are probably poorly diagnosed infectious agents in humans because of diagnostic limitations and detection difficulties. Treatment with albendazole is effective for initial therapy for intestinal and certain disseminated species of microsporidiosis while fumagillin is effective for treatment of *Enterocytozoon bieneusi*, the most prevalent species causing enteric infections, but has toxic side effects (Didier and Weiss 2011). There is a need for further research about persistence and reactivation as well as whether asymptomatic carriers of infection may transmit infections to those at risk.

Numerous reports of microsporidian infections in marine and freshwater crustacea include genera infecting marine crabs (Sprague and Couch 1971; Olson et al. 1994;

Stentiford and Bateman 2007; Stentiford et al. 2007), marine lobster, (Kiryu et al. 2009), freshwater crayfish (Freeman et al. 2010), marine shrimps (Lightner 1996; Azevedo 2000, 2001; Clotilde-Ba and Toguebaye 2001; Toubiana et al. 2004; Tourtip et al. 2009), brine shrimp (Martinez et al. 1992), Cladocerans (Refardt et al. 2002; Wolinska et al. 2009), ostracods (Diarra and Toguebaye 1996), amphipods (Dunn et al. 2001), and copepods (Sprague and Vávra 1977; Voronin 1996; Bronnvall and Larsson 2001; Freeman et al. 2003; Vávra et al. 2005).

Glugoides intestinalis, *Microsporidium* sp., *Ordospora colligata*, *Gurleya vavrai*, *Larssonia obtusa* and *Flabelliforma magnivora* are microsporidian parasites of the planktonic freshwater crustaceans *Daphnia* spp. (Refardt et al. 2002). *Giardia intestinalis* is related to two microsporidia infecting lepidopterans and to *Vittaforma corneae*, which has been described as a human pathogen. It is thought that *V. corneae* may have an invertebrate as its natural host (Refardt et al. 2002).

To date there is no report of microsporidia infecting fungi. Almost all described microsporidia infect animals while the rest infect protists which are themselves animal parasites (gregarines, myxosporidia, and certain ciliates). Such evidence suggests that microsporidia originated as animal parasites. Other targets of microsporidian infection are probably later, opportunistic developments (Kelling and Fast 2002). Microsporidia can affect host population densities and may even drive host population cycles (Kohler and Hoiland 2001).

Microsporidia also cause agriculturally important disease: 50%–94% of honey bees in the USA are infected with the microsporidian *Nosema ceranae*, which has been implicated in honey-bee colony collapse disorder (Johnson et al. 2009; Troemel 2011). The same species has been found infecting honey bees in Thailand (Chaimanee et al. 2010).

Pathogenic microsporidia have been isolated from various water sources including drinking water, wastewater, and recreational water (Dowd et al. 1998; Izquierdo et al. 2011) and they have been associated with waterborne outbreaks of disease worldwide (Cotte et al. 1999; Lobo et al. 2012). Drinking water treatment plants and wastewater treatment plants which comply with national regulatory standards on sanitary quality can still be contaminated with microsporidia and other pathogens (Galván et al. 2013). These researchers detected microsporidia in 49% of samples (109/223) from various water sources in Madrid, Spain. Human-pathogenic microsporidia were detected, including *Enterocytozoon bieneusi* (C, D, and D-like genotypes), *Encephalitozoon intestinalis*, *Encephalitozoon cuniculi* (genotypes I and III), and *Anncaliia algerae*. *Encephalitozoon intestinalis* was the most commonly detected species, with an annual prevalence of 26.6% in positive samples, followed by *Enterocytozoon bieneusi* (16.5%), *Encephalitozoon cuniculi* (11.9%), and *Anncaliia algerae* (1.8%). *Enterocytozoon bieneusi*, the more frequent microsporidia in humans, was detected in three of the four seasons and was most common in spring (27.7%), followed by winter (25%) and summer (9.7%). *Encephalitozoon intestinalis* was common in spring (58.3%) and winter (40%), and there were positive samples for *Enterocytozoon cuniculi* only in winter (65%).

Anncaliia algerae was found in one sample from spring and another from autumn. Undetermined species represented 59.6% (65 out of 109) of samples.

Such findings highlight the need to review the current controls and regulations for drinking water, waste and recreational waters to ensure risks to human health are removed or reduced. Because of their waterborne transmission potential, microsporidia (*Enterocytozoon bieneusi, Encephalitozoon intestinalis, Encephalitozoon cuniculi, Encephalitozoon hellem*, and *Vittaforma corneae*) have been included in the last two drinking water contaminant candidate lists [contaminant candidate list 2 (CCL2) and CCL3] of the U.S. Environmental Protection Agency (2005, 2009).

The detection of microsporidia along with some intestinal protozoa in commercial fresh fruit juices is especially noteworthy (Mossallam 2010). Results showed that 35.43% were contaminated with one or more of Cryptosporidia, Microsporidia, and Cyclospora, as well as *Giardia* spp. Strawberry was the most contaminated juice (54.28%), while orange was the slightest (22.86%). Cryptosporidia was the highest contaminant (61.29%), and Cyclospora was the least (14.52%). Microsporidia species were the most robust contaminants which retained its viability and infectivity in juices in which it was detected. Water is a major conduit for these parasites and contaminated water is an important source of human infection either by direct consumption or by the use of contaminated water during collection, processing and preparation of such fruit juices i.e. from "farm to the fork" (Slifko et al. 2000).

Hence consumers, especially high risk or immuno-compromised groups, are placed at hazard of contracting such intestinal infections. The risk to insect pollinators by some species of imported microsporidia via this source is also plausible, prompting a call for more stringent quarantine surveillance. The water/food connection for parasite zoonoses is complex (Slifko et al. 2000), with feces as a major vehicle for many environmental transmissive stages. However, the spores of some microsporidia (e.g. *Encephalitozoon cuniculi*) contaminate the environment through urine. The transmissible stages can contaminate water or foods directly, voided in feces, or indirectly (Slifko et al. 2000). The disposal of animal (and human) wastes remains a significant public health issue that has yet to be assessed or controlled in many countries.

7.2 Classification

Microsporidia have historically been classified, based on structural characters, with other intracellular parasites such as myxosporidia, actinomyxidia, haplosporidia, and sporozoa (e.g. Lom and Vávra 1966; Kudo 1966; Desportes and Nashed 1983). However, conventional eukaryotic classification, e.g. based on microtubule structures and mitochondria, is not possible in microsporidia because these microorganisms do not possess such structures but possess, instead, mitosome, thought to be a relic of mitochondria (Katinka 2001; Goldberg et al. 2008; Wu et al. 2009). In eukaryotic cells, energy is generated in the mitochondria; however, the cells of microsporidia reveal no such organelle and their exact mode of metabolism is not yet fully understood.

Microsporidia are obligate biotrophs and also lack motile structures such as flagella or cilia. They are very simplistic in form, harbouring highly reduced mitochondria or mitosomes, ribosomal RNAs, an unconventional Golgi apparatus and are unrelated to the Chytridiomycota (Corradi and Slamovits 2011). The challenges in phylogenetic classification of the Microsporidia are also related to their highly divergent sequences (Stiller and Hall 1999; Katinka et al. 2001). For example, in *Encephalitozoon cuniculi* (1,997 potential protein-coding genes) the strong host dependence is illustrated by the lack of genes for some biosynthetic pathways and for the tricarboxylic acid cycle while genome compaction is reflected by reduced intergenic spacers (Katinka et al. 2001). Microsporidia were first considered a deeply branching protist lineage that diverged before the endosymbiotic event that led to mitochondria. Such divergences could be due to rapid multiplication accompanied by a high mutation rate.

Earlier phylogenetic analyses, dependent on a small number of genes, led to caution in interpretation (e.g. Keeling et al. 2000, 2005) which prompted comparisons of many shared genes between Microsporidia and Fungi (Fischer and Palmer 2005; Vossbrinck and Debrunner-Vossbrinck 2005). The association gained further strength with earlier research into molecular phylogenies involving alpha- and beta-tubulins (Edlind et al. 1996; Keeling and Doolittle 1996). Macroscopic and imprecise molecular analyses placed the Microsporidia in the Archezoa (Keeling 2009).

Data published by Vossbrinck and Debrunner-Vossbrinck (2005) indicate the presence of five major clades of Microsporidia which are grouped according to habitat. These researchers presented three new classes of Microsporidia based on natural phylogenetic groupings as illustrated by the simple sequence repeats (ssr) DNA analysis: Aquasporidia, Marinosporidia and Terresporidia. The names of the proposed classes reflect the habitat of each group. The class Aquasporidia, found primarily in freshwater habitats, is a paraphyletic group consisting of three clades. The Marinosporidia are found in hosts of marine origin and the Terresporidia are primarily from terrestrial environments. For those genera known to infect aquatic hosts, ssrRNA gene phylogenetic data are available for 35 aquatic arthropod-infecting taxa, 17 fish-infecting taxa and six taxa associated with non-arthropod invertebrates (Stentiford et al. 2013b). To date, no sequence data exist for microsporidian parasites infecting protists or hyperparasites.

Many aquatic microsporidians infecting freshwater arthropods (e.g. crayfish) are more closely related phylogenetically to insect-infecting taxa than those infecting similar hosts in marine habitats (Stentiford et al. 2013b) thereby raising the potential for freshwater taxa to interact with both insect and crustaceans. Thus most freshwater taxa align with Clade 1 of the Aquasporidia. However, additional criteria such as site of infection, e.g. muscle, may help to differentiate phylogenetically similar but morphologically diverse microsporidian genera (Stentiford et al. 2010).

The challenges confronting biologists in trying to unravel the complexity of genetic analyses involving Microsporidia are illustrated by the research of Karpov et al. (2013) whose results suggest a close relationship of aphelids with rozellids and microsporidia. Since species of *Rozella* efficiently transfer carbon and energy from their hosts (primary consumers) to grazing zooplankton and other tertiary consumers (Gleason et al. 2012), unlike Microsporidia, without more quantitative data, it is difficult to understand the proposed phylogenetic relationship of Microsporidia and Rozella (Cryptomycota) (Gleason et al. 2012).

However, recent studies by James et al. (2013) showed a species of Cryptomycota (*Rozella allomycis*) is united with Microsporidia based on phylogenomics and shared genomic traits. The latter genomic elements include a nucleotide transporter that is also used by Microsporidia for accessing the host's ATP energy source. Other genomic features in common include chitin synthase genes. Showing that *Rozella* cysts actively produce a chitinous wall identifies a characteristic that unites fungi with Cryptomycota, microsporidia and possibly aphelids. However, unlike species of Microsporidia, those of Cryptomycota do possess mitochondria though somewhat degenerate in their capability. The *Rozella* genome has therefore not undergone major contraction as has the microsporidial genome (James et al. 2013).

Whilst Microsporidia have now been excluded from the *International Code of Botanical Nomenclature*, a proposal has been made to also exclude them from the *International Code of Zoological Nomenclature* (Redhead et al. 2009 a,b). A complication of microsporidia being identified with fungi and not the protista is that the majority of names given to microsporidial species are invalid because fungi are subject to botanical rules of nomenclature (Keeling 2009). Excluding the Microsporidia from the *Code* would resolve this dilemma (Redhead et al. 2006).

Morphological and life cycle data do not conclusively link microsporidia with a particular kind of fungus (Keeling and Fast 2002; Corradi and Keeling 2009). Nevertheless, the integrated use of available data including phylogenetics, ultrastructure, pathology, ecology and host type is now well accepted especially when attempting to classify novel aquatic pathogens within the phylum (Vossbrinck and Debrunner-Vossbrinck 2005; Stentiford et al. 2013a). These researchers demonstrated, however, that combining plasticity in morphological traits with phylogenetic data within individual aquatic taxa were unreliable metrics for classification purposes (Vossbrinck and Debrunner-Vossbrinck 2005; Stentiford et al. 2013a).

Not all microsporidia kill their host. Discoveries of natural parasites of the nematodes *Caenorhabditis elegans* and *Caenorhabditis briggsae* led to the classification of a new microsporidian genus (*Nematocida*) of two species (Troemel et al. 2008). Analyses of their genomes when compared with several other divergent microsporidia genomes have led to a better understanding of the evolutionary strategies leading to the emergence of this phylum of obligate intracellular biotrophs (Troemel et al. 2008).

7.3 Evolutionary origins

By sequencing beta-tubulins from three microsporidia, four chytrid fungi, and 12 zygomycete fungi, Keeling et al. (2000) showed beta-tubulin phylogeny supported the proposal that Microsporidia evolved from a fungus sometime after the divergence of chytrids, zoosporic fungi and zygosporic fungi (Hibbett et al. 2007; Blackwell 2011). This was further supported when 29 beta-tubulins were sequenced from microsporidia, zygomycetes, and chytrids (Keeling 2003). However, tubulin trees and chytrid tubulins were not without problems (Corradi and Keeling 2009).

A research team at Duke University Medical Centre used genetic studies to show that microsporidia are especially closely related to a zygomycete fungus. They found that of the 2,000 genes of zygomycetes, 33 genes were in common with microsporidia, but not with other fungi (Lee et al. 2008). This genomic evidence strongly suggests that microsporidia and zygomycetes have a common ancestor and their reduced genomes and gene structures, the result of gene loss and compaction, reveal a highly evolved system rather than a primitive one (Corradi and Keeling 2009).

The collective data suggested that the Microsporidia are a sister group to fungi but do not actually arise from within the group (Keeling and Doolittle 1996; Edlind et al. 1996; Hirt et al. 1999; Gill et al. 2010). Recent research by Capella-Gutiérrez et al. (2012) supports the position of Microsporidia in the Fungi Kingdom based on the combined analysis of 53 concatenated genes and was robust to filters controlling for rate heterogeneity, compositional bias, long-branch attraction and heterotachy.

While the fungal nature of microsporidia is now generally accepted by researchers investigating taxonomic and evolutionary schemes (Keeling et al. 1996, 1998; Cavalier-Smith 1998; Adl et al. 2005; Cuomo et al. 2012; James et al. 2013), their exact position in the fungal tree remains hotly debated (James et al. 2006, 2013; Gill et al. 2006; Lee et al. 2008).

The presence of a common gene repertoire among microsporidia with different genome sizes suggests that these parasites have evolved from a gene-sparing ancestor and that their current gene content may well represent the lower limit for a fully functional eukaryotic genome (Corradi and Slamovits 2011). Additional evidence indicates that the sexual cycle of microsporidia, in the infected host, is genetically controlled (Cuomo et al. 2012).

Whilst zygomycetes and their alleged sister taxon, the microsporidia, exclusively share the presence of a cluster of three genes encoding a sugar transporter, a high mobility group (HMG)-type transcription factor and an RNA helicase, claims are made that the presence of a gene cluster is not phylogenetically informative (Koestler and Ebersberger 2011). From their analysis, the latter researchers concluded that the phylogenetic placement of microsporidia as sister to the zygomycetes needs to be reconsidered (Koestler and Ebersberger 2011). Molecular data seem the only evidence to indicate microsporidia have undergone extraordinarily divergent and reductive evolution, losing most of the morphological and molecular clues that would establish the exact nature of their fungal heritage. Despite recent progress, many deep relationships within the fungi remain unclear (Ebersberger 2012).

Opportunities for horizontal and vertical transmission may be influenced by selective pressures among the Microsporidia. Whilst horizontal transmission is the main route among aquatic Microsporidia many of the species can adopt both horizontal and vertical transmission (Dunn et al. 2001; Smith 2001). In aquatic habitats, persistence and dispersal of the endoparasite may be important (Lucarotti and Andreadis 1995) as well as virulence. Horizontal transmission, when spores are released into water by the parasite, may select for virulence and high microsporidial replication (Terry et al. 2004) while vertical transmission dependent on host reproduction is likely to select for lower virulence (Dunn and Smith 2001).

7.4 Cell structure and spore significance

The life cycle of microsporidia includes a proliferative merogonic stage, followed by a sporogonic stage producing characteristically small (1–4 μm), environmentally resistant, infective spores. Being highly resistant, spores of microsporidia survive outside the host for many years and their morphological features can help to classify them to the genus level. In this form the parasite is most easily recognizable and identifiable to an individual species level.

Spore morphology ranges from predominantly pyriform or ovoid to spherical or rod-like. As with bacterial endospores, the resistance of spores of microsporidia is likely to be attributed to chemical composition and high internal pressure generated during spore development and contraction. The spore wall has three layers and encloses mainly two closely-associated nuclei forming a diplokaryon. The chitin-rich inner wall (Peuvel-Fanget et al. 2006) of the microsporidial spore also has proteins (Peuvel-Fanget et al. 2006; Wu et al. 2008). The formation of the spore wall is thought to be linked with a surface protein, chitin deacetylase (Brosson et al. 2005).

The spore has in the anterior half a unique spear-like ejectile organelle (Xu et al. 2005) located at the apex and affixed to the spore by way of an anchoring disk surrounded by a lamella of membranes. This "spear-head" is attached to a long coiled, thread-like, hollow polar filament, many times the length of the spore, located adjacent to a vacuole in the posterior half of the spore. A spore wall protein interacts with polar tube protein (Li et al. 2012). Most microsporidian infections are transmitted by oral ingestion of spores with the site of initial infection being the gastrointestinal tract (Li et al. 2012). While spore germination is initiated by a signal from the species' environment, e.g. gastrointestinal epithelium, the exact nature of these signals is not known for certain.

The polar tube distinguishes microsporidia from all other organisms and has an essential function in infectiousness by injecting the sporoplasm along with its nucleus into the cytoplasm of the target host cell (Bouzahzah et al. 2010). The entire germination process takes as little as two seconds to be completed. There appears to be more than one mode of infection. The entry of the parasite, propelled by the spore's internal pressure, is achieved after piercing the plasmalemma of the host cell or the membrane of the phagosomes containing the endocytosed spores (Couzinet et al. 2000; Franzen 2004).

In the latter situation, the fusion with the lysosome is inhibited thereby allowing the parasite (meront) to escape destruction and persist intracellularly. It has also been shown that microsporidia can be taken up by phagocytosis and then use the polar tube to escape from the phagosome vacuole (Franzen 2005). Amoeboid sporoplasm emerges from the polar tube in the infected cell where the spore activates to develop into a syncytial plasmodium and differentiate into infectious spores. This intimate relationship between parasite and host is unique. It allows the microsporidia to be highly exploitative of the host cell environment and cause such diverse effects as the induction of hypertrophied cells to harbour prolific spore development, host sex ratio distortion and host cell organelle and microtubule reorganization (Williams 2009).

This method of invading new host cells is one of the most sophisticated infection mechanisms in biology and ensures that the microsporidia enter the host cell unrecognized and protected from the host defense reactions. The extrusion apparatus of the microsporidia has obviously ensured the success of this phylum during evolution, resulting in a group of obligate intracellular biotrophs, capable of infecting almost any type of host and cell (Franzen 2005).

Intranuclear parasitism by microsporidians is uncommon but remarkable given that microsporidians lack mitochondria. Such is found in three genera: *Enterospora* in crabs, *Nucleospora* and *Paranucleospora* in fish and complete their life cycles within the nucleus of the host cell (Stentiford et al. 2013b).

7.5 Metabolism

The extreme compaction with reduction of most microsporidian genomes (Thomarat et al. 2004) has led to the loss of many metabolic pathways making these biotrophs extremely dependant on their host (Texier et al. 2010). Obligate intracellular microbes are completely dependent on intracellular resources from their hosts but how they thrive remains a puzzle. Some of the other infectious microbes of humans are also obligate intracellular pathogens such as species of Chlamydia. One hypothesis is that the ancestors of microsporidia resembled intracellular organelles like mitochondria (Keeling 2011). Their genome is only a fraction of that of eukaryotes and still they can proliferate dramatically in the host cell (Keeling and Corradi 2011). Obviously the rapid replication and development of the intracellular biotroph demands an uninterrupted supply of building blocks such as amino acids, lipids and nucleotides from the host cell which they usually disrupt. How this control by the biotroph is achieved is not well understood.

Dolgikh et al. (2011) described the presence of an alternative respiratory chain in spores of aquatic species which suggested their requirement for oxygen and some exchange with the water environment. The transition of the parasite to terrestrial life (the infecting of terrestrial hosts) and the long adaptation of spores to dry-air conditions might result in the loss of alternative respiration, slowing down spore metabolism (trehalase activity) and the enlargement of the ubiquinone pool by its redistribution from mitosomes to other spore membranes (Dolgikh et al. 2011).

7.6 Genome structure

When contrasting evolution of gene order and protein sequence in microsporidian genomes compared to other fungi, microsporidian genomes exhibit comparatively high degree of gene order conservation despite great divergence at the amino acid sequence level (Corradi and Slamovits 2011). Whether microsporidia are haploid or diploid is unclear, but are presumed to be haploid (Cuomo et al. 2012) although some microsporidia do undergo a sexual cycle (Becnel et al. 2005).

Of importance is that the gene order of the zygomycete mating locus is conserved in the microsporidia (Lee et al. 2008) suggesting microsporidia have a diploid stage and a mating cycle as in true fungi (Ni et al. 2011). Since mating appears to be important for the generation of diversity and virulence in other pathogenic fungi (Ni et al. 2011) an answer to whether they are diploid or haploid, as well as questions about mating and recombination, could provide insight into microsporidian speciation and pathogenic mechanisms (Cuomo et al. 2012).

Reductive mechanisms in microsporidia have included extensive gene losses, especially in biochemical pathways involved in metabolism, as well as the shrinking of virtually all intergenic regions and surrounding genes. The variable extent of gene losses has resulted in different levels of host reliance among Microsporidia (Corradi and Slamovits 2011).

7.7 Discussion and conclusion

Microsporidia are a group of intracellular parasites related to fungi with severely reduced metabolic capability thereby making them derive nutrients from the living tissues of another organism (its host) (James et al. 2013). As highly specialized parasites, of medical, agricultural and ecological significance, they are characterized by a number of unusual adaptations, many of which are manifested as extreme reduction at the molecular, biochemical and cellular levels. These biotrophs lack mitochondria, a Krebs cycle and a respiratory chain but possess genes encoding glycolysis enzymes, a glycerol phosphate shuttle, and ATP/ADP carriers to import host ATP (Dolgikh et al.

2011). Because these biotrophs lack any ATP-generating enzymes of mitochondrial origin, their mitosomes appear not to be involved in energy metabolism.

Without mitochondria, where do the microsporidia get their energy without having many metabolic pathways, such as oxidative phosporylation, electron transport, tricarboxylic acid cycle and when there are no genes for such metabolic pathways in microsporidia (Weider et al. 1999)? One important step toward becoming a successful obligate intracellular pathogen is to increase the biosynthetic output of the host (Cuomo et al. 2012). Evidence indicates that some parasitic species seems to import ATP from the host cell, while others import ATP into the relict mitosome (Tsaousis et al. 2008; Williams 2008). Furthermore, it has been reported that microsporidia infection causes depletion of host glycogen and rapid uptake of glucose (Meteier and Vivares 2001), events that could be promoted by delivery of the parasite hexokinase into host cells (Cuomo et al. 2012).

Whilst a view is held that the mitochondrion evolved from bacteria originating from within the bacterial phylum α-proteobacteria, new data continue to reshape our views regarding mitochondrial evolution, particularly raising the question of whether the mitochondrion originated after the eukaryotic cell arose, as assumed in the classical endosymbiont hypothesis, or whether this organelle had its beginning at the same time as the cell containing it (Gray 2012).

Cuomo et al. (2012) found that all microsporidia lost the tumor-suppressor gene retinoblastoma, which they speculated could accelerate the biotroph cell cycle and increase the mutation rate. These researchers also found that microsporidia-acquired transporters that could import nucleosides to fuel rapid growth. In addition, microsporidian hexokinases gained secretion signal sequence and in a functional assay these were found sufficient to export proteins out of the cell. Thus, the researchers concluded hexokinase may be targeted into the host cell to reprogram it toward biosynthesis. Cuomo et al. (2012) deduced that since similar molecular changes appear during formation of cancer cells, such may be evolutionary strategies adopted independently by microsporidia to proliferate rapidly within host cells. Finally, their analysis of genome polymorphisms revealed evidence for a sexual cycle that may provide genetic diversity to alleviate problems caused by clonal growth. Together these events may explain the emergence and success of these diverse intracellular biotrophs (Cuomo et al. 2012).

To meet these demands, the same researchers proposed that microsporidia hexokinase would be ideally situated to redirect host cells toward anabolic metabolism i.e. the host metabolic state sharing similarities with the Warburg effect, where cancer cells switch to glycolysis to help meet their proliferative demands for nucleotides, amino acids, and lipids (Vander et al. 2009). Noteworthy, the Warburg effect has been observed in host cells infected by Kaposi's sarcoma herpesvirus, another obligate intracellular pathogen (Delgado et al. 2010). The proposal (Cuomo et al. 2012) that rapid clonal proliferation using the cancer-like strategies together with a rare sexual cycle complementing each other to allow for rapid intracellular growth and successful spread of microsporidia throughout the animal kingdom is highly significant.

According to the phylogenetic system based on molecular analysis of the small subunit ribosomal DNA (rDNA) of 125 microsporidial species (Vossbrinck and Debrunner-Vossbrinck 2005), these biotrophs belong mainly to the phylogenetic clade IV (Terresporidia). Such adaptations are associated with genome compaction induced gene size reduction but also simplification of cellular processes such as transcription.

Microsporidioses are emerging and opportunistic infections in humans and animals (Didier 2005). The emergence of disease is more likely in freshwater habitats which interconnect with marine and terrestrial habitats under pressure from human activity (Didier and Weiss 2011).

Thus, microsporidia are excellent models for eukaryotic genome evolution and gene expression in the context of host-pathogen relationships. The suggestion that some of their adaptations are shared with the metabolic properties of invasive tumours has scope to unravel the challenging biochemical puzzles of these highly significant biotrophs in clinical application.

7.8 Further research avenues

- Have microsporidia arisen from host species in terrestrial or aquatic habitats?
- Is the chitinous wall needed to generate turgor pressure during the polarized movement of the protoplasm?
- Will full gene sequences for freshwater hosts and their microsporidian endoparasites facilitate investigation and an understanding of the emergence of this group of pathogens at all levels – individual, community, ecosystem and global distribution?
- More complete data on prevalence and severity of infections, with regard to fish age, geographic distribution and time of year, are required to determine potential mortality rates.
- Further research is needed to develop new and appropriate control and regulations for drinking, wastewater and recreational waters to avoid health risks from these pathogens.
- Which environmental factors may promote future adaptations of microsporidia to novel host species for emergent disease in freshwater?
- Research into phylogenetics to investigate intriguing links between some aquatic pathogens and those infecting humans.
- Which mechanisms govern high tissue/organ specificity in many freshwater microsporidian taxa?

References

Adl SM, Simpson AGB, Farmer MA, Andersen RA, Anderson OR, Barta JR, et al. The new higher level classification of Eukaryotes with emphasis on the taxonomy of Protists. J Eukar Microbiol 2005;52:399–451.

Azevedo C. Ultrastructural aspects of a new species, *Vavraia mediterranica* (Microsporidia, Pleistophoridae), parasite of the French Mediterranean shrimp, *Crangon crangon* (Crustacea, Decapoda). J Invert Path 2001;78:194–200.

Azevedo C, Corral L, Vivares CP. Ultrastructure of the microsporidian *Inodosporus octospora* (Thelohaniidae), a parasite of the shrimp *Palaemon serratus* (Crustacea, Decapoda). Dis Aquat Org 2000;41:151–158.

Becnel JJ, White SE, Shapiro AM. Review of microsporidia-mosquito relationships: from the simple to the complex. Folia Parasitol (Praha) 2005;52:41–50.

Blackwell M. The fungi: 1, 2, 3 ... 5.1 million species? Am J Bot 2011;98:426–438.

Bouzahzah B, Nagajyothi F, Ghosh K, Takvorian PM, Cali A, Tanowitz HB, Weiss LM. Interactions of *Encephalitozoon cuniculi* polar tube proteins. Infect Immun 2010;78:2745–2753.

Breitenmoser AC, Mathis A, Burgi E, Weber R, Deplazes P. High prevalence of *Enterocytozoon bieneusi* in swine with four genotypes that differ from those identified in humans. Parasitol 1999;118:447–453.

Bronnvall AM, Larsson JIR. Ultrastructure and light microscopic cytology of *Agglomerata lacrima* n. sp. (Microspora, Duboscqiidae) a microsporidian parasite of *Acanthocyclops vernalis* (Copepoda, Cyclopidae). Eur J Protistol 2001;37:89–101.

Brosson D, Kuhn L, Prensier G, Vivarès CP, Texier C. The putative chitin deacetylase of *Encephalitozoon cuniculi*: a surface protein implicated in microsporidian spore-wall formation. FEMS Microbiol Lett 2005;247:81–90.

Burri L, Williams BA, Bursac D, Lithgow T, Keeling PJ. Microsporidian mitosomes retain elements of the general mitochondrial targeting system. Proc Natl Acad Sci USA 2006;103:15916–15920.

Cali A, Kotler DP, Orenstein JANM. *Septata intestinalis* n. g., n.sp., an intestinal microsporidian associated with chronic diarrhea and dissemination in AIDS patients. J Eukaryot Microbiol 1993;40:101–112.

Capella-Gutiérrez S, Marcet-Houben M, Gabaldón T. Phylogenomics supports microsporidia as the earliest diverging clade of sequenced fungi. BMC Biol 2012;10:47–60.

Cavalier-Smith T. The kingdom Protozoa and its 18 phyla. Microbiol Rev 1993;57:953–994.

Cavalier-Smith T. A revised six-kingdom system of life. Biol Rev 1998;73:203–266.

Chaimanee V, Warrit N, Chantawannakul P. Infections of *Nosema ceranae* in four different honeybee species. J Invert Pathol 2010;105:207–210.

Clotilde-Ba FL, Toguebaye BS. Infection of *Penaeus monodon* (Fabricius, 1798) (Crustacea, Decapoda, Penaeidae) by *Agmasoma penaei* (Microspora, Thelohaniidae) in Senegal, West Africa. Bull Eur Ass Fish Pathol 2001;21:157–159.

Corradi N, Slamovits CH. The intriguing nature of microsporidian genomes. Brief Funct Genom 2011;10:115–124.

Corradi N, Keeling, P J. Microsporidia: a journey through radical taxonomic revisions. Fungal Biol Rev 2009;23:1–8.

Cotte L, Rabodonirina M, Chapuis F, Bailly F, Bissuel F, Raynal C, et al. Waterborne outbreak of intestinal microsporidiosis in persons with and without human immunodeficiency virus infection. J Infect Dis 1999;180:2003–2008.

Couzinet S, Cejas E, Schittny J, Deplazes P, Weber R, Zimmerli S. Phagocytic uptake of *Encephalitozoon cuniculi* by nonprofessional phagocytes. Infect Immun 2000;68:6939–6945.

Cuomo CA, Desjardins CA, Bakowski MA, Goldberg J, Ma AT, Becnel JJ, et al. Microsporidian genome analysis reveals evolutionary strategies for obligate intracellular growth. Genome Res 2012;22:2478–2488.
Delgado T, Carroll PA, Punjabi AS, Margineantu D, Hockenbery DM, Lagunoff M. Induction of the Warburg effect by Kaposi's sarcoma herpesvirus is required for the maintenance of latently infected endothelial cells. Proc Natl Acad Sci 2010;107:10696–10701.
Dengjel B, Zahler M, Hermanns W, Heinritzi K, Spillmann T, Thomschke A, et al. Zoonotic potential of *Enterocytozoon bieneusi*. J Clin Micro 2001;39:4495–4499.
Desportes I, Nashed NN. Ultrastructure of sporulation in *Minchinia dentali* (Arvy), an haplosporean parasite of *Dentalium entale* (Scaphopoda, Mollusca); taxonomic implications. Protistology 1983;19:435–460.
Diarra K, Toguebaye BS. Ultrastructure of *Nosema stenocypris* (Diarra & Toguebaye), 1994, a microsporidian parasite of *Stenocypris major* (Crustacea, Ostracoda, Cyprididae). Archiv Protisten 1996;146:363–367.
Didier ES. Microsporidiosis: an emerging and opportunistic infection in humans and animals. Acta Trop 2005;94: 61–76.
Didier ES, Maddry JA, Brindley PJ, Stovall ME, Didier PJ. Therapeutic strategies for human microsporidia infections. Expert Rev Anti Infect Ther 2005;3:419–434.
Didier ES, Weiss LM. Microsporidiosis: not just in AIDS patients. Curr Opin Infect Dis 2011;24:490–495.
Didier ES, Weiss LM. Microsporidiosis: current status. Curr Opin Infect Dis 2006; 19:485–492.
Dolgikh VV, Senderskiy IV, Pavlova OA, Naumov AM, Beznoussenko GV. Immunolocalization of an alternative respiratory chain in *Antonospora (Paranosema) locustae* spores: Mitosomes retain their role in microsporidial energy metabolism. Eukaryot Cell 2011;10:588–593.
Dowd SE, Gerba CP, Pepper IL. Confirmation of the human-pathogenic microsporidia *Enterocytozoon bieneusi*, *Encephalitozoon intestinalis*, and *Vittaforma corneae* in water. Appl Environ Microbiol 1998;64:3332–3335.
Dunn AM, Smith JE. Microsporidian life cycles and diversity: the relationship between virulence and transmission. Microbes Infect 2001;3:381–388.
Dunn AM, Terry RS, Smith JE. Transovarial transmission in the microsporidia. Adv Parasitol 2001;48:57–100.
Ebersberger I, de Matos Simoes R, Kupczok A, Gube M, Kothe E, Voigt K, von Haeseler A. A consistent phylogenetic backbone for the Fungi. Mol Biol Evol 2012;29:1319–1334.
Edlind TD, Li J, Visvesvara GS, Vodkin MH, Mclaughlin GL, Katiyar SK. Phylogenetic analysis of β-tubulin sequences from amitochondrial protozoa. Mol Phylogenet Evol 1996;5:359–367.
Fischer WM, Palmer JD. Evidence from small-subunit ribosomal RNA sequences for a fungal origin of Microsporidia. Mol Phylogenet Evol 2005;36:606–622.
Franzen C. Microsporidia: how can they invade other cells? Trends Parasitol 2004;20:275–279.
Franzen C. How do microsporidia invade cells? Folia Parasitol (Praha) 2005;52:36–40.
Freeman MA, Bell AS, Sommerville C. A hyperparasitic microsporidian infecting the salmon louse, *Lepeophtheirus salmonis*: an rDNA-based molecular phylogenetic study. J Fish Dis 2003;26:667–676.
Freeman MA, Turnbull JF, Yeomans WE, Bean CW. Prospects for management strategies of invasive crayfish populations with an emphasis on biological control. Aquat Conserv: Marine Freshw Ecostysm 2010;20:211–223.
Galván AL, Magnet A, Izquierdo F, Fenoy S, Rueda C, Vadillo CF, et al. Molecular characterization of human-pathogenic Microsporidia and *Cyclospora cayetanensis* isolated from various water sources in Spain: A year-long longitudinal study. Appl Environ Microbiol 2013;79:449–459.
Gill EE, Fast NM. Assessing the microsporidia-fungi relationship: combined phylogenetic analysis of eight genes. Gene 2006;375:103–109.

Gleason FH, Carney LT, Lilje O, Glockling S. Ecological potentials of species of Rozella (Cryptomycota). Fung Ecol 2012;5:651–656.

Goldberg AV, et al. Localization and functionality of microsporidian iron-sulphur cluster assembly proteins. Nature 2008;452:624–628.

Gill EE, Lee RC, Corradi N, Grisdale CJ, Limpright VO, Keeling PJ, et al. Splicing and transcription differ between spore and intracellular life stages in the parasitic microsporidia. Mol Biol Evol 2010;27:1579–1584.

Gray MW. Mitochondrial Evolution doi: 10.1101/cshperspect.a011403 Cold Spring Harbor Laboratory Press. 2012; Accessed 16.8.13 http://www.cshperspectives.net/content/4/9/a011403.full.

Hibbett DS, Binder M, Bischoff JF, et al. A higher-level phylogenetic classification of the Fungi. Mycol Res 2007;111:509–547.

Hirt RP, Logsdon Jr JM, Healy B, Dorey MW, Doolittle WF, Embley TM. Microsporidia are related to fungi: evidence from the largest subunit of RNA polymerase II and other proteins. Proc Natl Acad Sci USA 1999;96:580–855.

Izquierdo F, Castro Hermida JA, Fenoy S, Mezo M, Gonzalez-Warleta M, del Aguila C. Detection of microsporidia in drinking water, wastewater and recreational rivers. Water Res 2011;45:4837–4843.

James TY, Kauff F, Schoch CL, Matheny PB, Hofstetter V, Cox CJ, et al. Reconstructing the early evolution of Fungi using a six-gene phylogeny. Nature 2006;443:818–822.

James TY, Pelin A, Bonen L, Ahrendt S, Sain D, Corradi N, et al. Shared signatures of parasitism and phylogenomics unite Cryptomycota and Microsporidia. Curr Biol 2013;23: 1–6.

Johnson RM, Evans JD, Robinson GE, Berenbaum MR. Changes in transcript abundance relating to colony collapse disorder in honey bees (*Apis mellifera*). Proc Natl Acad Sci USA 2009;106:14790–14795.

Karpov SA, Mikhailov KV, Mirzaeva GS, Mirabdullaev IM, Mamkaeva KA, Titova NN, et al. Obligately phagotrophic aphelids turned out to branch with the earliest-diverging fungi. Protist 2013;164:195–205.

Katinka MD, Duprat S, Cornillot E, Metenier G, Thomarat F, Prensier G, et al. Genome sequence and gene compaction of the eukaryote parasite *Encephalitozoon cuniculi*. Nature 2001;414: 450–453.

Keeling PJ. Congruent evidence from alpha-tubulin and beta-tubulin gene phylogenies for a zygomycete origin of microsporidia. Fungal Genet Biol 2003;38:298–309.

Keeling PJ. Five things to know about Microsporidia. PLoS Pathog 2009;5:e1000489. doi:10.1371/journal.ppat.1000489. Accessed 16.8.13. http://www.plospathogens.org/article/info%3Adoi%2F10.1371%2Fjournal.ppat.1000489 http://genome.cshlp.org/cgi/ijlink?linkType=ABST&journalCode=pnas&resid=106/35/14790.

Keeling PJ. Endosymbiosis: Bacteria sharing the load. Curr Biol 2011;21:R623–R624.

Keeling PJ, Burger G, Durnford DG, Lang BF, Lee RW, Pearlman RE, Roger AJ, Gray MW. The tree of eukaryotes. Trends Ecol Evol 2005;20:670–676.

Keeling PJ, Corradi N. Shrink it or lose it: Balancing loss of function with shrinking genomes in the microsporidia. Virulence 2011;2:67–70.

Keeling PJ, Luker MA, Palmer JD. Evidence from beta-tubulin phylogeny that Microsporidia evolved from within the fungi. Mol Biol Evol 2000;17:23–31.

Keeling P J, McFadden GI. Origins of microsporidia. Trends Microbiol 1998;6:19–23.

Keeling PJ, Doolittle WF. Alpha-tubulin from early-diverging eukaryotic lineages and the evolution of the tubulin family. Mol Biol Evol 1996;13:1297–1305.

Keeling PJ, Fast NM. Microsporidia: Biology and evolution of highly reduced intracellular parasites. Ann Rev Microbiol 2002;56:93–116.

Keeling PJ, Williams BAP, Law J, Fast NM, Slamovits CH. Comparative genomics of microsporidia. Folia Parasitol 2005;52:8–14.

Kiryu Y, Behringer DC, Landsberg JH, Petty BD. Microsporidiosis in the Caribbean spiny lobster *Panulirus argus* from southeast Florida, USA. Dis Aquat Org 2009;83:237–242.
Koestler T, Ebersberge I. Zygomycetes, Microsporidia, and the evolutionary ancestry of sex determination. Genome Biol Evol 2011;3:186–194.
Kohler SL, Hoiland WK. Population regulation in an aquatic insect: the role of disease. Ecology 2001;82:2292–2305.
Kudo RR. *Protozoology*. 1966; 5th edition. Springfield, IL: Charles C. Thomas.
Lee SC, Corradi N, Byrnes EJ, Torres-Martinez S, Dietrich FS, Keeling PJ, et al. Microsporidia evolved from ancestral sexual fungi. Curr Biol 2008;18:1675–1679.
Levine ND, Corliss JO, Cox FE, Deroux G, Grain J, Honigberg BM, et al. A newly revised classification of the protozoa. J Protozool 1980;27:37–58.
Li Z, Pan G, Li T, Huang W, Chen J, Geng L, Yang D, et al. SWP5, a spore wall protein, interacts with polar tube proteins in the parasitic microsporidian *Nosema bombycis*. Eukaryotic Cell 2012;11:229–237.
Lightner DV. A handbook of shrimp pathology and diagnostic procedures for disease control of cultured penaeid shrimp. Baton Rouge, LA World Aquaculture Society. 1996.
Lobo ML, Xiao L, Antunes F, Matos O. Microsporidia as emerging pathogens and the implication for public health: a 10-year study on HIV-positive and -negative patients. Int J Parasitol 2012;42:197–205.
Lom J, Nilsen F. Fish microsporidia: fine structural diversity and phylogeny. Int J Parasitol 2003;33:107–127.
Lom J, Vávra J. A proposal to the classification within the subphylum Cnidospora. Syst Zool 1966; 11:172–175.
Lucarotti CJ, Andreadis TG. Reproductive strategies and adaptations for survival among obligatory microsporidian and fungal parasites of mosquitos – a comparative analysis of *Ambylospora* and *Coelomomyces*. J Am Mosq Control Assoc 1995;11:111–121.
Martinez MA, Vivares CP, Rocha RD, Fonseca AC, Andral B, Bouix G. Microsporidiosis on *Artemia* (Crustacea, Anostraca) Light and electron microscopy of *Vavraia anostraca* sp. nov. (Microsporidia: Pleistophoridae) in the Brazilian Solar Salterns. Aquaculture 1992;107: 229–237.
Matos O, Lobo ML, Xiao L. Epidemiology of *Enterocytozoon bieneusi* Infection in Humans. Parasitol Res. 2012; v.2012; PMC3469256. Accessed 16.8.13 http://www.ncbi.nlm.nih.gov/pmc/articles/PMC3469256/
Metenier G, Vivares CP. Molecular characteristics and physiology of microsporidia. Microbes Infect 2001;3:407–415.
Mori H, Mahittikorn A, Watthanakulpanich D, Komalamisra C, Sukthana Y. Zoonotic potential of *Enterocytozoon bieneusi* among children in rural communities in Thailand. Parasite 2013;20:14.
Mossallam SF. Detection of some intestinal protozoa in commercial fresh juices. J Egypt Soc Parasitol 2010;40:135–149.
Ni M, Feretzaki M, Sun S, Wang X, Heitman J. Sex in fungi. Annu Rev Genet 2011;45:405–430.
Nylund S, Andersen L, Saevareid I, Plarre H, Watanabe H, Arnesen CE, et al. Diseases of farmed Atlantic salmon *Salmo salar* associated with infections by the microsporidian *Paranucleospora theridon* Dis Aquat Org 2011;94:41–57.
Olson RE, Tiekotter KL, Reno PW. *Nadelspora canceri* n.g., n.s. an unusual microsporidian parasite of the Dungeness crab, *Cancer magister*. J Euk Microbiol 1994;41:349–359.
Peuvel-Fanget I, et al. EnP1 and EnP2, two proteins associated with the *Encephalitozoon cuniculi* endospore, the chitin-rich inner layer of the microsporidian spore wall. Int J Parasitol 2006;36:309–318.
Redhead SA, Kirk P, Keeling PJ, Weiss LM. Proposals to exclude the phylum Microsporidia from the Code. Taxon 2009a;58:10–11.

Redhead SA, Kirk PM, Keeling PJ, Weiss LM. Proposals 048-051 to amend the International Code of Botanical Nomenclature. Mycotaxon 2009b;108:505–507.

Redhead SA, Cushion MT, Frenkel JK. Proposals to exclude the phylum Microsporidia from the Code in J Eukaryot Microbiol 2006;53:8.

Refardt D, Canning EU, Mathis A, Cheney SA, Lafranchi-Tristem NJ, Ebert D. Small subunit ribosomal DNA phylogeny of microsporidia that infect *Daphnia* (Crustacea: Cladocera). Parasitologia 2002;124:381–389.

Slifko TR, Smith HV, Rose JB. Emerging parasite zoonoses associated with water and food. Int J Parasitol 2000;30:1379–1393.

Smith JE. The ecology and evolution of microsporidian parasites. Parasitol 2001;136:1901–1914.

Sprague V, Couch J. An annotated list of protozoan parasites, hyperparasites and commensals of decapod crustacea. J Protozool 1971;18:526–537.

Sprague V, Vávra J. Systematics of the microsporidia. *In*: (LA Bulla, CC Cheng, eds) Comparative Pathobiology. New York, NY: Pleum Press 1977;1 – 150.

Stark D, van Hal S, Barratt J, Ellis J, Marriott D, Harkness J. Limited genetic diversity among genotypes of *Enterocytozoon bieneusi* strains isolated from HIV-infected patients from Sydney, Australia. J Med Microbiol 2009;58:355–357.

Stentiford GD, Bateman KS. *Enterospora* sp., an intranuclear microsporidian infection of hermit crab *Eupagurus bernhardus*. Dis Aquat Org 2007;75:61–72.

Stentiford GD, Bateman KS, Longshaw M, Feist SW. *Enterospora canceri* n. gen., n. sp., intranuclear within the hepatopancreas of the European edible crab *Cancer pagurus*. Dis Aquat Org 2007;75:61–72.

Stentiford GD, Bateman KS, Small HJ, Moss J, Shields JD, Reece KS, et al. *Myospora metanephrops* (n.g.,n.sp.) from marine lobsters and a proposal for erection of a new order and family (Crustaceacida;Myosporidae) in the class Marinosporidia (Phylum Microsporidia). Int J Parasitol 2010;40:1433–1446.

Stentiford GD, Bateman KS, Feist SW, Chambers E, Stone DM. Plastic parasites: extreme dimorphism creates a taxonomic conundrum in the phylum Microsporidia. Int J Parasitol 2013a;43: 339–352.

Stentiford GD, Feist SW, Stone DM, Bateman KS, Dunn AM. Microsporidia: diverse, dynamic, and emergent pathogens in aquatic systems. Trends Parasitol 2013b;29:567–578.

Stiller JW, Hall BD. Long-branch attraction and the rDNA model of early eukaryotic evolution. Mol Biol Evol 1999;16:1270–1279.

Terry RS, Smith JE, Sharpe RG, Rigaud T, Littlewood Dt J, Ironside JE, et al. Widespread vertical transmission and associated host sex-ratio distortion within the eukaryotic phylum Microspora. Proc R Soc Lond Series B: Biol Sci 2004;271:1783–1789.

Texier C, Vidau C, Viguès B, El Alaoui H, Delbac F. Microsporidia: a model for minimal parasite-host interactions. Curr Opin Microbiol 2010;13:389–391.

Thomarat F, Vivares CP, Gouy M. Phylogenetic analysis of the complete genome sequence of *Encephalitozoon cuniculi* supports the fungal origin of microsporidia and reveals a high frequency of fast-evolving genes. J Mol Evol 2004;59:780–791.

Toubiana M, Guelorget O, Bouchereau JL, Lucien-Brun H, Marques A. Microsporidians in *penaeid* shrimp along the west coast of Madagascar. Dis Aquat Org 2004;58:79–82.

Tourtip S, Wongtripop S, Stentiford GD, Bateman KS, Sriurairatana S, Chavadej J, et al. *Enterocytozoon hepatopenaei* sp. nov. (Microsporida: Enterocytozoonidae), a parasite of the black tiger shrimp *Penaeus monodon* (Decapoda: Penaeidae): fine structure phylogenetic relationships. J Invert Path 2009;102:21–29.

Troemel ER, Felix MA, Whiteman NK, Barriere A, Ausubel FM. Microsporidia are natural intracellular parasites of the nematode *Caenorhabditis elegans*. PLoS Biol 2008;6:2736–2752.

Troemel ER. New models of microsporidiosis: Infections in zebrafish, *C. elegans*, and honey bee. PLoS Pathog. 2011; 7: e1001243. doi: 10.1371/journal.ppat.1001243. Accessed 16.8.13 http://www.plospathogens.org/article/info%3Adoi%2F10.1371%2Fjournal.ppat.1001243

Tsaousis AD, Kunji ER, Goldberg AV, Lucocq JM, Hirt RP, Embley TM. A novel route for ATP acquisition by the remnant mitochondria of *Encephalitozoon cuniculi*. Nature 2008;453:553–556.

US Environmental Protection Agency 2005. Contaminant candidate list 2 and regulatory determinations. US Environmental Protection Agency, Washington, DC: Accessed 16.8.13 http://water.epa.gov/scitech/drinkingwater/dws/ccl/ccl2.cfm.

US Environmental Protection Agency 2009. Contaminant candidate list 3. US Environmental Protection Agency, Washington, DC: Accessed 16.8.13 http://water.epa.gov/scitech/drinkingwater/dws/ccl/ccl3.cfm.

Vander Heiden MG, Cantley LC, Thompson CB. Understanding the Warburg effect: The metabolic requirements of cell proliferation. Science 2009;324:1029–1033.

Vávra J, Lukeš J. Microsporidia and "the art of living together". Adv Parasitol 2013;82:254–319.

Voronin VN. Ultrastructure and horizontal transmission of *Gurleya macrocyclopis* (Protozoa, Microspora) to *Macrocylops albidus* (Crustacea, Copepoda). J Invert Path 1996;67:105–107.

Vossbrinck CR, Debrunner-Vossbrinck BA. Molecular phylogeny of Microsporidia: ecological, ultrastructural and taxonomic considerations. Folio Parasitologica 2005;52:131–142.

Vávra J, Hyliš M, Obornik M, Vossbrink CR. Microsporidia in aquatic microcrustacea: the copepod Microsporidium *Marssoniella elegans* Lemmermann, 1990 revisited. Folia Parasitol 2005;52:163–172.

Weidner E, Findley AM, Dolgikh V, Sokolova J. Microsporidian biochemistry and physiology. *In*: (M Wittner LM Weiss, eds) *The microsporidia and microsporidiosis*. Washington (D. C.): ASM Press 1999.

Wittner M, Weiss LM. *The Microsporidia and Microsporidiosis*. 1999; Publisher: Amer Society for Microbiology ISBN-13: 978-1555811471 Accessed 16.8.13 http://www.amazon.com/The-Microsporidia-Microsporidiosis-Wittner-Weiss/dp/1555811477

Williams BA. Unique physiology of host-parasite interactions in microsporidia infections. Cell Microbiol 2009;11:1551–60.

Williams BA, Hirt RP, Lucocq JM, Embley TM. A mitochondrial remnant in the microsporidian *Trachipleistophora hominis*. Nature 2002;418:865–869.

Williams BA, Haferkamp I, Keeling PJ. An ADP/ATP-specific mitochondrial carrier protein in the microsporidian *Antonospora locustae*. J Mol Biol 2008;375:1249–1257.

Wolinska J, Giessler S, Koerner H. Molecular identification and hidden diversity of novel *Daphnia* parasites from European lakes. Appl Environ Microbiol 2009;75:7051–7059.

Wu Z, Li Y, Pan G, Zhou Z, Xiang Z. SWP25, a novel protein associated with the *Nosema bombycis* endospore. J. Eukaryot Microbiol 2009;56:113–118.

Wu Z, Li Y, Pan G, Tan X, Hu J, Zhou Z, Xiang Z. Proteomic analysis of spore wall proteins and identification of two spore wall proteins from *Nosema bombycis* (Microsporidia). Proteomics 2008;8:2447–2461.

Xu Y, Weiss LM. The microsporidian polar tube: a highly specialised invasion organelle. Int J Parasitol 2005;35:941–953.

Agostina V. Marano, Ana L. Jesus, Carmen L. A. Pires-Zottarelli, Timothy Y. James, Frank H. Gleason and Jose I. de Souza

8 Phylogenetic relationships of Pythiales and Peronosporales (Oomycetes, Straminipila) within the "peronosporalean galaxy"

8.1 Introduction

The Oomycetes, also called Peronosporomycetes (Dick 2001), are a heterogeneous assemblage of fungal-like eukaryotes that produce heterokont biflagellate zoospores and belong to phylum Oomycota, kingdom Straminipila (syn. Chromista *sensu* Cavalier-Smith and Chao 2006). These heterotrophic straminipiles are placed in the supergroup Chromalveolata (Cavalier-Smith 2004; Adl et al. 2005) together with other chromists (cryptomonads and haptophytes) and the alveolates (apicomplexans, ciliates, perkinsids and dinoflagellates) (Harper et al. 2005; Mangot et al. 2011; Marano et al. 2012).

Oomycetes have long been considered to be within the Kingdom Fungi, due to similarities in the production of complex hyphal systems, absorptive mode of nutrition and ecological niches (Richards et al. 2006). However, from the last part of the 20th century, molecular analyses have started to reveal strong evidence that Oomycetes are phylogenetically distant from members of the true fungi and are more closely related to brown algae and diatoms (Beakes et al. 2012). Major differences between Oomycetes and zoosporic true fungi are: 1) the state (ploidy) of vegetative mycelium, which is diploid in Oomycetes; 2) the zoospore ultrastructure and type of flagellation; 3) the composition of the cell wall, which is cellulosic in Oomycetes; and 4) the presence of tubular cristae in the mitochondria of Oomycetes (Rossman and Palm 2006). Recent findings suggest that lateral (horizontal) gene transfer (HGT), which occurred between the Oomycetes and true fungi (Richards et al. 2011) might have resulted in the acquisition of fungal-like characteristics, such as the fungal-like growth in Oomycetes (Richards et al. 2006), osmotrophy and pathogenesis (Soanes et al. 2007; Richards et al. 2011). Richards et al. (2011) have detected 34 high-confidence HGTs between Oomycetes and true fungi that together contributed up to ~8% of the secretome of *Phytophthora ramorum*. In addition, convergent mechanisms of sexual pheromone synthesis in the mucoralean fungi and *Phytophthora* have recently been discovered (Lee et al. 2012).

8.2 The monophyly of Chromalveolata and the relationships between heterotrophic straminipile lineages

The "supergroup" Chromalveolata (chromalveolates) contains a large fraction of the eukaryotic diversity. The monophyletic origin of this supergroup is currently controversial. Some authors (Cavalier-Smith 1999; Harper et al. 2005; Keeling 2009; Baurain et al. 2010) proposed a monophyletic origin based on the hypothesis of a flagellated cell with a red algal endosymbiont as the least common ancestor, which was originated by a single secondary endosymbiotic event. Consequently, plastid-lacking lineages within Chromalveolata, such as the Oomycetes, have lost their plastids secondarily and independently but retained some photosynthetic genes (Martens et al. 2008). Other authors (Burki et al. 2008; Hampl et al. 2009), on the contrary, support a polyphyletic origin of the supergroup by independent events of acquisition of secondary plastids among different lineages. These lineages, supported by molecular data, include the alveolates (apicomplexans, ciliates, dinoflagellates), which are also grouped by the presence of cortical alveoli underlying their plasma membranes (Hausmann et al. 2003), and the chromists (straminipiles, cryptomonads and haptophytes) (Baldauff 2008). The RAS group, which was recently recognized, includes the Rhizaria and Chromalveolata (alveolates + straminipiles) and is sister to the haptophytes (Burki et al. 2012). The cryptomonads contain a relict nucleus of the red-algal endosymbiont termed nucleomorph (Lane and Archivald 2006) and its phylogenetic placement is still not resolved (Harper et al. 2005). It has been recently suggested that inconsistencies in the systematics of chromalveolates can be due to acquisition of photosynthetic genes by lateral gene transfer instead of *via* endosymbiosis (Kemen and Jones 2012).

The straminipiles share the presence of flagella with rows of stiff, tripartite hairs (Baldauff 2008). However, absence of hairs on the flagellum has been documented in some of the basal lineages (Beakes et al. 2012). These eukaryotes are composed of members that present diverse ecological niches and lifestyles ranging from photosynthetic diatoms and brown algae to filamentous heterotrophic oomycetes, unicellular and colonial labyrinthulids and hyphochytrids. Heterotrophic lineages within the straminipiles include proteromonads, labyrinthulids, bicosoecids, oomycetes, perkinsids and the genus *incertae sedis Developayella* (Tong 1995; Mangot et al. 2011).

Although the whole straminipiles appear to be monophyletic, there is no consensus about relationships between the lineages of non-photosynthetic straminipiles, such as labyrinthulids and oomycetes (Tsui et al. 2009). Some studies based on rDNA phylogenies supported their monophyly, with independent loss of chloroplasts in oomycetes and labyrinthulids, while others postulated a single loss in the common ancestor of both groups (Oudot-Le Secq et al. 2006). Oomycetes and the photosynthetic Ochrophyta formed a well-supported monophyletic clade, which is sister to the earliest diverging Bigyra (bicoeceans + Labyrinthulomycetes) (Tsui et al. 2009). Independent loss of chloroplasts in the photosynthetic ochrophytes appears as a more plausible hypothesis than independent gains. If we consider that the straminipile ancestor was photosynthetic, independent losses of photosynthesis in Bigyra might have occurred first and subsequently in the oomycetes (Tsui et al. 2009).

8.3 Major lineages within the Oomycetes: the "galaxies"

The monophyletic Oomycetes (Beakes 1987; Riethmüller et al. 1999; Hudspeth et al. 2000) currently comprise ~956 species (Kirk et al. 2008), which can be assigned to two informally defined groupings termed "galaxies" (Sparrow 1976; Dick et al. 1984). These galaxies were formally defined as subclass rank (Saprolegniomycetidae and Peronosporomycetidae) by Dick et al. (1999). The "saprolegnian galaxy," traditionally comprised a single order, Saprolegniales, while the "peronosporalean galaxy," included the orders Leptomitales, Sclerosporales, Rhipidiales, Pythiales and Peronosporales. However, the Leptomitales and Sclerosporales were transferred to the "saprolegnian galaxy" based on DNA phylogenies (Klassen et al. 1988; Dick et al. 1989).

The existence of these two major lineages was more recently confirmed based on SSU and LSU rDNA phylogenies (Petersen and Rosendhal 2000; Lara and Belbahri 2011; Marano et al. 2012). Some physiological and morphological characteristics also support the existence of these subclasses: 1) Pythiales are not able to synthesize sterols *de novo* while the Leptomitales and Saprolegniales can; 2) K-bodies are absent in Peronosporomycetidae and 3) centripetal oosporogenesis results in oospores with persistent periplasm in Peronosporomycetidae (Beakes and Sekimoto 2009). A third subclass, the Rhipidiomycetidae, was proposed by Dick et al. (1989) and Dick (1995) to accommodate the members of the Rhipidiales, and was later confirmed by Riethmüller et al. (2002) based on SSU rDNA phylogeny. This subclass appears to be more closely related to the Saprolegniomycetidae (Riethmüller et al. 2002).

8.4 The "peronosporalean galaxy": a marine origin?

Current evidence suggests that Oomycetes have most likely evolved in a marine environment and migrated to land along with their hosts (Beakes and Sekimoto 2009). Most of the members of the earliest diverging clades (such as *Halodaphnea*, *Haliphthoros* and *Eurychasma dicksonii*), which are commonly called "basal Oomycetes", are predominantly obligate parasites of diverse marine microorganisms (Beakes and Sekimoto 2009). However, since some genera such as *Halophytophthora* and *Salisapilia* are exclusively saprophytes, it has been hypothesized that some marine lineages have reverted to a saprotrophic lifestyle (Hulvey et al. 2010) or they had reacclimatized to the marine or estuarine environment (Nakagiri 2002).

Dick (2001), on the contrary, proposed a freshwater or terrestrial origin since most of the "crown oomycetes" (*sensu* Beakes and Sekimoto 2009) are either freshwater (saprolegnian galaxy) or terrestrial (peronosporalean galaxy). Barr (1983) suggested that obligate parasites of terrestrial plants evolved from an aquatic saprolegnian saprobe (e.g. *Achlya* and *Saprolegnia*) through a primitive *Pythium* or *Phytophthora*. Terrestrial facultative soil-borne pathogens of plants developed from these "primitive" aquatic and saprotrophic genera and later obligate parasites of the aerial parts of plants, such as *Albugo* and *Bremia*, evolved. This is supported by the gradual reduction of zoospore stages and on the dependency on water for the release and dispersal of zoospores during evolution. In addition, Beakes (1987) proposed Saprolegniales as the ancestral group due to the similarity between *Saprolegnia* and the alga *Vaucheria*.

Obligate biotrophic genera, which are considered to have evolved from facultative pathogens or more specialized hemibiotrophs in *Phytophthora*, are currently considered to be placed in the Peronosporales while the facultative pathogens with broad host range are included within the order Pythiales (Hudspeth et al. 2003). Considering that Albuginales is basal to all members of the "peronosporalean galaxy" a reversal adaptation to the marine environment appears as the most plausible hypothesis for the marine representatives such as *Halophytophthora* (Thines et al. 2009).

8.5 Ecological and economical significance

The "peronosporalean galaxy" (~1,000 spp., Judelson 2012) is the most successful assemblage of Oomycetes in terms of number of recognized species and economic impact (Dick 2001). Their members are ubiquitous and usually found in a wide variety of terrestrial, freshwater, estuarine and marine habitats (Marano et al. 2011, 2012). The number of species that occur in freshwater is approximately 127 species: 87 in Pythiales and 40 in Peronosporales, which represents only 13% of the total number of species in this galaxy (www.mycobank.org; www.indexfungorum.org; Dick 2001). In freshwater microbial communities, they can be found as saprotrophs or parasites with great impact on plants, fishes, and amphibians of economical relevance (Marano et al. 2011). In the case of pathogens, lifestyles range from biotrophy to necrotrophy, obligate to facultative pathogenesis, and narrow to broad host ranges on plants or animals. Recently, some clinical findings have demonstrated that Oomycetes, particularly members of the genera *Pythium* and *Lagenidium*, represent an emerging but under-diagnosed group of human pathogens (Reinprayoon et al. 2013). *Pythium insidiosum* more frequently infects domestic animals such as dogs and horses (Mendoza 2009), but can also rarely infect humans, causing a chronic, pyogranulomatous disease, commonly called "pythiosis" (Phillips et al. 2008; See also Chapter 12), particularly when hemoglobinopathy syndromes are present (Thianprasit 1986). Unidentified species of *Lagenidium* have been responsible for corneal (Reinprayoon et al. 2013) and chronic soft tissue infections in humans (Grooters et al. 2004), diseases which are commonly called "lagenidiosis".

Members of *Phytophthora* are best known as pathogens of economically important crops and as invasive pathogens that cause extensive damage to natural forests (e.g. Brasier et al. 2005; Hansen 2007; Balci et al. 2010). Although *Phytophthora* assemblages are abundant and diverse in forested streams (Reeser et al. 2011), little is known about species that have not been associated with disease in nearby riparian ecosystems (Hansen et al. 2012). In the last few years, however, a considerable number of species from ITS clade 6 (Robideau et al. 2011), which are mostly aquatic, sterile and saprophytic but facultatively pathogenic to terrestrial plants, have been described (e.g. Brasier et al. 2003; Greslebin et al. 2005; Hansen et al. 2012; Oh et al. 2013).

The genus *Pythium* includes saprotrophic species in natural (such as grasslands, forests, freshwater courses and less frequently estuarine and marine habitats) and agricultural ecosystems, in which they are important soilborne plant pathogens (Schroeder et al. 2013). Most species are generalists and necrotrophic, causing damping-off and devastating mainly horticultural crops in early stages of development (Schroeder et al. 2013).

Obligate plant biotrophs in the Peronosporales ("downy mildews") and Albuginales ("white rusts") have been reported from a great variety of angiosperm hosts and can cause severe damage in several economically important plant families, such as the Brassicaceae (Thines and Volgmayr 2009). Members of both orders are mainly host family, host genus or host species specific (Thines et al. 2009). The obligate biotrophic dependence on angiosperms appears to have evolved independently in Peronosporaceae and Albuginales (Thines and Volgmayr 2009). In the "downy mildews", it appears that obligate biotrophy has evolved recently, because some of their basal members still exhibit characteristics commonly attributed to *Phytophthora* (Thines 2009).

8.6 The phylogeny of Pythiales and Peronosporales

In this chapter, we propose a reconstruction of the "peronosporalean galaxy" based on sequences of the complete ITS and partial LSU regions of the rDNA, particularly considering saprotrophic or facultatively-parasitic genera in the Pythiales and Peronosporales. For this purpose, we sequenced some members of the genera *Pythium, Phytophthora* and *Halophytophthora* and compared our data with published sequences from GenBank using BLASTn and the *Phytophthora* database (www.phytophthoradb.org) (Park et al. 2008). Original sequences and their best hits in both databases were used for phylogenetic reconstruction. In addition, each genus type species deposited in GenBank and species of *Pythium* from clades B, C, D, E, F, H and K (Robideau et al. 2011), *Phytophthora* from clades 1, 4, 6, 8 and 9 (Robideau et al. 2011; Kroon et al. 2012), *Halophytophthora* and other peronosporalean genera (*Lagenidium, Pythiogeton* and *Salisapilia*) were included. Members of the biotrophic Peronosporales and Albuginales were included only with the aim of testing the phylogenetic relatedness with saprotrophic and facultative parasitic Peronosporales-Pythiales.

Our ITS phylogeny showed that this galaxy is composed of three major clades (Fig. 8.1A): 1) the Albuginales clade, which is robustly supported (100% branch

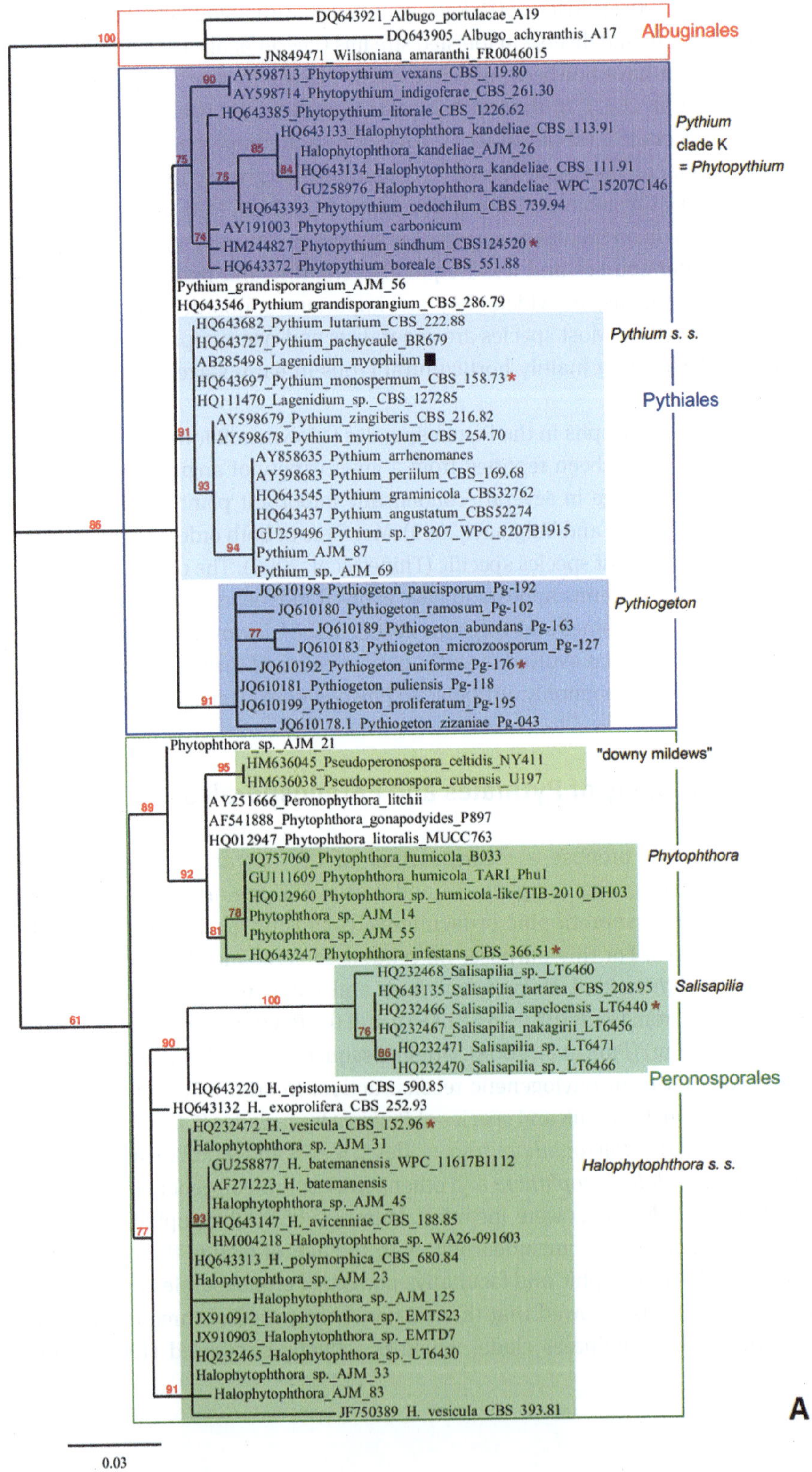

(Figure continued)

(*Figure continued*)

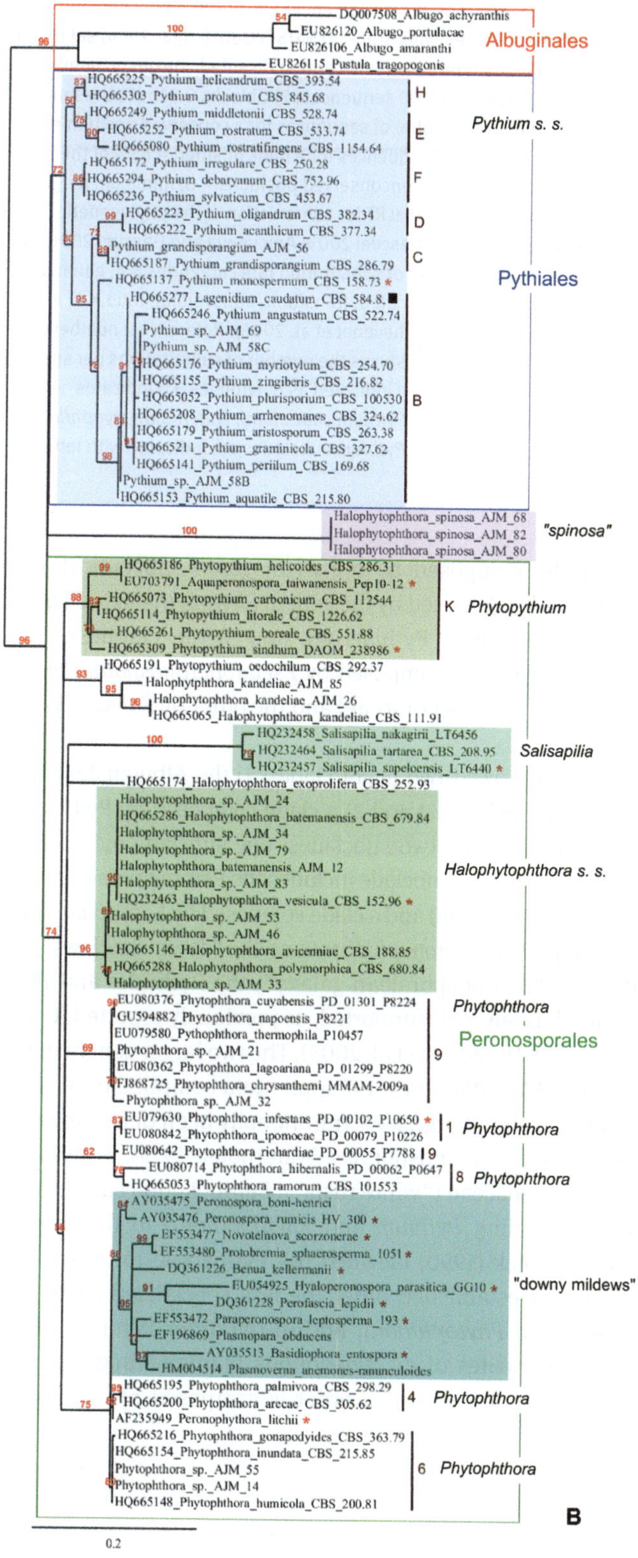

Fig. 8.1: Maximum likelihood tree inferred from ITS Fig. 8.1A and LSU Fig. 8.1B rDNA sequences of species in the "peronosporealean galaxy". The phylogeny.fr platform (Dereeper et al. 2008) on default mode ("One Click") was used. This mode comprises the following steps: 1) MUSCLE 3.7 (Edgar 2004) set to full processing mode for multiple sequence alignment, find diagonals option disabled, maximum number of iterations = 16, no duration limitation, no more than 200 sequences; 2) Gblocks 0.91b (Castresana 2000) for removal of ambiguous regions, minimum number of sequences for a conserved position = half the number of sequences + 1, minimum number of sequences for a flank position = 85% of the number of sequences, maximum number of contiguous nonconserved positions = 8, minimum length of a block = 10, allowed gap positions none; 3) PhyML 3.0 aLRT for construction of the phylogenetic trees by the maximum likelihood method (Guindon and Gascuel 2003; Anisimova and Gascuel 2006) set to substitution model HKY85, aLRT test = SH-like, number of substitution rate categories = 4, gamma parameter: estimated, proportion of invariable sites: estimated, transition/transversion ratio = 4; and 4) TreeDyn 198.3 for graphic representation of the trees (Chevenet et al. 2006). References: numbers next to branches indicate branch support (%) and the bar shows the number of substitutions per site. Branches with <50% of branch support were collapsed and indicated as polytomies. (*) indicates genus type species; (■) *Lagenidium caudatum* (currently *Pythium caudatum), Lagenidum myophilum* (currently *Salilagenidium myophilum*); *s. s: sensu stricto*. *Pythium* subclades are indicated with letters; *Phytophthora* subclades are indicated with numbers.

support), 2) the Pythiales clade, composed by the *Pythiogeton* subclade (family Pythiogetonaceae) and the subclade composed by the genera *Phytopythium* (= *Pythium* clade K) and *Pythium-Lagenidium* (family Pythiaceae), and 3) the Peronosporales clade, which includes two subclades, one composed by *Pseudoperonospora, Peronophythora* and *Phytophthora* (family Peronosporaceae) and the other by *Salisapilia* and *Halophytophthora sensu stricto*.

Our LSU phylogeny (Fig. 8.1B) showed two major clades: 1) the Albuginales clade, which is basal and also well-supported (96%) and 2) a clade composed by Pythiales + Peronosporales. Clade 2 is subdivided into two subclades; one subclade includes the Pythiales, represented by *Pythium*. Another subclade includes the Peronosporales, and is divided into *Pythium* clade K (= *Phytopythium*) and a clade composed by *Salisapilia, Halophytophthora sensu stricto, Phytophthora, Peronophythora* and the "downy mildews".

The main orders within the "peronosporalean galaxy", the Pythiales and Peronosporales, appeared to be moderately well supported, as previously reported (Cooke et al. 2000; Riethmüller et al. 2002; Hudspeth et al. 2003). The two larger genera in this galaxy, *Pythium* (~239 spp.) and *Phytophthora* (~157 spp.) (data compiled from www.indexfungorum.org and www.mycobank.org), have traditionally been placed together in the family Peronosporaceae, order Peronosporales (Waterhouse 1973). Waterhouse (1973) assigned the families Albuginaceae ("white rusts"), Peronosporaceae ("downy mildews"), and Pythiaceae (including *Pythium* and *Phytophthora*) to the Peronosporales. Dick et al. (1984) and Dick (1990) elevated the Pythiaceae to the level of order and included in this newly-established order the less specialized pathogens (facultative), such as *Pythium* and *Phytophthora*, and in the order Peronosporales all biotrophic genera that are parasites of dicotyledonous plants (Albuginaceae + Peronosporaceae). According to Dick (1990) this order is composed of one family, the Pythiaceae, which includes *Pythium, Phytophthora,* and *Halophytophthora.* Nevertheless, according to our results and some other studies (Cooke et al. 2000; Riethmüller et al. 2002; Kirk et al. 2008; Hulvey et al. 2010), *Phytophthora* and *Halophytophthora* should be still considered as part of the Peronosporales. Lara and Belbhari (2011) have previously suggested that *Phytophthora* should be removed from the Pythiales and placed in the Peronosporales because most species of *Phytophthora* included in their LSU rDNA phylogeny clustered with *Peronospora* rather than with *Pythium*, with a high support (100% bootstrap). Our ITS and LSU phylogenies (Fig. 8.1A and B) indicated that both *Halophytophthora* and *Phytophthora* belong to the order Peronosporales, but are placed in different subclades. Recently, the newly described genus *Salisapilia*, isolated from salt marshes, was placed in the Family Salisapiliaceae, order Peronosporales, based on similarities in oospore morphology, mycelium growth and sporangium formation with other members of this order (Hulvey et al. 2010). One of the three species included, *Salisapilia tartarea*, was transferred from *Halophytophthora*. Our results indicated that *Halophytophthora* and *Salisapilia* clustered together in the Peronosporales. The ITS phylogeny (Fig. 8.1A) suggested that they might belong to the same family. In this case, *Halophytophthora* should be included in the family Salisapiliaceae. Nevertheless, this result is not supported by our LSU phylogeny (Fig. 8.1B) and therefore, additional multigene phylogenies will be needed to confirm this assumption.

8.6.1 Clade 1: Albuginales

This group, commonly called "white rusts" is represented here by three genera (*Albugo*, *Wilsoniana* and *Pustula*) of non-culturable pathogens of angiosperms (Göker et al. 2007; Thines 2010). In our ITS and LSU phylogenies (Fig. 8.1A and B) it appears as a well-defined and supported clade (branch support 100% and 96%, respectively). Our LSU phylogeny (Fig. 8.1B) showed that this clade is ancestral to all members of the "peronosporalean galaxy", which justifies the circumscription of the order Albuginales, as previously demonstrated (Riethmüller et al. 2002; Thines and Spring 2005). Göker et al. (2007) suggested that biotrophy has evolved independently in the peronosporalean Oomycetes, such as the "white rusts" and "downy mildews". This assumption is supported by the basal phylogenetic position of the "white rusts", as compared to the "downy mildews", which are closely related to some facultative pathogens or hemibiotrophic members of *Phytophthora* (Fig. 8.1A and B).

8.6.2 Clade 2: Pythiales

In MycoBank (www.mycobank.org) the order Pythiales is composed of two families: Pythiaceae including the genera *Phytophthora, Pythium, Halophytophthora,*

Phytopythium, Peronophythora, Lagenidium and *Cystosiphon*; and Pythiogetonaceae with *Pythiogeton, Medusoides* and two genera *incertae sedis, Achlyopsis* and *Endosphaerium*. These last two genera, which are monotypic, have not been recorded again since described in 1896 and 1977, respectively, and their putative placement in the Pythiales is uncertain. In the case of the monotypic aero-aquatic genus *Medusoides* Voglmayr, since no asexual stage was found, the species *M. argyrocodium* could not be isolated into culture and there is absence of molecular information, its placement into the Pythiogetonaceae is speculative and more comprehensive studies need to be done before assigning this species to any known clade or family. *Cystosiphon* currently contains five species; one is the type species *C. pythioides*, and the rest were transferred by Dick (2001) from the genera *Pythium* (*C. dictyoporum*), *Lagenidium* (*C. canterae* and *C. closterii*) and *Lagenidiopsis* (*C. reductus*). Unfortunately, no molecular information is available to test its putative placement within the Pythiales.

8.6.2.1 *Pythiogeton*

This genus is composed of 16 species (www.mycobank.org), six of them recently described by Huang et al. (2013). Most species are saprophytic except *P. zeae* and *P. zizaniae*, which are pathogens of plants. Members of this genus have traditionally been placed in the family Pythiaceae because of the morphological similarities with *Pythium* (Hawksworth et al. 1995). However, zoospore discharge in *Pythiogeton* occurs through the formation of a very ephemeral vesicle, and posteriorly the naked protoplasma differentiates into zoospores (Sparrow 1960), in contrast to *Pythium*, and *Phytopythium*, in which differentiation into zoospores occurs inside a conspicuous vesicle. Because of these differences, Dick (2001) created a new family, Pythiogetonaceae to accommodate this genus and the newly-described genus *Medusoides*. Our results based on the ITS phylogeny showed that members of *Pythiogeton* formed a well-supported clade (91% branch support, Fig. 8.1A), which is sister to the clade containing *Pythium-Lagenidium* and *Phytopythium* (= *Pythium* clade K). Huang et al. (2013) also found that *Pythiogeton* is sister to a clade composed of *Pythium* species with filamentous to contiguous globose sporangia (*Pythium* clade B), as shown in Fig. 8.1B. These results support Dick's delimitation of the family Pythiogetonaceae.

8.6.2.2 *Pythium, Lagenidium* and *Phytopythium*

In our ITS phylogeny, species of *Pythium* are split into two different major clades: the *Pythium sensu stricto* clade (91%) and the *Pythium* clade K (75%) as shown in Fig. 8.1A. Therefore this genus appears to be paraphyletic, as previously reported (Beakes and Sekimoto 2009; Uzuhashi et al. 2010). In our LSU phylogeny, species in the core clade of *Pythium* clustered together according to the subclades traditionally delimited. In the case of most of the members of clade K (= *Phytopythium*), they form a well-supported clade (88% branch support), which is distant to the *Pythium sensu stricto* clade (Fig. 8.1B).

Primitive characteristics present in *Pythium*, such as external differentiation of zoospores in a vesicle rather than inside the zoosporangium and the less specialized parasitism, suggested that this genus is ancestral to *Phytophthora* (Gaümann 1952). Usually this characteristic is used to discriminate this genus from other genera morphologically similar, such as *Phytophthora* and *Halophytophthora* (van der Plaats-Niterink 1981). Our results showed that species in the genus *Pythium* (particularly from clade B [Fig. 8.1B] which produce filamentous sporangia) and *Lagenidium*, which share this ancestral feature, are grouped together in the same clade. This characteristic is also shared by the members of *Pythium* clade K (= *Phytopythium*). Based on these and previous results, the genus *Pythium sensu lato*, can be considered invalid as currently circumscribed. Uzuhashi et al. (2010) have proposed to split *Pythium* into five different genera based on the shape of the sporangium and on LSU rRNA gene and COX2-based phylogenies. One of these genera, *Ovatisporangium*, is currently invalid (www.mycobank.org) and corresponds actually to *Phytopythium* (= *Pythium* clade K), as suggested by Lévesque and de Cock on the basis of ITS sequences (Gloria Abad, personal communication). One species that was also recovered in this clade is *Halophytophthora kandeliae*, which forms zoospores within a vesicle and has papillate zoosporangia that proliferate internally. This species should be excluded from *Halophytophthora* and placed in *Phytopythium*, according to the SSU phylogeny (Lara and Belhari 2011) and our own results (Fig. 8.1A and B). According to Lévesque and de Cock (2004), the *Pythium* clade K is phylogenetically intermediate between the orders Pythiales and Peronosporales. Species of *Pythium* within this clade clustered with the type species (*Phytopythium sindhum*) of the recently established genus *Phytopythium* (Bala et al. 2010) and could justify the transference of the species in this clade to this new genus. Unfortunately, we did not reach a consensus between the phylogenetic placement of the *Phytopythium* (= *Pythium* clade K). In our ITS phylogeny, this clade appears as sister to the *Pythium* clade in the Pythiales, whereas in our LSU phylogeny, it is placed within the Peronosporales.

8.6.3 Clade 3: Peronosporales

According to Mycobank (www.mycobank.org), Peronosporales is composed of two families, Peronosporaceae ("downy mildews") and Salisapiliaceae, a monotypic family comprising *Salisapilia*. Currently, there is no consensus on which genera are included in this order. Hulvey et al. (2010) considered that the order is composed of three families: Pythiaceae, which includes *Pythium*; Peronosporaceae, including *Halophytophthora*, *Phytophthora* and *Phytopythium*; and Salisapiliaceae, with *Salisapilia*.

In Peronosporales, in contrast to Pythiales, zoospores are completely differentiated inside the sporangia and can be discharged through transient vesicles which rapidly rupture allowing the zoospores to swim away (Huang et al. 2013). In addition, some members have adapted to parasitism in terrestrial environment and as a consequence have lost the ability to produce zoospores (Hardham 2009). In this case, sporangia are produced in sporangiophores, which are dispersed directly by air currents.

8.6.3.1 Downy mildews

The term "downy mildews" is given to 22 genera (www.mycobank.org) that are pathogens of plants (biotrophs) and are included in the family Peronosporaceae (Göker et al. 2007). The "downy mildews" appeared as a monophyletic and well-supported group (88%) in our LSU phylogeny (Fig. 8.1B), as reported by Göker et al. (2007). Some studies (Riethmüller et al. 2002; Göker et al. 2007; Thines et al. 2009) that included both "downy mildews" and *Phytophthora* species have not resolved the relationship of the "downy mildews" with different clades of *Phytophthora* (Runge et al. 2011). Göker et al. (2007) based on ITS phylogeny postulated that the sister group to the "downy mildews" are the air-borne *Phytophthora* from clade 1, such as *Phytophthora infestans*, which are capable of forming haustoria, while Cooke et al. (2000) and Runge et al. (2011) postulated that members of *Phytophthora* from clade 4 are sister to the "downy mildews" clade. In our LSU phylogeny, this group formed a clade of moderate support (75%) with two species of *Phytophthora* from clade 4 (in this case, *Phytophthora arecae* and *P. palmivora*) and *Peronophytora litchii*, and some members of *Phytophthora* clade 6. Since the knowledge about the sister group of the "downy mildews" is essential for understanding the evolution of biotrophy (Runge et al. 2011), and we did not get resolution to confirm which *Phytophthora* clade is sister to the "downy mildews", a more comprehensive and multilocus phylogeny should be done to confirm previous results. It is currently widely accepted that biotrophy has more recently evolved in this group, as compared to the "white rusts" (Thines and Volgmayr 2009). The close relationship of the "downy mildews" and *Phytophthora* obtained in our phylogenies does not support the placement of the genera *Phytophthora* and *Pythium* in the family Pythiaceae as traditionally accepted (Waterhouse 1973; Dick 1984, 2001).

8.6.3.2 *Phytophthora* and *Peronophythora*

Until the late 90s there were approximately 60 valid species of *Phytophthora* (Erwin and Ribeiro 1996). This number has dramatically increased in the last

10 years (53 new species, compiled from www.indexfungorum.org), with approximately 113 species formally described (Hansen et al. 2012). Most of the species are facultative pathogens, especially in plants of economical importance (Erwin and Ribeiro 1996). Therefore, it has been hypothesized that rapid radiation within this genus has occurred due to increasingly specialized parasitism on plants which favored speciation (Seidl et al. 2012). The species included in our ITS phylogeny (except the type species *P. infestans*) belong to the clade 6 (Robideau et al. 2011), which is composed of mostly saprophytic species recovered from natural habitats, such as forests, riparian ecosystems and freshwater courses, even though under some circumstances some of the species included in this clade can be opportunistic and aggressive tree pathogens (Jung et al. 2011).

The phylogenetic position of the "downy mildews" in our LSU phylogeny (Fig. 8.1B) leads to the paraphyly of *Phytophthora*, as previously reported (Cooke et al. 2000; Göker et al. 2003; Runge et al. 2011). In our LSU phylogeny, species of *Phytophthora* clustered together according to the subclades traditionally delimited (Cooke et al. 2000). The only exceptions are *Phytophthora richardiae*, which clusters outside the subclade containing most members of subclade 9 and *Phytophthora thermophila*, which was originally described in clade 6 (Jung et al. 2011) but retrieved as a member of clade 9 (Fig. 8.1B), as previously obtained by Robideau et al. (2011).

Peronophythora litchii, the only species described in *Peronophythora*, is placed in the family Peronophythoraceae (Chen 1961). *Peronophythora litchii* is the causative agent of litchi fruit rot and has characteristics that are intermediate between Peronosporaceae (presence of sporangiophores) and Pythiaceae (characteristics of the sex organs) (Zhang et al. 2007). Chi et al. (1982) and Riethmüller et al. (2002) suggested that this species should be transferred to *Phytophthora*, due to morphological similarities (indeterminate growth of the sporangiophores) with this genus and its close ITS and LSU phylogenetic relationships with *Phytophthora arecae* and *P. palmivora*. Zhang et al. (2007), however, considered that molecular phylogenies do not support the transference of *P. litchii* to *Phytophthora* and also that substantial morphological differences support the validity of the genus *Peronophythora*. Göker et al. (2007) considered that there was no sound molecular and morphological evidence to maintain the genus *Peronophythora* and thus transferred *P. litchii* to *Phytophthora*. Nevertheless, the name *P. litchii* is still considered legitimate in Mycobank. In our LSU phylogeny (Fig. 8.1B), *P. litchii* clustered with considerable support (82%) within the *Phytophthora* clade 4 comprising *Phytophthora arecae* and *P. palmivora*, as previously registered (Riethmüller et al. 2002).

Recently, *Aquaperonospora* has been described as a monotypic genus in the family Peronophythoraceae (Ko et al. 2010) together with *Peronophythora*. Morphological similarities of *Aquaperonospora taiwanensis* with *Pythium* (= *Phytopythium*) *helicoides* (*cf.* van der Plaats-Niterink 1981) are very evident. In addition, our BLASTn analysis with its ITS and LSU sequences confirmed its maximum identity (99%) with

Phytopythium helicoides (AB108030 and HQ665186, respectively). Based on this evidence, and on our LSU phylogeny (Fig. 8.1B) we consider *A. taiwanensis* to be a synonym of *P. helicoides*. Nevertheless, the ex-types specimens of both species should be carefully examined before proposing this synonymy formally.

8.6.3.3 *Halophytophthora* and *Salisapilia*

The first *Phytophthora* described from marine and estuarine environments was *P. vesicula* (Anastasiou and Churchland 1969), which together with eight additional species posteriorly described, were transferred to the genus *Halophytophthora* (Ho and Jong 1990). It has been speculated that some species of *Phytophthora* have independently adapted to the marine/estuarine environment, as observed in our isolates AJM 14 and AJM 21, which were recovered from saline conditions (1.04–2.50% salinity). Based on this concept and as stated by Cooke et al. (2000), Nakagiri (2002), Hulvey et al. (2010) and Nigrelli and Thines (2013), the genus *Halophytophthora* might not represent a natural grouping, and therefore as currently defined is polyphyletic (Fig. 8.1A and B). Although most of the sequences branch in a well-supported *Halophytophthora sensu stricto* clade in both our ITS and LSU phylogenies (with 91 and 96% branch support, respectively) within the Peronosporales, some representatives fall into a clade outside of the core clade of *Halophytophthora*. For example, *Halophytophthora kandeliae* (isolates AJM 26 and AJM 85) branched in both ITS and LSU trees within the Pythiales, particularly within the *Phytopythium* = *Pythium* K clade (syn. *Ovatisporangium*). Other isolates of this species have been previously recovered in the *Phytopythium* clade (Nigrelli and Thines 2013) ex *Pythium* clade K (Bala et al. 2010). Therefore, the possibility of transferring *H. kandeliae* to the genus *Phytopythium* should be considered. Other species, like *Halophytophthora epistomia* and *H. exoprolifera* also clustered in different clades to the core *Halophytophthora*. Therefore, as stated by Hulvey et al. (2010), we believe that *H. epistomia* does not belong to the genus *Halophytophthora*. Based only on morphological characteristics (absence of vesicle during zoospore discharge and presence of a plug of material at the discharge pore), Hulvey et al. (2010) suggested that both species should be transferred to *Salisapilia*. Our ITS results showed that *H. epistomia* indeed appears to be phylogenetically closer but does not seem to belong to *Salisapilia*. Unfortunately, in the case of *H. exoprolifera*, we did not get strong branch support (<50%) in both ITS and LSU phylogenies (Fig. 8.1A and B) to putatively confirm the placement of this species in either of the two clades. The genus *Salisapilia* was erected on the basis of the absence of a vesicle during zoospore discharge, the presence of a plug of material at the apex of the discharge tube and oospore formation (Hulvey et al. 2010). The genus currently contains three species, one of them, *Salisapilia tartarea*, was transferred from *Halophytophthora* on the basis of oospore formation and its placement within the *Salisapilia* clade in both ITS and LSU phylogenies (Hulvey et al. 2010). Even *Salisapilia* appears as a well-supported monophyletic clade in ITS and LSU phylogenies (Hulvey et al. 2010; Nigrelli and Thines 2013,

Fig. 8.1A and B), morphological characters that circumscribe this genus are in need of revision. For example, oospore formation has been registered in other species of *Halophytophthora* when first described, such as in *H. epistomia* (Ho et al. 1990) and *H. exoprolifera* (Ho et al. 1992). In addition, in *Salisapilia nakagirii* (ex-type CBS 127.947), no zoosporangia were observed by Hulvey et al. (2010) and therefore, the absence of vesicle in this species could not be confirmed. However, in *S. nakagirii* CBS 127.947 we observed the absence of a plug of material at the discharge papilla and presence of a semi-persistent vesicle during zoospore discharge, when growing in half-strength seawater (Fig. 8.2).

Also, our LSU DNA phylogeny (Fig. 8.1B) showed that the three isolates identified as *Halophytophthora spinosa* from this study fall into a clade ("spinosa" clade),

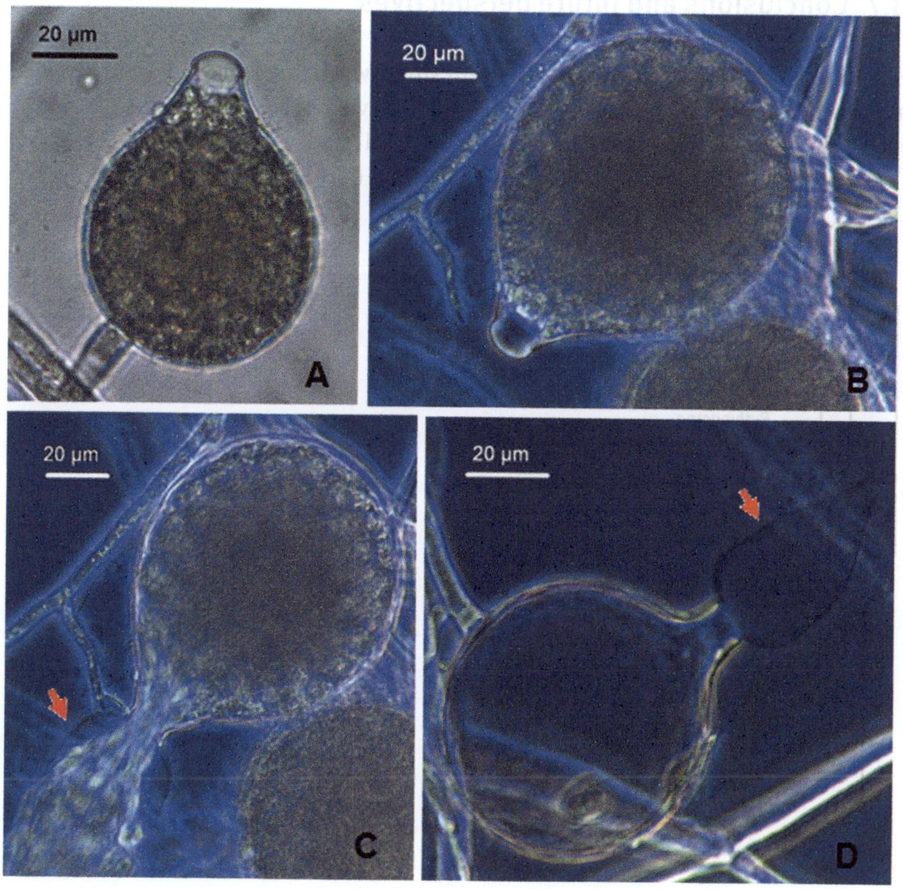

Fig. 8.2: *Salisapilia nakagirii* CBS 127.947. A. Detail of a papillate zoosporangium showing the absence of plug material at the discharge pore. B–C. Zoosporangium discharging zoospores in a conspicuous semi-persistent vesicle (⟶). D. Zoosporangium after the discharge of zoospores. Note the persistence (up to 1 h after zoospore discharge) of the vesicle (⟶).

phylogenetically distant from most of the species of *Halophytophthora sensu stricto* clade. This clade could not be putatively assigned to any of the two subclades (Pythiales or Peronosporales) due to low branch support (<50%) and appeared as a polytomy. Nakagiri (2002) has previously suggested that this species is not a member of the *Halophytophthora sensu stricto* and has a close relationship with *Sapromyces* (Rhipidiales), which has intermediate morphological features between the Saprolegniomycetidae and the Peronosporomycetidae (Dick 1969, 1995; Beakes 1987). However, it is more likely that this clade might represent a new genus, but additional multigene analysis is necessary to confirm this hypothesis and to putatively assign the members of the "spinosa" clade to any of the peronosporalean orders.

8.7 Conclusions and future perspectives

Based on the results obtained in the present study, which are summarized in Fig. 8.3, and on previous results obtained by other authors, we can confirm that:

1. Albuginales is a well-defined and supported clade, ancestral to the rest of members of the "peronosporalean galaxy" and deserves the category of order;

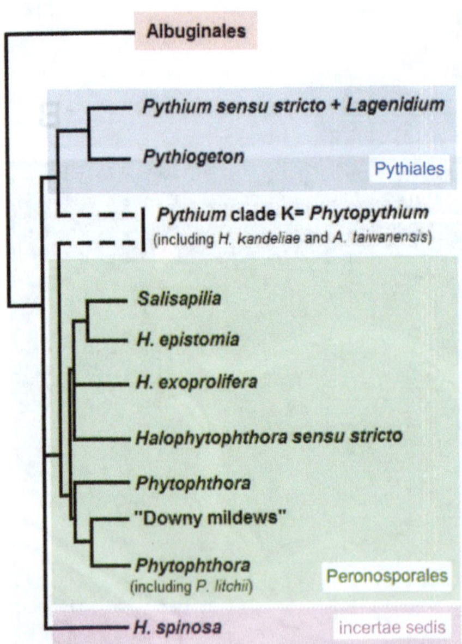

Fig. 8.3: Simplified phylogenetic tree for the "peronosporelean galaxy" based on our results of ITS and LSU phylogenies. Since the position of the *Pythium* clade K (= *Phytopythium*) cannot be putatively confirmed based on our results, it has been included in both orders and indicated in dashed lines. The degree of paraphyly in *Phytophthora* is herein simplified for easier comprehension.

2. Pythiales and Peronosporales are well-defined and moderately supported clades and therefore should be considered separate orders. Pythiales comprises *Pythiogeton, Pythium* and *Lagenidium* while Peronosporales includes *Salisapilia, Phytophthora, Halophytophthora* and the "downy mildews". Since the position of *Phytopythium* is controversial in our ITS and LSU phylogenies, additional multigene analysis should be performed to determinate its placement in one of these orders;
3. *Pythium* as currently established is paraphyletic. Members of *Pythium* clade K, which form a well-defined clade, clustered with the type species of the newly-described genus *Phytopythium*. As a consequence, members of this clade should be transferred to *Phytopythium*. Members of the rest of the subclades (A–J), which form a clade that includes the type species *Pythium monospermum* should be maintained in *Pythium* and thereafter the genus will be monophyletic;
4. *Pythiogeton*, appears to be a well-defined and monophyletic clade, which is sister to the clade containing *Pythium* and *Lagenidium* and deserves its placement in a separate family (Pythiogetonaceae);
5. *Halophytophthora* as currently established is polyphyletic because some of the species clustered outside the core *Halophytophthora* clade. This is particularly evident for *H. kandeliae, H. exoprolifera, H. epistomia*, and *H. spinosa*. In the case of *H. kandeliae*, its transference to the genus *Phytopythium*, based on morphological and molecular evidence should be considered. *Halophytophthora epistomia* is closely related to *Salisapilia* in our LSU phylogeny. Hulvey et al. (2010) has suggested that this species, together with *H. exoprolifera* might belong to *Salisapilia*, based on morphological characteristics. However, our results suggested that *H. epistomia*, although closely related, is not a member of *Salisapilia* but belongs to a different and yet undescribed genus. In the case of *H. exoprolifera* we did not get considerable support (>50%) to putatively place this species within any of the clades. Therefore, additional multigene analysis and a detailed re-evaluation of key morphological characteristics (e.g. zoospore discharge, presence of sexuality) are needed. In addition, it is necessary to carefully re-define the diagnose characters of the genus *Salisapilia*. The *H. spinosa* isolates clustered together as a separate clade which might correspond to a new genus, but unfortunately its putative placement could not be confirmed in our LSU analysis. Therefore, it is mandatory that more isolates and more genes should be sequenced and included in multigene phylogenies, before reaching any conclusion. Also it is necessary to include *Sapromyces* and other Rhipidiales in multigene phylogenies to test the hypothesis on its relatedness to the *H. spinosa* clade and clarify its phylogenetic position. Unfortunately, only one LSU sequence of *Sapromyces elongatus* is currently available in databases;
6. *Phytophthora* showed a high degree of paraphyly. Only the members of subclade 1, which includes the type species *Phytophthora infestans* should retain the generic name, but the introduction of additional generic names for each of the other subclades should be carefully considered;

7. *Peronophythora*, clustered in the *Phytophthora* subclade 4 and as previously suggested by other authors, we consider that this species should be transferred to *Phytophthora*;
8. *Aquaperonospora* has evident morphological similarities with *Pythium helicoides* (=*Phytopythium helicoides*). In addition, our results from BLASTn showed 99% identity with this species, and the LSU phylogeny showed that it is not closely related to *Peronophythora* (it was originally described in the family Peronophythoraceae) and instead clustered within the *Pythium* clade K. Therefore, this name should be considered a heterotypic synonym of *P. helicoides*;
9. The "downy mildews" appeared as a well-defined and supported clade, closely related with some members of *Phytophthora*. This demonstrated that biotrophy has appeared twice within the "peronosporalean galaxy", first in the Albuginales and more recently in the "downy mildews".

We believe that most of the diversity within several genera of Pythiales and Peronosporales remains poorly explored, as indicated by the considerable number of new taxa described in the last 10 years. For this reason, more isolates and particularly additional species need to be sampled, sequenced and deposited into public databases in order to perform multigene phylogenies that will help in delimiting genera and understanding the phylogenetic relationships of closely-related genera within the "peronosporalean galaxy".

Acknowledgments

The authors would like to thank FAPESP (Project 2012/50222-7) and CNPq (Process 304411/2012-4) for financial support. A. V. Marano is particularly indebted to CAPES ("Jovens Talentos" 006-2012, Science without Borders Program). We are grateful to CBS-KNAW Fungal Biodiversity Centre, The Netherlands, for providing the strain of *Salisapilia nakagirii* CBS 127.947 from the United States. The authors are indebted to Instituto Florestal for the permission given to collect samples at Parque Estadual da Ilha do Cardoso, São Paulo, SP.

References

Adl SM, Simpson AG, Farmer MA, Anderson RA, Anderson OR, Barta SR, et al. The new higher level classification of the eukaryotes with emphasis on the taxonomy of protists. J Eukaryot Microbiol 2005;52:399–451.

Anastasiou CJ, Churchland LM. Fungi on decaying leaves in marine habitats. Can J Bot 1969;47:251–257.

Anisimova M, Gascuel O. Approximate likelihood ratio test for branchs: a fast, accurate and powerful alternative. Syst Biol 2006;55:539–552.

Bala K, Robideau GP, Lévesque CA, de Cock AWAM, Abad ZG, Lodhi AM, et al. *Phytopythium* Abad, de Cock, Bala, Robideau, Lodhi & Lévesque, gen. nov. and *Phytopythium sindhum* Lodhi, Shahzad & Levésque, sp. nov. Persoonia 2010;24:127–139.
Balci Y, Long RP, Mansfield M, Balser D, MacDonald WL. Involvement of *Phytophthora* species in white oak (*Quercus alba*) decline in southern Ohio. Forest Pathol 2010;40:430–442.
Baldauff SL. An overview of the phylogeny and diversity of eukaryotes. J Syst Evol 2008;46:263–273.
Barr DJS. The zoosporic grouping of plant pathogens. Entity or non-entity? In: Buczacki ST, ed. Zoosporic Plant Pathogens, a Modern Perspective. London: Academic Press 1983:43–83.
Baurain D, Brinkmann H, Petersen J, Rodríguez-Ezpeleta N, Stechmann A, Demoulin V, et al. Phylogenomic evidence for separate acquisition of plastids in cryptophytes, haptophytes, and stramenopiles. Mol Biol Evol 2010;27:1698–1709.
Beakes GW, Glockling SL, Sekimoto S. The evolutionary phylogeny of the oomycete "fungi". Protoplasma 2012;249:3–19.
Beakes GW. Oomycete phylogeny: ultrastructural perspectives. In: Rayner ADM, Brasier CM, Moore DO, eds. Evolutionary Biology of the Fungi. Cambridge: Cambridge University Press 1987:405–421.
Beakes GW, Sekimoto S. The evolutionary phylogeny of oomycetes - insights gained from studies of holocarpic parasites of algae and invertebrates. In: Lamour K, Kamoun S, eds. Oomycete Genetics and Genomics. New York, NY: John Wiley & Sons 2009:1–24.
Brasier CM, Beales PA, Kirk SA, Denman S, Rose J. *Phytophthora kernoviae* sp. nov., an invasive pathogen causing bleeding stem lesions on forest trees and foliar necrosis of ornamentals in the UK. Mycol Res 2005;109:853–859.
Brasier CM, Cooke DEL, Duncan JM, Hansen EM. Multiple new phenotypic taxa from trees and riparian ecosystems in *Phytophthora gonapodydes-P. megasperma* ITS Clade 6, which tend to be high-temperature tolerant and either inbreeding or sterile. Mycol Res 2003;107:277–290.
Burki F, Okamoto N, Pombert JF, Keeling PJ. The evolutionary history of haptophytes and cryptophytes: phylogenomic evidence for separate origins. Proc R Soc B—Biol Sci 2012;279:2246–2254.
Burki F, Shalchian-Tabrizi K, Pawlowski J. Phylogenomics reveals a new "megagroup" including most photosynthetic eukaryotes. Biol Lett 2008;4:366–369.
Castresana J. Selection of conserved blocks from multiple alignments for their use in phylogenetic analysis. Mol Biol Evol 2000;17:540–552.
Cavalier-Smith T. Principles of protein and lipid targeting in secondary symbiogenesis: euglenoid, dinoflagellate, and sporozoan plastid origins and the eukaryote family tree. J Eukaryot Microbiol 1999;46:347–366.
Cavalier-Smith T. Chromalveolate diversity and cell megaevolution: interplay of membranes, genomes and cytoskeleton. In: Hirt RP, Horner D, eds. Organelles, Genomes and Eukaryotic Evolution. London: Taylor and Francis 2004:71–103.
Cavalier-Smith T, Chao EEY. Phylogeny and megasystematics of phagotrophic heterokonts (Kingdom Chromista). J Mol Evol 2006;62:388–420.
Chen CC. A species of *Peronophythora* gen. nov. parasitic on litchi fruit in Taiwan. Special Publication of College of Agriculture, National Taiwan University 1961;10:1–37.
Chevenet F, Brun C, Banuls AL, Jacq B, Chisten R. TreeDyn: toward dynamic graphics and annotations for analyses of trees. BMC Bioinform 2006;7:439.
Chi PK, Pang XP, Liu R. Studies on *Phytophthora* disease of litchi. I. Identification of causal agent. Proc Annual Meeting Chinese Phytopathol Soc 1982;72–73.
Cooke DEL, Drenth A, Duncan JM, Wagels G, Brasier CM. A molecular phylogeny of *Phytophthora* and related Oomycetes. Fungal Genet Biol 2000;30:17–32.
Dereeper A, Guignon V, Blanc G, Audic S, Buffet S, Chevenet F, et al. Phylogeny.fr: robust phylogenetic analysis for the non-specialist. Nucl Acids Res 2008;36:465–469.

Dick MW. Morphology and taxonomy of the Oomycetes, with special reference to Saprolegniaceae, Leptomitaceae and Pythiaceae. I. Sexual reproduction. New Phytol 1969;68:751–775.
Dick MW. Phylum Oomycota. In: Margulis L, Corliss JO, Melkonian M, Chapman D, eds. Handbook of Protoctista. Boston, MA: Jones & Bartlet 1990:661–685.
Dick MW. Sexual reproduction in the Peronosporomycetes (chromistan fungi). Can J Bot 1995;73:712–724.
Dick MW. Straminipilous Fungi: Systematics of the Peronosporomycetes including Accounts of the Marine Straminipilous Protists, the Plasmodiophorids and Similar Organisms. Dordrecht: Kluwer Academic, 2001.
Dick MW, Croft BJ, Magarey RC, de Cock AWAM, Clark G. A new genus of the Verrucalvaceae (Oomycetes). Bot J Linn Soc 1989;99:97–113.
Dick MW, Wong PTW, Clark G. The identity of the oomycete causing "Kikuyu Yellows", with a reclassification of the downy mildews. Bot J Linn Soc 1984;89:171–197.
Dick MW, Vick MC, Gibbings JG, Hedderson TA, Lopez-Lastra CC. 18S rDNA for species of *Leptolegnia* and other Peronosporomycetes: Justification for the subclass taxa Saprolegniomycetidae and Peronosporomycetidae and division of the Saprolegniaceae *sensu lato* into the Leptolegniaceae and Saprolegniaceae. Mycol Res 1999;103:1119–1125.
Dick MW. Straminipilous Fungi: Systematics of the Peronosporomycetes Including Accounts of the Marine Straminipilous Protists, the Plasmodiophorids and Similar Organisms. Dordrecht: Kluwer Academic Publishers 2001.
Edgar RC. MUSCLE: multiple sequence alignment with high accuracy and high throughput. Nucl Acids Res 2004;32:1792–1797.
Erwin DC, Ribeiro OK. *Phytophthora* Diseases Worldwide. St. Paul. MN: APS Press 1996.
Gaümann EA. The Fungi. A Description of their morphological Features and Evolutionary Development. New York and London: Hafner Publishing 1952.
Göker M, Voglmayr H, Riethmüller A, Weiß M, Oberwinkler F. Taxonomic aspects of Peronosporaceae inferred from Bayesian molecular phylogenetics. Can J Bot 2003;81:672–683.
Göker M, Voglmayr H, Riethmüller A, Oberwinkler F. How do obligate parasites evolve? A multi-gene phylogenetic analysis of downy mildews. Fungal Genet Biol 2007;44:105–122.
Greslebin AG, Hansen EM, Winton LM, Rajchenberg M. *Phytophthora* species from declining *Austrocedrus chilensis* forests in Patagonia, Argentina. Mycologia 2005;97:218–228.
Grooters AM, Proia LA, Sutton DA, Hodgin EC. Characterization of a Previously Undescribed *Lagenidium* Pathogen Associated with Soft Tissue Infection: Initial Description of a New Human Oomycosis. 14th Focus on Fungal Infection. New Orleans: LA 2004.
Guindon S, Gascuel O. A simple, fast and accurate algorithm to estimate large phylogenies by maximum likelihood. Syst Biol 2003;52:696–704.
Hampl V, Hug L, Leigh J, Dacks J, Lang B, Simpson A, et al. Phylogenomic analyses support the monophyly of Excavata and resolve relationships among eukaryotic "supergroups". Proc Natl Acad Sci USA 2009;106:3859–3864.
Hansen E. Alien forest pathogens: *Phytophthora* species are changing world forests. Boreal Environ Res 2007;13:33–41.
Hansen EM, Reeser PW, Sutton W. *Phytophthora* beyond agriculture. Annu Rev Phytopathol 2012;50:359–378.
Harper JT, Waanders E, Keeling PJ. On the monophyly of chromalveolates using a six-protein phylogeny of eukaryotes. Int J Syst Evol Microbiol 2005;55:487–496.
Hausmann K, Hülsmann N, Radek R. Protistology, 3rd ed. Stuttgart: E. Schweizerbart'sche Verlagsbuchhandlung 2003.
Hawksworth DL, Kirk PM, Sutton BC, Pegler DM. Ainsworth & Bisby's Dictionary of the Fungi, 8th ed. Egham: International Mycological Institute 1995.

Ho HH, Jong SC. *Halophytophthora*, gen. nov., a new member of the family Pythiaceae. Mycotaxon 1990;36:377–382.

Ho HH, Nakagiri A, Newell SY. A new species of *Halophytophthora* from Atlantic and Pacific subtropical islands. Mycologia 1992;84:548–554.

Huang JH, Chen CY, Lin YS, Ann PJ, Huang HC, Chung WH. Six new species of *Pythiogeton* in Taiwan, with an account of the molecular phylogeny of this genus. Mycoscience 2013;54:130–147.

Hudspeth DSS, Nadler SA, Hudspeth MES. A cox2 phylogeny of the Peronosporomycetes (Oomycetes). Mycologia 2000;92:674–684.

Hudspeth DSS, Stenger D, Hudspeth MES. A cox2 Phylogenetic Hypothesis for the Downy Mildews and White Rusts. Fungal Divers 2003;13:47–57.

Hulvey J, Telle S, Nigrelli L, Lamour K, Thines M. *Salisapiliaceae* – A new family of oomycetes from marsh grass litter of southeastern North America. Persoonia 2010;25:109–116.

Judelson HS. Dynamics and Innovations within Oomycete genomes: Insights into biology, pathology, and evolution. Eukaryot Cell 2012;11:1304–1312.

Jung T, Stukely MJC, Hardy GEStJ, White D, Paap T, Dunstan WA, et al. Multiple new *Phytophthora* species from ITS Clade 6 associated with natural ecosystems in Australia: evolutionary and ecological implications. Persoonia 2011;26:13–39.

Keeling PJ. Chromalveolates and the evolution of plastids by secondary endosymbiosis. J Eukaryot Microbiol 2009;56:1–8.

Kemen E, Jones JDG. Obligate biotroph parasitism: can we link genomes to lifestyles? Trends Plant Sci 2012;17:448–457.

Klassen GR, McNabb SA, Whitmore E, Gray M, Dick MW. Restriction sites in ribosomal DNA as characters for the evolutionary study of zoosporic fungi. Abstracts, 16th International Congress of Genetics 1988.

Kirk PM, Cannon PF, Minter DW, Stalpers JA. Dictionary of Fungi, 10th ed. Wallingford: CABI Bioscience 2008.

Kroon LPNM, Brouwer H, de Cock AWAM, Govers F. The genus *Phytophthora* anno 2012. Phytopathology 2012;102:348–364.

Ko WH, Lin MJ, Hu CY, Ann PJ. *Aquaperonospora taiwanensis* gen. *et* sp. nov. in Peronophythoraceae of Peronosporales. Bot Stud 2010;51:343–350.

Lane CE, Archibald JM. Novel nucleomorph genome architecture in the cryptomonad genus *Hemiselmis*. J Eukaryot Microbiol 2006;53:515–521.

Lara E, Lassaad B. SSU rRNA reveals major trends in oomycete evolution. Fungal Divers 2011;49:93–100.

Lee SC, Ristaino JB, Heitman J. Parallels in intercellular communication in oomycete and fungal pathogens of plants and humans. PLoS Pathog 2012 8:e1003028. doi:10.1371/journal.ppat.1003028, 2012.

Lévesque CA, de Cock AWAM. Molecular phylogeny and taxonomy of the genus *Pythium*. Mycol Res 2004;108:1363–1383.

Mangot JF, Debroas D, Domaizon I. Perkinsozoa, a well-known marine protozoan flagellate parasite group, newly identified in lacustrine systems: a review. Hydrobiologia 2011;659:37–48.

Marano AV, Pires-Zottarelli CLA, Gleason FH, Neuhauser S, Steciow MM. Assemblages of zoosporic true fungi, heterotrophic straminipiles and plasmodiophorids in freshwater ecosystems. In: Browne SA, ed. Aquatic Ecosystems Series. New York, NY: Nova Publishers 2011:1–56.

Marano AV, Pires-Zottarelli CLA, de Souza JI, Glockling S, Leaño E, Gachon CMM, et al. Chapter 11: Hyphochytriomycota, Oomycota and Perkinsozoa (Supergroup Chromalveolata). In: Jones EBG, Pang KL, eds. Marine Fungi and Fungal-like Organisms. Germany: Gruyter 2012:166–213.

Martens C, Vandepoele K, Van de Peer Y. Whole-genome analysis reveals molecular innovations and evolutionary transitions in chromalveolate species. Proc Natl Acad Sci USA 2008;105:3427–3432.

Mendoza L. *Pythium insidiosum* and mamalian hosts. In: Lamour K, Kamoun S, eds. Oomycete Genetics and Genomics: Biology, Interactions with Plants and Animals, and Toolbox. Hoboken, NJ: John Wiley 2009:387–405.

Nakagiri A. Diversity and phylogeny of *Halophytophthora* (Oomycetes). Abstracts 7th International Mycological Congress Oslo 2002;55:19.

Nigrelli L, Thines M. Tropical oomycetes in the German bight: climate warming or overlooked diversity? Fungal Ecol 2013;6:152–160.

Oh E, Gryzenhout M, Wingfield BD, Wingfield MJ, Burgess TI. Surveys of soil and water reveal a goldmine of *Phytophthora* diversity in South African natural ecosystems. IMA Fungus 2013;4:123–131.

Oudot-Le Secq MP, Loiseaux-de Goër S, Stam WT, Olsen J. Complete mitochondrial genomes of the three brown algae (Heterokonta: Phaeophyceae) *Dictyota dichotoma, Fucus vesiculosus* and *Desmarestia viridis*. Curr Genet 2006;49:47–58.

Park J, Park B, Veeraraghavan N, Jung K, Lee YH, Blair JE, et al. *Phytophthora* database: A forensic database supporting the identification and monitoring of *Phytophthora*. Plant Dis 2008;92:966–972.

Plaats-Niterink AJ van der. Monograph of the genus *Pythium*. Stud Mycol. 21. Baarn: Centraalbureau Voor Schimmelcultures 1981.

Petersen AB, Rosendahl S. Phylogeny of the Peronosporomycetes (Oomycota) based on partial sequences of the large ribosomal subunit (LSU rDNA). Mycol Res 2000;102:1295–1303.

Phillips AJ, Anderson VL, Robertson EJ, Secombes CJ, van West P. New insights into animal pathogenic oomycetes. Trends Microbiol 2008;16:1–19.

Reeser PW, Hansen EM, Sutton W, Remigi P, Adams GC. *Phytophthora* species in forest streams in Oregon and Alaska. Mycologia 2011;103:22–35.

Reinprayoon U, Permpalung N, Kasetsuwan N, Plongla R, Mendoza L, Chindamporn A. *Lagenidium* sp. ocular infection mimicking ocular pythiosis. J Clin Microbiol 2013;51:2778–2780.

Richards TA, Dacks JB, Jenkinson JM, Thornton CR, Talbot NJ. Evolution of filamentous plant pathogens: gene exchange across eukaryotic kingdoms. Curr Biol 2006;16:1857–1864.

Richards TA, Soanes DM, Jones MDM, Vasieva O, Leonard G, Paszkiewicz K, et al. Horizontal gene transfer facilitated the evolution of plant parasitic mechanisms in the Oomycetes. Proc Natl Acad Sci USA 2011;108:15258–15263.

Riethmüeller A, Weir M, Oberwinkler F. Phylogenetic studies of Saprolegniomycetidae and related groups based on nuclear large subunit ribosomal DNA sequences. Can J Bot 1999;77:1790–1800.

Riethmüller A, Voglmayr H, Göker M, Weiss M, Oberwinkler F. Phylogenetic relationships of the downy mildews (Peronosporales) and related groups based on nuclear large subunit ribosomal DNA sequences. Mycologia 2002;94:834–849.

Robideau GP, de Cock AWAM, Coffey MD, Voglmayr H, Brouwer H, Bala K, et al. DNA barcoding of oomycetes with cytochrome *c* oxidase subunit I and internal transcribed spacer. Mol Ecol Resour 2011;11:1002–1011.

Rossman AY, Palm ME. Why aren't *Phytophthora* and other Oomycota true Fungi? Outlooks Pest Manag 2006;17:217–219.

Runge F, Telle S, Ploch S, Savory E, Day B, Sharma R, et al. The inclusion of downy mildews in a multi-locus dataset and its reanalysis reveals a high degree of paraphyly in *Phytophthora*. IMA Fungus 2011;2:163–171.

Schroeder KL, Martin FN, de Cock AWAM, Lévesque CA, Spies CFJ, Okubara PA, et al. Molecular detection and quantification of *Pythium* species: evolving taxonomy, new tools, and challenges. Plant Dis 2013;97:4–20.

Seidl MF, Van den Ackerveken G, Govers F, Snel B. Reconstruction of oomycete genome evolution identifies differences in evolutionary trajectories leading to present-day large gene families. Genome Biol Evol 2012;4:199–211.

Soanes DM, Richards TA, Talbot NJ. Insights from sequencing fungal and oomycete genomes: what can we learn about plant disease and the evolution of pathogenicity? Plant Cell 2007;19:3318–3326.

Sparrow Jr FK. Aquatic Phycomycetes, 2nd ed. Ann Arbor, MI: University of Michigan Press 1960.

Sparrow FK. The present status of classification in biflagellate fungi. In: Jones EBG, ed. Recent Advances in Aquatic Mycology. London: Elek Science 1976:213–222.

Thianprasit M. Fungal infection in Thailand. Jpn J Dermatol 1986;96:1343–1345.

Thines M, Spring O. A revision of *Albugo* (Chromista, Peronosporomycetes). Mycotaxon 2005;92:443–458.

Thines M. Bridging the gulf: *Phytophthora* and downy mildews are connected by rare grass parasites. PLoS ONE 2009;4:e4790.

Thines M, Voglmayr H, Göker M. Taxonomy and phylogeny of the downy mildews (Peronosporaceae). In: Lamour K, Kamoun S, eds. Oomycete Genetics and Genomics: Biology, Interactions with Plants and Animals, and Toolbox. Hoboken: J Wiley 2009:47–75.

Thines M. Evolutionary history and diversity of white blister rusts (Albuginales). Polish Bot J 2010;55:259–264.

Tong SM. *Developayella elegans* nov. gen., nov. sp., a new type of heterotrophic flagellate from marine plankton. Eur J Protistol 1995;31:24–31.

Tsui CKM, Marshall W, Yokoyama R, Honda D, Lippmeier JC, Craven KD, et al. Labyrinthulomycetes phylogeny and its implications for the evolutionary loss of chloroplasts and gain of ectoplasmic gliding. Mol Phylogenet Evol 2009;50:129–140.

Uzuhashi S, Tojo M, Kakishima M. Phylogeny of the genus *Pythium* and description of new genera. Mycoscience 2010;51:337–365.

Waterhouse GM. Peronosporales. In: Ainsworth GC, Sparrow FK, Sussman AS, eds. The Fungi—An Advanced Treatise. New York, NY: Academic Press 1973:165–183.

Zhang ZG, Zheng XB, Wang YC, Ko WH. Evaluation of the rearrangement of taxonomic position of *Peronophythora litchii* based on partial DNA sequences. Bot Stud 2007;48:79–89.

Biodiversity of freshwater fungi

Sally L. Glockling, Wyth L. Marshall, Rodolphe E. Gozlan, Agostina V. Marano, Osu Lilje and Frank H. Gleason

9 The ecological and economic importance of zoosporic Mesomycetozoean (Dermocystida) parasites of freshwater fish

The Mesomycetozoea is a phylum of fungus-like organisms which includes many species which infect freshwater and marine bony fishes (Class Osteichthyes). Some of these pathogenic species can cause serious and fatal diseases in fish species which are of ecological and economic importance in natural freshwater ecosystems and in fish farms.

The phylum Mesomycetozoea comprises species of pathogens, parasites and mutualists which infect a wide range of invertebrates and vertebrates including mammals, amphibians, fish, arthropods and crustaceans. They are osmotrophs, absorbing nutrients from the host tissues through their cell walls. They are found in both freshwater and marine environments. Freshwater fishes that are known to succumb to infection by Mesomycetozoean parasites include several species of Salmonidae, Percae and Cyprinidae (see Tab. 9.1). Some parasites of fish in the order Dermocystida, such as *Sphaerothecum*, also known as the “rosette agent” and some species of *Dermocystidium*, produce zoospores. Members of the order Ichthyophonida are not zoosporic but the genus *Ichthyophonus* contains the important fish pathogen, *I. hoferi*. *Dermocystidium* contains 11 species that infect freshwater fish, whilst *Sphaerothecum* includes only *S. destruens*, which infects both salmonid and cyprinid fishes (see Tab. 9.1, Figs. 9.1, 9.2). These species cause dermocystidiosis in their hosts and can have a very detrimental effect on the health of fish populations in natural ecosystems and in aquaculture (Arkush et al. 1998; Kalesalu et al. 2000; Gozlan et al. 2005; Ramaiah 2006).

9.1 Phylogeny

The Mesomycetozoea are a group of fungus-like microorganisms which are considered to be basal to the animal lineage in the Opisthokont supergroup (Adl et al. 2012; Paps et al. 2012). The Mesomycetozoea is a monophyletic group which is divided into two orders, the Dermocystida and the Ichthyophonida, which are well supported in phylogenetic studies (Mendoza et al. 2002; Marshall et al. 2008; Lohr et al. 2010; Marshall and Berbee 2011). There are also consistent morphological and ecological differences between the two orders, as the Dermocystida and Ichthyophonida have different characteristics and life cycles (see below). The Dermocystida all infect

Tab. 9.1: Some Dermocystidan species which are parasites of fish.

Parasite	Host	Location	References
Dermocystidium anguillae	Eel, *Anguilla anguilla*	Europe	Wootten and McVical (1982)
Dermocystidium percae	Common perch, *Perca fluviatilis;* Ruff *Gymnocephalus cernuus*	Europe	Reichenbach-Klinke (1950); Pekkarinen and Lotman (2003)
Dermocystidium fenniculum	Perch, *Perca fluviatilis*	Europe	Pekkarinen et al. (2003)
Dermocystidium gasterostei	Three-spined stickleback, *Gasterosteus aculeatus*; Ninespine stickleback, *Pungitius pungitius*	Europe	Elkan (1962)
Dermocystidium granulosum	Humpback pufferfish, *Tetraodon palembangensis*	Asia	Sterba and Naumann (1970)
Dermocystidium koi	Common carp, *Cyprinus carpio*	Asia	Hoshina and Sahara (1950)
Dermocystidium macrophagi	Rainbow trout, *Oncorhynchus mykiss*	Europe	van de Moer et al. (1987)
Dermocystidium cyprini	Common carp, *Cyprinus carpio*	Europe	Červinka et al. (1974); Lotman et al. (2000)
Dermocystidium salmonis	Chinook salmon, *Oncorhynchus tshawytscha*; pink salmon, *Oncorhynchus gorbuscha;* common whitefish *Coregonus lavaretus*	North America	Davis (1947) Olson et al. (1991) Olson and Holt (1995)
Dermocystidium branchialis	Brown trout, *Salmo trutta*	Europe	Dunkerley (1914)
Dermocystidium vejdovskyi	Northern pike, *Esox lucius*	Europe	Jirovec (1939)
Sphaerothecum destruens	Atlantic salmon, *Salmo salar*; Chinook salmon *O. tshawytscha*; Coho salmon, *O. kisutch*, brown trout *S. trutta*, rainbow trout, *O. mykiss*, brook trout, *Salvelinus fontinalis*, Topmouth gudgeon, *Pseudorasbora parva*; sunbleak, *Leucaspius delineatus*, carp, *Cyprinus carpio*, bream *Abramis brama*	North America, Europe	Harrell et al. (1986); Hedrick et al. (1989); Arkush et al. (1998); Andreou et al. (2012)

vertebrate organisms whilst the Ichthyophonida are found in both vertebrates and invertebrates. The specific relationships within the Dermocystida have not been fully determined (Gonzalez-Hernandez et al. 2010), although *Dermocystidium* is considered to be a sister clade to *Rhinosporidium*, a genus containing mammalian pathogens (Herr et al. 1999; Mendoza et al. 2002; Pereira et al. 2005; Raffel et al. 2008; Gonzalez-Hernandez et al. 2010). *Dermocystidium percae* appears to be paraphyletic, complicating the phylogenetic relationships within the genus. *Sphaerothecum* was found to be a sister taxon to *Dermocystidium* and *Rhinosporidium* (Mendoza et al. 2002). *Ichthyophonus* has a well-supported relationship with species isolated from arthropods (Marshall et al. 2008; Marshall and Berbee 2011) and it has been speculated that the infective stage may be carried by a copepod (Sindermann and Scattergood 1994).

9.2 Life cycles

All species of Mesomycetozoea which are parasites of fish are considered to be obligate parasites or biotrophs and must grow in the tissues of their hosts in the natural environment. The life cycle of many members of the Mesomycetozoea has not been fully elucidated, and for several species it is not known how the host is infected or how the infective cells proliferate within the host. Studies of mesomycetozoean life cycles have often been limited to parasitic stages because axenic cultivation has not been successful for any species in the Dermocystida, although *Sphaerothecum* has been cultivated along with salmonid and cyprinid cell lines (Arkush et al. 1998, 2003; Andreou et al. 2012; Paley et al. 2012). More is known about the life cycles of some members of the Ichthyophonida as a total of six genera in this order, including the fish pathogen *I. hoferi*, have been successfully grown in pure culture (Whistler 1960; Okamoto et al. 1985; Jostensen et al. 2002; Marshall et al. 2008; Marshall and Berbee 2011). Despite cultivation of some species of Mesomycetozoea, sexual development has not been observed. Species in both orders of the Mesomycetozoea produce thick-walled endospores. In addition to these endospores, some members of the Dermocystida, such as the fish pathogen *Sphaerothecum destruens* and the *Dermocystidium* species *D. percae, D. fenniculum* and *D. salmonis*, produce zoospores. Zoospores have been shown to emerge from spherical intra-cytoplasmic spore stages when these are incubated in freshwater (Arkush 2003; Andreou et al. 2009). Some members of the Ichthyophonida, including the fish parasite *I. hoferi*, produce amoebae (Mendoza et al. 2002), although amoebae production in *I. hoferi in vitro* can be dependent upon pH (Okamoto et al. 1985). The life cycle of *Ichthyophonus* has been shown to be complex and the progression of life stages within the host is not fully understood (Kocan et al. 2013). These motile zoospores and amoebae found in the Mesomycetozoea may function as infective agents, although this has not been categorically proven. Evidence of a cryptic amoeboid stage in *Ichthyophonus*, which penetrates the gut epithelium has been reported (Kocan et al. 2013) and infection in *I. hoferi* may be transmitted via the fish skin or the urinary tract (Ramiah 2006) and can also be transferred from herring to salmon through the food chain in laboratory tests (Jones and Dawe 2002).

Mendoza et al. (2002) published a simplified diagrammatical life cycle comparison for the Dermocystida and Ichthyophonida. According to Mendoza et al. (2002), following infection, the zoospore of species in the Dermocystida encysts and enlarges to form a large, spherical, multinucleate walled cell (200–400 µm in diameter in *Dermocystidium*) inside the host. This develops into a sporangium. The cytoplasm of the sporangium cleaves into small spherical endospores which may be released within the host tissues. This developmental cycle is repeated inside the host so that thousands of endospores can be produced, which are eventually released into the environment.

In the Ichthyophonida, infection may be facilitated by the amoebae which also form sporangia inside the host. The sporangia containing the endospores are surrounded by thick cell walls or receptacles, which may give these structures longevity and protection from environmental extremes. The sporangia may be released from the receptacle inside or outside the host. Endospores may be released from the sporangia, which may develop into amoebae. The propagation of infection via amoebae in *Ichthyophonus* has also been suggested by Kocan et al. (2013). However, the *in vitro* life cycle of *I. hoferi* is very complex and the progression of life stages is not yet clearly understood (Okamoto et al. 1985; Kocan et al. 2013).

9.3 The zoospore

Some aspects of the physiology and ecology of zoospores of fungal and fungal-like organisms have been reviewed by Hickman and Ho (1966), Fuller (1977) and Gleason and Lilje (2009). Motile zoospores can be released from zoosporangia of parasitic species in large numbers. The relationships between zoospore movement and virulence are not understood in general. Zoospores of some parasitic species which have been studied in the laboratory, are known to be chemotactic, allowing them to navigate chemical gradients and swim for several hours. Moss et al. (2008) measured the attraction of zoospores of *Batrachochytrium dendrobatidis* (a uniflagellate fungal parasite of amphibians) to various chemical substances and El-Feki et al. (2003) conducted similar studies with *Saprolegnia parasitica* (an Oomycete fish parasite). The ability to sense chemical gradients (chemotaxis) allows parasites to efficiently locate and infect healthy host species in a watery environment. Alternatively, non-motile, thick-walled spores are carried by water or air currents and reach their hosts by passive means. The zoospores of the Dermocystida, where they are known, are posteriorly uniflagellate like other members of the Opisthokonta. It is not known whether the zoospores of *Dermocystidium* species and *Sphaerothecum* are chemotactic. The zoospores of these species can be very small; those of *S. destruens* are reported to be only 1–2 µm in diameter (Mendoza et al. 2002).

It has not been proven that zoospores are the infective agents in *Sphaerothecum*, and cursory laboratory experiments using immersion baths failed to induce infection in host fish species (Paley et al. 2012). Andreou et al. (2009) studied the influence of temperature on the production and longevity of zoospores of *S. destruens* at four temperatures in the laboratory: 4°C, 15°C, 25°C and 30°C. Both the healthy carrier host fish (the Asian cyprinid, topmouth gudgeon *Pseudorasbora parva*) and the pathogen have been observed at these temperatures in the wild. Results of this study showed that *S. destruens* could tolerate a broad range of temperature, although zoospores were produced later and had a longer period of viability at lower temperatures than at higher temperatures. The zoospores were able to survive at 30°C in laboratory studies. Zoospores of *D. salmonis* were also shown to survive at 4°C and, in this study, the zoospores were also shown to be the infective agents (Olsen et al. 1991).

9.4 Symptoms of disease

Species of *Dermocystidium* usually cause macroscopically visible infections of the gills, fins and skin (Olson et al. 1991; Olson and Holt 1995; Lotman et al. 2000; Pekkarinen et al. 2003; Pekkarinen and Lotman 2003). The infections appear as spherical, oval, elongate or dumbbell-shaped white cysts. Cysts of *Dermocystidium percae* have been found on the fins of perch and ruff in Finland (Pekkarinen and Lotman 2003). Perch can become infected in the summer of their first year of growth and the infection is thought to survive the winters in the form of resistant cysts. After 3 years of age, perch appear to develop some immunity to *D. percae*. Although the fins were the most common location of *D. percae* cysts, they also occurred in the gills in ruff (Pekkarinen and Lotman 2003). In the gills of cultured eels, *Dermocystidium* infection can lead to hyperplasia and fusion of the secondary lamellae with the potential to cause osmoregulatory problems (Wootten and McVicar 1982). *Dermocystidium* infection of the skin and fins is accompanied by a strong inflammatory response (Dykova and Lom 1992; Pekkarinen and Lotman 2003). *Dermocystidium* species, such as *D. salmonis*, infect and often subsequently kill fish either because infected fish are unable to take up oxygen due to the large numbers of cysts in the gills or because vital organs become infected. Systemic infections have been found to affect the kidney, liver, spleen and testes in rainbow trout *Oncorhynchus mykiss*, brown trout *Salmo trutta* and Atlantic salmon *Salmo salar* (Nash et al. 1989). Spores of *Dermocystidium* replicate asexually in the host through fission. They can infect epithelial, mesenchymal, and hematopoietic cells, and eventually cause cell death (Arkush et al. 1998). Following release, the spores can infect further tissues or they can be excreted in body fluids, such as bile, urine, seminal and ovarian fluids and also through the gut epithelium (Arkush et al. 2003). Fish infection in *Dermocystidium* is believed to occur through ingestion and gut penetration, or via attachment of spores to the gills or skin, or both (Arkush et al. 2003).

Sphaerothecum destruens (Figs. 9.1, 9.2) infects the visceral organs of fish, causing tissue degradation and cell death (Arkush et al. 1998), and internal lesions and disseminated forms have been described in the cyprinid, *Leucaspius delineatus* (Andreou et al. 2011). Internal symptoms have also been found in salmonid hosts, comprising white nodules on visceral organs (Arkush et al. 1998; Paley et al. 2012; Rowley et al. 2013). *Sphaerothecum* also causes anaemia and lymphocytosis in salmon as the macrophages of the liver and spleen are affected (Mendoza et al. 2002). In sunbleak, *S. destruens* causes necrosis of liver cells and a granulomatous response (Fig. 9.2a) and macrophages aggregate in infected intestine and connective tissues (Fig. 9.2b).

The fish parasite in the Ichthyophonida, *I. hoferi*, is also lethal to its hosts as thick-walled spherical cells invade internal organs such as the heart, liver, muscles, and spleen (Mendoza et al. 2002). *Ichthyophonus hoferi* produces thick-walled, multinucleate spherical cells in the host tissues (Franco-Sierra and Alvarez-Pellitero 1999; Mendoza et al. 2002).

Fig. 9.1: *Cyprinus carpio* (above) and *Leucaspius delineatus* (below) infected with *Sphaerothecum destruens*. Mature infections.

(A)

(B)

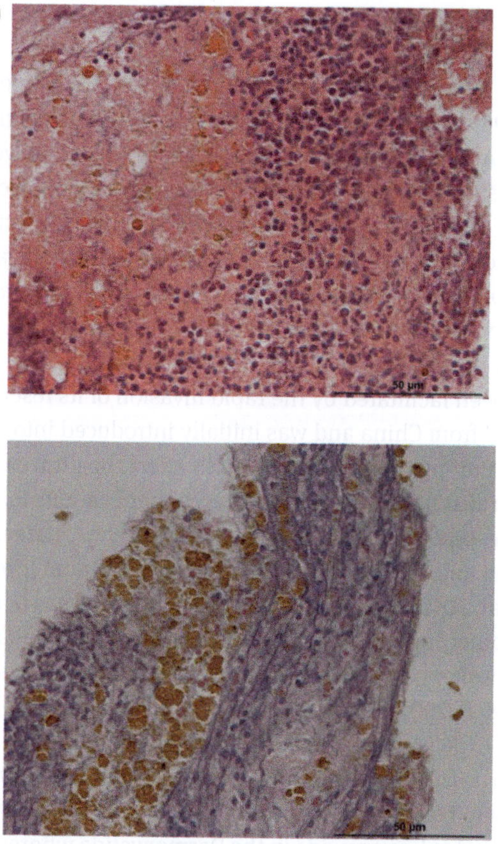

Fig. 9.2: *Sphaerothecum destruens* can cause systemic disease in sunbleak, with disseminated disease most often observed. Some granulomatous response is also observed. A. Liver with cell necrosis and some host response. 40 ×; hematoxylin and eosin stain. B. Connective tissue and intestine with macrophage aggregation. 40 ×; hematoxylin and eosin stain.

9.5 Ecological and economic significance

Although the Dermocystida remains a relatively poorly-studied order of fungus-like organism (Ragan et al. 1996), it includes important pathogens associated with high mortality in several species of fish (Allen et al. 1968; Olson et al. 1991; Rahimian and Thulin 1996; Arkush et al. 1998, 2003; Marty et al. 1998; Hershberger et al. 2002; Gozlan et al. 2005, 2009; Andreou et al. 2012; Paley et al. 2012). Allen et al. (1968) studied mortality in Chinook salmon caused by *Dermocystidium* and found that both fry and adults were infected. *Dermocystidium* was also frequently found on prespawned carcasses (Allen et al. 1968). Mortality in adult prespawning Chinook salmon was found to be associated with *Dermocystidium* infection of the gills in 25% of cases (Pauley 1967). Death was considered to be a result of anoxia due to inflammation of the gills and granulation of the tissues (Pauley 1967). Parasites in the Dermocystida infect many freshwater and anadromous fish belonging to a number of families (See Tab. 9.1).

Sphaerothecum destruens can infect and cause mortality in a range of salmonid (Harrel et al. 1986; Hendrick et al. 1989; Arkush et al. 1998; Paley et al. 2012) and cyprinid fish species (Gozlan et al. 2005; Andreou et al. 2011, 2012). These include several species of commercial importance such as Atlantic salmon (*S. salar*) and common carp (*Cyprinus carpio*). Figure 9.1 shows *C. carpio* and *L. delineatus* with advanced infection caused by *S. destruens*. However, hosts also include species of high conservation value such as the Chinook salmon *Oncorhynchus tshawytscha* and sunbleak *L. delineatus*. In addition, *S. destruens* has been found in the highly invasive healthy carrier, the topmouth gudgeon *Pseudorasbora parva* (Gozlan et al. 2010). The spread of *S. destruens* in Europe is believed to have been facilitated by the rapid invasion of its reservoir fish host *P. parva*, which originated from China and was initially introduced into carp aquaculture fish farms in Eastern Europe (Copp et al. 2010). In effect, healthy carrier hosts acting as reservoirs for generalist pathogens can have great implications for the pathogenicity of species like *S. destruens*. It is known that in single host parasites, the trade-off between transmission and virulence characterises the optimum level of a pathogen's virulence (Peeler et al. 2010). However, in generalist pathogens such as *S. destruens*, the introduction of other hosts with different susceptibility can change the virulence of the pathogen. This alters the initial transmission level and affects the outcome of infection (Woolhouse et al. 2001). This effect has been clearly illustrated by the pan-continental extinction of *L. delineatus* as a result of the co-introduction and spread of the reservoir host *P. parva*. The associated multi-host pathogen *S. destruens* now represents a significant threat to fish biodiversity worldwide (Gozlan et al. 2005).

The host specificity is considered to be very wide in the Dermocystida wherever it has been studied. Kalesalu et al. (2000) examined 12 fish farms and five natural freshwater systems in Estonia for the presence of *Dermocystidium* cysts on fish. They found cysts on several farmed fish species including carp *Cyprinus carpio*, rainbow trout *Oncorhynchus mykiss* and roach *Rutilus rutilus*. In natural water systems they found cysts on Northen pike *Esox lucius* and perch *Perca fluviatilis*. Infection of fish was found to be due to a total of four species of *Dermocystidium*. In the winters of four successive years it was found that *D. cyprini* infected between 18.4 and 24.6% of *C. carpio*, with an average of 18.7 cysts per fish (Kalesalu et al. 2000). Some Ichthyophonida also have wide host ranges. For example, *I. hoferi* is known to infect about 20 species of fish including six freshwater species (Mendoza et al. 2002). However, Rasmussen et al. (2010) found that in a molecular study, *Ichthyophonus* forms two separate clades. One clade corresponds to sequences from rainbow trout *O. mykiss* whilst the other corresponds to infections in anadromous American shad *Alosa sapidissima*, marine rockfish *Sebastes proriger* and Pacific herring *Clupea pallasi*. For this reason, Rasmussen et al. (2010) caution against using the species name. In the marine environment *Ichthyophonus* appears to have a broad host preference based on ribosomal sequencing (Halos et al. 2005).

There are a number of characteristics of Dermocystida that suggest that some species should be designated as emerging infectious diseases according to Fisher et al. (2012). These characteristics, which are typical of parasites capable of causing significant losses in host populations, include high virulence under certain conditions, infective propagules which are easily dispersed, long-lived resistant stages in the life cycle, a broad host range, and the potential for spread via international trade (Fisher et al. 2012).

9.6 Discussion and conclusion

In comparison with other diseases of fish, members of the Dermocystida have received remarkably little attention, despite causing declines in fish populations and mass mortalities since the early 1990s. Due to their interesting placement as a basal group in the animal lineage, close to where the fungal and animal kingdoms diverged in the Opisthokonta, there has been much interest in the Mesomycetozoea in recent years from a phylogenetic point of view (Adl et al. 2012; Paps et al. 2012). However, the ecology of these organisms still requires considerable research to elucidate their life cycles and infection strategies. There is still uncertainty as to whether the zoospore is the infective agent in *Dermocystidium* species and *Sphaerothecum*, and how the infection disperses within the host. Further work is also required to decipher exactly how infection is spread between conspecific host organisms and from one host species to another for species of Dermocystida and Ichthyphonida with broad host ranges. Until more is known about how these pathogenic organisms interact with their hosts, the diseases they cause cannot be properly understood or controlled. The inability to grow species of *Dermocystidium* and *Sphaerothecum* in pure culture has hindered physiological studies of these important pathogens and most of our knowledge of the species in the Dermocystida comes from field studies. Some long-term studies are helping to discover how diseases caused by the Dermocystida affect fish populations throughout the growth and reproductive development of the host (Allen et al. 1968; Pekkarinen and Lottman 2003), but further research is needed to establish the full ecological impact of fish diseases caused by Dermocystida species in the natural freshwater environment.

Dermocystidium has an extensive host range of animals in both freshwater and marine ecosystems. These hosts include fish species which are important in the aquaculture industry and also species of high conservation status (see Tab. 9.1). Many of the fish parasites in the Dermocystida can cause high rates of mortality in host populations. The economic losses due to dermocystidiosis in aquaculture and in natural freshwater ecosystems have not been comprehensively assessed. Appropriate controls against dermocystidiosis for fish farms await a more thorough understanding of the biology of both the pathogens and the diseases which they cause. However, there is a crucial need to evaluate the economic impact as a result of Dermocystida species in aquaculture. In addition, due to the close link between fish farms and natural ecosystems, the economic and ecological risks are closely linked (Gozlan et al. 2006).

Reports of disease in fishes caused by Dermocystida species have been increasing in recent years, but is not clear whether this apparent increase in disease is due to improvements in the effectiveness of detection techniques such as PCR, or whether environmental factors are involved. Some data indicate that the parasites in the Dermocystida are being spread along with their hosts by the international trade of animal species (Gozlan et al. 2005; Copp et al. 2010; Hershberger et al. 2010). The introduction of *P. parva* to carp fisheries may have spread *S. destruens* to susceptible host species (Copp et al. 2010). In addition, native salmonid fish populations in Columbia river systems could be affected by the introduction of *A. sapidissima* (Hershberger et al. 2010), which is a primary host of the Ichthophonidan pathogen, *I. hoferi*. It is imperative that the spread of species in the Dermocystida group be monitored carefully by governmental authorities that administer biosecurity programs at both the local and international levels. Although difficulties in pathological identification remain, the development of increasingly sensitive and relatively inexpensive molecular tools (e.g. PCR) would allow large scale screening of wild fish populations in freshwater ecosystems.

Acknowledgment

The authors thank Elayna Truszewski, Department of Biological Sciences, Macquarie University for her editorial assistance with preparation of the original manuscript.

References

Adl SM, Simpson AG, Lane CE, Lukeš J, Bass D, Bowser SS. The revised classification of eukaryotes. J Eukaryot Microbiol 2012;59:429–514.

Allen RL, Meekin TK, Pauley GB, Fujihara MP. Mortality among Chinook Salmon associated with the fungus *Dermocystidium*. J Fish Res Board Can 1968;25:2467–2475.

Andreou D, Gozlan RE, Paley R. Temperature influence on production and longevity of *Sphaerothecum destruens* zoospores. J Parasitol 2009;95:1539–1541.

Andreou D, Gozlan RE, Stone D, Martin P, Bateman K, Feist SW. *Sphaerothecum destruens* pathology in cyprinids. Dis Aquat Organ 2011;95:145–151.

Andreou D, Arkush KD, Guégan J-F, Gozlan RE. Introduced pathogens and native freshwater biodiversity: A case study of *Sphaerotheceum destruens*. PLoS ONE 2012;7:e36998.doi: 10.1371/journal.pone.0036998.

Arkush KD, Frasca S Jr, Hedrick RP. Pathology associated with the Rosette Agent, a systemic protist infecting salmonid fishes. J Aquat Anim Health 1998;10:1–11.

Arkush KD, Mendoza L, Adkinson MA, Hedrick RP. Observations on life stages of *Sphaerothecum destruens* n. g., n. sp., a Mesomycetozoean fish pathogen formerly referred to as the rosette agent. J Eukaryot Microbiol 2003;50:430–438.

Copp GH, Villizi L, Gozlan RE. Fish movements: the introduction pathway for topmouth gudgeon *Pseudorasbora parva* and other non-native fishes in the UK. Aquat Conserv: Mar Freshw Ecosyst 2010;20:269–273.

Dykova I, Lom J. New evidence of fungal nature of *Dermocystidium koi* Hoshina and Sahara. J Appl Ichthyol 1992;8:180–185.

El-Feki M, Hatai K, Hussein MMA. Chemotactic and chemokinetic activities of *Saprolegnia parasitica* toward different metabolites and fish tissue extracts. Mycoscience 2003;44:159–162.

Fisher MC, Henk DA, Briggs CJ, Brownstein JS, Madoff LC, McCraw SL, et al. Emerging fungal threats to animal, plant and ecosystem health. Nat 2012;484:186–194.

Franco-Sierra A, Alvarez-Pellitero P. The morphology of *Ichthyophonus* sp. in their mugilid hosts (Pisces: Teleostei) following cultivation *in vitro*. A light and electron microscopy study. Parasitol Res 1999;85:562–575.

Fuller MS. The zoospore, hallmark of the aquatic fungi. Mycol 1977;69:1–20.

Gleason FH, Lilje O. Structure and function of fungal zoospores: ecological implications. Fungal Ecol 2009;2:53–59.

González-Hernández M, Denoël M, Duffus AJ, Garner TW, Cunningham AA, Acevedo-Whitehouse K. Dermocystid infection and associated skin lesions in free-living palmate newts (*Lissotriton helveticus*) from Southern France. Parasitol Int 2010;59:344–350.

Gozlan RE, St-Hilaire S, Feist SW, Martin P, Kent ML. Biodiversity: disease threat to European fish. Nature 2005;435:1046.

Gozlan RE, St-Hilaire S, Feist SW, Longshaw M, Peeler EJ. The effect of microbial pathogens on the diversity of aquatic populations, notably in Europe. Microbes Infections 2006;8:1358–1364.

Gozlan RE, Whipps CM, Andreou D, Arkush KD. Identification of a rosette-like agent as *Sphaerothecum destruens*, a multi-host fish pathogen. Int J Parasitol 2009;39:1055–1058.

Gozlan RE, Andreou D, Asaeda T, Beyer K, Bouhadad R, Burnard D, et al. Pan-continental invasion of *Pseudorasbora parva*: towards a better understanding of freshwater fish invasions. Fish Fish 2010;11:315–340.

Halos D, Alexandra S, Hershberger P, Kocan R. *Ichthyophonus* in Puget sound Rockfish from the San Juan Islands Archipelago and Puget Sound Washington, USA. J Aquat Anim Health 2005;17:222–227.

Harrell LW, Elston RA, Scott TM, Wilkinson MT. A Significant new systemic-disease of net-pen reared Chinook salmon (*Oncorhynchus tshawytscha*) Brood Stock. Aquat 1986;55:249–262.

Hedrick RP, Friedman CS, Modin J. Systemic infection in Atlantic salmon *Salmo salar* with a *Dermocystidium*-like species. Dis Aquat Org 1989;7:171–177.

Herr AR, Akello L, Taylor JW, Arsecularatne SN, Mendoza L. Phylogenetic analysis of *Rhinosporidium seeberi*'s 18S among members of the protoctistant mesomycetozoan clade. J Clin Microbiol 1999;37:2750–2754.

Hershberger PK, Stick K, Bui B, Carroll C, Fall B, Mork C, et al. Incidence of *Ichthyophonus hoferi* in Puget Sound fishes and its increase with age of Pacific herring. J Aquat Anim Health 2002;14:50–56.

Hershberger PK, van der Leeuw BK, Gregg JL, Grady CA, Lujan KM, Gutenberger SK, et al. Amplification and transport of an endemic fish disease by an introduced species. Biol Invasions 2010;12:3665–3675.

Hickman CJ, Ho HH. Behaviour of zoospore in plant-pathogenic phycomycetes. Ann Rev Phytopathol 1966;4:195–214.

Jones SRM, Dawe SC. *Ichthyophonus hoferi* Plehn & Mulsow in British Columbia stocks of Pacific herring, *Clupea pallasi* Valenciennes, and its infectivity to Chinook salmon, *Oncorhynchus tshawytscha* (Walbaum). J Fish Dis 2002;25:415–421.

Jostensen JP, Sperstad S, Johansen S, Landfald B. Molecular phylogenetic, structural and biochemical features of a cold adapted marine ichthyosporean near the animal fungal divergence described from in vitro cultures. Eur J Protistol 2002;38:93–104.

Kalesalu J, Laius A, Lotman K. The occurrence and species composition of parasites of the genus *Dermocystidium* in Estonian fish farms and some natural water bodies. Agraarteadus 2000;11:205–212.

Kocan R, LaPatra S, Hershberger P. Evidence for an Amoeba-like infectious stage of *Ichthyophonus* sp. and description of a circulating blood stage: a probable mechanism for dispersal within the fish host. J Parasitol 2013;99:235–240.

Lohr JN, Laforsch C, Koerner H, Wolinska JA. *Daphnia* parasite (*Caullerya mesnili*) constitutes a new member of the Ichthyosporea, a group of protists near the animal-fungi divergence. J Eukaryot Microbiol 2010;57:328–336.

Lotman K, Pekkarinen M, Kasesalu J. Morphological observations on the life cycle of *Dermocystidium cyprini* Červinka and Lom, 1974, parasitic in carps (*Cyprinus carpio*). Acta Protozool 2000;39:125–134.

Marshall WL, Berbee ML. Facing unknowns: living cultures (*Pirum gemmata* gen. nov., sp. nov., and *Abeoforma whisleri*, gen. nov., sp. nov.) from invertebrate digestive tracts represent an undescribed clade within the unicellular Opisthokont lineage Ichthyosporea (Mesomycetozoa). Protist 2011;162:33–57.

Marshall WL, Celio G, McLaughlin DJ, Berbee ML. Multiple isolations of a culturable, motile Ichthyosporean (Mesomycetozoea, Opisthokonta), *Creolimax fragrantissima* n. gen., n. sp., from marine invertebrate digestive tracts. Protist 2008;159:415–433.

Marty GD, Freiberg EF, Meyers TR, Wilcock J, Farver TB, Hinton DE. Viral haemorrhagic septicemia virus, *Ichthyophonus hoferi*, and other causes of morbidity in Pacific Herring *Clupea pallasi* and its early appearance in wild Puget Sound herring. Dis Aquat Org 1998;35:23–29.

Mendoza L, Taylor JW, Ajello L. The Class Mesomycetozoea: A heterogeneous group of microorganisms at the animal-fungal boundary. Ann Rev Microbiol 2002;56:315–344.

Moss AS, Reddy NS, Dortaj IM, San Francisco MJ. Chemotaxis of the amphibian pathogen *Batrachochytrium dendrobatidis* and its response to a variety of attractants. Mycol 2008;100:1–5.

Nash G, Southgate P, Richards RH, Sochon E. A systemic protozoal disease of cultured salmonids. J Fish Dis 1989;12:157–173.

Okamoto N, Nakase K, Suzuki H, Nakai Y, Fujii K, Sano T. Life history and morphology of *Ichthyophonus hoferi in vitro*. Fish Pathol 1985;20:273–285.

Olson RE, Dungan CF, Holt RA. Water-borne transmission of *Dermocystidium salmonis* in the laboratory. Dis Aquat Org 1991;12:41–48.

Olson RE, Holt RA. The gill pathogen *Dermocystidium salmonis* in Oregon salmonids. J Aquat Anim Health 1995;7:111–117.

Paley RK, Andreou D, Bateman KS, Feist SW. Isolation and culture of *Sphaerothecum destruens* from sunbleak (*Leucaspius delineatus*) in the UK and pathogenicity experiments in Atlantic salmon (*Salmo salar*). Parasitol 2012;139:904–914.

Paps J, Medina-Chacon LA, Marshall W, Suga H, Ruiz-Trillo I. Molecular phylogeny of unikonts: new insights into the position of Apusomonads and Ancyromonads and the internal relationships of opisthokonts. Protist 2012;164:2e12.

Pauley GB. Prespawning adult salmon mortality associated with a fungus of the genus *Dermocystidium*. J Fish Res Board Can 1967;24:843–848.

Peeler EJ, Oidtmann BC, Midtlyng PJ, Miossec L, Gozlan RE. Non-native aquatic animals introductions have driven disease emergence. Eur Biol Invasions 2010;13:1291–1303.

Pekkarinen M, Lom J, Murphy CA, Ragan MA, Dykova I. Phylogenetic position and ultrastructure of two *Dermocystidium* species (Ichthyosporea) from the common perch (*Perca fluviatilis*). Acta Protozool 2003;42:287–307.

Pekkarinen M, Lotman K. Occurrence and life cycles of *Dermocystidium* species (Mesomycetozoea) in the perch (*Perca fluviatilis*) and ruff (*Gymnocephalus cernuus*) (Pisces: Perciformes) in Finland and Estonia. J Nat Hist 2003;37:1155–1172.

Pereira CN, Di Rosa I, Fagotti A, Simoncelli F, Pascolini R, Mendoza L. The pathogen of frogs *Amphibiocystidium ranae* is a member of the order Dermocystida in the class Mesomycetozoea. J Clin Microbiol 2005;43:192–198.

Raffel TR, Bommarito T, Barry DS, Witiak SM, Shackelton LA. Widespread infection of the Easter red-spotted new (*Notophthalmus viridescens*) by a new species of *Amphibiocystidium*, a genus of fungus-like mesomycetozoan parasites not previously reported in North America. Parasitol 2008;135:203–215.

Ragan MA, Goggins CL, Cawthorn RJ, Cerenius L, Jamieson AVC, Plourde SM, et al. A novel clade of protistan parasites near the animal-fungal divergence. Proc Natl Acad Sci USA 1996;93:11907–11912.

Rahimian H, Thulin J. Epizootiology of *Ichthyophonus hoferi* in herring populations off the Swedish west coast. Dis Aquat Org 1996;27:187–195.

Ramaiah N. A review on fungal diseases of algae, marine fishes, shrimps and corals. Indian J Mar Sci 2006;35:380–387.

Rasmussen C, Purcell MK, Gregg JL, LaPatra SE, Winton JR, Hershberger PK. Sequence analysis of the internal transcribed spacer (ITS) region reveals a novel clade of *Ichthyophonus* sp. from rainbow trout. Dis Aquat Org 2010;89:179–183.

Rowley JJL, Gleason FH, Andreou D, Wyth R, Marshall L, Lilje O, et al. Impact of mesomycetozoean parasites in amphibian and freshwater fish populations. Fungal Ecol 2013;27:100–111.

Sindermann C, Scattergood L. Diseases of fishes of the Western North Atlantic. II. *Icthyosporidium* disease of the sea herring (*Clupea harengus*). Res Bull Dep Sea Shore Fish Maine 1994;19:1–40.

Whistler HC. Pure culture of the Trichomycete, *Amoebidium parasiticum*. Nature 1960;186:732–733.

Woolhouse MEJ, Gowtage-Sequeria S. Host range and emerging and reemerging pathogens. Emerg Infect Dis 2005;11:1842–1847.

Wootten R, McVicar AH. *Dermocystidium* from cultured eels, *Anguilla anguilla* L., in Scotland. J Fish Dis 1982;5:215–222.

Mohammad N. Sarowar, Marcia Saraiva, Casey N. Jessop, Osu Lilje, Frank H. Gleason and Pieter van West

10 Infection strategies of pathogenic oomycetes in fish

10.1 Introduction

Studies on the evolutionary lineage of oomycetes show that these pathogens are capable of infecting different hosts, ranging from algae, plants, protists, fungi and arthropods to vertebrate animals (van West 2006; Phillips et al. 2008; Beakes et al. 2012). Animal pathogenic oomycetes are often neglected in spite of significant economic importance due to lack of study and scarcity of data (Diéguez-Uribeondo et al. 2009). Outbreaks of mycotic diseases, including saprolegniosis (causative agent *Saprolegnia* spp.) and epizootic ulcerative syndrome (EUS) (causative agent *Aphanomyces invadans*) can result in enormous economic losses in aquaculture threatening both production targets and even the long-term viability of the industry as well as natural ecosystems (van West 2006; Phillips et al. 2008; Sarowar et al. 2013; van den Berg et al. 2013). For example, saprolegniosis has resulted in heavy losses in salmonid hatcheries and "winter kill" in catfish aquaculture (Bangyeekhun et al. 2001; Torto-Alalibo et al. 2005). It has been estimated that 10% of all hatched salmon in the aquaculture industry succumb to *Saprolegnia* infections (van West 2006). The best-studied fish pathogenic oomycetes belong to the order Saprolegniales, which include the genera *Saprolegnia*, *Achlya* and *Aphanomyces*. Several other genera of Saprolegniales also have a significant impact on freshwater ecosystems by infecting different host organisms and their eggs (Tab. 10.1) (Kiesecker et al. 2001; Fernández-Benéitez et al. 2008; Ruthig 2009).

Shortly after 1905, and extending to about 1970, reports of saprolegniosis in the wild became more frequent and reflected increasing recognition while the cause of the disease was still in some dispute. Numerous reports support the presumption that all freshwater fish are susceptible to some oomycete infection under the right conditions (Srivastava 1980), since certain water moulds are consistently isolated from fish lesions. Most notable are *Saprolegnia*, *Achlya*, and *Aphanomyces* species from the order Saprolegniales and *Pythium undulatum* from the order Peronosporales (Noga 1993; El-Sharouny and Badran 1995). Most of our basic and applied knowledge of oomycete infections in freshwater fish is based on studies of salmonids due to their economic and ecological relevancy.

Oomycetes pathogenic to animals are zoosporic eukaryotic microorganisms – often referred to as water moulds (Coker 1923; Seymour 1970). They are fungal-like cosmopolitan organisms distributed in a variety of aquatic environments and ecological niches. Together with the brown algae and diatoms, they are classified as Stramenopiles (Heterokonts). The oomycetes however have lost their plastids in the evolutionary

Tab. 10.1: Oomycetes species reported from tissues of bony fishes (Osteichthyes) or their eggs (Adapted from Czeczuga et al. 2002; Boys et al. 2012; Czeczuga et al. 2013).

Oomycete family	Isolates	Known as pathogen (Y/N)	Observed on dead (D) or live (L) fish
Saprolegniaceae	*Achlya ambisexualis**	Y	D
	*A. americana**	Y	D
	*A. caroliniana**	Y	D & L
	A. crenulata	N	D
	A. debaryana	N	D
	*A. diffusa**	Y	D
	*A. dubia**	Y	D
	A. flagellata	–	L
	*A. intricate**	Y	D
	*A. klebsiana**	Y	D
	A. oblongata	N	D & L
	A. oligacantha	N	D
	*A. orion**	Y	D
	A. polyandra	N	D & L
	*A. prolifera**	Y	D & L
	*A. proliferoides**	Y	D
	*A. radioasa**	Y	D
	A. rodriqueziana	N	D
	A. treleaseana	N	D
	Saprolegnia ansiospora	–	L
	*S. diclina**	Y	D & L
	S. eccentrica	N	D & L
	*S. ferax**	Y	D & L
	S. hypogyna	N	D
	S. litoralis	N	D & L
	*S. mixta**	Y	D & L
	*S. monoica**	Y	D
	*S. parasitica**	Y	D & L
	S. pseudocrustosa	N	D
	*S. shikotsuensis**	Y	D & L
	*S. torulosa**	Y	D
	S. uliginosa	N	D
	Dictyuchus monosporus	Y	D
	D. pisci	–	L
	Isoachlya monilifera	Y	D & L
	Thraustotheca clavata	N	D
Leptolegniaceae	*Aphanomyces frigidophilus*	–	L
	*A. Invadans**	Y	L
	A. irregularis	N	D
	A. laevis	–	L
	A. stellatus	N	D
	Leptolegnia caudata	Y	D & L

(*continued*)

Tab. 10.1: (*continued*)

Oomycete family	Isolates	Known as pathogen (Y/N)	Observed on dead (D) or live (L) fish
Leptomitaceae	*Leptomitis lacteus**	Y	D & L
Pythiaceae	*Pythium afertile*	N	D
	P. aquatile	N	D
	P. arrhenomanes	N	D
	P. butleri	N	D
	P. diclinum	–	L
	P. dissotocum	N	D
	P. hemmianum	N	D
	P. intermedium	N	D
	P. myriotylum	N	D
	P. ostracodes	N	D
	P. periplocum	N	D
	P. proliferum	–	L
	P. tenue	N	D
	Zoophagus insidians	N	D

* Species known as pathogens; fish specimens include eggs of African catfish, *Clarias gariepinus*, muscle sections of monkey goby, *Neogobius fluviatilis*; racer goby, *Neogobius gymnotrachelus*; Chinese sleeper, *Perccottus glenii*; and stone moroko, *Pseudorasbora parva* and found on live fish bony herring, *Nematalosa erebi*; Golden Perch, *Macquaria ambigua*; Murray cod, *Maccullochella peelii*; and spangled perch, *Leiopotherapon*.

process (Baldauf 2003; Robideau et al. 2011). Water moulds are phylogenetically very distant from the Kingdom Eumycota, the true Fungi, unlike their initial appearance would suggest. For example, there are several clearly defined developmental stages in the life cycle that distinguish them from true fungi. Indeed the Stramenopiles are characterized by forming free-swimming spores, called zoospores that have two flagella of which one flagellum has tinsel hairs (Walker and van West 2007).

Common disease symptoms include visible white and gray cotton wool like patches of filamentous mycelium growing on the skin and gills of fish (Stueland et al. 2005). The rapid growing infections cause large wounds on the skin and head of the fish (Fig. 10.1), resulting in impaired osmoregulation and hemo-dilution, whereas extensive gill infections cause respiratory failure, which leads to mortality. Unlike *Saprolegnia* infection, *A. invadans* is capable of invading through fish muscle and other tissues, and induce a granulomatous host response characteristic of EUS lesions (Lilley and Roberts 1997). Fish are more prone to saprolegniosis and EUS in winter, when presumably lower water temperatures cause the fish to become immuno-compromised (Bly and Clem 1992; Bly et al. 1992; Oidtmann 2012).

Pathogenic oomycetes in the order Saprolegniales, such as *Saprolegnia* and *Aphanomyces* species, are responsible for infections on fish in aquaculture, wild population and even recreational fish tanks (Willoughby and Pickering 1977; Neish and Hughes

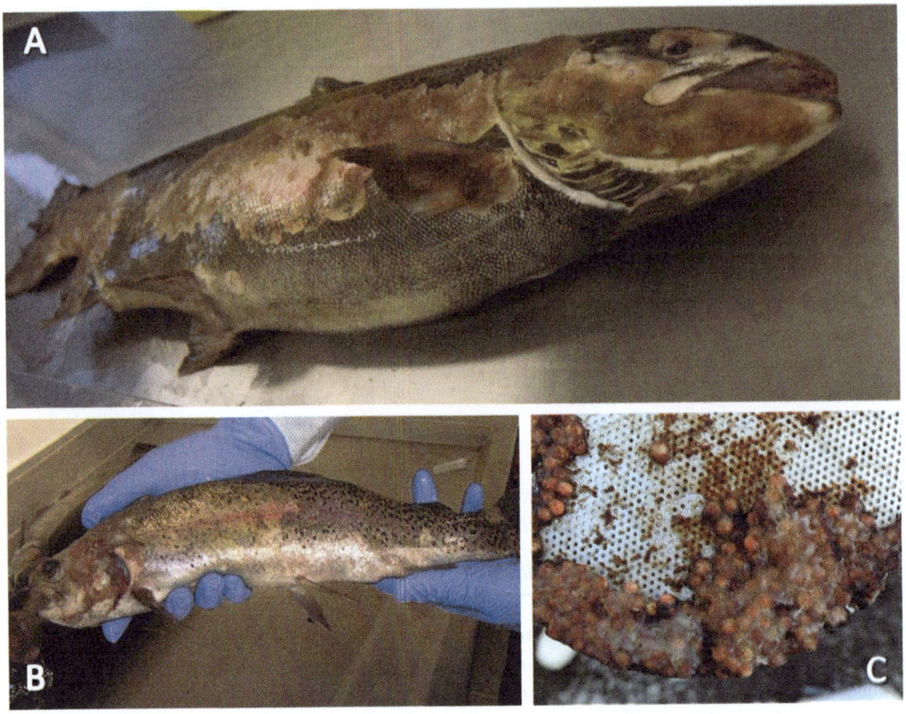

Fig. 10.1: Salmonid fish and eggs infected with Saprolegniosis. (A) Farmed salmon (*Salmo salar*), (B) farmed rainbow trout (*Onchorhynchus mykiss*). The white/gray patches on the skin, head and tail of the fish are signs of severe saprolegniosis. (C) Salmon eggs completely destroyed by saprolegniosis. The pictures were taken by Herbert van den Berg and Rodrigo Belmonte.

1980; Lilley and Roberts 1997; Bruno and Wood 1999; Hussein and Hatai 2002). One of the most destructive pathogens on fishes is *Saprolegnia parasitica*, which is endemic to all freshwater habitats around the world and is to some extent responsible for the decline of natural populations of salmonids and other freshwater fishes (Neitzel et al. 2004; van West 2006). The devastating infections are a major problem in the salmon aquaculture which includes hatcheries and farms in Scotland, Scandinavia, Chile, Japan, Canada, and the USA. Saprolegniosis costs the aquaculture industry tens of millions of pounds annually (Lilley and Roberts 1997; Hussein and Hatai 2002; Phillips et al. 2008). In the USA, "winter kill" in farmed catfish caused by *S. parasitica* results in financial losses of up to 50%, which represents an economic loss of about $40 million (Bruno and Wood 1999).

Epizootic ulcerative syndrome (EUS) is another devastating mycotic disease caused by the pathogen *A. invadans* affecting both wild and farmed fishes. It was first characterized and described in Japan in 1971 (Egusa and Masuda 1971). Since then *A. invadans* has been responsible for major epidemics in freshwater and estuarine fishes in Asia, Australia, North America and Africa. The rapid spread of the disease has substantially affected livelihoods of fish farmers and fishermen especially in Asia and Africa (Lilley et al. 2001, 2003). The infective life stage of *A. invadans* is the free-swimming zoospore that attaches to a fish host, encysts and germinates to develop vegetative hyphae invading and ramifying through host tissues (Lilley and Roberts 1997; Kiryu et al. 2003). The hyphae will invade the fish skin, muscular tissue and may reach the internal organs eventually killing the fish. The likely pathway of transmission of *S. parasitica* and *A. invadans* from infected to native fish is through the formation of a zoosporangium developing from the mycelium in fish tissue at the fish surface, followed by the formation of primary zoospores that are released from the zoosporangium.

This review focuses on the different aspects of the infection strategy of oomycetes that are pathogenic to fish. We review the physical strategy for infection with different life stages of the pathogens, which shaped their pathogenicity towards fish and molecular strategy in setting infection, which illustrates recent molecular work to support the notion that pathogenic oomycetes are capable of manipulating host immune responses in fish.

10.2 Taxonomy of oomycetes pathogenic to fish

Most oomycetes that are pathogenic to animals belong to the subclass Saprolegniomycetidae, which has two orders: Saprolegniales and Leptomitales (Phillips et al. 2008). Kützing (1843) formally erected the Saprolegniales but considered members of the order to be nonchlorophyllous algae. The first account (English) monograph of the family was published by Humphrey (1893) and revised by Coker (1923). Subsequently Moreau and Moreau (1935, 1938), Coker (1935), Dick (1969, 1972, 1978, 2001), Miller and Ristanovic (1975), Powell and Blackwell (1998) and Padgett et al. (2000) have made observations about systematic concepts in water moulds, most of which emphasize the significant morphological plasticity especially within the family Saprolegniaceae. The genera were distinguished by the structure and shape of asexual sporangia and mode of zoospore release (Seymour 1970; Daugherty et al. 1998). The species differentiation was based on morphological characteristics of their sexual structures, i.e. oogonia, oospores, and antheridia (Seymour 1970; Johnson et al. 2005). The traditional taxonomy often led to complicacy in identifying species, many species exhibit very similar or overlapping characters. For example, isolates of *Saprolegnia* often fail to produce sexual structures at all *in vitro* especially isolates of fish pathogenic *S. parasitica* and *A. invadans*. This has resulted in taxonomic confusion regarding isolates pathogenic to fish.

Advances in molecular biology have made it possible to perform complementary speciation at the taxon level by DNA sequencing. These developments not only provide more confidence but are also time-saving and are valuable in avoiding questionable or synonymous speciation (Robideau et al. 2011). The increasing availability of molecular sequence data of the members of Saprolegniales has guided studies into redrawing the phylogenetic tree. Analysis of different regions of the ribosomal RNA

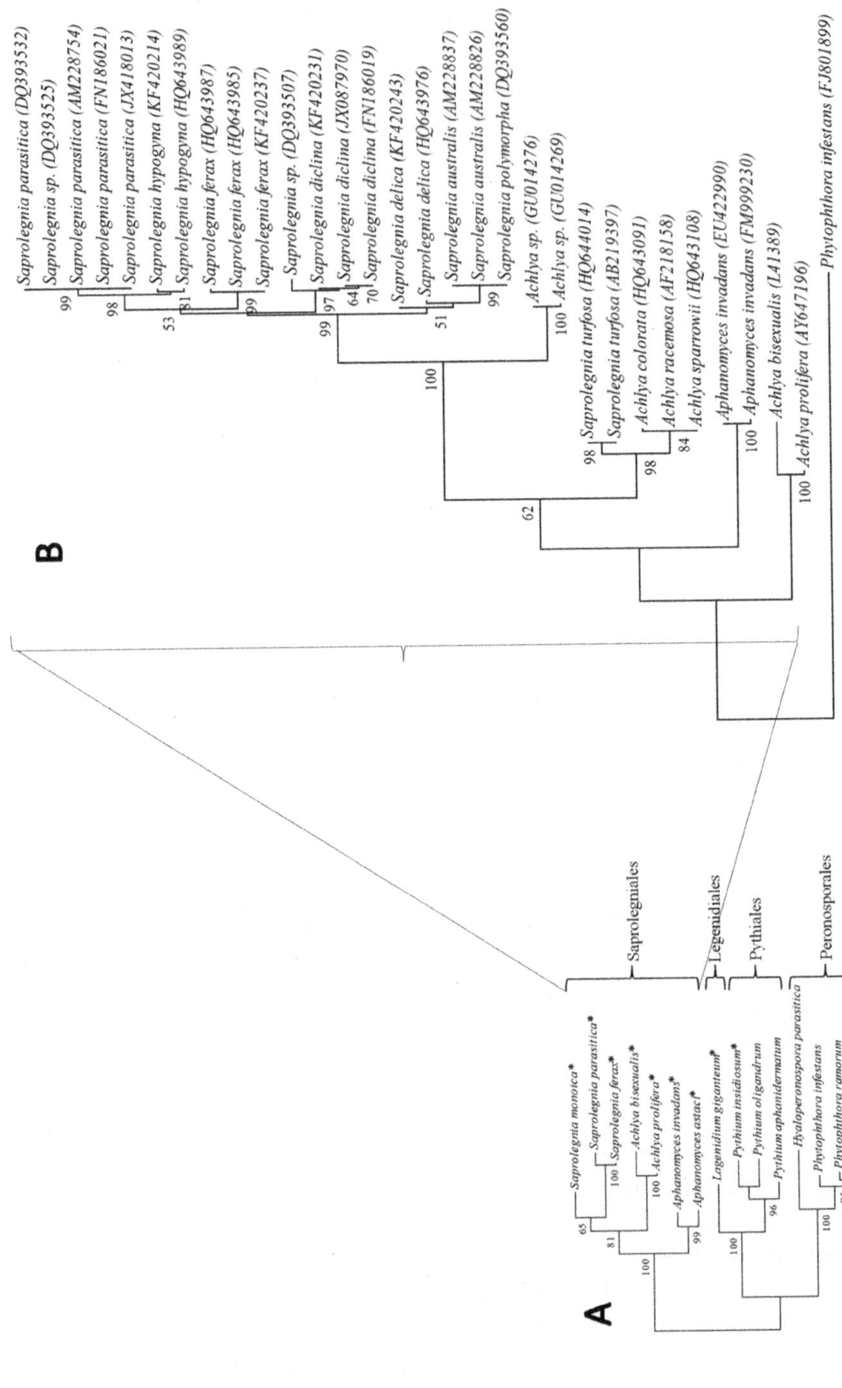
B
Saprolegnia parasitica (DQ393532)
Saprolegnia sp. (DQ393525)
Saprolegnia parasitica (AM228754)
Saprolegnia parasitica (FN186021)
Saprolegnia parasitica (JX418013)
Saprolegnia hypogyna (KF420214)
Saprolegnia hypogyna (HQ643989)
Saprolegnia ferax (HQ643987)
Saprolegnia ferax (HQ643985)
Saprolegnia ferax (KF420237)
Saprolegnia sp. (DQ393507)
Saprolegnia diclina (KF420231)
Saprolegnia diclina (JX087970)
Saprolegnia diclina (FN186019)
Saprolegnia delica (KF420243)
Saprolegnia delica (HQ643976)
Saprolegnia australis (AM228837)
Saprolegnia australis (AM228826)
Saprolegnia polymorpha (DQ393560)
Achlya sp. (GU014276)
Achlya sp. (GU014269)
Saprolegnia turfosa (HQ644014)
Saprolegnia turfosa (AB219397)
Achlya colorata (HQ643091)
Achlya racemosa (AF218158)
Achlya sparrowii (HQ643108)
Aphanomyces invadans (EU422990)
Aphanomyces invadans (FM999230)
Achlya bisexualis (L41389)
Achlya prolifera (AY647196)
Phytophthora infestans (FJ801899)
0.1
A
Saprolegnia monoica*
Saprolegnia parasitica*
Saprolegnia ferax*
Achlya bisexualis*
Achlya prolifera*
Aphanomyces invadans*
Aphanomyces astaci*
Lagenidium giganteum*
Pythium insidiosum*
Pythium oligandrum
Pythium aphanidermatum
Hyaloperonospora parasitica
Phytophthora infestans
Phytophthora ramorum
Phytophthora sojae
Saprolegniales
Legenidiales
Pythiales
Peronosporales
0.1

Fig. 10.2: Phylogeny of different orders of oomycetes. (A) Phylogenetic relationships of different orders of oomycetes. The maximum likelihood tree was constructed retrieving the internal transcribed spacer (ITS) regions sequence of the rRNA gene of the isolates from the GenBank. The animal pathogenic oomycetes are highlighted (*) and they are from a range of genera within the oomycetes. (B) The order Saprolegniales is extended into the maximum likelihood tree using the isolates that are most frequent from fish. Percentile bootstrap values for both trees based on 1,000 replications are indicated at the nodes (Adapted from Phillips et al. 2008).

(rRNA) gene, i.e. small ribosomal subunit (18S), internal transcribed spacer (including ITS1, 5.8S and ITS2) and large ribosomal subunit of rRNA gene (28S) determined phylogenetic data of the family Saprolegniaceae to be clustered in 10 robust clades (Hulvey et al. 2007). Interestingly most *Saprolegnia* isolates that are obtained from salmonid lesions fall into the same cluster based on ITS sequences of the isolates (Fig. 10.2) (Diéguez-Uribeondo et al. 2007). DNA sequence databases have been used as the main resource for identifying species with similar morphological features (Tautz et al. 2003; Hulvey et al. 2007). However, the major limitation of the databases is that there is a lack of biodiversity studies on the Saprolegniaceae. To confound this there are a number of incorrectly identified DNA sequences in the available public databases. Inevitably, this situation will result in incorrect identification of species (Bridge et al. 2003) and makes it difficult to understand the true extent of all possible oomycete species that are pathogenic to fish (Hulvey et al. 2007). Several studies have attempted to clarify the existing morphology based on traditional taxonomy using the full ITS sequences (Diéguez-Uribeondo et al. 2007; Nilsson et al. 2009).

10.3 Physical adaptation and strategy for infection: macroscopic infection, the face of infection on hosts

It has often been assumed that oomycetes pathogenic to fish set infection at the epidermis level, and probably, where the epithelial cells are ruptured (Rucker 1944). However, it has also been suggested that *Saprolegnia* species can gain entrance into susceptible animals by penetrating directly into the intact epidermis, or entering into natural body openings or wounds. The invading organism also may enter at points where the epidermis is not scaly (Willoughby 1971) or is thin. *Saprolegnia* species are known to infect gills, skin and fins of mature fish and also eggs from aquatic animals such as fish and amphibians (Hussein et al. 2001; van West 2006; Petrisko et al. 2008; Ghiasi et al. 2010; Thoen et al. 2011; Cao et al. 2012). Lesions associated with hyphae belonging to the Saprolegniaceae develop at the site of infection and are usually accompanied by necrosis, ulceration (Bootsma 1973) and inflammation (Vishniac and Nigrelli 1957). The hyphae of some water moulds can penetrate into the dermal areas (connective tissue) in fish and can also adhere closely to the stratified epithelial cells of the epidermis. According to Bootsma (1973), necrosis may be accompanied by cell proliferation.

In the late 1970s Neish (1977) observed the course and progression of saprolegniosis in a few Pacific salmons. Hyphae at first destroyed the multicellular stratified squamous epithelium followed by penetration into the basement membrane. Subsequently, as the infection proceeded, the hyphae grew into the *stratum spongiosum* of the dermis and then into the collagen bundles of the *stratum compactum*. From the dermis, hyphae can penetrate into the hypodermis, and then into the subtending musculature. The histopathology of the invasion of salmonid fry was investigated by Miyazaki et al. (1977). They found hyphal embolisms in the blood of infected fish, and mycelium throughout the abdominal cavity. Hyphae of *Saprolegnia* sp. was found to have penetrated into the stomach lumen at the pyloric region, and grew from sites in the stomach wall into the pancreas, pyloric caeca, spleen, abdominal musculature, and visceral blood vessels.

10.4 Oomycete zoospores, the first line of attack

One of the key characteristics that separate oomycetes from true fungi is the formation of free-swimming zoospores. The epidemic spread of the oomycete diseases is based on the very rapid dispersal by zoospores from host to host that are released from zoosporangia. The released zoospores can actively swim which probably increase the chance of finding new hosts. The number of zoospores released per zoosporangium varies from species to species. The zoospores of *Saprolegnia* species have a single nucleus and lack cell walls and can swim with the help of two flagella each of differing lengths; one tinsel (or anterior) flagellum, and one whiplash (or posterior) flagellum. The anterior flagellum is probably responsible for pulling the zoospore through the water whereas the posterior flagellum acts as a rudder for steering the cell (Judelson and Blanco 2005). Generally, zoospores are "bean" shaped, due to the ventral groove into which the two flagella are inserted, both in a lateral orientation (Dick 1973, 1990, 2000).

Zoospores are also essential for targeting the site of infection, which is often referred to as the homing response (van West et al. 2003). Oomycete zoospores have evolved intriguing homing responses. For example, they are able to make use of electro tactic as well as chemotactic clues (Carlile 1983; Donaldson and Deacon 1993; Morris and Gow 1993; van West et al. 2002, 2003). Once a suitable site, or a host, has been found, the zoospore undergoes encystment and becomes immobilized. This involves the rapid production of a cell wall surrounding the zoospore. The flagella are either discarded or retracted into the cell which depends on the species (Dick 2001). Subsequently the cyst attaches to the host.

Many pathogenic isolates of *Saprolegnia* and *Aphanomyces* species have two different types of motile zoospores. Initially a primary zoospore is released from the zoosporangium. It then encysts, and later a secondary zoospore can emerge from the cyst. Both primary and secondary zoospores have two flagella, an anterior and a posterior flagellum, although the lengths of the flagella in the secondary zoospores may be twice as long (Walker and van West 2007). The secondary zoospores are considered the better swimming spore and are more comparable to zoospores from the Peronosporales, such as, *Phytophthora* and *Pythium* (Beakes 1987). Secondary zoospores have a longer motile period and are considered crucial for successful infection.

The production of two generations of zoospores (diplanetism), often morphologically different, is one of the defining characteristics of the "saprolegnian line" or "saprolegnian galaxy" of oomycetes (Sparrow 1976; Barr and Désaulniers 1987; Dick 2001). Suppression of the "primary" (or first-formed) generation of zoospore occurs widely in the order Saprolegniales. However, non-motile aplanospores may be released from the sporangium, forming balls of cysts, as in the genera *Aphanomyces* and *Achlya*, or may be retained in the sporangium as in *Dictyuchus* and *Thraustotheca* (Beakes 1987; Dick 2001). Zoospores of oomycetes contain an array of peripheral vesicles, which upon encystment are discharged to form either a ventral "adhesive pad" or an outer cyst coat layer (Beakes 1987; Gubler and Hardham 1988; Lehnen and Powell 1989; Burr and Beakes 1994; Hardham 2005) which reflects evolutionary modification for better attachment to hosts.

10.5 Triggers for zoospore formation, waking up the beast

The asexual zoosporangial phase of fish pathogenic isolates of oomycetes is mainly responsible for dispersing free-swimming infectious propagules in order to find new hosts. The zoosporangia are formed at the end of the hyphae and are separated from the hyphae by basal septa. Specific triggers are known to set off zoospore release, e.g. the lack of nutrients (Fuller and Jaworski 1987) or a sudden drop in temperature. The presence and number of hosts might also bring about a queue in fish pathogenic oomycetes to initiate sporulation. In most cases, the release of zoospores is finely orchestrated with the host's physiology. The number of zoospores being produced and released can significantly increase with sudden drop of temperature (Bly et al. 1992). The latter also leaves a cold-blooded animal like fish compromised, which increases the chance of infection by pathogens (Bly et al. 1992). In case of *A. invadans*, infection in fish occurs mostly at water temperatures of 18–22°C or less (Bondad-Reantaso et al. 1992). These conditions also favor the sporulation of *A. invadans* (Lumanlan-Mayo et al. 1997) and temperatures of 17–19°C have been shown to delay the inflammatory response of some fish species to oomycete infection (Chinabut et al. 1995; Catap and Munday 1998).

10.6 Encystment and germination, one step closer to infection

Encystment and germination is considered to be vital for establishing a successful infection. Zoospores undergo the rapid process of encystment, which involves shedding the two flagella and the formation of a non-motile walled cyst (cytospore). This could help zoospores to thrive a bit longer in the absence of a suitable host. Contact of the zoospore with the host triggers encystment, which in turn initiates germination and eventually causes infection.

The triggers that lead to encystment under laboratory conditions include agitation by mechanical force and the presence of nutrients such as peptone (Dieguez-Uribeondo et al. 1994), dissolved fish mucus or nutrient solutions such as bovine serum albumin, haemoglobin and hempseeds (Willoughby 1986; Dieguez-Uribeondo et al. 1994). The environmental factors that lead to encystment and germination in zoospores may vary between species (Willoughby et al. 1983), but a single regulatory system is proposed to control both encystment and germination cycles (Andersson and Cerenius 2002). The encystment of zoospores has been found to be regulated by Ca^{2+} in *Aphanomyces astaci*, *Aphanomyces euteiches* and even in *Phytophthora parasitica* (Cerenius and Söderhäll 1984a, b, 1985; Warburton and Deacon 1998). It is thought that this may also be the case for other pathogenic oomycete species (Donaldson and Deacon 1992).

Cytological studies have identified key peripheral organelles associated with encystment of zoospores and cysts. One of these is the kinetosome-associated organelle (or K-body) (Holloway and Heath 1977; Beakes 1983; Lehnen and Powell 1989, 1991). The K-body is the first organelle to be discharged by fusing with the plasma membrane and forming a diffuse fibrillar pad (Lehnen and Powell 1989; Burr and Beakes 1994). Secondary encystment vesicles (SEVs) have been described in *S. diclina* and *S. parasitica* in association with encystment (Burr and Beakes 1994). Fully differentiated SEVs in *S. parasitica* contain boathook spines whereas in *S. diclina* the SEVs contain capsules shaped bodies (Beakes 1983). During encystment these vesicles break-up into smaller vesicles that appear to form bubbles on the surface of the secondary zoospores before discharging and releasing an electron dense surface in which, in the case of *S. parasitica*, the boathook spine bundles are embedded (Beakes 1983; Burr and Beakes 1994). The SEV discharge reacts with the lectin Concanavalin A (Con A), which indicates the presence of mannose and glucose sugars. The discharge becomes patchy in germlings and localised with the bundles of spines (Burr and Beakes 1994). Burr and Beakes (1994) infer that the entrapment of this discharge material in amongst the boathook spines may act as an adhesive complex that facilitates binding to fish. Con A induces strong encystment of *S. diclina* and *S. parasitica* secondary zoospores (Burr and Beakes 1994). The lectin wheat germ agglutinin (WGA) is known to react to N-acetyl glucosamine and binds to sites that correspond closely with K-body loci on primary and secondary zoospores. WGA induces the encystment of secondary zoospores of *S. diclina*, but not *S. parasitica*, which may correspond to its relatively poor expression on *S. parasitica* (Burr and Beakes 1994). The expression of N-acetyl glucosamine has also been observed on the adhesive pads of K_2-bodies of *S. ferax* spores attaching to a substrate (Lehnen and Powell 1989), hyphal adherence on plastic surfaces (Willoughby 1986) and *S. parasitica* germ tubes (Burr and Beakes 1994). Fibrillar type vesicles release fibrillar material but there is no evidence of spines (Burr and Beakes 1994). The lack of lectin reactive compounds in these vesicles, suggests the absence of glycoconjugates with accessible residues of sugars.

Evidence from *Phytophthora cinnamomi* suggests these vesicles are repositories for storage proteins (Gubler and Hardham 1988, 1990).

In several *Saprolegnia* species, a set of hairs or boathooks or spines is produced upon encystment on the surface. Although there is considerable variation both in the length and number of the long-hooked hairs associated with the secondary cysts (Pickering et al. 1979; Hatai et al. 1990; Fregeneda-Grandes et al. 2000), the size and structure of the hairs seem to be species specific, although more evidence is required for this characteristic to be used for species identification (Hatai et al. 1990; Diéguez-Uribeondo et al. 2007).

There is debate amongst scientists as to the function of hairs on the cysts. Some suggest that they may be required for attachment to a host. Others imply that it slows sedimentation when free flowing in a water column. These advantages would be consistent with increasing the ability of cysts to attach themselves to passing fish, as has been reported on several occasions (Willoughby and Pickering 1977; Willoughby et al. 1983; Burr and Beakes 1994; Quiniou et al. 1998). Long-hooked hairs of the cysts seem to favor contact and attachment to the surface of the host or substratum (Pickering and Duston 1983; Hallett and Dick 1986), and presence of bundles of long hairs appears to be often related to pathogenicity (Pickering and Willoughby 1982; Fregeneda-Grandes et al. 2000). Reports suggest that adaptation to parasitism in *Saprolegnia* might have occurred at cyst level by the development of long-hooked hairs to facilitate host attachment and selection of a retracting germination (Diéguez-Uribeondo et al. 2007). Diéguez-Uribeondo et al. (2007) found that most *Saprolegnia* isolates obtained from salmonid lesions contained cysts with bundles of long-hooked hairs and suggested to put them into the same cluster. Curiously, *Saprolegnia* isolates with the longest hairs (Fregeneda-Grandes et al. 2000, 2001) (Type I) appeared to be less aggressive pathogens or even non-pathogenic, at least towards salmonid fish, than those with shorter spines (Fregeneda-Grandes et al. 2000, 2001) (Type II) and these observations appear to be supported by other studies (Hatai et al. 1990; Hussein and Hatai 1999; Stueland et al. 2005). This could mean that the size and number of hairs might not be directly related to the pathogenicity of the species.

10.7 Repeated zoospore emergence, the back-up plan

The release of primary motile zoospores from sporangia serves a minimal level of dispersal from the parent colony. The released zoospores undergo encystment if a suitable host is found. In several fish pathogenic species, such as, *S. parasitica* and *A. invadans* the resulting cyst is capable of releasing a new zoospore. In several *Saprolegnia* species the secondary zoospores can undergo repeated cycles of encystment and zoospore release, which is known as "repeated zoospore emergence" or "polyplanetism" (Cerenius and Söderhäll 1984b; Hatai et al. 1990; Diéguez-Uribeondo et al. 1994, 2007). Repeated zoospore emergence may represent an adaptation to the parasitic mode of life in the Saprolegniales as it enables zoospores to make several attempts to find and colonise a suitable host (Cerenius and Söderhäll 1985; Diéguez-Uribeondo et al. 2007, 2009). *Aphanomyces invadans* isolates have been shown to produce one additional generation of motile zoospores from artificially encysted secondary zoospores, but no further generations could be induced. This is in contrast to *A. astaci*, where up to three successive generations can be induced (Cerenius and Söderhäll 1984b, 1985).

10.8 Chemotactic response of zoospores, the specialization

Some aspects of the physiological and ecological properties of zoospores in general have been reviewed by Hickman and Ho (1966), Chet and Mitchell (1976), Fuller (1977) and Gleason and Lilje (2009). Motile zoospores can be released from zoosporangia of many parasitic species in large numbers. The chemotactic properties of zoospores of a few species of various groups of fungi and fungus-like organisms have been studied in the laboratory. Some of the zoospores which infect plant and animal hosts are known to be chemotactic, sense exudates, navigate in chemical gradients and can swim for several hours. For example, Bimpong and Clerk (1970) found that zoospores of the plant parasite *Phytophthora palmivora* remain motile in distilled water for 84 h at 17°C, for 24 h at 27°C and for 1 h at 33°C. Zentmyer (1961) reported that zoospores of *Phytophthora cinnamomi* sense the presence of exudates from avocado roots, settle on the surface of the roots and begin to encyst and germinate within 30 to 60 min in the laboratory.

Much of the early research on chemotaxis in zoospores of oomycetes focused on parasites of vascular plants, particularly species of *Phytophthora*. For example, Khew and Zentmyer (1973) tested the responses of zoospores of *P. cactorum, P. capsici, P. cinnamomi, P. citrophthora* and *P. palmivora* to 12 amino acids (alanine, arginine, aspartate, asparagine, glutamate, glutamine, histidine, leucine, methionine, phenylalanine, proline and serine). Zoospores of all five species were attracted to all amino acids, but there was considerable variation in the magnitude of the responses to amino acids by different species. Halsall (1975) tested the responses of zoospores of *P. drechsleri, P. cryptogea, P. cinnamomi, P. nicotianae* var. *parasitica* and *P. citricola* to four amino acids (asparagine, glutamine, aspartate and glutamate) and to sucrose and ethanol. Zoospores of all five species were attracted to all six compounds.

More recently the properties of zoospores have been studied in parasites of animals. For example, El-Feki et al. (2003) tested the responses of zoospores of *S. parasitica* to metabolites (aspartate, alanine, leucine, tyrosine, histidine, arginine, glucose and sucrose). Zoospores were attracted to all eight chemical compounds.

It is clear from these studies that zoospores of facultative parasites of both plant and animal species are attracted to a large variety of exudates of metabolites from tissues of their hosts. The chemotactic responses of zoospores, particularly attraction to amino acids, expedite quick arrival at uninfected and susceptible animal hosts which are rich in proteins. However, studies providing data on positive chemotaxis of zoospores and metabolism capacities by parasites of amino acids and sugars also suggest that it is also possible that some facultative parasites might be able to utilize protein and carbohydrate compounds in both plant and animal substrates.

10.9 Proteins and amino acids as substrates for growth

Proteins from many sources, e.g. casein and keratin, are known to be good substrates for the isolation of many species of oomycetes, particularly facultative parasites of animals, into pure culture and for their subsequent growth in liquid media (Sparrow 1960). Living animal hosts and parts of dead animals, in particular, are protein rich environments. Proteins must be digested outside the cell and the amino acids produced must be transported into the cell prior to their catabolism. Smith et al. (1994) demonstrated proteolytic activity of *S. diclina*, *S. ferax* and *S. parasitica* by observing the clearing of casein in solid media. Jiang et al. (2013) documented the presence of genes for serine, metallo- and cysteine proteases and genes for amino acid transporters in the complete sequence of the genome of *S. parasitica*.

Once the protein substrates have been cleaved into their component amino acids by proteases, the amino acid molecules must be catabolized and the energy in their C-C bonds converted into high energy compounds, such as ATP. Some amino acid molecules can be used directly for the synthesis of proteins or converted into nucleotides for the synthesis of DNA in intermediary metabolism. Some saprotrophic isolates of *Saprolegnia*, *Achlya*, *Dictyuchus*, *Leptolegnia*, *Aphanomyces*, *Apodachlya*, *Leptomitus* and *Pythium* grow rapidly on many but not all amino acids as sole sources of carbon and nitrogen in liquid media (Tab. 10.2) (Gleason 1968; Gleason et al. 1970a, b). Both saprotrophic and facultatively parasitic isolates of *Saprolegnia* can grow rapidly on mixtures of amino acids, and these isolates are capable of removing the molecules of the amino acids from a mixture in the growth medium (Gleason 1973; Nolan 1976). These data indicate that many oomycetes have the capacity for the digestion of proteins, the subsequent uptake of amino acids from a mixture and their subsequent catabolism, and that carbohydrates do not need to be present in the growth medium although many oomycetes can use carbohydrates as well as proteins as carbon sources. Therefore, we would expect many species of oomycetes to grow in protein rich environments such as dead or living animal hosts as saprotrophs and facultative parasites. For example, Czechuga et al. (2002) isolated many species of oomycetes from pieces of fish muscle tissue placed in lakes in Poland.

Sarowar et al. (2013) conducted challenge experiments to see whether *Saprolegnia* spp. isolates from insects and amphipods were able to infect and kill stonefly nymphs, salmon eggs and frog embyos and found that *S. diclina* and *S. ferax* were pathogenic to all three hosts whereas *S. parasitica* was less aggressive. They hypothesised that *S. parasitica* may have evolved more specific affinity to live fish compared to *S. diclina* and *S. ferax* which seem to be more saprotrophic and facultative parasites

Tab. 10.2: The growth of ten species of oomycetes on single amino acid compounds as sole sources of carbon and nitrogen. The oomycetes were grown in 50 ml of growth media with only amino acid containing fixed Carbon amount of 4 g/l (pH 6.6 continuously) in a 125 ml Erlenmeyer flasks placed on a rotary shaker. Dry weight indicates the amount of growth after various times of incubation (days). The conditions for growth and the composition of the growth media are described in detail by Gleason (1968) and Gleason et al. (1970a, b). The blanks indicate that the amino acid was not tested.

Compound	Dry weight (mg/flask) and days (in parenthesis)									
	Saprolegnia sp.	*Achlya ambisexualis*	*Leptolegnia eccentrica*	*Dictyuchus sterile*	*Leptomitus lacteus**	*Apodachlya punctata*	*Apodachlya brachynema*	*Apodachlya minima*	*Aphanomyces laevis*	*Pythium ultimum*
Alanine	67 (6)	83 (6)	44 (7)	78 (7)	50 (11)	88 (9)	67 (10)	83 (16)	72 (9)	52 (9)
Glutamate	71 (6)	72 (11)	70 (14)	71 (12)	63 (14)	65 (14)	70 (8)	87 (11)	73 (15)	65 (9)
Aspartate	36 (21)				22 (22)	43 (17)	45 (21)	35 (20)	0 (14)	43 (8)
Proline	99 (9)	112 (6)	94 (10)	93 (9)	77 (16)	79 (18)	107 (21)	0 (14)	0 (14)	25 (21)
Leucine	59 (21)	73 (15)	26 (21)	92 (12)	97 (14)	86 (21)	83 (12)	0 (14)	0 (14)	0 (21)
Hydroxyproline	0 (21)				57 (21)	0 (16)				
Arginine	72 (6)	73 (8)	31 (16)	22 (120)	0 (16)		0 (14)	0 (14)	8 (21)	2 (21)
Phenylalanine	81 (9)	70 (9)	34 (18)	1 (16)	0 (16)				0 (14)	0 (21)
Serine	33 (21)				0 (16)				0 (14)	25 (21)
Lysine	20 (21)				0 (16)					
Glucose	108 (3)	109 (5)	70 (7)	91 (8)	0 (16)	88 (7)	78 (10)	100 (10)	76 (9)	77 (5)

* The growth of *Leptomitus lacteus* was measured after incubation for 21 days on a mixture of amino acids (asparagine, arginine, phenylalanine, glycine, serine, threonine, isoleucine and valine). The dry weight was 59 mg. No growth was reported after 16 days with each of these amino acids as single sources of carbon.

that adapt to seasonal variations in the presence of potential hosts (Sarowar et al. 2013). Whether the generalist nature of the latter species correlates with wider distribution in the environment needs to be determined.

10.10 Sexual reproduction, seeing through the bad times

Sexual reproduction in oomycetes is oogamous and fertilization is achieved by the direct transfer of a non-motile gametic nucleus from an accompanying antheridium into the egg in the oogonium (Sparrow 1960, 1976; Karling 1981; Dick 2001). Oogamous sexual reproduction is clearly one of the major evolutionary developments that define more advanced oomycetes (Beakes et al. 2012). The sexual phase is thought to enable survival under harsh conditions such as dryness and extreme temperature until conditions become more favorable for germination. Sexual reproduction in *Saprolegnia* begins with the formation of an oogonium and an antheridium. Fertilization occurs when the antheridium grows towards and its nucleus fuses with the egg nucleus in the oogonium. Oospores are released as enclosed eggs characterized by a thick wall that enables survival under harsh conditions and germinate when conditions improve (Beakes and Bartinicki-Garcia 1989; Diéguez-Uribeondo et al. 1996). Sexually produced oospores differ in number, size and shape from species to species allowing this feature to be one of the key tools for taxonomic speciation (Coker 1923; Beakes et al. 2012). Under laboratory conditions, sexual reproduction of several *Saprolegnia* species can be initiated by growth on nutrient-poor substrates or by reducing the temperature. As sexual reproduction appears to be difficult to initiate *in vivo*, and is sometimes impossible *in vitro*, it is not always possible to identify species based on morphological characteristics of their sexual structures, especially the fish pathogenic isolates (Beakes et al. 1994; Diéguez-Uribeondo et al. 2007, 2009). This severely hampers the identification of *Saprolegnia* species, especially the isolates from salmonid lesions. Sexual reproductive characters are fundamental to the identification of the Peronosporomycetes (Dick 2001), but such structures have not been observed in cultures of *A. invadans*.

10.11 Molecular adaptation and strategy in setting infection: microscopic infection

In order to successfully establish an infection, oomycetes secrete effector proteins. Effector molecules act as “ligands” that can regulate enzyme activity, gene expression, or cell signaling. Once inside or within the host cell surroundings the effector proteins can alter the host defences by modulating host immune responses or by inhibiting host cell functions, such as phagocytosis, proinflammatory responses, apoptosis, and intracellular trafficking allowing the infection to progress (Dou 2008; Wawra et al. 2012a, b). Depending on where the effectors fulfil their function, they are classified as extracellular or apoplastic effectors when it acts on the host cell surroundings, or intracellular effectors if they are able to translocate into host cells where they can interfere with host defence mechanism (enzymatic machinery and/or immune system) (Stassen and Van den Ackerveken 2011; Wawra et al. 2012a). Usually effector molecules that mediate invasion of host cells are also secreted from mycelia. Like plant cells, oomycetes secrete a number of hydrolytic proteins into the extracellular space. For example, glycosyl hydrolases have been found in *S. parasitica* (Torto-Alalibo et al. 2005), *P. infestans* (McLeod et al. 2003) and *A. euteiches* (Gaulin et al. 2008) and will aid in breaking down cell wall components and thus allow entry into host tissues.

With the publication of the first complete genome sequence of *S. parasitica*, several genes encoding serine proteases, metalloproteases and cysteine proteases, were found to be largely extended in numbers (76, 69 and 85 genes, respectively) when compared to other oomycete sequences, like *P. sojae* or *P. infestans* (Jiang et al. 2013). Interestingly, an extended group of cysteine proteases (48 out of 85 genes) was found of which the papain-like peptidase C1 proteases were most prominent. Cysteine proteins are mostly found in fruits. They seem to play several roles in plant physiology and development. In addition, they are involved in signalling pathways and in the response to biotic and abiotic stresses. In humans cysteine proteases are responsible for MHC class II immune responses, apoptosis, hormone processing, and extracellular matrix remodeling being important for bone development (Grudkowska and Zagdanska 2004). In humans and livestock, fruits like papaya or pineapple that are rich in cysteine proteases are used against intestinal worm infections in homeopathic medicine (Behnke et al. 2008). In the particular case of papain-like peptidase C1 proteases it is known that it plays a role in defence against herbivorous insects (van der Hoorn and Jones 2004). Nevertheless nothing is known about the functions of cysteine proteases in *S. parasitica*. Notably, *S. parasitica* possesses one of the largest repertoires of proteases (270 in total) among eukaryotes that are deployed in waves at different points during infection as determined from RNA-Seq data (Jiang et al. 2013).

A signal that might direct some secreted proteins for uptake by the host cell has been found in plant pathogenic oomycetes. A short, conserved amino acid sequence signature, named “RxLR” (arginine – any amino acid – leucine – arginine sequence), occurs within 30 residues of the N-terminal signal peptide and is common to the known oomycete avirulence proteins as well as to other secreted proteins of unknown function (Ellis and Dodds 2011; Wawra et al. 2012b). In 2010, a host target protein, SpHtp1, a putative effector from *S. parasitica*, was found to be delivered *in vitro* by the pathogen into cells of a rainbow trout cell line (RTG-2). Although SpHtp1 contains a RHLR sequence within the characteristic position for an RxLR motif, it is not considered as a true RxLR effector due to the lack of a W-Y-sequence element and the gene is not under diversifying selection as is usually the case for *Phytophthora* RxLR effectors (Wawra et al. 2012a, b). SpHtp1 is translocated specifically into fish cells independently of the pathogen.

Since 2005, more genome sequences of several oomycete pathogens have become available. Computational analysis showed an enrichment of the RxLR-motif within the secreted proteins of species only from the order Peronosporales with up to several hundred putative RxLR-effectors (Tyler et al. 2006; Whisson et al. 2007; Haas et al. 2009; Baxter et al. 2010) whereas other oomycetes, e.g. *S. parasitica* show no enrichment or could have different conserved motifs (Lévesque et al. 2010; Schornack et al. 2010; Kemen et al. 2011; Links et al. 2011; Jiang et al. 2013). Nevertheless, several genes encoding host-targeting proteins (including SpHtp1) were identified in the genome of *S. parasitica*. Most of these candidates have no homology to known proteins, and their functions are currently unknown (Jiang et al. 2013).

Oomycetes evolved the ability to infect plants and animals independently of other eukaryotic microbes, due to the development of unique mechanisms of pathogenicity. With the availability of a respectable molecular toolbox and a multitude of gene sequences, significant progress has been made in understanding the molecular basis of infection by oomycetes. However, most of our knowledge remains limited to economically important species in the *Phytophthora* genus, and little is known about infection by other pathogenic oomycetes (Kamoun 2003).

10.12 Host responses to oomycete infections

There is a lack of knowledge about animal responses to the presence of oomycete pathogens. In the case of *A. astaci*, germination frequently takes place on the joints or between the abdominal segments of the crayfish exoskeleton. The infection can be seen as black spots due to heavy melanisation. This phenomenon occurs around the hyphae of *A. astaci* within the cuticle of the crayfish cells. The melanization process occurs due to the active form of the enzyme prophenoloxidase (proPO). Cerenius et al. (2003) revealed that crayfish species that are resistant to *A. astaci* produce proPO transcripts constitutively, whereas in susceptible crayfish the expression of this gene is significantly lower. The authors suggest an adaptation of the resistant (North American) crayfish to the presence of *A. astaci* by producing proPo continually, enhancing their defence mechanism against *A. astaci*. A similar mechanism can be found in insect larvae exposed to the pathogen *Lagenidium giganteum* (Golkar et al. 1993). In this case, the larvae of *Anopheles gambiae* induce an intense melanization which encapsulates the germ tubes of the pathogen.

It has been assumed, but never proven, that small wounds on the epidermal layer of the fish are possibly the point of entry for *Saprolegnia*. This would thus suggest that the intact epidermis plays a key role in fish protection against infection by *Saprolegnia*. The mucosal layer present on the epidermis acts as a protective shield and contains several enzymes, such as proteases, that seem to help its protective role (Pickering

1994). Furthermore, Chivers et al. (2007) demonstrated that the epidermis of challenged fish with cysts of either *S. ferax* or *S. parasitica* had an enhanced number of club cells when compared to the epidermis of uninfected (control) fish. Club cells have an interesting role in protection against pathogens. These cells are present in the surface layers of the epidermis and when attacked by a pathogen it releases chemical alarm cues. Interestingly, fish that had been immune-suppressed after exposure to the heavy-metal cadmium (Cd) were not able to increase production of club cells in response to *Saprolegnia* challenges (Chivers et al. 2007). This is very important once immune-compromised fish have a higher incidence of saprolegniosis (Bly et al. 1994). The infection outcome can vary greatly not only due to factors that predispose fish to disease, but also to difference in virulence of individual strains of *Saprolegnia* (Stueland et al. 2005). Saprolegniosis can progress rapidly and often inflammatory responses in the fish do not occur (López-Dóriga and Martínez 1998).

10.13 The animal trade is responsible for the spread of pathogens into novel and wild ecosystems

The oomycetes impact a wide array of taxa including; fish, insects, zooplankton, nematodes, crayfish and amphibians (Phillips et al. 2008; Beakes et al. 2012). However, they are most commonly known for their detrimental impact on aquatic taxa (finfish, crustaceans and molluscs) (Arthur and Bondad-Reantaso 2002), as several outbreaks involving the oomycetes have caused substantial economic damage and mass mortalities in both novel ecosystems and in the wild (van West 2006). *Saprolegnia parasitica* has caused global losses worth tens of millions of dollars in salmon and catfish farms impacting Scotland, Scandinavia, Chile, Japan, Canada and the USA (van West 2006). Additionally, *S. parasitica* could possibly partly explain the global declines in wild salmonids, having been reported to cause 22% mortality in the USA (Neitzel et al. 2004).

One major contributor that can be directly linked to oomycete outbreaks is the global transport of aquatic taxa (Arthur and Bondad-Reantaso 2002). Live aquatic taxa are transported for a number of reasons; 1) intentional introduction or transfer into aquaculture facilities or natural waters, 2) intentional introduction of ornamental fish, and 3) unintentional introduction via ballast water of ships. In addition to the transport and introduction of non-native aquatic taxa, oomycetes can be transported and introduced along with resistant non-native hosts, accidentally introducing diseases into new local ecosystems (Phillips et al. 2008). While there are many benefits to the global trade of aquatic taxa including the increase in aquaculture production, recreational fishing and biological control (Stewart 1991; Arthur and Bondad-Reantaso 2002; Fisher and Garner 2007), oomcyetes can be highly lethal to native hosts and can cause 100% mortality (Fisher et al. 2012). *A. astaci*, commonly referred to as the crayfish plague, is responsible for eradicating crayfish *Astacus astacus, A. leptodactylus* and *Austropotamobius pallipes* populations in the UK. Native species in the UK were exposed to this Oomycetes when resistant

North American species *Procambarus clarkii, Orconectes limosus* and *Pacifastacus leniusculus* were introduced (Phillips 2008). *Aphanomyces invadans* which causes epizootic ulcerative syndrome (EUS) also commonly known as red spot in over a hundred freshwater and estuarine fish species worldwide has been isolated in Asia, Australia and the USA (Callinan et al. 1995; Lilley and Roberts 1997; Chinabut 1998; Boys et al. 2012; Huchzermeyer and van der Waal 2012). In 2006, the pathogen was also found in the Zambezi River System (ZRS) in Africa and by 2011 had spread further along the ZRS, invading several new ecosystems (Huchzermeyer and van der Waal 2012; Nsonga et al. 2013). The low host specificity of *Aphanomcyes* spp. increases its prevalence amongst a range of species, increasing the chance of disease outbreaks (Huchzermeyer and van der Waal, 2012). This raises concerns about the potential release of *Aphanomyces* spp. into new habitats as a result of global trade of aquatic animals. Especially since global trade has increased from 25% to 38% (57 million tonnes) in 1976 and 2010, respectively (The State of Food and Agriculture 1998; The 2012 State of World Fisheries and Aquaculture). This reflects how significant the impact of the transport and introduction of oomycete infected individuals could be for novel and wild ecosystems at both the local and the pan-continental scale.

10.14 Future perspectives

Oomycetes that are pathogens of animals are currently receiving substantially more attention than in the past, which is primarily due to their roles in causing serious economic damage in aquaculture. However, besides causing problems in fish farms, these water moulds are also causing major problems in the environment. Remarkably more than 150 years after their discovery, we still know very little about the infection process both microscopically and at the molecular level. In fact we even don't have an answer to one of the most basic questions: "do oomycetes only infect wounded or immune compromised animals, or can they also infect healthy animals?"

Recent developments regarding the basic molecular processes of *Saprolegnia* species pathogenic to fish, the nature of the interactions with their hosts, and the identification of genes and proteins involved in these processes and host immune responses have increased our understanding of some aspects of the infection process of the pathogen. The high level of genome plasticity and ever expanding reservoirs of effectors proteins disclosed from the genome sequence of *S. parasitica*, will hopefully help us to locate the most essential targets for genetic resistance breeding or chemical controls. The unique metabolic features of oomycetes and common effector translocation pathways as reviewed in this chapter may offer key treatment targets.

Efforts to elucidate the molecular mechanisms by which *Saprolegnia* infect its host and studying the host's immune responses, have mainly focused on *S. parasitica* infections on salmon and trout. New genomic approaches to gain insight into the transcriptome of *S. parasitica* and the functional characterization of the virulent genes will undoubtedly aid in our understanding of how the pathogen is able to cause disease and henceforth leading to a novel control strategies to increase fish health, reduce losses due to mycotic disease and increase production.

Acknowledgments

The authors would like to acknowledge the Biotechnology and Biological Sciences Research Council (BBSRC, BB/G012075/1, PvW); Natural Environment Research Council (NERC, NE/F012578/1, PvW); the University of Aberdeen (PvW); The Total Foundation (PvW); and Commonwealth Scholarship Commission (BDCS-2010-42, MNS), a Marie Curie Initial Training Networks with the SAPRO (Sustainable approaches to reduce Oomycete (*Saprolegnia*) infections in aquacultures) grant PITN-GA-2009-238550 (MS and PvW) for supporting our works. The authors thank Elayna Truszewski, Department of Biological Sciences, Macquarie University for her editorial assistance with preparation of the original manuscript.

References

Andersson MG, Cerenius L. Pumilio homologue from *Saprolegnia parasitica* specifically expressed in undifferentiated spore cysts. Eukaryot Cell 2002;1:105–111.

Arthur JR, Bondad-Reantaso MG. Capacity and awareness building on import risk analysis (IRA) for aquatic animals. In: Proceedings of the Workshops held 2002. Network of Aquaculture Centres in Asia-Pacific, Bangkok, Thailand 1–6 April 2002;1–6.

Baldauf SL. The deep roots of eukaryotes. Science 2003;300:1703–1706.

Bangyeekhun E, Quiniou SMA, Bly JE, Cerenius L. Characterisation of *Saprolegnia* sp. isolates from channel catfish. Dis Aquat Org 2001;45:53–59.

Barr DJS, Désaulniers NL. Ultrastructure of the *Lagena radicola* zoospore, including a comparison with the primary and secondary *Saprolegnia* zoospores. Can J Bot 1987;65:2161–2176.

Baxter L, Tripathy S, Ishaque N, Boot N, Cabral A, Kemen E, et al. Signatures of Adaptation to Obligate Biotrophy in the *Hyaloperonospora arabidopsidis* Genome. Science 2010;330:1549–1551.

Beakes GW. A comparative account of cyst coat ontogeny in saprophytic and fish-lesions (pathologenic) isolates of *Saprolegnia diclina-parasitica* complex. Can J Bot 1983;61:603–625.

Beakes GW. Oomycete phylogeny: Ultrastructural perspectives. In: Rayner ADM, Brasier CM, Moore D, eds. Evolutionary biology of the fungi. Cambridge: Cambridge University Press 1987:405–422.

Beakes GW, Bartinicki-Garcia S. Ultrastructure of mature oogonium-oospore wall complexs in *Phytophthora megasperma*: a comparison of in vivo and in vitro dissolution of the oospore wall. Mycol Res 1989;93:321–334.

Beakes GW, Glockling SL, Sekimoto S. The evolutionary phylogeny of the oomycete "fungi". Protoplasma 2012;249:3–19.

Beakes GW, Wood SE, Burr AW. Features which characterize *Saprolegnia* isolates from salmon fish lesions. A Review. In: Mueller GJ. ed. Salmon *Saprolegniasis*. U.S. Department of Energy, Portland, OR 1994:33–66.

Behnke J, Buttle D, Stepek G, Lowe A, Duce I. Developing novel anthelmintics from plant cysteine proteinases. Parasite Vect 2008;1:29.

Bimpong CE, Clerk GC. Motility and chemotaxis in zoospores of *Phytophthora palmivora* (Butl.) Butl Ann Bot 1970;34:617–624.
Bly JE, Clem LW. Temperature and teleost immune functions. Fish Shellfish Immunol 1992;2:159–171.
Bly JE, Lawson LA, Abdel-Aziz E, Clem LW. Channel catfish, *Ictalurus punctatus*, immunity to *Saprolegnia* sp. J Appl Aquac 1994;3:35–50.
Bly JE, Lawson LA, Dale DJ, Szalai AJ, Durborow RM, Clem LW. Winter saprolegniosis in channel catfish. Dis Aquat Org 1992;13:155–164.
Bondad-Reantaso MG, Lumanlan SC, Natividad JM, Phillips MJ. Environmental monitoring of the epizootic ulcerative syndrome (EUS) in fish from Munoz, Nueva Ecija in the Philippines. In: Shariff M, Subasinghe RP, Arthur JR, eds. Diseases in Asian Aquaculture 1. Manila: Asian Fisheries Society 1992:475–490.
Bootsma R. Infections with *Saprolegnia* in pike culture (*Esox lucius* L.). Aquac 1973;2:385–394.
Boys CA, Rowland SJ, Gabor M, Gabor L, Marsh IB, HumS, et al. Emergence of Epizootic Ulcerative Syndrome in native fish of the Murray-Darling River System, Australia: Hosts, distribution and possible vectors. PLoS One. 2012; 7: e35568. doi:10.1371/journal.pone.0035568
Bruno DW, Wood BP. *Saprolegnia* and other oomycetes. Viral Bact Fungal Infect 1999;3:599–659
Bridge PD, Roberts PJ, Spooner BM, Panchal G. On the unreliability of published DNA sequences. New Phytol 2003;160:43–48.
Burr AW, Beakes GW. Characterization of zoospore and cyst surface structure in saprophytic and fish pathogenic *Saprolegnia* species (oomycete fungal protists). Protoplasma 1994;181:142–163.
Cao H, Zheng W, Xu J, Ou R, He S, Yang X. Identification of an isolate of *Saprolegnia ferax* as the causal agent of saprolegniosis of Yellow catfish (*Pelteobagrus fulvidraco*) eggs. Vet Res Commun 2012;36:239–244.
Callinan RB, Paclibare JO, Bondad-Reantaso MG, Chin JC, Gogolewski RP. *Aphanomyces* species associated with epizootic ulcerative syndrome (EUS) in the Philippines and red spot disease (RSD) in Australia: preliminary comparative studies. Dis Aquat Org 1995;21:233–238.
Carlile MJ. Motility, taxis and tropism in *Phytophthora*. In: Erwin DC, Bartnicki-Garcia S, Tsao PH, eds. *Phytophthora*. Its Biology, Taxonomy, Ecology and Pathology. American Phytopathological Society, St Paul, MN 1983:95–107.
Catap ES, Munday BL. Effects of variation of water temperature and dietary lipids on the expression of experimental epizootic ulcerative syndrome (EUS) in sand whiting, *Sillago ciliata*. Fish Pathol 1998;33:327–335.
Cerenius L, Söderhäll K. Chemotaxis in *Aphanomyces astaci*, an arthropod-parasitic fungus. J Invertebr Pathol 1984a;43:278–281.
Cerenius L, Söderhäll K. Repeated zoospore emergence from isolated spore cysts of *Aphanomyces astaci*. Exp Mycol 1984b;8:370–377.
Cerenius L, Söderhäll K. Repeated zoospore emergence as a possible adaptation to parasitism in *Aphanomyces*. Exp Mycol 985;9:259–263.
Cerenius L, Bangyeekhun E, Keyser P, Söderhäll I, Söderhäll K. Host prophenoloxidase expression in freshwater crayfish is linked to increased resistance to the crayfish plague fungus, *Aphanomyces astaci*. Cell Microbiol 2003;5:353–357.
Chet I, Mitchell R. Ecological aspects of microbial chemotactic behaviour. Annu Rev Microbiol 1976;30:221–239.
Chinabut S, Roberts RJ, Willoughby GR, Pearson MD. Histopathology of snakehead, *Channa striatus* (Bloch), experimentally infected with the specific *Aphanomyces* fungus associated with epizootic ulcerative syndrome (EUS) at different temperatures. J Fish Dis 1995;18:41–47.
Chivers DP, Wisenden BD, Hindman CJ, Michalak TA, Kusch RC, Kaminskyj SGW, et al. Epidermal "alarm substance" cells of fishes maintained by non-alarm functions: Possible defence against pathogens, parasites and UVB radiation. Proc R Soc B: Biol Sci 2007;274:2611–2619.

Coker WC. Interrelationships of the Saprolegniales. Proceedings of the 6th International Botanical Congress. Amsterdam (Zesde Internationaal Botanisch Congres) 1935;1:268–270.
Coker WC. The Saprolegniaceae with Notes on Other Water Molds. Chapel Hill, NC: University of North Carolina Press 1923.
Czeczuga B, Kiziewicz B, Danilkiewicz Z. Zoosporic fungi growing on thespecimens of certain fish species recently introduced to Polish waters. Acta Ichthyol Piscat 2002;32:117–125.
Daugherty J, Evans TM, Skillom T, Watson LE, Money NP. Evolution of spore release mechanisms in the Saprolegniaceae (oomycetes): Evidence from a phylogenetic analysis of internal transcribed spacer sequences. Fungal Genet Biol 1998;24:354–363.
Dick MW. Morphology and taxonomy of the oomycetes, with special reference to Saprolegniaceae, Leptomitaceae and Pythiaceae I. Sexual reproduction. New Phytol 1969;68:751–775.
Dick MW. Morphology and taxonomy of the oomycetes, with special reference to Saprolegniaceae, Leptomitaceae and Pythiaceae II. Cytogenetic systems. New Phytol 1972;71:1151–1159.
Dick MW. Saprolegniales. In: Ainsworth GC, Sparrow FK, Sussman AL, eds. The Fungi, An Advanced Treatise, IVB. New York, NY: Academic Press 1973:113–144.
Dick MW. Systematics, taxonomy and phylogeny in the oomycetes. In: Subramanian CV, ed. Taxonomy of Fungi. Proceedings of the International Symposium on Taxonomy of Fungi. University of Madras, Madras 1978:82–87.
Dick MW. Phylum Oomycota. In: Margulis L, Corliss JO, Melkonian M, Chapman D, eds. Handbook of Protoctista. Boston, MA: Jones and Bartlett 1990:661–685.
Dick MW. The Peronosporomycetes. In: McLaughlin DJ, McLaughlin EF, Lemke PA, eds. The Mycota, Vol. 7, Part A. Systematics and Evolution. Berlin: Springer-Verlag 2000:39–72.
Dick MW. Straminipilous Fungi. Systematics of the Peronosporomycetes Including Accounts of the Marine Straminipilous Protists, the Plasmodiophorids and Similar Organisms. Dordrecht: Kluwer Academic Publishers 2001:670.
Diéguez-Uribeondo J, Cerenius L, Soderhall K. Repeated zoospore emergence in *Saprolegnia parasitica*. Mycol Res 1994a;98:810–815.
Diéguez-Uribeondo J, Cerenius L, Söderhäll K. Physiological characterization of *Saprolegnia parasitica* isolates from brown trout. Aquaculture 1996;140:247–257.
Diéguez-Uribeondo J, Fregeneda-Grandes JM, Cerenius L, Pérez-Iniesta E, Aller-Gancedo JM, Tellería MT, et al. Re-evaluation of the enigmatic species complex *Saprolegnia diclina-Saprolegnia parasitica* based on morphological, physiological and molecular data. Fungal Genet Biol 2007;44:585–601.
Diéguez-Uribeondo J, García MA, Cerenius L, Kozubíková E, Ballesteros I, Windels C, et al. Phylogenetic relationships among plant and animal parasites, and saprotrophs in *Aphanomyces* (oomycetes). Fungal Genet Biol 2009;46:365–376.
Donaldson SP, Deacon JW. Effects of amino acids and sugars on zoospore taxis, encystment and cyst germination in *Pythium aphanidermatum* (Edson) Fitzp., *P. catenulatum* Matthews and *P. dissotocum* Drechs. New Phytol 1993;123:289–295.
Donaldson SP, Deacon JW. Role of calcium in adhesion and germination of zoospore cysts of *Pythium*: A model to explain infection of host plants. J Gen Microbiol 1992;138:2051–2059.
Dou D. Conserved C-terminal motifs required for avirulence and suppression of cell death by *Phytophthora sojae* effector Avr1b. Plant Cell 2008;20:1118–1133.
Egusa S, Masuda N. A new fungal disease of *Plecoglossus altivelis*. Fish Pathol 1971;6:41–46. (in Japanese).
El-Feki M, Hatai K, Hussein MA. Chemotactic and chemokinetic activities of *Saprolegnia parasitica* toward different metabolites and fish tissue extracts. Mycoscience 2003;44:159–162.
Ellis JG, Dodds PN. Showdown at the RXLR motif: Serious differences of opinion in how effector proteins from filamentous eukaryotic pathogens enter plant cells. Proc Natl Acad Sci USA 2011;108:14381–14382.

El-Sharouny HM, Badran RAM. Experimental transmission and pathogenicity of some zoosporic fungi to Tilapia fish. Mycopathologia 1995;132:95–103.

Fernández-Benéitez MJ, Ortiz-Santaliestra ME, Lizana M, Diéguez-Uribeondo J. *Saprolegnia diclina*: Another species responsible for the emergent disease "Saprolegnia infections" in amphibians. Microbiol Lett 2008;279:23–29.

Fisher MC, Garner TW. The relationship between the emergence of *Batrachochytrium dendrobatidis*, the international trade in amphibians and introduced amphibian species. Fungal Biol Rev 2007;21;2–9.

Fisher MC, Henk DA, Briggs CJ, Brownstein JS, Madoff LC, McCraw SL, et al. Emerging fungal threats to animal, plant and ecosystem health. Nature 2012;484:186–194.

Fregeneda-Grandes JM, Fernández Díez M, Aller Gancedo JM. Experimental pathogenicity in rainbow trout, *Oncorhyncus mykiss* (Walbaum) of two distinct morphotypes of long-spined *Saprolegnia* isolates obtained from wild brown trout, *Salmo trutta* L., and river water. J Fish Dis 2001;24:351–359.

Fregeneda-Grandes JM, Fernández Díez M, Aller Gancedo JM. Ultrastructural analysis of *Saprolegnia* secondary zoospore cyst ornamentation from infected wild brown trout, *Salmo trutta* L., and river water indicates two distinct morphotypes amongst long-spined isolates. J Fish Dis 2000;23:147–160.

Fuller MS. The zoospore, hallmark of the aquatic fungi. Mycol 1977;69:1–20.

Fuller MS, Jaworski A. Zoosporic Fungi in Teaching and Research. Athens, GA: Southeastern Publishing 1987.

Gaulin E, Madoui MA, Bottin A, Jacquet C, Mathe C, Couloux A, et al. Transcriptome of *Aphanomyces euteiches*: new oomycete putative pathogenicity factors and metabolic pathways. PLoS One 2008; 3:e1723. doi:10.1371/journal.pone.0001723.

Ghiasi M, Khosravi AR, Soltani M, Binaii M, Shokri H, Tootian Z, et al. Characterization of *Saprolegnia* isolates from Persian sturgeon (*Acipencer persicus*) eggs based on physiological and molecular data. J Med Mycol 2010;20:1–7.

Gleason FH. Nutritional comparisons in the Leptomitales. Am J Bot 1968;55:1003–1010.

Gleason FH. Uptake of amino acids by *Saprolegnia*. Mycologia 1973;65:465–468.

Gleason FH, Lilje O. Structure and function of fungal zoospores: ecological implications. Fungal Ecol 2009;2:53–59.

Gleason FH, Rudolph CR, Price JS. Growth of certain aquatic oomycetes on amino acids I. *Saprolegnia, Achlya, Leptolegnia*, and *Dictyuchus*. Physiol Plant 1970a;23:513–516.

Gleason FH, Stuart TD, Price JS, Nelbach WT. Growth of certain aquatic oomycetes on amino acids II. *Apodachlya, Aphanomyces*, and *Pythium*. Physiol Plant 1970b;23:769–774.

Golkar L, Lebrun RA, Ohayon H, Gounon P, Papierok B, Brey PT. Variation of larval susceptibility to *Lagenidium giganteum* in three Mosquito species. J Invertebr Pathol 1993;62:1–8.

Grudkowska M, Zagdanska B. Multifunctional role of plant cysteine proteinases. Acta Biochim Polon 2004;51:609–624.

Gubler F, Hardham AR. Secretion of adhesive material during encystment of *Phytophthora cinnamomi* zoospores characterized by immunogold labeling with monoclonal antibodies to components of peripheral vescicles. J Cell Sci 1988;90:225–235.

Gubler F, Hardham AR, Protein storage in large peripheral vesicles in *Phytophthora* zoospores and its breakdown after cyst germination. Exp Mycol 1990;14:393–404.

Haas BJ, Kamoun S, Zody MC, Jiang RHY, Handsaker RE, Cano LM, et al. Genome sequence and analysis of the Irish potato famine pathogen *Phytophthora infestans*. Nature 2009;461:393–398.

Humphrey JE. The Saprolegniaceae of the United States, with notes on other species. Trans Am Phil Soc 1893;17:63–148.

Hallett IC, Dick MW. Fine structure of zoospore cyst ornamentation in the Saprolegniaceae and Pythiaceae. Trans Br Mycol Soc 1986;86:457–463.

Halsall DM. Zoospore chemotaxis in Australian isolates of *Phytophthra* species. Can J Microbiol 1976;22:409–422.
Hardham AR. *Phytophthora cinnamomi.* Mol Plant Pathol 2005;6:589–604.
Hatai K, Willoughby LG, Beakes GW. Some characteristics of *Saprolegnia* obtained from fish hatcheries in Japan. Mycol Res 1990;94:182–190.
Hickman CJ, Ho HH. Behaviour of zoospores in plant-pathogenic phycomycetes. Annu Rev Phytopathol 1966;4:195–214.
Huchzermeyer KDA, van der Waal BCW. Epizootic ulcerative syndrome: exotic fish disease threatens Africa's aquatic ecosystems. J South African Vet Assoc 2012;83:1–6.
Holloway SA, Heath IB. An ultrastructural analysis of the changes in organelle arrangement and structure between the various spore types of *Saprolegnia*. Can J Bot 1977;55:1328–1339.
Hulvey JP, Padgett DE, Bailey JC. Species boundaries within *Saprolegnia* (Saprolegniales, Oomycota) based on morphological and DNA sequence data. Mycol 2007;99:421–429.
Hussein MMA, Hatai K. Pathogenicity of *Saprolegnia* species associated with outbreaks of salmonid saprolegniosis in Japan. Fish Sci 2002;68:1067–1072.
Hussein MMA, Hatai K. *Saprolegnia salmonis* sp. nov. isolated from sockeye salmon, *Onchrhynchus nerka*. Mycoscience 1999;40:387–391.
Hussein MMA, Hatai K, Nomura T. Saprolegniosis in salmonids and their eggs in Japan. J Wildl Dis 2001;37:204–207.
Jiang RHY, Bruijn Id, Haas BJ, Belmonte R, Löbach L, Christie J, et al. Distinctive expansion of potential virulence genes in the genome of the oomycete fish pathogen *Saprolegnia parasitica*. PLoS Genet.2013; 9: e1003272. doi:10.1371/journal.pgen.1003272.
Johnson Jr. TW, Seymour RL, Padgett DE. Systematics of the Saprolegniaceae: New combinations. Mycotaxon 2005;92:11–32.
Judelson HS, Blanco FA. The spores of *Phytophthora*: weapons of the plant destroyer. Nat Rev Microbiol 2005;3:47–58.
Kamoun S. Molecular Genetics of pathogenic oomycetes. Eukaryot Cell 2003;2:191–199.
Karling JS. Predominantly holocarpic and ucarpic imple iflagellate Phycomycetes, 2nd ed. J. Cramer, Liechtenstein:Vaduz 1981.
Kemen E, Gardiner A, Schultz-Larsen T, Kemen AC, Balmuth AL, Robert-Seilaniantz A, et al. Gene gain and loss during evolution of obligate parasitism in the white rust pathogen of *Arabidopsis thaliana*. PLoS Biology2011;9: e1001094. doi:10.1371/journal.pbio.1001094.
Khew KL, Zentmyer GA. Chemotactis response of zoospores of five species of *Phytophthora*. Phytopathologia 1973;63:1511–1517.
Kiesecker JM, Blaustein AR, Miller CL. Transfer of a pathogen from fish to amphibians. Conserv Biol 2001;15:1064–1070.
Kiryu Y, Shields JD, Vogelbein WK, Kator H, Blazer VS. Infectivity and pathogenicity of the oomycete *Aphanomyces invadans* in Atlantic menhaden *Brevoortia tyrannus*. Dis Aquat Org 2003;54:135–146.
Kützing FT. Phycologia generalis oder natomie, hysiologie und ystemkunde der Tange. F.A. Brockhaus, Leipzig 1843;458.
Lehnen Jr LP, Powell MJ. The role of kinetosome-associated organelles in the attachment of encysting secondary zoospores of *Saprolegnia ferax* to substrates. Protoplasma 1989; 149:163–174.
Lehnen LP, Powell MJ. Formation of K_2-bodies in primary cysts of *Saprolegnia ferax*. Mycologia 1991;83:163–179.
Lévesque CA, Brouwer H, Cano L, Hamilton JP, Holt C, Huitema E, et al. Genome sequence of the necrotrophic plant pathogen *Pythium ultimum* reveals original pathogenicity mechanisms and effector repertoire. Genome Biol 2010; 11: R73. doi:10.1186/gb-2010-11-7-r73.
Lilley JH, Beakes GW, Hetherington CS. Characterization of *Aphanomyces invadans* isolates using pyrolysis mass spectrometry (PyMS). Mycoses 2001;44:383–389.

Lilley JH, Hart D, Panyawachira V, Kanchanakhan S, Chinabut S, Söderhäll K, et al. Molecular characterization of the fish-pathogenic fungus *Aphanomyces invadans*. J Fish Dis 2003;26:263–275.
Lilley JH, Roberts RJ. Pathogenicity and culture studies comparing the *Aphanomyces* involved in epizootic ulcerative syndrome (EUS) with other similar fungi. J Fish Dis 1997;20:135–144.
Links MG, Holub E, Jiang RHY, Sharpe AG, Hegedus D, Beynon E, et al. De novo sequence assembly of *Albugo candida* reveals a small genome relative to other biotrophic oomycetes. BMC Genomics 2011;12:503.
López-Dóriga MV, Martínez JL. Ultrastructure of fish cells involved in cellular defences against *Saprolegnia* infections: Evidence of non-leucocytic nature. Dis Aquat Org 1998;32:111–117.
Lumanlan-Mayo SC, Callinan RB, Paclibare JO, Catap ES, Fraser GC. Epizootic ulcerative syndrome (EUS) in rice-fish culture systems: an overview of field experiments 1993–1995. In: Flegel TW, MacRae IH, eds. Diseases in Asian Aquaculture III. Manila: Asian Fisheries Society 1997:129–138.
McLeod A, Smart CD, Fry WE. Characterization of 1,3-beta-glucanase and 1,3;1,4-beta-glucanase genes from *Phytophthora infestans*. Fungal Genet Biol 2003;38:250–63.
Miller CE, Ristanović B. Systematic studies on some taxa of the Saprolegniaceae. Milrobiol Acta Biol Iugoslavica 1975;12:47–58.
Miyazaki T, Kubota SS, Tashiro F. Studies on visceral mycosis of salmonid fry—I. Histopathology. Fish Pathol 1977;11:183–186.
Moreau F, Moreau M. Recherches sur les Saprolégniées 11e sér. Ann Sci Nat Bot 1938;1:221–358.
Moreau F, Moreau M. Remarques sur la systématique des Saprolégniacées. Bull Soc Bot Fr 1935;82,354–358.
Morris BM, Gow NAR. Mechanism of electrotaxis of zoospores of phytopathogenic fungi. Phytopathol 1993;8:877–882.
Neish GA. Observations on saprolegniasis of adult sockeye salmon, *Oncorhynchus nerka* (Walbaum). J Fish Biol 1977;10:513–522.
Neish GA, Hughes GC. Fungal Diseases of Fishes: Book 6. Neptune City: T.F.H. Publications 1980.
Neitzel DA, Abernethy CS, Elston RA, Pacific NL. Prevention of respawning mortality: a use of almon eadburns and ranial esions. U.S. Department of Energy, Portland, OR. 2004.
Nilsson RH, Ryberg M, Abarenkov K, Sjökvist E, Kristiansson E. The ITS region as a target for characterization of fungal communities using emerging sequencing technologies. FEMS Microbiol Lett 2009;296:97–101.
Nolan RA. Physiological studies on an Isolate of *Saprolegnia ferax* from the larval gut of the blackfly *Simulium vittatum*. Mycologia 1976;68:523–540.
Noga EJ. Water mold infections of freshwater fish: Recent advances. Annu Rev Fish Dis 1993;3:291–304.
Nsonga A, Mfitilodze W, Samui KL, Sikawa D. Epidemiology of epizootic ulcerative syndrome in the zambezi river system. A case study for Zambia. HVM Bioflux 2013;5:1–8.
Oidtmann B. Review of Biological factors relevant to import risk assessments for epizootic ulcerative syndrome (*Aphanomyces invadans*). Transbound Emerg Dis 2012;59:26–39.
Padgett DE, Puckett JC, Strickland WM. Taxonomic implications of unusual zoospore discharge in an isolate of *Saprolegnia unispora*. Mycotaxon 2000;76:171–174.
Petrisko JE, Pearl CA, Pilliod DS, Sheridan PP, Williams CF, Peterson CR, et al. Saprolegniaceae identified on amphibian eggs throughout the Pacific Northwest, USA, by internal transcribed spacer sequences and phylogenetic analysis. Mycologia 2008;100:171–180.
Phillips AJ, Anderson VL, Robertson EJ, Secombes CJ, van West P. New insights into animal pathogenic oomycetes. Trends Microbiol 2008;16:13–19.
Pickering AD. Factors Which Predispose Salmonid Fish to Saprolegniasis. U.S. Department of Energy, Portland, Or, 1994.

Pickering AD, Willoughby LG. *Saprolegnia* infections of salmonid fish. In: Roberts RJ, ed. Microbial Diseases of Fish. London: Academic Press 1982:271–297.
Pickering AD, Willoughby LG, McGrory CB. Fine structure of secondary zoospore cyst cases of *Saprolegnia* isolates from infected fish. Trans Br Mycol Soc 1979;72:427–436.
Powell MJ, Blackwell WH. Phenetic analysis of genera of the Saprolegniaceae (oomycetes). Mycotaxon 1998;68:505–516.
Quiniou SM, Bigler S, Clem LW, Bly JE. Effects of water temperature on mucous cell distribution in channel catfish epidermis: A factor in winter saprolegniasis. Fish Shellfish Immunol 1998;8:1–11.
Robideau GP, De Cock AWAM, Coffey MD, Voglmayr H, Brouwer H, Bala K, et al. DNA barcoding of oomycetes with cytochrome c oxidase subunit I and internal transcribed spacer. Mol Ecol Resour 2011;11:1002–1011.
Rucker RR. A study of *Saprolegnia* infections among fish 1944. PhD thesis, University of Washington, Seattle, 92pp.
Ruthig GR. Water molds of the genera *Saprolegnia* and *Leptolegnia* are pathogenic to the north American frogs *Rana catesbeiana* and *Pseudacris crucifer*, respectively. Dis Aquat Org 2009;84:173–178.
Sarowar MN, van den Berg AH, McLaggan D, Young MR, van West P. *Saprolegnia* strains isolated from river insects and amphipods are broad spectrum pathogens. Fungal Biol 2013;117:752–763.
Seymour RL. The genus *Saprolegnia*. Nova Hedwig 1970;19:1–124.
Smith SN, Chohan R, Howitt SG, Armstrong RA. Proteolytic activity amongst selected *Saprolegnia* species. Mycol Res 1994;98:389–395.
Sparrow Jr FK. Aquatic Phycomycetes, 2nd ed. Ann Arbor, MI: The University of Michigan Press 1960.
Sparrow Jr FK. The present status of classification in biflagellate fungi. In: Jones EBG. ed. Recent Advances in Aquatic Mycology. London: Elek Science 1976:213–222.
Schornack S, Van Damme M, Bozkurt TO, Cano LM, Smoker M, Thines M, et al. Ancient class of translocated oomycete effectors targets the host nucleus. Proc Natl Acad Sci USA 2010;107:17421–17426.
Srivastava RC. Fungal parasites of certain fresh water fishes of India. Aquac 1980;21:387–392.
Stassen JH, Van den Ackerveken G. How do oomycete effectors interfere with plant life? Curr Opin Plant Biol 2011;14:407–414.
Stewart JE. Introductions as factors in diseases of fish and aquatic invertebrates. Can J Fish Aquat Sci 1991;48:110–117.
Stueland S, Hatai K, Skaar I. Morphological and physiological characteristics of *Saprolegnia* spp. strains pathogenic to Atlantic salmon, *Salmo salar* L. J. Fish Dis 2005;28:445–453.
Tautz D, Arctander P, Minelli A, Thomas RH, Vogler AP. A plea for DNA taxonomy. Trends Ecol Evol 2003;18:70–74.
The 2012 State of World Fisheries and Aquaculture (http://www.fao.org/docrep/016/i2727e/i2727e00.htm) Site visited on the 24th November 2013.
The State of Food and Agriculture 1998 (http://www.fao.org/docrep/w9500e/w9500e00.htm) Site visited on the 24th November 2013.
Thoen E, Evensen Ø, Skaar I. Pathogenicity of *Saprolegnia* spp. to Atlantic salmon, *Salmo salar* L., eggs. J Fish Dis 2011;34:601–608.
Torto-Alalibo T, Tian M, Gajendran K, Waugh ME, Van West P, Kamoun S. Expressed sequence tags from the oomycete fish pathogen *Saprolegnia parasitica* reveal putative virulence factors. BMC Microbiol 2005;5:46.
Tyler BM, Tripathy S, Zhang X, Dehal P, Jiang RHY, Aerts A, et al. *Phytophthora* genome sequences uncover evolutionary origins and mechanisms of pathogenesis. Science 2006;313:1261–1266.
van den Berg AH, McLaggan D, Diéguez-Uribeondo J, van West P. The impact of the water moulds *Saprolegnia diclina* and *Saprolegnia parasitica* on natural ecosystems and the aquaculture industry. Fungal Biol Rev 2013;27:33–42.

van der Hoorn RAL, Jones JDG. The plant proteolytic machinery and its role in defence. Curr Opin Plant Biol 2004;7:400–407.
van West P. *Saprolegnia parasitica*, an oomycete pathogen with a fishy appetite: new challenges for an old problem. Mycol 2006;20:99–104.
van West P, Appiah AA, Gow NAR. Advances in research on oomycete root pathogens. Physiol Molec Plant Pathol 2003;62:99–113.
van West P, Morris BM, Reid B, Appiah AA, Osborne MC, Campbell TA, et al. Oomycete plant pathogens use electric fields to target roots. Mol Plant-Microbe Interact 2002;15:790–798.
Vishniac HS, Nigrelli RF. The ability of saprolegniaceae to parasitise Platyfish. Zool 1957;42:131–134.
Walker CA, van West P. Zoospore development in the oomycetes. Fungal Biol Rev 2007;21:10–18.
Warburton AJ, Deacon JW. Transmembrane Ca^{2+} fluxes associated with zoospore encystment and cyst germination by the phytopathogen *Phytophthora parasitica*. Fungal Genet Biol 1998;25:54–62.
Wawra S, Bain J, Durward E, de Bruijn I, Minor KL, Matena A, et al. Host-targeting protein 1 (SpHtp1) from the oomycete *Saprolegnia parasitica* translocates specifically into fish cells in a tyrosine-O-sulphate-dependent manner. Proc Natl Acad Sci 2012a;109:2096–2101.
Wawra S, Belmonte R, Löbach L, Saraiva M, Willems A, van West P. Secretion, delivery and function of oomycete effector proteins. Curr Opin Microbiol 2012b;15:685–691.
Whisson SC, Boevink PC, Moleleki L, Avrova AO, Morales JG, Gilroy EM, et al. A translocation signal for delivery of oomycete effector proteins into host plant cells. Nat 2007;450:115–118.
Willoughby LG. An ecological study of water as the medium for growth and reproduction of the *Saprolegnia* from salmonid fish. Trans Br Mycol Soc 1986;87:493–502.
Willoughby LG. Rapid preliminary screening of *Saprolegnia* on fish. J Fish Dis 1985;8:473–476.
Willoughby LG. Observations on fungal parasites of Lake district salmonids. Salmon Trout Mag 1971; 192:152–158.
Willoughby LG, McGrory CB, Pickering AD. Zoospore germination of *Saprolegnia* pathogenic to fish. Trans Br Mycol Soc 1983;80:421–435.
Willoughby LG, Pickering AD. Viable saprolegniaceae spores on the epidermis of the salmonid fish *Salmo trutta* and *Salvelinus alpinus*. Trans Br Mycol Soc 1977;68:91–95.
Zentmyer GA. Chemotaxis of zoospores for root exudates. Science 1961;133:1595–1596.

Frank H. Gleason, Jodi L. Rowley, Casey N. Jessop and Osu Lilje

11 Zoosporic parasites of amphibians

The amphibian chytrid, *Batrachochytrium dendrobatidis*, is the most frequently reported zoosporic pathogen causing diseases in amphibians. Nevertheless, it is becoming increasingly obvious that the spectrum of pathogens infecting amphibians and causing declines in population sizes is much wider than previously recognized (Pessier 2008). In addition to *B. dendrobatidis*, a number of other groups of zoosporic parasites are likely to be important in precipitating amphibian population declines, particularly in the face of climate change or other stressors (Alford et al. 2007; Di Rosa et al. 2007; Pounds et al. 2007; Rowley et al. 2013). Furthermore, in addition to zoosporic parasites, many other groups of parasites are associated with population declines in amphibians. These include for example viruses, bacteria, fungi, platyhelminthes, mesozoa, nematoda, annelida and other groups (Raffel et al. 2008; Berger et al. 2009; Hartigan et al. 2013). This chapter focuses on primary infections by zoosporic parasites, although secondary infections by a variety of microorganisms may follow primary infections and cause mortality. We discuss zoosporic parasites in four phyla: Chytridiomycota, Mesomycetozoea, Oomycota and Perkinsozoa. Species of parasites in all of these groups may infect different stages of the life cycles of individuals within populations of hosts. Any one individual may be infected by more than one parasite simultaneously or at different times. Population declines are also affected by environmental factors, such as temperature, salinity and pH, which are discussed briefly here. Many of these parasites fit into the model for emerging infectious diseases described by Fisher et al. (2012). Some significant characteristics and strategies for infection are also discussed. Finally the possible roles of the animal trade in introduction of new species are briefly discussed.

11.1 Chytridiomycota

The most well-known and extensively-studied amphibian parasite, *Batrachochytrium dendrobatidis*, belongs to the Chytridiomycota. *Batrachochytrium dendrobatidis* is responsible for the amphibian disease chytridiomycosis (Longcore et al. 1999), infects species in all major lineages of amphibians, including Anura (frogs and toads), Caudata (salamanders) and Gymnophiona (caecilians) (Gower et al. 2013) and has been involved in mass mortalities, population declines, and extinctions of amphibian populations and species around the world (Berger et al. 1998; Lips 1999; Longcore et al. 1999; Bosch et al. 2001; Weldon and Du Preez 2004; Lips et al. 2006; Rachowicz et al. 2006). In many cases, declines in frog populations attributed to

B. dendrobatidis have been dramatic, resulting in the rapid local extinction of 50% or more of the anuran species at particular sites (Lips 1999; Lips et al. 2006) and large reductions in the abundance of remaining species (Lips and Donnelly 2002; Lips et al. 2006).

This pathogen infects keratinized tissues in the epidermis of metamorphosed amphibians (Longcore et al. 1999) and the mouthparts of larval anurans (Fellers et al. 2001; Rachowicz and Vredenburg 2004). Chytridiomycosis can cause rapid mortality (Nichols et al. 2001), with frogs of susceptible species dying less than 3 weeks after experimental infection in the laboratory (Berger et al. 1998, 2004, 2005), with epidermal damage caused by *B. dendrobatidis* disrupting critical cutaneous functioning and causing fatal disruptions in fluid and electrolyte balance (Voyles et al. 2009; Voyles et al. 2012). Clinical signs of severe chytridiomycosis include lethargy, irregular skin sloughing, abnormal posture (hind legs abducted), and loss of righting reflex (Berger et al. 2009; Voyles et al. 2009).

The only known mechanism of transmission among hosts, and of intrahost increase in pathogen load, is by infection and re-infection with motile, waterborne zoospores (Longcore et al. 1999). This parasite infects individuals and spreads within and among populations of amphibians via the release of these propagules (Nichols et al. 2001). As such, interspecific variation in behavior (with respect to contact with water and conspecifics) may affect the transmission of chytridiomycosis (Rowley and Alford 2007).

It is well known that environmental temperature can determine the progress and outcome of *B. dendrobatidis* infections in the laboratory. Piotrowski et al. (2004) incubated three isolates at 10°C, 17°C, 23°C, 25°C and 28°C in growth media in the laboratory. In culture, the isolates of *B. dendrobatidis* grew best between 17°C and 25°C (Piotrowski et al. 2004). All cultures between 10°C and 25°C contained live zoospores. No growth or live zoospores was observed at 28°C. In the laboratory, low and/or fluctuating temperatures can retard the pathogen's growth, and elevated body temperatures can clear frogs of *B. dendrobatidis* infection (Woodhams et al. 2003). In nature, population-level infection prevalence and host mortality rates are often correlated with ambient environmental conditions; they are highest during cooler months (Woodhams and Alford 2005; Alford 2010; Puschendorf et al. 2011) and at higher elevations (Woodhams and Alford 2005). Individual histories of body temperature can explain a substantial fraction of these seasonal, elevational, and interspecific patterns of prevalence of infection in *B. dendrobatidis* that have been observed in nature (Rowley and Alford 2013).

Environmental pH also impacts the growth of *B. dendrobatidis* in the laboratory. Piotrowski et al. (2004) tested the growth of three isolates in liquid culture between pH 4 and 8. Maximum growth occurred at pH 6–7 with less growth at pH 8 and minimal growth at pH 4 and 5. Zoospores were produced and remained motile at all of the pH values tested. Acid precipitation can result in a decrease in pH in aquatic ecosystems. Vatnick et al. (2006) documented the effect of exposure of *Lithobates pipiens* to an acid environment in the laboratory. The inflammatory response in adult *L. pipiens* was disrupted at pH 5.5. More nonphagocytic leukocytes and fewer highly phagocytic leukocytes were found at pH 5.5 than at pH 7.0. Exposure of *L. pipiens* to pH 5.5 resulted in 70% mortality within 10 days. Vatnick et al. (2006) and Brodkin et al. (2003) have suggested that acid stress and exposure to low temperatures may result in weakening of natural defence mechanisms in *L. pipiens* and therefore contribute to population declines in northern United States. Although these studies focused on the roles of bacteria in systemic infections in frogs, they indicated that *B. dendrobatidis* can also grow at low temperatures and pH values, but there is little data on the effects of low temperature and pH on infection of amphibians by this fungus.

Salinity is also an important environmental factor. High osmolarities may cause severe stress to osmoregulation in freshwater species, but low osmolarities may be tolerated by the host species and at the same time inhibit growth of the pathogen. Sodium chloride is known to be an antifungal agent which acts by altering osmotic gradients (Stockwell et al. 2012). *Batrachochytrium dendrobatidis* grows well and produces zoospores on media supplemented with 0.5% NaCl and grows slowly on media supplemented with 1% NaCl (Piotrowski et al. 2004). Stockwell et al. (2012) tested the effects of 0 to 5 ppt (or 0 to 0.5%) NaCl on growth, zoospore motility and survival of *B. dendrobatidis* in liquid culture media incubated for an 11-day period and on survivorship of infection of *Litoria peronii* in captivity for a 2-month period. NaCl at a concentration of 5 ppt does not negatively affect survival of the host or the pathogen, but concentrations greater than 3 ppt significantly reduced growth and zoospore motility of *B. dendrobatidis*. This fungus is extremely sensitive to salinity.

Recently, a second species of *Batrachochytrium* has been described, *B. salamandrivorans* (Martel et al. 2013). This species is known to cause chytridiomycosis in salamanders and is implicated in dramatic population declines in fire salamanders (*Salamandra salamandra*) (Martel et al. 2013). Martel et al. (2013) recently studied the effect of temperature on growth and sporulation of *B. salamandrivorans*. Interestingly, *B. salamandrivorans* grew well at 5°C, and the optimal temperature for growth was between 10°C and 15°C. Higher temperatures inhibited growth. Motile zoospores were present at temperatures between 5°C and 20°C but not at higher temperatures. Martel et al. (2013) concluded that *B. salamandrivorans* has a markedly lower thermal preference than *B. dendrobatidis*.

11.2 Mesomycetozoea

Some parasites in the Mesomycetozoea are capable of infecting amphibians (Pereira et al. 2005; Gozlan et al. 2009; Glockling et al. 2013; Rowley et al. 2013) (Tabs. 11.1, 11.2). Amphibian disease associated with infection by mesomycetozoeans has been reported in Italy (Pascolini et al. 2003), France (Pérez 1907; González-Hernádez

Tab. 11.1: Oomycete parasites of amphibians (Kiesecker et al. 2001; Fernández-Beneítez et al. 2007; Petrisko et al. 2008; Ruthig 2008; Romansic et al. 2009; Fernández-Beneítez et al. 2011).

Genera	Species	Observed on eggs (E), larvae (L) amphibian specimens
Achlya	*A. americana*	E
	A. aquatica	E
	A. colorata	E
	A. intricata	E
	A. oligacantha	E
	A. papillosa	E
	A. racemosa	E
Saprolegnia	*S. anomalies*	E
	S. bulbosa	E
	*S. diclina**	E
	*S. ferax**	E, L
	S. hypogyna	E
	S. longicaulis	E
	S. oliviae	E
	S. parasitica	E
	S. salmonis	E
	S. semihypogyna	E
Leptolegnia	NA*	E
Aphanomyces	*Aphanomyces* sp.	L

*Species known as pathogens.
Amphibian specimens include eggs of western toad, *Bufo boreas*; long toe salamander, *Ambystoma macrodactylum*; cascades frog, *Rana cascadae*; Columbia spotted frog, *Rana luteiventris*; Oregon spotted frog, *Rana pretiosa*; spring peeper, *Pseudacris crucifer*; natterjack toad, *Bufo Calamita*; western spadefoot toad, *Pelobates cultripes* and the eggs and larvae of Pacific tree frog, *Pseudacris regilla* and northern red-legged frog, *Rana aurora*.

2010), Switzerland (Guyénot and Naville 1922), Czechoslovakia (Broz and Privora 1952), on the Isle of Rum, Scotland (Grey 2008) and in the Scottish Highlands (Duffus and Cunningham 2010; Duffus, unpublished observations), and the USA (Raffel et al. 2008; Duffus and Cunningham 2010). Mortality and malaise have been reported in populations of infected amphibians (Moral 1913). Infected newts have been shown to have a significantly lower survival rate in captivity than uninfected newts (Raffel et al. 2008). Little is known about the extent or pathogenicity of mesomycetozoean parasites in amphibian populations in natural ecosystems. The only evidence of a significant impact on populations of amphibians is provided by a study in central Italy where declines in populations of the Pool Frog, *Rana lessonae*, have been attributed, at least in part, to an outbreak of the mesomycetozoean parasite *Amphibiocystidium ranae* (Di Rosa et al. 2007).

Tab. 11.2: Dermocystidian parasites of amphibians.

Parasite	Host	Location	References
Amphibiocystidium armoriacus	Palmate newt, *Lissotriton helveticus*	Europe	Poisson (1937)
Amphibiocystidium beccarii	Common newt, *Lissotriton vulgaris*	Europe	Granata (1919)
Amphibiocystidium granulosum	Common frog, *Rana temporaria*	Europe	Broz and Privora (1952)
Amphibiocystidium hyalarum	Red snouted treefrog, *Scinax ruber*	South America	Carini (1940)
Amphibiocystidium pusula	Marbled newt, *Triturus marmoratus;* crested newt, *Triturus cristatus*	Europe	Pérez (1907, 1913)
Amphibiocystidium ranae	Water frog, *Pelophylax esculentus* complex; Common Frog, *Rana temporaria;* palmate newt, *Lissotriton helveticus;* common midwife toad, *Alytes obstetricans*	Europe	Guyénot and Naville (1922); Broz and Privora (1952); Broz and Kulda (1954); Pascolini et al. (2003); Pereira et al. (2005); González-Hernández et al. (2010)
Amphibiocystidium viridescens	eastern redspotted Newt, *Notophthalmus viridescens*	North America	Raffel et al. (2008)
Amphibiothecum penneri	American toad, *Anaxyrus americanus*	North America	Jay and Pohley (1981)
Histocystidium ranae	Bullfrogs, *Lithobates catesbeianus*	North America	Goodchild (1953)

Andreou et al. (2009) studied the influence of temperature on the production and longevity of zoospores of *Sphaerothecum destruens* (rosette agent) at four temperatures in the laboratory: 4°C, 15°C, 25°C and 30°C. Both the host and parasite have been observed within this temperature range in the field. After infection with this multi-host intracellular parasite, high mortalities within the host fish populations can result. Their results indicate a wide temperature tolerance with a later onset of the production of zoospores and a longer life span of zoospores at lower temperatures. Although these zoospores are somewhat temperature sensitive, they still survive and spread disease at 30°C in laboratory studies. The host for the isolate of *Sphaerothecum destruens* used in the study by Andreou et al. (2009) is *Leucaspius delineates* (European cyprid). Andreou et al. (2009) did not study this phenomenon in zoospores of amphibian parasites, and as far as we know data on the effect of temperature on stability of zoospores in amphibians is not yet available.

11.3 Oomycota (oomycetes or water moulds)

Water moulds are a very large group of heterotrophic stramenopiles that are commonly observed growing on detritus in many freshwater and soil habitats (Sparrow 1960; Marano et al. 2011). Some species are well known as economically and ecologically important parasites of plants, amphibians, fish, crustaceans and many other freshwater animals (Sparrow 1960; Czeczuga et al. 1999; Czeczuga and Muszyńska 1999; van den Berg et al. 2013). For example, some species of *Saprolegnia* cause saprolegniasis which is a common disease of fish especially in hatcheries (Kiesecker et al. 2001; van den Berg et al. 2013). In this chapter we discuss primarily water moulds which infect the eggs and tissues of the larval stages of amphibians. These include in particular many species in the genera *Achlya*, *Aphanomyces*, *Leptolegnia* and *Saprolegnia* (Tab. 11.3). These and other genera of water moulds are thought to be facultative pathogens. Some of the recent research on these pathogens has been reviewed by Berger et al. (2001), Romansic et al. (2009), Sarowar et al. (2013) and van den Berg et al. (2013).

Saprolegnia ferax is known to infect amphibian embryos causing mass mortality (Romansic et al. 2009). Romansic et al. (2009) tested the infectivity of one isolate of *S. ferax* against larvae of three species of frogs and one salamander. This isolate killed larvae of two species of frogs. Berger et al. (2001) observed a species of *Aphanomyces* growing on the skin of tadpoles of *Bufo marinus*. Petrisko et al. (2008) assessed the diversity and phylogeny of Saprolegniaceae growing on amphibian eggs of six species in the Pacific Northwest. A wide diversity of these parasites was recorded.

Some virulent ecotypes of water moulds may grow in the soil as saprotrophs (Kiesecker and Skelly 2001). Ruthig and Provost-Javier (2012) isolated six saprotrophic genotypes of water moulds from dead arthropods, vertebrates, plant leaves and frogs

Tab. 11.3: Ichthyophonid parasites of amphibians.

Parasite	Host	Location	References
Anurofeca richardsi	Tadpoles: common frog, *Rana temporaria;* common toad, *Bufo bufo*, natterjack toad, *Epidalea calamita.*	Europe	Wong and Beebee (1994); Baker et al. (1999)
Ichthyophonus sp.	Redspotted newt, *Notophthalmus viridescens;* spotted salamander, *Ambystoma maculatum;* wood frog, *Lithobates sylvaticus;* green frog, *Lithobates clamitans*, bullfrogs, *Lithobates catesbeianu:* pickerel frog, *Lithobates palustris;* spring peeper, *Pseudacris crucifer*	North America	Herman (1984); Mikaelian et al.(2000); Raffel et al. (2006); Ware et al. (2008)

eggs from a small pond in Virginia. The ITS (rDNA) sequences of these genotypes clustered near sequences of *Saprolegnia, Leptolegnia, Achlya, Aphanomyces* and *Pythium* from GenBank. This study indicated that all of these genotypes infected eggs of the bullfrog *Lithobates catesbeianus* in the laboratory and colonized many other substrates as well. Furthermore, some isolates of *Saprolegnia* are capable of infecting a taxonomically wide range of hosts other than amphibians. For example, four isolates from insects could also infect Atlantic salmon eggs and embryos of the African clawed frog, and one isolate from Atlantic salmon eggs could infect nymphs of stone fly larvae and embryos of the African clawed frog (Sarowar et al. 2013).

The maximum temperature for growth of some parasites of insects and fishes in the Oomycota has been estimated. Nolan (1976) tested an isolate of *Saprolegnia ferax* from larvae of *Simulium vittatum*. The minimum was 10°C, the optimum was 25°C and the maximum was 30°C. Pacheco-Marino et al. (2011) tested one isolate of *Saprolegnia ferax* and one of *Achlya racemosa* both from *Odontesthes bonariensis* at 5°C, 15°C and 25°C. The growth rate was highest at 25°C for both species. Hatai et al. (1990) compared growth rates of 13 isolates from diseased fish at a hatchery in Japan at 30°C. The species of fish included *Oncorhynchus kisutch, Odonthestes bonariensis, Salmo gairdneri* and *S. trutta*, and the *Saprolegnia* species included *S. parasitica, S diclina* and *S. hypogyna*. All isolates grew at 30°C but at different rates and none grew at 35°C. All isolates could survive exposure to 35°C for 4 h. *Aphanomyces frigidophilus*, isolated from eggs of *Salvelinius leucomaenis*, grew at a maximum rate at 25°C and at less than half the maximum rate at 30°C and did not grow at 35°C (Kitancharoen and Hatai 1997). Kitancharoen et al. (1996) also tested other isolates of *S. diclina* and *S. parasitica* from species of *Oncorhynchus, Odonthestes* and *Plecoglossus* and from pond water for growth at 30°C. Seven isolates of *S. diclina* grew rapidly and eight isolates of *S. parasitica* had little or no growth. The data on temperature and pH effects presented in this chapter comes from studies with fish, but many of these species also infect amphibians. Kitancharoen et al. (1996) found that all 15 isolates grew well at pH 4.8. The seven isolates of *S. diclina* grew at pH 3.5, but the eight isolates of *S. parasitica* could not grow at pH 3.5. The optimum pH for growth of *S. ferax* is 5.6 (Nolan 1976).

11.4 Perkinsozoa

A previously unknown pathogen, originally described as belonging to Alveolata, has been linked to new infections causing mass mortalities in amphibian populations throughout the USA (Green et al. 2002). This pathogen appeared similar in morphology to the species of *Perkinsus* which are known pathogens of oysters in the ocean, but instead it infected both adult and larval ranid frogs in fresh water ecosystems (Green et al. 2002; Davis et al. 2007). In 2007, following a massive mortality of the frog *Lithobates sphenocephala*, Davis et al. (2007) performed histological analyses of infected organisms, and described the presence of hundreds of round single-cells infecting preferentially the frog liver. The morphology of these pathogens appeared similar to that of non-photosynthetic dinoflagellates. The *Perkinsus*-like pathogens of frogs have not been thoroughly studied yet.

Perkinsus marinus causes dermo disease in *Crassostrea virginica* (eastern oyster) along the East Coast of North America. Cook et al. (1998) found a positive correlation between the increasing sea-surface temperature and the northward spread of this parasite. The disease is prevalent during the summer months and early autumn when the water temperature can reach above 20°C, but becomes dormant during the winter and spring (Chu and Greene 1989). Temperatures less than 15°C cause a decrease in the incidence and intensity of disease in the field. In the laboratory temperatures greater than 18°C are required for development of sporangia, and the optimum temperature for sporulation and survival of zoospores was 28°C.

11.5 The Fisher concept of emerging infectious diseases (EIDs)

Fisher et al. (2012) have defined significant properties for identifying emerging infectious diseases (EID's) in aquatic and terrestrial ecosystems caused by fungi and fungus-like organisms. Some groups of protists can also be recognized as EIDs. The principals discussed by Fisher et al. (2012) apply to the ecology of disease in amphibians. One example of an EID discussed by Fisher et al. (2012) was chytridiomycosis which is caused by *B. dendrobatidis*. Furthermore, several other groups of zoosporic parasites of amphibians have properties described in the concept of EIDs and which allow them to be characterized as causing EIDs. Although there is insufficient evidence for all of the characteristics listed by Fisher (2012) in most of the virulent parasites of amphibians, we have summarized our knowledge at present in Tab. 11.4.

11.6 Host switching by parasites

In general, there is considerable evidence that many zoosporic parasites can switch between hosts, that some hosts can function as reservoirs and that resistant hosts can act as carriers. We present several specific examples here.

There has been much speculation about the possibility of alternative hosts for *B. dendrobatidis*, but until recently, none have been found. Tadpoles can be infected with *B. dendrobatidis* but chytridiomycosis does not usually cause death in resistant species. Some species of tadpoles can persist in the environment for multiple seasons, therefore tadpoles can be carriers of this disease spreading it into other stages of the life cycle of the same or different species. For example, the North American bull frog, *Lithobates catesbeianus*, can be infected but is highly resistant to chytridiomycosis (Schloegel et al. 2012).

Tab. 11.4: Properties of emerging infectious diseases.

Properties of Emerging Infectious Diseases	***Chytridiomycota***	***Mesomycetozoea***	***Oomycota***	***Perkinsozoa***
Propagules for rapid transmission	Uniflagellate zoospores	Uniflagellate zoospores	Biflagellate zoospores	Biflagellate zoospores
Threshold host population size for disease persistence	Probably none	Unknown	None	Unknown
Host species resistant to disease	Some host species	Some host species	Probably none	Unknown
Ecotypes with high virulence	Yes	Yes	Yes	Unknown
Long-lived Environmental stages	Resistant sporangia	Endospores	Oospores	Unknown
Generalist and opportunist pathogens	Yes	Yes	Yes	Unknown
Sexual life cycle	Rare	Unknown	Common	Unknown
Transport of infected species	amphibian trade	amphibian trade	amphibian and fish trade	Unknown

Very recently, field studies in Louisiana and Colorado have revealed *B. dendrobatidis* zoosporangia within the gastro-intestinal tract of crayfish with a prevalence as high as 29% (McMahon et al. 2013). The presence of chytridiomycosis in crayfish was a positive predictor of infection co-occurring in amphibians but the roles of crayfish as hosts, reservoirs or vectors need further research. Tadpoles and adults of resistant species of frogs and crayfish have been proposed as reservoirs and vectors for chytridiomycosis in amphibians (McMahon et al. 2013). However, quantitative data are not available.

Kiesecker et al. (2001) provided evidence that *Saprolegnia ferax* can be transferred from fish (*Oncorhynchus mykiss*) to toads (*Anaxyrus boreas*) and to unknown substrates in the soil. This raises the possibility that introduced (non-native) fish may act as vectors and reservoirs for virulent ecotypes of Oomycetes which can infect amphibians, and that populations of virulent ecotypes may grow in the soil as saprotrophs. Several saprotrophic species of Oomycetes isolated from non-living substrates in aquatic environments can infect bullfrog eggs in the laboratory (Ruthig and Provost-Javier 2012).

Many species in the Oomycota are capable of growth on a large variety of substrates including live species of animals and plants as well as decaying and dead animal and plant debris in freshwater ecosystems (Sparrow 1960; Czeczuga et al. 2002). Host species include crustacean, insects, molluscs (gastropods and bivalves), leeches, and eggs, larval stages and adult stages of amphibians and fishes (Sparrow 1960; Czeczuga and Muszynska 1999; Czeczuga et al. 1999, 2003, 2008, 2013; Czeczuga 2000). Many commonly introduced fishes, such as, *Salmo* sp., *Salvelinus* sp. and *Oncorbynchus* sp. are common carriers of species of *Saprolegnia*, including *Saprolegnia ferax* (Peeler et al. 2011), and it is possible that these pathogens can switch to amphibian hosts (Ruthig and Provost-Javier 2012; Sarowar et al. 2013). Parts of dead animals can be used to bait for these facultative parasites (Czeczuga et al. 2002). This suggests that both dead and live animals may serve as reservoirs in the benthic and pelagic zones of freshwater ecosystems. Unfortunately, quantitative data on the relationships between population sizes of pathogens and prevalence of infection of hosts in the field are not currently available for Oomycota.

11.7 Genetic variation in parasite populations

Zoosporic parasites often display a high degree of genetic variation. Fisher et al. (2009) showed that 11 isolates of *B. dendrobatidis* from different geographical regions can be profiled by their genotypic and proteomic characteristics, as well as the size of their sporangia. Rosenblum et al. (2013) sequenced the genomes of 29 isolates of *B. dendrobatidis* collected from a wide geographical area and found considerably more genetic variation than previously thought possible. Further genetic data may reveal that *B. dendrobatidis* is in fact a cluster of multiple species. DNA sequence data from the various isolates of *B. dendrobatidis* has revealed considerable genetic variation (Fisher et al. 2009; Joneson et al. 2011; Bataille et al. 2013; Rosenblum et al. 2013).

Analysis of ITS (rDNA) indicate that oomycete parasites of amphibians fall into at least five genera (Ruthig and Provost-Javier 2012). It is possible for amphibian eggs to be infected by more than one genotype of oomycete simultaneously. Jiang et al. (2013) have highlighted the wide genetic variation in oomycete genomes and the expansion of potential virulence genes. Genetic variations among the other groups of zoosporic parasite have not been as intensively studied. The relationship of variation in rDNA and other genes to virulence of pathogens is presently under investigation (Jiang et al. 2013).

An important source of genetic variation is sexual reproduction and recombination. Many species of Oomycota undergo sexual reproduction as a normal part of their life cycles with the production of resistant structures (oospores) which can survive extreme environmental conditions (van den Berg et al. 2013). Sexual reproduction has not yet been observed in the life cycle of *B. dendrobatidis* with the microscope but evidence is growing which indicates some form of genetic recombination (Rosenblum et al. 2013). Since the complete genomic sequence is currently available for *B. dendrobatidis* and *S. parasitica*, studies on the expansion of genes for pathogenicity in these species are in progress (Joneson et al. 2011; Jiang et al. 2013; Rosenblum et al. 2013).

11.8 Proteases

Dead and live animals are known to be protein rich substrates. Proteases are extracellular enzymes which are necessary for the digestion of proteins by pathogens in tissues of living host animals. For example, proteases are released by *B. dendrobatidis* during the digestion of the tissues of amphibian hosts (Piotrowski et al. 2004). Piotrowski et al. (2004) measured the effects of pH and temperature on the activity of extracellular proteases. The supernatant concentrated 10× from the cultures contained the enzyme activity, azocasein was used as substrate, the reaction mixtures were incubated for 36 h and the absorbance was measured at 440 nm. Five pH values were tested at 23°C: pH 6.0, 6.5, 7.0, 7.5 and 8.0, and five temperatures were tested at pH 7.5: 6°C, 15°C, 23°C, 30°C and 37°C. The culture supernatant was most active in degrading azocasein at pH 6–8 and 23°C–30°C.

The complete genomic sequence for *B. dendrobatidis* and *S. parasitica*, provides an opportunity to mine for protease genes (Joneson et al. 2011; Jiang et al. 2013; Rosenblum et al. 2013). Joneson et al. (2011) describe significant lineage-specific expansions in three families of proteases (metallo, serine and aspartyl proteases) in isolates of *B. dendrobatidis*. Duplication of protease genes facilitates the production of more protease enzymes. Jiang et al. (2013) documented the presence of genes for three families of proteases (serine, metallo- and cysteine proteases) and genes for amino acid transporters in *Saprolegnia parasitica*. Serine protease genes have been characterized in *Perkinsus marinus*, the oyster parasite (Perkinsozoa) (Villalba et al. 2004). We expect that genes for proteases are present in freshwater species as well. Furthermore, we expect that other groups of zoosporic parasites also use a variety of proteases in different families to digest the tissues in their animal hosts.

11.9 International animal trade

The global trade in amphibians results in a huge number of species being translocated around the world every year (Fisher and Garner 2007; Jenkins et al. 2007). Along with the potential spread of non-native amphibian species (Fisher and Garner 2007), there is also the potential release of non-native parasites into novel environments (Daszak et al. 2000; Gozlan et al. 2006; Peeler et al. 2011; McKenzie and Peterson 2012). The emergence of *B. dendrobatidis*, as a result of the trade of the American bullfrog *L. catesbeianus* for human consumption (Farrer et al. 2011; Schloegel et al. 2012), has been implicated as a significant factor in the decline of amphibians. It is thought that highly virulent genotypes of chytridiomycetes have been introduced into new environments along with their resistant hosts (Schloegel et al. 2012). As discussed previously, there are a number of mesomycetozoean parasites that are carried by amphibians (Tab. 11.2). The heavily traded *L. catesbeianus* is also a known carrier of *Histocystidium ranae* (Goodchild 1953). The invasive amphibian species may also act as a reservoir or vector for the parasite and therefore magnify it and the parasites impact on the novel environment (Rowley et al. 2013). Currently there is limited surveillance of diseases in the wildlife trade (Smith et al. 2008, 2009). Unregulated introduction of introduced species can only be controlled by effective monitoring and quarantine where appropriate. Unless this is addressed, this "pathogen pollution" will have a significant role in the emergence of numerous economically and ecological diseases (Daszak et al. 2000; Rowley et al. 2013).

11.10 Discussion and conclusion

A large number of zoosporic parasites infect the different stages of the life cycles of amphibians. Therefore, a wide range of parasites and interactions between them must be considered when attempting to explain declines in population sizes. Understanding and ultimately managing the impacts of zoosporic parasites of amphibians requires understanding the factors determining the susceptibility of individuals to disease, and the susceptibility of populations to epidemic outbreaks. Individual susceptibility to disease is often considered to be largely a function of the immune system (Morales and Robert 2007; Woodhams et al. 2007), and the susceptibility of populations to epidemic outbreaks is thought to be related to individual susceptibility and to factors affecting transmission rates (de Castro and Bolker 2005). Micro-environmental selection, frequency of contact with other amphibians and environmental reservoirs, as well as the movement patterns of hosts may be particularly important in determining susceptibility to zoosporic parasites of amphibians (Rowley and Alford 2009).

All of the parasites discussed in this review appear to be well adapted to infecting amphibian and fish hosts in freshwater environments which are slightly acidic and at approximately 20°C. If species of mesomycetozoean, perkinsozoan or oomycota can grow at higher temperatures than *B. dendrobatidis*, it is possible that some of these species could replace *B. dendrobatidis* as the dominant parasite of frog populations with increasing water temperatures due to global warming. There is no quantitative evidence that this is occurring at present, but this is possibility needs further investigation. It is more likely that changes in physical factors with climate change (such as temperature, pH and salinity) could result in small but significant changes in the composition of food webs and the ecological services provided. These changes could be cumulative and possible result in cascades which would have major ecological consequences.

The continued expansion of the global wildlife trade, which involves billions of animals per year, has and will continue to have major impacts on the global translocation and release of pathogens. The trade in amphibians alone is vast. Not only is there the potential of the species becoming invasive, but they have the potential to be reservoirs or vectors of many of the pathogens discussed above. Many species in the Chytridiomycota, Mesomycetozoea, Perkinsozoa and Oomycota are parasite of amphibians and some of these species have characteristics which according to the concepts published by Fisher et al. (2012) suggest that they should be classified as EIDs.

Acknowledgments

The authors thank the University of Sydney Library for the use of its resources and Elayna Truszewski, Department of Biological Sciences, Macquarie University for her editorial assistance with preparation of the original manuscript.

References

Alford RA, Bradfield KS, Richards SJ. Ecology – global warming and amphibian losses. Nature 2007;447:E3–E4.

Alford RA. Declines and the Global Status of Amphibians. In: Sparling DW, Linder G, Bishop CA, Krest S, eds. Ecotoxicology of Amphibians and Reptiles, 2nd ed. Pensacola, FL: SETAC Press. 2010:13–46.

Andreou D, Gozlan RE, Paley R. Temperature influence on production and longevity of *Sphaerothecum destruens* zoospores. J Parasitol 2009;95:1539–1541.

Baker GC, Beebee TJ, Ragan MA. *Prototheca richardsi*, a pathogen of anuran larvae, is related to a clade of protistan parasites near the animal-fungal divergence. Microbiology 1999;145:1777–1784.

Bataille A, Fong JJ, Cha M, Wogan GOU, Baek HJ, Lee H, et al. Genetic evidence for a high diversity and wide distribution of endemic strains of the pathogenic chytrid fungus *Batrachochytrium dendrobatidis* in wild Asian amphibians. Mol Ecol 2013;22:4196–4209.

Berger L, Marantelli G, Skerratt LF, Speare R. Virulence of the amphibian chytrid fungus *Batrachochytrium dendrobatidis* varies with the strain. Dis Aquat Org 2005;68:47–50.

Berger L, Speare R, Daszak P, Green DE, Cunningham AA, Goggin CL, et al. Chytridiomycosis causes amphibian mortality associated with population declines in the rainforests of Australia and Central America. Proc Natl Acad Sci USA 1998;95:9031–9036.

Berger L, Speare R, Hines HB, Marantelli G, Hyatt AD, McDonald KR, et al. Effect of season and temperature on mortality in amphibians due to chytridiomycosis. Aust Vet J 2004;82:434–439.

Berger L, Speare R, Marantelli G, Skerratt LF. A zoospore inhibition technique to evaluate the activity of antifungal compounds against *Batrachochytrium dendrobatidis* and unsuccessful treatment of experimentally infected green tree frogs (*Litoria caerulea*) by fluconazole and benzalkonium chloride. Res Vet Sci 2009;87:106–110.

Berger L, Speare R, Thomas A, Hyatt A. Mucocutaneous fungal disease in tadpoles of *Bufo marimus* in Australia. J Herpetol 2001;35:330–335.

Berger L, Longcore JE, Speare R, Hyatt A, Skerratt LF, eds. Amphibian Biology vol. 8, Australia: Surrey Beatty & Sons 2009:2986–3052.

Bosch I, Martinez-Solano I, Garcia-Paris M. Evidence of a chytrid fungus infection involved in the decline of the common midwife toad (*Alytes obstetricans*) in protected areas of central Spain. Biol Conserv 2001;97:331–337.

Brodkin M, Vatnick I, Simon M, Hopey H, Butler-Holston K, Leonard M. Effects of acid stress in adult *Rana pipiens*. J Expt Zoo 2003;298A:16–22.

Broz O, Kulda J. *Dermosporidium multigranulare* n.sp parasit z kuze *Rana esculenta*. Acta Soc Zool Bohem 1952;18:91–97.
Broz O, Privora M. 2 skin parasites of *Rana-temporaria* - *Dermacystidium ranae* guyenot and Naville and *Dermosporidium granulosum* N-SP. Parasitology 1952;42:65–69.
Carini A. Sobre um parasito semelhante ao 'Rhinosporidium' encontrado em quistos da pele de uma 'Hyla'. Arq Inst Biol 1940;11:93–98.
Chu FLE, Greene KH. Effect of temperature and salinity on in vitro culture of the oyster pathogen, Perkinsus-Marinus (Apicomplexa-Perkinsea). J Invertebr Pathol 1989;53:260–268.
Cook T, Folli M, Klinck J, Ford S, Miller J. The relationship between increasing sea-surface temperature and the northward spread of *Perkinsus marinus* (Dermo) disease epizootics in oysters. Estuar Coast Shelf Sci 1998;46:587–597.
Czeczuga B, Godlewska A, Semeniuk J, Czeczuga-Semeniuk E, Kosielinski P. Distribution of fungi and straminipiles in different stem parts of submerged aquatic plants. Nova Hedwig 2013;97:239–250.
Czeczuga B, Kiziewicz B, Danilkiewicz Z. Zoosporic fungi growing on the specimens of certain fish species recently introduced to Polish waters. Acta Ichthyol Piscat 2002;32:117–125.
Czeczuga B, Kiziewicz B, Godlewska A. Zoosporic fungi growing on leeches (Hirudinea). Pol J Environ Stud 2003;12:361–369.
Czeczuga B, Kozlowskaz M, Godlewska A, Velu SC. *Moina macrocopa* (Straus): a plankton crustacean as a vector for fungus-like fish parasites. Turk J Zool 2008;32:19–26.
Czeczuga B, Muszynska E, Pankiewicz A. Zoosporic fungi in the ice of some water reservoirs. Acta Soc Bot Pol 1999;68:143–147.
Czeczuga B, Muszyńska E. Aquatic fungi growing on percid fish eggs (Persidae) in Poland. Pol J Environ Stud 1999;8:31–34.
Czeczuga B. Zoosporic fungi growing on freshwater molluscs. Pol J Environ Stud 2000;9:151–156.
Daszak P, Cunningham AA, Hyatt AD. Emerging infectious diseases of wildlife-threats to biodiversity and human health. Science 2000;287:443–449.
Davis AK, Yabsley MJ, Keel MK, Maerz JC. Discovery of a novel alveolate pathogen affecting southern leopard frogs in Georgia: Description of the disease and host effects. Ecohealth 2007;4:310–317.
De Castro FF, Balker B. Mechanisms of disease-induced extinction. Ecol Lett 2005;8:117–126.
Di Rosa I, Simoncelli F, Fagotti A, Pascolini R. Ecology – the proximate cause of frog declines? Nat 2007;447:E4–E5.
Duffus ALJ, Cunningham AA. Major disease threats to European amphibians. Herpetol J 2010;20:117–127.
Farrer RA, Weinert LA, Bielby J, Garner TW, Balloux F, Clare F, et al. Multiple emergences of genetically diverse amphibian-infecting chytrids include a globalized hypervirulent recombinant lineage. Proc Natl Acad Sci USA 2011;108:18732–18736.
Fellers GM, Green DE, Longcore JE. Oral chytridiomycosis in the mountain yellow-legged frog (*Rana mucosa*). Copeia 2001;2001:943–953.
Fisher MC, Bosch J, Yin Z, Stead DA, Walker J, Selway L, et al. Proteomic and phenotypic profiling of the amphibian pathogen *Batrachochytrium dendrobatidis* shows that genotype is linked to virulence. Mol Ecol 2009;18:415–429.
Fisher MC, Garner TW. The relationship between the emergence of *Batrachochytrium dendrobatidis*, the international trade in amphibians and introduced amphibian species. Fungal Biol Rev 2007;21:2–9.
Fisher MC, Henk DA, Briggs CJ, Brownstein JS, Madoff LC, McCraw SL, et al. Emerging fungal threats to animal, plant and ecosystem health. Nature 2012;484:186–194.
Glockling SL, Marshall WL, Gleason FH. Phylogenetic interpretations and ecological potentials of the Mesomycetozoea (Ichthyosporea). Fungal Ecol 2013;6:237–247.

González-Hernández M, Denoël M, Duffus AJ, Garner TW, Cunningham AA, Acevedo-Whitehouse K. Dermocystid infection and associated skin lesions in free-living palmate newts (*Lissotriton helveticus*) from Southern France. Parasitol Int 2010;59:344–350.

Goodchild CG. A subcutaneous, cyst-parasite of Bullfrogs: *Histocystidium ranae*, n. g, n. sp. J Parasitol 1953;39:395–405.

Gower DJ, Doherty-Bone T, Loader SP, Wilkinson M, Kouete MT, Tapley B, et al. *Batrachochytrium dendrobatidis* infection and lethal chytridiomycosis in caecilian amphibians (Gymnophiona), EcoHealth 2013;10:173–183.

Gozlan RE, Peeler EJ, Longshaw M, St-Hilairee S, Feist SW. Effect of microbial pathogens on the diversity of aquatic populations, notably in Europe. Microbes Infect 2006;8:1358–1364.

Gozlan RE, Whipps CM, Andreou D, Arkush KD. Identification of a rosette-like agent as *Sphaerothecum destruens*, a multi-host fish pathogen. Int J Parasitol 2009;39:1055–1058.

Granata L. *Dermomycoides beccarii* n.g n.sp nuovo enigmatico parassita di Molge vulgaris L. Monit. Zool Ital 1919;20:153–160.

Green DE, Converse KA, Schrader AK. Epizootiology of sixty-four amphibian morbidity and mortality events in the USA, 1996–2001. Ann NY Acad Sci 2002;969:323–339.

Grey AM. Infection of the palmate newt (*Triturus helveticus*) by a novel species of *Amphibiocystidium* on the Isle of Rum, Scotland. MSc thesis. London Institute of Zoology, Zoological Society of London and Royal Veterinary College, University of London 2008.

Guyénot E, Naville A. Un nouveau protiste, du genre *Dermocystidium*, parasite de la grenouille. *Dermocystidium ranae* nov. spec. Rev Suisse Zool 1922;29:133–145.

Hartigan A, Phalen DN, Šlapeta J. Myxosporean parasites in Australian frogs: importance, implications and future directions. Int J Parasitol: Parasites and Wildl 2013;2:62–68.

Hatai K, Willoughby LG, Beakes GW. Some characteristics of *Saprolegnia* obtained from fish hatcheries in Japan. Mycol Res 1990;94:182–190.

Herman RL. *Ichthyophonus*-like infection in newts (*Notophthalmus viridescens* Rafinesque). J Wildl Dis 1984;20:55–56.

Jay JM, Pohley WJ. *Dermosporidium penneri* sp n from the skin of the American toad *Bufo americanus* (Amphibia: Bufonidae). J Parasitol 1981;67:108–110.

Jenkins PT, Genovese K, Ruffler H. Broken Screens: The Regulation of Live Animal Imports in the United States. Washington DC: Defenders of Wildlife 2007.

Jiang RHY, de Bruijn I, Haas BJ, Belmonte R, Lobach L, Christie J, et al. Distinctive expansion of potential virulence genes in the genome of the oomycete fish pathogen *Saprolegnia parasitica*. PLoS Genet 2013;9:e1003272. doi: 10.1371/journal.pgen.1003272.

Joneson S, Stajich JE, Shiu SH, Rosenblum EB. Genomic transition to pathogenicity in Chytrid fungi. PLoS Pathog 2011;7:e1002338. doi: 10.1371/journal.ppat.1002338.

Kiesecker JM, Blaustein AR, Miller CL. Transfer of a pathogen from fish to amphibians. Conserv Biol 2001;15:1064–1070.

Kiesecker JM, Skelly DK. Effects of disease and pond drying on gray tree frog growth, development, and survival. Ecologia 2001;82:1956–1963.

Kitancharoen N, Hatai K. *Aphanomyces frigidophilus* sp. nov. from eggs of Japanese char, *Salvelinus leucomaenis*. Mycoscience 1997;38:135–140.

Kitancharoen N, Yuasa K, Hatai H. Effects of pH and temperature on growth of *Saprolegnia diclina* and *S. parasitica* isolated from various sources. Mycoscience 1996;37:385–390.

Lips KR, Brem F, Brenes R, Reeve JD, Alford RA, Voyles J, et al. Emerging infectious disease and the loss of biodiversity in a Neotropical amphibian community. Proc Natl Acad Sci USA 2006;103:3165–3170.

Lips KR, Donnelly MA. What the tropics can tell us about declining amphibian populations: current patterns and future prospects. In: Lannoo LJ, ed. North American Amphibians: Status and Conservation. Berkeley, CA: University of California Press 2002:388–406.

Lips KR. Mass mortality and population declines of anurans at an upland site in western Panama. Conserv Biol 1999;13:117–125.
Longcore JE, Pessier AP, Nichols DK. *Batrachochytrium dendrobatidis* gen. *et* sp. nov., a chytrid pathogenic to amphibians. Mycologia 1999;91:219–227.
Marano AV, Pires-Zottarelli CLA, Gleason FH, Neuhauser S, Steciow MW. Assemblage of zoosporic true fungi, hetyerotrophic stramenopiles and plasmodiophorids in freshwater ecosystems. In: Brown SA, ed. Aquatic Ecosystems. Hauppauge, NY: Nova Science Publishers 2011:1–56.
Martel A, Spitzen-van der Sluijs A, Blooi M, Bert W, Ducatelle R, Fisher MC, et al. *Batrachochytrium salamandrivorans* sp. nov causes lethal chytridiomycosis in amphibians. Proc Natl Acad Sci USA 2013;110:15325–15329.
McKenzie VJ, Peterson AC. Pathogen pollution and the emergence of a deadly amphibian pathogen. Mol Ecol 2012;21:5151–5154.
McMahon TA, Brannelly LA, Chatfield MWH, Johnson PTJ, Joseph MB, McKenzie VJ, et al. Chytrid fungus *Batrachochytrium dendrobatidis* has nonamphibian hosts and releases chemicals that cause pathology in the absence of infection. Proc Natl Adad Sci USA 2013;110:210–215.
Mikaelian I, Ouellet M, Pauli B, Rodrigue J, Harshbarger JC, Green DM. *Ichthyophonus*-like infection in wild amphibians from Quebec, Canada. Dis Aquat Org 2000;40:195–201.
Moral H. Concerning the appearance of *Dermocystidium pusula* (Perez), a single cell parasite in the skin of *Triton cristatus* newts. Arch Mikrosk Anat 1913;81:381–393.
Morales HD, Robert J. Characterization of primary and memory CDS T-cell response against ranavims (FV3) in *Xenopus laevis*. J Virol 2007;81:2240–2248.
Nichols DK, Lamirande EW, Pessier AP, Longcore JE. Experimental transmission of cutaneous chytridiomycosis in dendrobatid frogs. J Wildl Dis 2001;37:1–11.
Nolan RA. Physiological studies on an isolate of *Saprolegnia ferax* from larval gut of blackfly *Simulium vittatum*. Mycologia 1976;68:523–540.
Pacheco-Marino S, Steciow M, Paul B. Culture media and temperature influence on growth and sexual reproduction of the fish pathogens *Achlya racemosa* and *Saprolegnia ferax*. Nova Hedwig 2011;92:273–282.
Pascolini R, Daszak P, Cunningham AA, Tei S, Vagnetti D, Bucci S, et al. Parasitism by *Dermocystidium ranae* in a population of *Rana esculenta* complex in Central Italy and description of *Amphibiocystidium* n. gen. Dis Aquat Org 2003;56:65–74.
Peeler EJ, Oidtmann BC, Midtlyng PJ, Miossec L, Gozlan RE. Non-native aquatic animals introductions have driven disease emergence in Europe. Biol Invasions 2011;13:1291–1303.
Pereira CN, Di Rosa I, Fagotti A, Simoncelli F, Pascolin R, Mendoza L. The pathogen of frogs *Amphibiocystidium ranae* is a member of the order dermocystida in the class Mesomycetozoea. J Clin Microbiol 2005;43:192–198.
Pérez C. *Dermocystis pusula,* organisme nouveau parasite de la peau des tritons. Comptes Rendus Soc Biol 1907;63:445–447.
Pérez C. *Dermocystis pusula*, parasite de la peau des tritons. Arch Zool Exp Gén 1913;52: 343–57.
Pessier AP. Management of disease as a threat to amphibian conservation. Int Zoo Yearb 2008;42:30–39.
Petrisko JE, Pearl CA, Pilliod DS, Sheridan PP, Williams CF, Peterson CR, et al. Saprolegniaceae identified on amphibian eggs throughout the Pacific Northwest, USA, by internal transcribed spacer sequences and phylogenetic analysis. Mycologia 2008;100:171–180.
Piotrowski JS, Annis SL, Longcore JE. Physiology of *Batrachochytrium dendrobatidis*, a chytrid pathogen of amphibians. Mycologia 2004;96:9–15.
Poisson R. Sur une nouvelle espèce du genre Dermomycoides Granata 1919: *Dermomycoides armoriacus* Poisson 1936 parasite cutané de *Triturus palmatus* (Schneider): genèse *et* structure de la zoospore. Bull Biol 1937;71:91–116.

Pounds JA, Bustamante MR, Coloma LA, Consuegra JA, Fogden MPL, Foster PN, et al. Ecology. Pounds et al. reply. Nature 2007;447:E5–E6.

Puschendorf R, Hoskin CJ, Cashins SD, McDonald K, Skerratt LF, Vanderwal J, et al. Environmental refuge from disease-driven amphibian extinction. Conserv Biol 2011;25:956–964.

Rachowicz LJ, Knapp RA, Morgan JAT, Stice MJ, Vredenburg VT, Parker JM, et al. Emerging infectious disease as a proximate cause of amphibian mass mortality. Ecologia 2006;87:1671–1683.

Rachowicz LJ, Vredenburg VT. Transmission of *Batrachochytrium dendrobatidis* within and between amphibian life stages. Dis Aquat Org 2004;61:75–82

Raffel TR, Dillard JR, Hudson PJ. Field evidence for leech-borne transmission of amphibian *Ichthyophonus*. J Parasitol 2007;92:1256–1264.

Raffel TR, Martin LB, Rohr JR. Parasites as predators: unifying natural enemy ecology. Trends Ecol Evol 2008;23:610–618.

Raffel TR. Causes and consequences of seasonal dynamics in the parasite community of red-spotted newts (*Notophthalmus viridescens*). (Doctoral dissertation, College of Science). Pennsylvania State University, PA, USA 2006.

Romansic JM, Diez KA, Higashi EM, Johnson JE, Blaustein AR. Effects of the pathogenic water mold *Saprolegnia ferax* on survival of amphibian larvae. Dis Aquat Org 2009;83:187–193.

Rosenblum EB, James TY, Zamudio KR, Poorten TJ, Ilut D, Rodriguez DE, et al. Complex history of the amphibian-killing chytrid fungus revealed with genome resequencing data. Proc Natl Acad Sci USA 2013;110:9385–9390.

Rowley JJL, Alford RA. Behaviour of Australian rainforest stream frogs may affect the transmission of chytridiomycosis. Dis Aquat Org 2007;77:1–9.

Rowley JJL, Alford RA. Factors affecting interspecific variation in susceptibility to disease in amphibians. Amphib Biol 2009;8:3053–3066.

Rowley JJL, Alford RA. Hot bodies protect amphibians against chytrid infection in nature. Sci Rep 2013;3:doi: 10.1038/srep01515.

Rowley JJL, Gleason FH, Andreou D, Marshall W, Lilje O, Gozlan R. Impacts of mesomycetozoean parasites on amphibian and freshwater fish populations. Fungal Biol Rev 2013;27:100–111.

Ruthig GR, Provost-Javier KN. Multihost saprobes are facultative pathogens of bullfrog *Lithobates catesbeianus* eggs. Dis Aquat Org 2012;101:13–21.

Sarowar MN, van den Berg AH, Mclaggan D, Young MR, van West P. *Saprolegnia* strains isolated from river insects and amphipods are broad spectrum pathogens. Fungal Biol 2013;117:752–763.

Schloegel LM, Toledo LF, Longcore JE, Greenspan SE, Vieira CA, Lee M, et al. Novel, panzootic and hybrid genotypes of amphibian chytridiomycosis associated with the bullfrog trade. Mol Ecol 2012;21:5162–5177.

Smith KF, Behrens MB, Schloegel LM, Marano N, Burgiel S, Daszak P. Reducing the risks of the wildlife trade. Science 2009;324:595.

Smith KF, Behrens MD, Max LM, Daszak P. U. S. drowning in unidentified fishes: scope, implications and regulation of live fish import. Conserv Lett 2008;1:103–109.

Sparrow Jr FK. Aquatic Phycomycetes, 2nd ed. Ann Arbor, MI: The University of Michigan Press 1960.

Stockwell MP, Clulow J, Mahony MJ. Sodium Chloride inhibits the growth and infective capacity of the amphibian Chytrid fungus and increases host survival rates. PLoS One 2012;7:e36942. doi: 10.1371/journal.pone.0036942.

Van den Berg AH, Mclaggan D, Diéguez-Uribeondo J, van West P. The impact of the water moulds *Saprolegnia diclina* and *Saprolegnia parasitica* on natural ecosystems and the aquaculture industry. Fungal Biol Rev 2013;27:33–41.

Vatnick I, Andrews J, Colombo M, Madhoun H, Rameswaran M, Brodkin MA. Acid exposure is an immune disruptor in adult *Rana pipiens*. Environ Toxicol Chem 2006;25:199–202.

Villalba A, Reece KS, Camino Ordás M, Casa SM, Figueras A. Perkinsosis in molluscs: a review. Aquat Living Res 2004;17:411–432.

Voyles J, Vredenburg VT, Tunstall TS, Parker JM, Briggs CJ, Rosenblum EB. Pathophysiology in mountain yellow-legged frogs (*Rana muscosa*) during a chytridiomycosis outbreak. PloS One 2012;7:e35374.

Voyles J, Young S, Berger L, Campbell C, Voyles WF, Dinudom A, et al. Pathogenesis of chytridiomycosis, a cause of catastrophic amphibian declines. Science 2009;326:582–585.

Ware JL, Viverette C, Kleopfer JD, Pletcher L, Massey D, Wright A. Infection of spotted salamanders (*Ambystoma maculatum*) with Ichthyophonus-like organisms in Virginia. J Wildl Dis 2008;44:174–176.

Weldon C, Du Preez L. Decline of the Kihansi Spray Toad, *Nectophrynoides asperginis*, from the Udzungwa Mountains, Tanzania. Froglog 2004;62:2–3.

Wong A, Beebee T. Identification of a unicellular, non-pigmented alga that mediates growth inhibition in anuran tadpoles: a new species of the genus *Prototheca* (Chlorophyceae: Chlorococcales). Hydrobiologia 1994;277:85–96.

Woodhams DC, Alford RA, Marantelli G. Emerging disease of amphibians cured by elevated body temperature. Dis Aquat Organ 2003;55:65–67.

Woodhams DC, Alford RA. The ecology of chytridiomycosis in rainforest stream frog assemblages of tropical Queensland. Conserv Biol 2005;19:1449–1459.

Woodhams DC, Ardipradja K, Alford RA, Marantelli G, Reinert LK, Rollins-Smith LA. Resistance to chytridiomycosis varies among amphibian species and is correlated with skin peptide defenses. Anim Conserv 2007;10:409–417.

Angkana Chaiprasert and Theerapong Krajaejun

12 Pythiosis

12.1 History

Pythium insidiosum (de Cock et al. 1987) is the only pathogenic species in the genus that causes disease in mammals, called pythiosis insidiosi. Most cases of pythiosis have been reported in horses, dogs and humans and rarely in calves (Perez et al. 2005), cats (Miller et al. 1985; Rakish et al. 2005), sheep (Miller et al. 1985; Santurio et al. 2008) and birds (Pesavento et al. 2008). Human pythiosis is considered endemic in Thailand and is a life threatening disease with high rates of morbidity and mortality especially when lack of tools for early diagnosis and effective treatment (Laohapensang et al. 2009).

Pythiosis was described in 1884 by British veterinarians working with horses in India (Smith 1884). The etiologic agent was named *Hyphomyces destruens equi* and firstly isolated in 1901 by Dutch scientists working with horses in Indonesia (de Haan and Hoogkamer 1901) and re-isolated again in 1924 by Witkamp (Witkamp 1924). The disease has many names, such as, bursattee or bursatte (from Indian word burus or bursator which means rainy season), espundia, equine phycomycosis, granular dermatitidis, hyphomycosis destruens equi, leeches, swamp cancer and summer sores (Fish 1895; Witkamp 1924; Gonzales and Ruiz 1975; Ichitani and Amemiya 1980).

12.2 Biology

The fungus-like nature of the causative agent of pythiosis was firstly reported by Smith (1884) and Drouin (1896) without naming it due to lack of sporulation. Bridges and Emmons (1961) named the etiologic agent as *Hyphomyces destruens* from a disease name described by de Haan and Hoogkamer (1901). However, this name was not valid due to lack of publication with Latin description. Bridges and Emmons also proposed the name phycomycosis for the disease and *Mortierella* in zygomycete for the organism. Austwick and Copland (1974) observed zoosporogenesis of the causative agent when grown in aqueous medium and concluded that the agent belonged to the Oomycete genus *Pythium* and the name pythiosis was proposed by Chandler et al. (1980). Ichitani and Amemiya isolated a *Pythium* sp. from a diseased horse in 1980 and proposed that it should be *P. gracile* (Amemiya 1982). The causative Oomycete was finally described as *P. insidiosum* species *novum* by de Cock et al. (1987) after successful observation of sexual sporulation. The other binomial *P. destruens* was proposed almost simultaneously for the isolate obtained from infected horse in

Australia by Shipton (1987) and later placed in synonym with *P. insidiosum* (Mendoza and Martin 1989).

Pythium insidiosum can be isolated from standing inland waters and occasionally in soil (Mendoza et al. 1993, 1996) and also natural water reservoirs from canals and irrigation water in northern and north-eastern parts of Thailand (Supabandhu et al. 2008). The organism usually occurs in two forms; branching non-septate hyphae on solid medium and biflagellate zoospores in aquatic environments (Alexopoulos et al. 1996).

It grows well on different artificial media such as Sabouraud agar, vegetable extract agar, and peptone yeast glucose agar, with optimum temperature at 25–32°C but can also grow at human body temperature (de Cock et al. 1987; Shipton 1987; Grooters et al. 2002). On Sabouraud dextrose agar, it forms glabrous white to yellowish-white colony consisting of coenocytic hyphae with a diameter of 4–12 µm and branching perpendicular to the main hyphae. It is not a true fungus because of following reasons; its cell wall does not contain chitin but cellulose and β-glucans, ergosterol is absent from its cell membrane, the thallus is diploid and sexual process is oogamous and it develops biflagellate zoospores in wet environments (Alexopoulos et al. 1996). *Pythium* spp. belongs to the Kingdom Chromista (Mendoza et al. 1996; See also Chapter 8).

Borisooth (1993) studied 16 clinical isolates of *P. insidiosum* from Thai patients (9 from arteritis form and 7 from pythium keratitis cases) and showed that they could be identified by an exoantigen test and all strains produced zoospores. The size of vegetative hyphal tips ranged from 2.9–5.5 µm, reproductive tips 4.6–8.9 µm, immature zoosporangia 22–40 µm and encysted zoospores 8–10 µm, similar to the report from de Cock et al. (1987). There was no difference among strains causing pythium arteritis and keratitis in term of morphological characteristics. Moreover, the strains could be molecular typed based on 16 rDNA restriction fragment length polymorphism with *Eco* RI and *Pst* I or ribotyping into four and seven types, respectively. Again there was no difference in molecular types between isolates from the two clinical manifestations. The growth of two *P. insidiosum* strains obtained from arteritis cases revealed no significant difference either in iron-free or iron-containing media and also no difference when grown in media supplemented with either normal human sera or thalassemic patients' sera and let the mystery there about the risk factors prone thalassemic patients to get pythiosis (Borisooth 1993). Zoospores developed only in water cultures after stimulation by the presence of ions such as K^+, Ca^{2+}, Mg^{2+} and chemically attracted by plant material, hairs and animal tissues (Austwick and Copeland 1974; Shipton 1987; Mendoza and Prendas 1988; Chaiprasert et al. 1990; Mendoza et al. 1993). The single nucleated zoospores without cell wall with two flagella (one anterior tinsel and one posterior whiplash) are considered to be the infective cells which show chemotaxis and become encysted when they make contact to injured plant tissue or animal tissue (Mendoza et al. 1993, 1996). A glycoprotein secreted on the surface of the encysted zoospores allows adhesion to both the injured tissue (Estrad-Garcia et al. 1990;

Mendoza et al. 1996) and normal hair and epithelial cell lines (Monkongtanakit et al. 2011). Oospores are rarely seen in *P. insidiosum* isolates obtained from human or environmentally, and maintained in the laboratory.

12.3 Molecular typing

The first report showing that *P. insidiosum* is a unique oomycetous pathogen of mammals came from a molecular phylogenetic study of several *Pythium* species (Martin 2000). The phylogenetic tree based on *COX II* gene sequences divided *Pythium* species into at least 10 different phylogenetic groups (designated A to K). *P. insidiosum* is found in group C together with *P. grandisporangium*. Schurko et al. (2003a, b) showed heterogeneity in ribosomal DNA sequences (intergenic spacer or IGS and internal transcribed spacer or ITS) among 23 isolates of *P. insidiosum* which correlated to geographic origin, but not to the host. *Pythium insidiosum* isolates grouped into three clades: clade I composed of isolates from North, Central and South America; clade II consists of isolates from Asia and Australia, with clade III containing isolates from Thailand and the USA. Clade I and II are closely related but clade III is significantly different from the other two and a new species was proposed. Supabandhu et al. (2008) investigated ITS sequences of 59 environmental isolates of *P. insidiosum* obtained from many regions in northern Thailand, 20 *P. insidiosum* DNA sequences available in the database and 11 other *Pythium* species. The phylogenetic tree revealed three distinct groups A, B and C. Clade A is equal to clade I included the American strains, seven equine isolates from Costa Rica, two canine and one feline strains from USA, one equine from Brazil and one human isolate from Haiti. Clade B or clade II included 41 of 59 environmental isolates, four human isolates from Thailand, two strains from equine in Japan and Papua New Guinea and one strain from an Indian mosquito. Clade C or clade III contained 18 of 59 environmental isolates as well as clinical human isolates from Thailand and the USA, and one isolate obtained from spectacled bear in a zoo in the USA. It is apparent that none of the Thai human and environmental isolates fell into clade A or clade I. Chaiprasert et al. (2009), Kammarnjesdakul et al. (2011) and Arthithanyaroj et al. (2013) analyzed the correlation between phylogenetic group, geographic distribution and host specificity of 113 rDNA ITS sequences and *COX II* gene sequences of 92 *P. insidiosum* isolates from different regions of the world. The results confirmed that *P. insidiosum* could be divided into three clades in accordance with geographic distribution as previously reported by Schurko et al. (2003a, b). Thai isolates, both from the environment and human cases, grouped only in clade II and III in the ratio of 2:1. The *COX II* gene sequences provided higher discrimination power and separated *P. insidiosum* strains into three clades and seven subgroups namely I–III, IIA–D and IIIA–C. The majority of strains from both the environment and human pythiosis cases are in clade II. Interestingly subgroup IID and IIIC contained only environmental strains and whether this phenomenon depends on the virulence of the strains needs to be investigated in the future. In the study of Arthithanyaroj et al. (2013), subgroup IIIA contained more keratitis than arteritis strains with a ratio of 3:1 and the subgroup IIIB had more arteritis than corneal keratitis with a ratio of 2:1. This study indicated that *COX II* gene polymorphism can be used to determine genetic relationship of *P. insidiosum* human isolates and showed some correlation with clinical syndromes.

12.4 Epidemiology

Pythiosis in humans has been mostly reported from all over Thailand, while the disease state in animals (i.e. horses, dogs, cats, and cattle) has been found in tropical areas worldwide (Sathapatayavongs et al. 1989; Tanphaichitra 1989; Chetchotisakd et al. 1992; Wanachiwanawin et al. 1993; Mendoza et al. 1996; Thianprasit et al. 1996; Thitithanyanont et al. 1998; Prasertwitayakit et al. 2003; Krajaejun et al. 2004; Wanachiwanawin et al. 2004; Krajaejun et al. 2006; Pupaibool et al. 2006). This observation suggests that the pathogen is ubiquitously dispersed in nature. *P. insidiosum* prefers inhabiting swampy areas (i.e. paddy field, canal, and water reservoir), as demonstrated by Miller (1983) and Supabandhu et al. (2008), who have isolated the pathogen in water samples from various locations in Queensland, Australia and in northern regions of Thailand. Epidemiological analysis of demographic information of Thai patients with pythiosis shows that ~90% of the cases are in working ages ranging from 20 to 60 years, ~70% are male, and ~80 are agriculturists (Krajaejun et al. 2006). An individual with either of these epidemiological characteristics have a higher risk to directly contact *P. insidiosum* and develop pythiosis. Additionally, it has been found human pythiosis coexists with hematological underlying diseases, mostly thalassemia hemoglobinopathies (Sathapatayavongs et al. 1989; Wanachiwanawin et al. 1993; Krajaejun et al. 2006). Even though no pythiosis cases has been reported from most countries in Southeast Asia, existence of the disease is expected in these countries due to similar ethnical group of people, and geographic and climate conditions which are suitable for *P. insidiosum* growth.

12.5 Pathogenesis

Direct contact to host is a route of *P. insidiosum* entry. As a water mold, *P. insidiosum* hyphae colonizes water plants (Miller 1983; Mendoza et al. 1993, 1996). Under optimal conditions, the organism can produce asexual biflagellate zoospores, which can swim to attach to the host surface, and initiate hyphal invasion into host tissue (Mendoza et al. 1993; Ravishankar et al. 2001; Krajaejun et al. 2006). Thus, the zoospore is considered as an infective unit. Miller (1983) and Mendoza et al. (1993) demonstrate that the *P. insidiosum* zoospores move towards plant and mammalian tissues by chemotaxis. It is likely that the zoospore produces amorphous substances on the surface that enable attachment to the host (Mendoza et al. 1993, 1996; Banyong 2010; Monkongtanakit 2011). Ravishankar et al. (2001) proposed that hyphal invasion of *P. insidiosum* requires secreted proteolytic enzymes to weaken host barrier which was later supported by the studies of Davis et al. (2006) and Banyong (2010). They demonstrated the presence of secreted serine proteases and metalloprotease from clinical strains of *P. insidiosum* grown on gelatin containing medium (Davis et al. 2006; Banyong 2010). *Pythium insidiosum* prefers to infect arterial and ocular tissues in human patients, and cutaneous and gastrointestinal tissues in other animal hosts (Mendoza et al. 1996; Krajaejun et al. 2006). Some other tissue types, i.e. brain, sinus, bone, spleen, and liver, have been reported as less common targets of the infection (Miller and Campbell 1983; Mendoza et al. 1996; Krajaejun et al. 2006).

Temperature resistance is an important virulence factor for a successful pathogen. It has been demonstrated that *P. insidiosum* can grow well at body temperature of a mammalian host (37°C) (Krajaejun et al. 2010), whereas most of other *Pythium* species lack this capability (Van de Plaats-Niterink 1981). The strong association between *P. insidiosum* infection of non-ocular tissues and the thalassemia underlying disease suggests that *P. insidiosum* requires a thalassemia-related condition (i.e. impaired host immunity, increased iron availability, limited oxygen content from chronic hypoxia) that contributes to increased pathogen virulence or host susceptibility (Wanachiwanawin 2000; Krajaejun et al. 2006; Vento et al. 2006). In ocular pythiosis, most patients are healthy and do not have any underlying disease, indicating that *P. insidiosum* can infect healthy individuals as well (Krajaejun et al. 2006).

In vascular pythiosis, while venous vessels are not affected, arterial occlusion (by fibrosis and thrombi) or aneurysm of aorta and main arteries of lower extremities are main features of this type of the disease (Krajaejun et al. 2006). *Pythium insidiosum* gradually and centrally progresses from a more distal infection site. In ocular infection, the pathogen resides in corneal stroma. Histological examination of deep affected tissues (i.e. arteries and subcutaneous tissue) usually demonstrates *P. insidiosum* hyphal elements, chronic inflammation, and eosinophilia (Thianprasit et al. 1996; Krajaejun et al. 2006).

12.6 Clinical features

12.6.1 Human pythiosis

Various clinical features of pythiosis in humans can be observed. The most common forms of *P. insidiosum* infection are vascular and ocular pythiosis, which accounts for ~60% and ~30% of the reported cases, respectively (Krajaejun et al. 2006). For vascular pythiosis, patients usually come with arterial insufficiency syndrome of lower extremities, including intermittent claudication, rest pain, and gangrenous ulcers (Krajaejun et al. 2006). The disease can occur in both legs. Some other symptoms and signs can be found: i.e. fever, vesicle, cellulitis, absence of arterial pulse, and palpable mass (from aneurysm) (Krajaejun et al. 2006). Duration of the disease, from first recognition of symptoms to hospitalization, is several months. History of contact with swampy areas prior to the disease is obtained from some patients. Radiography reveals dilatation or occlusion of affected arteries. Histologically, *P. insidiosum* hyphal elements are normally found in arterial walls and thrombi in the arterial lumen. It should be noted that the pathogen does not infect venous tissue. In advanced infection, ruptured aneurysm is often a result and can lead to death.

For ocular pythiosis, patients have corneal ulcer, keratitis, eye pain, irritation, poor visual acuity, conjunctivitis, and hypopyon (Krajaejun et al. 2004, 2006). Histological examination shows the organism resides in the corneal stroma. Endophthalmitis can be observed in cases with advanced infection. The infection is usually confined to one eye. Duration of recognized symptoms before hospitalization varies from days to weeks. Like vascular pythiosis, history of direct contact to *P. insidiosum* habitat or history of eye trauma is usually obtained.

Not only arteries and eyes, *P. insidiosum* can also infect other tissue types. Patients with cutaneous pythiosis have acute or chronic infection, which manifests

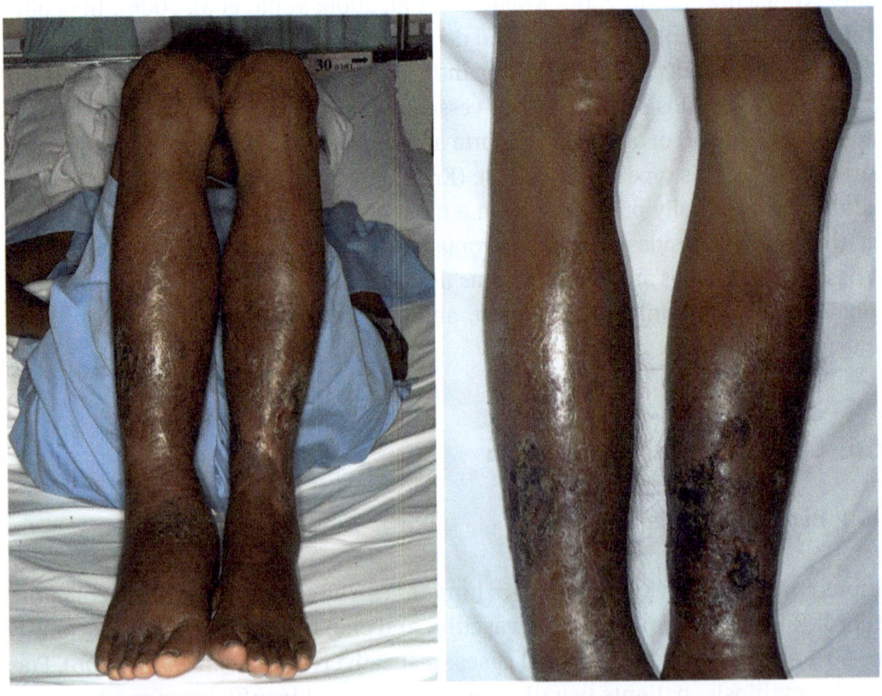

Fig. 12.1: Photograph of patient with advanced stages of pythiosis of both legs.

as cellulitis, swelling, painful subcutaneous lumps and ulcer, on the face or extremities (Fig. 12.1). Histological examination of infected tissue shows acute or chronic inflammation with eosinophilia (Thianprasit et al. 1996; Krajaejun et al. 2006). Other pathological sites of the infection include gastrointestinal tract, brain, and sinus (Krajaejun et al. 2006). It should be noted that gastrointestinal tract is a common infection site of pythiosis in animals (Mendoza et al. 1996), but it is a rare infection site in humans. Drinking water contaminated with *P. insidiosum* hyphae or zoospores could initiate the infection. How *P. insidiosum* enters brain tissue and sinus is unclear. *P. insidiosum* might enter the infection sites by direct invasion through the nasal cavity.

12.6.2 Animal pythiosis

Pythiosis in animals is more prevalent than the disease in humans. Many animals can be infected by *P. insidiosum*, including horse, dog, cattle, cat, sheep, camel, tiger, jaguar and bear (Miller and Campbell 1982; Miller and Campbell 1984; Miller et al. 1985; Mendoza and Alfaro 1986; Brown and Roberts 1988; Bentinck-Smith et al. 1989; Meireles et al. 1993; Mendoza et al. 1996; Patton et al. 1996; Dykstra et al. 1999; Reis et al. 2003; Camus et al. 2004; Tabosa et al. 2004; Wellehan et al. 2004; Rakich et al. 2005; Rivierre et al. 2005; Santurio et al. 2005; Buergelt et al. 2006; Berryessa et al. 2008). Among animals, horse and dog are most affected. Like humans, pythiosis, has a high rate of morbidity and mortality. Clinical features of animal pythiosis are various, but cutaneous tissues (i.e. of a limb, chest, abdomen, and face) or gastrointestinal tract (i.e. esophagus, stomach, and intestines) are the most common pathological sites of *P. insidiosum* infection. Disseminated infection to other internal organs can be observed as well. Animals with cutaneous pythiosis usually have an ulcer and serosanguineous discharge. The lesion often looks like a tumor, as it is called "swamp cancer". Animals with gastrointestinal tract infection often come with anorexia, vomiting, weight loss, and diarrhea, abdominal mass and mucosal ulceration. Granulomatous tissue with eosinophilia is histologically observed at the infection sites. "Kunker", an eosinophilic coral-like mass containing *P. insidiosum* hyphae, can be found in animals with pythiosis, but not in human patients.

12.7 Diagnosis

Diagnosis of *P. insidiosum* infection can be obtained by examination of the agent in 10% KOH (Fig. 12.2), followed by culturing, or detection of antibodies against *P. insidiosum* using serological assays or detection of DNA of the infectious agent in the infected tissue by PCR and sequencing (Miller 1983; Imwidthya and Srimuang 1989;

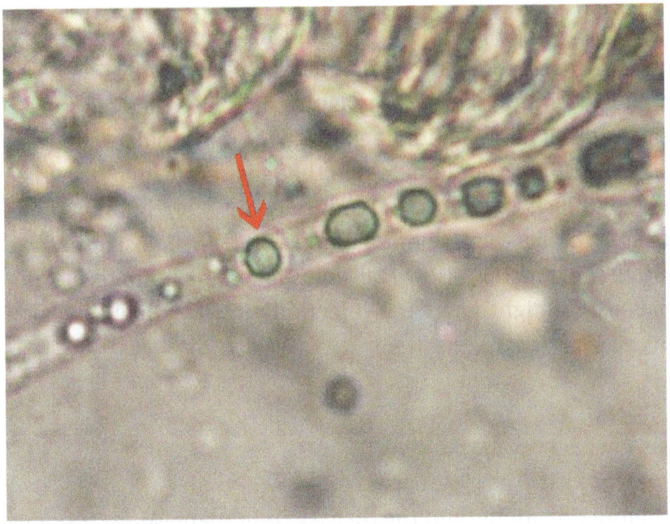

Fig. 12.2: Arrow shows a hypha running across the photo as demonstrated in a KOH preparation from left anterior tibial artery of pythium arteritis patient (400 × magnification).

Badenoch et al. 2001; Mendoza 2005; Grooters 2007; Krajaejun et al. 2009). Thrombus and infected tissue are specimens of choice for examination of the etiologic agent and isolation by culture. The samples should be washed with sterile distilled water or a saline solution with streptomycin or ampicillin or chloramphenicol or tetracycline during transportation at room temperature and before cultivation (Grooters et al. 2002b). Storing the samples at 4°C (shipping on ice) inhibits growth of *P. insidiosum* for about 20% of the clinical samples (Gaastra et al. 2010). These samples could also be used for DNA detection by PCR and sequencing (Grooters and Gee 2002; Vanittanakom et al. 2004; Thongsri et al. 2013). There are many serological tests developed for rapid diagnosis of pythiosis based on detection of antibodies against *P. insidiosum*, both from sera and plasma. An immunodiffusion (ID) test (Fig. 12.3) is very specific but has low sensitivity (Mendoza et al. 1986; Imwidthya and Srimuang 1989; Pracharktam et al. 1991). Other tests like enzyme linked immuno-sorbent assay (ELISA), immunochromatographic (ICT) assay or Western blot were developed later for increased sensitivity and specificity (Mendoza et al. 1997; Grooters et al. 2002a; Krajaejun et al. 2002; Chindamporn et al. 2009; Krajaejun et al. 2009). Jindayok et al. (2009) developed a simple haemagglutination (HA) test for serodiagnosis of vascular and cutaneous pythiosis. Recently, Krajaejun et al. (2013) compared the diagnostic performance of ID, ELISA, ICT and HA using sera from 37 pythiosis patients and 248 control subjects. They found that ICT and ELISA showed best diagnostic results of 100% for sensitivity, specificity, positive predictive value and negative predictive value. ICT was both rapid and user friendly and could promote earlier diagnosis and

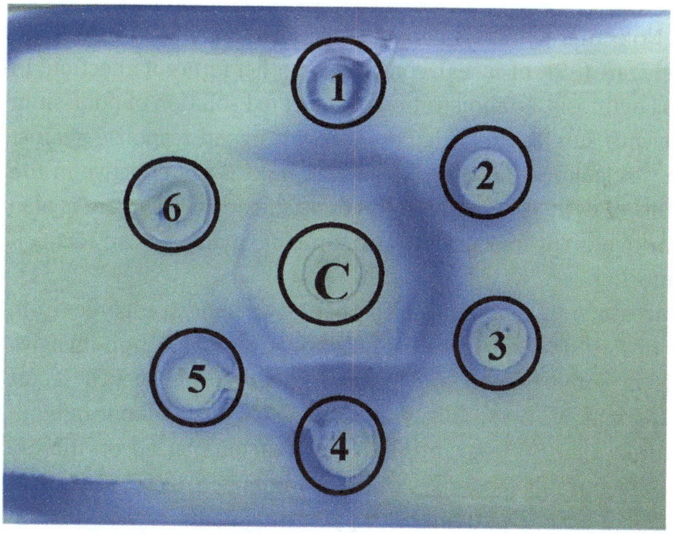

Fig. 12.3: Positive immunodiffusion test: serum from infected patient. 1, 4 = reference Ab, 2 = undiluted, 3 = dilution 1:2, 5 = dilution 1:4, 6 = dilution 1:8, C = reference Ag.

improved patients' care for pythiosis (Krajaejun et al. 2013). *P. insidiosum* can also be identified in fixed tissues by immunofluorescence (Mendoza et al. 1987) or by the immunoperoxidase staining technique (Brown and Roberts 1988).

12.8 Management

In most cases, conventional antifungal drugs are ineffective in controlling *P. insidiosum* infection. This could be due to the fact that oomycetes, including *Pythium* spp., lack the drug-target ergosterol biosynthesis pathway (Krajaejun et al. 2006). However, there is an exceptional pythiosis case, reported from USA, which has been cured by administration of a combination of itraconazole and terbinafine (Shenep et al. 1998).

Depending on the site of infection and disease progression, radical excision of infected tissue is required as the main treatment option for cure. In vascular pythiosis, resection of all infected artery or limb amputation (i.e. below-knee or above-knee amputation) has been performed in majority (~80%) of cases (Krajaejun et al. 2006). Removal of infective thrombi from arterial lumen tends to spread infection, and thus, is not recommended. With intensive medical care, ~60% of vascular patients were cured, but with physical disability (losing leg), whereas the rest (~40%) die from the advanced infection. In ocular pythiosis, eye removal (by enucleation or evisceration) is necessary, if keratectomy or keratoplasty failed to control the *P. insidiosum* infection. As a treatment outcome (Krajaejun et al. 2006), ~80% of ocular pythiosis patients undergo eye removal, while for the rest (~20%), their infected eyes can be saved by keratoplasty. No ocular pythiosis patients died from the infection. Cutaneous pythiosis is relatively rare (~5% of cases), compared to other forms of infection. In patients with cutaneous pythiosis, administration of saturated solution of potassium iodide (SSKI), together with surgical debridement, provides good response against *P. insidiosum* infection (Krajaejun et al. 2006). However, SSKI is not effective in the treatments of other forms of pythiosis. In animal pythiosis, extensive surgery is also considered as the main treatment for cure. After treatment, pythiosis can recur, if there is a remnant of infection.

As a conjunctive treatment, an immunotherapeutic vaccine (prepared from crude protein extract of *P. insidiosum*) has been administered as a non-invasive treatment of pythiosis in some human and animal patients, as the last resort (Mendoza et al. 1992; Dixon et al. 1998; Thitithanyanont et al. 1998; Mendoza et al. 2003; Wanachiwanawin et al. 2004). According to the outcome based on limited data, the vaccine shows favourable responses in 60% of horses, 97% of cattle, 33% of dogs, and at least 42% of human patients with pythiosis (Mendoza and Newton 2005; Krajaejun et al. 2006).

12.9 Research direction

Research is needed to provide insights into *P. insidiosum*, and lead to discovery of novel methods for disease prevention and control. Research direction should focus on the biology, pathogenesis, host-pathogen interaction and evolutionary aspect, using modern biochemical, cellular, molecular, and bioinformatics techniques. Antifungal drugs are not generally effective for treatment of pythiosis, because *Pythium* species may lack the drug-target ergosterol biosynthesis pathway (Schlosser and Gottlieb 1966; Krajaejun et al. 2008). Finding a new drug for controlling the infection caused by *P. insidiosum* is urgently needed. β-glucan is a major cell wall component of oomycetes, including *P. insidiosum* (Kwon-Chung 1994). Thus, using β-glucan inhibitors, a new group of antifungal agents, might be beneficial to the treatment of pythiosis. However, recent study shows that such drugs (i.e. caspofungin), which have been designed for fungal pathogens, has limited efficacy against *P. insidiosum* (Pereira et al. 2007). This might be due to the drug target structures are different between oomycetes and fungi. There are attempts to search for bioactive compound that have anti-*P. insidiosum* effect. B23 is a compound isolated from the endophytic fungus *Muscodor crispans*, which has a broad anti-microbial spectrum against bacteria, fungi, and oomycetes (Krajaejun et al. 2012). Interestingly, B23 has showed a potent *in vitro* anti-*P. insidiosum* effect. B23 could be a hope for new effective drug for treatment of pythiosis, but it requires further studies, i.e. toxicity and *in vivo* analyses.

The *P. insidiosum* immunotherapeutic vaccine is considered a potential non-invasive treatment option, and has shown favorable treatment outcome in some human and animal patients with pythiosis (Mendoza et al. 1992; Dixon et al. 1998; Thitithanyanont et al. 1998; Mendoza et al. 2003; Wanachiwanawin et al. 2004; Krajaejun et al. 2006). Nevertheless, the underlying mechanism of the vaccine is unclear and should be investigated, because it could provide a clue as to how to improve its efficacy. Components of the immunotherapeutic vaccine, which is prepared from crude extract of *P. insidiosum* culture, should be elucidated. There are a few studies that explore antigenic components of *P. insidiosum* (Krajaejun et al. 2006, 2010), that could potentially be a diagnostic marker and a vaccine candidate. Such studies use Western blot analysis and identify the 74-kDa immunodominant antigens of *P. insidiosum*, which is then characterized as an exo-1,3-β glucanase. In depth study of this protein might also lead to discovery of a new therapeutic target for pythiosis.

Human pythiosis has an obvious association with thalassemia (Krajaejun et al. 2006). Iron overload, chronic anemia, and hemolysis, which are pathological changes found in thalassemic patients, might increase host susceptibility or promote infectivity of the pathogen. Better understanding of how exactly thalassemia is a predisposing factor for pythiosis will help in finding a way to prevent and control the disease. Thorough studies are needed to elucidate this pythiosis-thalassemia association.

Genetic information, i.e. genome and transcript sequences, is an important resource for understanding biology, pathogenesis, and evolution of a pathogen. Genetic comparison of *P. insidiosum* to non-pathogenic oomycetes will help in explaining why *P. insidiosum* is the only oomycete species that infects both humans and animals. So far, only a few hundreds of *P. insidiosum* genes, identified by conventional Sanger-based sequencing of cDNA library (also known as: transcriptome or expressed sequence tags), has been reported (Krajaejun et al. 2011). This genetic data is small, and can only provide a limited picture of *P. insidiosum* biology. Recently, new technology for DNA sequencing, called "next generation sequencing (NGS)", is available. NGS is a high-throughput sequencing method, and it is relatively less expensive compared to the conventional Sanger-based sequencing. Thus, NGS should be used to generate complete *de novo* genetic information of a non-model organism like *P. insidiosum*, which will be a significant step for the study of *P. insidiosum*, and could lead to discovery of new therapeutic options (i.e. drug and vaccine candidates) for pythiosis.

Acknowledgment

I would like to dedicate this chapter to commemorate the 90th birthday of Prof. Dr. Emeritus Karl Esser, my advisor for the Dr. rer. nat study. With his, and his staff kind support, in my training, educating and giving me experiences during my study at Ruhr University Bochum, Germany, has made this chapter contribution possible.

References

Alexopoulos CJ, Mims CW, Blackwell M. Introductory Mycology. 4th ed. New York, NY: John Wiley & Sons, Inc. 1996;61–85:683–737.

Amemiya J. Granular dermatitis in the horse, caused by *Pythium gracile*. Bull Fac Agric Kagoshima Univ 1982;32:141–147.

Austwick PKC, Copland JW. Swamp cancer. Nature 1974;250:84.

Badenoch PR, Coster DJ, Wetherall BL, Brettig HT, Rozenbilds MA, Drenth A, et al. *Pythium insidiosum* keratitis confirmed by DNA sequence analysis. Br J Ophthalmol 2001;85:502–503.

Banyong R. Determination of protease activities in *Pythium insidiosum* human isolates. A thesis submitted in partial fulfillment the requirement for the degree of Master of Science in Microbiology, Faculty of Graduate Studies, Mahidol University, Bangkok, Thailand 2010:1–74.

Bentinck-Smith J, Padhye AA, Maslin WR, Hamilton C, McDonald RK, Woody BJ. Canine pythiosis–isolation and identification of *Pythium insidiosum*. J Vet Diagn Investig Off Publ Am Assoc Vet Lab Diagn Inc 1989;1:295–298.

Berryessa NA, Marks SL, Pesavento PA, Krasnansky T, Yoshimoto SK, Johnson EG, et al. Gastrointestinal pythiosis in 10 dogs from California. J Vet Intern Med Am Coll Vet Intern Med 2008;22:1065–1069.

Borisooth R. Biological and molecular characterization of *Pythium insidiosum* the etiologic agent of human pythiosis. A thesis submitted in partial fulfillment the requirement for the degree of Master of Science in Microbiology, Faculty of Graduate Studies, Mahidol University, Bangkok, Thailand 1993:1–105.

Brown CC, Roberts ED. Intestinal pythiosis in a horse. Aust Vet J 1988;65:88–89.

Buergelt C, Powe J, White T. Abdominal pythiosis in a Bengal tiger (*Panthera tigris tigris*). J Zoo Wildl Med Off Publ Am Assoc Zoo Vet 2006;37:186–189.

Camus AC, Grooters AM, Aquilar RE. Granulomatous pneumonia caused by *Pythium insidiosum* in a central American jaguar, *Panthera onca*. J Vet Diagn Investig Off Publ Am Assoc Vet Lab Diagn Inc. 2004;16:567–571.

Chaiprasert A, Samerpitak S, Wanachiwanawin W, Thasnakorn P. Induction of zoospore formation in Thai isolates of *Pythium insidiosum*. Mycoses 1990;33:317–323.

Chandler FWW, Kaplan W, Ajello L. A Color Atlas and Textbook of the Histopathology of Mycotic Diseases. Chicago, IL: Year Book Medical Publishers Inc. 1980:104–105.

Chareonsirisuthigul T, Khositnithikul R, Intaramat A, Inkomlue R, Sriwanichrak K, Piromsontikorn S, et al. Performance comparison of immunodiffusion, enzyme-link immunosorbent assay, immunochromatography and hemagglutination for serodiagnosis of human pythiosis. Diag Microbiol Infect Dis 2013;76:42–45.

Chetchotisakd P, Pairojkul C, Porntaveevudhi O, Sathapatayavongs B, Mairiang P, Nuntirooj K, et al. Human pythiosis in Srinagarind Hospital: one year's experience. J Med Assoc Thai 1992;75:248–254.

Chindamporn A, Vilela R, Hoag KA, Mendoza L. Antibodies in the sera of host species with pythiosis recognize a variety of unique immunogens in geographically divergent *Pythium insidiosum* strains. Clin Vac Immunol 2009;16:330–336.

Davis DJ, Lanter K, Makselan S, Bonati C, Asbrock P, Ravishankar JP, et al. Relationship between temperature optima and secreted protease activities of three *Pythium* species and pathogenicity toward plant and animal hosts. Mycol Res 2006;110:96–103.

De Cock AW, Mendoza L, Padhye A, Ajello L, Kaufman L. *Pythium insidiosum* sp. nov., the etiologic agent of pythiosis. J Clin Microbiol 1987;25:344–349.
De Haan J, Hoogkamer LJ. Hypho-mycosis destruens. Veeartsen-nijkundige Bld voor Ned Indië 1901;13:350–374.
Dick MW. Straminipilous Fungi: Systematics of the Peronosporomycetes Including Accounts of the Marine Straminipilous Protist, the Plasmodiophorids and Similar Organisms. London: Kluwer Academic Publishers 2001:67.
Dixon DM, Casadevall A, Klein B, Mendoza L, Travassos L, Deepe GS. Development of vaccines and their use in the prevention of fungal infections. Med Mycol Off Publ Int Soc Hum Anim Mycol 1998;36 Suppl 1:57–67.
Drouin V. Sur une nouvelle mycose du cheval. Recl Med Vet 1896;3:337–344.
Dykstra MJ, Sharp NJ, Olivry T, Hillier A, Murphy KM, Kaufman L, et al. A description of cutaneous-subcutaneous pythiosis in fifteen dogs. Med Mycol Off Publ Int Soc Hum Anim Mycol 1999;37:427–433.
Estrada-Garcia MT, Callow JA, Green JR. *Monoclonal antibodies* to the adhesive cell coat secreted by *Pythium aphanidermatum* zoospores recognize 200 × 300 M glycoproteins stored within large peripheral vesicles. J Cell Sci 1990;95:199–206.
Fish PA. Leeches: a histological investigation of two cases of equine mycosis with a historical account of a supposed similar disease, called bursatte, occurring in India. In: 12th and 13th Annual Report of the Bureau of Animal Industry. Italy: Ulan Press 1923:229–259.
Gonzalez HE, Ruiz A. Espundia equine: etiologiay patogenesisi de una ficomicosis. Revista ICA Bogota, Colombia 1975;10:175.
Gooters AM, Whittington A, Lopez MK, Boroughs MN, Roy AF. Evaluation of microbial culture techniques for the isolation of *Pythium insidiosum* from equine tissues. J Vet Diagn Invest 2002;14:288–294.
Grooters AM, Gee MK. Development of a nested polymerase chain reaction assay for the detection and identification of *Pythium insidiosum*. J Vet Intern Med 2002;16:147–152.
Grooters AM. Pythiosis and zygomycosis. In: Sellon D, Long M, eds. Equine Infectious Diseases. St Louis, MO: Saunders Elsevier 2007:412–419.
Hnilica KA. Difficult dermatologic diagnosis. Pythiosis. J Am Vet Med Assoc 1998;212:1192–1193.
Ichitani T, Amemiya J. *Pythium gracile* isolated from the foci of granular dermatitis in the horse (*Equus caballus*). Trans Mycol Soc Jpn 1980;21:263–265.
Imwidthaya P, Srimuang S. Immunodiffusion test for diagnosing human pythiosis. Mycopathologia 1989;106:109–112.
Jindayok T, Piromsontikorn S, Srimuang S, Khupulsup K, Krajaejun T. Hemagglutination test for rapid serodiagnosis of human pythiosis. Clin Vac Immunol 2009;16:1047–1051.
Keoprasom N, Chularojanamontri L, Chayakulkeeree M, Chaiprasert A, Wanachiwanawin W, Ruangsetakit C. Vascular pythiosis in a thalassmic patient presenting as bilateral leg ulcers. Med Mycol Case Rep 2013;51:25–28.
Krajaejun T, Chongtrakool P, Angkananukul K, Brandhorst TT. Effect of temperature on growth of the pathogenic oomycete *Pythium insidiosum*. Southeast Asian J Trop Med Public Health 2010;41:1462–1466.
Krajaejun T, Imkhieo S, Intaramat A, Ratanabanangkoon K. Development of an immunochromatographic test for rapid serodiagnosis of human pythiosis. Clin Vac Immunol 2009;16:506–509.
Krajaejun T, Keeratijarut A, Sriwanichrak K, Lowhnoo T, Rujirawat T, Petchthong T, et al. The 74-kilodalton immunodominant antigen of the pathogenic oomycete *Pythium insidiosum* is a putative exo-1,3-beta-glucanase. Clin Vaccine Immunol Cvi 2010;17:1203–1210.
Krajaejun T, Khositnithikul R, Lerksuthirat T, Lowhnoo T, Rujirawat T, Petchthong T, et al. Expressed sequence tags reveal genetic diversity and putative virulence factors of the pathogenic oomycete *Pythium insidiosum*. Fungal Biol 2011;115:683–696.

Krajaejun T, Kunakorn M, Niemhom S, Chongtrakool P, Pracharktam R. Development and evaluation of an in-house enzyme-linked immunosorbent assay for early diagnosis and monitoring of human pythiosis. Clin Diagn Lab Immunol 2002;9:378–382.
Krajaejun T, Kunakorn M, Pracharktam R, Chongtrakool P, Sathapatayavongs B, Chaiprasert A, et al. Identification of a novel 74-kiloDalton immunodominant antigen of *Pythium insidiosum* recognized by sera from human patients with pythiosis. J Clin Microbiol 2006;44:1674–1680.
Krajaejun T, Lowhnoo T, Yingyong W, Rujirawat T, Fucharoen S, Strobel GA. *In vitro* antimicrobial activity of volatile organic compounds from *Muscodor crispans* against the pathogenic oomycete *Pythium insidiosum*. Southeast Asian J Trop Med Public Health 2012;43:1474–1483.
Krajaejun T, Pracharktam R, Wongwaisayawan S, Rochanawutinon M, Kunakorn M, Kunavisarut S. Ocular pythiosis: is it under-diagnosed? Am J Ophthalmol 2004;137:370–372.
Krajaejun T, Sathapatayavongs B, Pracharktam R, Nitiyanant P, Leelachaikul P, Wanachiwanawin W, et al. Clinical and epidemiological analyses of human pythiosis in Thailand. Clin Infect Dis Off Publ Infect Dis Soc Am 2006;43:569–576.
Kwon-Chung KJ. Phylogenetic spectrum of fungi that are pathogenic to humans. Clin Infect Dis Off Publ Infect Dis Soc Am 1994;19 Suppl 1:S1–7.
Laohapensang K, Rutherford RB, Supabandhu J, Vanittanakorn N. Vascular pythisosis in a thalassemic patient. Vascular 2009;17:234–238.
Martin FN. Phylogenetic relationships among some *Pythium* species inferred from sequence anaylsis of the mitochondrially encoded cytochrome oxidase II gene. Mycologia 2000;92:711–727.
Meireles MC, Riet-Correa F, Fischman O, Zambrano AF, Zambrano MS, Ribeiro GA. Cutaneous pythiosis in horses from Brazil. Mycoses 1993;36:139–142.
Mendoza L, Ajello L, McGinnis MR. Infections caused by the oomycetous pathogen *Pythium insidiosum*. J Mycol Med 1996;6:151–164.
Mendoza L, Alfaro AA. Equine pythiosis in Costa Rica: report of 39 cases. Mycopathologia 1986;94:123–129.
Mendoza L, Hernandez F, Ajello L. Life cycle of the human and animal oomycete pathogen *Pythium insidiosum*. J Clin Microbiol 1993;31:2967–2973.
Mendoza L, Kaufman L, Mandy W, Glass R. Serodiagnosis of human and animal pythiosis using an enzyme-linked immunosor-bent assay. Clin Diagn Lab Immunol 1997;4:715–718.
Mendoza L, Kaufman L, Standard P. Antigenic relationship between the animal and human pathogen *Pythium insidiosum* and nonpathogenic *Pythium* spp. J Clin Microbiol 1987;25:2159–2162.
Mendoza L, Mandy W, Glass R. An improved *Pythium insidiosum*-vaccine formulation with enhanced immunotherapeutic properties in horses and dogs with pythiosis. Vaccine 2003;21:2797–2804.
Mendoza L, Marin G. Antigenic relationship between *Pythium insidiosum* de Cock et al. 1987 and its synonym *Pythium destruens* Shipton 1987. Mycoses 1989;32:73–77.
Mendoza L, Newton JC. Immunology and immunotherapy of the infections caused by *Pythium indidiosum*. Med Mycol 2005;43:477–486.
Mendoza L, Prendas J. A method to obtain rapid zoosporogenesis of *Pythium insidiosum*. Mycopathologia 1988;104:59–62.
Mendoza L, Villalobos J, Calleja CE, Solis A. Evaluation of two vaccines for the treatment of pythiosis insidiosi in horses. Mycopathologia 1992;119:89–95.
Miller RI, Campbell RS. Clinical observations on equine phycomycosis. Aust Vet J 1982;58:221–226.
Miller RI, Campbell RS. Experimental pythiosis in rabbits. Sabouraudia 1983;21:331–41.
Miller RI, Campbell RS. The comparative pathology of equine cutaneous phycomycosis. Vet Pathol 1984;21:325–332.
Miller RI, Olcott BM, Archer M. Cutaneous pythiosis in beef calves. J Am Vet Med Assoc 1985;186:984–986.

Miller RI. Granulomatous and neoplastic disease of equine skin. Veterinary Reviews and Monographs No.6 James Cook University of North Queensland, Townsville, Australia 1983:1–92.
Miller RI. Investigations into the biology of three "phycomycotic" agents pathogenic for horses in Australia. Mycopathologia 1983;81:23–28.
Monkongtanakit O. Determination of adhesive materials secreted from zoospores of *Pythium insidiosum* on different host tissue. A thesis submitted in partial fulfillment the requirement for the degree of Master of Science in Microbiology, Faculty of Graduate Studies, Mahidol University, Bangkok, Thailand 2011:1–98.
Patton CS, Hake R, Newton J, Toal RL. Esophagitis due to *Pythium insidiosum* infection in two dogs. J Vet Intern Med Am Coll Vet Intern Med 1996;10:139–142.
Pereira DI, Santurio JM, Alves SH, Argenta JS, Pötter L, Spanamberg A, et al. Caspofungin *in vitro* and *in vivo* activity against Brazilian *Pythium insidiosum* strains isolated from animals. J Antimicrob Chemother 2007;60:1168–1171.
Pérez R, Luis-León JJ, Vivas JL, Mendoza L. Epizootic cutaneous pythiosis in beef calves. Vet Microbiol 2005;109:121–128.
Pesavento PA, Barr B, Riggs SM, Eigenheer AL, Pamma R, Walker RI. Cutaneous pythiosis in a nestling white faced ibis. Vet Pathol 2008;45:538–541.
Pracharktam R, Changtrakool P, Sathapatayavongs B, Jayanetra P, Ajello L. Immunodiffusion test for diagnosis and monitoring of human pythiosis insidiosi. J Clin Microbiol 1991;29:138–141.
Prasertwitayakij N, Louthrenoo W, Kasitanon N, Thamprasert K, Vanittanakom N. Human pythiosis, a rare cause of arteritis: case report and literature review. Semin Arthritis Rheum 2003;33:204–214.
Pupaibool J, Chindamporn A, Patrakul K, Suankratay C, Sindhuphak W, Kulwichit W. Human pythiosis. Emerg Infect Dis 2006;12:517–518.
Rakich PM, Grooters AM, Tang K-N, Gastrointestinal pythiosis in two cats. J Vet Diagn Invest 2005;17:262–269.
Ravishankar JP, Davis CM, Davis DJ, MacDonald E, Makselan SD, Millward L, et al. Mechanics of solid tissue invasion by the mammalian pathogen *Pythium insidiosum*. Fungal Genet Biol 2001;34:167–175.
Reis JL, de Carvalho ECQ, Nogueira RHG, Lemos LS, Mendoza L. Disseminated pythiosis in three horses. Vet Microbiol 2003;96:289–295.
Rivierre C, Laprie C, Guiard-Marigny O, Bergeaud P, Berthelemy M, Guillot J. Pythiosis in Africa. Emerg Infect Dis 2005;11:479–481.
Santurio JM, Argenta JA, Schwendler SE, Cavalheiro AS, Pereira DIB, Zanette RA, et al. Granulomatous rthinitis associated with *Pythium insidiosum* infection in sheep. Vet Rec 2008;163:276–277.
Santurio JM, Monteiro AB, Leal AT, Kommers GD, de Sousa RS, Catto JB. Cutaneous pythiosis insidiosi in calves from the Pantanal region of Brazil. Mycopathologia 1998;141:123–125.
Sathapatayavongs B, Leelachaikul P, Prachaktam R, Atichartakarn V, Sriphojanart S, Trairatvorakul P, et al. Human pythiosis associated with thalassemia hemoglobinopathy syndrome. J Infect Dis 1989;159:274–280.
Schlosser E, Gottlieb D. Sterols and the sensitivity of *Pythium* species to filipin. J Bacteriol 1996;91:1080–1084.
Schurko AM, Mendoza L, de Cock AW, Klassen GR. Evidence for geographic clusters: molecular genetic differences among strains of *Pythium insidiosum* from Asia. Australia and the Americas are explored. Mycologia 2003a;95:200–208.
Schurko AM, Mendoza L, Lévesque CA, Désaulniers NL, de Cock AW, Klassen GR. A molecular phylogeny of *Pythium insidiosum*. Mycol Res 2003b;107:537–544.

Shenep JL, English BK, Kaufman L, Pearson TA, Thompson JW, Kaufman RA, et al. Successful medical therapy for deeply invasive facial infection due to *Pythium insidiosum* in a child. Clin Infect Dis Off Publ Infect Dis Soc A 1998;27:1388–1393.

Shipton WA. *Pythium destruens* sp. nov, an agent of equine pythiosis. J Med Vet Mycol 1987;25:137–151.

Smith F. The pathology of bursattee. Vet J 1884;19:16–17.

Supabandhu J, Mathew CF, Mendoza L, Vanittanakom N. Isolation and identification of the human pathogen *Pythium insidiosum* from environmental samples collected in Thai agricultural areas. Med Mycol 2008;46:41–52.

Tabosa IM, Riet-Correa F, Nobre VMT, Azevedo EO, Reis-Júnior JL, Medeiros RMT. Outbreaks of pythiosis in two flocks of sheep in northeastern Brazil. Vet Pathol 2004;41:412–415.

Tanphaichitra D. Tropical disease in the immunocompromised host: melioidosis and pythiosis. Rev Infect Dis 1989;11:S1629–S1643.

Thianprasit M, Chaiprasert A, Imwidthaya P. Human pythiosis. Curr Top Med Mycol 1996;7:43–54.

Thitithanyanont A, Mendoza L, Chuansumrit A, Pracharktam R, Laothamatas J, Sathapatayavongs B, et al. Use of an immunotherapeutic vaccine to treat a life-threatening human arteritic infection caused by *Pythium insidiosum*. Clin Infect Dis Off Publ Infect Dis Soc Am 1998;27:1394–1400.

Thongsri Y, Wonglakorn L, Chaiprasert A, Svobodava L, Hamal P, Pakarasang M, et al. Evaluation for the clinical diagnosis of *Pythium insidiosum* a single-tube nested PCR. Mycopathologia 2013;176:369–376.

Van der Plaats-Niterink AJ. Monograph of the genus *Pythium*. Stud Mycol 1981;21:1–242.

Vanittanakorn N, Supabandhu J, Khamwan C, Praparattanapan J, Thirach S, Prasetwikayakij N, et al. Identification of emerging human-pathogenic *Pythium insidiosum* by serological and molecular assay based methods. J Clin Microbiol 2004;42:3970–3974.

Vento S, Cainelli F, Cesario F. Infections and thalassaemia. Lancet Infect Dis 2006;6:226–233.

Wanachiwanawin W, Mendoza L, Visuthisakchai S, Mutsikapan P, Sathapatayavongs B, Chaiprasert A, et al. Efficacy of immunotherapy using antigens of *Pythium insidiosum* in the treatment of vascular pythiosis in humans. Vaccine 2004;22:3613–3621.

Wanachiwanawin W, Thianprasit M, Fucharoen S, Chaiprasert A, Sudasna N, Ayudhya N, et al. Fatal arteritis due to *Pythium insidiosum* infection in patients with thalassaemia. Trans R Soc Trop Med Hyg 1993;87:296–298.

Wanachiwanawin W. Infections in E-beta thalassemia. J Pediatr Hematol Off J Am Soc Pediatr Hematol 2000;22:581–587.

Wellehan JF, Farina LL, Keoughan CG, Lafortune M, Grooters AM, Mendoza L, et al. Pythiosis in a dromedary camel (*Camelus dromedarius*). J Zoo Wildl Med Off Publ Am Assoc Zoo Vet 2004;35:564–568.

Withamp J. Bijdrage tot de kennis van de *Hyphomycosis destruens*. Ned Ind Blad voor Diergeneeskd en Dierenteelt 1924;36:229–245.

Frank H. Gleason, Sergey A. Karpov, Osu Lilje, Deborah J. Macarthur, Floris F. van Otgen and Telesphore Sime-Ngando*

13 Zoosporic parasites of phytoplankton

Sparrow (1960) described many species of aquatic phycomycetes which are parasites of phytoplankton and which are propagated by zoospores. At that time the Class Phycomycetes were considered to be a valid taxonomic group within the Kingdom Fungi, but this concept is no longer accepted by mycologists. The aquatic phycomycetes, as a group, still may be important as an ecological concept however. Based on data from sequencing genes, particularly rDNA sequences, these parasites have been reassigned to different branches or supergroups within the tree of life by Baldauf (2003), namely the opisthokonts, cercozoa, alveolates and heterokonts. One cluster of phylogenetic clades, the zoosporic true fungi, is still currently considered acceptable as a valid taxonomic group within the opisthokont supergroup. The zoosporic true fungi include parasitic and saprotrophic species in the Chytridiomycota, Blastocladiomycota, Neocallimastigomycota, Monoblepharomycota and Olpidium clade. Cryptomycota and Aphelida are not fungi and belong to superphylum Opisthosporidia (Karpov et al. 2014a). All true fungi are thought to have chitin in the cell walls of hyphae, sporangia, and zoospore cysts. Interestingly, *Rozella* (Cryptomycota) has four chitin synthetase genes (James et al. 2013).

Many zoosporic parasites produce infective, motile propagules which are usually uniflagellate or heterokont zoospores. Unikont zoospores have one posteriorly directed whiplash flagellum, whereas heterokont zoospores have one posteriorly directed whiplash flagellum and one anteriorly directed flagellum with mastigonemes, although there are some variations of this structure. A few species produce amoebae with pseudopods as infective propagules.

In the present review eukaryotic parasites with motile propagules, typically zoospores, but sometimes amoebae, are called zoosporic parasites. Most zoosporic parasites and parasitoids of freshwater phytoplankton are included in the phyla Chytridiomycota, Blastocladiomycota and Aphelida. A few species are placed in the Olpidium clade which is related to the polyphyletic Zygomycetes (Sekimoto et al. 2011). Some examples of parasites in these phyla are discussed in this chapter. Some species in the Oomycota, Perkinsozoa and Labyrinthulomycota have frequently been reported as zoosporic parasites of marine phytoplankton and therefore are not discussed in the present chapter.

The hosts for the parasites and parasitoids considered in this chapter are all phytoplankton which are microscopic photosynthetic organisms (autotrophs) inhabiting

* All authors contributed significantly to the preparation of this manuscript.

euphotic zones of marine and freshwater ecosystems. As a group the phytoplankton is at the base of producer-based food webs and produces much of earth's oxygen. In this chapter hosts in five eukaryotic and one prokaryotic phyla will be considered: Chlorophyta, Euglenophyta, Xanthophyta, Chrysophyta, Bacillariophyta and Cyanobacteria.

13.1 The main groups of zoosporic parasites and parasitoids of phytoplankton

13.1.1 Aphelida

Three phyla Aphelida, Cryptomycota and Microsporidia at the base of Holomycota form a superphylum Opisthosporidia, which is sister to all the other fungi (Karpov et al. 2014a). The Class Aphelidea includes two freshwater genera (*Aphelidium* and *Amoeboaphelidium*) and one marine genus (*Pseudaphelidium*). Phylum Cryptomycota contains at present only two genera *Rozella* and *Paramicrosporidium*, and a number of uncultured organisms designated by environmental sequences (Jones et al. 2011; Corsaro et al. 2014). *Paramicrosporidium* is a parasitoid of benthic and periphytic amoeba *Vannella* (Corsaro et al. 2014). All known species of *Rozella* (Class Rozellidea) are holocarpic, intracellular parasitoids of zoosporic true fungi and heterotrophic stramenopiles, while the aphelids (class Aphelidea) are holocarpic, intracellular parasites of freshwater and marine phytoplankton (Held 1981; Gromov 2000; Gleason et al. 2012; Karpov et al. 2014a). Like other Opisthokonts, *Rozella* and the aphelids have uniflagellate zoospores with posteriorly directed flagella and flat/tubular mitochondrial cristae at the stages of the life cycle (Karpov et al. 2014a). Nevertheless, because species of *Rozella* are not known to be parasitoids of phytoplankton, the discussion of these species is limited to comparisons with aphelids in this chapter.

In freshwater, aphelids infect many species of unicellular eukaryotic algae in the Chlorophyta, Xanthophyta and Bacillariophyta (Gromov 2000). However, the most common hosts for the aphelids are the chlorococcous algae (Fott 1957; Gromov and Mamkaeva 1968; Gromov 2000), and *Tribonema gayanum* as well. The latter is often used as a host for the aphelid cultures (Gromov 1972). Host specificity has been investigated in only a few species: *Rozella allomycis* (Held 1981) and four ecotypes of *Amoeboaphelidium* (Gromov and Mamkaeva 1968). However, both *Rozella* and aphelids appear to be very common parasitoids in many aquatic ecosystems (Held 1981; Gromov 2000; Jones et al. 2011). Recently, Letcher et al. (2013) highlighted the roles of aphelids in causing epidemics in outdoor ponds used for production of algae in the biofuel industry. Livermore and Mattes (2013) detected DNA sequences belonging to novel Cryptomycota in an aquifer in Iowa and in data sets in GenBank from previous collections from freshwater and marine environments. The identity of these operational taxonomic units is unknown, but it is possible that some of these could be aphelids.

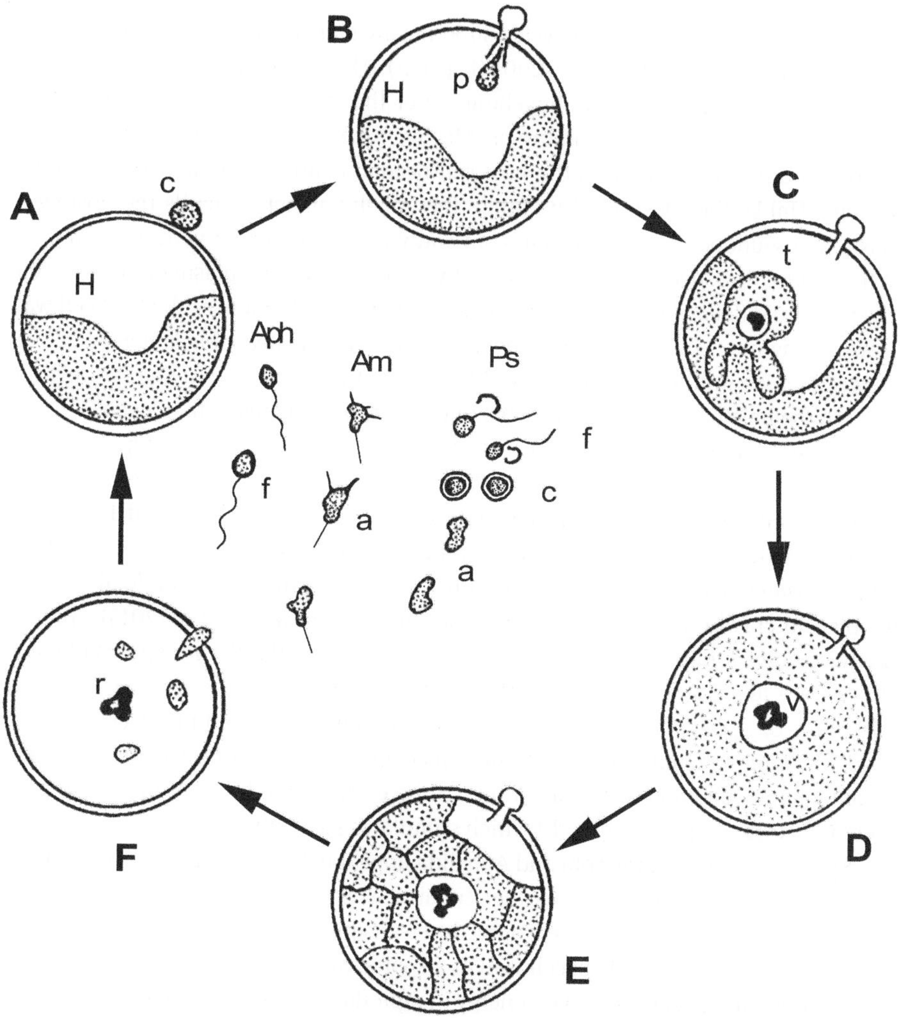

Fig. 13.1: Life cycle of aphelids: *Aphelidium* (Aph), *Amoeboaphelidium* (Am) and *Pseudaphelidium* (Ps), distinguishing by zoospore structure and development. Modified after: Gromov (2000). A, Zoospore encystment on the host surface; B, propagule penetration into the host; C, trophic amoeba engulfs host cytoplasm; D, parasite totally replaced the host, contains central vacuole with residual body; E, immature sporangium: parasitoid divides producing uninuclear cells, F, mature zoospores release the sporangium. Abbreviations: a, amoeba, with posterior pseudocilium in *Amoeboaphelidium*, c, cyst; f, flagellate; p, propagule; t, trophont, r, residual body as food remnants in central vacuole (v), H, host.

Aphelids have complex life cycles superficially similar to those of many species in the Chytridiomycota (Gromov 2000) (Fig. 13.1). The zoospore attaches to the host alga and encysts discarding its flagellum. Later the cyst germinates and penetrates into the host with its pseudopodium or infective tube, and the whole contents of the parasitoid are transported into the host cell becoming an intracellular phagotrophic amoeba. Unlike the zoosporic true fungi, the aphelid amoeba engulfs the host cytoplasm by pseudopodia, grows forming an endobiotic sporangium with residual body in the central digestive vacuole, and it totally consumes the cytoplasm of the host cell. The parasitoid does not form its own sporangium wall, rather it uses the host cell wall as a sporangium. After maturation, the amoeboid cells (*Amoeboaphelidium* and *Pseudaphelidium*) or uniflagellate zoospores (*Aphelidium*) are released from the sporangium through the hole made previously by the infection tube and infect other algae.

Another difference between species of *Rozella* and aphelids is the type of motile propagules produced. All known *Rozella* species produce uniflagellate zoospores. Aphelids can produce either uniflagellate zoospores, amoebae without flagella or amoebae with flagella reduced in size (pseudocilia). Zoospores are adapted for swimming in water while amoebae (even with a posterior immobile pseudocilium) are adapted to crawl on surfaces (Gleason and Lilje 2009; Karpov et al. 2013). Interestingly, in *Aphelidium* having zoospores with normal flagella, the zoospores of some species are able to crawl like amoebae. For example, according to Gromov (1972) zoospores of *Aphelidium tribonemae* can swim using the flagellum or crawl using filopodia while the immobile flagellum trails behind the zoospore. The physiological flagellate-amoebae-flagellate transition can be easily shifted in culture by environmental factors such as bright light which stimulates swimming.

Species in the Cryptomycota and Aphelida appear to be very efficient in converting energy from one trophic level to another, because a relatively high percentage of energy is likely to be transferred between trophic levels. As previously stated both *Rozella* and the aphelid genera use the host cell wall as a sporangium. Therefore, we would expect these endobiotic species to use less energy in reproduction than epibiotic species since they do not produce an extra structure for zoospore release. Since *Rozella* and aphelid species have small thalli and most of the cytoplasm goes into zoospores, reproduction is energy efficient. Because of their small sizes (Held 1981; Gromov 2000) zoospores produced by *Rozella* and aphelids are likely to be easily eaten by ciliates and filter feeding zooplankton and metazoans at the third or fourth trophic level (Gleason et al. 2008). Thus, the nutrients of the host plankton are recycled by the microbial loop while they remain in the euphotic zones.

The host range has been estimated in some *Amoeboaphelidium* species. Gromov and Mamkaeva (1968) measured the sensitivity of 226 different strains of green and yellow green algae to infection by four strains of *Amoeboaphelidium* (x-1 through x-4 CALU). Some cultures were sensitive to infection by strains x-1 and x-4 (*A. protococcarum*), while others were resistant. Strain x-2 (*A. chlorellavorum*) infected *Chlorella* strains only.

Gromov et al. (2002) investigated distribution of species of *Aphelidium* and *Rhizophydium* (Chridiomycota) in samples from ten stations in Ladoga Lake and adjacent water bodies in August 2000, and found *Aphelidium* at five and *Rhizophydium* at 10 stations. The number of infectious units (zoospores or infected algal cells) for *Aphelidium* varied from 0.01 to 0.92 per 300 ml of water, while for *Rhizophydium* it was from 0.1 to 1.6. These data suggest that rates of infection (prevalence) by *Aphelidium* are comparable with those by *Rhizophydium*, the most common representative of the Chytridiomycota.

13.1.2 Chytridiomycota

The Phylum Chytidiomycota includes a very large group of morphologically and genetically diverse fungi (Sparrow 1960; Shearer et al. 2007). Many species are saprotrophs, while others are parasites of phytoplankton (Sparrow 1960; Kagami et al. 2007; Sime-Ngando 2012; Voigt et al. 2013). During the last decade several new orders have been established in the class Chytridiomycetes (Phylum Chytridiomycota; See also Chapter 6) (Letcher et al. 2006; Mozley-Standridge et al. 2009; Simmons et al. 2009; Longcore and Simmons 2012; Karpov et al. 2014b). The genus *Rhizophydium* is an especially large group of species and contains many parasites of phytoplankton and a few species of parasites of protists (zooplankton) (Sparrow 1960; Kagami et al. 2007; Letcher and Powell 2012). In this chapter we focus on parasites which have been described since this topic was reviewed by Kagami et al. (2007).

Most recently on the basis of the ultrastructure and molecular phylogeny of *Gromochytrium mamkaevae* (a parasite of the alga *Tribonema gayanum*), and the molecular phylogenetic analysis of the *Mesochytrium* cluster, two new orders were introduced: Gromochytriales and Mesochytriales (Karpov et al. 2014b). Detailed investigation of *Gromochytrium mamkaevae* revealed unique characters of zoospore ultrastructure which was not found in any other order of Chytridiomycetes: posterior ribosomal core unbounded by the endoplasmic reticulum and detached from nucleus or microbody-lipid complex (MLC), and kinetosome composed of microtubular doublets. 18S and 28S rRNA gene sequences of *G. mamkaevae*, together with few environmental sequences form a branch, the Gromochytriales, which is sister to the *Mesochytrium* cluster (Mesochytriales).

Characteristics of Mesochytriales as illustrated by *Mesochytrium penetrans* (Karpov et al. 2010) include: zoospores with a unique ultrastructural organization (ribosomes dispersed through the cytoplasm; mitochondrion and MLC surrounded by rough endoplasmic reticulum) and a sessile zoosporangium, partially penetrating the host cell.

Mesochytrium penetrans and *Gromochytrium mamkaevae* have thus far not been found in environmental studies with DNA sequences. This indicates that these species may not be prevalent on hosts in the sampled ecosystems, and thus much broader sampling is needed for a more exhaustive analysis of fungal diversity. At the same time, some undescribed Mesochytriales species, represented by almost identical clones, were repeatedly recovered in several environmental samples (Lefèvre et al. 2008; Lepère et al. 2008; Jobard et al. 2012).

The DNA sequences from *G. mamkaevae* group together with environmental sequences are not as abundant as the *Mesochytrium* clade (Genitsaris et al. 2009; Monchy et al. 2011; Karpov et al. 2014b). The diversity and abundance of rDNA clones from these environmental samples suggest that members of the Mesochytriales may play important roles as regulators of the size and diversity of phytoplankton populations (Karpov et al. 2014b).

Some of the documented parasites of prokaryotic and eukaryotic algae (producers) are listed here in Tabs. 13.1 and 13.2. During the seasons when algal blooms are common, most zoosporic parasites reproduce quickly by asexual reproduction (r-strategy) because uninfected hosts are plentiful. Some species can reproduce by sexual reproduction as well (s-strategy) when conditions for growth are sub-optimal. A good example of a genus with a complete lifecycle is *Zygorhizidium* (Order *incertae sedis*), a parasite of diatoms in the genus *Synedra* (Doggett and Porter 1995, 1996) (Fig. 13.2). Gerphagnon et al. (2013) divided the asexual life cycles of two species of *Rhizosiphon*, which are parasites of *Anabaena macrospora* (Cyanobacteria) into three phases and seven stages (Fig. 13.3). *R. crissum* infects both vegetative cells and akinetes, while *R. akinetum* infects only akinetes. These categories facilitate comparison of the infection process between different species.

Host range has been estimated for some species. For example, Barr and Hickman (1967) studied the parasitism of isolates of *Spirogyra* by *Rhizophydium sphaerocarpum* with both field and laboratory-based research. *R. sphaerocarpum* is a monocentric fungus with an epibiotic sessile sporangium and endobiotic rhizoids. Both the parasite and the hosts could be grown easily in pure culture. Although ecotypes of *R. sphaerocarpum* have been reported to infect a wide range of species in the genera *Spirogyra*, *Oedogonium*, and *Mougeotia*, the isolate that the authors chose infected only a very restricted range of hosts. Sixteen isolates of *Spirogyra* were tested: three were very susceptible, four were moderately susceptible, and nine were immune to infection. Doggett and Porter (1995) studied the host range for *Zygorhizidium planktonicum*, an obligate, epibiotic parasite of the freshwater diatom *Synedra acus*. Host range studies revealed that germination and sporangium development were confined to three host species, *Synedra* sp., *S. acus*, and *S. ulna*. Sønstebø and Rohrlack (2011) observed that infection of *Planktothrix* hosts (cyanobacteria) by fungal parasites was chemotype-dependent. The existence of different *Planktothrix* chemotypes and of zoosporic parasites with chemotype preferences reveal narrow host ranges. According to Sønstebø and Rohrlack (2011) the data indicating narrow host ranges for these parasites is compatible with general ecological concepts such as the "kill-the-winner" concept and Red Queen hypothesis.

Tab. 13.1: Some species of Chytridiomycota and Aphelida which are putative parasites of freshwater eukaryotic phytoplankton in five phyla.

Host phyla	Host genera	Parasite genera	Reference
Charophyta	*Arthrodesmus*	*Rhizophydium*	Letcher and Powell (2012)
	Chara	*Diplophlyctis*, *Entophlyctis*	Kagami et al. (2007)
	Closterium	*Ancylistes*, *Entophlyctis*, *Micromyces*, *Micromycopsis*, *Myzocytium*, *Olpidium*, *Phlyctochytirium*, *Rhizophydium*	Kagami et al. (2007); Letcher and Powell (2012)
	Coleochaete	*Aphelidium*, *Phlyctidium*, *Rhizophydium*	Sparrow (1960); Gromov (2000); Letcher and Powell (2012)
	Cosmarium	*Myzocytium*, *Olpidium*, *Phlyctochytirium*, *Rhizophydium*	Karling (1965); Kagami et al. (2007)
	Cylindrocystis	*Endodesmidium*, *Rhizophydium*	Kagami et al. (2007); Letcher and Powell (2012)
	Desmidium	*Myzocytium*	Kagami et al. (2007)
	Docidium	*Mitochytridium*	Sparrow (1960)
	Elakatothrix	*Zygorhizidium*	Kagami et al. (2007)
	Euastrum	*Olpidium*	Sparrow (1960)
	Gonatozygon	*Myzocytium*	Kagami et al. (2007)
	Hyalotheca	*Olpidium*, *Micromyces*	Sparrow (1960); Kagami et al. (2007)
	Mesotaenium	*Zygorhizidium*	Sparrow (1960)
	Micrasterias	*Olpidium*	Sparrow (1960)
	Mougeotia	*Chytridium*, *Micromyces*, *Olpidium*, *Phlycidium*, *Rhizophydium*, *Zygorhizidium*	Kagami et al. (2007); Letcher and Powell (2012)
	Netrium	*Endodesmidium*, *Rhizophydium*	Kagami et al. (2007); Letcher and Powell (2012)
	Nitella	*Chytridium*, *Diplophlyctis*, *Entophlyctis*	Kagami et al. (2007)
	Penium	*Myzocytium*, *Olpidium*, *Rhizophydium*	Sparrow (1960); Kagami et al. (2007); Letcher and Powell (2012)
	Phycastrum	*Rhizophydium*	Letcher and Powell (2012)
	Pleurotaenium	*Olpidium*	Sparrow (1960)

(continued)

Tab. 13.1: (*continued*)

Host phyla	Host genera	Parasite genera	Reference
	Spirogyra	*Chytridium, Entophlyctis, Lagenidium, Micromyces, Phlyctidium, Phlyctochytirium, Rhizidium, Rhizophydium*	Sparrow (1960); Kagami et al. (2007)
	Spirotaenia	*Rhizophydium*	Kagami et al. (2007)
	Spondylosium	*Phlyctochytirium, Rhizophydium*	Kagami et al. (2007)
	Staurastrum	*Chytridium, Myzocytium, Phlyctidium, Phlyctochytirium, Rhizophydium, Zygorhizidium*	Kagami et al. (2007)
	Staurodesmus	*Chytridium, Phlyctochytirium, Rhizophydium*	Kagami et al. (2007)
	Tetmemorus	*Olpidium*	Sparrow (1960)
	Xanthidium	*Myzocytium*	Kagami et al. (2007)
	Zygnema	*Micromyces, Micromycopsis, Olpidium, Phlyctochytrium, Rhizophydium, Scherffeliomyces*	Sparrow (1960); Kagami et al. (2007)
Chlorophyta	*Ankistrodesmus*	*Amoeboaphelidium, Rhizophydium, Zygorhizidium*	Gromov et al. (1999); Gromov (2000); Kagami et al. (2007)
	Ankyra	*Hapalopera, Rhizophydium*	Holfeld (2000); Rasconi et al. (2012)
	Apiocystis	*Rhizophydium*	Kagami et al. (2007)
	Asterococcus	*Phlyctochytirium*	Kagami et al. (2007)
	Botryococcus	*Chytridium*	Kagami et al. (2007)
	Bulbochaete	*Chytridium, Phlyctidium*	Sparrow (1960); Kagami et al. (2007)
	Chaetophora	*Phlyctidium, Rhizophydium*	Letcher and Powell (2012)
	Characium	*Chytridium, Phlyctidium, Rhizophydium*	Sparrow (1960); Kagami et al. (2007)
	Cladophora	*Olpidium, Phlyctidium, Phlyctochytirium, Rhizophydium*	Sparrow (1960); Kagami et al. (2007)
	Chlamydocapsa	*Zygorhizidium*	Kagami et al. (2007)
	Chlamydomonas	*Dangeardia, Entophlyctis, Polyphagus, Phlyctochytirium, Rhizophydium, Scherffeliomyces*	Kagami et al. (2007)
	Chlorella	*Amoeboaphelidium, Rhizophydium*	Gromov et al. (1999); Gromov (2000)

(*continued*)

Tab. 13.1: (*continued*)

Host phyla	Host genera	Parasite genera	Reference
	Chlorococcum	*Amoeboaphelidium*, *Mesochytrium*, *Rhizophydium*	Gromov et al. (1999); Gromov (2000); Kagami et al. (2007)
	Chlorogonium	*Phlyctidium*	Kagami et al. (2007)
	Chodatella	*Rhizophydium*	Rasconi et al. (2012)
	Cladophora	*Chytridium*, *Olpidium*, *Phlyctochytirium*, *Rhizophydium*	Sparrow (1960); Kagami et al. (2007)
	Coelastrum	*Zygorhizidium*	Kagami et al. (2007)
	Draparnaldia	*Sporophlyctis*	Sparrow (1960)
	Dictyosphaerium	*Rhizophydium*, *Zygorhizidium*	Kagami et al. (2007)
	Eremosphaera	*Olpidium*, *Rhizophydium*	Kagami et al. (2007)
	Eudorina	*Dangeardia*, *Endocoenobium*, *Polyphagus*, *Phlyctidium*, *Rhizophydium*, *Zygorhizidium*	Sparrow (1960); Kagami et al. (2007)
	Gemellicystis	*Rhizidium*, *Rhizophydium*	Kagami et al. (2007)
	Kirchneriella	*Amoeboaphelidium*, *Rhizophydium*	Gromov (2000); Kagami et al. (2007)
	Oedogonium	*Achlyogeton*, *Chytridium*, *Micromyces*, *Plasmophagus*, *Rhizophydium*	Sparrow (1960); Kagami et al. (2007)
	Oocystis	*Chytridium*, *Diplochytridium*, *Lagenidium*, *Rhizophydium*, *Zygorhizidium*	Kagami et al. (2007); Letcher and Powell (2012)
	Palmodictyon	*Rhizophydium*	Letcher and Powell (2012)
	Pandorina	*Dangenardia*, *Polyphagus*, *Rhizophydium*, *Zygorhizidium*	Kagami et al. (2007); Letcher and Powell (2012)
	Pediastrum	*Phlyctidium*, *Zygorhizidium*	Kagami et al. (2007)
	Pseudosphaerocystis	*Rhizophydium*	Kagami et al. (2007)
	Scenedesmus	*Amoeboaphelidium*, *Aphelidium*, *Chytridium*, *Phlyctidium*, *Rhizophydium*	Gromov et al. (1999); Gromov (2000); Kagami et al. (2007)
	Sphaerella	*Rhizidium*	Sparrow (1960)

(*continued*)

Tab. 13.1: (*continued*)

Host phyla	Host genera	Parasite genera	Reference
	Sphaerocystis	*Rhizophydium, Zygorhizidium*	Kagami et al. (2007)
	Sphaeroplea	*Rhizophydium*	Letcher and Powell (2012)
	Stigeoclonium	*Phlyctidium, Rhizophydium*	Sparrow (1960); Letcher and Powell (2012)
	Stylosphaeridium	*Rhizophydium*	Kagami et al. (2007)
	Ulothrix	*Rhizophydium*	Kagami et al. (2007)
	Volvox	*Loborhiza, Polyphagus*	Sparrow (1960); Kagami et al. (2007)
	Volvulina	*Polyphagus*	Kagami et al. (2007)
Cryptophyta	*Cryptomonas*	*Rhizophydium*	Kagami et al. (2007)
Dinophyta	*Ceratium*	*Amphicypellus, Rhizophydium, Zygorhizidium*	Canter and Heaney (1984); Kagami et al. (2007)
	Glenodinium	*Rhizophydium*	Letcher and Powell (2012)
	Peridinium	*Amphicypellus, Phlyctochytirium*	Kagami et al. (2007)
Euglenophyta	*Colacium*	*Rhizophydium*	Letcher and Powell (2012)
	Euglena	*Chytridium, Entophlyctis, Olpidium, Phlyctochytrium, Polyphagus, Psuedosphaerita, Scherffeliomyces, Sphaerita*	Sparrow (1960); Karling, (1965); Kagami et al. (2007)
Heterokontophyta	*Achnanthes*	*Amoeboaphelidium, Rhizophydium*	Gromov (2000); Letcher and Powell (2012)
	Amphora	*Physorhizophidium*	Sparrow (1960)
	Asterionella	*Chytridium, Chytriomyces, Rhizophydium, Septosperma, Zygorhizidium*	Kagami et al. (2007)
	Botrydiopsis	*Aphelidium*	Gromov (2000)
	Chaetoceros	*Rhizophydium*	Kagami et al. (2007)
	Chlorobotrys	*Chytridium*	Sparrow (1960)
	Chrysocapsa	*Dangeardia*	Kagami et al. (2007)
	Chrysochaete	*Dangeardia*	Kagami et al. (2007)
	Cocconeis	*Chytridium, Rhizophydium*	Kagami et al. (2007); Letcher and Powell (2012)
	Cyclotella	*Rhizophydium, Zygorhizidium*	Kagami et al. (2007; Rasconi et al. (2012)

(*continued*)

Tab. 13.1: *(continued)*

Host phyla	Host genera	Parasite genera	Reference
	Cymbella	*Chytridium, Rhizophydium*	Sparrow (1960); Letcher and Powell (2012)
	Dinobryon	*Rhizophydium, Zygorhizidium*	Kagami et al. (2007)
	Fragilaria	*Chytridium, Podochytrium, Rhizophydium, Zygorhizidium*	Kagami et al. (2007)
	Gonyostomum	*Sphaerita*	Sparrow (1960)
	Gomphonema	*Phlyctidium, Rhizophydium*	Sparrow (1960); Letcher and Powell (2012)
	Hantzschia	*Phlyctidium, Rhizophydium*	Sparrow (1960); Letcher and Powell (2012)
	Hyalobryon	*Rhizophydium*	Kagami et al. (2007)
	Mallomonas	*Chytridium, Pseudopileum, Septosperma*	Kagami et al. (2007)
	Melosira	*Aphelidium, Chytridium, Podochytrium, Rhyzidiopsis, Rhizophydium, Zygorhizidium*	Gromov (2000); Kagami et al. (2007)
	Navicula	*Rhizophydium*	Letcher and Powell (2012)
	Nitzschia	*Chytridium*	Kagami et al. (2007)
	Ochromonas	*Chytridiium*	Kagami et al. (2007)
	Ophiocytium	*Rhizophydium*	Letcher and Powell (2012)
	Pinnularia	*Podochytrium, Rhizophydium*	Kagami et al.(2007); Letcher and Powell (2012)
	Rhizosolenia	*Rhizophydium, Zygorhizidium*	Kagami et al. (2007); Letcher and Powell (2012)
	Stephanodiscus	*Rhizophydium, Zygorhizidium*	Holfeld (1998); Kagami et al. (2007)
	Surirella	*Rhizophydium*	Letcher and Powell (2012)
	Synedra	*Rhizophydium, Septolpidium, Zygorhizidium*	Kagami et al. (2007); Rasconi et al. (2012)
	Tabellaria	*Chytridium, Chytriomyces*	Kagami et al. (2007)
	Tribonema	*Aphelidium, Chytridium, Plasmophagus, Polyphagus, Rhizophydium*	Sparrow (1960); Gromov (2000); Kagami et al. (2007)
	Uroglena	*Zygorhizidium*	Kagami et al. (2007)
	Vaucheria	Entophlyctis, Rhizophydium	Kagami et al. (2007); Letcher and Powell (2012)

Tab. 13.2: Some species of Chytridiomycota which are putative parasites of freshwater prokaryotic phytoplankton in the Cyanobacteria.

Cyanobacteria	Putative chytrid parasites	Reference
Anabaena affanis var. *intermedia*	*Rhizosiphon akinetum*	Canter (1954)
	Rhizosiphon anabaenae (Phlyctidium anabaenae)	Canter (1953)
	Rhizosiphon crassum	Canter (1953)
Anabaena circinalis	*Blastocladiella anabaenae*	Canter and Willoughby (1964)
	Rhizophydium sp.	Gleason and Macarthur (2008)
	Rhizosiphon anabaenae (Phlyctidium anabaenae)	Canter (1972)
	Rhizosiphon crassum	Canter (1953)
	Chytridium cornutum	Braun (1856)
Anabaena flos-aquae	*Blastocladiella anabaenae*	Canter and Willoughby (1964); Canter (1968)
	Phlyctidium megastomum (Rhizophydium sphaerocarpum)	Raitschenko (1902)
	Rhizosiphon crassum	Rasconi et al. (2009)
	Unidentified chytrids	Sigee et al. (2007)
Anabaena macrospora	*Rhizosiphon akinetum*	Canter (1954)
	Rhizosiphon anabaenae (Phlyctidium anabaenae)	Seligo (1909); Skuja (1948); Fott (1951)
Anabaena planktonica	*Rhizosiphon anabaenae (Phlyctidium anabaenae)*	Paterson (1960)
Anabaena smithii	Unidentified chytrid 1	Takano et al. (2008)
	Unidentified chytrid 2	Takano et al. (2008)
Anabaena solitaria	*Blastocladiella anabaenae*	Canter (1968)
	Rhizophydium ubiquetum	Canter (1968)
	Rhizosiphon crassum	Canter (1968)
Anabaena sp.	*Rhizosiphon crassum*	Canter (1951); Sparrow et al. (1965)
Anabaena sp.	*Scherffeliomyces* sp.	Canter (1972)
Anabaena spherica	*Rhizosiphon anabaenae (Phlyctidium anabaenae)*	Skuja (1948)
Anabaena spiroides	*Rhizosiphon anabaenae (Phlyctidium anabaenae)*	Skuja (1948); Canter (1953)
	Rhizosiphon crassum	Canter (1953)
Aphanizomenon flos-aquae	*Blastocladiella anabaenae*	Canter (1968)
	Chytridium cornutum	Canter (1963)
	Phlyctidium globosum	Skuja (1956)
	Rhizophydium megarrhizum	Paterson (1958)
	Unidentified chytrid	Seligo (1909)
Aphanizomenon gracile	*Rhizophydium subangulosum*	Fott (1951); Fjerdingstad (1966)
Chroococcus turgidus	*Rhizophydium agile*	Sparrow (1960)
Cyanobacteria spp.	Rhizidiaceae (possibly)	Holfeld (1998)

(continued)

Tab. 13.2: (*continued*)

Cyanobacteria	Putative chytrid parasites	Reference
Cylindrospermopsis raciborskii	Unidentified chytrids	Fabbro and Duivenvoorden (1996)
Filarszkya sp.	*Rhizosiphon crassum*	Sparrow (1960)
Gomphosphaeria naegeliana	Unidentified chytrid	Canter (1954)
Gomphosphaeria sp.	Unidentified chytrid	Rasconi et al. (2009)
Lyngbya sp.	*Rhizophydium megarrhizum*	Paterson (1958)
Microcystis aeruginosa	*Chytridium microcystidis*	Skuja (1948); Szklarczyk (1956); Canter (1972); Van Wichelen et al. (2010)
	Rhizidium microcystidis	Canter (1972); Sen (1988)
	Rhizophydium sp.	Gleason and Macarthur (2008)
Microcystis flosaquae	*Chytridium microcystidis*	Skuja (1948); Canter (1972)
Microcystis sp.	Unidentified chytrid	Rasconi et al. (2009)
Microcystis sp.	Unidentified chytrid	Paterson (1958)
Microcystis viridis	*Chytridium microcystidis*	Skuja (1948)
Oscillatoria agardhii	*Rhizophydium megarrhizum*	Skuja (1956)
	Rhizophydium subangulosum	Fott (1951)
Oscillatoria agardhii var. *isothrix*	*Rhizophydium megarrhizum*	Canter and Lund (1951); Canter (1972)
Oscillatoria rubescens	*Rhizophydium deformans*	Jaag and Nipkow (1951)
	Rhizophydium megarrhizum	Canter and Lund (1951)
	Rhizophydium oscillatoriae-rubescentis	Jaag and Nipkow (1951)
Oscillatoria sp.	*Rhizophydium megarrhizum*	Canter and Lund (1951)
Oscillatoria sp.	*Rhizophydium megarrhizum*	Canter (1953)
Oscillatoria sp.	*Rhizophydium megarrhizum* (*Rhizophydium subangulosum)*	de Wildeman (1890)
Oscillatoria sp.	*Rhizophydium* sp.	Paterson (1958)
Oscillatoria tenuis	*Rhizophydium subangulosum*	Berczi (1940)
Oscillatoria tenuis var. *subfusca*	*Rhizophydium subangulosum* (*Chytridium subangulosum*)	Braun (1856)
Planktothrix agardhii	*Rhizophydium megarrhizum* (possibly)	Davis et al. (2003)
Planktothrix rubescens	*Rhizophydium megarrhizum* (possibly)	Davis et al. (2003)
Planktothrix sp.	Unidentified chytrid from the Angulomycetaceae family (Chy-Kol2008)	Sønstebø and Rohrlack (2011)
Planktothrix sp.	Unidentified chytrid from the Angulomycetaceae family (Chy-Lys2009)	Sønstebø and Rohrlack (2011)

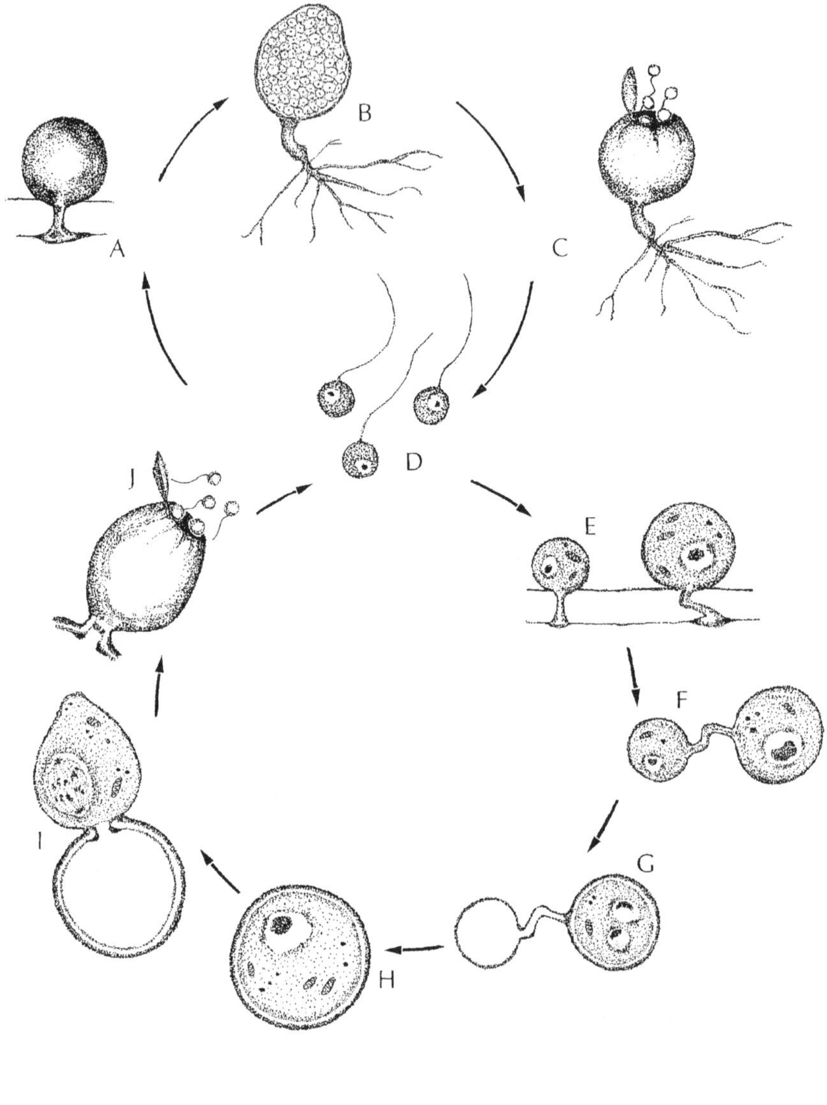

Fig. 13.2: Life cycle of parasites of eukaryotic phytoplankton in the Chytridiomycota. The confirmed life cycle of *Zygorhizidium planktonicum*. The asexual, infection phase (A–D) is characterized by uninucleate, haploid thalli (A) that differentiate into operculate zoosporangia (B). Upon release (C), the uniflagellate zoospores (D) give rise to either asexual thalli or incipient gametangia (E). Sexual reproduction is evidenced by heterothallic gametangia (E–G) which fuse by means of a single, epibiotic conjugation tube. Plasmogamy immediately follows wall fusion and the donor protoplast migrates into the recipient gametangium, which in turn becomes the thick walled resting spore (G). Following wall deposition, the nuclei fuse and the zygotic nucleus remains undifferentiated throughout spore dormancy (H). Meiosis occurs within the newly-formed meiosporangium following resting spore germination (I). The haploid zoospores are released by an operculate meiosporangium (J). Reproduced from Doggett and Porter (1996) with permission.

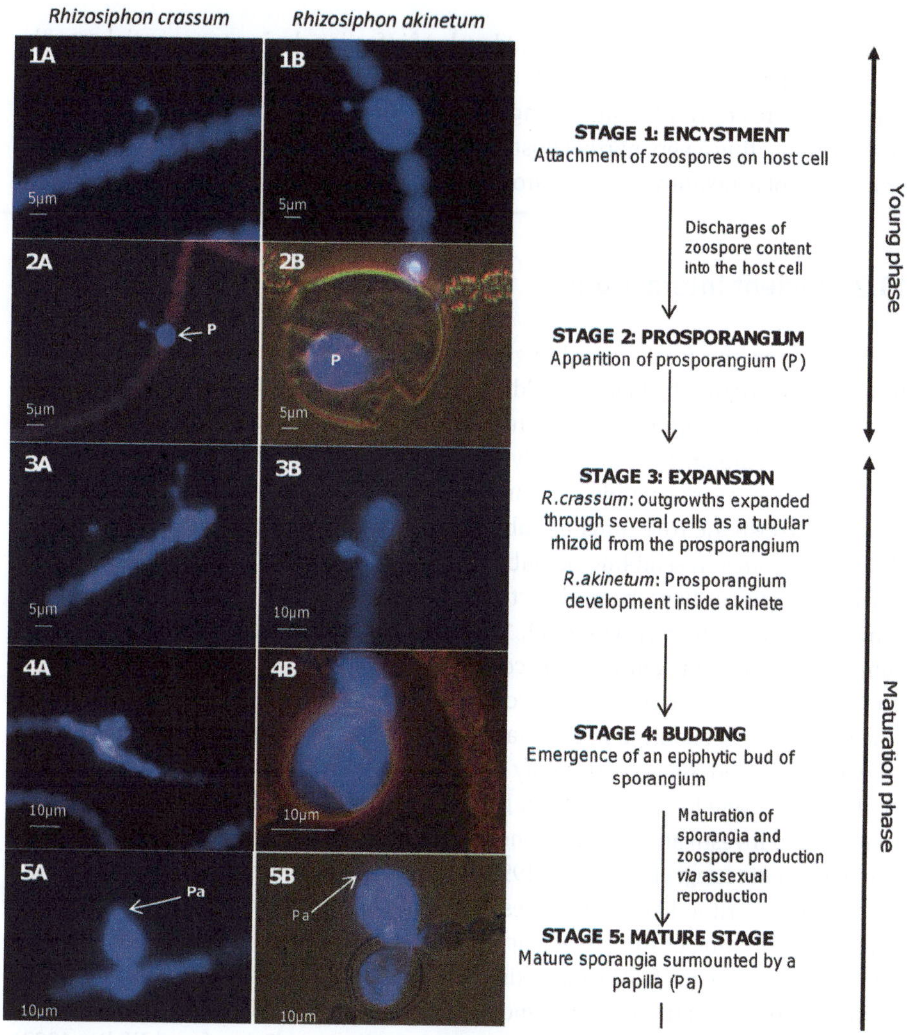

Fig. 13.3: Stages in the life cycle of parasites of cyanobacteria (Chytridiomycota). Reproduced from Gerphagnon et al. (2013).

13.1.3 Blastocladiomycota

Paraphysoderma sedebokerensis is a parasite of the green alga, *Haematococcus puvialis* (Gutman et al. 2009). The infective propagules of this species are amoebae. Gutman et al. (2009) tested the specificity of *Paraphysoderma sedebokerensis* to 13 species of algae belonging to the Chlorophyta: *Chlamydomonas nivalis, Chlorella emersonii, Chlorella vulgaris, Chlorococcum* sp., *Chlorogonium elongatum, Monoraphidium braunii, Muriella zofigiensis, Scenedesmus obliquus, Scenedesmus vacuolatus* (2 strains), *Scotiellopsis oocystiformis* and *H. pluvialis*. Most hosts were either resistant to infection, or only experienced slight levels of infection during exponential growth. Infections developed rapidly (in 3–4 days) and intensively (100% cells infected) only in cultures of *H. pluvialis* during both the logarithmic and stationary stages of growth, indicating a very narrow host range. *H. pluvialis* is used commercially for the production of astaxanthin, a ketocarotenoid.

13.2 Ancient interactions

Evidence from the fossil record suggests that parasitism by zoosporic true fungi has ancient origins. In fact, the oldest fossil fungi are zoosporic true fungi from the Vendian strata from the late Precambrian (~600 Ma) (Burzin 1993). Microbes in older Precambian fossils, initially identified as fungi, are either empty sheaths of filamentous cyanobacteria, or cannot properly be assigned (Hass et al. 1994). Fossils of zoosporic true fungi are abundant in Devonian Rhynie Chert (~400 Ma). Rhynie parasites resembling members of the Blastocladiales, Olpidiaceae and Spizellomycetales, have been detected, indicating the diversity of zoosporic true fungi at this early date (Taylor et al. 1992a,b). For example, fungi, with morphology similar to the modern genus *Allomyces* (Blastocladiales), were probably parasitic on the extinct vascular plants, the rhyniophytes (Taylor et al. 1994). Species of zoosporic true fungi have been found in association with all of the Rhynie chert plants, as well as with fungi and algae, with many associations appearing to be parasitic (Taylor et al. 2004). Some fossils have been found in such good condition, that researchers were able to investigate host relationships, life histories and palaeoecology of these ancient aquatic fungi (Taylor et al. 1992a,b; Taylor and Osborn 1996).

Cyanobacterium-like prokaryotes are thought to be present and widespread since the beginning of the Proteozoic era (~2.5 Ga), however identification is almost impossible for the few fossils that exist from this era (Schopf and Walter 1982). There is little to observe in these fossils, and apart from incompletely preserved cells; there is just the structure and form of stromatolitic sediments (Schopf and Walter 1982). Stronger evidence comes from the Late Archaean (~ 2.8–2.5 Ga) from the interpretation of the internal structure of stromatolites as well as filamentous fossils, similar to *Oscillatoria* and *Lyngbya* within the stromatolites (Schopf and Walter 1982). Extant species of both *Oscillatoria* and *Lyngbya* are hosts to parasitic zoosporic true fungi (Braun 1856; de Wildeman 1890; Berczi 1940; Canter and Lund 1951; Fott 1951; Jaag and Nipkow 1951; Canter 1953; Skuja 1956; Paterson 1958; Canter 1972).

In general, parasitism of cyanobacteria by zoosporic true fungi is difficult to establish by investigating fossil evidence. For example, cyanobacteria have rarely been described from the Rhynie chert, despite evidence that cyanobacteria are common in virtually all aquatic environments (Schopf and Walter 1982). Studies where adequate preservation of the structure of the organism has allowed identification are rare (Schopf and Walter 1982). Recently *Croftalania venusta* (Oscillatoriales), a non-marine aquatic filamentous-colonial fossil cyanobacterium from the Early Devonian Rhynie chert, was described (Krings et al. 2007). *Croftalania venusta* occurs in the Rhynie chert in different growth forms and is associated with the formation of microbial mats, where it coexists with other microorganisms, including zoosporic true fungi. Also samples of *C. venusta* were found to contain chytrid-like structures (Krings et al. 2007). Recent advances in technology, such as Fourier transform infrared spectroscopy have allowed the identification of biochemical signatures of microorganisms in ancient deposits (Preston and Genge 2010), however the valuable equipment required for these techniques is not available for use by the average laboratory.

In conclusion, it appears it is extremely unlikely we will ever determine the actual nature of ancient interactions between zoosporic true fungi and cyanobacteria. Evidence suggests that both were present in ancient ecosystems in vast quantities, and given current knowledge of interactions, fossil evidence may at least promote awareness into the evolution of parasitism of cyanobacteria by zoosporic true fungi.

13.3 Novel food webs

13.3.1 *Vorticella* communities attached to cyanobacterial filaments

Zoospores of fungi are also thought to be consumed by ciliates (Gleason and Lilje 2009; Sime-Ngando 2012) but feeding rates have not been measured. Species of *Vorticella* often attach to colonies and filaments of cyanobacteria, especially to species of *Anabaena* (Canter et al. 1992). It is possible that fungal zoospores released by the zoosporic parasites on the cyanobacteria may be an important food source for *Vorticella* (Fig. 13.4). These species of cyanobacteria have gas vacuoles which keep them floating in the euphotic zones. Long filaments have enough buoyancy to keep the attached *Vorticella* colonies and thalli of zoosporic true fungi near the surface as well. *Vorticella* cannot consume the long filaments. Parasitism by zoosporic true fungi can cause the long filaments to break up into short filaments (Sime-Ngando 2012). These shorter filaments can be eaten more easily by grazing zooplankton, but they are also more likely to sink to the bottom. If the filaments carry resistant sporangia of zoosporic true fungi and akinetes of *Anabaena*, they provide a large inoculum which can spread both the hosts and the parasites along spatial and temporal scales within the freshwater ecosystem.

13.3.2 Communities involving other protists

Zoosporic fungal parasites of other protists are also found in freshwater ecosystems. For example, vampyrellid amoebae are well known predators of soil fungi and phytoplankton (Chakraborty and Old 1982; Hess et al. 2012). *Rhizophydium vampyrella*

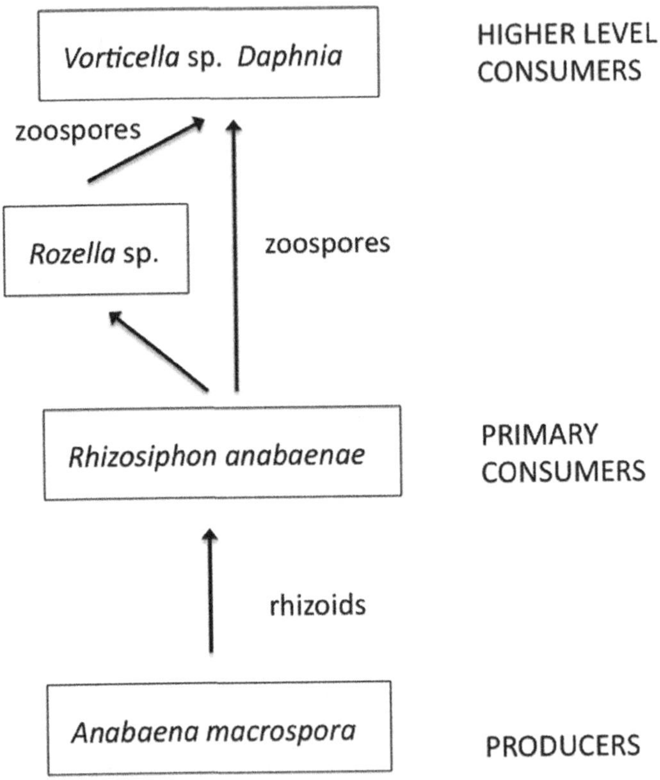

Fig. 13.4: Additions to food webs by *Vorticella* and *Rozella*.

is a parasite of cysts of *Vampyrella* sp. (Letcher and Powell 2012). This amoeba is a predator of *Gloeocystis vesiculosa*. This food chain therefore provides another link for carbon flow in freshwater ecosystems.

13.4 Host parasite dynamics

To understand and predict changes in dynamic aquatic food webs, chemical, physical and biological factors need to be considered (Stewardson and Webb 2010). Up until now the emphasis has largely been placed on physico-chemical data to provide predictions of changes in biological systems such as phytoplankton blooms. Recent advances in molecular techniques have allowed more substantial gathering of empirical biological data. This should allow more holistic conceptual models of these complex dynamic systems to be developed.

There has been increasing emphasis on the relationships between trophic levels in a food web and how these relationships influence the overall state of the ecosystem. The presence of trophic links between the micro, meso and macroscales is a strong indicator of the health of the ecosystem (Stewardson and Webb 2010). Blooms of cyanobacteria or diatoms are most likely examples of temporal disruptions of these trophic links. Recent research on aquatic systems has supported the theory that parasitism and predation play important roles in re-establishing these trophic links and reducing blooms of cyanobacteria and other groups of phytoplankton (Fabbro and Duivenvoorden 1996; Jobard et al. 2010; Gleason et al. 2011; Grami et al. 2011; Kagami et al. 2011; Miki et al. 2011; Rasconi et al. 2011; Sønstebø and Rohrlack 2011). The parasitism of phytoplankton has recently been recognized as a potential source of energy for higher trophic levels that would otherwise be inaccessible (Sparrow 1960, 1968; Canter 1972; Fabbro and Duivenvoorden 1996; Jobard et al. 2010; Gleason et al. 2011; Grami et al. 2011; Miki et al. 2011; Rasconi et al. 2011; Sime-Ngando et al. 2011; Sønstebø and Rohrlack 2011).

The relationships between host and parasite are dependent on host susceptibility, genotypes, chemotactic responses and severity of infection. The dynamics of host-parasite relations is deserved of some attention as parasitism may play a major role in host population size regulation, host evolution, parasite evolution, and act as a food source. Early empirical work by Canter and Lund (1951) has shown that zoosporic true fungal parasites may regulate phytoplankton sizes. Field measurements and laboratory studies, summarised by van Donk (1989), clearly show the effect of temperature, host cell exudate, nutrients, and light on host-parasite relations. Furthermore, host size may also play a role in the susceptibility of the host to parasite attack (Kagami et al. 2007). Turbulence may also affect host parasite dynamics (Kühn and Hofmann 1999). Gerphagnon et al. (2013) observed that the parasitic chytrid *Rhizosiphon crassum* caused relatively high infection in low *Anabaena macrospora* cell densities compared to higher densities. They suggest that chemotactic responses of new *R. crassum* zoospores to nearby healthy uninfected A. *macrospora* may be stronger than to those further away or those intermingled with infected filaments. This would result in localized prevalence of parasitic infection during post-bloom conditions where *A. macrospora* populations are declining. Zoosporic true fungal and other eukaryotic parasites are likely to play a more significant part in the “top-down” model of phytoplankton dynamics. Their role in the turn-over and release of nutrients to higher trophic levels of the food web would in turn encourage and accelerate phytoplankton reduction during post-bloom conditions.

Furthermore, the resting spores of cyanobacteria, i.e. akinetes, were found to be highly infected by parasites in temperate lakes (Gerphagnon, pers com). Akinetes are the only cell types which overcome unfavorable conditions and lead to the colonization of the water column by cyanobacteria when favorable seasonal conditions for growth return (Baker and Bellifemine 2000). Kravchuk et al. (2011) concluded that the size of the inoculum (akinete density in sediment) is one biotic factor which determines the proportion of *Anabaena flos-aquae* in the total phytoplankton biomass in blooms. Additionally, Tsujimura et al. (2000) have suggested that the start of bloom formation was related to the quantity of *Microcystis* sp. colonies in the sediment. By impacting akinetes, fungal parasitism could be responsible for an important loss of the inoculum size. Consequently, zoosporic true fungi could delay or be responsible of a decrease in the competitive ability of cyanobacterium host populations in subsequent seasons. In addition, by infecting akinetes, zoosporic true fungi could severely affect the genetic structure of cyanobacterial populations from year to year. Indeed, as akinetes are differentiated from clonal vegetative cells, their genetic structure reflects the genetic structure of the cyanobacterial bloom and they constitute the "genetic seed bank" for cyanobacterial population. Because the genetic structure of host population is a major factor determining its susceptibility to and resistance against infection by parasites (De Bruin et al. 2004), the parasitism of this "genetic seed bank" may strongly impact on the cyanobacterial population responses to infections by parasites during successive seasons.

Many aquatic fungal parasites may be host specific or have a narrow range of hosts, this provides opportunity for driving change in species types based on susceptibility to parasites (Ibelings et al. 2004). It has been argued that this host specificity may drive evolution in the host and potentially in the parasites (coevolution, red queen hypothesis). For example, host chemotype, which may in fact attract specific parasite zoospores, may shift as a result parasitic pressure (Sønstebø and Rohrlack 2011). To further add to the complexity, parasites may also be subject to parasitism (hyper parasitism) by other species such as *Rozella* sp. (Ibelings et al. 2004; Gleason et al. 2012).

Early work by Hassell and May (1973) and Anderson and May (1978) developed various mathematical models that capture the dynamics of observed host parasite relations. These models have aided understanding of population dynamics and equilibrium states in the host-parasite populations. For example, using models, Anderson and May (1978) identified a number of categories critical in maintaining equilibrium in host-parasite relations. The relationship between parasite induced host death and parasite burden (the number of parasites per host) is one of these. Bruning et al. (1992) developed a population-dynamic model specifically for chytrid parasites on phytoplankton in a lake system. As well as physical and chemical properties of the water, the model required four parameters; the number of host cells infected, time for sporangia to develop, growth rate of an uninfected host and the difference between the mortality rate of an uninfected host and an infected host.

13.5 Conclusion

Many species in three phyla Aphelida, Chytridiomycota and Blastocladiomycota are most common parasites of prokaryotic and eukaryotic phytoplankton in freshwater ecosystems. Many species of these parasites have been described, and environmental surveys using molecular methods suggest that many undescribed species are also present.

Zoosporic parasites play major roles in control of sizes of phytoplankton populations and therefore succession and ecosystem complexity. These parasites also increase the number of food chains in freshwater food webs. The problem with the lack of quantitative data remains because the ecology of these parasites has been so poorly studied.

The motile propagules of zoosporic parasites are usually zoospores but can be amoebae in some species. In fact a few species produce only amoebae. In some cases centrioles are present in amoebae that lack flagella. Karpov et al. (2014b) found pseudopods on zoospores of one species in the Chytridiomycota. This suggests that amoeboid movement may be more common in zoosporic true fungi than previously thought.

The Cryptomycota and Aphelida are very ancient microorganisms which presumably evolved near the point of divergence of animals and fungi and prior to the divergence of Chytridiomycota from other groups of fungi. Fossil evidence indicates that microorganisms with morphologies similar to zoosporic true fungi were parasites of cyanobacteria. Presumably, before the appearance of multicellular plants and animals, the only hosts available were phytoplankton and zooplankton. At present these hosts are still susceptible to infection.

Ibelings et al. (2011) considered the effect of global warming on the composition and complexity of phytoplankton populations which were subject to parasitism. There are very few studies and very little data available on this topic. With accelerated rates of climate change this topic needs to be thoroughly addressed in the immediate future. It is possible that global climate change may significantly alter the composition of phytoplankton populations and the ecological services provided by these species in freshwater ecosystems in the future. Also accelerated global warming could cause destructive cascade effects which could cause the disappearance of key species, at least in localized environments.

Acknowledgments

The authors thank the Russian Foundation for Basic Research (projects No 12-04-01486), the program "Problems of life origin and biosphere development", launched by the Presidium of the Russian Academy of Sciences, "*Research in Paris* grant from the Mairie de Paris" and the University of Sydney for providing library resources. The ANR Project ROME (ANR 12 BSV7 0019 02, France) is a contributor of this study. The authors thank Elayna Truszewski, Department of Biological Sciences, Macquarie University for her editorial assistance with preparation of the original manuscript. The authors thank the University of Sydney library for providing access to literature.

References

Anderson RM, May RM. Regulation and stability of host-parasite population interactions: I. Regulatory processes. J Anim Ecol 1978;47:219–247.

Baker PD, Bellifemine D. Environmental influences on akinete germination of *Anabaena circinalis* and implications for management of cyanobacterial blooms. Hydrobiol 2000;427:65–73.

Baldauf SL. The deep roots of Eukaryotes. Science 2003;300:1703–1706.

Barr DJS, Hickman CJ. Chytrids and algae I. Host-substrate range, and morphological variation of species of *Rhizophydium*. Can J Bot 1967;45:423–430.

Berczi L. Adatok Magyarorszag vfzigomba vegeta-ciqjahoz. Vizigombakkal vegzett mesterseges fertozesi kfserletek. Acta Geobot Hung 1940;3:79–99.

Braun A. Über *Chytridium*, eine gattung einzelliger schmarotzergewächse auf algen und infusorium. Gedruckt in der Druckerei der Königl, Berlin 1856.

Bruning K, Lingeman R, Ringelberg J. Estimating the impact of fungal parasites on phytoplankton populations. Limnol Oceanogr 1992;37:252–260.

Burzin M. The oldest chytrid (Mycota, Chytridiomycetes *Incertae sedis*) from the Upper Vendian of the East European Platform. In: Sokolov, ed. Fauna and Ecosystems of the Geological Past. Moscow: Nauka 1993:21–33.

Canter HM. Fungal parasites of the phytoplankton. II (Studies on British Chytrids, XII). 1951. Ann Bot 1951;15:129–156.

Canter HM. Annotated list of British aquatic chytrids. Trans Br Mycol Soc 1953;36:278–301.

Canter HM. Fungal parasites of the phytoplankton. III. Trans Br Mycol Soc 1954;37:111–133.

Canter HM. Concerning *Chytridium cornutum* Braun. Trans Br Mycol Soc 1963;46:208–212.

Canter HM. Studies on British chytrids: XXVII. *Rhizophydium fugax* sp. nov., a parasite of planktonic cryptomonads with additional notes and records of planktonic fungi. Trans Br Mycol Soc 1968;51:699–705.

Canter HM. A guide to the fungi occurring on phytoplanktonic blue-green algae. In: Desikachary TV, ed. Taxonomy and Biology of Blue-Green Algae. India: University of Madras 1972:145–159.

Canter HM, Heaney SI. Observations on zoosporic fungi of *Ceratium* spp. in lakes of the English Lake District; importance for phytoplankton population dynamics. New Phytol 1984;97:601–612.

Canter HM, Lund JWG. Studies on plankton parasites. Ann Bot 1951;15:359–371.

Canter HM, Walsby AE, Kinsman R, Ibelings BW. The effect of attached vorticellids on the buoyancy of the colonial cyanobacterium *Anabaena lemmermannii*. Br Phycol J 1992;27:65–74.

Canter HM, Willoughby LA. A parasitic *Blastocladiella* from Windermere plankton. J R Micros Soc 1964;83:365–372.

Chakraborty S, Old KM. Mycophagous soil amoeba: interactions with three plant pathogenic fungi. Soil Biol Biochem 1982;14:247–255.

Corsaro D, Walochnik J, Venditti D, Steinmann J, Müller KD, Michel R. Microsporidia-like parasites of amoebae belong to the early fungal lineage Rozellomycota. Parasitol Res 2014; DOI:10.1007/s00436-014-3838-4

Davis PA, Dent M, Parker J, Reynolds CS, Walsby AE. The annual cycle of growth rate and biomass change in *Planktothrix* spp. in Blelham Tarn, English Lake District. Freshwater Biol 2003;48:852–867.

De Bruin A, Ibelings BW, Rijkeboer M, Brehm M, van Donk E. Genetic variation in *Asterionella formosa* (Bacillariophyceae): is it linked to frequent epidemics of host specific parasitic fungi? J Phycol 2004;40:823–830.

De Wildeman E. Chytridiacees de Belgique. Ann Soc Belg Microsc (Mem.) 1890;14:5–28.

Doggett MS, Porter D. Further Evidence for host-specific variants in *Zygorhizidium planktonicum*. Mycologia 1995;87:161–171.

Doggett MS, Porter D. Sexual reproduction in the fungal parasite, *Zygorhizidium planktonicum*. Mycologia 1996;88:720–732.

Fabbro LD, Duivenvoorden LJ. Profile of a bloom of cyanobacterium *Cylindrospermopsis raciborskii* (Woloszynska) Seenaya and Subba Raju in the Fitzroy River in tropical central Queensland. Mar Freshw Res 1996;47:685–694.

Fjerdingstad E. Cell dimensions of some members of *Aphanizomenon* Morren (Cyanophyceae). Aquat Sci 1966;28:133–147.

Fott B. New chytrids parasitizing on algae. Mem Soc R Sci Bohème 1951;4:1–10.

Fott B. *Aphelidium chlorococcarum* spec. nova, ein neuer Parasit in Grunalgen. "Biologica". Univ Carol 1957;3:229–237.

Genitsaris S, Konstantinos AK, Moustaka-Gouni M. Microscopic eukaryotes living in a dying lake (Lake Koronia, Greece). FEMS Microbiol Ecol 2009;69:75–83.

Gerphagnon M, Latour D, Colombet J, Sime-Ngando T. Fungal parastitism: life cycle, dynamics and impact on cyanobacterial blooms. PLoS One 2013;8:1–10.

Gleason FH, Carney LT, Lilje O, Glockling SL. Ecological potentials of species of *Rozella* (Cryptomycota). Fungal Ecol 2012;5:651–656.

Gleason FH, Kagami M, Lefevre E, Sime-Ngando T. The ecology of chytrids in aquatic ecosystems: roles in food web dynamics. Fungal Biol Rev 2008;22:17–25.

Gleason FH, Kupper FC, Amon JP, Picard K, Gachon CMM, Marano AV, et al. Zoosporic true fungi in marine ecosystems: a review. Mar Freshw Res 2011;62:383–393.

Gleason FH, Lilje O. Structure and function of fungal zoospores: ecological implications. Fungal Ecol 2009;2:53–59.

Gleason FH, Macarthur DJ. The chytrid epidemic revisited. Inoculum 2008;59:1–3.

Grami B, Rasconi S, Niquil N, Jobard M, Saint-Béat B, Sime-Ngando T. Functional effects of parasites on food web properties during the spring diatom bloom in Lake Pavin: A linear inverse modeling analysis. PLoS ONE 2011;6:1–14.

Gromov BV. *Aphelidium tribonemae* Scherffel parasitizing yellow green algae. Mikopologia Phytopathol 1972;6:443–445.

Gromov BV. Algal parasites of the genera *Aphelidium*, *Amoeboaphelidium* and *Pseudoaphelidium* from the Cienkovski's "Monadea" group as representatives of new class. Zool Zhurnal 2000;79:517–525.

Gromov BV, Mamkaeva KA. Sensitivity of different *Scenedesmus* strains to the endoparasitic microorganism *Amoeboaphelidium*. Phycologia 1968;7:19–23.

Gromov BV, Pljusch AV, Mamkaeva KA. Morphology and possible host range of *Rhizophydium algavorum* sp. nov. (Chytridiales) – an obligate parasite of algae. Protistology 1999;1:62–65.

Gromov BV, Mamkaeva KA, Pliusch AV, Voloshko LN, Mamkaeva MA. Estimation of the occurrence of certain algal parasites in Lake Ladoga and adjacent water bodies. Algological Stud 2002;107: 173–180.

Gutman J, Zarka A, Boussiba S. The host-range of *Paraphysoderma sedebokerensis*, a chytrid that infects *Haematococcus fluvialis*. Eur J Phycol 2009;44:509–514.

Hass H, Taylor TN, Remy W. Fungi from the Lower Devonian Rhynie chert: mycoparasitism. Am J Bot 1994;81:29–37.

Hassell MP, May RM. Stability in insect host-parasite models. J Anim Ecol 1973;42:693–726.

Held AA. *Rozella* and *Rozellopsis*: Naked endoparasitic fungi which dress-up as their hosts. Bot Rev 1981;47:451–515.

Hess S, Sausen N, Melkonian M. Shedding light on vampires: The phylogeny of Vampyrellid amoebae revisited. PLoS ONE 2012;7:e31165. doi:10.1371/journal.pone.0031165.

Holfeld H. Fungal infections of the phytoplankton: seasonality, minimal host density, and specificity in a mesotrophic lake. New Phytol 1998;138:507–517.

Holfeld H. Relative abundance, rate of increase, and fungal infections of freshwater phytoplankton. J Plankton Res 2000;22:987–995.

Ibelings BW, de Bruin A, Kagami M, Rijkeboer M, Brehm M, van Donk E. Host parasite interactions between freshwater phytoplankton and chytrid fungi (Chytridiomycota). J Phycol 2004;40:437–453.

Ibelings BW, Gsell AS, Mooij WM, van Donk E, van Wyngaert S, de Senerpont Domis LN. Chytrid infections and diatom spring blooms: paradoxical effects of climate warming on fungal epidemic in Lakes. Freshw Biol 2011;56:754–766.

Jaag O, Nipkow F. Neue and wenig bekannte parasitische pilze auf planktonorganismen Schweizerischer Gewisser I. Berichte Schweiz Bot Ges 1951;61:478–498.

James TY, Pelin A, Bonen L, Ahrendt S, Sain D, Corradi N, et al. Shared signatures of parasitism and phylogenomics unite cryptomycota and microsporidia. Curr Biol 2013;23:1–6.

Jobard M, Rasconi S, Sime-Ngando T. Diversity and functions of microscopic fungi: a missing component in pelagic food webs. Aquat Sci 2010;72:255–268.

Jobard M, Rasconi S, Solinhac L, Cauchie MH, Sime-Ngando T. Molecular and morphological diversity of fungi and the associated functions in three European nearby lakes. Env Microbiol 2012;14:2480–2494.

Jones MDM, Forn I, Gadelha C, Egan M, Bass D, Massana R, et al. Discovery of novel intermediate forms redefines the fungal tree of life. Nature 2011;474:200–203.

Kagami M, de Bruin A, Ibelings BW, van Donk E. Parasitic chytrids: their effects on phytoplankton communities and food-web dynamics. Hydrobiol 2007;578:113–129.

Kagami M, Helmsing NR, van Donk E. Parasitic chytrids could promote copepod survival by mediating material transfer from inedible diatoms. Hydrobiologia 2011;659:49–54.

Karling JS. Some Zoosporic Fungi of New Zealand. I. Sydowia 1965;19:213–226.

Karpov SA, Kobzeva AA, Mamkaeva MA, Mamkaeva KA, Mikhailov KV, Mirzaeva GS, et al. *Gromochytrium mamkaevi* gen. *et* sp. nov. and two new orders: *Gromochytriales* and *Mesochytriales* (*Chytridiomycetes*). Persoonia 2014b;32:115–126.

Karpov SA, Letcher PM, Mamkaeva MA, Mamkaeva KA. Phylogenetic position of the genus *Mesochytrium* (Chytridiomycota) based on zoospore ultrastructure and 18S and 28S rRNA gene sequences. Nova Hedwig 2010;90:81–94.

Karpov SA, Mamkaeva MA, Aleoshin VV, Nassonova E, Lilje O, Gleason FH. Morphology, phylogeny, and ecology of the aphelids (Aphelidea, Opisthokonta) and proposal for the new superphylum Opisthosporidia. Front Microbiol 2014a;5:112. Hyperlink http://dx.doi.org/10.3389/fmicb.2014.00112; doi: 10.3389/fmicb.2014.00112

Karpov SA, Mikhailov KV, Mirzaeva GS, Mirabdullaev IM, Mamkaeva KA, Titova NN, et al. Obligately phagotrophic aphelids turned out to branch with the earliest-diverging fungi. Protist 2013;164:195–205.

Kravchuk ES, Ivanova EA, Gladyshev MI. Spatial distribution of resting stages (akinetes) of the cyanobacteria *Anabaena flos-aquae* in sediments and its influence on pelagic populations. Mar Freshw Res 2011;62:450–461.

Krings M, Kerp H, Hass H, Taylor TN, Dotzler N. A filamentous cyanobacterium showing structured colonial growth from the Early Devonian Rhynie chert. Rev Palaeobot Palynol 2007;146:265–276.

Kühn SF, Hofmann M. Infection of *Coscinodiscus granii* by the parasitoid nanoflagellate *Pirsonia diadema*: III. Effects of turbulence on the incidence of infection. J Plankton Res 1999;21:2323–2340.

Lefèvre E, Roussel B, Amblard C, Sime-Ngando T.The molecular diversity of freshwater picoeukaryotes reveals high occurrence of putative parasitoids in the plankton. PLoS ONE 2008;3:e2324. doi:10.1371/journal.pone.0002324

Lepère C, Domaizon I, Debroas D. Unexpected importance of potential parasites in the composition of the freshwater small eukaryotes community. Appl Environ Microbiol 2008;74:2940–2949.

Letcher PM, Lopez S, Schmieder R, Lee PA, Behnke C, Powell MJ, et al. Characterization of *Amoeboaphelidium protococcarum*, an algal parasite new to the Cryptomycota isolated from an outdoor algal pond used for the production of biofuel. PLoS ONE 2013;6:e562342. doi: 10.1371/journal.pone.0056232.

Letcher PM, Powell MJ. A taxonomic summary and revision of *Rhizophydium* (Rhizophydiales, Chytridiomycota). University of Alabama, Tuscaloosa 2012:1–222.

Letcher PM, Powell MJ, Churchill PE, Chambers JG. Ultrastructural and molecular phylogenetic delineation of a new order, the Rhizophydiales (Chytridiomycota). Mycol Res 2006;110:898–915.

Livermore JA, Mattes TE, Phylogenetic detection of novel Cryptomycota in an Iowa (United States) aquifer and from previously collected marine and freshwater targeted high-throughput sequencing sets. Env Microbiol 2013;15:2333–2341.

Longcore JE, Simmons DR. The Polychytriales ord. nov. contains chitinophilic members of the rhizophlyctoid alliance. Mycologia 2012;104:276–194. doi: 10.3852/11-193.

Miki T, Takimoto G, Kagami M. Roles of parasitic fungi in aquatic food webs: a theoretical approach. Freshw Biol 2011;56:1173–1183.

Monchy S, Jobard M, Rasconi S, Gerphagnon M, Chabe M, Cian A, et al. Exploring and quantifying fungal diversity in freshwater lake ecosystems using rDNA cloning/sequencing and SSU tag. Env Microbiol 2011;13:1433–1453.

Mozley-Standridge SE, Letcher PM, Longcore JE, Porter D, Simmons DR. Cladochytriales – a new order in Chytridiomycota. Mycol Res 2009;113:498–507.

Paterson RA. On the planktonic chytrids *Zygorhizidium melosirae* Canter and *Z. planktonicum* Canter. Trans Br Mycol Soc 1958;41:457–460.

Paterson RA. Parasitic and saprophytic phycomycetes which invade planktonic organisms. II. A new species of *Dangeardia* with notes on other Lacustrine fungi. Mycologia 1958;50:453–468.

Paterson RA. Infestation of chytridiaceous fungi on phytoplankton in relation to certain environmental factors. Ecology 1960;41:416–424.

Preston LJ, Genge MJ. The Rhynie Chert, Scotland, and the search for life on Mars. Astrobiology 2010;10:549–560.

Raitschenko AA. Über eine Chytridiacee: *Rhizophidium sphaerocarpum* (Zopf) Fischer. Bulletin du Jardin impérial botanique de St.-Pétersbourg 1902;2:119–126.

Rasconi S, Jobard M, Jouve L, Sime-Ngando T. Use of calcofluor white for detection, identification, and quantification of phytoplanktonic fungal parasites. Appl Environ Microbiol 2009;75:2545–2553.

Rasconi S, Jobard M, Sime-Ngando T. Parasitic fungi of phytoplankton: ecological roles and implications for microbial food webs. Aquat Microbiol Ecol 2011;62:123–137.

Rasconi S, Niquil N, Sime-Ngando T. Phytoplankton chytridiomycosis: community structure and infectivity of fungal parasites in aquatic ecosystems. Environ Microbiol 2012;14:2151–2170.

Schopf JW, Walter MR. Origin and early evolution of cyanobacteria: the geological evidence. In: Carr NG, Witton BA, eds. The Biology of Cyanobacteria. Los Angeles, CA: University of California Press 1982:543–564.

Sekimoto S, Rochon DA, Long JE, Dee JM, Berbee ML. A multigene phylogeny of *Olpidium* and its implications for early fungal evolution. BMC Evol Biol 2011;11:331. doi:10.1186/1471-2148-11-331.

Seligo T. Tiere und pflanzen des seenplanktons. Stuttgart Franckhische Verlag 1909.

Sen B. Fungal parasitism of planktonic algae in Shearwater. IV. Parasitic occurrence of a new chytrid species on the blue-green alga *Microcystis aeruginosa* Kuetz. Emend. Elenkin. Archiv Hydrobio Suppl 1988:79:177–184.

Shearer CA, Descals E, Kohlmeyer B, Kohlmeyer J, Marvanová L, Padgett D, et al. Fungal biodiversity in aquatic habitats. Biodivers Conserv 2007;16:49–67.
Sigee DC, Selwyn A, Gallois P, Dean AP. Patterns of cell death in freshwater colonial cyanobacteria during the late summer bloom. Phycologia 2007;46:284–292.
Sime-Ngando T. Phytoplankton chytridiomycosis: fungal parasites of phytoplankton and their impacts on the food web dynamics. Front Microbiol 2012;14:2151–2170.
Sime-Ngando T, Lefèvre E, Gleason FH. Hidden diversity among aquatic heterotrophic flagellates:ecological potentials of zoosporic fungi. Hydrobiology 2011;659:5–22.
Simmons DR, James TY, Meyer AF, Longcore JE. Lobulomycetales, a new order in the Chytridiomycota. Mycol Res 2009;113:450–460.
Skuja H. Taxonomie des phytoplanktons einiger seen in Uppland, Schweden. Uppland: Lundequististaka bokhandeln 1948.
Skuja H. Taxonomische und biologische studien über das phytoplankton Schwedischer binnengewässer. Nova Acta Reg Soc Sci Sci Ups Ser 4 1956;16:1–404.
Sønstebø JH, Rohrlack T. Possible implications of chytrid parasitism for population subdivision in freshwater cyanobacteria of the genus *Planktothrix*. Appl Env Microb 2011;77:1344–1351.
Sparrow Jr FK. Aquatic Phycomycetes, 2nd ed. Ann Arbor, MI: The University of Michigan Press 1960.
Sparrow FK, Paterson RA, Johns RM. Additions to the phycomycete flora of the Douglas Lake Region. V. New or interesting fungi. Papers of the Michigan Academy of Science, Arts and Letters 1965;50:115–123.
Sparrow Jr FK. Ecology of freshwater fungi. In: Ainsworth GC, Sussman AS, eds. The Fungal Population. The fungi, An Advanced Treatise. Vol. 3. New York (NY): Academic Press 1968:41–93.
Stewardson M, Webb A. Modelling ecological responses to flow alteration: making the most of existing data and knowledge. In: Saintilan N, Overton I, eds. Ecosystem response modeling in the Murray-Darling Basin. Victoria: CSIRO Publishing 2010:37–49.
Szklarczyk C. Phytoplankton of the dam reservoir at Kzlowa Gora in the years 1951–1953. Acta Soc Bot Pol 1956;25:538–578.
Takano K, Ishikawa Y, Mikami H, Igarashi S, Hino S, Yoshioka T. Fungal infection for cyanobacterium *Anabaena smithii* by two chytrids in eutrophic region of large reservoir Lake Shumarinai, Hokkaido, Japan. Limnology 2008;9:213–218.
Taylor TN, Hass H, Remy W. Devonian fungi: Interactions with the green alga *Palaeonitella*. Mycologia 1992a;84:901–910
Taylor TN, Klavins SD, Krings M, Taylor EL, Kerp H, Hass H. Fungi from the Rhynie chert: a view from the dark side. Trans R Soc Edinb: Earth Sci 2004;94:457–473.
Taylor TN, Osborn JM. The importance of fungi in shaping the paleoecosystem. Rev Palaeobot Palynol 1996;90:249–262.
Taylor TN, Remy W, Hass H. Fungi from the Lower Devonian Rhynie Chert: Chytridiomycetes. Am J Bot 1992b;79:1233–1241.
Taylor TN, Remy W, Hass H. *Allomyces* in the Devonian. Nature 1994;367:601–601.
Tsujimura S, Tsukada H, Nakahara H, Nakajima T, Nishino M. Seasonal variations of *Microcystis* populations in sediments of Lake Biwa, Japan. Hydrobiology 2000;434:183–192.
Van Donk E. The role of fungal parasites in phytoplankton succession. In: Brock TD, ed. Plankton Ecology. Berlin: Springer 1989:171–194.
Van Wichelen J, Van Gremberghe I, Vanormelingen P, Debeer AE, Leporcq B, Menzel D, et al. Strong effects of amoebae grazing on the biomass and genetic structure of a *Microcystis* bloom (Cyanobacteria). Environ Microbiol 2010;12:2797–2813.
Voigt K, Marano AV, Gleason FH. Ecological and economical importance of parasitic zoosporic true fungi. In: Kempken F, ed. The Mycota Vol XI, Agricultural Applications, 2nd ed. Berlin: Springer Verlag 2013:243–270.

Sally L. Glockling, Agostina V. Marano, Osu Lilje
and Frank H. Gleason

14 Zoosporic parasites of freshwater invertebrates

In freshwater ecosystems, algal tissues and cells of unicellular phytoplankton (producers) are broken down into particles by small grazing and filter feeding invertebrates. Some algae may also be parasitized by fungi and fungal-like organisms which assimilate their nutrients and produce zoospores. Small freshwater invertebrates (consumers), many of which comprise the zooplankton, feed on the particulate matter from the algae and phytoplankton and also consume the zoospores produced by their parasites (Sime-Ngando 2013; Rasconi et al. 2014). Freshwater zooplankton comprises of a wide diversity of microscopic invertebrates from many different taxonomic groups. Larger invertebrates feed on both the phytoplankton and the zooplankton. This results in complex food webs that can be assessed to some extent using various analytical models (Rasconi et al. 2014). It has been calculated that chytrid algal parasites channel nearly 20% of primary production (Rasconi et al. 2014).

Many freshwater invertebrates are parasitized by zoosporic fungi and fungal-like organisms, including members of the Blastocladiomycota, Chytridiomycota, Oomycota and Mesomycetozoea. This chapter will describe the parasites involved at these higher trophic levels in both a taxonomic context and in relation to the host. Some parasites of freshwater invertebrates are host-dependent or obligate parasites, whilst others are facultative parasites with the ability to live saprotrophically outside the host. Some are host-specific, infecting a single host or only a few related host taxa. Others are generalist parasites, infecting several species, often from different trophic levels and different phyla. Parasitoids kill the host and holocarpic parasites fill the host body with the developing sporangia. The sporangial content cleaves and zoospores are released into the water. Other species may live symbiotically, usually in the host gut, absorbing nutrients without apparently causing harm to their host.

Pathogens and parasites may cause disease and substantial losses in economically important species, such as those that are commercially exploited for food. Crayfish plague, caused by the oomycete, *Aphanomyces astaci*, threatens to destroy populations of European crayfish, and has devastated the crayfish industry in many areas (Alderman 1986, 1996). Other important parasites infect and kill vectors of disease. For example, the oomycete *Lagenidium giganteum*, can cause high mortality in larvae of mosquito species, some of which are vectors of human malaria (Kerwin 2007). Parasites like *L. giganteum*, which kill organisms that are harmful to human health, may be effective as biological control agents (Federici 1981; Scholte et al. 2004; Kerwin 2007).

The majority of zoosporic parasites infecting freshwater invertebrates have no commercial importance and, with a few exceptions, are not well- known or extensively studied. Recently a few long-term ecological studies have been carried out, which have shown that zoosporic parasites of invertebrates are important components of freshwater ecosystems, often regulating host population sizes (Wolinska et al. 2007; Johnson et al. 2009). The infective propagule is usually the zoospore, which requires a watery habitat to navigate environmental gradients and actively locate a suitable host, sometimes by chemotaxis (Cerenius and Söderhäll 1984a; Deacon and Saxena 1997). The zoospore is usually uniflagellate in the zoosporic true fungi (formerly known as chytrids) or biflagellate (heterokont and anisokont) in the Oomycota (fungal-like organisms).

Invertebrate hosts infected by zoosporic parasites in freshwater ecosystems primarily include crustaceans, insect larvae and rotifers. Free-living nematodes are associated with a large number of fungal and fungal-like predators and parasites, but these susceptible bactivorous nematodes are more common in wet terrestrial habitats such as soil, compost and dung, than in freshwater ecosystems (Drechsler 1940; Barron 1976; Glockling 1994; Glockling and Beakes 2000). Plant-parasitic cyst nematodes are also host to some zoosporic true fungi and oomycete parasites (Tribe 1977; Kerry 1980; Kerry and Crump 1980). A few genera of zoosporic parasites of nematodes also include species that are host specific for loricate rotifers. Some loricate rotifers are found in both wet terrestrial and aquatic environments, whilst other species are planktonic. Rotifers and nematodes comprise the Aschelminths, a cluster of roundworm phyla. Many aquatic invertebrate hosts for zoosporic parasites are common species inhabiting the shallow, plant rich waters around the margins of ponds and lakes and in narrow water courses, whilst others are part of the zooplankton in lakes and rivers.

14.1 Parasites in the Blastocladiomycota and Chytridiomycota

Zoosporic true fungi are known to be an early evolved lineage of true fungi which mostly release zoospores with a single posterior whiplash flagellum (James et al. 2006). A separate lineage of uniflagellate parasites, comprising the environmental sequences of the Cryptomycota, together with the parasitic genus *Rozella* and three genera of aphelids, is considered to be either basal to the Chytridiomycota or a separate lineage at the animal-fungi divergence (James et al. 2006; Jones et al. 2011; Gleason et al. 2012; James and Berbee 2012; Lara et al. 2012; Karpov et al. 2013; Letcher et al. 2013). *Rozella* species are not known to infect freshwater invertebrates but most environmental sequences for the Cryptomycota have originated from freshwater environments (James and Berbee 2012).

Within the zoosporic true fungi there are three phyla: Blastocladiomycota, Chytridiomycota and Neocallimastigomycota, and also the *Olpidium* clade (James et al. 2006). Most of the zoosporic true fungal parasites of freshwater invertebrates belong to the Blastocladiomycota (order Blastocladiales). Gleason et al. (2010) have listed six blastocladialean genera with invertebrate hosts that include mosquitoes and other dipteran larvae, rotifers, nematodes, crustaceans and tardigrades. Species of zoosporic true fungi which infect Aschelminthes (rotifers and nematodes), insects and protists have been reviewed by Dick (2003), along with oomycete parasites.

The well-known chytrid parasite lethal to frogs, *Bactrachochytrium dendrobatatis*, also known as Bd, has been found in the form of zoosporangia in the gastrointestinal tracts of crayfish (McMahon et al. 2013).

The life cycle in zoosporic true fungal parasites can be very complex, with an alternation of diploid and haploid generations requiring two host taxa in genera such as *Coelomomyces* (Federici 1981). Other holocarpic zoosporic parasites in the Chytridiomycota have a much simpler life cycle and are associated with a single host.

The genus *Coelomomyces* contains many parasites of mosquito larvae (Couch and Bland 1985) and *Coelomycidium* is a common parasite of blackflies (McCreadie and Alder 1999). *Catenaria* species (Fig. 14.1a–c) are known to infect midge eggs (Martin 1975, 1978), bactivorous nematodes (Sorokine 1876) and cyst nematodes (Tribe 1977). *Phlyctochytrium* (Fig. 14.2a–b) parasitizes adult nematodes (Karling 1946a) and *Rhizophydium* species parasitize nematodes, adult rotifers and their eggs (Karling 1946b; Dick 2003). Two species of *Endochytrium* have been recorded from protists and from eggs of rotifers and nematodes (Karling 1937). *Callimastix* species have been isolated from various copepods (Vavra and Joyon 1966) and *Polycaryum* is a common parasite of the water flea, *Daphnia* (Johnson et al. 2006). *Sorochytrium* species are parasites of tardigrades, commonly called water bears. *Pseudosphaerita* and *Sphaerita* spp. have been described as parasites of protozoans such as amoebae, euglenids and cryptomonads. However, descriptions of these parasites are mostly from the early 1900s, with no subsequent reports, and their taxonomic placement is considered dubious (Dick 2003).

The genus *Olpidium* (Figs. 14.3, 14.4, 14.5a–c), which in addition to plant parasites, contains parasites of nematodes, protists and adult rotifers and their eggs, was previously considered to be a zoosporic true fungus as it produces uniflagellate zoospores. However, multigene phylogenies have shown that *Olpidium* species cluster with the zygomycetes (James et al. 2006; Sekimoto et al. 2011). *Olpidium nematodae* was recorded from adult nematodes (Skvortzow 1927), and *O. vermicola* (Fig. 14.5a–c) is a common parasite of rhabditid nematode eggs (Barron and Szijarto 1986, Glockling 1994). Some *Olpidium* species, such as *O. rotiferum*, *O. gregarium*, *O. granulatum* and *O. paradoxum* infect adults or eggs of loricate rotifers (Karling 1946; Glockling 1998; Dick 2003).

14.2 Parasites in the Oomycota

Recent phylogenetic studies have shown that morphologically simple, holocarpic parasites of invertebrates and algae evolved early (Beakes et al. 2006; Beakes and Sekimoto 2009; Lara and Belbahri 2011). Studies from both freshwater and marine environments have shown that the genus *Haptoglossa*, which infects free-living nematodes and bdelloid rotifers, occupies one of the most basal positions in phylogenetic trees along with the marine algal parasite *Eurychasma* (Beakes et al. 2006; Beakes and Sekimoto 2009). Several other genera of invertebrate parasites are considered to have evolved before the oomycetes branched into the main peronosporalean

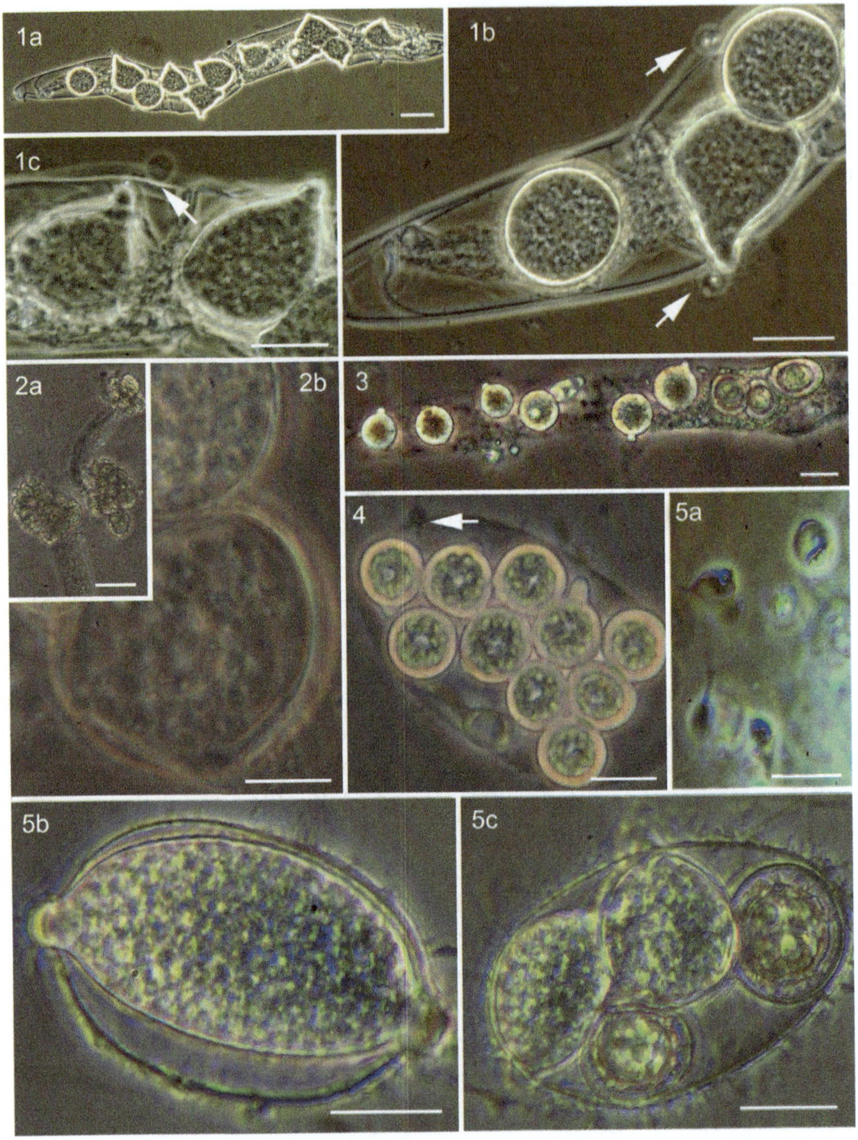

Figs. 14.1–14.5: Fig. 14.1a–c *Catenaria anguillulae*. a) General morphology of the thallus inside the body of an unidentified nematode. Scale = 10 µm. b) Sporangia inside the head of nematode, showing two encysted zoospores (arrows). Scale = 10 µm. c) Detail of the sporangia inside the nematode, and encysted zoospore with a germ tube on the nematode cuticle (arrow). Scale = 10 µm. **Fig. 14.2a–b** *Phlyctochytrium nematodae*. a) Nematode with several bundles of external sporangia. Scale = 20 µm. Scale = 10 µm. b) Sporangia containing zoospores. Scale = 10 µm. **Fig. 14.3** *Olpidium* sp. in adult nematode body with sporangia and resting spores. Scale = 10 µm. **Fig. 14.4** *Olpidium* sp in nematode egg with thick-walled resting spores, and encysted zoospore penetrating the egg (arrow). Scale = 10 µm. **Fig. 14.5a–c** *Olpidium vermicola*. a) Zoospores. Scale = 10 µm. b) Nematode egg containing single sporangium with two exit papillae. Scale = 10 µm. c) Nematode egg containing two sporangia and two resting spores. Scale = 10 µm.

and saprolegnialean lineages. These include the nematode and rotifer-infecting genus *Chlamydomyzium* (Glockling and Dick 1997; Beakes et al. 2006; Glockling and Beakes 2006; Beakes et al. 2014b) and the marine crustacean parasite, *Atkinsiella. Crypticola*, a genus that contains a mosquito parasite, has recently been placed in the Atkinsiellales (Beakes et al. 2014a). Oomycete parasites of invertebrates also fall in basal positions in both the saprolegnialean and peronosporalean lineages and are considered to be imperative to understanding oomycete evolution (Beakes and Sekimoto 2009; Beakes et al. 2012).

Oomycetes reproduce asexually and the heterokont, principal-form zoospore is usually the infective agent in parasitic species, although some species produce adhesive aplanospores with a passive mode of infection (Glockling and Beakes 2000). Most Oomycetes also reproduce sexually but in some of the basal parasites, only asexual reproduction is known.

Whilst genera such as *Saprolegnia* (Figs. 14.9a–b, 14.10a–b) are predominantly found in freshwater as saprophytes or as parasites of fish, other genera such as *Aphanomyces* (Figs. 14.6a–b, 14.8a) are more diverse, comprising of saprophytes and parasites of both animals and plants, and occur in a wide range of environmental habitats. *Aphanomyces* was shown to be the earliest evolved genus in the Saprolegniales (Peterson and Rosendhal 2000; Diéguez-Uribeondo et al. 2009).

Nematode and rotifer infecting parasites include holocarpic genera such as *Myzocytiopsis* (Figs. 14.7a–b, 14.13) and *Chlamydomyzium*. Unlike *Haptoglossa*, in which some species infect wet terrestrial bdelloid rotifers (Barron 1990), rotifer-attacking *Myzocytiopsis* (Fig. 14.7a–b) and *Chlamydomyzium* mainly infect aquatic loricate rotifers. Although both genera were previously considered to be closely related (Dick 1997), it has been shown that *Myzocytiopsis* is in the Peronosporales, close to *Lagenidium* and *Pythium* (Beakes et al. 2006; Beakes and Sekimoto 2009), whilst *Chlamydomyzium* is an early evolved genus, probably diverging before the split into the two main oomycete lineages, and has been placed in the Leptomitales (Beakes et al. 2014a, b). The core *Myzocytiopsis* species form a separate clade close to *Lagenidium* in both LSU and SSU trees (Lévesque, Spies and Glockling, unpublished). Sequencing of 18S rRNA of the planktonic rotifer parasites in *Aquastella* indicated that these species are most closely related to *Aphanomyces* (Molloy et al. 2014). Another rotifer parasite, *Sommerstorffia spinosa*, appears to mainly infect loricate rotifers in rivers and ponds (Karling 1952).

Myzocytiopsis and *Chlamydomyzium* species appear to have a restricted host range, often infecting a single host species or a few closely related taxa. Some species may infect both nematodes and rotifers. *Chlamydomyzium sphaericum* was reported to occasionally infect tardigrades in addition to nematodes (Glockling 1994; Dick 2001). *Myzocytiopsis parthenospora* mainly infects *Lecane* and *Philodina* rotifers and their eggs, but also reportedly infects the adult cyst nematode, *Heterodera*, and eggs of the gastrotrich protist, *Chaetonotus larus* (Dick 2003). Some species in other oomycete genera, such as *Saprolegnia, Aphanomyces* and *Lagenidium*, have been shown to infect a wide diversity of distantly related host organisms of both vertebrates and invertebrates (Wolinska et al. 2009).

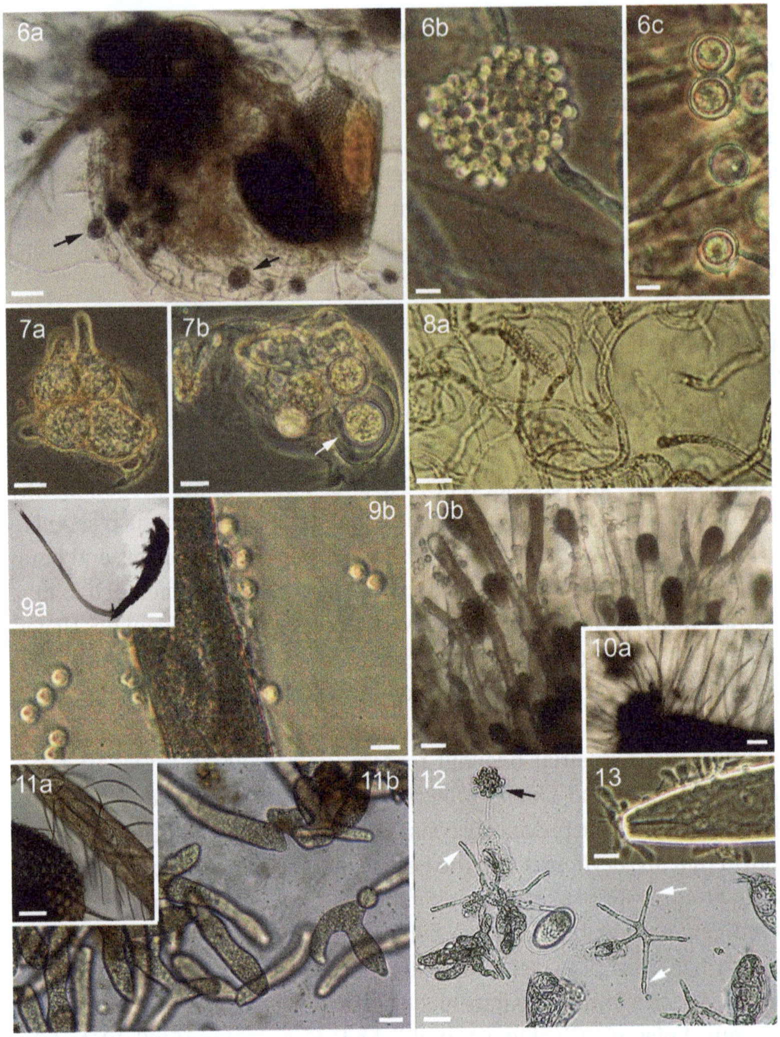

Figs. 14.6–14.13: Fig. 14.6a–c *Aphanomyces daphniae*. a) *Daphnia* with sporulating infection, showing bundles of encysted spores (arrows). Scale = 50 μm. b) Encysted spores at the tip of the sporangium. Scale = 20 μm. c) Oospores in pure culture. Scale = 20 μm. **Fig. 14.7a–b** *Myzocytiopsis microspora*. a) Body of aquatic rotifer containing several sporangia with short exit tubes. b) Rotifer body containing smooth-walled oospores inside the oogonial wall (arrow). Scales = 10 μm. **Fig. 14.8** *Aphanomyces astaci*. Hyphae growing inside the cuticle of the noble crayfish *Astacus astacus*. Scale = 20 μm. **Fig. 14.9a–b** Saprolegniaceous species in caddisfly larva (Tricoptera). a) Single large, broad sporangium growing from abdomen of larva. Scale = 50 μm. b) Spore cysts around the sporangium. Scale = 20 μm. **Fig. 14.10a–b** *Saprolegnia ferax*. a) Sporangia growing out from body of a chironomid midge. Scale = 40 μm. b) Mature sporangia and cysts. Scale = 40 μm. **Fig. 14.11a–b** *Lagenidium giganteum*. a) Thalli inside the leg of a mosquito. Scale = 20 μm. b) Excised thalli from the mosquito body. Scale = 10 μm. **Fig. 14.12** *Sommerstorffia spinosa* showing predacious structures on which rotifers can be captured (white arrows). A bundle of discharged spores have encysted at the tip of an exit tube (black arrow). Scale = 20 μm. **Fig. 14.13** *Myzocytiopsis humicola*. Head of a rhabditid nematode with several adhered aplanospores. Scale = 5 μm.

The holocarpic genus *Lagenidium* used to contain many species parasitic to a wide range of hosts including animals and algae (Karling 1944; Sparrow 1960). Some of the nematophagous species were misplaced in *Lagenidium* and transferred to *Myzocytiopsis* or *Chlamydomyzium* (Dick 2001). Parasites of other animals and of algae were placed in other genera while the mosquito parasite, *L. giganteum* (Fig. 14.9a–b), remained as the sole representative of *Lagenidium* (Dick 2001). However, the separation of the former species of *Lagenidium* based on morphology and host type has not been universally accepted (Kiziewicz 2004; Beakes and Sekimoto 2009). Recent systematics of Straminipila has discontinued the use of the order Lagenidiales and the genus *Lagenidium* is placed in the Peronosporales (Beakes et al. 2013, 2014a). The large and phylogenetically complex genus *Pythium* is mainly comprised of plant parasites, but some species have been reported to parasitize freshwater invertebrates such as mosquito larvae (Saunders et al. 1988; Su et al. 2001) and planktonic rotifers (Thomas et al. 2011). Also in the Pythiales, *Endosphaerium funiculatum* was described as a symbiont in the gills of the freshwater bivalve mollusc (D'Eliscu 1977).

14.3 Parasites in the Mesomycetozoea

The Mesomycetozoea are a group of fungal-like protists, which are considered to have diverged early in the animal lineage, near to the animal-fungi divergence (Marshall et al. 2008; Lohr et al. 2010). This group has been reviewed by Mendoza et al. (2002) and more recently by Glockling et al. (2013) in the light of new discoveries and phylogenetic studies. The Mesomycetozoea are divided into two orders; the Dermocystida and the Ichthyophonida, which may produce uniflagellate zoospores or amoebae respectively (Mendoza et al. 2002). Species in both orders produce walled endospores. Parasites of freshwater invertebrates are only known in the order Ichthyophonida. The Mesomycetozoeans are osmotrophs and some are parasitic or symbiotic in marine and freshwater invertebrates. The group also includes parasites of fish, amphibians and mammals (Mendoza et al. 2002; Glockling et al. 2012, 2013). Several species infect freshwater crustaceans such as crayfish, freshwater shrimps (*Gammarus*) and water fleas (*Daphnia*). Other species infect insect larvae (Mendoza et al. 2002; Kim and Alder 2005; Marshall et al. 2008; Glockling et al. 2013). Members of the Amoebidiales are all symbionts in freshwater invertebrates such as small crustaceans and aquatic stages of insects. They were formerly considered to be trichomycete fungi (Zygomycotina) (Lichtwardt 1986), but were later found to belong to the Mesomycetozoea following phylogenetic studies (Benny and O'Donnel 2000). Similarly, the Eccrinales, which are mainly symbionts of marine and terrestrial invertebrates, were also formerly classified as trichomycetes, but were found to belong to the Mesomycetozoea (Cafaro 2005). Only a few freshwater invertebrates (such as the freshwater shrimp *Gammarus*, which is host to *Astreptonema gammari*) are host to eccrinid species, although these symbionts have a broad range of invertebrate hosts in terrestrial and marine environments. Eccrinids inhabit the hindgut of crustaceans, diplopods, copepods, isopods and insects. Although some members of the Mesomycetozoea have been cultivated (Marshall et al. 2008), including some members of the Amoebidiales (Whistler 1962), the Eccrinales and many other species are considered unculturable (Cafaro 2005). It is not generally known whether mesomycetozoean symbionts have any beneficial or detrimental effect on the host although some mutualistic species such as the anaerobic fungi in the rumen gut have long been known to assist in providing nutrients to the herbivore. Some mesomycetozoeans, especially those which infect fish, may be pathogenic and lethal (Mendoza et al. 2002).

14.4 Parasites of crayfish

These include the well-documented oomycete, *Aphanomyces astaci* (Fig. 14.8), which causes crayfish plague (Oidtmann 2012) and the Mesomycetozoean, *Psorospermium haekeli* (Mendoza et al. 2002). Söderhäll et al. (1989) isolated the oomycete, *Saprolegnia parasitica*, normally a parasite of fish, from the narrow-clawed crayfish *Astacus leptodactylus*. In addition, the oomycete *Aphanomyces frigidophilus* is able to infect crayfish (Ballesteros et al. 2006). Field sampling of crayfish species, *Procambarus alleni* and *P. clarkii*, in America revealed that up to 30% of crayfish had zoosporangia of the frog parasite, *Batrachochytrium dendrobatidis*, embedded in their gastrointestinal tracts (McMahon et al. 2013). This study concluded that crayfish may be a reservoir host for *B. dendrobatidis*, and showed that infected crayfish could transmit the infection to amphibians under laboratory conditions (McMahon et al. 2013).

14.4.1 Crayfish plague

Probably the most infamous pathogenic oomycete in freshwater invertebrates is *Aphanomyces astaci*, the causal agent of "krebspest" or crayfish plague, which has dramatically reduced European and Asian crayfish populations (Alderman 1996; Egerton et al. 2004). As an economically important parasite, *A. astaci* has been extensively studied (Alderman 1996, 1990; Söderhäll and Cerenius 1999; Oidtmann et al. 2006) and has recently been comprehensively reviewed (Oidtmann 2012).

Four groups of strains of *A. astaci* have been identified by PCR (Huang et al. 1994). One group is thought to have been in Europe for a long time without having any serious impact on indigenous crayfish, but the introduction of North American crayfish into European waters has caused massive losses of European crayfish which were highly susceptible to disease from virulent strains of *A. astaci* (Alderman 1986, 1990; Dick 2003). Disease in crayfish has been reported since the early 1900s (Shikora 1906). The North American signal crayfish (*Pacifastacus leniusculus*) is thought to have spread crayfish plague to European species via river systems throughout Europe and into Russia, Scandinavia and Asia Minor (Cerenius et al. 1988; Alderman et al. 1990; Dick 2003; Egerton et al. 2004).

North American signal crayfish and Louisiana swamp crayfish (*Procambarus clarkia*) are carriers of the disease and, except for superficial lesions on the exoskeleton, these hosts do not usually show any clinical signs of disease and infection is not fatal (Oidtmann et al. 2006). The North American crayfish responds to hyphal infections of *A. astaci* from encysted zoospores by quickly producing melanin, which arrests the growth of the pathogen. Native European crayfish species, however, such as the Noble crayfish (*Astacus astacus*) and the white-clawed crayfish (*Austropotamobius pallipes*), do not have a strong melanin immune response (Unestam and Weiss 1970; Söderhäll and Cerenius 1992). This renders them very susceptible to crayfish plague, and the disease spreads into the haemocoel, causing high rates of mortality very quickly. These susceptible hosts are infected by biflagellate zoospores of *A. astaci* that are attracted to the host cuticle by chemotaxis (Cerenius and Söderhäll 1984a). The zoospores encyst on the cuticle, often infecting the soft cuticle of the ventral abdomen (Fig. 14.8) and areas around the joints. Following penetration, hyphae grow into the body where the connective tissue and haemal sinuses, as well as other tissues, can become infected (Oidtmann 2012). The pathogen forms sporangia which release amoeboid primary zoospores that encyst at the tip of the exit tube before the infective secondary or principal-form zoospores emerge. The infective period in *A. astaci* is considered to be extended by repeated encystment and emergence of principal-form zoospores (Cerenius and Söderhäll 1984b), a trait that is considered to be typical of a parasitic lifestyle (Beakes and Sekimoto 2009).

14.4.2 *Psorospermium haekeli*

Another parasite of freshwater crayfish is the mesomycetozoean, *Psorospermium haekeli*, which has been known to cause disease in crayfish since the mid 1800s. This species belongs to the order Ichthyophonida and produces amoebae as infective propagules. It mainly affects the connective tissues around the gut and is thought to have several morphotypes (Mendoza et al. 2002). *Psorospermium haekeli* has been classified incorrectly in the past with several groups of organisms including other protist groups, algae and fungi.

14.5 Parasites of mosquitoes, blackflies and midges

Parasites of aquatic larval stages of mosquitoes, blackflies and midges can be found in the Blastocladiomycota, the Oomycota and the Mesomycetozoea. Within the zoosporic true fungi, these dipteran parasites occur in the order Blastocladiales (Gleason et al. 2010). Larval blackflies are considered to be model organisms for studying symbiotic associations because they are hosts to several symbionts from a range of phyla and they are also ubiquitous in freshwater streams and rivers (Kim and Alder 2005). The blastocladialean genus, *Coelomomyces* contains several species which infect mosquito larvae and midge eggs (Couch and Bland 1985; Dick 2003). Species of *Catenaria* have been described infecting the eggs of chironomid midges, including species of *Chironomus, Endochironomus, Glyptotendipes,* and *Dicrottendipes* (Martin 1975, 1978, 1981b, 1991; Dick 2003). In addition, a few species in the genus *Coelomycidium* infect the larvae of blackflies (McCreadie and Adler 1999). Oomycetes are also known to infect mosquito larvae and sometimes adults. These include the peronosporalean species, *Lagenidium giganteum* (Fig. 14.11a–b), the saprolegnialean species, *Leptolegnia caudata* and *L. chapmanii* and *Crypticola clavulifera* in the Atkinsiellales. Laboratory cultures of mosquito larvae are also reported to be susceptible to infection by *Aphanomyces* species (Seymour and Briggs 1985). Zoosporic parasites that are specific to mosquitoes have been reviewed by Scholte et al. (2004) in relation to their effectiveness as potential biocontrol organisms. Mosquitoes can transmit several diseases to humans including malaria, dengue fever and filariasis, and mosquitoes are developing resistance to chemical forms of control.

Scholte et al. (2004) listed 27 zoosporic parasites of mosquitoes, together with their host species from both laboratory and field investigations. Many of the species which parasitize mosquito larvae are also able to infect midge larvae. *Couchia complexa, C. amphora* and *C. limnophora* in the Saprolegniales (Martin 1981a, 2000), *Crypticola entomophaga* (Martin 1977, Dick 1998) and *Aphanomycopsis sexualis* in the Leptomitales (Martin 1975) are parasites of midge eggs. *Saprolegnia ferax* has been observed to infect and kill larvae of the chironomids (Fig. 14.10a–b) (Glockling, Spies and Lévesque, unpublished).

Blackfly larvae are associated with several mesomycetozoean species, such as *Amoebidium* and *Paramoebidium* that live as symbionts in the gut. They do not produce zoospores but have amoebae as their motile stage.

14.5.1 *Coelomomyces*

Dick (2003) listed a total of 64 *Coelomomyces* species which are all host dependent in aquatic stages of dipteran larvae, including six genera of mosquitoes. Scholte et al. (2004) listed 16 species of *Coelomomyces* that have been recorded from mosquito larvae. An overview of *Coelomomyces* and their effectiveness in controlling mosquito populations is given in Couch and Bland (1985). This genus has a complicated life cycle requiring two mosquito generations and also an intermediate copepod (Crustacea) host, which may be a cyclops or an ostracod (Federici 1981; Whistler et al. 1999). One species, *Coelomomyces indicus* has been studied more than most because it infects 16 species of mosquito including the malaria carrying species *Anopheles gambiae*. Mortality rates in mosquitoes infected with *Coelomomyces* in laboratory studies can be very high and it can persist and cause epizootic infections periodically for several years (Whistler et al. 1999). Field trials have been conducted to study the effect of *Coelomomyces* on populations of *Anopheles quadrimacutus* (Umphlett 1969). However, field trials generally render lower infection rates than laboratory tests and infection decreases over time, possibly due to limited copepod populations. The obligatory parasitic nature of *Coelomomyces* together with its complex life cycle makes it impractical for use as a biological control against mosquitoes. *Coelomomyces* species have never been successfully grown on artificial media.

14.5.2 *Lagenidium giganteum*

Lagenidium giganteum (Fig. 14.11a–b) is probably the best known oomycete parasite of mosquito larvae and has been extensively studied (Kerwin and Washino 1986; Kerwin and Peterson 1997; Kerwin 2007). Couch (1935) initially isolated *L. giganteum* from a collection of mosquito larvae and copepods, but it has been isolated repeatedly since then from mosquito larvae across several continents (Scholte et al. 2004). As well as infecting mosquito larvae, *L. giganteum* has been reported to infect *Daphnia* and other cladocerans, copepods, chironomids and midge larvae (Couch 1935; Frances et al. 1989; Nestrud and Anderson 1994; Wright and Easton 1996). It has also been isolated from nematodes (Glockling and James, unpublished), but most freshwater aquatic invertebrates are not susceptible to infection (Nestrud and Anderson 1994). Genera of mosquito larvae that may be killed by *L. giganteum* include *Culex, Ochlerotatus, Anopheles, Aedes,* and *Mansonia.* Many laboratory experiments have been performed to ascertain percentage mortality in mosquito species subjected to *L. giganteum* and, under optimal controlled laboratory conditions, 100% mortality can be achieved against *Culex* and *Aedes* larvae (McCray et al. 1973). *Lagenidium giganteum* is a freshwater facultative parasite as it can grow both saprotrophically in water and pathogenically in association with the host (Scholte et al. 2004). Infection by *L. giganteum* is via the biflagellate zoospores which occur near the surface of the water and infect mosquito larvae by encysting on the cuticle. Mosquitoes have an immune melanization response to *L. giganteum*, which is strong in some species, making them less susceptible to infection and death (Golkar et al. 1993). As zoospores are required for infection, a zoospore suspension has sometimes been used to test effectiveness against mosquito species. In the laboratory, *L. giganteum* can be grown on either solid or liquid media for asexual reproduction, but spore production has been shown to require sterols (Domnas et al. 1977). The use of vegetative hyphae or zoospore suspensions for biological control is problematic as they are not very durable and zoospores do not remain viable for long. In contrast, sexually produced oospores are much more resistant and have much greater longevity than zoospores, but cultivation of large quantities of oospores is difficult and germination is not easily achievable. In depth studies to elucidate the chemical regulation of oosporogenesis in *L. giganteum* have found that the sexual stages in the life cycle from antheridial formation through to oospore maturation are controlled by calcium-dependent events (Kerwin and Washino 1986).

Pilot studies have been made to test the efficiency of *L. giganteum* in rice fields and in natural freshwater environments (Washino and Laird 1983; Kerwin et al. 1994).

This species is currently the only zoosporic fungal-like organism registered with USEPA as a biological control for mosquitoes. It is able to persist in the environment and to infect multiple generations of mosquitoes.

14.5.3 *Pythium*

Some *Pythium* species have been reported to parasitize mosquito larvae. *Pythium flevoense* was found to infect larvae of the tree hole mosquito, *Aedes* (*Ochlerotatus*) *sierrensis* in natural (Saunders et al. 1988) and laboratory conditions (Washburn et al. 1988), while *P. carolinianum* was found to infect larvae of *Culex* and *Aedes* mosquitoes (Su et al. 2001) and *P. sierrensis* infects seven species of mosquito larvae (Scholte et al. 2004). Other *Pythium* species found to infect mosquitoes have not been formally described or identified with known species. *Pythium* species are not considered suitable for biological control as conditions for sporulation are difficult to achieve and mortality rates in laboratory tests have been very variable (Scholte et al. 2004).

14.5.4 *Leptolegnia*

In the order Saprolegniales two species of *Leptolegnia, L. caudata* and *L. chapmanii,* infect mosquitoes. *Leptolegnia* species grow readily on artificial media and *L. chapmanii* appears to be host specific for mosquito larvae (McInnis and Schimmel 1985; López Lastra et al. 1999). Bisht et al. (1996) isolated *L. caudata* from the larva of the malaria carrying mosquito, *Anopheles culicifacies,* and the parasite produced 100% mortality in these larvae in laboratory tests. *Leptolegnia chapmanii* was isolated from the mosquito *Ochlerotatus triseriatus* (Seymour 1984), but subsequent laboratory tests found it killed 100% of *Aedes egyptii* larvae within 24 h of exposure, and was also effective against *Culex* and *Anopheles* species (McInnis and Zattau 1982; Nnakumusana 1986). However, *Leptolegnia* spp. are not considered suitable for biological control as infectivity decreases with prolonged culture and zoospores are not produced in saline conditions (Scholte et al. 2004).

14.5.5 *Crypticola*

Crypticola clavulifera, now considered to be in the Atkinsiellales (Beakes et al. 2014a), was isolated from the midge *Forcipomyia marksae* (Frances et al. 1989), but has also been shown to be effective against larvae of several species of mosquito and the chironomid *Chironomus tepperi* in laboratory tests (Frances et al. 1991). Another species, *C. entomophaga,* is parasitic in dipteran eggs (Martin 1977; Dick 1998).

14.5.6 *Amoebidium* and *Paramoebidium*

Parasites of blackfly larvae include the Mesomycetozoean genera *Amoebidium* and *Paramobidium*, which were formerly included in the trichomycete fungi (Lichtwardt 1986; Cafaro 2005). These two genera make up the order Amoebidales. Whilst *Amoebidium* species attach to the exoskeleton of the host, *Paramoebidium* species attach to the lining of the hindgut (Cafaro 2005). *Amoebidium parasiticum*, which has been isolated from blackfly larvae including *Simulium vandalicum*, was the first trichomycete to be artificially cultivated (Whistler 1960) and therefore many studies were made of its biology (Whistler 1962, 1968). Another species, *A. appalenchense*, was isolated from the larva of a chironomid midge in the genus *Chironomus*. *Paramoebidium chattoni* infects the blackflies *Simulium furculatum*, *S. vandalicum*, *S. innoxium*, *Cnephia ornithophilum* and *Nemura variegate* (Glockling et al. 2013) and has also been isolated from larvae of Ephemeropterans and Plecopterans. In a study of multiple symbionts in blackfly larvae, Kim and Alder (2005) found that *P. chattoni* was often in physical association with a trichomycete species in the hindgut of the blackfly larvae, suggesting mutual dependence. *Paramoebidium chattoni* infected 100% of isolated larvae of *Simulium innoxium* compared to 39% of *S. furculatum* and 0%–4% of *Cnephia ornithophilum* (Kim and Alder 2005).

14.6 Parasites of *Daphnia*

The water flea, *Daphnia*, is a small crustacean in the Cladocera, which is known to be parasitized by species of zoosporic true fungi, oomycetes and mesomycetozoeans. *Daphnia* has been shown to be a model organism for long-term ecological studies of their parasites (Johnson et al. 2006, 2009; Wolinska et al. 2009).

Within the Blastocladiomycota, *Polycaryum laeve* infects *Daphnia* species such as *D. pulicaria* (Johnson et al. 2009). A study of the effects of *P. laeve* on host *Daphnia* populations in lakes found that *Daphnia* populations decreased by as much as 99% during epizootic infections (Johnson et al. 2006; Gleason et al. 2010). In one of the very few long-term studies of parasitism by zoosporic true fungi, Johnson et al. (2009) made a 15-year study of population dynamics in relation to *Polycaryum* infection of *Daphnia*. They found that *Daphnia* populations decreased by up to 50% due to *Polycaryum* infection, depending on various factors including reproductive rates, host population sizes and prevalence of infection.

Oomycete parasites of *Daphnia* include species in the genus *Aphanomyces* (Fig. 14.6a–c) such as *A. daphniae*, which was isolated from *Daphnia hyalina* (Prowse 1954b) and *A. patersonii* (Scott 1961). Wolinska et al. (2009) examined *Daphnia* from 19 European lakes and attempted to identify parasites using both microscopy and molecular sequencing. In 147 infected *Daphnia*, a total of nine oomycete taxa were identified by their SSU and ITS rRNA gene regions. Some sequences showed proximity to the mosquito parasite, *Leptolegnia caudata*, whilst others fell close to the fish parasites, *Saprolegnia parasitica*, *Aphanomyces invadans* (=*piscida*) and *A. frigidophilus*. It was concluded that some oomycete species are highly virulent to *Daphnia*, but that some *Daphnia* parasites are also generalists, infecting species from a variety of trophic levels (Wolinska et al. 2009). However, it appears from sequence data that there are several parasites of *Daphnia* which have not been morphologically or formally described. *Daphnia* is also infected by several species of *Saprolegnia* including *S. diclina*, *S. ferax* and *S. australis* (Wollinska et al. 2008). Wolinska et al. (2008, 2009) have also reported *Pythium* infections in *Daphnia* from the USA and from Europe.

Another species recorded from *Daphnia* includes *Aphanomycopsis sexualis*, in the Leptolegniellaceae, initially described as a midge parasite (Martin 1975).

Caullerya mesnili, a member of the Mesomycetozoea, is a common parasite of *Daphnia*, which was discovered over a hundred years ago (Lohr et al. 2010; Glockling et al. 2013). For a long time it was mistakenly classified as a member of the Haplosporidia until recent gene sequencing showed it to be in the Mesomycetozoea and clustering in phylogenies with environmental sequences in the order Ichthyophonida (Lohr et al. 2010). *Caullerya mesnili* infects the gut epithelium of *Daphnia*, where it forms clusters of ovoid spores, which are later released into the gut lumen. No motile stage has been observed in *C. mesnili* although some members of the Icthyophonida produce amoebae. Many species of Mesomycetozoea appear to have lost the motile stage altogether (Marshall et al. 2008). It is thought that infection in *Daphnia* is probably via ingestion of *C. mesnili* spores, although the infection process has not been clarified and it is still not known how the infection spreads within the host or whether there is another spore type involved (Lohr et al. 2010). Epidemics of *C. mesnili* in *Daphnia* populations in permanent European lakes can affect 40% of the *Daphnia* population and, although the parasite does not kill the host, infected individuals are debilitated by reduced fecundity and shorter life span (Wolinska et al. 2006, 2007; Lohr et al. 2010).

Amoebidium parasiticum, also in the Ichthyophonida, parasitizes *Daphnia* in addition to blackfly larvae, and this genus, like some others in the order, produces amoebae.

14.7 Parasites of rotifers and nematodes

In freshwater ecosystems, several zoosporic members of the Blastocladiomycota, Chytridiomycota and the Oomycota are known to parasitize adult nematodes and rotifers and their eggs. Mesomycetozoeans have not been recorded from Aschelminthes. Species of *Catenaria* (Fig. 14.1a–c) in the Blastocladiomycota and *Phlyctochytrium* (Fig. 14.2a–b) in the Chytridiomycota have been isolated from aquatic habitats, as have species of *Olpidium* (Figs. 14.3, 14.4, 14.5a–c). But the majority of *Olpidium* species and zoosporic true fungal parasites in nematodes and rotifers have been isolated from rotting organic debris from terrestrial environments. *Endochytrium oophilum* and *E. operculatum* have been reported to parasitize the eggs of rotifers and nematodes (Karling 1937). Parasites of nematodes in the Oomycetes, such as *Haptoglossa, Gonimochaete, Myzocytiopsis* and *Chlamydomyzium* are also mainly found in terrestrial habitats (Barron 1976, 1989; Glockling 1994; Beakes et al. 2014b). Wet terrestrial bdelloid rotifers are infected by some *Haptoglossa* species (Barron 1990) and the host range of *Haptoglossa* in laboratory tests has been shown to include both bdelloid rotifers and nematodes for several species (Barron 1988). The genera *Myzocytiopsis* and *Chlamydomyzium* contain parasites of aquatic loricate rotifers in addition to parasites of bactivorous nematodes, so they can be found in both terrestrial and aquatic habitats.

Myzocytiopsis is a holocarpic genus and some species infect rhabditid nematodes in terrestrial habitats, such as decaying vegetation and herbivore dung (Barron 1976; Glockling 1994). Many *Myzocytiopsis* species reproduce sexually to form oospores which often have a thick, reticulate outer wall. Nematode and rotifer hosts are infected by zoospores or adhesive aplanospores (Fig. 14.13) which attach to the body and penetrate with a narrow germ tube to produce an internal thallus. The thallus usually divides into segments, each of which becomes a sporangium and develops a discharge tube through which the spores are released. Zoospores are often released into a vesicle that forms at the tip of the exit tube as the sporangial content is released. This vesiculate mode of spore release is similar to that of *Lagenidium*. Most *Myzocytiopsis* species described from aquatic loricate rotifers or their eggs, for instance *M. zoophthora, M. parthenospora, M. distylae, M. fijiensis* and *M. microspora* (Fig. 14.7a–b), have infected *Lecane* (*Distyla*) rotifers, which are common littoral rotifers in ponds and streams (Sparrow 1936; Karling 1944, 1969). Another species, *M. oophila*, was described from rotifer eggs in aquatic river debris (Sparrow 1939) while *M. indica*, an unusual species with narrow tubes connecting the thalli, was described from unidentified loricate rotifers (Singh 1973).

A few *Chlamydomyzium* species such as *C. septatum, C. aplanosporum* and. *C. internum*, infect aquatic loricate rotifers or their eggs (Karling 1969; Glockling and Dick 1997), although other members of the genus infect bactivorous nematodes (Dick 2001; Beakes et al. 2014b). *Chlamydomyzium oviparasiticum* was initially described as a parasite of bdelloid rotifer eggs (Barron 1989), but was later isolated from adult *Rhabditis* nematodes (Glockling and Beakes 2006). *Chlamydomyzium* does not have a male gametangium but forms resting spores, which are either chlamydospores or parthenogenetically produced oospores. The resting spores are often very thick-walled when mature. This genus has recently been reviewed (Beakes et al. 2014b) and accommodated in the Leptomitales (Beakes et al. 2014a).

The genus *Aquastella* contains two species *A. acicularis* and *A. attenuata* that were found infecting planktonic rotifers in a remote North American lake. *Aquastella* produces unusual sporangial outgrowth which projects outside the dead host.

The genus *Hydatinophagus* (1931a, b) was described for *H. apsteinii*, a parasite of rotifers belonging to the genus *Hydatina* (now *Epiphanes*), a semi-planktonic or planktonic species. Another species, *H. americanus*, which infected *Lecane* (*Monostyla*) rotifers, was later added, although asexual reproduction in this species was not observed (Bartsch and Wolf 1938). *Hydatinophagus* was considered to differ from

Aphanomyces because infection was thought to be via ingested spores (Bartsch and Wolf 1938). However, Scott (1961) transferred members of *Hydatinophagus* to *Aphanomyces* along with an earlier species described from rotifers, *Aphanomyces gordajeverae* (Skvortzow 1925).

Within the genus *Pythium*, *P. caudatum* is currently the only described member of the genus which is host-specific for nematodes (Dick 2001, 2003). Recently, a species considered to belong to *Pythium*, based on morphology and on molecular sequencing, was found infecting the planktonic rotifer, *Asplanchna girodi* (Thomas et al. 2011). The species has yet to be formally named. Previously, a *Pythium*-like infection was reported to infect *A. pridonta* in a lake in Russia (Gorbunov and Kosova 2001).

14.7.1 *Sommerstorffia spinosa*

Sommerstorffia spinosa is probably the most well known and most extraordinary oomycete infecting rotifers as it is a predator which captures aquatic rotifers on specialized structures (Fig. 14.12). *Sommerstorffia* captures species of *Lecane, Colurella* and *Lepadella* rotifers, which are all littoral or periphytic loricate species. Karling (1952) found it also captured the protozoan, *Entosiphon ovatum*, but did not capture bdelloid rotifer species in laboratory tests.

Sommerstorffia has been isolated from several countries including Bulgaria (Arnaudow 1923a), the USA (Sparrow 1929; Karling 1952), UK (Prowse 1954a), Poland (Czeczuga and Próba 1980) and Japan (Saikawa and Hoshino 1986; Akamatsu and Saikawa 2005). Arnaudow (1923a, b) described the formation of bullate oospores, but sexual reproduction has only been reported once since from a Polish isolate (Czeczuga and Próba 1980). This genus was placed in the Saprolegniales, and was considered to be close to *Aphanomyces* based on the mode of spore release and zoospore development (Arnaudow 1923a, b; Karling 1952). The primary spores encyst in a bundle at the mouth of the exit tube (Fig. 14.10, white arrow). *Sommerstorffia* has been studied ultrastructurally (Saikawa and Hoshino 1986), and its rDNA has recently been sequenced. Sequencing of the LSU region demonstrated that it fell closest to *Aphanomyces* in the Saprolegniales (Spies and Lévesque, unpublished) and supported the initial classification. *Sommerstorffia* is an unusual species because, although it is largely a predator, capturing rotifers on predacious hyphae, rotifers are also infected by specialized spores, termed sporelings. The adhesive sporelings can attach to passing rotifers and stick to them before penetrating to initiate infection. The sporelings of *S. spinosa* may also initiate infection when a rotifer tries to feed on them (Karling 1952). The predacious hyphae and the apices of the sporelings were observed to be optically refractile, and Karling (1952) suggested they may contain an adhesive substance or mucilage. Transmission electron microscopy (TEM) showed the trapping structures were densely packed with vesicles, which were considered to contain adhesive material (Saikawa and Hoshino 1986).

14.7.2 *Aquastella*

A new oomycete genus, *Aquastella*, containing two species, *A. attenuata* and *A. acicularis* has recently been described in the Saprolegniales (Molloy et al. 2014). *Aquastella* is unusual as the thalli produces elongate outgrowths which project through the rotifer body wall (Molloy et al. 2014). These *Aquastella* species were found to be host specific in planktonic rotifers. *Aquastella attenuata* infected a single loricate species, *Keratella taurocephala*, whereas *A. acicularis* infected both a non loricate species, *Polyarthra vulgaris* and a loricate species, *Ploesoma truncatum*. Molloy et al. (2014) conducted an ecological study of parasitism by *Aquastella* over the course of 7 years of sampling in a North American lake. Molloy et al. (2014) found that there was a correlation between peak host population densities and the occurrence of parasitism. It was hypothesized that the unique sporangial outgrowths may slow descent of the dead hosts and also deter their predation.

14.8 Parasites of protozoans

Few descriptions of zoosporic parasitic infections in freshwater protozoans exist, and these are mainly poorly or incompletely described. In addition, many of these records are quite old and, in some cases, morphological criteria that are essential for classification, such as the number of flagella of the zoospore, are not detailed. Thus, they may be considered taxonomically dubious (Dick 2003). In the oomycetes, some parasites appear to have lost their ability for sexual reproduction, so the oospore, which is a defining characteristic, may be lacking. Dick (2003) listed the taxa of zoosporic true fungi and oomycetes that have been described as parasites of protozoans. These include 24 species of zoosporic true fungi, including six species of *Olpidium*, and nine species of Oomycetes. *Olpidium* species have been described as parasites of zoocysts of various species of protozoans including *Pseudospora, Difflugia, Peptophrys, Pseudospora* and *Vampyrella* (Dick 2003). Bartsch and Wolf (1938) described *Aphanomyces acinetophagus*, a parasite of the protozoan *Acineta flava* from a river in the USA. This parasite reproduced asexually with primary spores encysting at the mouth of numerous long exit tubes and secondary zoospores encysting after several hours. *Lambornella clarki*, the parasitic ciliate of the treehole mosquito *Aedes sierriensis*, was infected experimentally by *Pythium flevoense* (Washburn et al. 1988). Most of the ciliate cysts were killed 72 h after inoculation (Washburn et al. 1988).

14.9 Discussion

Analytical models are proving to be useful in assessing and quantifying aspects of the complex food webs found in freshwater ecosystems. It has been shown that parasitic forms represents about half of the zooplankton diet when inedible algal blooms occur (Rasconi et al. 2014).

Freshwater ecosystems and the associations between fungal and fungal-like parasites and invertebrate hosts have, in general, been poorly studied by mycologists. In contrast, certain species of commercial importance or with the potential for biological control have been extensively documented. In particular, large lakes have been poorly sampled for invertebrates and parasites. Many zoosporic parasites of invertebrates are considered difficult to find and require special techniques for collection of hosts. They are also difficult to study because they are holocarpic and host-specific and cannot be artificially cultured. Furthermore, a multidisciplinary approach is needed to establish the exact taxonomic identification of the host. However, several researchers from the early part of the last century helped to highlight the variety of zoosporic parasites and invertebrate host species found in ponds, streams and rivers. Many early records were incomplete and some of the reported parasites are only known from their original description. In particular, further sampling and research of the zoosporic parasites of protozoans is necessary to better understand their biology. Improved sampling methods for freshwater systems may allow for re-examination of species which are currently considered dubious due to lack of morphological details (Dick 2003). Species such as *Sommerstorffia*, which has been found by several researchers since its initial discovery, has subsequently been recorded in great detail. Repeated findings of a ubiquitous species can lead to a much more comprehensive description, with advances in technology enabling biochemical, ultrastructural and phylogenetic studies.

Although all the members of a genus may, in general, parasitize the same type of host organism in some cases, such as *Coelomomyces* in Dipterans, other genera, such as *Aphanomyces*, are known to be much more versatile. Classification of pathogenic and parasitic genera based on host type is not always being borne out by phylogenetic data, particularly of those in the oomycetes. As well as *Aphanomyces*, genera such as *Pythium, Lagenidium* and *Saprolegnia* are proving to have a broader host range than was formerly considered. Indications are that there is a much wider diversity of oomycete parasites of freshwater invertebrates in these genera than is currently reflected by the number of morphologically described species. This hidden diversity is indicated by recent descriptions of new species, reports of undescribed parasites, and also by environmental sequences which do not match any formally described taxa.

The majority of environmental sequences of Cryptomycota have come from freshwater habitats (James and Berbee 2012) and, although they are not known morphologically, the possibility of some species infecting freshwater invertebrates cannot be ruled out.

Many oomycete holocarpic parasites of freshwater invertebrates such as *Myzocytiopsis* and *Chlamydomyzium* exhibit a high degree of host specificity in their natural environment. Some fish parasites in *Saprolegnia* and in *Aphanomyces*, however, appear to be opportunistic parasites of freshwater invertebrates, sometimes infecting crayfish, *Daphnia*, or midge larvae. It is not clear whether these invertebrates are only susceptible to infection when they are compromised by poor health or laboratory conditions.

Recent studies compiling phylogenies of members of the Chytridiomycota, Cryptomycota (James et al. 2006; Lara et al. 2010; James and Berbee 2012; Jones et al. 2012), Oomycota (Beakes and Sekimoto 2009; Dieguez-Uribeondo et al. 2009; Beakes et al. 2006, 2012, 2014a) and Mesomycetozoea (Cafaro 2005; Marshall et al. 2008), which include holocarpic parasites, are clarifying phylogenetic relationships within and between these groups. These studies have shown that similarities in morphological traits can be misleading and some parasites, particularly some of those in the Mesomycetozoea, were incorrectly classified for many years. Other taxa with similar morphologies, now known to be more distantly related, include the Oomycetes *Myzocytiopsis* and *Chlamydomyzium*. More surprisingly, the holocarpic and uniflagellate parasite *Olpidium* was shown not to belong to the Chytridiomycota but to the zygomycetes (Sekimoto et al. 2011). Many zoosporic parasites of freshwater invertebrates may still currently be misclassified, and subsequent findings, together with further morphological study and sequencing of their DNA, may be necessary to reclassify them. *Olpidium paradoxum* produces chains of spiny resting spores in culture (Glockling 1998), which are reminiscent of *Catenaria* (Couch 1935, 1945). Some members of *Myzocytiopsis*, particularly those which are easily cultivated and produce smooth-walled oospores or oospores not formed by gametangial copulation, may also currently be misplaced and may belong to other oomycete genera. Several species are described only from asexual features, which may show variation in different host species such as nematodes and rotifers, and lead to separate classifications of conspecific taxa. Details of sexual reproduction need to be clarified for these species. Ongoing phylogenetic studies may eventually resolve these issues, especially now that specific primers have allowed for sequencing of unculturable species within the host. Beakes and Sekimoto (2009) and Beakes et al. (2014a) have stressed the importance of including parasites in phylogenetic studies as a means of understanding their evolution, as many morphologically simple holocarpic taxa fall into key basal positions in phylogenetic trees (Beakes et al. 2006, 2014a; Beakes and Sekimoto 2009).

As yet, little is known about host-parasite interactions. However, recent ecological studies of model host species such as blackflies (McCreadie and Alder 1999; Kim and Alder 2005) and *Daphnia* (Johnson et al. 2006; Wolinska 2009), in their natural habitats, are clarifying aspects of the long-term effects of parasites on host populations. Ecological studies over several consecutive years, such as that for the rotifer parasites in *Aqustella*, suggest that host population levels correlate with the occurrence of parasitism by *Aquastella* species (Molloy et al. 2014). In addition, these ecological studies show how the relationships between parasite and host are affected by external environmental factors. With accelerated rates of climate change and environmental deterioration, changes in the composition of assemblages of parasites as well as their hosts in aquatic ecosystems need to be monitored.

Acknowledgments

SLG is grateful to Birgit Oidtmann, CEFAS, for contributing Fig. 14.8 and Paul G. Davison, University of North Alabama, USA for contributing Fig. 14.12. AVM is grateful to Carmen L. A. Pires-Zottarelli for contributing Fig. 14.11.

References

Alderman DJ. Fungi as pathogens of non-insect invertebrates. In: Samson RA, Vlak JM, Peters D, eds. Fundamental and Applied Aspects of Invertebrate Pathology, Wageningen, The Netherlands. Foundation of the Fourth International Colloquium of Invertebrate Pathology. The Netherlands: Weldhoven 1986;354–355.

Alderman DJ. Geographical spread of bacterial and fungal diseases of crustaceans. Rev Sci Tech Off Int Epiz 1996;15:603–632.

Alderman DJ, Holdich D, Reeve I. Signal crayfish as vectors in crayfish plague in Britain. Aquaculture 1990;86:3–6.

Akamatsu Y, Saikawa M. Giant mycelium by *Sommerstorffia spinosa*. Proc Ann Meet Mycolog Soc Jap 2005;49:79.

Arnaudow N. Untersuchungen über *Sommerstorffia spinosa* nov. gen., nov. sp. Jahrbuch der Sofianer Universität 1923a;19, Heft 2, Abt 1a:161–196.

Arnaudow N. Ein neuer Rädertiere (Rotaria) fangender Pilz. (*Sommerstorffia spinosa*, nov. gen., nov. sp.). Flora 1923b;116:109–113.

Ballesteros I, Martin MP, Dieguez-Uribeondo J. First isolation of *Aphanomyces frigidophilus* (*Saprolegniales*) in Europe. Mycotaxon 2006;95:335–340.

Barron GL. Nematophagous fungi: Three new species of *Myzocytium*. Can J Microbiol 1976;22:752–762.

Barron GL. Host range studies for *Haptoglossa* and a new species, *Haptoglossa intermedia*. Can J Bot 1988;67:1645–1648.

Barron GL. New species and new records of fungi that attack microscopic animals. Can J Bot 1989;67:267–271.

Barron GL. A new and unusual species of *Haptoglossa*. Can J Bot 1990;68:435–438.

Barron GL, Szijarto E. A new species of *Olpidium* parasitic in nematode eggs. Mycologia 1986;78:972–975.

Bartsch AF, Wolf FT. Two new saprolegniaceous fungi. Am J Bot 1938;25:392–395.

Beakes GW, Sekimoto S. The evolutionary phylogeny of oomycetes – insights gained from studies of holocarpic parasites of algae and invertebrates. In: Lamour K, Kamoun S, eds. Oomycete Genetics and Genomics: Diversity, Interactions and Research Tools. New York, NY: John Wiley and Sons Inc. 2009:1–24.

Beakes GW, Glockling SL, James TY. The diversity of oomycete pathogens of nematodes and its implications to our understanding of oomycete phylogeny. In: Meyer W, Pearce C, eds. Proceedings of the Eighth International Mycological Congress, August 20–25, Cairns, Australia. Medimond, Italy, 2006:7–12.

Beakes, GW, Glockling SL, Sekimoto S. The evolutionary phlyogeny of the oomycete "fungi". Protoplasma 2012;249:3–19.
Beakes GW, Honda D, Thines M. Systematics of the Straminipila: Labyrinthulomycota, Hyphochytriomycota and Oomycota. In: McLaughlin DJ, Spatafora JW, eds. The Mycota VIIA (2nd Edition) Heidelberg, New York: Springer 2014a. (In press)
Beakes GW, Glockling SL, James TY. A new Oomycete species parasitic in nematodes, *Chlamydomyzium dictyuchoides* sp. nov.: developmental biology and phylogenetic studies. Fung Biol 2014b (In press). http://dx.doi.org/10.1016/j.funbio.2014.04.005.
Benny GL, O'Donnel K. *Amoebidium parasiticum* is a protozoan, not a Trichomycete. Mycologia 2000;92:1133–1137.
Bisht GH, Joshi C, Khulbe RD. Watermolds: Potential biological control agents of malaria vector *Anopheles culicifacies*. Curr Sci 1996;70:393–395.
Cafaro MJ. Eccrinales (Trichomycetes) are not fungi, but a clade of protists at the early divergence of animals and fungi. Mol Phyl Evol 2005;35:21–34.
Cerenius L, Söderhäll K. Chemotaxis in *Aphanomyces astaci*, an arthropod parasitic fungus. J Invert Pathol 1984a;43:278–281.
Cerenius L, Söderhäll K. Repeated zoospore emergence from isolated spore cysts of *Aphanomyces astaci*. Exp Mycol 1984b;8:370–377.
Cerenius L, Söderhäll K, Persson M, Ajaxon R. The crayfish plague fungus *Aphanomyces astaci* – diagnosis, isolation and pathobiology. Freshw Crayfish 1988;7:131–144.
Couch JN. A new saprophytic species of *Lagenidium*, with notes on other species. Mycologia 1935;27:376–387.
Couch JN. Observations on the genus *Catenaria*. Mycologia 1945;37:163–193.
Couch JN, Bland CE. The genus *Coelomomyces*, 1st edition. Orlando, FL: Academic Press 1985.
Czeczuga B, Próba D. The characteristics of the environment of *Sommerstorffia spinosa* (Oomycetes: Saprolegniales), a parasite of certain rotifers. Mycologia 1980;72:702–707.
Deacon JW, Saxena G. Orientated zoospore attachment and cyst germination in *Catenaria anguillulae*, a facultative endoparasite of nematodes. Mycol Res 1997;101:513–522.
D'Eliscu PN. *Endosphaerium funiculatum*, gen. nov., sp. nov., a new predaceous fungus mutualistic in the gills of freshwater pelecypods. J Invert Pathol 1977;30:418–421.
Dick MW. The Myzocytiopsidaceae. Mycol Res 1997;101:878–882.
Dick MW. The species and systematic position of *Crypticola* in the Peronosporomyces, and new names for the genus *Halocrusticida* and species therein. Mycol Res 1998;102:1062–1066.
Dick MW. Straminipilous Fungi. Systematics of the Peronosporomycetes including accounts of the marine straminipilous protists, and the plasmodiophorids and similar organisms. Dortrecht: Kluwer Academic Publishers 2001.
Dick MW. The Peronosporales and other flagellate fungi. In: Howard DH, ed. Pathogenic Fungi in Humans and Animals, 2nd edn. New York, NY: Marcel Dekker Inc 2003:17–66.
Diéguez-Uribeondo J, Garcia MA, Cerenius L, Kozubikova E, Ballesteros I, Windels C, et al. Phylogenetic relationships among plant and animal parasites, and saprotrophs in *Aphanomyces* (oomycetes). Fung Gen Biol 2009;46:365–376.
Domnas AJ, Srebo JP, Hicks BF. Sterol requirement for zoospore formation in the mosquito-parasitizing fungus *Lagenidium giganteum*. Mycologia 1977;69:875–886.
Drechsler C. Three fungi destructive to free-living terricolous nematodes. J Wash Acad Sci 1940;30:240–254.
Edgerton BF, Henttonen P, Jussila, J, Mannonen A, Paasonen P, Taugbil T, Edsman L, Souty-Grosset C. Understanding the cause of disease in European freshwater crayfish. Conservat Biol 2004;18:1466–1474.

Federici B. Mosquito control by the fungi *Culicinomyces, Lagenidium* and *Coelomomyces*. In: Burges H, ed. Microbial Control of Pests and Plant Diseases 1970–1980. New York, NY: Academic Press 1981:555–572.
Frances SP. Pathogenicity, host range and temperature tolerance of *Crypticola clavulifera* (Oomycetes: Lagenidiales) in the laboratory. J Am Mosquito Control Assoc 1989;7:504–506.
Frances SP, Sweeney AW, Humber RA. *Crypticola clavulifera* gen. *et* sp. nov. and *Lagenidium giganteum*: Oomycetes pathogenic for Dipterans infesting leaf axils in an Australian rain forest. J Invert Pathol 1991;54:103–111.
Gleason FH, Marano AV, Johnson P, Martin WW. Blastocladian parasites of invertebrates. Fung Biol Rev 2010;24:56–67.
Gleason FH, Carney LT, Lilje O, Glockling SL. Ecological potentials of species of *Rozella* (Cryptomycota). Fung Ecol 2012;5:651–655.
Glockling SL. Predacious and parasitoidal fungi in association with herbivore dung in deciduous woodlands. PhD Thesis 1994. University of Reading.
Glockling SL. *Zoophagus cornus*: a new species from Japan. Mycol Res 1997;101:1179–1182.
Glockling SL, Dick MW. New species of *Chlamydomyzium* from Japan and pure culture of *Myzocytiopsis* species. Mycol Res 1997;101:883–896.
Glockling SL, Beakes GW. A review of the biology and infection strategies of biflagellate zoosporic parasites of nematodes. Fungal Divers 2000;4:1–20.
Glockling SL, Beakes GW. Structural and developmental studies of *Chlamydomyzium oviparasiticum* from *Rhabditis* nematodes and in culture. Mycol Res 2006;110:1119–1126.
Glockling SL, Jones DM, Gleason FH. Cryptomycota (Rozellida) and Mesomycetozoea (Icthyosporea) the Super-group Opisthokonta. In: Jones EBG, Ka-Lai Pang K-L, eds. Marine Fungi. Berlin: Walter De Gruyter 2012:153–165.
Glockling SL, Marshall WH, Gleason FH. Phylogenetic interpretations and ecological potentials of Mesomycetozoea (Icthyosporea). Fung Ecol 2013;6:237–247.
Golkar L, LeBrun RA, Ohayon H, Gounon P, Papierok B, Brey PT. Variation of larval susceptibility to *Lagenidium giganteum* in three mosquito species. J Invert Pathol 1993;62:1–8.
Gorbunov AK, Kosova AA. Parasites in rotifers from the Volga delta. Hydrobiol 2001;446:51–55.
Huang TS, Cerenius L, Söderhäll K. Analysis of genetic diversity in the crayfish plague fungus, *Aphanomyces astaci,* by random amplification of polymorphic DNA. Aquaculture 1994;126:1–10.
James TY, Berbee ML. No jacket required – new fungal lineage defies dress code. Bioassays 2012;34:94–102.
James TY, Letcher PM, Longcore JE, Mozely-Stanridge SE, Porter D, Powell MJ, et al. A molecular phylogeny of the flagellated fungi (Chytridiomycota) and description of a new phylum (Blastocladiomycota). Mycologia 2006;98:860–871.
Johnson PTJ, Longcore JE, Stanton DE, Carnegie RB, Shields JD, Preu ER. Chytrid infections of *Daphnia pulicaria*: development, ecology, pathology and phylogeny of *Polycaryum laeve*. Freshw Biol 2006;51:634–648.
Johnson PTJ, Ives AR, Lathrop RC, Carpenter SR. Longterm disease dynamics in lakes: causes and consequences of chytrid infections in *Daphnia* populations. Ecology 2009;90:132–144.
Jones MDM, Forn I, Gadelha C, Egan M, Bass D, Massana R, et al. Discovery of novel intermediate forms redefines the fungal tree of life. Nature 2011;474:200–203.
Karling JS. The structure, development and identity of *Endochytrium*. Am J Bot 1937;24:352–364.
Karling JS. Simple Holocarpic Biflagellate Phycomycetes. New York, NY: Columbia Academic Press 1942.
Karling JS. New lagenidiaceous parasites of rotifers from Brazil. Lloydia 1944;7:328–342.
Karling JS. Brazilian chytrids. VIII. Additional parasites of rotifers and nematodes. Lloydia 1946a;9:1–12.
Karling, J.S. Brazilian chytrids. IX. Species of *Rhizophidium*. Am J Bot 1946b;33:328–334

Karling JS. *Sommerstorffia spinosa* Arnaudow. Mycologia 1952;44:387–412.
Karling JS. Zoosporic fungi of oceania. IV. Species with biflagellate zoospores. Mycopathol Mycolog Appl 1969;37:161–170.
Karpov SA, Mikhailov KV, Mirzaeva GS, Mirabdullaev IM, Mamkaeva KA, Titova NN, et al. Obligately phagotrophic aphelids turned out to branch with the earliest-diverging fungi. Protist 2013;164:195–205.
Kerry B. Biocontrol: Fungal parasites of female cyst nematodes. J Nematol 1980;12:253–259.
Kerry B, Crump DH. Two fungi parasitic on females of cyst nematodes (*Heterodera* spp.). Trans Br Mycol Soc 1980;74:119–125.
Kerwin JL. Oomycetes: *Lagenidium giganteum*. J Am Mosquito Contr 2007;23 (2 Suppl.):50–57.
Kerwin JL, Washino RK. Sterol induction of sexual reproduction in *Lagenidium giganteum*. Exp Mycol 1986;7:109–115.
Kerwin JL, Peterson EE. Fungi: Oomycetes and Chytridiomycetes. In: Lacey LA, ed. Manual of Techniques in Insect Pathology. New York, NY: Academic Press 1997:251–268.
Kerwin JL, Dritz DA, Washino RK. Pilot scale production and application in wild-life ponds of *Lagenidium giganteum* (Oomycetes: Lagenidiales) to mammals. J Am Mosquito Contr 1994;10:451–455.
Kim S-K, Alder PH. Multiple symbiotes in larval black flies (Diptera: Simuliidae): associations of trichomycete fungi with ichthyosporeans, mermithid nematodes, and microsporidia. J Invert Pathol 2005;88:247–253.
Kiziewicz B. Aquatic fungi and fungus-like organisms in the bathing sites of the river Supraśl in Poland. Mycolog Balcanica 2004;1:77–83.
Lara E, Moriera D, Lopez-Garcia P. The environmental clade LKM11 and *Rozella* from the deepest branching clade of the fungi. Protist 2010;161:116–121.
Lara E, Belbahri L. SSU rRNA reveals major trends in oomycete evolution. Fungal Divers 2011;49:93–100.
Letcher PM, Lopez S, Schmieder R, Lee PA, Behnke C, Powell MJ, et al. Characterization of *Amoeboaphelidium protococcarum*, an algal parasite new to the Cryptomycota isolated from an outdoor algal pond used for the production of biofuel. PLoS ONE 2013;8:e56232. doi:10.1371/journal.pone.0056232.
Lichtwardt RW. The Trichomycetes, Fungal Associates of Arthropods. New York, NY: Springer-Verlag 1986.
Lohr JN, Laforsch C, Koerner H, Wolinska JA. Daphnia parasite (*Caullerya mesnili*) constitutes a new member of the Ichthyosporea, a group of protists near the animal-fungi divergence. J Eukaryot Microbiol 2010;57:328–336.
Lopez Lastra CC, Steciow MM, Garcia JJ. Registro más austral del hongo *Leptolegnia chapmanii* (Oomycetes: Saprolegniales) como patógeno de larvas de mosquitos (Diptera: Culicidae). Rev Iberoamericana Micol 1999;16:143–145.
Marshall WL, Celio G, McLaughlin DJ, Berbee ML. Multiple isolations of a culturable, motile Ichthyosporean (Mesomycetozoa, Opisthokonta), *Creolimax fragrantissima* n. gen., n. sp., from marine invertebrate digestive tracts. Protist 2008;159:415–433.
Martin WW. A new species of *Catenaria* parasitic in midge eggs. Mycologia 1975;67:264–72.
Martin WW. *Aphanomycopsis sexualis*, a new parasite of midge eggs. Mycologia 1975;67:923–933.
Martin WW. The development and possible relationships of a new *Atkinsiella* parasitic in insect eggs. Am J Bot 1977;64:760–769.
Martin WW. Two additional species of *Catenaria* (Chytriodiomycetes: Blastocladiales) parasitic in midge eggs. Mycologia 1978;70:461–467.
Martin WW. *Couchia complexa*, a water mold parasitic in midge eggs. Mycologia 1981a;73:1143–1157.
Martin WW. The natural regulation of midge populations by aquatic fungi in Virginia. J Elisha Mitch Sci Soc 1981b;97:162–170.

Martin WW. Egg parasitism by zoosporic fungi in a littoral chironomid community. J N Am Benthol Soc 1991;10:455–462.
Martin WW. Two new species of *Couchia* parasitic in midge eggs. Mycologia 2000;92: 1149–1154.
McCray Jr. EM, Umphlett CJ, Fay RW. Laboratory studies on a new fungal pathogen of mosquitoes. Mosquito News 1973;33:54–60.
McCreadie JW, Adler PH. Parasites of blackflies (Diptera: Simuliidae) and environmental factors associated with their distribution. Invert Biol 1999;118:310–318.
McInnis TM, Schimmel LE. Host range studies on a new fungus *Leptolegnia*, a parasite of mosquito larvae (Diptera: Culicidae). J Med Entomol 1985;22:226–227.
McInnis Jr. T, Zattau WC. Experimental infection of mosquito larvae by a species of the aquatic fungus *Leptolegnia*. J Invert Pathol 1982;39:98–104.
McMahon TA, Branelly LA, Chatfield MWH, Johnson PTJ, Joseph MB, McKenzie VJ, et al. Chytrid fungus *Batrachochytrium dendrobatidis* has nonamphibian hosts and releases chemicals that cause pathology in the absence of infection. PNAS 2013;110:210–215.
Mendoza L, Taylor JW, Ajello L. The class Mesomycetozoea: a heterogeneous group of microorganisms at the animal fungal boundary. Ann Rev Microbiol 2002;56:315–344.
Molloy DP, Glockling SL, Siegfried CA, Beakes GW, James TY, Mastitsky SE, et al. *Aquastella* gen nov.: A new genus of saprolegniacious oomycete rotifer parasites related to *Aphanomyces*, with unique sporangial outgrowths. Fung Biol 2014 (In press). http://dx.doi.org/10.1016/j.funbio.2014.01.007.
Nestrud LB, Anderson RL. Aquatic safety of *Lagenidium giganteum*: effects on freshwater fish and invertebrates. J Invert Pathol 1994;64:228–233.
Nnakumusana ES. Histopathological studies on the progress of infection of *Leptolegnia* sp. (SC-1) in *Anopheles gambiae* larvae exposed to zoospores in the laboratory. Curr Sci 1986;55:633–636.
Oidtmann B. Crayfish Plague (*Aphanomyces astaci*). Chapter 2.2.1. In: Manual of Diagnostic Tests for Aquatic Animals 2013; http://www.oie.int/international-standard-setting/aquatic-manual/access-online/;accessed (15th Oct 2012).
Oidtmann B, Geiger S, Steinbauer P, Culas A, Hoffmann RW. Detection of *Aphanomyces astaci* in North American crayfish by polymerase chain reaction. Dis Aquat Org 2006;72:53–64.
Petersen AB, Rosendhal S. Phylogeny of the Peronosporomycetes (Oomycota) based on partial sequences of the large ribosomal subunit (LSU rDNA). Mycol Res 2000;104:1295–1303.
Prowse GA. *Sommerstorffia spinosa* and *Zoophagus insidians* predacious on rotifers, and *Rozellopsis inflata* the endoparasite of *Zoophagus*. Trans Br Mycol Soc 1954a;36:134–150.
Prowse GA. *Aphanomyces daphniae* sp. nov., parasitic on *Daphnia hyalina*. Trans Br Mycol Soc 1954b;37:22–28.
Rasconi S, Grami B, Niquil N, Jobard M, Sime-Ngando T. Parasitic chytrids sustain zooplankton growth during inedible algal blooms. Frontiers Microbiol 2014;5:229. doi:10.3389/fmicb.2014.00229.
Saikawa M, Hoshino J. Electron microscopy on *Sommerstorffia spinosa*, a watermold parasitic on rotifers. Mycologia 1986;78:554–561.
Saunders GA, Washburn JO, Egerter DE, Anderson JR. Pathogenicity of fungi isolated from field collected larvae of the western treehole mosquito, *Aedes sierrensis* (Diptera: Culicidae). J Invert Pathol 1988;52;360–363.
Scholte E-J, Knols BGJ, Samson RA, Takken W. Entomopathogenic fungi for mosquito control: a review. J Insect Sci 2004;4:19.
Scott WW. A monograph of the genus *Aphanomyces*. Virginia Agric Exp Station Tech Bull 1961;151:1–95.

Sekimoto S, Rochon D, Long JE, Dee, JM, Berbee ML. A multigene phylogeny of *Olpidium* and its implications for early fungal evolution. BMC Evol Biol 2011;11:331.

Seymour RL. *Leptolegnia chapmanii*, an oomycete pathogen of mosquito larvae. Mycologia 1984;76:670–674.

Seymour RL, Briggs JD. Occurrence and control of *Aphanomyces* (Saprolegniales: Fungi) infections in laboratory colonies of larval *Anopheles*. J Am Mosquito Contr 1985;1:100–102.

Shikora F. Die Krebspest. Ficherei Zeitung 1906;9:529–583.

Sime-Ngando T. Phytoplankton chytridiomycosis: fungal parasites of phytoplankton and their imprints on the food web dynamics. Frontiers Microbiol 2012;3:1–13.

Singh UP. A new species of *Myzocytium* from India. Hydrobiol 1973;42:445–450.

Skvortzow BW. Zur Kenntnis der Phycomyceten aus der Nordmandshurei, China. Archiv Protistenkunde 1925;51:428–433.

Skvortzow BW. Über ein Phycomyceten aus China. Archiv für Protistenkunde 1927;57:204–206.

Söderhäll K and Cerenius L. Crustacean immunity. Ann Rev Fish Dis 1992;2:3–23.

Söderhäll K, Cerenius, L. The crayfish plague fungus: history and recent advances. Freshw Crayfish 1999;12:11–35.

Söderhäll K, Dick MW, Clarke G, Fürst M. Isolation of *Saprolegnia parasitica* from the crayfish *Astacus leptodactylus*. Aquaculture 1989;92:121–125.

Sorokine N. Les végeteaux parasites des Anguillulae. Annals des Sciences Naturelles, Botanique, Série V1. 1876;4:62–71.

Sparrow Jr FK. A note on the occurrence of two rotifer-capturing Phycomycetes. Mycologia 1929;21:90–96.

Sparrow FK. A contribution to our knowledge of the aquatic Phycomycetes of Great Britain. J Linn Soc Bot 1936;5:417–478.

Sparrow FK. A new species of *Lagenidium* parasitic on rotifer eggs. Mycologia 1939;31:527–532.

Sparrow FK. Aquatic Phycomycetes. 2nd edn. Ann Arbor, MI: Michigan University Press 1960.

Su X, Zou F, Guo Q, Huang J, Chen TX. A report on a mosquito-killing fungus, *Pythium carolinianum*. Fungal Divers 2001;7:129–133.

Thomas SH, Housley JM, Reynolds AN, Penczykowski RM, Kenline KH, Hardegree N, et al. The ecology and phylogeny of oomycete infections in *Asplanchna* rotifers. Freshw Biol 2011;56:384–394.

Tribe HT. A parasite of white cysts of *Heterodera: Catenaria auxillaris*. Trans Br Mycol Soc 1977;69:367–376.

Umphlet CJ. Infection levels in *Coelomyces punctatus*, an aquatic fungus parasite, in a natural population of the common malaria mosquito *Anopheles quadrimacutus*. J Invert Pathol 1969;15:299–305.

Unestam T, Weiss DW. The host-parasite relationship between freshwater crayfish and the crayfish disease fungus, *Aphanomyces astaci*: Responses to infection by a susceptible and a resistant species. J Gen Microbiol 1970;60:77–90.

Valkanov A. Über Morphologie und Systematik der rotatorienbefallendden Pilze. Arch Protestenk 1931a;74;5–17.

Valkanov A. Über Morphologie und Systematik der rotatorienbefallendden Oomyceten (Bulgarian German summary). Jahrb Inov Sofia 1931b;27:215–134.

Vavra L, Joyon L. Étude sur le morphologie, the cycle evolutif et la position systematique de *Callimastix cyclops* Weisenberg 1912. Protistologica 1966;2:5–16.

Washburn JO, Egerter DE, Anderson JR, Saunders GA. Density reduction in larval mosquito (Diptera: Culicidae) populations by interactions between a parasitic ciliate (Ciliophora: Tetrahymenidae) and an opportunistic fungal (oomycetes: Pythiaceae) parasite. J Med Entomol 1988;25: 307–314.

Washino RK, Laird M. Biocontrol of Mosquitoes Associated With Californian Rice Fields With Special Reference to the Recycling of *Lagenidium giganteum* Couch and Other Microbial Agents: Biocontrol of Medical and Veterinary Pests. New York, NY: Praeger 1983:122–139.
Whistler HC. Pure culture of the Trichomycete, *Amoebidium parasiticum*. Nature 1960;186:732–733.
Whistler HC. Culture and nutrition of *Amoebidium parasiticum*. Am J Bot 1962;49:193–199.
Whistler HC. Developmental control of *Amoebidium parasiticum*. Dev Biol 1968;17:562–570.
Whistler HC, Gabriel PB, Chanpaisaeng J, Zebold SL, Padua LE. Observations on the life cycle of *Coelomomyces indicus* (Blastocladiales: Coelomycetaceae) in Anopheline mosquitoes from the Phillipines and Thailand. J Med Entomol 1999;36:695–701.
Wolinska J, Bittner K, Ebert D, Spaak P. The coexistence of hybrid and parental *Daphnia*: the roles of parasites. Proc R Soc London 2006;273:1977–1983.
Wolinska J, Keller B, Manca M, Spaak P. Parasite survey of a *Daphnia* hybrid complex: host-specificity and environment determine infection. J Animal Ecol 2007;76:191–200.
Wolinska J, King KC, Vigneux F, Lively CM. Virulence, cultivating conditions, and phylogenetic analyses of oomycete parasites in *Daphnia*. Parasitology 2008;135:1667–1678.
Wolinska J, Giessler S, Koerner H. Molecular identification and hidden diversity of novel *Daphnia* parasites from European lakes. Appl Environ Microbiol 2009;75:7051–7059.

Ecology

Holger Thüs, André Aptroot and Mark R. D. Seaward

15 Freshwater lichens

Lichens are symbiotic systems consisting of a dominating fungus, one or several photosynthetically active algae or cyanobacteria and co-inhabitants such as heterotrophic bacteria and other fungi. Their balanced composition of two to many different organisms makes lichens miniaturized ecosystems in themselves (Seaward 1988). They are distinguished from other fungal-algal associations by the dominating role of the fungus, the exhabitant, which provides the overall form and structure of the lichen, over the photobiont, the inhabitant surrounded by the fungal hyphae (Hawksworth and Honegger 1994). Furthermore, fungi and heterotrophic bacteria can co-inhabit the symbiotic structure as parasites or saprobes (Hodkinson and Lutzoni 2009; Bates et al. 2011; Bjelland et al. 2011). Lichenization is a successful mode of nutrition for circa 40% of all known ascomycetes, but only for a small percentage of basidiomycetes. Hawksworth (2000), mostly based on morphological characters, estimated there to be circa 13,500 lichenized fungi, but the rapidly increasing number of taxonomic revisions via molecular markers has revealed a large number of new cryptic and semi-cryptic species (Crespo and Lumbsch 2010). Recent examples of such speciation in freshwater lichens were demonstrated for *Peltigera* (Lendemer and O'Brien 2011) and *Hydropunctaria* (Orange 2012). Based on a case study on the family Lobariaceae, the number of phylogenetic species of lichenized fungi may be exceptionally four to five times higher than previous estimates for some genera (Moncada et al. 2013).

Lichenization, mainly in the Ascomycota, has independently evolved on several occasions (Lutzoni et al. 2001), but there is no clear evidence if or how often this occurred on land or in aquatic environments. Today the vast majority of lichenized fungi are restricted to terrestrial environments and their thalli disintegrate or are quickly taken over by non-symbiotic fungi if they are fully hydrated for extended periods of time. Among the Ascomycota, the Verrucariales and Lichinales stand out as the orders which contribute the highest numbers of species to freshwater habitats and include most of the dominating lichens in the permanently submerged zone (Tab. 15.1). A few lineages of fully aquatic lichens have also evolved independently several times in the Peltigerales (e.g. *Collema dichotomum, Peltigera hydrothyria*). Further taxa in the Pertusariales, Peltigerales and Ostropales can be important in sub-aquatic habitats, but most other orders are represented by only a few species or are entirely absent in aquatic and sub-aquatic environments. The Lecanorales, the largest order of mostly lichenized Ascomycota, is remarkable for the poor representation of its members in freshwater habitats.

The number of truly aquatic lichens in permanently submerged habitats is small, with less than ten species in temperate areas of the Northern Hemisphere confirmed as capable of surviving several years of submersion (e.g. *Hydropunctaria rheithrophila*, Fig. 15.1). About 170 species worldwide are more or less restricted to freshwater

Tab. 15.1: Global numbers of lichens and lichenicolous fungi from freshwater habitats (taxa which are either specific for or at least regularly reported from this habitat).

Class	Order	Lichen genera	Lichen species	Lichenicolous genera	Lichenicolous species
Arthoniomycetes	Arthoniales	0	0	1	1
Ascomycota, families *incertae sedis*		3	7	0	0
Chaetothyriomycetidae, families *incertae sedis*		1	3	1	1
Dothideomycetes	Dothideomycetes, families *incertae sedis*	1	1	7	10
Eurotiomycetes	Pyrenulales	1	5	0	0
Eurotiomycetes	Verrucariales	9	110	5	13
Lecanoromycetes	Baeomycetales	1	1	0	0
Lecanoromycetes	Lecanorales	3	5	0	0
Lecanoromycetes	Ostropales	3	18	0	0
Lecanoromycetes	Peltigerales	4	20	0	0
Lecanoromycetes	Pertusariales	2	5	0	0
Lecanoromycetes	Teloschistales	3	9	0	0
Lecanoromycetidae, families *incertae sedis*		5	10	0	0
Lichinomycetes	Lichinales	16	43	0	0
Ostropomycetidae, families *incertae sedis*		3	7	0	0
Sordariomycetes	Hypocreales	0	0	1	1
	total	55	244	15	26

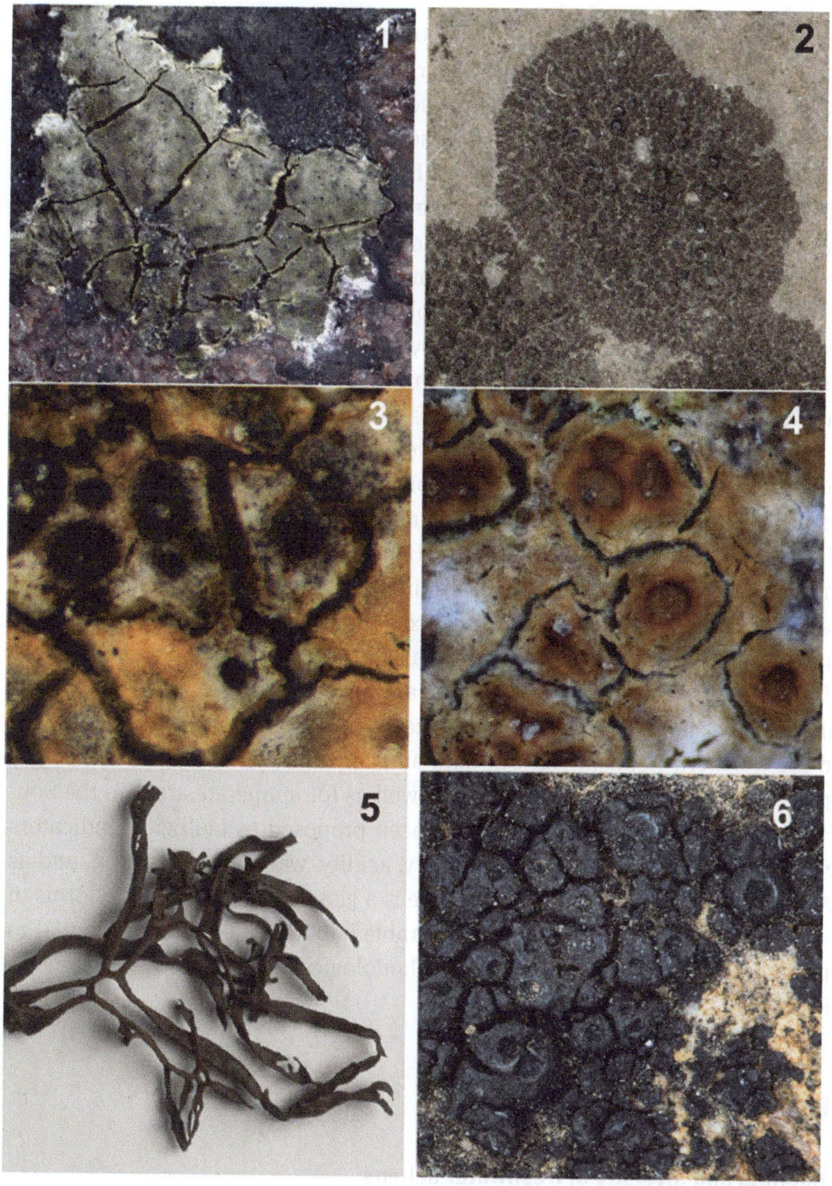

Figs. 15.1–15.6: Fig. 15.1 *Hydropunctaria rheithrophila* (Verrucariaceae) from permanently submerged rocks in the river Anger (Bergisches Land, Germany, Thüs & Zimmermann W0935). **Fig. 15.2** *Staurothele drumondii* (Verrucariaceae), a typical species of temporarily inundated sites in North America (Kingston, Canada, Holotype, AT Drummond). **Fig. 15.3** *Verrucaria conturmatula* (Verrucariaceae), parasitic on *Ionaspis lacustris* (as illustrated in Fig. 15.4). **Fig. 15.4** *Ionaspis lacustris* (Hymeneliaceae), thallus with apothecia from moist steep rock face (Wales, United Kingdom). **Fig. 15.5** *Jenmania goebelii* (Lichinaceae), submerged in the Mazaruni River, Marshall Falls (British Guyana, K.Kato 103). **Fig. 15.6** *Pterygiopsis australiensis* (Lichinaceae) from submerged stones in the river Iris, Tasmania (Australia, Kantvilas 466/84).

habitats, but these are typically at most temporarily immersed, being best characterized as amphibious or sub-aquatic. However, together with other taxa with wider ecological amplitude but regularly present in semi-aquatic habitats, the total number of described freshwater lichens in this wide sense (Keller 2000) comes close to 250 species. Furthermore, freshwater lichens are host to more than 20 species of lichenicolous fungi, a remarkably high percentage of which belong to the family Verrucariaceae (Tab. 15.1), probably due to physiological adaptations of the fungi in this family to a pronounced gas diffusion-limited micro-environment. It could also reflect a facilitated interaction with the dominating photobionts (*Diplosphaera, Dilabifilum, Heterococcus*) in freshwater Verrucariaceae by parasites which have evolved in the same family and which may still retain relevant physiological traits to access these unusual algal groups as a food source. This fascinating hypothesis requires further studies since these parasitic fungi could have evolved from ancestors which have used the same algae as their current hosts. Convergent evolution of similar morphologies is common in freshwater lichens and traditional species delimitations are in need of revision based on molecular markers. Hence, all estimates of species diversity in freshwater habitats are still highly uncertain. Specialized aquatic lichens in temperate areas are often associated with algal taxa which are not found in lichens from other habitats, but little is known of the algal partners of aquatic lichens in other parts of the world. Decline and recovery of aquatic and amphibious lichens as a result of environmental changes have been analyzed in various countries of temperate areas in the Northern Hemisphere, but data and conservation strategies are largely lacking for subtropical and tropical areas, as well as for temperate areas in the Southern Hemisphere. Freshwater lichens have been proposed as biological indicators of water level fluctuations, stream-bed stability, acidity, water quality, silting, and as passive monitors for heavy metals. Since there is a general similarity of life-forms of freshwater lichens across the world in comparable habitats, there is the potential to develop standards for globally applicable methodologies using freshwater lichens as bioindicators.

15.1 Ecology

15.1.1 Habitats and diversity of freshwater lichens

Lichens occur in most types of permanent and temporary freshwater habitats if they have a water temperature which remains low, support stable enough substrata, the light supply is guaranteed, and where competition with other autotrophic organisms is not too strong. Most freshwater lichens live on rocks and pebbles, but a few species appear to be restricted to temporarily inundated bark (Zschacke 1927; Aptroot and Thüs 2011), including some saxicolous species where the substratum is thinly covered with mineral sediment particles (Motiejunaite 2003; Hadchułka 2011). Hard surfaces are host to a larger number of lichens than soft ones, and the number of freshwater species on siliceous rocks is much larger than that on limestone or chalk, with tufa deposition resulting in particularly species-poor lichen vegetation. Specialized freshwater species (never occurring in terrestrial habitats) make up a higher percentage of the species pool in acidic streams compared to those in limestone streams, where species with a wide ecological amplitude dominate (Gilbert 1996; Gilbert and Giavarini 1997). This difference may be partially linked to the higher number of streams which can dry-up during summer droughts in catchment areas with calcareous bedrock, but similar results are also noted for permanent springs in the Southern Alps (Nascimbene et al. 2007).

All freshwater lichens are more or less sensitive to silting and large amounts of fine sediment cover, particularly organic deposits, such as in organically polluted streams or due to sediment-rich agricultural run-off, which soon destroys all lichen growth (Thüs 2002). Elevated nitrate levels have been shown to cause direct damage to *Peltigera hydrothyria* in laboratory experiments (Davis et al. 2000). The combined factors of stream-bed stability, scouring, silting and eutrophication often result in a characteristic pattern of diversity along a watercourse, the richest sections of acidic streams in Britain being within the middle reaches of the stream system, where there is a rich diversity of alternating structural features, such as rapids, waterfalls and gorges, and a permanent water supply is guaranteed (Gilbert 2000).

Typical habitats for freshwater lichens are large permanent water bodies fed by rain or groundwater, but more short-lived or small-scale water bodies, such as more or less continuously wet seepage tracks, can support characteristic lichen communities which overlap with those found in streams and lakes. In arid to sub-arid areas, occasional rainfalls can produce small and short-lived water-filled basins with a specific lichen community dominated by the genus *Peltula*. Its species are mainly characterized by their ability to withstand months of desiccation rather than their inundation tolerance (Büdel 1987; Wessels and Büdel 1989), but there is some overlap with the species from river banks in subtropical streams with large water level fluctuations. In particular, the euryoecous *P. euploca* has been found as a component in temporary rainwater rock pools in South Africa, as well as providing a locally dominant component in temporarily inundated lichen communities on the banks of the Rio Dayman in Uruguay (Thüs, unpubl. observations based on collections communicated by G. Mazzetta). Short-lived freshwater habitats can occur in water-filled tree axils, bromeliad reservoirs or leaf axils of other epiphytic seed plants and ferns in the tropics, but they appear to support no lichens at all or only a sporadic occurrence of generalists.

The number of lichens which not only survive but flourish in fully submerged conditions over several years is difficult to estimate, due to the scarcity of long-term observations of freshwater lichen habitats. Confirmed observations for survival in streams and lakes over several years of uninterrupted submergence have only been undertaken in temperate areas of the Northern Hemisphere for less than ten species, but on a global scale can be estimated to be circa 50. Many more lichens are specialized amphibious taxa which are found in at least occasionally inundated or wetted banks of streams and large rivers or the shores of lakes, but neither tolerate full submersion over most of the year, nor are they known from terrestrial habitats. There are circa 100 described species in this group; this is roughly the same number of species classified as regular facultative freshwater lichens which become locally characteristic elements in communities temporarily inundated or influenced by splash-water that also exist in much drier habitats. Examples of this group are *Porpidia* species, which can be visually dominant elements of streams in more or less shaded habitats, and *Caloplaca flavovirescens* which forms extensive populations on subtropical river banks in Brazil and also on the dykes surrounding freshwater lakes in The Netherlands.

15.1.2 Collecting and identifying freshwater lichens

Most freshwater lichens can be spotted in the field, but small species are easily overlooked when light conditions are not ideal or when they are submerged. However, even for those lichens easily observed in the field, identification to species level often relies on anatomical characters, and modern revisions reveal that some taxa, such as members of the Verrucariaceae, can only be identified unambiguously by using molecular markers (Orange 2012). Collecting and preserving voucher specimens is therefore essential for all work involving freshwater lichens, such as species inventories, ecological and physiological studies. However, legal obligations may limit the size of the sample taken, particularly in the case of rare species. For analyses of communities, Coste (2005) recommends rock slices of 150–200 cm^2, but this is not justifiable at protected sites and technically impossible to achieve with hammer and chisel at other sites, especially if a large sample could eradicate most or all of a population of a rare species. Collection of some taxa, such as *Collema dichotomum* in the United Kingdom, is prohibited without a permit from the respective conservation authority. In an area where the species pool is well known or where larger samples cannot be obtained, the collection of smaller pieces of thalli and individual perithecia from thalli with a razor blade, and photographing the thallus from which the fragments have been extracted, are recommended.

For each specimen collected, its vertical distance to the waterline (and an estimate of its relation to the current waterline and to the low- and high-water levels if visible marks can be traced), the type of substratum, the degree of shading, and, if relevant, the occurrence of splash-water should preferably be noted, ideally in a field notebook and on an annotation slip stored with the sample in the collecting bag. Wet samples can be stored in plastic bags for a short period of time in order to transport them to a field station, laboratory or herbarium, but need to be dried as soon as possible to avoid the risk of mould and the degradation of DNA, particularly in warm climates. If plastic bags have to be used, a pencil-written annotation slip with the collection number should be inserted into the bag and also written on the outside of the bag with a waterproof permanent marker. Wherever possible, samples should be air-dried directly at the collecting site and then stored in paper packets or cardboard boxes. If the collection consists of several chips of rock, wrapping them individually in tissue paper helps to prevent damage of the lichen surfaces during transport. The storage and management of a collection of freshwater lichens follows the same principles as for lichens from other habitat types (See Obermayer 2002). On site measurements of water, pH and conductivity are particularly important, but for amphibious lichens they should be complemented in the laboratory by measurements of the surface pH of the substratum using a flat tip sensor.

In order to achieve an approximation of the full diversity of freshwater lichens, all sufficiently hard substrate types (e.g. rocks, roots) should be inspected. The highest diversity can be expected on the largest and most stabilized rocks, particularly if they are in a semi-shaded position, but both shaded and fully exposed areas need to be examined to capture the full diversity. Wood can host freshwater lichens if it is firmly fixed and the surface is hard and stable. Loose pebbles can be colonized by pioneer lichens and add species otherwise not found on other substrata, but sampling this habitat alone will give a strongly distorted image of the lichen diversity at a site. Methods for plot based and plotless randomized sampling strategies can be found in Douglas (2008), Wolseley and James (2002), and Aptroot and Sparrius (2002). Patchiness of occurrences and small-scale mosaics of different freshwater lichen communities in small streams, rapids or waterfalls limits the applicability of commonly used methods. Flexibility and case-specific solutions for ecological studies on freshwater lichens are needed for individual areas and habitat types. (See also Sections 15.5 on zonation and 15.7 on communities.)

15.2 Physiological challenges for freshwater lichens

15.2.1 Water saturation and diffusion resistance

The frequency and length of periods with full immersion in water, either by splash-water or as a result of raising water levels, is critical for the survival of all aquatic lichens. Laboratory experiments with the specialized aquatic lichen *Verrucaria funckii* showed irreversible damage following even short-term desiccation (Ried 1960b), but populations of this species exist at sites which are not permanently inundated (Keller and Scheidegger 1994; Gilbert and Giavarini 1997; Thüs 2002). Most species which typically occur in the permanently inundated parts of a stream can be found occasionally in the splash-water zone or in seepages if the micro-habitat is shaded (Ried 1960a); the water-holding capacity of the substratum can also largely influence the desiccation tolerance of aquatic lichens (Coste 2005). Generally, desiccation tolerance may depend partly on the conditions under which a lichen thallus has established and developed at a particular site. Water temperatures also affect inundation tolerance of lichens (Ried 1960b) and species such as *Phaeophyscia sciastra*, called "terrestrial" elsewhere, tolerate long periods of inundation (up to 3 months continuous submersion in standing water with only minor vitality loss) in arctic areas of Canada (Marsh and Timoney 2005). In warmer areas, this species is absent from regularly inundated sites, and only occasionally colonizes the splash water zone (e.g. Wirth 1972).

The productivity of lichens as a miniaturized ecosystem is limited by the burden of a large proportion of purely heterotrophic fungal cells which have to be supported by the algal or cyanobacterial partner. This puts the lichen symbiosis at a disadvantage compared to competing free-living algae, cyanobacteria, mosses or liverworts, at least in habitats under light stress. Consequently, whilst many aquatic lichens can become dominant in well-lit or half-shaded stretches of water bodies, and are often absent from deeply shaded areas (Thüs 2002), some specialized species (e.g. *Dermatocarpon luridum*, *Porina guentheri*, *P. grandis*) are better developed in shaded stretches.

Lichens are generally absent from deep water, mainly due to light limitation (Hawksworth 2000), but their vertical distribution limits seem to be strongly influenced by the transparency of the water. In clear lakes and rivers, lichens can be found several metres below the water surface; e.g., Luther (1954) found *Hydropunctaria rheithrophila* in a Finnish river at a depth of 2.6 m and an unidentified *Verrucaria* sp. was found at more than 7 m depth in the extremely clear waters of Lake Baikal (Kulikowa et al. 2008). This is a range similar to the only truly aquatic lichen in salt water, *Wahlenbergiella serpuloides*, which has been recorded down to 10 m below the low tide in Antarctic waters (Lamb 1973; Øvstedal and Smith 2001). It is remarkable that Santesson (1939) in his pioneer study on the lichens of the Aneboda lake system in Sweden did not record any lichens at a considerable depth and concluded that "aquatic" lichens did not exist and freshwater lichens should at best be called "amphibious". However, although species numbers are small, there is clear evidence that the majority of the known populations of species such as *Collema dichotomum*, *C. ramenskii*, *Peltigera hydrothyria*, *Hydropunctaria rheithrophila*, *Verrucaria aquatilis* and *V. funckii* remain fully submerged for at least several uninterrupted years (Hawksworth 2000; Thüs 2002).

The lack of sufficient light is discussed as one of the reasons why eutrophication in lakes and slow-flowing rivers appears to have a stronger effect on freshwater lichens compared to sites in fast-flowing smaller streams. Erichsen (1957) found fully aquatic lichens in several lakes in Northern Germany, but only *Verrucaria aquatilis* was still found in shallow parts of a single lake in a recent re-survey of this area (Thüs 2002); due to increased eutrophication and plankton blooms in the other lakes, all other fully aquatic lichens had vanished. Sub-aquatic lichens coped better with eutrophication of the German lakes and *Verrucaria praetermissa* could still be found on the shores of several lakes. Similarly, in permanently submerged parts of large rivers with a high sediment load, such as in the lower stretches of the Rhine and Elbe in Germany, lichens are entirely absent; however, in the splash-water zone of these rivers, where their sediment cover is removed by rainfall and wave action, amphibious lichen communities are well developed (Heibel and Mies 1996; Thüs 2006; Thüs and Schultz 2008).

While specialized aquatic lichens are typically sensitive to high desiccation, for most sub-aquatic lichens desiccation is more easily tolerated. Inundation is a stress factor for sub-aquatic lichens which limits their vertical distribution on stream banks and lake shores (Ried 1960a, b). Various strategies have evolved which reduce the risks from "drowning" following the blocked diffusion of gases through the lichen system. Many lichens employ anatomical and morphological structures to reduce and channel the influx of excess water and extend the length of periods when the water content is at a suitable level for net carbon uptake (Büdel and Scheidegger 2008). Routes for water influx and their containment can be regulated by a highly complex mosaic of hydrophilic and hydrophobic structures in the lichen thallus (Hamlet et al. 2011). Most lichens have hydrophobic coatings of excreted proteins (hydrophobins) surrounding fungal and algal cells in the thallus; a high hydrophobicity has been demonstrated for the surface of *Bacidia inundata* (Hauck et al. 2008), but this sub-aquatic species is not restricted to freshwater habitats and is possibly only inundation tolerant to a certain degree. While periods of over-saturation can be efficiently reduced with hydrophobic surfaces, this strategy does not work under permanent submersion, where sooner or later the entire thallus becomes water-filled. In such cases, maintaining empty spaces between hyphae and algae and the production of hydrophobins is probably superfluous. Fully aquatic lichens typically have no free spaces between fungal hyphae and the photobiont cells, and sub-aquatic lichens can easily be distinguished from these species by the presence of these empty spaces in the thallus (Zschacke 1927).

Lichens can resist extreme temperatures (Lange 1953), but only in a state of desiccation where respiratory losses are kept to a minimum. The combination of high temperatures and high water levels results in quickly rising respiration rates which cannot be compensated by photosynthesis and the lichen starves due to carbon depletion, as occurs in the understorey of tropical lowland rainforests. The lichen biota in this habitat are largely specific to this environment and are mostly composed of thin crustose growth forms which facilitate gas exchange and a have a high photobiont density to enhance the balance between respiration loss and photosynthetic gains (Lakatos et al. 2006). Similar morphological traits are found in aquatic lichens, but remarkable differences exist in the identity of the associated photobionts. Typical for the understorey of the tropical lowland rainforest are the genera *Trentepohlia* and *Phycopeltis*, the latter only on living leaves. Both genera, known to be free-living and lichenized, cope well with high temperatures and become active even in high air humidity, but are rarely found free-living in aquatic environments.

The fungal components of lichens in aquatic habitats often invest heavily in hydrophobic coatings of the hyphae in order to prevent diffusion limitation during the periods of photosynthetic activity of its photobiont. In sub-aquatic environments,

Trentepohlia occurs in some species but *Phycopeltis* seems to be entirely absent. Unusual algal partners instead dominate in truly aquatic lichens, such as the genera *Dilabifilum* (Ulvophyceae, Chlorophyta) and *Heterococcus* (Heteropediaceae, Chromista) which are unknown from lichens in other habitats but are the commonest photobionts for specialized aquatic and amphibious species in the family Verrucariaceae (Tschermak-Woess 1988; Thüs et al. 2011). Only in temporarily inundated zones and areas wetted mainly by splash-water does the diversity of freshwater lichen algae increase with *Trentepohlia*, *Asterochloris* and *Trebouxia* all being present, particularly in cool areas where the low water temperatures ensure a high solubility of gaseous CO_2 during periods of inundation. Many Verrucariaceae in sub-aquatic (amphibious) habitats are associated with representatives of the genus *Diplosphaera*. This genus appears to consist of taxa with a wide tolerance to drought and (limited) periods of full submersion. *Diplosphaera* strains in freshwater lichens, and those from drier habitats, are not separated in molecular studies (Thüs et al. 2011).

Most work on the phylogenetic position of aquatic or sub-aquatic lichens has focussed on the eukaryotic photobionts associated with fungi of the family Verrucariaceae. The molecular identities of the symbiotic green algae and cyanobacteria in other freshwater lichens remain to be explored. However, based on their morphology, the genera *Nostoc*, *Stigonema*, *Scytonema* and representatives of the family Rivulariaceae have been identified (Tschermak-Woess 1988). Cyanobacteria possess a highly efficient CO_2 concentration mechanism and the eukaryotic algae involved in the freshwater lichen symbiosis are prospective candidates for equally efficient pathways to make use of particularly low concentrations of carbon sources in the strongly diffusion limited micro-environment of the constantly hydrated aquatic lichen thallus.

The Verrucariaceae (Figs. 15.1–15.3), the fungal family which has the highest number of amphibious species (Hawksworth 2000), contains taxa with an aquatic lifestyle spread over many different clades (Gueidan et al. 2007), although some clades seem to contain mostly or exclusively freshwater species, and others marine or terrestrial taxa (Gueidan et al. 2009; Thüs et al. 2011). Most species have a clear preference for either saline or freshwater conditions (Orange 2012), but there are exceptions, and the recently described *Hydropunctaria oleae* has been found in the intertidal zone as well as in a coastal freshwater stream (Orange 2013). Very few lichen species are known to have their ecological optimum in brackish water. *Verrucaria ditmarsica* is primarily a coastal species which is particularly frequent in estuaries with slightly brackish water. It was recently collected in large quantities from the pebbles of a raised beach on the Isles of Scilly (UK) which was constantly percolated by dripping water from a higher freshwater aquifer.

There is no clear evidence whether the pH or the total amount of solutes in the water is more relevant for the distribution patterns of lichens since these factors are often correlated in field studies on freshwater lichens (Nascimbene et al. 2007). In British inland lakes, Gilbert and Giavarini (2000) observed major differences between the lichen communities of lakes with hard and soft water, but they omit pH values.

However, water acidity had already been recognized as a major factor determining the lichen community composition of streams and rivers there (Pentecost 1977; Gilbert and Giavarini 1997). In German streams, Thüs (2002) found that differences in the surface pH of the rock substrata were more important than the pH of the stream water itself for the lichen species composition in temporarily inundated areas or the splash-water zone of small streams. Local small-scale differences can result from the interaction of the rock surface, airborne (e.g. acidic) pollutants and the stream water which can have a buffering and alkalising effect, which gradually reduces with decreasing wetting frequency and duration. A strong impact of alkaline deposits along the lichen trimline was also suggested as the reason for the significant proportion of calciphiles on an otherwise strongly acidic granitic substratum at lake margins of the Peace-Athabasca Delta in Canada (Timoney and Marsh 2005). Similar concentrations of calciphiles above the water line have been observed on river banks of rivers over acidic bedrock in subtropic parts of South America.

15.3 Freshwater lichens as a food source for other organisms

Many lichens are hosts to more or less specific parasitic (lichenicolous) fungi; of the 20 or so species observed on freshwater lichens, about half are specific to such hosts. *Ionaspis lacustris* (Fig. 15.4) and *Koerberiella wimmeriana* are by far the hosts to the highest number of parasite species (Rambold et al. 1999; Orange 2002, 2009).

Few field observations and no experimental work have been carried out on the role of freshwater lichens as a food source, their role as shelter for animals, and as a substratum for other algae and cyanobacteria. Similarly, no studies have as yet been carried out on associated heterotrophic bacteria. Large cushions of *Dermatocarpon* spp. can provide shelter for small invertebrates, but it is still unknown whether the presence of foliose freshwater lichens has a significant effect for maintaining local invertebrate populations and whether they are also using the lichen as a food source. Studies on coastal rocks in the Baltic Sea area suggest that foliose lichens can host large numbers of Collembola, while mites inhabit crustose lichens (Søchting and Gjelstrup 1985). Interestingly, there are no observations of mites from amphibious or aquatic freshwater lichens. Occasional observations have been made of snails and Trichoptera larvae grazing on submerged lichens. Gastropoda and Collembola have been found on sub-aquatic species and taxon-specific grazing marks were observed on amphibious lichens, although no major damage by herbivores has been observed in fully aquatic lichens (Thüs 2002). It is possible that the main food source for the observed grazers is epiphytic algae and detritus from the lichen surface. Contamination by epiphytic algae (e.g. *Stichococcus*, *Trebouxia*) is a significant problem for molecular approaches to characterise the symbiotic algae of freshwater lichens and microscopically checked culturing of algae from the inside of thallus fragments is recommended in critical cases (Thüs et al. 2011). The subgelatinous thalli of many aquatic Verrucariaceae have never been observed with grazing damage when growing submerged, and epiphytic algae establish less frequently on them, compared to non-subgelatinous lichens from sub-aquatic areas (Thüs 2002). If the smooth thallus surface and the subgelatinous consistency of the aquatic Verrucariaceae represent an efficient anti-fouling mechanism that holds back both invertebrate grazing and reduces algal growth, or if the reduced grazing frequency is due to the lack of a dense cover by epiphytic algae, merits further investigation.

15.4 Biogeography of freshwater lichens

Saxicolous freshwater lichens occur in all parts of the world wherever stable substrata are present, water chemistry is not too extreme, and silting is limited. Species numbers are generally higher in cooler water bodies, but are lower in polar and alpine environments (Thomson 1997; Øvstedal and Smith 2001; Nascimbene and Nimis 2006). In the tropics, amphibious species are common in cool mountain streams, but also locally abundant on the banks of lowland watercourses. However, entirely aquatic lichens are rare in lowlands, but exceptions exist, such as the endemic *Jenmania goebelii* in rapids and waterfalls at rather low elevations of the Guiana Shield (Fig. 15.5). For epiphytic freshwater lichens, restriction to cool areas seems to be more pronounced: in Europe they are common, e.g., in the countries surrounding the eastern part of the Baltic Sea (Motiejunaite 2003; Pykälä and Breuss 2009; Hadchułka 2011) and in parts of the Alps (Thüs and Schultz 2008), but only a few species are known from lowland areas of western Europe (Zschacke 1927; Aptroot and Thüs 2011).

The numbers of known freshwater species indicate a diversity centre in Europe, but this is mostly likely due to the longer history of exploration of such lichens there. There is no reason to assume that, e.g. North America has a lower diversity of freshwater lichens. In other continents the exploration of freshwater lichens has often focussed on taxonomic work in Verrucariaceae, but knowledge of other taxa frequently found in freshwater habitats, such as the Lichinaceae and species of *Aspicilia* and *Ionaspis*, is still rudimentary for most parts of the world. Little is known of the freshwater lichens of permanent water bodies in Africa, except for some scattered reports from Morocco (mostly by R.G. Werner, compiled in Egea 1996). In Central and South America some freshwater lichens were described as new and endemic (e.g. *Ephebe brasiliensis*, *Jenmannia goebeli*, *J. osorioii*, *Rinodina rivularis*, *Staurothele diffractella*, *Verrucaria laevatoides*), but no ecological studies or surveys specifically dedicated to freshwater lichens have been carried out and species numbers are certainly much higher than currently known. In Asia important taxonomic accounts, ecological notes and biogeographic information can be found for Japan (Yoshimura 1963; Harada 1992, 1993, 1996a, b) and for Yunnan in China (Harada and Wang 1996, 2004, 2006a, b, 2008). Subtropical parts of Japan have a high number of endemics and other species not found in temperate areas, but as one moves towards the humid tropics, the percentage of species found in other climatic zones drops further. Of the ten species identified by Zahlbruckner (1934) from Indonesian collections at elevations of 500–1700 m, only two are reported from other tropical countries and none from temperate areas. Freshwater lichens in Australia have been studied in more detail and a total of 35 species are currently listed as typical for its creeks and rivers (McCarthy 2013), circa 50% of which are restricted to either Australia, or Australia and Melanesia. Most of the more widespread taxa are reported from the highest elevations in mainland Australia, and from Tasmania which has a temperate to alpine climate in its mountainous parts (McCarthy 1995, 2002; Kantvilas and McCarthy 2003). All freshwater lichens known from Papua New Guinea are also reported from Australia (Aptroot 2010). In New Zealand four species are regarded as endemic (Galloway 2007), while the rest are shared with temperate areas of Australia (Johnson and McCarthy 1997; McCarthy and Johnson 1997, 1998). In the polar regions, Arctic freshwater lichens appear to be a reduced set of species which are widespread in the temperate zone (Thomson 1997), but species in the melt-water streams and seepages of Antarctica are mostly endemics (Lamb 1948; Øvstedal and Smith 2001).

Determination of global distribution patterns for freshwater lichens are hampered by the absence of molecular data for collections from tropical and Southern Hemisphere areas, and it can be expected that cryptic speciation at one site and high morphological plasticity on another may significantly distort the current picture. Genetically, the specialized freshwater lichen *Dermatocarpon luridum* var. *luridum* is remarkably uniform across populations of Europe and North America, while the amphibious taxa in the *D. miniatum* species complex consist of poorly understood semi-cryptic and cryptic species, including some varieties of *D. luridum* (Fontaine et al. 2012). Similarly, the exclusively aquatic to sub-aquatic genus *Hydropunctaria* includes taxa which can only be distinguished with certainty by DNA sequencing (Orange 2012). Based on these results, it has to be concluded that most species circumscriptions in Verrucariaceae will need to be reassessed by molecular markers before a realistic understanding of the biogeographic patterns for individual species can be achieved.

15.5 Zonation

The co-occurrence of the same lichen species under specific ecological conditions has been observed in the study of freshwater habitats for many years (Frey 1922). These lichen communities can occupy large continuous areas, such as lake shores (Santesson 1939; Gilbert and Giavarini 2000) and river banks (Rosentreter 1984; Aptroot and Seaward 2003). On banks and shores subjected to fairly regular patterns of water level fluctuations, such communities can typically occur in vertically arranged "belts" (zones) which are often dominated by differently colored taxa. These zonations can be observed on the shores of large water bodies worldwide, but they are often indistinct or absent in small streams and in situations with intensive turbidity where wetting by splash water or frequent water level fluctuations are not conducive to the formation of distinctive zonations (Wirth 1972; Thüs 2002).

Nevertheless, the sequence of life-forms in horizontal zonations is surprisingly constant even across climatic gradients worldwide, with crustose pyrenocarpous lichens, usually dark colored, in the most frequently submerged zones (e.g. *Hydropunctaria rheitrophila*, Fig. 15.1, and *Pterygiopsis australiensis*, Fig. 15.6), followed by an intermediate band which usually includes specialized freshwater lichens with perithecia or apothecia, often with lighter colors (e.g. *Ionaspis lacustris*, Fig. 15.4), and finally an area of terrestrial lichens, dominated by apothecioid crusts, often associated with more complex foliose and fruticose growth forms. At the species level, these bands can contain different taxa, depending on the climatic conditions of the area, the chemistry of the water and the composition of the substratum. A detailed analysis of these patterns in various parts of the world is currently not yet possible due to the different methodologies adopted for ecological studies and the various species circumscriptions adopted for the frequently dominating *Verrucaria* taxa. The zonations are best described in terms of their relation to the inundation frequency where this factor can be demonstrated to be of a dominating nature (Gilbert 1996) (Figs. 15.7–15.10).

The *submerged zone* is inundated for most or all of the year. This is the typical habitat for aquatic lichens (*sensu* Thüs and Schultz 2008) that are highly desiccation sensitive specialist taxa forming extensive colonies here and able to grow for at least several years in permanently submerged conditions. They can be accompanied by some sub-aquatic lichens, particularly in cold springs and headwaters (Nascimbene et al. 2007), but usually only in small patches. Species numbers in this zone generally increase with decreasing water temperatures, being, e.g. more diverse in upland areas and cold springs of the temperate areas, but are not known from this habitat in tropical lowland water bodies.

The *fluvial mesic zone* is the area immediately above the low water level and hosts the highest diversity of specialized freshwater lichens. All aquatic lichens (with the possible exception of *Collema elenkii*) have been found at least occasionally in this zone, but more freshwater lichens are specific for this habitat where lichens face frequent changes between extended periods of desiccation and inundation.

The *fluvial xeric zone* is rarely inundated or splashed, being dominated by lichens which can also occur in other habitats. Sub-aquatic freshwater lichens can extend into this zone but contribute only small cover percentages.

The *upper terrestrial zone* usually lacks inundation and freshwater lichens are absent, but due to increased air humidity and lower air temperatures the species composition on rocks and trees can differ considerably from other sites further away from the water body. Lichens which have their ecological optimum in this habitat or

Figs. 15.7–15.10: Fig. 15.7 Cold mountain stream in the Talamanca Mountains (Costa Rica) with rich populations of aquatic lichens. **Fig. 15.8** River Trettach (Northern Alps, Germany) where fully aquatic lichens are absent but stabilized boulders on the river bank are covered in the sub-aquatic *Verrucaria submersella*. **Fig. 15.9** Sun exposed stretches with rapids and large rock outcrops supporting freshwater lichens in a tropical lowland river (Sabah, Malaysia). **Fig. 15.10** Habitat of the facultative sub-aquatic lichens *Ephebe lanata* and *Ionaspis lacustris* below a waterfall (Isle of Man, UK).

are a characteristic element here are often referred to as "riparian" lichens (Thüs and Schultz 2008).

15.6 Lichen trimlines

The so-called "trimline" is defined as the mark created at the lower limit to which typical terrestrial lichens with very limited inundation tolerance occur and where they become replaced by (often mostly crustose) freshwater communities. In cool temperate areas the trimline is often marked by a sharp transition from communities dominated by crustose lichens to those with large foliose species (e.g. Blomgren and Naumann 1925; Santesson 1939; Beschel 1954). Common foliose lichens above the trimline are often representatives of the family Parmeliaceae with generally more inundation sensitive photobionts of the genus *Trebouxia*. The trimline can be obscured by the occurrence of some specialized freshwater taxa in the foliose genus *Dermatocarpon* with particularly leathery and firmly attached thalli and an inundation tolerant photobiont (*Diplosphaera* sp.). These lichens typically grow around the low water level or even beyond and should not to be taken into account for localising the trimline.

In lowland streams of the wet tropics, a trimline is often clearly visible, but marked by a change in the composition of crustose taxa instead of a switch from crustose to foliose species. Using calibrated growth rate data obtained from an analysis of lichens growing on dated tombstones of nearby cemeteries, the trimlines can be dated, allowing a detailed analysis of high floods over periods of time between several years and decades (Gregory 1976; Timoney and Marsh 2004). In northern Canada, the form and structure of the trimline can be used as an indicator of physical disturbance; the absence of a distinct trimline is as a sign of scouring of ice blocks during the winter months (Timoney and Marsh 2005). In lowland areas without rocks or with only unstable pebble banks, a "lichen line" can be observed on trees bordering watercourses or lakes. Similar to the trimlines on rocks, these lichen lines have been proposed as markers of exceptional high water levels, but here they are usually not marking a transition from terrestrial to amphibious lichen communities, but the lower limit of all lichens (Beckelhimer and Weaks 1984; Hale 1984). The absence of amphibious lichens below this lichen line is due not only to the large amount of fine sediments deposited on the bark during the flood period of many lowland watercourses, but also the possible effects from contamination with pesticides (Beckelhimer and Weaks 1986). Currently there is a lack of robust data for a comparative analysis of the patterns of epiphytic lichens in lowland and mountainous areas with different levels of sediment load and pollutants. Ideally such studies should be carried out in cool-temperate areas where epiphytic freshwater lichens occur in order to determine which factors affect either the transition from terrestrial to amphibious or from terrestrial to a total absence of lichens on trees in the temporarily inundated habitats. Ecophysiological studies in the field and under laboratory conditions could complement this approach to provide a robust response model for individual lichen species.

15.7 Freshwater lichen communities

Lichens are generally long-lived organisms and due to their association with photoautotrophic symbionts have a permanent presence in vegetation cover, the different habitat types characterized by typical sets of co-occurring species, the resulting communities of which have been described and named in a similar way as plant-dominated vegetation units. Frey (1922) informally described and named freshwater lichen communities in the Swiss Alps, but only through Santesson (1939) did a formalized system for freshwater lichen communities start to emerge. He distinguished five different communities ("Sozietäten") in the Swedish Aneboda lake system. The classification follows the Scandinavian tradition of phytosociology with an emphasis on dominating species. Attempts to create a single hierarchically structured classification of lichen dominated communities for terrestrial and aquatic habitats have mostly followed modifications of the phytosociological system of Braun-Blanquet (1951). In Japan, Yoshimura (1963) described an amphibious community which he named as *Endocarpeto-Dermatocarpetum minitati*, but his description is based on studies of sites in only a single gorge, and the name of the community does not follow the rules of phytosociological nomenclature (Weber et al. 2000) and requires validation. In Central Europe, Klement (1955) created the first synopsis and provided synthetic tables, summarizing earlier relevés by other authors (e.g. Frey 1922), but created a hybrid of data derived by various different methodologies.

The apparent heterogeneity of the sample areas for some of the documented amphibious communities was criticized, e.g. by Wirth (1972) who based a revised system on original relevés for siliceous saxicolous communities of the non-alpine areas of Central Europe obtained by a single methodology. Further relevés following this principle were added by Thüs (2002) and integrated into a phytosociological framework which took into account changes in species circumscriptions. Both publications focussed on saxicolous communities on siliceous rocks, except the description of the *Verrucario-Staurotheletum cataleptae* (Wirth 1972) which describes communities from large rivers with calcareous substrata in the Alps. Communities in limestone streams and rivers in France are described by Asta et al. (1973), and observations from streams on both carbonaceous and siliceous bedrock in Spain were contributed by Pereira (1992). Freshwater lichens on siliceous rocks are also treated in an overview of the lichen communities in the British Isles (James et al. 1977), but there are no original relevés published from this area and it is unclear which methodology was followed to recognize the named communities.

For freshwater streams with siliceous rocks, analyses of phytosociological data following a modified Braun-Blanquet system have been published for the French Département of Tarne (Coste 2005). A combination of these data with new relevées from Central and SW France was analyzed by multivariate statistics and confirmed previous results obtained from manual sorting (Coste 2009). Both studies are based on a "total sampling" approach where entire slabs of rock are sampled in the field and examination of the community is carried out in the laboratory. The sizes of the slabs chosen were in the range of 150–200 cm^2; this is based on the assumption that the same size of the minimum area for terrestrial saxicolous lichen communities (Roux 1990) is also satisfactory for freshwater habitats, but this has yet to be tested. Once the minimum area for freshwater lichen communities is established, the methodology would have greater appeal given the extent of taxonomic changes in freshwater lichens, and the opportunity to eventually study all associated organisms including microbial communities, which can only be analyzed by molecular markers. In protected areas, however, where sampling has to be limited to a necessary minimum, the recommended choice will remain, namely the collection of vouchers of selected representatives for each "morphotaxon". However, this method is a highly invasive sampling technique and unsuitable for permanent plots and long-term monitoring studies. In streams with very hard rock types and smooth surfaces very often it is simply impossible to obtain a slab of the recommended size, and in small streams a different methodology has to be applied where pebbles and rocks are so small that a minimum area of 150 cm^2 already exceeds the total upper surface of the substrata. Outside Central and SW Europe, the focus of studies on freshwater lichens has been mostly based on more case-specific and hypothesis-driven classifications, developed for the individual task and without the need for a hierarchical naming system.

15.8 Freshwater lichens as bioindicators

Freshwater lichens are typical elements of benthic communities in pristine water bodies and their diversity often decreases with the intensity of landscape transformation by human activities. For this reason, lichens continue to be discussed as potential indicators of water quality, patterns of water level fluctuations, the impact of acidic pollution, silting intensity and stream-bed stability (Thüs 2002; Gob et al. 2003; Nascimbene et al. 2013). Comparisons with other organisms recognized as bioindicators for freshwater habitats (e.g. diatoms, green algae, bryophytes) have shown that lichen diversity and response patterns to environmental change cannot be predicted from analyses of other organisms (Nascimbene et al. 2011). Consequently, lichen communities can be used to assess niches and aspects of environments which are insufficiently covered in current standard procedures for environmental assessments. Methodological problems remain due to taxonomic problems with semi-cryptic and cryptic species, the scarcity of experts for the identification of critical taxa, and the limited number of publications using statistically robust methods for randomized sampling and analysing data from freshwater lichen assessments (Nascimbene et al. 2013). For these reasons there are many papers proposing freshwater lichens as bioindicators, but only a few where the results can be directly applied in landscape management (Beschel 1954; Chatenet and Bonieau 2001; Timoney and Marsh 2004, 2005). A larger number of studies has made use of the lowermost growth points of terrestrial lichens on trees as indicators of water levels. (See Section 15.6 on trimlines.) Although these studies often do not include any specialized freshwater lichens, they are important for freshwater ecological studies as tools for the assessment of the magnitude and dating of former high water levels.

15.9 Water quality

Freshwater lichens are generally poor competitors and for this reason become easily replaced by faster growing organisms in nutrient-rich water bodies (Thüs 2002). Exceptions in polluted streams can be stretches with fast water-flow, where mats of cyanobacteria and filamentous green algae may be swept away, but some crustose lichen species such as *Verrucaria aquatilis* and *V. praetermissa* survive (Thüs 2002, 2006). Experimental evidence has shown, e.g., that high concentrations of more than 40 mmol of nitrate are detrimental to the growth rate of *Peltigera hydrothyria*, but at lower levels growth was actually enhanced (Davis et al. 2000). Generalized conclusions on lichens as indicators of water quality must take these facts into account. While it is correct to say that freshwater lichens are absent wherever stronger competitors can make more efficient use of elevated nutrient levels, data on the mechanisms underlying the widespread assumption that water pollutants directly damage particular freshwater species are limited.

The role of pesticides attached to fine sediment deposits from run-off water in agricultural areas has been discussed as a possible reason for the absence of lichens on the temporarily inundated bark of lower tree trunk zones downstream of intensively used agricultural areas, but no experimental evidence for this hypothesis has been provided (Beckelhimer and Weaks 1986). In the absence of a methodological framework, specific responses of most freshwater lichens to particular pollutants, and their interaction with other environmental factors, have not been applied to map water quality. Hence, Thüs (2002) questioned the suitability of lichens as bioindicators for this purpose unless more specific aspects of water quality are defined and the applicability of lichen-based methodologies for these aspects is demonstrated for a specific area. Most proposals to use lichens as bioindicators are based on passive monitoring approaches, with the exception of transplants of *Dermatocarpon luridum* as active accumulation indicators for heavy metal deposition in a stream in France (Chatenet and Botineau 2001).

15.10 Conservation

Classifications of the threat to a large number of freshwater lichen species have been carried out mainly in European countries (e.g. Liška et al. 2008; Wirth et al. 2011; Woods and Coppins 2012). Traditionally these have been based on fact-gathering on the changes to habitat conditions which are assumed to have affected particular species in a defined area. These assessments are usually based on the personal experience of the respective authors, but are rarely supported by quantified published data to document these changes. Based on field observations and laboratory experiments, the following anthropogenic activities have been identified as relevant for the threat to freshwater lichens: disturbance in the continuity of water supply, reduction of flow velocity due to dam construction, extraction of stable substrata, reduced transparency of the water, increased shading, direct and indirect acidification of water, increase in water temperature, and elevated levels of nitrogen compounds (Davis et al. 2000, 2003; Thüs 2002).

The effects of some of these changes on freshwater lichens at specific sites have been documented. In Germany, historic reports of freshwater lichen occurrences were compared with the results of recent surveys at the same sites by Thüs (2002), and in the UK, individual populations of *Collema dichotomum* were analyzed and recommendations for site management and further research were summarized in a Biodiversity Action Plan (JNCC 2010). Methodologies for quantitative assessments of individual populations are described by Douglas (2008). At the landscape level, sampling and data analysis strategies for conservation assessments of freshwater lichens, based on a combination of randomized sampling across large areas and targeted sampling at known sites, are outlined in Glavich (2009). Some freshwater lichen species respond surprisingly quickly after improvement of habitat conditions. Successful recolonization of freshwater habitats has been documented for areas in Germany and Italy after improvements in water and air quality (Thüs 2002, 2006) or the opening and reconstruction of previously closed watercourses (Nascimbene et al. 2009), in the latter case, recolonization occurring within 2–3 years.

Although populations of other European freshwater lichens, such as *Collema dichotoma*, continue to decline for unknown reasons, changes in water-flow resulting from the erection of small-scale hydroelectric facilities has been identified as a growing concern for this species (JNCC 2010). The loss of water supply due to abstraction for drinking water has also eradicated all aquatic lichens in a small stream in the Taunus Mountains north of Frankfurt (Thüs 2002). In North America, *Peltigera hydrothyria* is declining in parts of its distribution range where extended drought periods have caused the small streams and springs which make up its habitat to dry out during summer (D.H.S. Richardson, personal communication). Hence, there is an urgent need to catch up with our knowledge of diversity and distribution patterns in other countries with particularly high rates of change in landscape use, such as tropical and sub-tropical hills and mountains where deforestation and conversion to agricultural land is accelerating. Collecting data on freshwater lichens now can help to identify, e.g., streams with clean, relatively cold water and high stream bed stability; such desirable attributes, from a sustainable land use perspective, can provide the baseline against which changes in the stability of freshwater ecosystems can be effectively monitored.

Acknowledgments

The authors would like to thank Prof. D.H.S. Richardson and Dr G. Mazzetta for communicating unpublished observations and voucher samples.

References

Aptroot A. Macrolichen keys and checklist for New Guinea lichens. Zschackia 2010;7:1–78.

Aptroot A, Seaward MRD. Freshwater lichens. In: Tsui CKM, Hyde KD, eds. Freshwater Mycology. Fungal Diversity Research. Hong Kong: Fungal Diversity Press 2003:101–110.

Aptroot A, Sparrius L. Monitoring red-listed lichens using permanent plots. In: Nimis PL, Scheidegger C, Wolseley PA, eds. Monitoring with Lichens – Monitoring Lichens. Dordrecht: Kluver 2002:395–398.

Aptroot A, Thüs H. *Verrucaria rhizicola*. In: Lumbsch HT, et al., eds. One Hundred New Species of Lichenized Fungi: a Signature of Undiscovered Global Diversity. Phytotaxa 2011;18:112.

Asta J, Clauzade G, Roux C. Etude de quelques groupements licheniques saxicoles et calcicoles du parc national de la Vanoise. Trav Sci Parc Natl Vanoise 1973;3:73–104.

Bates ST, Cropsey GW, Caporaso JG, Knight R, Fierer N. Bacterial communities associated with the lichen symbiosis. Appl Environ Microbiol 2011;77:1309–1314.

Beckelhimer SL, Weaks TE. The effects of periodic inundation and sedimentation on lichens occurring on *Acer saccharinum* L. Bryologist 1984;87:193–196.

Beckelhimer SL, Weaks TE. Effects of water transported sediment on corticolous lichen communities. Lichenologist 1986;18:339–347.

Beschel R. Die Stufung der Flechtenvegetation an den Inn-Ufermauern in Insbruck. Phyton 1954;5:247–266.

Blomgren N, Naumann, E. Untersuchungen über die höhere Vegetation des Sees Stråken bei Aneboda. Lunds Univers Årsskr N F. Abt 2 1925;21:1–52.

Bjelland T, Grube M, Hoem S, Jorgensen SL, Daae FL, Thorset IH, et al. Microbial metacommunities in the lichen-rock habitat. Environ Microbiol Rep 2011;3:434–442.

Braun-Blanquet J. Pflanzensoziologie. 2nd edition. Wien: Springer 1951.

Büdel B. Zur Biologie und Systematik der Flechtengattungen *Heppia* und *Peltula* im südlichen Afrika. Biblioth Lichenol 1987;23:1–105.

Büdel B, Scheidegger C. Thallus morphology and anatomy. In: Nash TH, ed. Lichen Biology. Cambridge: Cambridge University Press 2008:40–68.

Chatenet P, Botineau M. Utilisation des lichens dans la mise en évidence des éléments traces présents dans les cours d'eau. Crypto Mycol 2001;22:225–237.

Coste C. Lichens et communautés hydrophiles et subhydrophiles saxicoles-calcifuges du Départment du Tarn (France). Bull Soc Bot Centre-Ouest, n.s 2005;36:577–596.

Coste C. New ecology and new classification for phytosociology of hydrophilic lichens in acid watercourses in France. Acte du colloque des 3émes recontres Naturalistes de Midi-Pyrénées. Toulouse 2009:157–168.

Crespo A, Lumbsch HT. Cryptic species in lichen-forming fungi. IMA Fungus 2010;1:167–170.

Davis WC, Gries C, Nash TH. The ecophysiological response of the aquatic lichen *Hydrothyria ventosa* to nitrate in terms of weight and photosynthesis over long periods of time. Biblioth Lichenol 2000;75:201–208.

Davis WC, Gries C, Nash TH. The influence of temperature on the weight and net photosynthesis of the aquatic lichen *Peltigera hydrothyria* over long periods of time. Biblioth Lichenol 2003;86:233–242.

Douglas J. River lichens in Scotland: an introduction to the ecology, identification, management and monitoringog River Jelly Lichen and other species. Plantlife Scotland. URL: http://www.plantlife.org.uk/uploads/documents/PLINKS%20River%20Jelly%20Lichen%20Course%20notes%20 2008%20FINAL.pdf. Accessed on 16.8.2013.

Egea JM. Catalogue of lichenized and lichenicolous fungi of Morocco. Bocconea 1996;6:19–114.

Erichsen CFE. Flechten von Nordwestdeutschland. Stuttgart: Gustav Fischer Verlag 1957.

Fontaine KM, Beck A, Stocker-Wörgötter E, Piercey-Normore MD. Photobiont relationships and phylogenetic history of *Dermatocarpon luridum* var. *luridum* and related *Dermatocarpon* species. Plants 2012;1:39–60.

Frey E. Die Vegetationsverhältnisse der Grimselgegend im Gebiet der künftigen Stauseen. Mitt Naturforsch Gesell Bern 1922;6:85–281.

Galloway D. Flora of New Zealand Lichens. 2nd edition. Lincoln: Manaaki Whenua 2007.

Gilbert OL. The lichen vegetation of chalk and limestone streams in Britain. Lichenologist 1996;28:145–159.

Gilbert OL. Aquatic lichens. In: Seaward MRD, ed. Lichen Atlas of the British Isles. Fasc.5. London: British Lichen Society 2000.

Gilbert OL, Giavarini VJ. The lichen vegetation of acid watercourses in England. Lichenologist 1997;29:347–367.

Gilbert OL, Giavarini VJ. The lichen vegetation of lake margins in Britain. Lichenologist 2000;32:365–386.

Glavich DA. Distribution, rarity and habitats of three aquatic lichens on federal land in the U.S. Pacific Northwest. Bryologist 2009;112:54–72.

Gob F, Petit F, Bravard JP, Ozer A. Lichenometric application to historical and subrecent dynamics and sediment transport of a Corsican stream (Figarella River-France). Quaternary Sci Rev 2003;22:2111–2124.

Gregory KJ. Lichens and the determination of river channel capacity. Earth Surface Processes 1976;1:273–285.

Gueidan C, Roux C, Lutzoni F. Using a multigene phylogenetic analysis to assess generic delineation and character evolution in *Verrucaria* (Verrucariales, Ascomycota). Mycol Res 2007;111: 1145–1168.

Gueidan C, Savić S, Thüs H, Roux C, Keller C, Tibell L, et al. Generic classification of the Verrucariaceae (Ascomycota) based on molecular and morphological evidence: recent progress and remaining challenges. Taxon 2009;58:184–208.

Hadchułka M. Freshwater lichens on submerged stones and alder roots in the Polish lowland. Acta Mycol 2011;46:233–244.

Hale ME. The lichen line and high water levels in a freshwater stream in Florida. Bryologist 1984;87:261–265.

Hamlet CAE, Shirtcliffe NJ, Pyatt FB, Newton MI, McHale G, Koch K. Passive water control at the surface of a superhydrophobic lichen. Planta 2011;234:1267–1274.

Harada H. A taxonomic study on the lichen genus *Staurothele* (Verrucariaceae) in Japan. Nat Hist Res 1992;2:39–42.

Harada H. A taxonomic study on *Dermatocarpon* and its allied genera (Lichenes, Verrucariaceae) in Japan. Nat Hist Res 1993;2:113–152.

Harada H. Taxonomic notes on the lichen family Verrucariaceae in Japan (IX). *Verrucaria rheitrophila* Zsch., new to Japan. J Japan Bot 1996a;71:317–322.

Harada H. *Verrucaria igii*, a new freshwater species of Verrucariaceae (lichenized Ascomycotina) from Japan. Bryologist 1996b;99:343–344.

Harada H. *Pyrenocollema japonicum*, a new freshwater species of pyrenocarpous lichen from Japan. Bryologist 1999;102:50–52.

Harada H, Wang LS. Taxonomic study on the freshwater species of Verrucariaceae (lichenised Ascomycota) of Yunnan, China (1). *Thelidium yunnanum* sp.nov. Lichenology 2004;3:47–50.

Harada H, Wang LS. Taxonomic study on the freshwater species of Verrucariaceae (lichenised Ascomycota) of Yunnan, China (2). Genus *Staurothele*. Lichenology 2006a;5:13–22.

Harada H, Wang LS. Taxonomic study on the freshwater species of Verrucariaceae (lichenized Ascomycota) of Yunnan, China (3). Genus *Thelidium*. Lichenology 2006b;5:23–30.

Harada H, Wang LS. Taxonomic study on the freshwater species of Verrucariaceae (lichenised Ascomycota) of Yunnan, China (4). Genus *Verrucaria*. Lichenology 2008;7:1–24.

Hauck M, Jürgens SR, Brinkmann M, Herminghaus S. Surface hydrophobicity causes SO_2 tolerance in lichens. Ann Bot 2008;101:531–539.

Hawksworth DL. Freshwater and marine lichen-forming fungi. In: Hyde KD, Ho WH, Pointing SB, eds. Aquatic Mycology across the Millenium. Fungal Divers 2000;5:1–7.

Hawskworth DL, Honegger R. The lichen thallus: a symbiotic phenotype of nutritionally specialised fungi and its response to gall producers. In: Williams MAJ, ed. Plant Galls: Organisms, Interactions, Populations. Clarendon: Fairlawn 1994:77–98.

Heibel E, Mies BA. *Staurothele frustulenta* Vainio – Verbreitung und Ökologie einer Flechte entlang des Niederrheines. Decheniana 1996;150:87–90.

Hodkinson BP, Lutzoni F. A microbiotic survey of lichen-associated bacteria reveals a new lineage from the Rhizobiales. Symbiosis 2009;49:163–180.

James PW, Hawksworth DL, Rose F. Lichen communities in the British Isles: a preliminary conspectus. In: Seaward MRD, ed. Lichen Ecology. London: Academic Press 1977:295–413.

Johnson PN, McCarthy PM. Aquatic pyrenolichens in New Zealand 1, distribution and habitats. Australasian Lichenology 1997;41:10–15.

Joint Nature Conservation Committee. UK priority species data collation – *Collema dichotomum*, 2010:1–15. URL: http://jncc.defra.gov.uk/_speciespages/234.pdf. Accessed on 15.8.2013.

Kantvilas G, McCarthy PM. *Hueidea* (Fuscideaceae), a new lichen genus from alpine Australia. Lichenologist 2003;35:397–407.

Keller C. Die Wasserflechten der Teigitsch zwischen der Langmannsperre und dem Kraftwerk Arnstein (Steiermark, Österreich). Herzogia 2000;14:49–58.

Keller C, Scheidegger C. Zur Verbreitung von Wasserflechten in Abhängigkeit zur jährlichen Überflutungsdauer im Flüelatal (Schweiz, Kanton Graubünden). Herzogia 1994;10:99–114.

Klement O. Prodromus der mitteleuropaischen Flechtengesellschaften. Feddes Repert Beih 1955;135:5–194.

Kulikova NN, Suturin AN, Boiko SM, Lishtva AV, Paradina LF, Saibatalova EV, et al. Original data on the diversity, ecology and chemical composition of aquatic and subaquatic lichens (Lichenes) of the stony littoral of Lake Baikal. Siberian J Ecol 2008;15:399–340.

Lakatos M, Rascher U, Büdel B. Functional characteristics of corticolous lichens in the understorey of a tropical lowland rain forest. New Phytol 2006;172:679–695.

Lamb IM. Antarctic pyrenocarp lichens. Disc Rep 1948;25:1–30.

Lamb IM. Further observations on *Verrucaria serpuloides* M. Lamb, the only known permanently submerged marine lichen. Occas Papers Farlow Herb Cryptog Bot. Harvard Univ 1973;6:1–5.
Lange OL. Hitze- und Trockenresistenz der Flechten in Beziehung zu ihrer Verbreitung. Flora 1953;140:39–97.
Lendemer JC, O'Brien HE. How do you reconcile molecular and non-molecular datasets? A case study where new molecular data prompts a revision of *Peltigera hydrothyria* s.l. in North America and the recognition of two species. Opuscula Philolichenum 2011;9:99–110.
Liška J, Palice Z, Slavíková S. Checklist and Red List of lichens in the Czech Republic. Preslia 2008;80:151–182.
Luther H. Über Krustenbewuchs an Steinen fliessender Gewässer, speziell in Süd-Finland. Acta Bot Fennica 1954;55:1–61.
Lutzoni F, Pagel M, Reeb V. Major fungal lineages are derived from lichen symbiotic ancestors. Nature 2001;411:937–940.
Marsh JE, Timoney KP. How long must northern saxicolous lichens be immersed to form a waterbody trimline? Wetlands 2005;25:495–499.
McCarthy PM. Aquatic species of *Verrucaria* in Eastern Australia. Lichenologist 1995;27:105–126.
McCarthy PM. *Porina otagensis* (Trichotheliaceae), a new semi-aquatic lichen from southern New Zealand. Lichenologist 1999;31:337–340.
McCarthy PM. A new aquatic species of *Verrucaria* from alpine Australia. Lichenologist 2002;34:207–210.
McCarthy PM. Checklist of the Lichens of Australia and its Island Territories. Australian Biological Resources Study, Canberra. Version 17 September 2013. http://www.anbg.gov.au/abrs/lichenlist/introduction.html
McCarthy PM, Johnson PN. *Verrucaria amnica*, a new aquatic lichen species from New Zealand. Lichenologist 1997;29:385–393.
Moncada B, Lücking R, Suárez A. Molecular phylogeny of the genus *Sticta* in Colombia. Fungal Divers 2014;64:205–231.
Motiejunaite J. Aquatic lichens in Lithuania. Lichens on submerged alder roots. Herzogia 2003;16:113–121.
Nascimbene J, Nimis PL, Thüs H. Lichens as bioindicators in freshwater ecosystems – challenges and perspectives. Ann Bot 2013;3:45–50.
Nascimbene J, Nimis PL. Freshwater lichens in the Italian Alps: a review. Ann Limnol – Int J Lim 2006;42:27–32.
Nascimbene J, Thüs H, Marini L, Nimis PL. Freshwater lichens in springs of the eastern Italian Alps: floristics, ecology and potential for bioindication. Ann Limnol – Int J Lim 2007;43:285–292.
Nascimbene J, Thüs H, Marini L, Nimis PL. Early recolonization of stone by freshwater lichens of restored habitats: a case study in northern Italy. Sci Total Env 2009;407:5001–5006.
Nascimbene J, Spitale D, Thüs H, Cantonati M. Congruencies between photoautotrophic groups in springs of the Italian Alps: Implications for conservation strategies. J Limnol 2011;70(Suppl. 1): 3–8.
Obermayer H. Management of a lichen herbarium. In*:* Kranner I, Beckett RP, Varma AK, eds. Protocols in Lichenology. Berlin: Springer 2002:507–523.
Orange A. Lichenicolous fungi on *Ionaspis lacustris*. Mycotaxon 2002;81:265–279.
Orange A. Two parasitic species of *Placopyrenium* (Verrucariaceae) from freshwater habitats in north-west Europe. Lichenologist 2009;41:131–139.
Orange A. Semi-cryptic marine species of *Hydropunctaria* (Verrucariaceae, Ascomycota) from north-west Europe. Lichenologist 2012;44:299–320.
Orange A. Four new species of *Verrucaria* (Verrucariaceae, Ascomycota) from freshwater habitats in Europe. Lichenologist 2013;45:305–322.

Øvstedal DO, Smith RI. Lichens of Antarctica and South Georgia. Studies in Polar Resarch. Cambridge: Cambridge University Press 2001.

Pentecost A. A comparison of the lichens of two mountain streams in Gwynedd. Lichenologist 1977;9:107–111.

Pereira I. Flora, Vegetacion y ecologia de los liquenes acuaticos de Espana. Doctoral thesis, University of Barcelona 1992.

Pykälä J, Breuss O. Six rare *Verrucaria* species new to Finland. Österr Z Pilzkunde 2009;18:123–127.

Rambold G, Hertel H, Triebel D. *Koerberiella wimmeriana* (Lecanorales, Porpidiaceae) and its lichenicolous fungi. Lichenologist 1990;22:225–240.

Ried A. Stoffwechsel und Verbreitungsgrenzen von Flechten I. Flechtenzonierungen an Bachufern und ihre Beziehungen zur jährlichen Überflutungsdauer und zum Mikroklima. Flora 1960a;148:612–638.

Ried, A. Stoffwechsel und Verbreitungsgrenzen von Flechten II. Wasser- und Assimilationshaushalt, Entquellungs- und Subermsionsresistenz von Krustenflechten benachbarter Standorte. Flora 1960b;149:345–385.

Rosentreter R. The zonation of mosses and lichens along the Salmon River in Idaho. NW Sci 1984;58:108–117.

Roux C. Échantillonage de la végétation lichénique et approche critique des méthodes de relevés. Cryptogamie Bryol Lichénol 1990;11:95–108.

Santesson R. Über die Zonationsverhältnisse einiger Seen im Anebodagebiet. Meddel Lunds Univ Limnol Inst 1939;1:1–70.

Seaward MRD. Contribution of lichens to ecosytems. In: Galun M, ed. CRC Handbook of Lichenology. Vol.2. Boca Raton: CRC Press 1988:107–129.

Søchting U, Gjelstrup P. Lichen communities and the associated fauna on a rocky sea shore on Bornholm in the Baltic. Holarct Ecol 1985;8:66–75.

Thomson JW. American Arctic Lichens 2. The Microlichens. Madison, WI: University of Wisconsin Press 1997.

Thüs H. Taxonomie, Verbreitung und Ökologie silicoler Süßwasserflechten im außeralpinen Mitteleuropa. Biblioth Lichenol 2002;83:1–214.

Thüs H. Neufunde und Bemerkungen zur Gefährdung amphibischer Flechten aus dem Saarland und Rheinland-Pfalz. Delattinia 2006;32:127–140.

Thüs H, Muggia L, Pérez-Ortega S, Favero-Longo SE, Joneson S, O'Brien H, et al. Revisiting photobiont diversity in the lichen family Verrucariaceae (Ascomycota). Eur J Phycol 2011;46:399–415.

Thüs H, Schultz M. Freshwater Flora of Central Europe. Vol. 21, *Fungi*, 1st Part *Lichens*. Heidelberg: Spektrum 2008.

Timoney KP, Marsh J. Lichen trimlines in northern Alberta: establishment, growth rates, and historic water levels. Bryologist 2004;107:429–440.

Timoney KP, Marsh J. Lichen trimlines in the Peace-Athabasca Delta: variations in flora, form and disturbance regime. Canad Field-Nat 2005;119:76–81.

Tschermak-Woess E. The algal partner. In: Galun M, ed. CRC Handbook of Lichenology, Vol.1. Boca Raton: CRC Press 1988:39–92.

Weber HE, Moravec J, Theurillat JW. International Code of Phytosociological Nomenclature. 3rd edition. J Veg Sci 2000;11:739–768.

Wessels DCJ, Büdel B. A rock pool lichen community in northern Transvaal, South Africa: composition and distribution patterns. Lichenologist 1989;21:259–277.

Wirth V. Die Silikatflechten-Gemeinschaften im ausseralpinen Zentraleuropa. Dissertationes Botanicae 1972;17:1–306.

Wirth V, Hauck M, vonBrackel W, Cezanne R, deBruyn U, Dürrhammer O, et al. und Artenverzeichnis der Flechten und flechtenbewohnenden Pilze Deutschlands. NaBiV 2011;70:7–122.

Wolseley PA, James PW. Asssessing changes in density and condition of lichens for species recovery programmes. In: Nimis PL, Scheidegger C, Wolseley PA, eds. Monitoring with Lichens – Monitoring Lichens. Dordrecht: Kluver 2002:391–393.
Woods RG, Coppins BJ. A Conservation Evaluation of British Lichens and Lichenicolous Fungi. Species Status 13. Joint Nature Conservation Committee, Peterborough 2012.
Yoshimura I. On the *Dermatocarpon miniatum* community of the Kuroson Gorge, Kochi Pref. J Jap Bot 1963;38:33–41.
Zahlbruckner A. Die von der Deutschen Limnologischen Sunda-Expedition gesammelten Flechten. Arch f Hydrobiol Suppl Bd 1934;12:730–738.
Zschacke H. Die mitteleuropäischen Verrucariaceen V. Hedwigia 1927;67:45–85.

Robert W. Lichtwardt

16 Aquatic Trichomycetes

16.1 Trichomycetes, an ecological group

Most described trichomycetes, commonly known as gut fungi, are aquatic. The class Trichomycetes consists of two orders of fungi, the Harpellales, currently with 176 described species and 44 genera, all of which live in aquatic insect larvae, with one exception, and are found in the hindgut or attached to the peritrophic matrix of the midgut. In the lesser studied Asellariales, half of the 14 described species (3 genera) live in freshwater arthropod hosts, the remainder being found in either marine or terrestrial hosts. Mention is made here of the Eccrinales, previously classified in the Trichomycetes, whose species resemble fungi, being filamentous and producing sporangiospores, but are now known to be protozoans in the Mesomycetozoa group placed low in the tree of life where it is believed fungi and animals diverged (Cafaro 2005). A fourth order, the aquatic Amoebidiales, long suspected not to be fungi (Lichtwardt 1986), have also been shown to be protozoans (Benny and O'Donnell 2000; Cafaro 2005), and form a sister group with the Eccrinales in the Mesomycetozoa. Because all four orders are studied by mycologists and entomologists, trichomycetes with a lower case "t" often is used to designate an ecological group of organisms, all living obligately as symbionts of arthropods. Emphasis in this article will be on the Harpellales, because it has the most known species, and because many isolates of more than 35 species from many parts of the world have been cultured axenically, thus providing cultures for experimental purposes and DNA sequencing studies.

16.2 Phylogenetic considerations

It is of interest to understand better how such unusual and specialized organisms like Trichomycetes evolved both within their taxonomic limits and their relationships with other fungi. Morphological characters have some value, but like many fungi more precise understanding of relationships has come from molecular techniques.

Figure 16.1 is an abbreviated phylogenetic tree, prepared by Eric Tretter, Boise State University, ID, USA from published and original data, showing the relationship of the Harpellales and Asellariales to the Kickxellomycotina, their closest relatives. Two unusual genera of gut fungi, *Orphella* and *Barbatospora*, once classified in the Harpellales, are now known to fall within the Kickxellomycotina (White et al. 2006; Tretter et al. 2013).

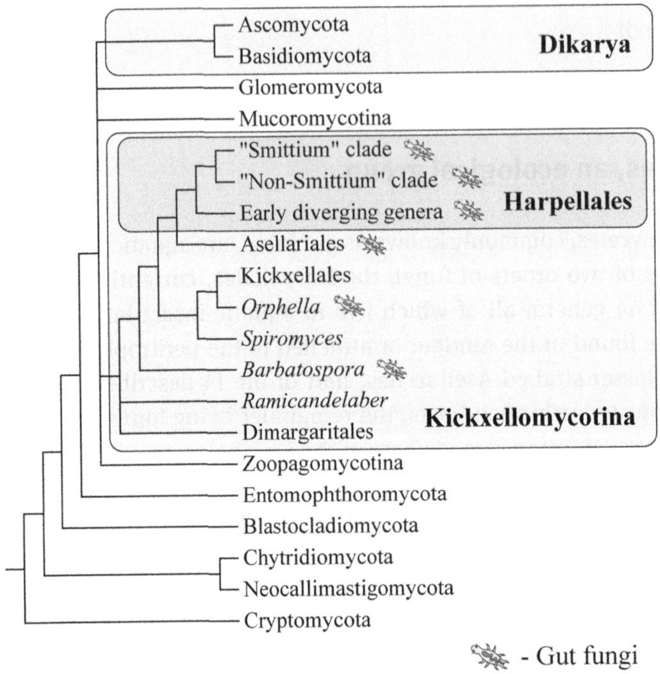

Fig. 16.1: An abbreviated tree showing the relationships of Harpellales and Asellariales to other fungi. See text.

Asexual spores of Harpellales (Figs. 16.2–16.7), commonly called trichospores (basically they are monosporous sporangia) are unlike asexual spores of other fungi, most producing one or more non motile appendages that help to retain the spore in the vicinity of feeding insect larvae in freshwater habitats. Like other fungi, many Harpellales are known to produce a sexual stage, zygospores (Figs. 16.8–16.11), that are morphologically distinct from other zygospore-producing fungi. A typical life cycle of a species of branched Harpellales is illustrated in Fig. 16.12.

Species of the fungal order Asellariales inhabit the hindgut of freshwater, marine, or terrestrial arthropods. They reproduce by arthrospore-like cells that break away from their branched thalli. Sexual reproduction may be present, but needs further study. Representative species of two genera are illustrated in Figs. 16.13–16.17.

Eccrinales, one of two orders now known to be protozoans in the ecological group trichomycetes (Cafaro 2005), currently has 17 known genera and 66 species that live in the hindguts of their hosts, 26 of which are species of *Enterobryus*, a common genus of unbranched fungus-like organisms that produce several types of sporangiospores, and exist worldwide mostly in millipedes (Diplopoda). Other species of Eccrinales are adapted to living in Crustacea and Insecta hosts. Two representatives are illustrated in Figs. 16.18–16.21.

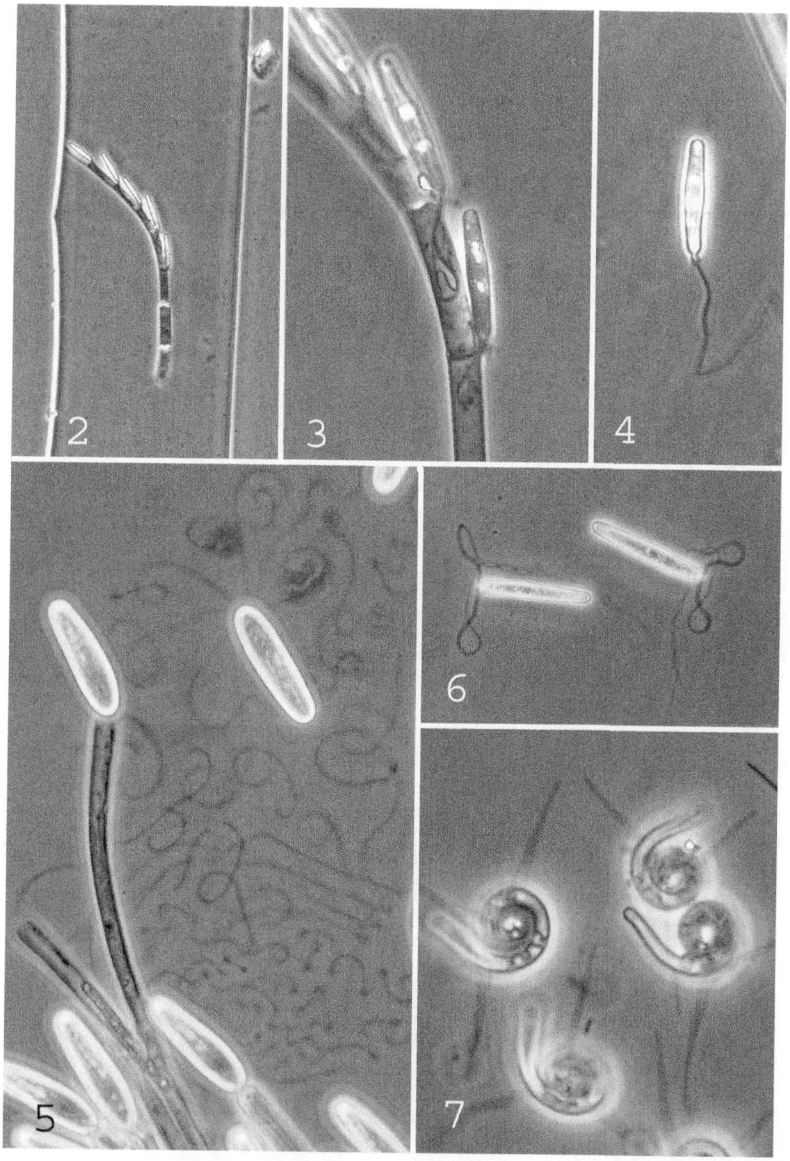

Figs. 16.2–16.7: Representative asexual spores (trichospores) produced by different genera of Harpellales. **Figs. 16.2–16.4** *Stahylina gravicaudata* thallus within the transparent peritrophic matrix of a nonbiting midge larva (Chironomidae) collected from a metal-polluted stream in southern Kansas, USA, the thallus producing trichospores with one appendage (Ferrington et al. 2000). **Fig. 16.5** *Genistellospora homothallica* with released trichospores each bearing six fine appendages, from a black fly larva (Simuliidae) collected in southern Chile (Lichtwardt and Arenas 1996). **Fig. 16.6** *Legerioides tumidus* with two released trichospores each bearing two broad appendages, removed from the hindgut of an aquatic isopod collected in Massachusetts, USA; this is the only Harpellales known to inhabit isopods (White 1999). **Fig. 16.7** *Pseudoharpella arcolamilica* whose coiled trichospores each produce three appendages, from a meniscus midge larva (Dixidae) in Minnesota, USA (Ferrington et al. 2003).

Figs. 16.8–16.11: Four basic types of sexual spores (zygospores) produced in Harpellales based on their method of attachment to the zygosporophore. **Fig. 16.8** *Simuliomyces spica* zygospores produced after hyphal conjugations, from a black fly larva (Simuliidae) collected in Arkansas, USA. **Fig. 16.9** *Trichozygospora chironomidarum* released zygospore bearing a collar and multiple fine appendages from a chironomid larva in the Rocky Mountains of Colorado, USA. **Fig. 16.10** *Genistellospora homothallica* mature zygospore from a black fly larva (Simuliidae) collected in the high Rocky Mountains of Wyoming, USA. **Fig. 16.11** *Carouxella coemeteriensis* elongate zygospores attached at one end arising from conjugated hyphae taken from a biting midge larva, *Dasyhelia morbida* (Ceratopogonidae), from a breeding colony in a cemetery vase in the La Plata, Argentina (from Lichtwardt et al. 1999).

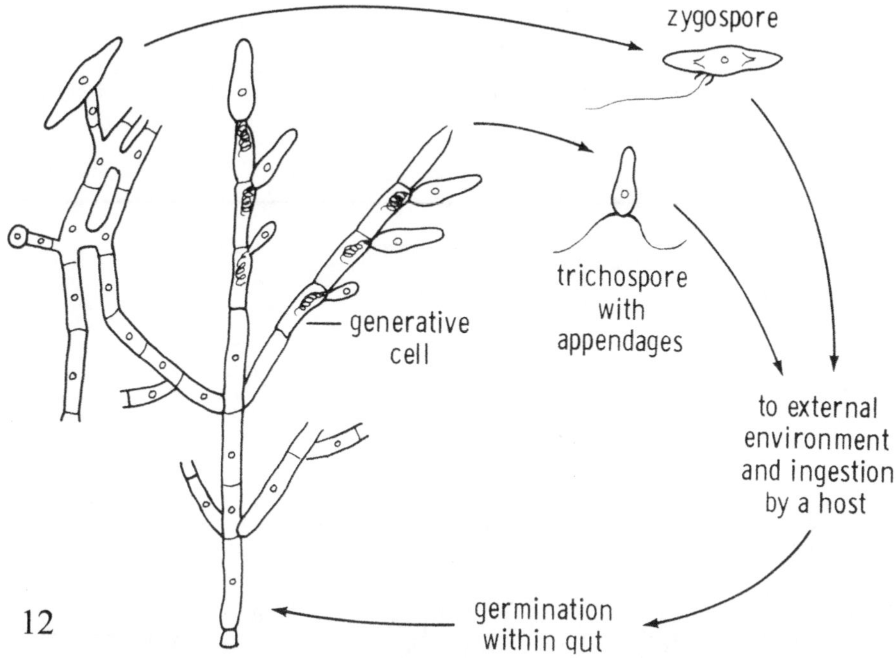

Fig. 16.12: Life cycle of a typical Harpellales producing both trichospores and zygospores. From Lichtwardt (1986).

The other protozoan order, Amoebidiales (Figs. 16.22–16.27), though a sister group of Eccrinales (Cafaro 2005), are unusual in producing amoeboid cells that move about for a short period of time, encyst, and produce cystospores that can infest another individual. There are two genera and 18 species, though it is quite evident that many more remain to be described. The five described *Amoebidium* species grow on the external surfaces of their freshwater Crustacea or Insecta hosts. In addition, *Amoebidium* species produce sporangiospores that can attach to passing hosts and develop into new thalli. *Amoebidium parasiticum* has been studied intensively by Whisler (1962, 1968), and several other articles Lichtwardt et al. (2001). He was the first to obtain *A. parasiticum* in pure culture (despite the specific epithet the species is a benign ectocommensal), and to discover how to induce amoeboid cells to form in culture. The unculturable 13 *Paramoebidium* species, on the other hand, are common within the hindgut of their freshwater hosts that include larvae of black flies (Simuliidae), stoneflies (Plecoptera), mayflies (Ephemeroptera), among other aquatic larvae. Many *Paramoebidium* species that have been found are not described and named for lack of good differentiating morphological features, but current DNA sequencing studies will help to rectify that problem. *Paramoebidium* species differ from species of *Amoebidium*, in addition to being located in the host hindgut, by not producing

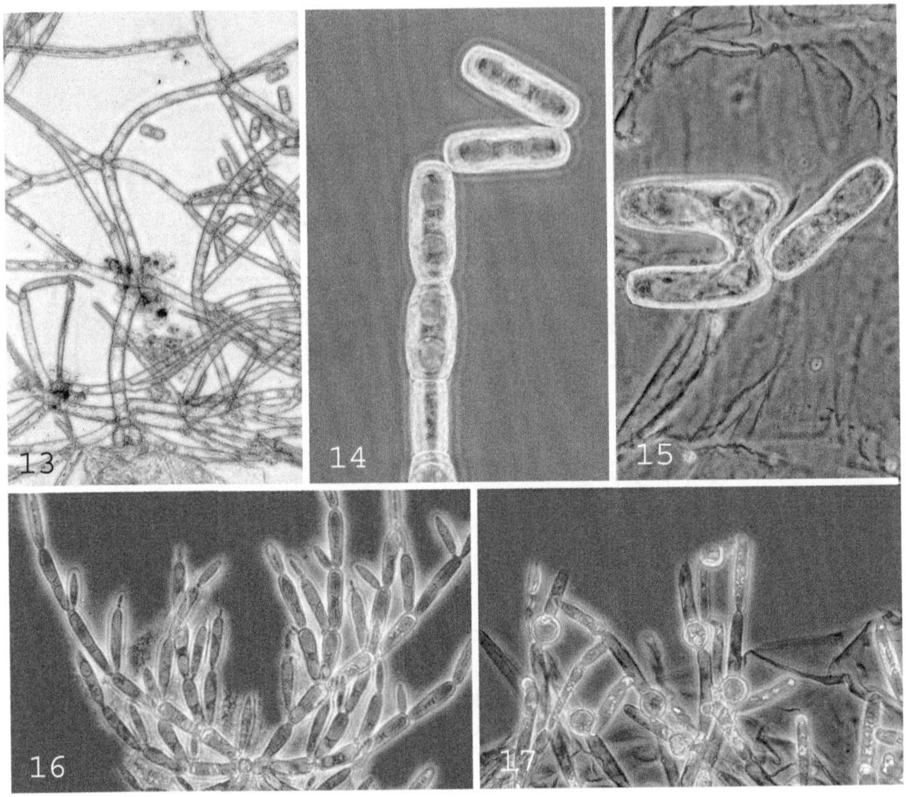

Figs. 16.13–16.17: Representatives of Asellariales. **Figs. 16.13–16.15** *Asellaria ligiae* from an intertidal isopod (*Ligia* sp.) showing a branched thallus producing cylindrical arthrospores, two of which are conjugating but without any further development of a sexual structure. **Figs. 16.16–16.17** Thalli of *Orchesellaria mauguioi* removed from the hindguts of springtails (Collembola) that were skating on freshwater in a pond, one producing thick-walled round cells believed to be a sexual stage.

sporangiospores. They remain unreproductive in the gut until the host begins to molt, at which time they form and release large numbers of amoeboid cells. Many other trichomycetes show integration of their development with their hosts' cycles (Lichtwardt et al. 2001).

16.3 Distribution and success of Trichomycetes

Many genera and some species of trichomycetes have a worldwide distribution, and are found wherever suitable hosts occur (Lichtwardt et al. 2001). New species are constantly being described whenever new sites are explored. The very wide distribution of taxa and the intimate symbiotic relationships that have evolved with different kinds of particular hosts speak to the success of the group. Though lacking fossil

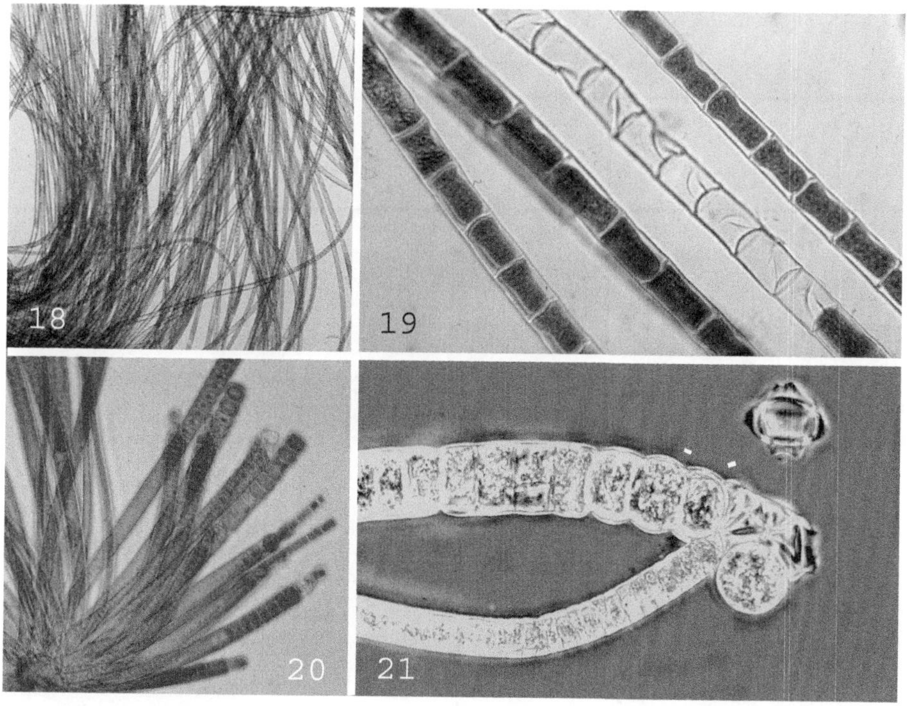

Figs. 16.18–16.21: Representatives of the trichomycete protozoan order Eccrinales. **Figs. 16.18–16.19** Mass of thalli of *Arundinula orconectis* removed from the hindgut of a freshwater crayfish, on the right with sporangiospores and empty sporangia. **Figs. 16.20–16.21** *Enteromyces callianassae* that infests both freshwater and marine crabs (Crustacea) seen here producing sporangiospores.

evidence, it has been suggested that these organisms have been evolving for a very long time, perhaps during the early stages of their hosts' evolution, and that their current patterns of distribution on different continents are attributable in some cases to tectonic events (Lichtwardt 2012). Harpellales and other trichomycetes have evolved to live in some of the most primitive arthropods that include Nematocera (the most primitive group of Diptera), stoneflies (Plecoptera), mayflies (Ephemeroptera), caddis worms (Trichoptera), isopods, springtails (Collembola), various kinds of crabs and crablike Crustaceans, and millipedes (Diplopoda), as examples. Many of the aquatic arthropods with gut fungi are food sources for fish, some birds, and predacious aquatic insects, and consequently are important players in such ecological systems (Merritt and Cummins 1996). A few aquatic larvae such as large, predacious stoneflies are used as bait by fishermen.

16.4 Variations in symbiotic associations

In many circumstances the gut fungi seem to be innocuous to the host, but evidence indicates that under particular conditions they may function as mutualists or even

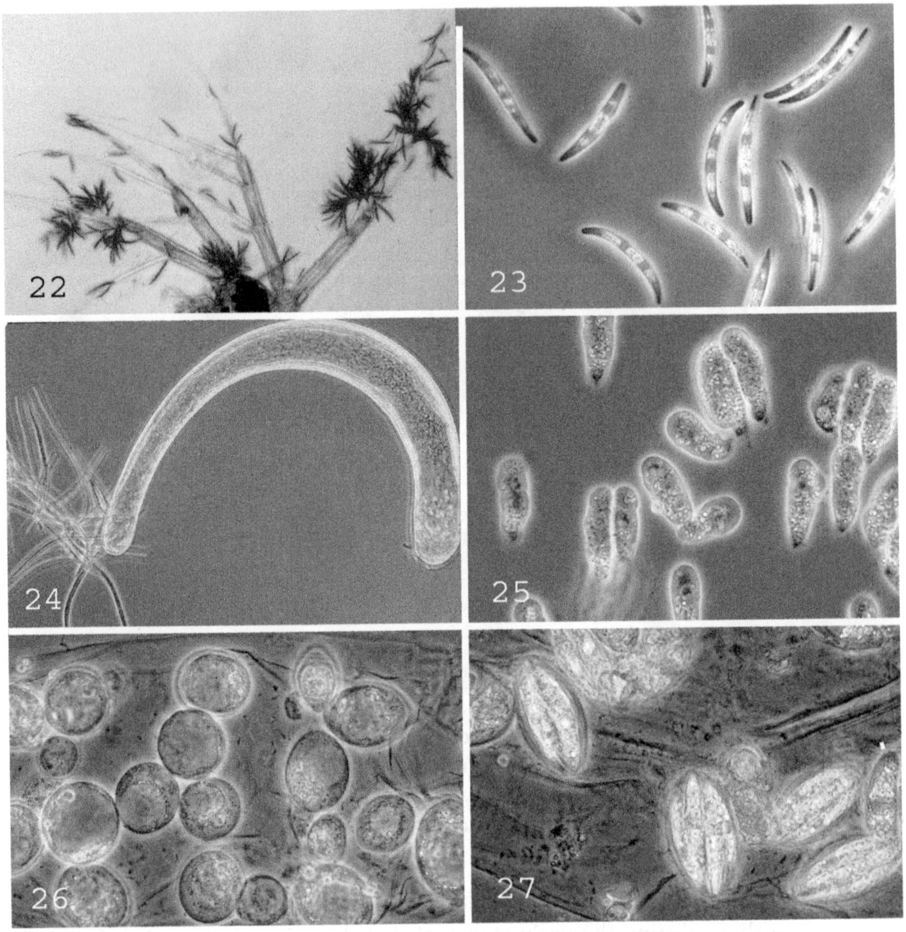

Figs. 16.22–16.27: Representatives of the trichomycete protozoan order Amoebidiales. **Figs. 16.22–16.23** *Amoebidium parasiticum* thalli attached to the antennae of a freshwater daphnia (Crustacea) extruding sporangiospores, and released lunate sporangiospores; this species is found worldwide attached to many kinds of small freshwater crustaceans and insect larvae (from Lichtwardt 1986). **Figs. 16.24–16.27** *Paramoebidium grande* thallus removed from the hindgut of a black fly larva (Simuliidae) in southern Chile, released migrating amoebae, encysted amoebae, and sporangiospores produced in the cysts (from Lichtwardt and Arenas 1996).

as pathogens. Horn and Lichtwardt (1981) were able to use a culturable Harpellales, *Zancudomyces culisetae* [formerly *Smittium culisetae* (Wang et al. 2013)], and the yellow fever mosquito *Aedes aegypti* larvae raised in the laboratory on artificial media with essential nutrients such as sterol and B-vitamins lacking individually in sets of larvae with and without the intestinal fungus. Larvae with the fungus developed through more instars and more rapidly than larvae without the fungus, and it was demonstrated that the gut fungus could provide sterol and other nutrients that led to better larval development. In nature, one can visualize conditions where very dense populations of mosquito larvae could lead to some essential nutrients becoming suboptimal which might then be provided by the gut fungus. How widespread this might be among Harpellales is difficult to determine, because experimental evidence requires both a culturable fungus and a host that can be raised axenically in the laboratory without bacteria or any other gut organisms other than the fungus fed to the larvae.

Harpellales infesting larvae in streams and other waters would have difficulty in dispersing to other streams or lentic bodies of water, activities essential to long-term survival, were it not for a parasitic stage discovered some years ago whereby flying adults can serve that function. There is no evidence that aquatic harpellids can disperse over distances by themselves beyond spreading from larva to larva within local populations. First noticed in adult black flies from Guatemala were fungal bodies in the ovaries that were at first misidentified (Garms 1975), and later studied in more detail in black flies by Yeboah et al. (1984) in Canada, and identified as fungal cysts. At certain times of the year under circumstances not yet understood, fungi begin to grow from the gut and penetrate the developing ovaries of the larva, pass through the pupal stage, ending in adults either completely or partially sterilized by hundreds of very small fungal cysts that replace eggs (Labeyrie et al. 1996; Lichtwardt 2008). These cysts, when "oviposited" by the female, germinate in water and produce asexual trichospores that can be ingested by larvae in the vicinity. Up to now only a few species of Harpellales producing such cysts have been identified. Based on trichospore morphology, *Harpella melusinae* and *Genistellospora homothallica* have been identified (Labeyrie et al. 1996), but using molecular methods (18S and 28S rDNA) other harpellid species have been identified, including *Pennella simulii* and several species of *Smittium* (White et al. 2006). Although relatively few Harpellales have been shown to produce ovarian cysts, we hypothesize that probably all extant Harpellales must have evolved this mechanism for essential dispersal, or perhaps something similar, which allows flying adults to transport the fungi to new sites (Lichtwardt 2012).

16.5 Medical implications

The hosts of Harpellales include several important types of insects that can be serious biting pests and/or carriers of human and animal pathogens. As a consequence, any survival effects gut fungi may have on these insects, whether positive or negative, are of interest. The insects include mosquitoes (Culicidae) that can carry the sporozoites of malaria, the viruses of yellow fever and dengue fever, the parasitic worms of elephantiasis, and the bacteria that cause tularemia. Black flies (Simuliidae) can transmit the parasitic nematodes that cause the serious disease called river blindness (onchocerciacis). In addition, the bites of these insects can be considerably annoying, as well as those of adult biting midges (Ceratopogonidae), also known as sand flies, some of whose larvae contain species of Harpellales (Lichtwardt et al. 2001). Adults of at least one species of non-biting midges (Chironomidae), a very common harpellid host family, have been known to cause serious allergies in humans (L.C. Ferrington Jr., personal communication).

Acknowledgments

This material is based upon work supported by the National Science Foundation under Grant No. DEB-0918169. The author thanks Eric Tretter, Boise State University, Boise, Idaho, USA for providing the tree in Fig. 16.1.

References

Benny GL, O'Donnell K. *Amoebidium parasiticum* is a protozoan, not a Trichomycete. Mycologia 2000;92:1133–1137.

Cafaro MJ. Eccrinales (Trichomycetes) are not fungi, but a clade of protists at the early divergence of animals and fungi. Molec Phylog Evol 2005;35:21–34.

Ferrington LC Jr, Lichtwardt R, Hayford B. *Smittium gravimetalum* (Trichomycetes: Harpellales), a new species of gut fungus from *Dicrotendipes fumidus* (Johannsen) (Diptera: Chironomidae) in a metal-polluted stream. In: Hoffrichter J, ed., Late 20th Century Research on Chironomidae: an Anthology from the 13th International Symposium on Chironomidae Shaker Verlag 2000:253–257.

Ferrington LC Jr, White MM, Lichtwardt RW. A new genus of Trichomycetes (*Pseudoharpella arcolamylica*) from *Dixa fluvica* Peters (Diptera: Dixidae). Aquat Insects 2003;25:85–94.

Garms R. Observations on filarial infections and parous rates of anthropophilic blackflies in Guatemala, with reference to the transmission of *Onchocerca volvulus*. Tropenmed Parasitol 1975;26:169–182.

Horn BW, Lichtwardt RW. Studies on the nutritional relationship of larval *Aedes aegypti* (Diptera: Culicidae) with *Smittium culisetae* (Trichomycetes). Mycologia 1981;73:724–740.

Labeyrie ES, Molloy DP, Lichtwardt RW. An investigation of Harpellales (Trichomycetes) in New York State blackflies (Diptera: Simuliidae). J Invert Pathol 1996;68:293–298.

Lichtwardt RW. The Trichomycetes: Fungal Associates of Arthropods. New York: Springer-Verlag 1986:1–343.

Lichtwardt RW. Trichomycetes and the arthropod gut. In: Brakhage AA, Zipfel PF, eds. The Mycota, Human and Animal Relations, 2nd ed. Berlin: Springer-Verlag 2008:3–19.

Lichtwardt RW. Evolution of Trichomycetes. In: Misra JK, Tewari JP, Deshmukh SK, eds. Systematics and Evolution of Fungi. Enfield, NH: Science Publishers 2012:107–114.

Lichtwardt RW, Arenas J. Trichomycetes in aquatic insects from southern Chile. Mycologia 1996;88:844–857.

Lichtwardt RW, Cafaro MJ, White MM. The Trichomycetes, fungal associates of arthropods. Revised edition. Published on the Internet: 2001:www.nhm.ku.edu/~fungi.

Lichtwardt RW, Ferrington LC Jr, López Lastra C. Trichomycetes in Argentinean aquatic insect larvae. Mycologia 1999;91:1060–1082.

Merritt RW, Cummins KW, eds. An Introduction to the Aquatic Insects of North America 3rd ed. Kendall: Hunt Publishing Co 1996:1–862.

Tretter ED, Johnson EM, Wang Y, Kandel P, White MM. Examining new phylogenetic markers to uncover the evolutionary history of early-diverging fungi: comparing MCM7, TSR1 and rRNA genes for single- and multi-gene analyses of the Kickxellomycotina. Persoonia 2013;30:106–125.

Wang Y, Tretter, ED, Lichtwardt RW, White MM. Overview of 75 years of *Smittium* research, establishing a new genus for *Smittium culisetae*, and prospects for future revisions of the '*Smittium*' clade. Mycologia 2013;105:90–111.

Whisler H. Pure culture of the Trichomycete, *Amoebidium parasiticum*. Nature 1960;186:732–733.

Whisler HC. Developmental control of *Amoebidium parasiticum*. Develop Biol 1968;17:562–570.

White MM. *Legerioides*, a new genus of Harpellales in isopods and other Trichomycetes from New England, USA. Mycologia 1999;91:1021–1030.

White MM, Lichtwardt RW, Colbo MH. Confirmation and identification of parasitic stages of obligate endobionts (Harpellales) in blackflies (Simuliidae) by means of rRNA sequence data. Mycol Res 2006;110:1070–1079.

Yeboah DO, Undeen AH, Colbo MH. Phycomycetes parasitizing the ovaries of blackflies (Simuliidae). J Invert Pathol 1984;43:363–373.

Umpava Pinruan, Aom Pinnoi, Kevin D. Hyde and E. B. Gareth Jones

17 Tropical peat swamp fungi with special reference to palms

Peat swamp forests are unique ecosystems that are extremely important globally because of their rich species biodiversity. Peat swamps in Asia can be found in Brunei, Indonesia, Malaysia, Papua New Guinea and southern Thailand (Jackson et al. 2009) and account for most of the worlds peat lands (60%). They also exist in a few areas in Africa and parts of Central America (Mohd Azmi et al. 2009). Draining of peat swamps for planting oil palm and rubber, for agriculture, for logging, for industrial and residental development, and for extraction of peat for fuel and gardening, threatens their existence (Ng and Shamsudin 2001). These activities involves digging canals to drain off water, leading to drying out of the peat and the potential danger of fires, especially in the dry season (Ng and Shamsudin 2001). Miettinen et al. (2012) have shown that peat swamp forest cover in Sumatran peatlands dropped from 75% to 28% between 1990 and 2010, and that at this rate of deforestation, peat swamp forests in Southeast Asia may disappear by 2030. Estimates of peatlands are given in Tab. 17.1.

Peat swamps are formed by long-term accumulation of plant material under acidic water-logged conditions, and account for one of the largest terrestrial organic sinks of carbon dioxide, storing 30 times more carbon dioxide than rainforests (http:/ www.panda.org/borneo; Jusoff et al. 2007). They are generally waterlogged even in the dry seasons with peat composed of more than 65% organic matter. The habitat is characterized by waterlogged soils, that is anaerobic, with a thick layer of leaf litter, low pH (2.9–4), low nutrients and dissolved oxygen, the water colored red, dark brown to black due to tannins that leach from fallen leaves, and the peat may be up to 20 m deep. Tropical peat swamps may be dominated by palms and *Pandanus* sp. (subcanopy), trees (many with buttresses and pneumatophores to enable oxygen uptake), including such species as *Alstonia scholaris*, *Terminalia* spp., *Pometia pinnata* and *Octomeles sumatrana*.

A tropical peat swamp forest is an unique ecosystem because of the conditions listed above, and suprisingly supports a rich diversity of plants (400+ flowering plants), and animals (200+), however, few studies have documented its rich microbial diversity. For example, Ng and Shamsudin (2001) list 216 plants, 62 mammals, 192 birds, 56 fish, 17 frog species and 21 others for the south-east Pahang peat swamp forest, Malaysia, but no fungus was mentioned. Yule and Gomez (2009) opine that while decomposition of most forest trees is slow, due to the presence of compounds that impede microbial growth, the leaves of other trees were almost completely degraded. Thus the DOC from leaf decomposition of this tree enabled the transfer of energy to other trophic levels. The phylogenetic diversity of Pru Toh Daeng peat swamp, Thailand, was surveyed by the pyrosequencing of environmental DNA and analysis of

Tab. 17.1: Estimates of peat lands (after Parish et al. 2012).

Country	Ha of peatlands
Brunei	90,000
Cambodia	4,580
Indonesia	20,695,000
Lao	19,000
Malaysia	2,588,900
Myanmar	122,800
Philippines	64,500
Singapore	50
Thailand	63,800
Vietnam	53,300

16S rRNA gene library, resulted in the documentation of a community dominated by aerobic microbes including Acidobacteria and Proteobacteria (chiefly), e.g. Alphaproteobacteria (Kanokratana et al. 2010).

Hawksworth (1991) estimated there are 1.5 million fungi on earth. However, there have been other estimates of global fungal numbers ranging from less than one million to more than nine million (Cannon 1997). The current (10th) edition of the Dictionary of the Fungi gives a figure of 98,128 species (Kirk et al. 2008). Thai fungal diversity has been poorly studied with *ca* 1.5% of the total known 90,000 worldwide fungal species reported (Hywel-Jones, unpublished), while Jones and Hyde (2004) indicate that 6,000 fungi are documented for Thailand. A wide range of palm fungi have been extensively studied in tropical locations including Australia, Brunei, Ecuador, Hong Kong and Thailand (Fröhlich and Hyde 1994, 2000; Hyde 1994; 1996a, b, c; Hyde et al. 1997a; Yanna et al. 2001a, b; Taylor and Hyde 2003; Pinnoi et al. 2006; Pinruan et al. 2007). The results indicate a rich fungal diversity and numerous new species (Hyde and Alias 1998; Hyde et al. 1998; Fröhlich and Hyde 2000; Yanna et al. 2001b; Hidayat et al. 2006). *Anthostomella*, *Astrosphaeriella*, *Capsulospora*, *Linocarpon*, *Neolinocarpon* and *Oxydothis* are common genera on terrestrial palm material (Fröhlich and Hyde 2000; Yanna, 2001a, b; Taylor and Hyde 2003). Ascomycetes are the dominant group of palm fungi (Hyde et al. 1997; Fröhlich and Hyde 2000; Pinnoi et al. 2006; Pinruan et al. 2007). Fungi are abundant on petioles and rachides, but few occur on leaves (Hyde et al. 1997; Pinnoi et al. 2006; Pinruan et al. 2007).

Few studies of palm fungi in tropical peat swamp forests have been undertaken, and this chapter will document our current knowledge. Aramsiriujiwet (1996) collected 29 hyphomycetes from terrestrial palms in Southern Thailand: *Borassus flabellifer*, *Caryota* sp., *Cocos nucifera*, *Cyrostachys lakka*, *Corypha lecomtei*, *Elaeis guineensis*, and *Roystonea regin*, while Sarapat (2003) recorded fungi from twelve palm species sampled in Sirindhorn peat swamp forest. The 340 samples were divided into three parts: palm fronds, leaves and fruits and yielded 111 taxa with 73 (57 genera) asexual fungi and 38 (22 genera) ascomycetes. Dominant fungi were *Astrosphaeriella* sp., *Circinoconis paradoxa*, and *Linocarpon* sp.

The objectives of this study are 1) to document the diversity of fungi occurring on four palms in Pru Toh Daeng, Sirindhorn Peat swamp forest, Narathiwat Province, Thailand, and 2) to compare the fungal community on the different palms.

17.1 Material and methods

17.1.1 Sample collection

Eleiodoxa conferta, *Licuala longicalycata*, *Metroxylon sagu* and *Nenga pumila* were collected at Sirindhorn Research and Nature Study Center (Pru Toh Daeng, Sirindhorn Peat Swamp Forest), Narathiwat Province, Thailand (06°12'N 101°57'E, elevation is 0–4 m). These unique habitats support a large variety of flora and fauna. Some 470 plant species in 124 families have been recorded; 217 bird, 52 reptile, 62 fish, 19 amphibian, 106 butterfly and 60 mammal species have been identified (Phlengklai et al. 1991; Vijarnsorn 1996; Hankaew 2003). The surrounding areas are privately owned and used for agricultural purposes (rice, rubber and fruit plantations) and settlements (Apiwathnasorn et al. 2009). The area has a constant mean temperature of 26–27°C with an annual rainfall of 2,560 mm and an average of 171 days of rain per year with the heaviest rainfall in November–December. There is no distinct dry season; the average humidity is 77–83% (Phlengklai et al. 1989). The peat water is acidic (pH level about 4.5–6.0). Collections of *Eleiodoxa conferta*, *L. longicalycata* and *M. sagu* were divided into three parts: leaves, rachides, and petioles while the *Nenga pumila* collections were divided into four parts: leaves, rachides, petioles and sheaths, all from three microhabitats: wet (constantly submerged), damp (moist and on the surface of the soil), and dry (aerial part) microhabitats and then placed in plastic bags. Collections were made in 2001–2004. Samples were returned to the laboratory where the material was incubated in plastic boxes on sterile moist tissue. The material was kept moist and examined periodically for fungal fruiting structures. Fungi collected were identified and isolated into axenic culture where possible using a single spore technique (Choi et al. 1999).

17.2 Results

17.2.1 Abundance of fungi on four palms (*Eleiodoxa conferta*, *Licuala longicalycata*, *Metroxylon sagu* and *Nenga pumila*)

One thousand and eighty-four fungal records (290 taxa, 129 genera) were found on four palms. One hundred and eighteen were fully identified, 110 were partially identified, and 62 were unidentified. The most common species on four palms were

Microthyrium sp. (16.5% of all records), *Cancellidium applanatum* (14.4%), *Diplococcium stoveri* (13.6%), *Xylomyces aquaticus* (13.2%), *Phaeoisaria clematidis* (12%), *Astrosphaeriella* sp. (10.7%), *Stilbohypoxylon elaeicola* (10.7%), *Jahnula appendiculata* (10.3%), *Lophiostoma frondisubmersa* (9.9%), and *Morenoina palmicola* (8.6%) (Tab. 17.2). The most common genera were *Astrosphaeriella* (6% of all records), *Microthyrium* (3.6%), *Stilbohypoxylon* (3.6%), *Cancellidium* (3.4%), *Diplococcium* (3.2%), *Oxydothis* (3.1%), *Xylomyces* (2.9%), *Lophiostoma* (2.8%), *Phaeoisaria* (2.7%) and *Annulatascus* (2.6%). However, 117 taxa (50% of total taxa) were represented by only one record and can be regarded as infrequent or rare.

Tab. 17.2: Distribution of fungi on four peat swamp palms.

Fungus	*Eleiodoxa conferta*	*Licuala longicalycata*	*Metroxylon sagu*	*Nenga pumila*
Acrocalymma medicaginis	1	–	–	–
Acrogenospora sphaerocephala	–	–	2	–
Acrospeira-like	–	1	–	–
Annulatascus aquaticus	–	3	–	–
Annulatascus palmae	–	3	–	–
Annulatascus sp. 1	8	–	–	–
**Annulatascus* sp. 2	1	–	–	–
Annulatascus velatispora	3	10	–	–
**Anthostomella* sp.1	–	–	5	–
**Anthostomella* sp.2	1	–	–	–
Anthostomella palmaria	–	2	–	–
Apioclypea apiosporioides	1	–	–	–
**Apioclypea* sp	–	–	3	–
Apiospora sp.	–	–	3	–
Arecomyces bruneiensis	–	1	–	–
Arecomyces frondicola	–	1	–	–
Arecomyces epigenus	–	1	–	13
Arecophila striatispora	–	1	1	–
Arthrinium arundinis	–	2	–	–
Arthrobotrys oligospora	–	2	–	–
Arthrobotrys sp.	2	–	–	–
Ascominuta lignicola	–	2	–	1
Ascominuta sp.	–	–	1	–
Aspergillus sp.	–	1	–	–
Astrocystis rachidis	–	1	2	–
Astrocystis sp.	–	1	–	–
Astrocystis eleiodoxae	14	4	–	–
Astrosphaeriella angustispora	1	4	–	–
Astrosphaeriella aquatica	7	6	–	–
Astrosphaeriella aquatica-like	26	–	–	–
Astrosphaereilla floridana	–	1	–	–

(*continued*)

Tab. 17.2: (*continued*)

Fungus	***Eleiodoxa conferta***	***Licuala longicalycata***	***Metroxylon sagu***	***Nenga pumila***
Astrosphaeriella fronsicola	–	2	–	–
Astrosphaeriella lophiostomopsis	–	2	1	–
Astrosphaeriella malayensis	–	6	–	–
Astrosphaeriella papillata	–	2	–	–
Astrosphaeriella tornata	–	–	1	–
Astrosphaeriella sp. 1	1	–	–	–
Astrosphaeriella sp. 2	1	–	–	–
Astrosphaeriella sp. 3	2	–	–	–
**Astrosphaeriella* sp. 4	–	1	–	–
Bactrodesmium sp.	–	2	1	–
**Baipadisphaeria spathulospora*	–	3	–	–
Beltraniella sp.	–	–	–	1
Berkleasmium micronesicum	–	–	–	4
Berkleasmium typhae	10	3	–	–
Bionectria sp.	3	–	–	–
Brachysporiella gayana	4	–	–	–
Cancellidium applanatum	32	3	–	–
Cancellidium-like 1	1	–	–	–
Cancellidium-like 2	1	–	–	–
Candelabrum brocchiatum	–	6	–	–
Canalisporium caribense	–	1	–	–
Canalisporium variable	–	2	–	–
Capnodiastrum sp./ *Kamatia* sp.	9	–	–	–
Capsulospora frondicola	2	–	2	–
Carinispora nypae	–	1	–	–
**Caryospora* sp.	–	1	–	–
Chaetopsina sp.	1	–	–	1
Chaetospermum camelliae	–	5	–	–
Chaetosphaeria hiugensis	–	–	–	1
Chaetosphaeria sp.	–	1	–	3
**Chalara siamense*	1	2	–	–
Chlamydospore 1	1	–	–	–
Chlamydospore 2	2	–	–	–
Chloridium sp.	2	–	–	–
Ciliciopodium-like	–	1	–	–
Ciliciopodium sp.	–	–	–	1
Clasterosporium sp.	–	–	–	1
Coleodictyospora micronesica	8	–	–	–
**Craspedodidymum licualae*	–	3	–	–
**Craspedodidymum microsporum*	–	3	–	–
**Craspedodidymum siamense*	–	1	–	–
Cryptophiale sp.	–	–	–	1
Cryptophialoidea sp.	–	–	–	5

(*continued*)

Tab. 17.2: (*continued*)

Fungus	*Eleiodoxa conferta*	*Licuala longicalycata*	*Metroxylon sagu*	*Nenga pumila*
**Custingophora undulatistipes*	3	–	–	–
Cylindrocladium sp. 1	–	–	2	–
Cylindrocladium sp. 2	–	–	1	–
Dactylaria hemibeltranioidea	–	1	–	–
Dactylaria flammulicornuta	3	–	–	2
Dactylaria palmae	–	–	–	6
Dactylaria uliginicola	1	–	–	–
Dactylella sp. 1	3	–	–	–
Dactylella sp. 2	1	–	–	–
Delortia palmicola	5	5	–	–
Diaporthe sp.	1	–	–	–
Dictyochaeta gyrosetula	–	3	–	–
Dictyochaeta ramuloseula	–	–	2	–
Dictyochaeta sp. 2	–	–	–	3
Dictyosporium alatum	–	–	2	–
Dictyosporium digitatum	–	2	–	–
**Dictyosporium* sp.1	–	1	–	–
Didymobotryum sp.	6	–	–	–
Didymosphaeria bisphaerica	–	1	–	1
Dinemasporium lanatum	–	–	3	–
Dinemasporium sp.	–	–	–	15
Diplococcium asperum	2	–	–	–
Diplococcium stoveri	–	–	–	33
Dischloridium sp.	2	–	–	–
Endocalyx melanoxanthus	–	1	–	–
**Flammispora bioteca*	–	2	1	–
Fluviatispora reticulata	3	–	–	–
Fluviatispora sp.	–	–	1	1
Gliocladium sp.	–	3	–	–
Glomerella sp.	–	2	–	–
Gnomonia sp.	1	–	–	–
**Goidanichiella fusiformis*	1	–	–	–
Gonytrichum macrocladium	1	4	–	–
Gaeumannomyces sp.	12	–	–	–
Guignadia manokwaria	–	2	–	–
Haplographium state of *Hyaloscypha dematiicola*	1	–	–	–
Helicoubisia coronata	1	–	–	–
Helicoma-like 2	–	2	–	–
Helicoma sp. 1	1	7	–	–
Helicomyces roseus	3	–	–	–
Helicosporium gigasporum	–	2	–	–
Helicosporium sp.	1	–	–	–

(*continued*)

Tab. 17.2: (*continued*)

Fungus	*Eleiodoxa conferta*	*Licuala longicalycata*	*Metroxylon sagu*	*Nenga pumila*
Heteroconium sp.	1	–	–	–
**Jahnula appendiculata*	13	6	–	6
Koorchaloma bambusae	–	1	–	–
Lanceispora amphibia	–	1	–	–
Lasiodiplodia theobromae	–	5	–	–
Lasiodiplodia sp.	–	3	–	–
Lasiosphaeria-like	–	–	2	–
Linocarpon elaeidis	–	2	–	–
Linocarpon livistonae	–	1	–	–
Linocarpon pandani	–	3	–	–
Linocarpon sp. 2	–	3	–	–
Linocarpon sp. 4	2	–	–	12
Linocarpon sp. 5	–	–	–	1
Lophiostoma frondisubmersa	23	1	–	–
Lophiostoma-like	–	1	–	–
Lophiostoma sp.	–	–	–	6
Lophodermium licualae	–	2	–	–
Lophodermium sp.	1	–	–	9
Massarina bipolaris	–	9	3	–
Massarina-like	2	–	–	1
Massarina corticola	–	3	–	–
**Massarina* sp.	–	2	–	–
Melanographium citri	1	–	1	–
Melanographium palmicola	–	1	–	–
Monotosporella rhizoidea	1	1	–	–
Morenoina palmicola	21	–	–	–
**Munkovalsaria* sp.	1	–	–	–
Microthyrium sp.	23	10	2	5
Mycocalia denudata	4	–	–	–
Myelosperma tumidum	–	3	–	–
Nawawia filiformis	9	2	5	–
Nectria sp. 1	–	7	–	–
Nectria sp. 2	–	2	–	–
Niesslia sp.	–	1	–	–
Neolinocarpon sp.	–	–	2	–
Ophioceras sp.	–	1	–	–
Ophiostoma sp.	1	–	–	–
Orbilia sp.	1	1	–	–
Ornatispora sp.	1	–	3	–
Oxydothis angustispora	–	2	–	–
Oxydothis frondicola	–	1	–	–
Oxydothis grisea	–	3	–	–
Oxydothis hoehnelii	–	3	–	–

(*continued*)

Tab. 17.2: (*continued*)

Fungus	*Eleiodoxa conferta*	*Licuala longicalycata*	*Metroxylon sagu*	*Nenga pumila*
Oxydothis licualae	–	2	–	–
Oxydothis livistonae	–	2	–	–
Oxydothis livistonica	–	1	–	–
Oxydothis oraniopsis	–	1	–	–
Oxydothis rattanicola	6	–	–	–
**Oxydothis* sp. 1	–	2	–	–
Oxydothis sp. 8	–	–	–	6
Oxydothis sp. 9	–	–	–	1
Oxydothis-like	–	–	4	–
Penicillium sp.	2	1	–	–
Pestalosphaeria austroamericana	2	–	–	–
Petrakiopsis sp.	–	5	–	–
Phaeodothis sp.	–	1	–	–
Phaeoisaria clematidis	19	10	–	–
Phialogeniculata sp.	2	–	–	–
Phomatospora berkelyi	–	2	–	–
**Phruensis brunneispora*	–	9	–	–
Pleurophragmium sp. 1	2	–	–	–
Pleurophragmium sp. 2	1	–	–	–
Pseudohalonectria sp.	–	–	–	1
Pseudorobillarda sojae	–	3	–	–
Rhytisma sp.	–	–	–	1
Rosellinia corticum	–	1	–	–
Sarocladium-like	–	1	–	–
Septomyrothecium sp. 1	5	–	–	–
Septomyrothecium sp. 2	2	–	–	–
Septomyrothecium sp. 3	1	–	–	–
Solheimia costispora	–	8	–	1
Solheimia sp.	–	–	2	–
**Spadicoides* sp.	–	–	–	5
Spadicoides obovatum	–	3	–	–
Spadicoides klotzchii	–	1	–	–
Spadicoides sp. 1	–	1	–	–
Spadicoides sp. 4	–	–	–	6
Spiropes sp.	–	–	1	–
Sporidesmiella oraniopsis	–	2	–	–
Sporidesmiella sp.	–	–	–	3
Sporidesmium sp. 1	5	–	–	–
Sporidesmium sp. 2	1	–	–	–
Sporidesmium sp. 3	1	–	–	1
Sporidesmium sp. 4	–	–	–	2
Sporidesmium sp. 5	–	–	–	2
Sporidesmium sp. 6	–	–	1	–
Sporidesmium sp. 7	–	–	1	–

(*continued*)

Tab. 17.2: (*continued*)

Fungus	*Eleiodoxa conferta*	*Licuala longicalycata*	*Metroxylon sagu*	*Nenga pumila*
Sporidesmium sp. 8	–	–	–	4
Sporidesmium-like (AOM128)	2	–	–	–
Sporoschisma sp.	–	–	2	–
Stachybotrys albipes	3	–	1	–
Stachybotrys bambusicola	–	2	–	–
**Stachybotrys palmae*	–	2	–	–
Stictis sp.	5	–	–	–
Stilbohypoxylon elaeicola	24	3	–	–
Submersisphaeria aquatica	–	5	–	–
**Submersisphaeria palmae*	7	4	–	3
Tetraploa aristata	–	–	3	–
Thailandiomyces bisetulosus	–	8	–	–
Thozetella nivea	3	6	–	–
Thozetella radicata	–	3	–	–
Trichoderma harzianum	–	4	–	–
Trichoderma sp.	3	–	–	5
Tubeufia claspisphaeria	1	6	–	1
Unidentified asexual morphs	16 (15 taxa)	10 (8 taxa)	3 (2 taxa)	6 (4 taxa)
Unidentified ascomycetes	14 (7 taxa)	14 (12 taxa)	9 (8 taxa)	1 (1 taxa)
Unidentified basidiomycetes	1	6 (3 taxa)	–	–
**Unisetosphaeria penguinoides*	1	–	–	–
**Vanakripa minutiellipsoidea*	1	–	–	–
Vanakripa sp.	–	–	1	–
Verticillium sp.	1	1	–	–
Wiesneriomyces-like	1	–	–	–
Wiesneriomyces javanicus	–	4	–	1
Xylomyces aquaticus	27	5	–	–
Zanclospora sp.	–	–	–	1
Zalerion-like	–	1	–	–
Zalerion sp.	–	–	2	–
Total collections (1,084)	462	358	82	184
Total species (290)	111	137	45	47

* New species or probable new species.

17.2.1.1 *Eleiodoxa conferta*

Four hundred and sixty-two fungal records were reported, with 111 taxa collected and identified including 44 ascomycetes, 66 asexual morphs and one basidiomycete (*Mycocalia denudata*). Forty species (36% of total species) were represented by only one record and can be regarded as infrequent or rare.

Percentage abundance of fungi on different parts of the *E. conferta* was: dry material supported 17%, damp material 34.5%, while the wet material supported the most fungi 48.5%, with *Xylomyces aquaticus* (19 records), *Microthyrium* sp. (18), *Astrosphaeriella* sp. (15), and *Jahnula appendiculata* (12) being the most common taxa (Pinnoi et al. 2006). The percentage coverage of fungi on different parts of *E. conferta* was petioles 53%, leaves 17%, and rachides 30%. The following species appeared exclusively or primarily on the petioles of the palm: *Stilbohypoxylon moelleri* (12 records), *Morenoina palmicola* (10), *Nemania eleiodoxae* (9), *Astrosphaeriella* sp. 4 (6), *Capnodiastrum*/*Kamatia* sp. (6), *Gaeumannomyces* sp. (6) *Coleodictyospora micronesica* (5), *Delortia palmicola* (4) and *Nawawia filiformis* (4). Fewer fungi were recorded on the palm rachides, and on leaf material and diversity was low (Pinnoi et al. 2006).

17.2.1.2 *Licuala longicalycata*

Three hundred and fifty-eight fungal identifications were made yielding 137 taxa, representing 70 ascomycetes, 64 asexual fungi, and three basidiomycetes. The ten most common fungi were *Annulatascus velatisporus* (2.8%), *Microthyrium* sp. (2.8%), *Phaeoisaria clematidis* (2.8%), *Massarina bipolaris* (2.5%), *Phruensis brunneispora* (2.5%), *Solheimia costaspora* (2.2%), *Thailandiomyces bisetulosus* (2.2%), *Helicoma* sp. 1 (2%), *Nectria* sp. 1 (2%), and *Astrosphaeriella aquatica* (1.7%). However, 66 species (19.8%) were represented by only one record, including 38 ascomycetes (58%), 27 hyphomycetes (41%) and one basidiomycete (1%). Dry material supported the greatest diversity (40%) and the most frequent fungi included: *Solheimia costaspora* (8 records), *Phaeoisaria clematidis* (7 records), *Astrosphaeriella malayensis* (6 records) and *Thailandiomyces bisetulosus* (6 records). Submerged material accounted for 32% of the diversity with damp material supporting the least number of fungi (28%) (Pinruan et al. 2007). Three tissue types were collected in this study including petioles, trunks, and leaves. Fungi were identified with a percentage occurrence of 61%, 24%, and 15%, respectively; petioles supported the greatest number of fungal records and diversity (Pinruan et al. 2007).

17.2.1.3 *Metroxylon sagu*

Eighty-two records were reported with 45 taxa including 21 ascomycetes, and 24 asexual morphs (Tab. 17.2). The percentage of fungi occurring on different parts of the palm was 47% 34% and 19%, respectively, on dry, damp and submerged material. The percentage of fungi recorded on petioles, leaves and rachides, was 47%, 19%, and 34% respectively (data not shown). The most common fungi were *Nawawia fuliformis* (6%), *Anthostomella bipapillispora* (6%), *Oxydothis*-like (4.9%), *Massarina bipolaris* (3.6%), *Ornatispora* sp. (3.6%), *Apioclypea eccentricospora* (3.6%), *Apiospora* sp. (3.6%), *Dinemasporium lanatum* (3.6%), and *Tetraploa aristata* (3.6%). Fourteen taxa (31%) were represented by only one record (Tab. 17.2).

17.2.1.4 *Nenga pumila*

One hundred and eighty-four records were reported yielding 47 taxa, with 28 asexual morphs taxa and 19 ascomycetes. The ten most common fungi were *Diplococcium stoveri* (17.9%), *Dinemasporium* sp. (8.2%), *Arecomyces epigeni* (7.1%), *Linocarpon* sp. 4 (6.5%), *Lophodermium* sp. (4.9%), *Jahnula appendiculata* (3.3%), *Dactylaria palmae* (3.3%), *Lophiostoma* sp. (3.3%), *Oxydothis* sp. 8 (3.3%), and *Spadicoides* sp. 4 (3.3%) while 20 taxa (10.9%) were found once (Tab. 17.2). Dry material supported high fungal diversity (49.5%), with 31.5% and 19% on damp and submerged material, respectively (data not shown). Percentage fungal diversity on different parts of the palm: petioles, leaves, rachides and sheaths were 22.8%, 2.7%, 25% and 49.5%, respectively.

17.2.2 Fungal diversity

This study is the first account to detail the diversity of fungi in a peat swamp, and in particular on senescent palm material. These fungi include taxa also reported from submerged wood, e.g. *Candelabrum brocchiatum, Dictyochaeta gyrosetula, Thozetella nivea* (Sivichai et al. 2000). However, the percentage overlap between the two communities is low. Many of the fungi encountered are common to those known from terrestrial palms, especially the genera *Astrosphaeriella, Linocarpon, Massarina* and *Oxydothis* (Hyde 1994; Hyde et al. 1997; Taylor and Hyde 2003). A significant number of taxa are unique to the peat swamp and have not been documented from any other habitat (Tab. 17.3).

Tab. 17.3: New fungi described from various substrata collected in Thai peat swamp forests.

New genera	New species	Reference
	Astrocystis eleiodoxae	Pinnoi et al. (2010)
Baipadisphaeria	*B. spathulospora*	Pinruan et al. (2010)
	Berkleasmium crunisia	Pinnoi et al. (2007)
	Chalara siamense	McKenzie et al. (2002)
	Craspedodidymum siamense	Pinruan et al. (2004a)
	C. licualae	Pinruan et al. (2004a)
	C. microsporum	Pinruan et al. (2004a)
	Custingophora undulatistipes	Pinnoi et al. (2003a)
	Dactylaria flammulicornuta	Pinnoi et al. (2003b)
	D. palmae	Pinnoi et al. (2003b)
	D. uliginicola	Pinnoi et al. (2003b)
Flammispora	*F. bioteca*	Pinruan et al. (2004c)
	Goidanichiella fusiforma	Hyde et al. (2002)
	Jahnula appendiculata	Pinruan et al. (2002)
Phruensis	*P. brunneispora*	Pinruan et al. (2004b)
	Stachybotrys palmae	Pinruan et al. (2004d)
	Submersisphaeria palmae	Pinnoi et al. (2004)
Thailandiomyces	*T. bisetulosus*	Pinruan et al. (2008)
Unisetosphaeria	*U.penguinoides*	Pinnoi et al. (2003b)
	Vanakripa minutellipsoidea	Pinnoi et al. (2003a)

Tab. 17.4: Ten most common species for each palm.

Eleiodoxa conferta	*Licuala longicalycata*	*Metroxylon sagu*	*Nenga pumila*
Cancellidium applanatum	*Microthyrium* sp.	*Anthostomella bipapillispora*	*Diplococcium stoveri*
Xylomyces aquaticus	*Phaeoisaria clematidis*	*Nawawia filiformis*	*Dinemasporium* sp.
Astrosphaeriella sp.	*Annulatascus velatispora*	*Oxydothis*-like	*Arecomyces epigeni*
Stilbohypoxylon elaeicola	*Massarina bipolaris*	*Apioclypea eccentricospora*	*Linocarpon* sp. 4
Lophiostoma frondisubmersa	*Phruensis brunneispora*	*Apiospora* sp.	*Lophodermium* sp.
Microthyrium sp.	*Solheimia costaspora*	*Dinemasporium lanatum*	*Dactylaria palmae*
Morenoina palmicola	*Thailandiomyces bisetulosus*	*Tetraploa aristata*	*Lophiostoma* sp.
Phaeoisaria clematidis	*Nectria* sp. 1	*Ornatispora* sp.	*Oxydothis* sp. 8
Stilbohypoxylon eleiodoxae	*Helicoma* sp. 1	*Massarina bipolaris*	*Spadicoides* sp. 4
Jahnula appendiculata	*Astrosphaeriella malayensis*	*Acrogenospora sphaerocephala*	*Jahnula appendiculata*

17.2.3 Percentage overlap in fungal diversity between the four palms

The ten most common species on each palm are listed in Tab. 17.4. No species appears to be common to the four palms, each supporting its own fungal community. Similarly, when the ten most common genera on the four palms are compared there is no single genus common to all palms (Tab. 17.5). The overlap in fungal species between the four palms was less than 1% (Tab. 17.6, Fig. 17.1), while overlap between three and

Tab. 17.5: Ten most common genera for each palm.

Eleiodoxa conferta	*Licuala longicalycata*	*Metroxylon sagu*	*Nenga pumila*
Astrosphaeriella	*Astrosphaeriella*	*Nawawia*	*Diplococcium*
Stilbohypoxylon	*Oxydothis*	*Anthostomella*	*Dinemasporium*
Cancellidium	*Annulatascus*	*Oxydothis*	*Linocarpon*
Xylomyces	*Massarina*	*Apiospora*	*Arecomyces*
Lophiostoma	*Microthyrium*	*Cylindrocladium*	*Spadicoides*
Microthyrium	*Phaeoisaria*	*Dinemasporium*	*Lophodermium*
Morenoina	*Nectria*	*Tetraploa*	*Sporidesmium*
Phaeoisaria	*Phruensis*	*Apioclypea*	*Dactylaria*
Jahnula	*Submersisphaeria*	*Ornatispora*	*Oxydothis*
Annulatascus	*Thozetella*	*Massarina*	*Jahnula*

Tab. 17.6: Percentage of fungal overlapping.

Eleiodoxa conferta & *Licuala longicalycata* = (19 species/315 species) $\times$ 100 = 6%
E. conferta & *Metroxylon sagu* = (5 species/315 species) $\times$ 100 = 1.6%
E. conferta & *Nenga pumila* = (8 species/315 species) $\times$ 100 = 2.5%
L. longicalycata & *M. sagu* = (9 species/315 species) $\times$ 100 = 2.8%
L. longicalycata & *N. pumila* = (15 species/315 species) $\times$ 100 = 4.8%
M. sagu & *N. pumila* = (2 species/315 species) $\times$ 100 = 0.6%
E. conferta & *L. longicalycata* & *M. sagu* = (2 species/315 species) $\times$ 100 = 0.6%
E. conferta & *L. longicalycata* & *N. pumila* = (6 species/315 species) $\times$ 100 = 1.9%
L. longicalycata & *M. sagu* & *N. pumila* = (1 species/315 species) $\times$ 100 = 0.3%
E. conferta & *M. sagu* & *N. pumila* = (1 species/315 species) $\times$ 100 = 0.3%
E. conferta & *L. longicalycata* & *M. sagu* & *N. pumila* = (1 species/315 species) $\times$ 100 = 0.3%

two palms are less than 2 and 6%, respectively (Figs. 17.2, 17.3), although all grew in close proximity within the forest.

17.3 Conclusion

Previously, many fungal taxa have been described from tropical and subtropical palms (Hyde et al. 1998; Fröhlich and Hyde 1999; Yanna et al. 2001a, b), an environment that supports a high fungal diversity. Taylor and Hyde (2003) report on 288 species (in 153 genera) of microfungi on the palms *Cocos nucifera* and *Archontophoneix alexandrae*, including 159 ascomycetes, two basidiomycetes and 127 asexual fungi. This resulted in the description of 34 new species. However, there has been no systematic study of fungi occurring on palms in a peat swamp.

The fungi collected on *Eleiodoxa conferta* yielded 111 taxa, and 138 taxa collected on *Licuala longicalycata*. These figures are considerably higher than those reported by Fröhlich and Hyde (2000) from palms in Australia 89 taxa, Brunei 70 taxa, Hong Kong 63 taxa and Ecuador 31 taxa. It is also higher than data reported by Yanna et al. (2001a) from five terrestrial palms; *Oraniopsis appendiculatum* (56 species), *Livistona australis* (32 species), *Oncosperma horridium* (69 species), *Salacca affinis* (65 species) and *Phoenix haceana* (87 species); while Taylor and Hyde (2003) reported fungi from: *Archonotophoenix alexandrae* (151 species), *Cocos nucifera* (77 species), and *Trachycarpus fortunei* (144 species). The high number of fungi reported on *E. conferta* and *L. longicalycata* may be attributed to the peat swamp habitat with its moist humid conditions and high rainfall, while fungi from another two palms, *Metroxylon sagu* and *Nenga pumila* were just a preliminary study (only 2 collecting times) and showed 45 and 47 taxa, respectively.

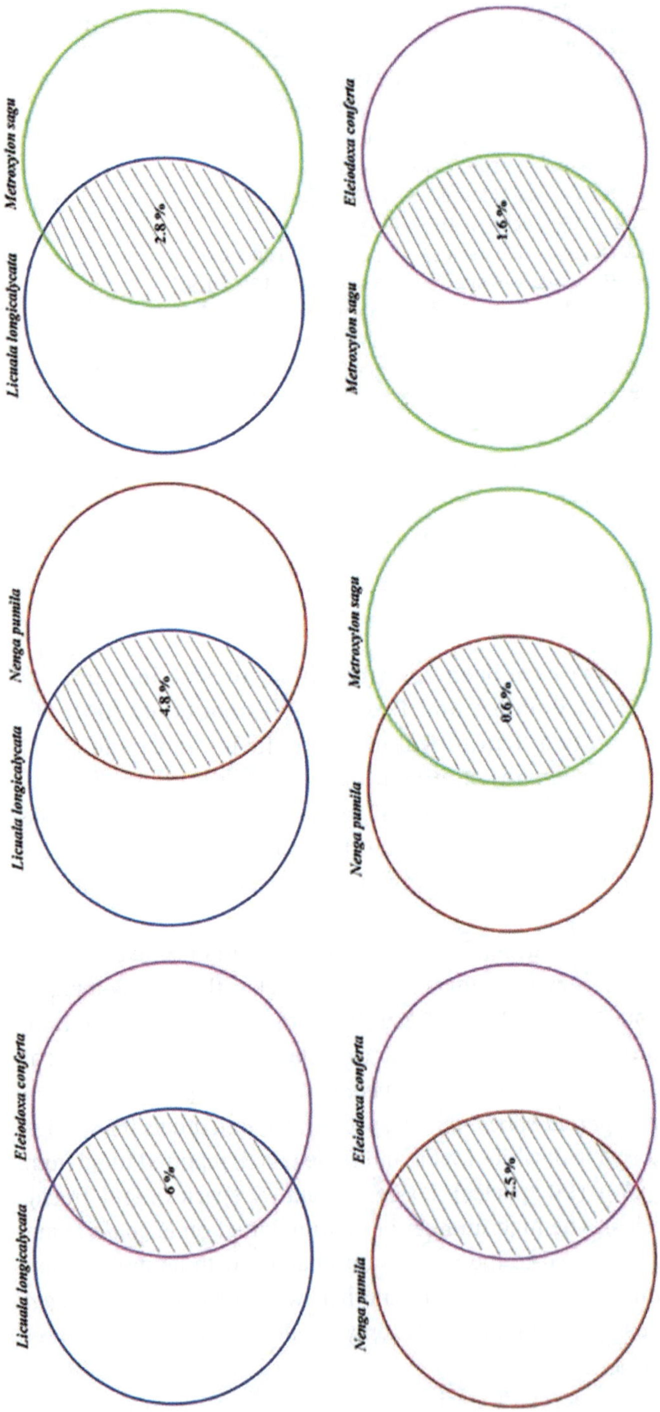

Fig. 17.1: Percentage of fungal overlapping between two palms.

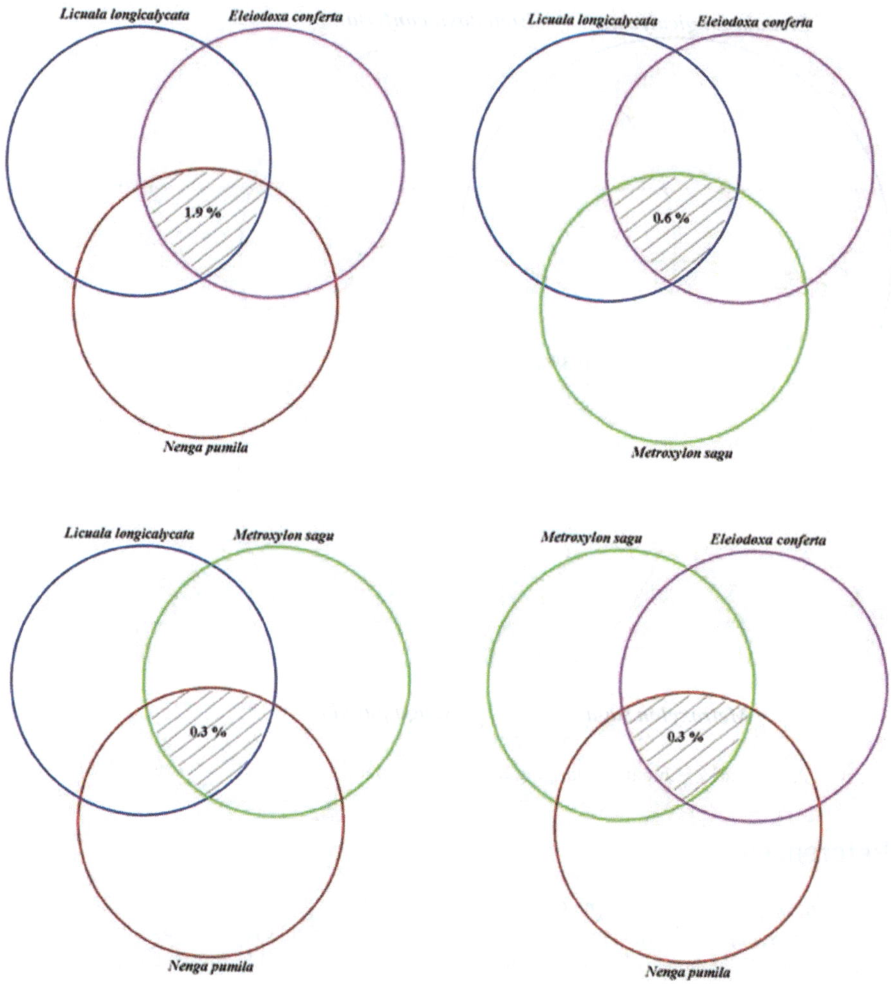

Fig. 17.2: Percentage of fungal overlapping between three palms.

Acknowledgments

This project is supported by research grant BRT R_145008 (Biodiversity Research and Training Program). We are grateful to Kanyawim Kirtikara and Morakot Tanticharoen for continued support, to Manetr Boonyanant and his staff for research facilities at the Sirindhorn Field and Nature Study Center, Narathiwat. Gareth Jones is supported by the Distinguished Scientist Fellowship Program (DSFP), King Saud University, Saudi Arabia. We are grateful to Dr Lai Pang for useful comments on this chapter.

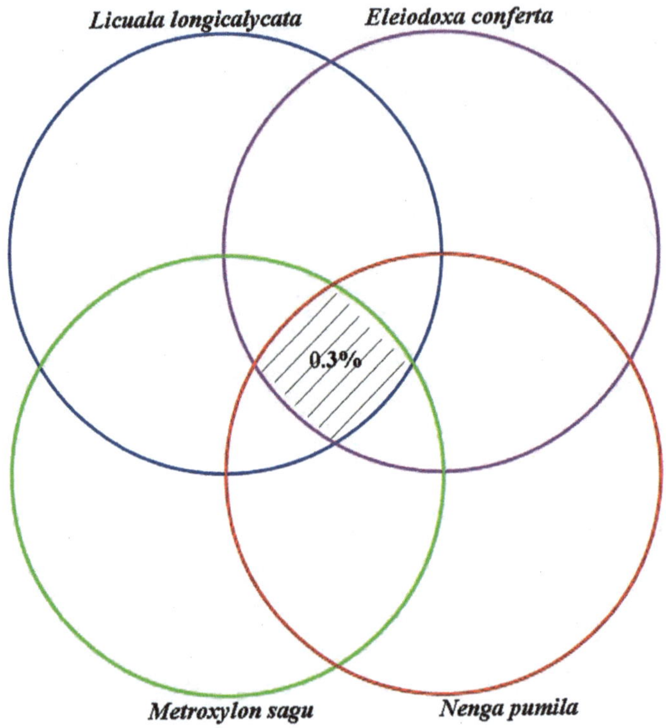

Fig. 17.3: Percentage of fungal overlapping between four palms.

References

Apiwathnasorn C, Samung Y, Prummongkol S, Panasoponkul C, Loymek S. Mosquito fauna of "Toh Daeng" swamp forest, Thailand. Southeast Asian J Trop Med Public Health 2009;40:720–726.

Aramsiriujiwet Y. Hyphomycetes on bamboo and Palmae in Thailand. M.S. Thesis. Bangkok, Thailand: Kasetsart University 1996.

Cannon PF. Diversity of the Phyllachoraceae with special reference to the tropics. In: Hyde KD, ed. Biodiversity of Tropical Microfungi. Hong Kong: Hong Kong University Press 1997:255–278.

Choi Y, Hyde KD, Ho WH. Single spore isolation of fungi. Fungal Divers 1999;3:29–38.

Fröhlich J, Hyde KD. New *Oxydothis* species associated with palm leaf spots in north Queensland. Mycol Res 1994;98:213–218.

Fröhlich J, Hyde KD. Biodiversity of palm fungi in the tropics: are global fungal diversity estimates realistic. Biodiver Conser 1999;8:977–1004.

Fröhlich J, Hyde KD. Palm Microfungi. Fungal Diversity Research Series: 3. Hong Kong: Fungal Diversity Press 2000.

Hawksworth DL. The fungal dimension of biodiversity: magnitude, significance, and conservation. Mycol Res 1991; 95:641–655.

Hawksworth DL. The magnitude of fungal diversity: the 1.5 million species estimate revisited. Mycol Res 2001;105:1422–1432.

Hankaew C. Integrated management plan on peat swamp forests, Southern Thailand. Workshop on approaches for integrated and wise use of peatlands in Thailand. Royal Princess Hotel, Narathiwat, Thailand. 9–11 April,2003;9pp.

Hidayat I, Jeewon R, To-anun C, Hyde KD. The genus *Oxydothis*: New palmicolous taxa and phylogenetic relationships within Xylariales. Fungal Divers 2006;23:367–406.

Hyde KD. Fungi from palms. XIII. The genus *Oxydothis*, a revision. Sydowia 1994;46:265–314.

Hyde KD. Fungi from palms XXIX. The genus *Anthostomella*, with ten new species. Nova Hedwig 1996a;62:273–340.

Hyde KD. Fungi from palms XXVI. *Arecomyces* gen. nov., with seven new species. Sydowia 1996b;48:224–240.

Hyde KD. Fungi from palms XXVII. *Capsulospora* gen. nov., with three new species. Sydowia 1996c;48:273–340.

Hyde KD, Alias SA. *Linocarpon angustatum* sp. nov., and *Neolinocarpon nypicola* sp. nov from petioles of *Nypa frusticans*, and a list of fungi from aerial parts of this host. Mycoscience 1999;440:145–149.

Hyde KD, Fröhlich, J, Taylor JE. Diversity of ascomycetes on plams in the tropics. In: Hyde KD, ed. Biodiversity of Tropical Microfungi. Hong Kong: Hong Kong Univ Press 2007:141–156.

Hyde KD, Yanna, Pinnoi A, Jones EBG. *Goidanichiella fusiforma* sp. nov. from palm fronds in Brunei and Thailand. Fungal Divers 2002;11:119–122.

Hyde KD, Taylor, JE, Fröhlich J. Fungi from palms XXXIV. The genus *Neolinocarpon* with five new species and one new combination. Fungal Divers 1998;1:115–131.

Jackson CR, Liew KC, Yule CM. Structural and functional changes with depth in microbial communities in a tropical Malaysian peat swamp forest. Microb Ecol 2009;57:402–412.

Jones EBG, Hyde KD. Introduction to Thai fungal diversity. In: Jones EBG, Tanticharoen M, Hyde KD, eds. Thai Fungal Diversity. BIOTEC: Thailand 2004:7–35.

Jusoff K, Hassan C H, Hamzah KA. Tropical peat swamp forest ecosystem and floristic diversity in Pahang, Malaysia. Inter J Systems Applications Eng Development 2007;3:41–44.

Kanokratana P, Urngwetwanit T, Rattanachomsri U, Bunterngsook B, Nimchua T, Tangphatsornuang et al. Insights into the phylogeny and metabolic potential of a primary tropical peat swamp forest microbial community by metagenomic analysis. Microb Ecol 2010;61:518–528.

Kirk PM, Cannon PF, Minter DW, Stalpers JA. Dictionary of the Fungi. Wallingford (UK): CABI Publishing 2008.

Magurran AE, Ecological Diversity and its Measurement. Princeton, NJ University Press 1988:1–179.

McKenzie EHC, Pinnoi A, Wong, MKM, Hyde KD, Jones EBG. Two new hyaline *Chalara* and key to species described since 1975. Fungal Divers 2002;11:129–139.

Miettinen J, Shi C, Liew SC. Two decades of destruction in Southeast Asia's peat swamps. Front Ecol Environ 2012;10:124–128.

Mohd Azmi MI, Cullen R, Bigsby H, Awang Noor AG. The existence value of peat swamp forest in Peninsula Malaysia. Paper presented at New Zealand Agriculture and Resource Economics Society (NZARES) Conference 2009. 27–28 August 2009. Tahuna Beach Resort, Nelson, New Zealand 2009:1–22.

Ng TP, Shamsudin I. Common trees in peat swamp forests of peninsular Malaysia. For Res Instit Malaysia 2001:1–97.

Parish F, Yiew CT, Yun CS. Integrated management planning of peatland. Regional Project Agency. Workshop on sustainable management of peatland forests in Southeast Asia. Bogor, 2012.

Phlengklai C, Niyomdham C, Premrasmi A, Chirathanakorn K, Phuma R. Peat swamp forest of Thailand. Thai Forest Bull 1989;18:1–42.

Phlengklai C, Niyomdham C, Ueachirakan W. Flora in Peat Swamp Areas of Narathiwat, Thailand. Phikun Thong Study Centre. Bangkok: Sombun Press 1991.

Pinnoi A, Jeewon R, Sakayaroj J, Hyde KD, Jones EBG. *Berkleasmium crunisia* sp. nov. and its teleomorphic affinities to the Pleosporales based on 18S, 28S and ITS-5.8S rDNA sequence analyses. Mycologia 2007;99:378–384.

Pinnoi A, Lumyong S, Hyde KD, Jones EBG. Biodiversity of fungi on the palm *Eleiodoxa conferta* in Sirindhorn peat swamp forest, Narathiwat, Thailand. Fungal Divers 2006;22:205–208.

Pinnoi A, McKenzie EHC, Jones EBG, Hyde KD. Palm fungi from Thailand: *Custingophora undulatistipes* sp. nov. and *Vanakripa minutellipsoidea* sp. nov. Nova Hedwig 2003a;77:213–219.
Pinnoi A, McKenzie EHC, Jones EBG, Hyde KD. Aquatic fungi from peat swamp palms: *Unisetosphaeria penguinoides* gen. et sp. nov., and three new *Dactylaria* species. Mycoscience 2003b;44:377–382.
Pinnoi A, Phongpaichit S, Jeewon R, Tang A, Hyde KD, Jones EBG. Phylogenetic relationships of *Astrocystis eleiodoxae* sp. nov. (Xylariaceae). Mycosphere 2010;1:l–9.
Pinnoi A, Pinruan U, Hyde KD, Lumyong S. *Submersisphaeria palmae* sp. nov. and key to the genus and notes on *Helicoubisia*. Sydowia 2004;56:72–78.
Pinruan U, Jones EBG, Hyde KD. Aquatic fungi from peat swamp palms: *Jahnula appendiculata* sp. nov. Sydowia 2002;54:242–247.
Pinruan U, Lumyong S, Hyde KD, Jones EBG. Occurrence of fungi on tissues of the peat swamp palm *Licuala longecalycata*. Fungal Divers 2007;25:157–273.
Pinruan U, Lumyong S, McKenzie EHC, Jones EBG, Hyde KD. Three new species of *Craspedodidymum* from palm in Thailand. Mycoscience 2004a;45:177–180.
Pinruan U, McKenzie EHC, Jones EBG, Hyde KD. Two new species of *Stachybotrys*, and a key to the genus. Fungal Divers 2004d;17:145–157.
Pinruan U, Rungjindamai N, Sakayaroj J, Lumyong S, Hyde KD, Jones EBG. *Baipadisphaeria* gen. nov., a freshwater ascomycete (Hypocreales, Sordariomycetes) from decaying palm leaves in Thailand. Mycosphere 2010;1:53–63.
Pinruan U, Sakayaroj J, Hyde KD, Jones EBG. Aquatic fungi from peat swamp palm: *Phruensis brunneispora* and its hyphomycete anamorph. Mycologia 2004b;96:1161–1168.
Pinruan U, Sakayaroj J, Hyde KD, Jones EBG. *Flammispora* gen. nov., a new freshwater ascomycete from decaying palm leaves. Stud Mycol 2004c;50:381–386.
Pinruan U, Sakayaroj J, Hyde KD, Jones EBG. *Thailandiomyces bisetulosus* gen. *et* sp. nov. (Diaporthales, Sordariomycetidae) and its anamorph *Craspedodidymum*, is described based on nuclear SSU and LSU rDNA sequences. Fungal Divers 2008;29:89–98.
Sarapat T. Diversity of Ascomycetes and Mitosporic fungi on Palmae in Sirindhorn Swamp Forest Ecosystem, Narathiwat Province, Thailand. M.S. Thesis. Songkla, Thailand: Prince of Songkla University 2003.
Sivichai S, Jones EBG, Hywel-Jones NL. Fungal colonization of wood in a freshwater stream, at Khao Yai National Park. Fungal Divers 2000;5:71–88.
Taylor J, Hyde KD. Microfungi of Tropical and Temperate Palms. Fungal Diversity Research Series 12. Hong Kong: Fungal Diversity Press 2003.
Vijarnaorn P. Peatlands in Southeast Asia: a regional perspective. In: Maltby E, Immirzi CP, Safford RJ, eds. Tropical lowland peatlands of Southeast Asia. Gland, Switzerland: ICUN.
Voglmayr HM, Yule CM. *Polyancora globosa* gen. *et* sp. nov., an aeroaquatic fungus from Malaysian peat swamp forest. Mycol Res 2006;110:1242–1252.
Yanna. Biodiversity, ecology and taxonomy of saprobic fungi on palm fronds. Ph.D. Thesis. The University of Hong Kong, Hong Kong, 2001.
Yanna, Ho WH, Hyde KD. Fungal communities on decaying palm founds in Australia, Brunei, and Hong Kong. Mycol Res 2001a;105:1458–1471.
Yanna, Hyde KD, Goh TK. *Koorchaloma novojornalis* sp. nov., a new sporodochial fungus from Hong Kong. Fungal Divers 1998;1:193–197.
Yanna, Ho WH, Hyde KD, Goh TK. Occurrence of fungi on tisues of *Livistonia chinensis*. Fungal Divers 2011b;6:167–180.
Yule CM, Gomez LN. Leaf litter decomposition in a tropical peat swamp forest in Peninsular Malaysia. Wetlands Ecol Manage 2009;17:231–241.

Verónica Ferreira, Vladislav Gulis, Cláudia Pascoal and
Manuel A. S. Graça

18 Stream pollution and fungi

18.1 The importance of aquatic hyphomycetes in woodland streams

Aquatic hyphomycetes are a group of freshwater fungi that includes asexual stages of ascomycetes and basidiomycetes. Even though aquatic hyphomycetes are a phyllogenetically heterogeneous group, there are common characteristics to most species: (i) they are saprotrophs, using submerged leaves and wood as substrates and sources of carbon (Gulis et al. 2006b), (ii) they can obtain nutrients from both the water column and from organic substrates (Suberkropp 1995), (iii) they inhabit mostly well aerated and turbulent waters (Medeiros et al. 2009), and (iv) they produce large numbers of conidia (asexual spores) with distinct shapes and sizes (Gessner and Chauvet 1994; Gulis and Suberkropp 2003c; Ferreira et al. 2006a) (Fig. 18.1).

Aquatic hyphomycetes are important players in small woodland streams, where they facilitate a fundamental ecosystem-level process—the decomposition of allochthonous plant litter entering streams from the riparian zone, mostly as leaves (Webster et al. 1997). Once in water, leaves leach soluble compounds, particularly during the first 24–48 h of submersion and are colonized by a range of microbes. In temperate streams, aquatic hyphomycetes are often the main microbial colonizers of submerged litter contributing >90% to the total microbial biomass and production (Weyers and Suberkropp 1996; Hieber and Gessner 2002; Pascoal and Cássio 2004). After attachment to the leaf litter surface, aquatic hyphomycete conidia germinate (Treton et al. 2004) and mycelia penetrate the leaf mesophyll (Canhoto and Graça 1999). Aquatic hyphomycetes macerate the leaf matrix with extracellular enzymes (Chamier and Dixon 1982; Arsuffi and Suberkropp 1984, 1988), mineralize organic carbon (Gulis et al. 2006b), or convert it into fungal biomass (Gessner and Chauvet 1994; Gulis et al. 2006b), and release large numbers of conidia (Gessner and Chauvet 1994; Gulis et al. 2006b; Ferreira et al. 2006a). The activities of aquatic hyphomycetes on leaf litter have been shown to correlate well with litter mass loss (Gessner and Chauvet 1994; Niyogi et al. 2003), and they can directly account for the loss of up to 39% of initial litter mass under field conditions (Hieber and Gessner 2002; Pascoal and Cássio 2004). The accumulation of fungal biomass and the maceration of litter also increase litter quality and palatability to invertebrate detritivores whose feeding activities lead to further litter mass loss (Canhoto and Graça 2008). Also, conidia released into the water column are used as food by the filter feeders (Bärlocher and Brendelberger

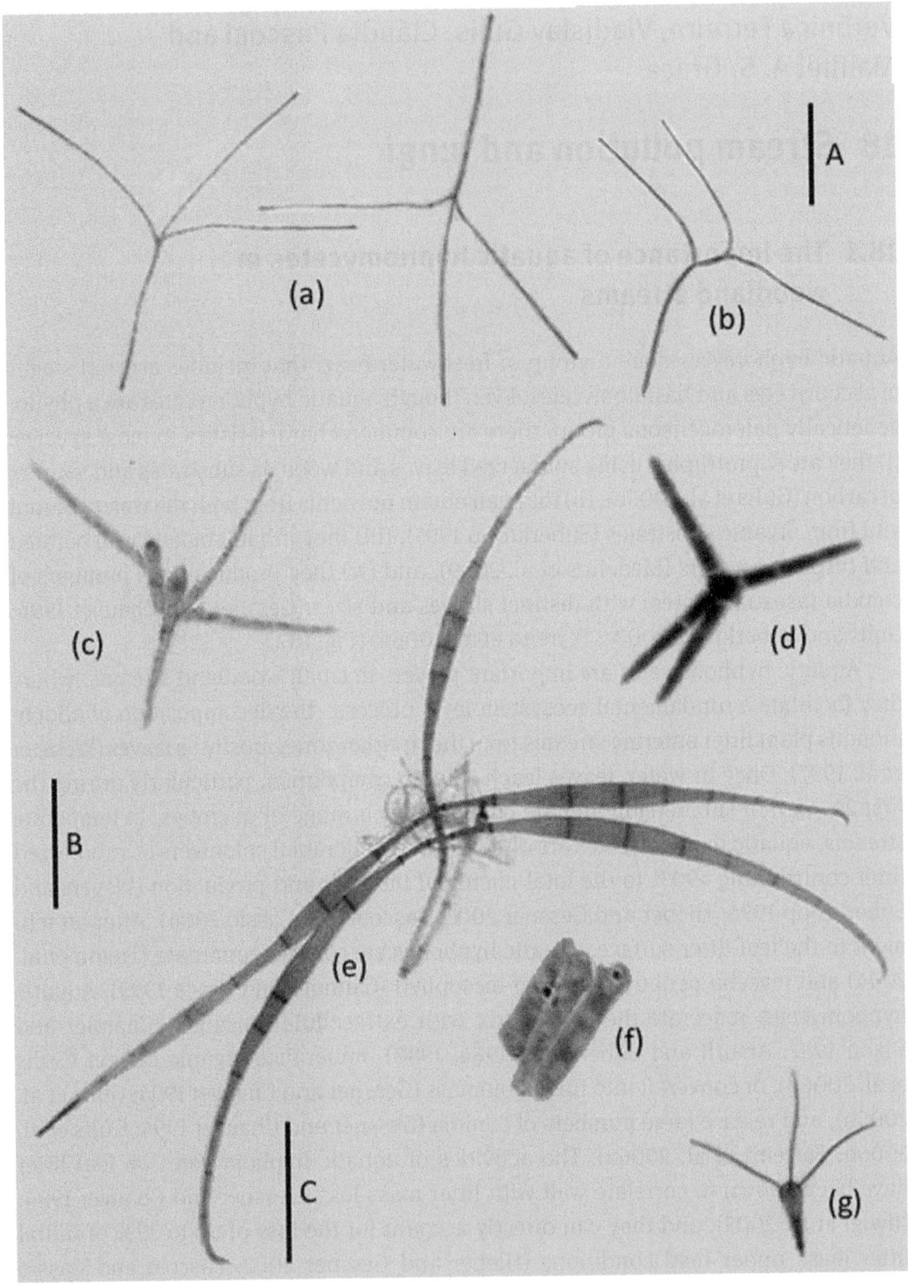

Fig. 18.1: Conidia of aquatic hyphomycetes. (a) *Tetrachaetum elegans*, (b) *Tricladium chaetocladium*, (c) *Tetracladium marchalianum*, (d) *Lemonniera terrestris*, (e) *Casaresia sphagnorum*, (f) *Heliscus lugdunensis* and (g) *Clavariopsis aquatica*. Scale bar A (species a, b, and g) = 50 µm, B (species c, d, and f) = 25 µm, C (species e) = 100 µm.

2004). Thus, aquatic hyphomycetes establish a link between dead organic matter and secondary invertebrate production.

Fungal community structure (species richness and identities) may also affect litter decomposition and invertebrate feeding (Arsuffi and Suberkropp 1984; Lecerf et al. 2005; Pérez et al. 2011). This can be attributed to species-specific differences in enzymatic activities (Zemek et al. 1985; Arsuffi and Suberkropp 1988) or elemental composition (Danger and Chauvet 2013), which suggests that litter colonized by certain fungal species will have distinct nutritive quality and could be preferentially consumed by detritivores.

Aquatic hyphomycete assemblages and activities are controlled by environmental conditions (e.g. nutrient availability, pH, temperature; Boyero et al. 2011; Cornut et al. 2012; Woodward et al. 2012). Therefore, human-induced changes in environmental conditions may affect fungal activity and community structure, fungi-driven litter decomposition and secondary production of invertebrates, and thus the functioning of woodland streams. Here, we address the effects of increases in nutrients, heavy metals, nanoparticle and xenobiotic concentrations, and water temperature, resulting from numerous human activities, on aquatic fungi.

18.2 Effects of nutrient enrichment on stream fungi

Pollution of streams and rivers due to agricultural or urban activities is often accompanied by increases in the concentration of dissolved organic and inorganic (nitrogen and phosphorus) nutrients, known as eutrophication. In addition to agricultural and urban runoff, atmospheric deposition of mostly nitrogen in highly industrialized regions of the world resulted in significant shifts from pristine oligotrophic conditions. Increased nutrient availability promotes the excessive growth of algae and heterotrophic organisms, such as bacteria and aquatic fungi (Vitousek et al. 1997; Carpenter et al. 1998). However, the majority of basic and applied studies dealt with the effects of nutrients on primary producers, while the heterotrophic pathways, including the effects of nutrients on fungal decomposers, received less attention (Dodds and Cole 2007).

The direct and indirect effects of dissolved organic nutrients on fungi are difficult to separate. Even though fungal growth in streams can be directly stimulated by readily available dissolved organic carbon (DOC) (Wilcox et al. 2005), prokaryotes generally outcompete fungi for dissolved labile C. The major indirect effects of organic pollution in streams and rivers are related to decreases in oxygen availability brought about by stimulation of bacterial respiration. Aquatic hyphomycetes are sensitive to dissolved oxygen concentration, and their growth, reproduction and diversity are negatively affected by oxygen deficiency (Medeiros et al. 2009). Several field studies have reported changes in fungal community structure in organically polluted reaches of streams or rivers, including increases in the relative abundances of terrestrial fungi (Au et al. 1992b; Raviraja et al. 1998; Tsui et al. 2001). The activity of fungal decomposers and decomposition rate of plant litter are less sensitive to organic pollution than fungal diversity (Suberkropp et al. 1988; Raviraja et al. 1998), but can be negatively affected (Au et al. 1992a). However, the extent of organic pollution and the presence of other accompanying pollutants complicate the interpretation (Pascoal et al. 2003, 2005).

The effect of dissolved inorganic nitrogen (N) and phosphorus (P) on litter-associated fungi in streams has received considerable attention in the recent years. Aquatic hyphomycetes are capable of obtaining N and P from both the plant litter and from the water column (Suberkropp 1995). From the perspective of stoichiometric imbalance, since fungal biomass C:N and C:P ratios are considerably lower than that of plant litter (Danger and Chauvet 2013; Gulis et al. unpublished), fungi have to either retain N and P from the organic substrate more efficiently than C or immobilize inorganic nutrients from stream water. In addition, mining N and P from organic substrates requires considerable energy expenditures to produce extracellular enzymes to attack complex, often recalcitrant, organic molecules and eventually cleave amino or phosphate groups. Thus, based on microbial energetics, fungi should preferentially use dissolved inorganic nutrients from the water column. Indeed, increased availability of dissolved N and P in laboratory studies, whole-stream nutrient addition experiments or due to anthropogenic activities stimulates fungal activity and lead to nutrient immobilization (e.g. Gulis and Suberkropp 2003a; Ferreira et al. 2006a; Gulis et al. 2006a). The effects of dissolved nutrients on fungi are also greatest for low quality plant litter (i.e. low N and P, high lignin concentrations; e.g. rhododendron or oak leaves, wood) than for high-nutrient litter such as alder or maple leaves, as, for the former substrates, nutrient or carbon limitation of fungal growth can be alleviated or reduced by external inorganic nutrients (Stelzer et al. 2003; Gulis et al. 2004, 2008; Ferreira et al. 2006a).

Experiments in laboratory microcosms simulating stream conditions (i.e. using plant litter as a substrate, providing aeration and realistic nutrient concentrations) (e.g. Sridhar and Bärlocher 1997; Suberkropp 1998; Gulis and Suberkropp 2003a, b; Ferreira and Graça 2007; Ferreira and Chauvet 2011) showed that even moderate increases in dissolved nutrient concentrations typically stimulate fungal biomass accrual, fungal production, sporulation, respiration, growth efficiency and, ultimately, litter decomposition. Fungal sporulation rate or total reproductive output are most sensitive to nutrient availability and often increase by one to two orders of magnitude under elevated nutrient conditions.

Whole-stream nutrient addition experiments provide additional realism to the interpretation of the effects of dissolved nutrients on fungi. Short-term studies in temperate, Arctic and Mediterranean streams (Stelzer et al. 2003; Benstead et al. 2005; Ferreira et al. 2006a) demonstrated that N or N + P additions increased fungal biomass accrual, sporulation and decomposition of leaf litter and wood veneers. A 5-year whole-stream nutrient addition (N + P) experiment in southern Appalachians

(USA) showed that increased dissolved nutrient availability stimulated fungal biomass accrual, fungal growth rate and production, microbial (mostly fungal) respiration and sporulation rate of aquatic hyphomycetes associated with decomposing leaf litter and wood, and resulted in increases in concentration of fungal spores in transport (Gulis and Suberkropp 2003c, 2004; Gulis et al. 2004, 2008; Suberkropp et al. 2010). In addition, the direct effect of dissolved nutrients on fungi led to important ecosystem-level consequences. Nutrient enrichment altered storage and fluxes of detrital carbon by accelerating organic matter decomposition and carbon export as CO_2 and as fine particulate organic matter downstream, and decreasing carbon storage (Benstead et al. 2009). This affected higher trophic levels by changing production and biomass of detritivorous invertebrates and the flow of microbially immobilized N and P to these consumers (Cross et al. 2006, 2007; Greenwood et al. 2007).

Increases in dissolved inorganic nutrients in whole-stream nutrient enrichment experiments also affected fungal community structure. For example, Gulis and Suberkropp (2003c, 2004) reported changes in dominance patterns and relative abundances of aquatic hyphomycetes associated with decomposing leaf litter. Specifically, following nutrient addition, species that produced relatively large conidia (*Anguillospora filiformis*, *Terachaetum elegans*, *Tricladium chaetocladium*) assumed dominance, whereas species with smaller conidia dominated before the nutrient addition and in the reference stream. A somewhat similar pattern has been observed by Artigas et al. (2008). This trend implies that production of conidia with higher biovolume requires additional N and P supply and that these additional nutrients may be indeed shunted largely to reproduction (Suberkropp 1998).

Correlative studies based on natural or anthropogenic gradients of dissolved inorganic nutrients showed mixed results. Studies performed in streams across a natural (geological) nutrient gradient (e.g. Suberkropp and Chauvet 1995; Rosemond et al. 2002) or affected by moderate agricultural land use (e.g. Niyogi et al. 2003; Gulis et al. 2006a; Bergfur et al. 2007) demonstrated stimulation of fungal biomass accrual, sporulation of aquatic hyphomycetes and/or microbial respiration associated with decomposing plant litter by dissolved nutrients. However, in streams that experience high levels of eutrophication, factors other than nutrients, e.g. oxygen deficiency, accumulation of fine sediments or toxic pollutants (pesticides, heavy metals, etc.), may override the positive effects of nutrients on fungal activity. As a result, fungal activity and microbially mediated plant litter decomposition may be inhibited in highly polluted streams despite high nutrient availability (Pascoal and Cássio 2004; Duarte et al. 2009; Woodward et al. 2012). Likewise, even though some studies reported increased fungal diversity in moderately eutrophied streams, it decreased in highly eutrophied streams (Pascoal et al. 2005; Duarte et al. 2009; Sridhar et al. 2009); the latter effect is probably attributable to other accompanying pollutants rather than elevated dissolved inorganic nutrients.

Whole-stream nutrient addition experiments and correlative studies in streams suggest that fungal nutrient demands are met at relatively low concentrations of dissolved inorganic N and P, and the relationship between nutrient availability and fungal activity and/or fungi-driven plant litter decomposition often follows asymptotic saturation-type models (e.g. Michaelis-Menten) (Rosemond et al. 2002; Ferreira et al. 2006a; Gulis et al. 2006a, b). The half saturation constants (i.e. concentration at which half of the maximum rate of activity is achieved) for nitrate-N and soluble reactive phosphorus (SRP) were estimated at 16–303 and 7–21 μg/L, respectively. This type of data together with the information on the responses of primary producers (Dodds 2007) can be useful for establishing nutrient criteria in streams. As mentioned earlier, due to potentially negative effects of other pollutants on fungi in highly impacted streams, the relationship between dissolved nutrient availability and microbial-driven decomposition on a continental scale (Europe) is rather unimodal (bell shaped) (Woodward et al. 2012).

18.3 Effects of heavy metals and acidification on stream fungi

Humans have been mining for metals for millennia. For instance, ancient mining in Rio Tinto, Spain, dating from circa 3000 BC, has led to a river with pH of 2.2 and red colour ("tinto" in Spanish). Worldwide mining has resulted in thousands of active and abandoned mines. The exposure of sulphide minerals to weathering results in the formation of sulphuric acid and the release of metal ions. Some metals such as Al and Fe may form hydroxides that precipitate and give sediments an orange color. Drainage waters from mines commonly have low pH and high concentration of metals, whereas benthic substrates in impacted streams are often coated with metal hydroxides (e.g. Hogsden and Harding 2013). Acidification of streams can also be caused by acid rain. The burning of fossil fuels results in the formation of sulphate and nitrous oxides that end up in streams *via* atmospheric deposition, with the consequent decrease in water pH (Allan and Castillo 2007). Other sources of heavy metals in the environment are the industry and agricultural practices (e.g. Corbi et al. 2006; Duan and Tan 2013). Because heavy metals are highly reactive and retained by living organisms, they are highly toxic (e.g. Nagajyoti et al. 2010). The biological effects of acid waters and metals on the biota include reduced biodiversity with the predominance of few tolerant species, decreased production and shortened food webs (Gray 1998; Hogsden and Harding 2013).

Increased concentrations of heavy metals reduce growth and sporulation of aquatic hyphomycetes under laboratory conditions (Miersch et al. 2001; Azevedo and Cássio 2010; Medeiros et al. 2010), with sporulation often being more sensitive to heavy metals than fungal growth (Abel and Bärlocher 1984). Low levels of heavy metals and acidity may lead to decreases in sporulation with no effect on fungal biomass (Duarte et al. 2004; Baudoin et al. 2008; Clivot et al. 2013). Suppression of reproduction or growth results in organisms channelling energy for maintenance, if stress conditions do not affect energy intake. This may explain why Gonçalves et al.

(2011) found higher microbial respiration rates on leaves from an uranium contaminated site than on leaves from a reference site. However, if environmental stress interferes with energy intake or expenditure, respiration rates are likely to decrease (Bermingham 1996; Dangles et al. 2004; Niyogi et al. 2009, 2013).

Heavy metals and acidification may also affect fungal community structure and ecosystem functioning. Stress conditions can cause changes in community structure (e.g. relative abundances of species) with no major changes in species richness on plant litter exposed to Zn (Duarte et al. 2004). In other studies, decreases in biodiversity with no changes in plant litter decomposition rates (Medeiros et al. 2010) or decreases in biodiversity and decomposition (Bermingham 1996; Baudoin et al. 2008; Niyogi et al. 2009; Clivot et al. 2013) have been reported. Decomposition rates were also reported to decrease with increased acidity (Dangles et al. 2004; Simon et al. 2009; Cornut et al. 2012). Hydroxide precipitation in streams affected by acid mine drainage may create a physical barrier for microbial colonization of plant litter and negatively affect decomposition (Schlief and Mutz 2005; Niyogi et al. 2013).

Several species of aquatic hyphomycetes have been often found in highly polluted streams (Sridhar et al. 2000, 2001; Krauss et al. 2001), which suggests some degree of tolerance or adaptation to high metal concentrations. Evidence to support this assumption comes from reports of higher tolerance to metals in fungal strains isolated from contaminated sites compared to conspecific strains from unpolluted streams (Chamier and Tipping 1997; Miersch et al. 1997; Baldrian and Gabriel 2002; Fernandes et al. 2011), and evidence of morphologic (Ferreira et al. 2010) and enzymatic differences between strains (Braha et al. 2007).

The effects of metals and acidification on aquatic hyphomycetes can have repercussions for higher trophic levels given the dependence of invertebrates on fungi to access the energy and nutrients locked in plant litter (Suberkropp 1992; Graça 1993; Canhoto and Graça 2008). Gonçalves et al. (2011) reported that although there were no differences in fungal biomass accrual between leaves incubated in uranium polluted and clean waters, the shredder *Sericostoma vittatum* discriminated between leaf types and preferentially consumed leaves from the non-polluted site. Similar pattern was found by Medeiros et al. (2008) in a stream receiving waters from an abandoned gold mine; fungal biomass and decomposition rates were similar between polluted and reference sites, but the shredder *Allogamus ligonifer* preferentially fed on leaves from the reference site.

18.4 Ecological and toxicological effects of engineered nanoparticles on stream fungi

The advances in nanotechnology have raised concerns about the potential impacts of the enormous amounts of nanomaterials that were produced over the past decade. Indeed, engineered nanoparticles (ENPs) have become part of our daily life in the form of cosmetics (Perugini et al. 2002), antimicrobial paints (Kaegi et al. 2008), textile fabrics (Zhang et al. 2009) and electronic devices (Luechinger et al. 2008). Natural waters are likely to serve as a final repository of natural and engineered nanoparticles *via* waste release, soil runoff and atmospheric deposition (Klaine et al. 2008).

Nanomaterials are generally defined as materials that are < 100 nm in at least one dimension. A common difficulty in assessing their properties and behavior arises because nanomaterials can exist in many forms and sizes (Bernhardt et al. 2010). The surface properties of ENPs are of essential importance for their aggregation behaviour, and thus for their mobility and interaction with biota. The aggregation into larger particles and deposition in aquatic systems are determined by the ENP surface properties, which are mainly dependent on parameters such as temperature, ionic strength, pH, particle concentration and size (Navarro et al. 2008).

A major gap that influences the research of the environmental behavior of nanoparticles is the lack of analytical methods for their analysis at low and environmentally relevant concentrations (Hassellöv et al. 2008). Although ENPs have been already detected in aquatic environments (e.g. TiO_2 ENPs leached from the paint of house facades into a stream; Kaegi et al. 2008), their potential impacts on ecosystem health are largely unknown. The vast majority of ecotoxicological studies of ENP effects to date have looked at model organism responses to pure solutions/suspensions of ENPs. Such studies, though essential, are insufficient to predict the ecological impacts of the much wider range of likely exposure scenarios (Bernhardt et al. 2010).

Given the critical role of microorganisms in organic matter and nutrient cycling in aquatic ecosystems, ENPs can potentially alter ecosystem productivity and biogeochemistry by affecting aquatic fungi. Indeed, the exposure to nanoparticulate silver (nano Ag) or copper oxide nanoparticles (nano CuO) led to a reduction in fungal sporulation and leaf decomposition (Pradhan et al. 2011). These effects have been accompanied by shifts in the composition of leaf-associated fungal assemblages based on fungal spore morphology and DNA fingerprints. The dominant fungal species *Articulospora tetracladia* was stimulated, while the co-dominant species *Flagellospora* sp. was inhibited by exposure to nano or ionic metals, being replaced by *Heliscus lugdunensis* (Pradhan et al. 2011). *Articulospora tetracladia* and *Heliscus lugdunensis* are reported to occur in metal contaminated streams (Jaeckel et al. 2005; Pascoal et al. 2005) and some strains of these species were found to be resistant to high concentrations of metals in their ionic (Braha et al. 2007) or nano form (Pradhan et al. 2013). Results for aquatic microorganisms are in agreement with those obtained for terrestrial microbial assemblages. Exposure to nano Ag decreased the biomass of soil microbes in concert with changes in the activity of both leucine aminopeptidase and phosphatase (Coleman et al. 2013), which are extracellular enzymes often used as indicators of the microbial potential to decompose organic matter.

Effects on microbial decomposers are expected to translate to higher trophic levels. Uptake or incorporation of metal ENPs into microbial biomass suggests the potential for trophic transfer of these contaminants (Coleman et al. 2013). A significant inhibition of leaf consumption rate (up to 47%) and growth rate of invertebrate detritivores (up to 46%) was observed when invertebrates were exposed to sublethal concentration of nano CuO (75 mg/L) through either contaminated stream water or pre-contaminated food (Pradhan et al. 2012), which is possibly related to the food avoidance behavior. Moreover, the exposure to increased concentration of nano CuO via water or pre-contaminated food led to accumulation of Cu in the larval body (Pradhan et al. 2012). Thus, the stress induced by nanoparticles may affect the invertebrates directly or indirectly through its effects on microbes.

The toxicity of ENPs can be attributed to physical characteristics, metal ions released by dissolution of metal ENPs, or production of reactive oxygen species. For instance, Ag antimicrobial activity is most often attributed to the dissolved cation rather than to the high surface area and low solubility of nonionic metallic nanoparticle (Klaine et al. 2008). Consistently, effects of metal nanoparticles are generally less pronounced than those of the respective metal ions (Pradhan et al. 2011, 2012). However, some studies reported that leached metal ions are insufficient in explaining the toxicity of nanoparticles due to low solubility of some metal ions (e.g. Griffitt et al. 2008). A differential response of aquatic microbial communities to nanoparticulate form of metals and their respective ionic form suggests different modes of action.

In a study with pure cultures of fungi, nano CuO expressed its toxicity by inhibiting fungal biomass production and altering the mycelium morphology in a dose- and time-dependent manner (Pradhan et al. 2013). Nano CuO was more toxic to fungi isolated from non-polluted than from metal-polluted streams (Pradhan et al. 2013), suggesting adaptive mechanisms underlying the tolerance/resistance against metal nanoparticles as found for several metal ions in aquatic fungi (Jaeckel et al. 2005; Azevedo et al. 2007; Guimarães-Soares et al. 2007; Krauss et al. 2011).

Overall, the data obtained so far point to the toxicity of nanoparticles to freshwater organisms which potentially places at risk fungal diversity and stream ecosystem processes. However, further studies are needed to identify the bioavailable forms of nanoparticles and effects of different metals. Moreover, ENPs themselves may serve as pollutant carriers because of interactions of nanoparticles with other pollutants (Bernhardt et al. 2010), further complicating predictions of their behavior in the environment. Furthermore, new ENPs are being developed, such as carbon-based nanoparticles, and their effects on fungi and fungi-driven ecological processes are yet unknown.

18.5 Effects of organic xenobiotics on stream fungi

Organic xenobiotics are man-made compounds that are not naturally found in the environment. Their effects depend on the nature of the compound, its mode of action and the target organism (Proia et al. 2012). Effects can be acute or chronic depending on the exposure time and concentration. There is no direct evidence of the mineralization of organic xenobiotic pollutants by freshwater fungi. However, aquatic hyphomycetes can be affected by or bioaccumulate xenobiotics or their metabolites. For instance, dichloro-diphenyl-trichloroethane (DDT, an insecticide) at concentrations between 0.1 and 60 μg/mL enhanced the growth of several fungal species (Hodkinson 1976). Within 2 weeks, *Heliscus submersus* was able to convert 27% and 8% of DDT into 1,1-dichloro-2,2-bis(4′-chlorophenyl)ethane (DDD) and 1,1-dichloro-2,2-bis(4′-chlorophenyl)ethylene (DDE), respectively, whereas *Clavariopsis aquatica* transformed 6% of DDT into these metabolites (Hodkinson 1976). The fungicides mancozeb and carbendazim did not inhibit fungal growth or reproduction at concentrations up to 5 mg/L or conidial germination up to 1 mg/L (Chandrashekar and Kaveriappa 1994). In contrast, Bärlocher and Premdas (1988) reported a linear negative relationship between conidium production and concentration of the fungicide pentachlorophenol (0.0001–10 mg/L).

The contamination of alder leaves with the herbicide mecoprop reduced the biomass of *Heliscus lugdunensis* and inhibited leaf consumption by the shredder *Gammarus pseudolimnaeus* (Bermingham et al. 1998). The major factor affecting food choice was the concentration of mecoprop that the leaves were exposed to, not the mecoprop-mediated effects on fungal biomass. On the other hand, antibiotic mixtures, at concentrations typical of streams receiving wastewater treatment plant effluents, had no effects on bacterial assemblages but led to stimulation of fungal biomass on leaves, which increased leaf palatability to shredders (Bundschuh et al. 2009).

Some freshwater fungi, including *Clavariopsis aquatica*, have shown the ability to degrade xenobiotics, such as xenoestrogen nonylphenol (Junghanns et al. 2005), polycyclic musks (Martin et al. 2007) and synthetic azo and anthraquinone dyes (Junghanns et al. 2008). This has been related to the activity of extracellular laccases that are involved in biodegradation of polyphenols and lignin found in plant tissues. Also, *Heliscus lugdunensis* showed the ability to metabolize polycyclic aromatic hydrocarbons such as naphtol (Augustin et al. 2006). *Heliscus lugdunensis* was consistently found among the top-ranked fungal species identified at different highly polluted sites (Krauss et al. 2001; Sridhar et al. 2005), which suggests a potential for some aquatic hyphomycetes to affect the fate of pollutants in aquatic ecosystems.

18.6 Effects of thermal pollution on stream fungi

Temperature is a major environmental factor shaping the activity of individual organisms, species distribution (Parmesan and Yohe 2003; Brown et al. 2004; Thackeray et al. 2010) and, therefore, affecting key ecosystem processes (Brown et al. 2004; Traill et al. 2010). Alterations of the thermal regimes can have serious effects on aquatic biota since most organisms have limited optimal thermal ranges (Ward and Stanford 1984; Lehtonen 1996). Alterations of the thermal regime of streams and rivers can result from human activities such as urbanization, forest clearance, thermal discharges, and flow modification, which presently affect many streams worldwide (Poole and Berman 2001; Cassie 2006). These activities will likely intensify over this century as a result of human population growth and concomitant needs for housing, food, water and energy (Vörösmarty et al. 2000). Thus, the number of streams and rivers affected by changes in thermal regime will probably increase in the future. In addition to these local or regional effects, global warming is predicted to affect thermal regimes of streams and rivers worldwide. Current climate models considering a doubling in atmospheric carbon dioxide concentration predict increases in mean global air temperature of 1.1°C–6.4°C by 2100 (IPCC 2007), which may translate into increases in stream water temperature of up to 0.6°C–0.9°C for each degree in air temperature (Pilgrim et al. 1998; Morrill et al. 2005). Further changes to the thermal regime of streams and rivers will result from the interaction between stressors (Stefan and Sinokrot 1993). Laboratory experiments have shown that responses of fungal growth and reproduction to increases in temperature depend on species identity, background temperature, and their interaction (Tab. 18.1). Fungal growth on solid media in the laboratory is stimulated by increases in temperature until the optimum of 10°C–30°C (species specific) is reached; any further increase in temperature leads to lower rates of colony extension (Koske and Duncan 1974; Webster et al. 1976; Graça and Ferreira 1995; Chauvet and Suberkropp 1998; Rajashekhar and Kaveriappa 2000; Dang et al. 2009; Duarte et al. 2013) (Tab. 18.1). For any given species, reported thermal optimum range is relatively narrow (<10°C), and the precision of the estimate depends on the number of temperature treatments and strain chosen. The increase in temperature also stimulates the reproductive activity in microcosm experiments up to the thermal optimum, which differs depending on species (Webster et al. 1976; Chauvet and Suberkropp 1998; Ferreira et al. 2012). For instance, *Lunulospora curvula* exhibits high spore production at high temperatures (25°C) and is therefore known as a warm-water species, while *Flagellospora curvula* has the highest spore production at lower temperatures (15°C) and is known as a cold-water species (Tab. 18.1). The different temperature optima of individual species partially determine seasonal patterns of conidia abundance in individual streams (Suberkropp 1984; Thomas et al. 1989; Gessner et al. 1993; Nikolcheva and Bärlocher 2005) and biogeographic distribution (Wood-Eggenschwiler and Bärlocher 1985; Ferreira et al. 2006b).

In experiments with fungal assemblages in laboratory microcosms, the increase in water temperature from low background levels (<15°C) stimulated biomass build-up and fungal sporulation on decomposing leaf litter (Ferreira and Chauvet 2011a, b; Bärlocher et al. 2013), while increases in water temperature from higher background levels (>15°C) generally had no effect or inhibited fungal growth and/or reproductive activity (Fernandes et al. 2009, 2012; Batista et al. 2012; Geraldes et al. 2012; Bärlocher et al. 2013). This apparent temperature threshold for fungal assemblages around 15°C is somewhat lower than the optimal temperatures for many species when grown in pure cultures (Tab. 18.1). Indeed, Webster et al. (1976) have shown that the sensitivity of individual species to warming depends on the presence of other species. They

Tab. 18.1: Effect of temperature on colony growth and spore production in 29 species of aquatic hyphomycetes (only studies that tested individual species using at least three temperatures were included).

Fungal species	Temperatures (°C) tested	Optimal temperature (°C) for colony growth[a]	Optimal temperature (°C) for spore production[a]	Reference
Alatospora acuminata	2–24 (17 temperatures)	20–24		Dang et al. (2009)
Alatospora acuminata	5–25 (at 5° intervals)	20		Koske and Duncan (1974)
Anguillospora filiformis	15, 20, and 25	25	20 = 25	Chauvet and Suberkropp (1998)
Anguillospora longissima	15, 20, and 25	25	20	Chauvet and Suberkropp (1998)
Articulospora tetracladia	15, 20, and 25	20	20	Chauvet and Suberkropp (1998)
Articulospora tetracladia	11, 13, 16, 21, and 27	13 = 16		Duarte et al. (2013)
Articulospora tetracladia	5–30 (at 5° intervals)	10–15		Graça and Ferreira (1995)
Articulospora tetracladia	5–25 (at 5° intervals)	20		Koske and Duncan (1974)
Dactylella aquatica	5–35 (at 5° intervals)	25		Rajashekhar and Kaveriappa (2000)
Flagellospora curvula	15, 20, and 25	20	15	Chauvet and Suberkropp (1998)
Flagellospora penicillioides	5–35 (at 5° intervals)	25		Rajashekhar and Kaveriappa (2000)
Flagellospora saccata	5–35 (at 5° intervals)	30		Rajashekhar and Kaveriappa (2000)
Gyoerffyella craginiformis	5–25 (at 5° intervals)	20		Koske and Duncan (1974)
Helicomyces sp.	5–35 (at 5° intervals)	25		Rajashekhar and Kaveriappa (2000)
Heliscus lugdunensis	2–24 (17 temperatures)	20–24		Dang et al. (2009)
Heliscus lugdunensis	5–30 (at 5° intervals)	20–25		Graça and Ferreira (1995)
Heliscus submersus	11, 13, 16, 21, and 27	27		Duarte et al. (2013)
Lemonniera aquatica	5–30 (at 5° intervals)	20–25		Graça and Ferreira (1995)
Lemonniera terrestris	5–30 (at 5° intervals)	10		Graça and Ferreira (1995)
Lunulospora curvula	15, 20, and 25	25	25	Chauvet and Suberkropp (1998)
Lunulospora curvula	11, 13, 16, 21, and 27	27		Duarte et al. (2013)

(*continued*)

Tab. 18.1: (*continued*)

Fungal species	Temperatures (°C) tested	Optimal temperature (°C) for colony growth[a]	Optimal temperature (°C) for spore production[a]	Reference
Lunulospora curvula	5–35 (at 5° intervals)	25		Rajashekhar and Kaveriappa (2000)
Lunulospora curvula	5–30 (at 5° intervals)	20	25	Webster et al. (1976)
Margaritispora aquatica	5–25 (at 5° intervals)	25		Koske and Duncan (1974)
Phalangispora constricta	5–35 (at 5° intervals)	25		Rajashekhar and Kaveriappa (2000)
Tetrachaetum elegans	15, 20, and 25	20	20 = 25	Chauvet and Suberkropp (1998)
Tetracladium marchalianum	15, 20, and 25	20	25	Chauvet and Suberkropp (1998)
Tetracladium marchalianum	2–24 (17 temperatures)	20		Dang et al. (2009)
Tetracladium setigerum	5–25 (at 5° intervals)	20		Koske and Duncan (1974)
Tetracladium setigerum	5–35 (at 5° intervals)	25		Rajashekhar and Kaveriappa (2000)
Tricellula aquatica	5–25 (at 5° intervals)	20		Koske and Duncan (1974)
Tricellula aurantiaca	5–25 (at 5° intervals)	20		Koske and Duncan (1974)
Tricellula curvatis	5–25 (at 5° intervals)	20		Koske and Duncan (1974)
Tricellula inaequalis	5–25 (at 5° intervals)	25		Koske and Duncan (1974)
Tricladium chaetocladium	15, 20, and 25	25	15 = 20 = 25	Chauvet and Suberkropp (1998)
Tricladium chaetocladium	5–30 (at 5° intervals)	15	25	Webster et al. (1976)
Tricladium splendens	5–25 (at 5° intervals)	20		Koske and Duncan (1974)
Varicosporium elodeae	11, 13, 16, 21, and 27	27		Duarte et al. (2013)
Vermispora cauveriana	5–35 (at 5° intervals)	30		Rajashekhar and Kaveriappa (2000)
Wiesneriomyces laurinus	5–35 (at 5° intervals)	30		Rajashekhar and Kaveriappa (2000)

[a] indicates no difference between temperatures

demonstrated that the optimum temperature for reproduction of *Lunulospora curvula* and *Tricladium chaetocladium* was 15° and 10° lower, respectively, when both species were incubated together compared to monocultures (Webster et al. 1976). This suggests that the optimum temperatures estimated for individual species in the laboratory are likely overestimates compared to those under field conditions, where species form diverse communities. The magnitude of the stimulation of fungal growth and reproduction by warming is also modulated by other factors such as substrate type, stage of decomposition, structure of the fungal assemblage (Bärlocher et al. 2013), metal contamination (Batista et al. 2012), and nutrient availability (Ferreira and Chauvet 2011a).

Increases in water temperature also affect the relative contribution of individual fungal species to the overall spore output (Dang et al. 2009; Fernandes et al. 2009, 2012; Ferreira and Chauvet 2011a, b; Batista et al. 2012; Geraldes et al. 2012), and in some cases this translates into shifts in the dominance pattern (Dang et al. 2009; Ferreira and Chauvet 2011a, b; Fernandes et al. 2012). Generally, the contribution of cold water species (e.g. *Articulospora tetracladia*, *Flagelospora curvula*) to the cumulative spore output decreases while that of warm-water species (e.g. *Lunulospora curvula*, *Lemonniera aquatica*) increases under elevated temperature (Fernandes et al. 2012).

Fungal respiration rates, as a surrogate for overall metabolism, are stimulated by warming both in experiments with monocultures (Ferreira et al. 2012) and multi-species assemblages (Ferreira and Chauvet 2011a, b) in laboratory microcosms. This indicates that increases in temperature stimulate carbon mineralization with the probable consequence of increased litter mass loss. In fact, fungi-driven litter decomposition is generally stimulated by temperature increases in the laboratory (Fernandes et al. 2009, 2012; Ferreira and Chauvet 2011a, b; Batista et al. 2012; Ferreira et al. 2012; Geraldes et al. 2012).

Taking advantage of the experimental warming of the hyporheic zone of a 1st order stream, Bärlocher et al. (2008) compared the number of litter fragments buried in the sediment and their fungal colonization. The average 4.3°C warming resulted in a decrease in the number of leaf fragments suggesting faster decomposition. Fungal species richness on leaf litter, but not on other litter types (e.g. grass, wood), was lower and the amount of fungal spores produced was higher under warming than at ambient temperature (Bärlocher et al. 2008). This agrees with laboratory results suggesting that warming may alter fungal diversity and community structure while promoting reproductive activity and fungi-driven litter decomposition. In another field experiment, the decomposition of oak leaf litter was followed in both halves of a small mountain stream monthly over 1 year, one stream half being at ambient temperature and the other being continually warmed by ~3°C above ambient temperature (Ferreira and Canhoto 2013). Results show stimulated microbially-mediated litter decomposition rates in the coldest, but not the warmest, months. This can be partially attributed to higher temperature sensitivity of enzymatic activity at low background temperature (Brown et al. 2004), and together with the results from laboratory experiments suggests higher sensitivity of cold water than warm water streams to increases in temperature.

Litter breakdown rates are also higher in warmer than colder streams across geographical thermal gradients (Irons et al. 1994; Fabre and Chauvet 1998; Friberg et al. 2009; Boyero et al. 2011), which can be attributed to enhanced microbial activity at warmer sites (Irons et al. 1994; Fabre and Chauvet 1998). This suggests that increases in temperature stimulate fungal activity, which translates into changes in ecosystem functioning. However, Gonçalves et al. (2006, 2007) found faster fungi-driven litter decomposition in a temperate stream than in warmer Mediterranean and tropical Cerrado streams. This may have resulted from higher fungal colonization of litter in the temperate stream, as suggested by the higher fungal biomass accumulation and sporulation rates in the temperate compared to the Mediterranean and tropical streams (Gonçalves et al. 2006, 2007). Ferreira et al. (2012a) also reported faster fungi-driven litter decomposition in a temperate stream than in a tropical rainforest stream, which was also explained by higher fungal colonization of litter in the temperate stream. This suggests that the contribution of fungi to litter decomposition in different climate zones is shaped not only by temperature but other factors as well (e.g. biogeography, oxygen availability), and that when fungal activity is low and/or background temperature is already high, further increases in temperature may not necessarily translate into measurable stimulation of litter decomposition.

When seasonal temperature gradients are considered, fungi-driven litter decomposition rates and associated fungal biomass and reproductive activity per gram of plant litter are higher in the warm spring/summer than in the cold autumn/winter (Nikolcheva and Bärlocher 2005; Ferreira et al. 2013). However, at an areal basis (whole stream scale), higher standing stock of fungal biomass, higher fungal production and spore output are found in autumn/winter in temperate streams due to higher standing crop of submerged litter resulting from the litter fall (Suberkropp 1997; Suberkropp et al. 2010). Thus, the increase in temperature may have stronger ecosystem-level effects on fungal parameters and fungi-driven litter decomposition during the colder season when litter and fungal standing stocks are greatest.

18.7 Effects of the interaction among factors on stream fungi

The stimulating effect of increased water temperature on fungal activity (provided that it is still below the thermal optimum) may be modified by the concomitant changes in other environmental factors. For instance, the increase in water temperature resulting from water abstraction for agriculture or from forest clearance may be accompanied by an increase in nutrient concentrations, which may interact either synergistically to stimulate (Ferreira and Chauvet 2011a) or antagonistically to reduce (Piggott et al. 2012) fungal activity and fungi-driven litter decomposition compared to that expected from the effect of each individual factor. Nutrient enrichment is often accompanied by decreases in dissolved oxygen concentration; in addition, oxygen solubility decreases as water temperature rises. As aquatic hyphomycetes prefer well aerated waters (Medeiros et al. 2009), oxygen depletion may counteract the potentially stimulating effects of nutrient enrichment or warming on fungal activity.

The increase in water temperature and nutrient availability in an agricultural landscape may also be accompanied by increases in the concentration of pesticides, which may be toxic to aquatic fungi (Bärlocher and Premda 1988) and can interact antagonistically with warming and nutrients to reduce fungal activities. The release of thermal effluents may be accompanied by the release of biocides used to control biofouling, which can pose additional stress on aquatic communities (Cooke and Schreer 2001). Increases in water temperature may exacerbate the toxicity of pollutants to fungi, while heavy metal pollution may in turn increase the sensitivity of fungi to warming (Batista et al. 2012). The relationship between temperature and fungi-driven litter decomposition in metal polluted streams can be further complicated if we consider that fungal strains in metal polluted streams may have adapted to these harsh conditions at the cost of reduced performance and therefore may have a reduced ability to cope with additional stressors (Ferreira et al. 2010, 2012a).

18.8 Conclusions

Aquatic hyphomycetes occupy a key position in aquatic food webs by establishing the link between terrestrial-derived organic matter and higher trophic levels. Therefore, changes in fungal community structure and/or activity can have strong impacts on ecosystem functioning (e.g. carbon and nutrient cycling), secondary production of invertebrates and fishes, and ecosystem services (e.g. water quality, food provision). The available literature shows that aquatic hyphomycetes are highly sensitive to environmental change, and several authors have suggested their potential utility as bioindicators of ecosystem health (Castela et al. 2008; Lecerf and Chauvet 2008; Solé et al. 2008). However, significant gaps still exist in our understanding of the effects of new pollutants (e.g. nanoparticles, pharmaceuticals) and interactions of multiple stressors affecting aquatic fungi at organismal, community and, especially, ecosystem scales.

Acknowledgments

The authors are grateful to FEDER-POFC-COMPETE and the Portuguese Foundation for Science and Technology – FCT (PEst-C/BIA/UI4050/2011, PTDC/AAC-AMB/121650/2010, and PTDC/AAC-AMB/117068/2010). Financial support by FCT to VF (postdoctoral fellowship SFRH/BPD/76482/2011, program POPH/FSE) and NSF DEB 0919054 to VG is also acknowledged.

References

Abel TH, Bärlocher F. Effects of cadmium on aquatic hyphomycetes. Appl Environ Microbiol 1984;48:245–251.

Allan D, Castillo M. Stream Ecology – Structure and Function of Running Waters. 2nd edition. Dordrecht: Springer 2007:1–436.

Arsuffi TL, Suberkropp K. Leaf processing capabilities of aquatic hyphomycetes: interspecific differences and influence on shredder preferences. Oikos 1984;42:144–154.

Arsuffi TL, Suberkropp K. Effects of fungal mycelia and enzymatically degraded leaves on feeding and performance of caddis fly (Trichoptera) larvae. J N Am Benthol Soc 1988;7:205–211.

Artigas J, Romaní AM, Sabater S. Effect of nutrients on the sporulation and diversity of aquatic hyphomycetes on submerged substrata in a Mediterranean stream. Aq Bot 2008;88:32–28.

Au DWT, Hodgkiss IJ, Vrijmoed LLP. Decomposition of *Bauhinia purpurea* leaf litter in a polluted and unpolluted Hong-Kong waterway. Can J Bot 1992a;70:1061–1070.

Au DWT, Hodgkiss IJ, Vrijmoed LLP. Fungi and cellulolytic activity associated with decomposition of *Bauhinia purpurea* leaf litter in a polluted and unpolluted Hong Kong waterway. Can J Bot 1992b;70:1071–1079.

Augustin T, Schlosser D, Baumbach R, Schmidt J, Grancharov K, Krauss G, et al. Biotransformation of 1-naphthol by a strictly aquatic fungus. Curr Microbiol 2006;52:216–220.

Azevedo MM, Cássio F. Effects of metals on growth and sporulation of aquatic fungi. Drug Chem Toxicol 2010;33:269–278.

Azevedo MM, Carvalho A, Pascoal C, Rodrigues F, Cássio F. Responses of antioxidant defenses to Cu and Zn stress in two aquatic fungi. Sci Total Environ 2007;377:233–243.

Baldrian P, Gabriel J. Intraspecific variability in growth response to cadmium of the wood-rotting fungus *Piptoporus betulinus*. Mycologia 2002;94:428–436.

Bärlocher F, Brendelberger H. Clearance of aquatic hyphomycete spores by a benthic suspension feeder. Limnol Oceanogr 2004;49:2292–2296.

Bärlocher F, Kebede YK, Gonçalves AL, Canhoto C. Incubation temperature and substrate quality modulate sporulation by aquatic hyphomycetes. Microb Ecol 2013;66:30–39.

Bärlocher F, Premdas PD. Effects of pentachlorophenol on aquatic hyphomycetes. Mycologia 1988;80:135–137.

Bärlocher F, Seena S, Wilson KP, Williams DD. Raised water temperature lowers diversity of hyporheic aquatic hyphomycetes. Freshwat Biol 2008;53:368–379.

Batista D, Pascoal C, Cássio F. Impacts of warming on aquatic decomposers along a gradient of cadmium stress. Environ Poll 2012;169:35–41.

Baudoin J, Guérold F, Felten V, Chauvet E, Wagner P, Rousselle, et al. aluminium concentration in acidified headwater streams lowers aquatic hyphomycete diversity and impairs leaf-litter breakdown. Microb Ecol 2008;56:260–269.

Benstead JP, Rosemond AD, Cross WF, Wallace J, Eggert SL, Suberkropp K, et al. Nutrient enrichment alters storage and fluxes of detritus in a headwater stream ecosystem. Ecology 2009;90:2556–2566.

Benstead JP, Deegan LA, Peterson BJ, Huryn AD, Bowden WB, Suberkropp K, et al. Responses of a beaded Arctic stream to short-term N and P fertilisation. Freshwat Biol 2005;50:277–290.

Bergfur J, Johnson RK, Sandin L, Goedkoop W, Nygren K. Effects of nutrient enrichment on boreal streams: invertebrates, fungi and leaf-litter breakdown. Freshwat Biol 2007; 52:1618–1633.

Bermingham S. The effects of pollutants on aquatic hyphomycetes colonising leaf material in freshwaters. In: Frankland JC, Magan N, Gadd GM, eds. Fungi and Environmental Change. Cambridge: Cambridge University Press 1996:201–216.

Bermingham S, Fisher PJ, Martin A, Marriott M, Lappin-Scott H. The effect of the herbicide mecoprop on *Heliscus lugdunensis* and its influence on the preferential feeding of *Gammarus pseudolimnaeus*. Microb Ecol 1998;35:199–204.

Bernhardt ES, Colman BP, Hochella Jr MF, Cardinale BJ, Nisbet RM, Richardson CJ, et al. An ecological perspective on nanomaterial impacts in the environment. J Environ Qual 2010;39:1–12.

Boyero L, Pearson RG, Gessner MO, Barmuta LA, Ferreira V, Graça MAS, et al. A global experiment suggests climate warming will not accelerate litter decomposition in streams but may reduce carbon sequestration. Ecol Lett 2011;14:289–294.

Braha B, Tintemann H, Krauss G, Ehrman J, Bärlocher F, Krauss GJ. Stress response in two strains of the aquatic hyphomycete *Heliscus lugdunensis* after exposure to cadmium and copper ions. BioMetals 2007;20:93–105.

Brown JH, Gillooly JF, Allen AP, Savage VM, West GB. Toward a metabolic theory of ecology. Ecology 2004;85:1771–1789.

Bundschuh M, Hahn T, Gessner MO, Schulz R. Antibiotics as a chemical stressor affecting an aquatic decomposer-detritivore system. Environ Toxicol Chem 2009;28:197–203.

Canhoto C, Graça MAS. Interactions between fungi and stream invertebrates:back to the future. In: Sridhar KR, Bärlocher F, Hyde KD, eds. Novel Techniques and Ideas in Mycology. Hong Kong: Hong Kong Univ Press 2008:305–325.

Canhoto C, Graça MAS. Leaf barriers to fungal colonization and shredders (*Tipula lateralis*) consumption of decomposing *Eucalyptus globulus*. Microb Ecol 1999;37:163–172.

Carpenter SR, Cole JJ, Kitchell JF, Pace ML. Impact of dissolved organic carbon, phosphorus, and grazing on phytoplankton biomass and production in experimental lakes. Limnol Oceanogr 1998;43:73–80.

Cassie D. The thermal regime of rivers: a review. Freshwat Biol 2006;51:1389–1406.

Castela J, Ferreira V, Graça MAS. Evaluation of stream ecological integrity using litter decomposition and benthic invertebrates. Environ Poll 2008;153:440–449.

Chamier A-C, Dixon PA. Pectinases in leaf degradation by aquatic hyphomycetes: the enzymes and leaf maceration. J Gen Microbiol 1982;128:2469–2483.

Chamier A-C, Tipping E. Effects of aluminium in acid streams on growth and sporulation of aquatic hyphomycetes. Environ Poll 1997;96:289–298.

Chandrashekar KR, Kaveriappa KM. Effects of pesticides on sporulation and germination of conidia of aquatic hyphomycetes. J Environ Biol 1994;15:315–324.

Chauvet E, Suberkropp K. Temperature and sporulation of aquatic hyphomycetes. App Environ Microbiol 1998;64:1522–1525.

Clivot H, Danger M, Pagnout C, Wagner P, Rousselle P, Poupin P, et al. Impaired leaf litter processing in acidified streams: learning from microbial enzyme activities. Microb Ecol 2013;65:1–11.

Coleman BP, Arnaout CL, Anciaux S, Gunsch CK, Hochella MF Jr, Kim B, et al. Low Concentrations of silver nanoparticles in biosolids cause adverse ecosystem responses under realistic field scenario. PLoS ONE 2013;8:e57189.

Cooke SJ, Schreer JF. Additive effects of chlorinated biocides and water temperature on fish in thermal effluents with emphasis on the Great Lakes. Rev Fish Sci 2001;9:69–113.

Corbi JJ, Strixino ST, do Santos A, Del Grande M. Environmental diagnostic of metals and organo-chlorinated compounds in streams near sugar cane plantations activity (Sao Paulo State, Brazil). Quimica Nova 2006;29:61–65.

Cornut J, Clivot H, Chauvet E, Elger A, Pagnout C, Guérold F. Effect of acidification on leaf litter decomposition in benthic and hyporheic zones of woodland streams. Water Res 2012;46:6430–6444.

Cross WF, Wallace JB, Rosemond AD, Eggert SL. Whole-system nutrient enrichment increases secondary production in a detritus-based ecosystem. Ecology 2006;87:1556–1565.

Cross WF, Wallace JB, Rosemond AD. Nutrient enrichment reduces constraints on material flows in a detritus-based food web. Ecology 2007;88:2563–2575.
Dang K, Schindler M, Chauvet E, Gessner MO. Temperature oscillation coupled with fungal community shifts can modulate warming effects on litter decomposition. Ecology 2009;90:122–131.
Danger M, Chauvet E. Elemental composition and degree of homeostasis of fungi: are aquatic hyphomycetes more like metazoans, bacteria or plants? Fungal Ecol 2013;6:453–457.
Dangles O, Gessner MO, Guérold F, Chauvet E. Impacts of stream acidification on litter breakdown: implications for assessing ecosystem functioning. J Appl Ecol 2004;41:365–378.
Dodds W, Cole J. Expanding the concept of trophic state in aquatic ecosystems: it's not just the autotrophs. Aq Sci 2007;69:427–439.
Dodds WK. Trophic state, eutrophication and nutrient criteria in streams. Trends Ecol Evol 2007;22:669–676.
Duan JC, Tan JH. Atmospheric heavy metals and Arsenic in China: situation, sources and control policies. Atmos Environ 2013;74:93–101.
Duarte D, Fernandes I, Nogueira MJ, Cássio F, Pascoal C. Temperature alters interspecific relationships among aquatic fungi. Fungal Ecol 2013;6:187–191.
Duarte S, Pascoal C, Cássio F. Effects of zinc on leaf decomposition by fungi in streams: studies in microcosms. Microb Ecol 2004;48:366–374.
Duarte S, Pascoal C, Garabetian F, Cássio F, Charcosset J-Y. Microbial decomposer communities are mainly structured by trophic status in circumneutral and alkaline streams. Appl Environ Microbiol 2009;75:6211–6221.
Fabre E, Chauvet E. Leaf breakdown along an altitudinal stream gradient. Arch Hydrobiol 1998;141:167–179.
Fernandes I, Pascoal C, Cássio F. Intraspecific traits change biodiversity effects on ecosystem functioning under metal stress. Oecologia 2011;166:1019–1028.
Fernandes I, Pascoal C, Guimarães H, Pinto R, Sousa I, Cássio F. Higher temperature reduces the effects of litter quality on decomposition by aquatic fungi. Freshwat Biol 2012;57:2306–2317.
Fernandes I, Uzun B, Pascoal C, Cássio F. Responses of aquatic fungal communities on leaf litter to temperature-change events. Internat Rev Hydrobiol 2009;94:410–418.
Ferreira V, Canhoto C. Effect of experimental and seasonal warming on litter decomposition in a temperate stream. Aq Sci 2013;doi:10.1007/s00027-013-0322-7.
Ferreira V, Chauvet E. Synergistic effects of water temperature and dissolved nutrients on litter decomposition and associated fungi. Global Change Biol 2011a;17:551–564.
Ferreira V, Chauvet E. Future increase in temperature more than decrease in litter quality can affect microbial litter decomposition in streams. Oecologia 2011b;167:279–291.
Ferreira V, Elosegi A, Gulis V, Pozo J, Graça MAS. Eucalyptus plantations affect fungal communities associated with leaf litter decomposition in Iberian streams. Arch Hydrobiol 2006b;166:467–490.
Ferreira V, Gonçalves AL, Canhoto C. Aquatic hyphomycete strains from metal-contaminated and reference streams might respond differently to future increases in temperature. Mycologia 2012;104:613–622.
Ferreira V, Gonçalves AL, Pratas J, Canhoto C. Contamination by uranium mine drainages may affect fungal growth and interactions between fungal species and strains. Mycologia 2010;102:1004–1011.
Ferreira V, Graça MAS. Fungal activity associated with decomposing wood is affected by nitrogen concentration in water. Internat Rev Hydrobiol 2007;92:1–8.
Ferreira V, Gulis V, Graça MAS. Whole-stream nitrate addition affects litter decomposition and associated fungi but not invertebrates. Oecologia 2006a;149:718–729.

Ferreira V, Lírio AV, Rosa J, Canhoto C. Annual organic matter dynamics in a small temperate mountain stream. Ann Limnol Internat J Limnol 2013;49:13–19.
Friberg N, Dybkjaer JB, Olafsson JS, Gislason GM, Larsen SE, Lauridsen TL. Relationships between structure and function in streams contrasting in temperature. Freshwat Biol 2009;54:2051–2068.
Geraldes P, Pascoal C, Cássio F. Effects of increased temperature and aquatic fungal diversity on litter decomposition. Fungal Ecol 2012;6:734–740.
Gessner MO, Chauvet E. Importance of stream microfungi in controlling breakdown rates of leaf litter. Ecology 1994;75:1807–1817.
Gessner MO, Thomas M, Jean-Louis A-M, Chauvet E. Stable successional patterns of aquatic hyphomycetes on leaves decaying in a summer cool stream. Mycol Res 1993;97:163–172.
Gonçalves AL, Lírio AV, Pratas J, Canhoto C. Uranium contaminated water does not affect microbial activity but decreases feeding by the shredder *Sericostoma vittatum*. Fund Appl Limnol Arch Hydrobiol 2011;179:17–25.
Gonçalves JF Jr, Graça MAS, Callisto M. Leaf-litter breakdown in 3 streams in temperate, Mediterranean, and tropical Cerrado climates. J N Am Benthol Soc 2006;25:344–355.
Gonçalves JF Jr, Graça MAS, Callisto M. Litter decomposition in a Cerrado savannah stream is retarded by leaf toughness, low dissolved nutrients and a low density of shredders. Freshwat Biol 2007;52:1440–1451.
Graça MAS. Patterns and processes in detritus-based stream systems. Limnologica 1993;23:107–114.
Graça MAS, Cressa C, Gessner MO, Feio MJ, Callies KA, Barrios C. Food quality, feeding preferences, survival and growth of shredders from temperate and tropical streams. Freshwat Biol 2001;46:947–957.
Graça MAS, Ferreira R. The ability of selected aquatic and terrestrial fungi to decompose leaves in freshwater. Sydowia 1995;47:167–179.
Gray NF. Acid mine drainage composition and the implications for its impact on lotic systems. Water Res 1998;32:2122–2134.
Greenwood JL, Rosemond AD, Wallace JB, Cross WF, Weyers HS. Nutrients stimulate leaf breakdown rates and detritivore biomass: bottom-up effects via heterotrophic pathways. Oecologia 2007;151:637–649.
Griffitt RJ, Luo J, Bonzongo JC, Barber DS. Effects of particle composition and species on toxicity of metallic nanomaterials in aquatic organisms. Environ Toxicol Chem 2008;27:1972–1978.
Guimarães-Soares L, Pascoal C, Cássio F. Effects of heavy metals on the production of thiol-compounds by the aquatic fungi *Fontanospora fusiramosa* and *Flagellospora curta*. Ecotoxicol Environ Safe 2007;66:36–43.
Gulis V, Ferreira V, Graça MAS. Stimulation of leaf litter decomposition and associated fungi and invertebrates by moderate eutrophication: implications for stream assessment. Freshwat Biol 2006a;51:1655–1669.
Gulis V, Kuehn KA, Suberkropp K. The role of fungi in carbon and nitrogen cycles in freshwater ecosystems. In: Gadd GM, ed. Fungi in Biogeochemical Cycles. Cambridge: Cambridge University Press 2006b:404–435.
Gulis V, Rosemond AD, Suberkropp K, Weyers HS, Benstead JP. Effects of nutrient enrichment on the decomposition of wood and associated microbial activity in streams. Freshwat Biol 2004;49:1437–1447.
Gulis V, Suberkropp K. Effect of inorganic nutrients on relative contributions of fungi and bacteria to carbon flow from submerged decomposing leaf litter. Microb Ecol 2003a;45:11–19.
Gulis V, Suberkropp K. Interactions between stream fungi and bacteria associated with decomposing leaf litter at different levels of nutrient availability. Aq Microb Ecol 2003b;30:149–157.
Gulis V, Suberkropp K. Leaf litter decomposition and microbial activity in nutrient-enriched and unaltered reaches of a headwater stream. Freshwat Biol 2003c;48:123–134.

Gulis V, Suberkropp K. Effects of whole-stream nutrient enrichment on the concentration and abundance of aquatic hyphomycete conidia in transport. Mycologia 2004;96:57–65.

Gulis V, Suberkropp K, Rosemond AD. Comparison of fungal activities on wood and leaf litter in unaltered and nutrient-enriched headwater streams. Appl Environ Microbiol 2008;74:1094–1101.

Hassellöv M, Readman JW, Ranville JF, Tiede K. Nanoparticle analysis and characterization methodologies in environmental risk assessment of engineered nanoparticles. Ecotoxicology 2008;17:344–361.

Hieber M, Gessner MO. Contribution of stream detritivores, fungi, and bacteria to leaf breakdown based on biomass estimates. Ecology 2002;83:1026–1038.

Hodkinson M. Interactions between aquatic fungi and DDT. In: Jones EBG, ed. Recent Advances in Aquatic Mycology. New York: Wiley 1976:447–467.

Hogsden KL, Harding JS. Consequences of acid mine drainage for the structure and function of benthic stream communities: a review. Freshwat Sci 2013;31:108–120.

Imberger SJ, Walsh CJ, Grace MR. More microbial activity, not abrasive flow or shredder abundance, accelerates breakdown of labile leaf litter in urban streams. J N Am Benthol Soc 2008;27:549–561.

IPCC (Intergovernmental Panel on Climate Change) *Climate change 2007: The physical science basis.* Contribution of the working group I to the Fourth Assessment report of the Intergovernmental panel on climate change. Cambridge: Cambridge University Press 2007:pp 996.

Irons JG, Oswood MW, Stout RJ, Pringle CM. Latitudinal patterns in leaf litter breakdown: is temperature really important? Freshwat Biol 1994;32:401–411.

Jaeckel P, Krauss GJ, Krauss G. Cadmium and zinc response of the fungi *Heliscus lugdunensis* and *Verticillium* cf. *alboatrum* isolated from highly polluted water. Sci Total Environ 2005;346:274–279.

Junghanns C, Krauss G, Schlosser D. Potential of fungi derived from diverse freshwater environments to decolourise synthetic azo and anthraquinone dyes. Bioresour Technol 2008;99:1225–1235.

Junghanns C, Möder M, Krauss G, Martin C, Schlosser D. Degradation of the xenoestrogen nonylphenol by aquatic fungi and their laccases. Microbiol 2005;151:45–57.

Kaegi R, Ulrich A, Sinnet B, Vonbank R, Wichser A, Zuleeg S, et al. Synthetic TiO_2 nano-particle emission from exterior facades into the aquatic environment. Environ Pollut 2008;156:233–239.

Klaine SJ, Alvarez PJJ, Batley GE, Fernandes TF, Handy RD, Lyon DY, et al. Nanomaterials in the environment: behavior, fate, bioavailability, and effects. Environ Toxicol Chem 2008;27:1825–1851.

Koske RE, Duncan IW. Temperature effects on growth, sporulation, and germination of some 'aquatic' hyphomycetes. Can J Bot 1974;52:1387–1391.

Krauss G, Bärlocher F, Schreck P, Wennrich R, Glasser W, Krauss G-J. Aquatic hyphomycetes occur in hyperpolluted waters in Central Germany. Nova Hedwig 2001;72:419–428.

Krauss GJ, Solé M, Krauss G, Schlosser D, Wesenberg D, Bärlocher F. Fungi in freshwaters: ecology, physiology and biochemical potential. FEMS Microbiol Rev 2011;35:620–51.

Lecerf A, Chauvet E. Diversity and functions of leaf-decaying fungi in human-altered streams. Freshwat Biol 2008;53:1658–1672.

Lecerf A, Dobson M, Dang C, Chauvet E. Riparian plant species loss alters trophic dynamics in detritus-based stream ecosystems. Oecologia 2005;146:432–442.

Lehtonen H. Potential effects of global warming on northern European freshwater fish and fisheries. Fish Manag Ecol 1996;3:59–71.

Luechinger NA, Athanassiou EK, Stark WJ. Graphene-stabilized copper nanoparticles as an air-stable substitute for silver and gold in low-cost ink-jet printable electronics. Nanotechnology 2008;19:445–201.

Martin C, Moeder M, Daniel X, Krauss G, Schlosser D. Biotransformation of the polycyclic musks HHCB and AHTN and metabolite formation by fungi occurring in freshwater environments. Environ Sci Technol 2007;41:5395–5402.
Medeiros AO, Duarte S, Pascoal C, Cássio F, Graça MAS. Effects of Zn, Fe and Mn on leaf litter breakdown by aquatic fungi:a microcosm study. Internat Rev Hydrobiol 2010;95:1–11.
Medeiros AO, Pascoal C, Graça MAS. Diversity and activity of aquatic fungi under low oxygen conditions. Freshwat Biol 2009;54:142–149.
Medeiros AO, Rocha P, Rosa C, Graça MAS. Litter breakdown in a stream affected by drainage from a gold mine. Fund Appl Limnol 2008;172:59–70.
Miersch J, Bärlocher F, Bruns I, Krauss G-J. Effects of cadmium, copper, and zinc on growth and thiol content of aquatic hyphomycetes. Hydrobiol 1997;346:77–84.
Miersch J, Tschimedbalshir M, Bärlocher F, Grams Y, Pierau B, Schierhorn A, et al. Heavy metals and thiol compouds in *Mucor racemosus* and *Articulospora tetracladia*. Mycol Res 2001;105:883–889.
Min JS, Kim KS, Kim SW, Jung JH, Lamsal K, Kim SB, et al. Effects of colloidal silver nanoparticles on sclerotium-forming phytopathogenic fungi. Plant Pathol J 2009;25:376–380.
Nagajyoti PC, Lee KD, Sreekanth TVM. Heavy metals, occurrence and toxicity for plants:a review. Environ Chem Lett 2010;8:199–216.
Nikolcheva LG, Bärlocher F. Seasonal and substrate preferences of fungi colonizing leaves in streams: traditional versus molecular evidence. Environ Microbiol 2005;7:270–280.
Niyogi DK, Cheatham CA, Thompson WH, Christiansen JM. Litter breakdown and fungal diversity in a stream affected by mine drainage Fund Appl Limnol 2009;175:39–48.
Niyogi DK, Harding JS, Simon KS. Organic matter breakdown as a measure of stream health in New Zealand streams affected by acid mine drainage. Ecol Indic 2013;24:510–517.
Niyogi DK, Simon KS, Townsend CR. Breakdown of tussock grass in streams along a gradient of agricultural development in New Zealand. Freshwat Biol 2003;48:1698–1708.
Panácek A, Kolár M, Vecerová R, Prucek R, Soukupová J, Krystof V, et al. Antifungal activity of silver nanoparticles against *Candida* spp. Biomaterials 2009;30:6333–6340.
Parmesan C, Yohe G. A globally coherent fingerprint of climate change impacts across natural systems. Nature 2003;421:37–42.
Pascoal C, Cássio F. Contribution of fungi and bacteria to leaf litter decomposition in a polluted river. Appl Environ Microbiol 2004;70:5266–5273.
Pascoal C, Cássio F, Marvanová L. Anthropogenic stress may affect aquatic hyphomycete diversity more than leaf decomposition in a low-order stream. Arch Hydrobiol 2005;162:481–496.
Pascoal C, Marvanová L, Cássio F. Aquatic hyphomycete diversity in streams of Northwest Portugal. Fungal Divers 2005;19:109–128.
Pascoal C, Pinho M, Cássio F, Gomes P. Assessing structural and functional ecosystem condition using leaf breakdown: studies on a polluted river. Freshwat Biol 2003;48:2033–2044.
Pérez J, Menéndez M, Larrañaga S, Pozo J. Inter- and intra-regional variability of leaf litter breakdown in reference headwater streams of Northern Spain: Atlantic versus Mediterranean streams. Internat Rev Hydrobiol 2011;96:105–117.
Perugini P, Simeoni S, Scalia S, Genta I, Modena T, Conti B, et al. Effect of nanoparticle encapsulation on the photostability of the sunscreen agent, 2-ethylhexyl-p-methoxycinnamate. Int J Pharm 2002;246:37–45.
Piggott JJ, Lange K, Townsend CR, Matthaei CD. Multiple stressors in agricultural streams: a mesocosm study of interactions among raised water temperature, sediment addition and nutrient enrichment. PlosOne 2012;7:e49873.
Poole GC, Berman CH. An ecological perspective on in-stream temperature: natural heat dynamics and mechanisms of human-caused thermal degradation. Environ Manag 2001;27:787–802.

Pradhan A, Seena S, Dobritzsch D, Helm S, Gerth K, Dobritzsch M, et al. Physiological responses to nanoCuO in fungi from non-polluted and metal-polluted streams. Science Total Environ 2013;466:556–563.
Pradhan A, Seena S, Pascoal C, Cássio F. Can metal nanoparticles be a threat to microbial decomposers of plant litter in streams? Microb Ecol 2011;62:58–68.
Pradhan A, Seena S, Pascoal C, Cássio F. Copper oxide nanoparticles can induce toxicity to the freshwater shredder *Allogamus ligonifer*. Chemosphere 2012;89:1142–1150.
Proia L, Cássio F, Pascoal C, Tili A, Romani A. The use of attached microbial communities to assess ecological risks of pollutants in aquatic ecosystems. The role of heterotrophs. In: Guasch H, Ginebreda A, Geiszinger A, eds. Emerging and priority pollutants in Rivers: Bringing Science into River Management Plans. Germany: Springer 2012:55–84.
Rajashekhar M, Kaveriappa KM. Effects of temperature and light on growth and sporulation of aquatic hyphomycetes. Hydrobiologia 2000;441:149–153.
Raviraja NS, Sridhar R, Bärlocher F. Breakdown of *Ficus* and *Eucalyptus* leaves in an organically polluted river in India: fungal diversity and ecological functions. Freshwat Biol 1998;39:537–545.
Rosemond AD, Pringle CM, Ramírez A, Paul MJ, Meyer JL. Landscape variation in phosphorus concentration and effects on detritus-based tropical streams. Limnol Oceanogr 2002;47:278–289.
Schlief J, Mutz M. Long-term leaf litter decomposition and associated microbial processes in extremely acidic (pH < 3) mining waters. Arch Hydrobiol 2005;164:53–68.
Simon KS, Simon MA, Benfield EF. Variation in ecosystem function in Appalachian streams along an acidity gradient. Ecol Appl 2009;19:1147–1160.
Solé M, Fetzer I, Wennrich R, Sridhar KR, Harms H, Krauss G. Aquatic hyphomycetes communities as potential bioindicators for assessing anthropogenic stress. Sci Total Environ 2008;389:557–565.
Sridhar KR, Bärlocher F. Water chemistry and sporulation by aquatic hyphomycetes. Mycol Res 1997;101:591–596.
Sridhar KR, Bärlocher F, Krauss GJ, Krauss G. Response of aquatic hyphomycete communities to changes in heavy metal exposure. Internat Rev Hydrobiol 2005;90:21–32.
Sridhar KR, Duarte S, Cássio F, Pascoal C. The role of early fungal colonizers in leaf-litter decomposition in Portuguese streams impacted by agricultural runoff. Internat Rev Hydrobiol 2009; 94:399–409.
Sridhar KR, Krauss G, Bärlocher F, Raviraja NS, Wennrich R, Baumbach R, et al. Decomposition of alder leaves in two heavy metal-polluted streams in central Germany. Aq Microb Ecol 2001;26:73–80.
Sridhar KR, Krauss G, Bärlocher F, Wennrich R, Krauss G-J. Fungal diversity in heavy metal polluted waters in central Germany. Fungal Divers 2000;5:119–129.
Stefan HG, Sinokrot BA. Projected global climate change impact on water temperatures in five north central U.S. streams. Climate Change 1993;24:353–381.
Stelzer RS, Heffernan J, Likens GE. The influence of dissolved nutrients and particulate organic matter quality on microbial respiration and biomass in a forest stream. Freshwat Biol 2003;48:1925–1937.
Suberkropp K. Effect of temperature on seasonal occurrence of aquatic hyphomycetes. Trans Br Mycol Soc 1984;82:53–62.
Suberkropp K. Interactions with invertebrates. In: Bärlocher F, ed. The Ecology of Aquatic Hyphomycetes. Berlin/Heidelberg: Springer 1992:118–134.
Suberkropp K. The influence of nutrients on fungal growth, productivity, and sporulation during leaf breakdown in streams. Can J Bot 1995;73:S1361–S1369.
Suberkropp K. Annual production of leaf-decaying fungi in a woodland stream. Freshwat Biol 1997;38:169–178.
Suberkropp K. Effect of dissolved nutrients on two aquatic hyphomycetes growing on leaf litter. Mycol Res 1998;102:998–1002.

Suberkropp K, Chauvet E. Regulation of leaf breakdown by fungi in streams: influences of water chemistry. Ecology 1995;76:1433–1445.
Suberkropp K, Gulis V, Rosemond AD, Benstead JP. Ecosystem and physiological scales of microbial responses to nutrients in a detritus-based stream: results of a 5-year continuous enrichment. Limnol Oceanogr 2010;55:149–160.
Suberkropp K, Michelis A, Lorch HJ, Ottow JCG. Effect of sewage treatment plant effluents on the distribution of aquatic hyphomycetes in the River Erms, Schwabische Alb, FRG. Aq Bot 1988;32:141–153.
Thackeray SJ, Sparks TH, Frederiksen M, Burthe S, Bacon PJ, Bell JR, et al. Trophic level asynchrony in rates of phenological change for marine, freshwater and terrestrial environments. Global Change Biol 2010;16:3304–3313.
Thomas K, Chilvers GA, Norris RH. Seasonal occurrence of conidia of aquatic hyphomycetes (Fungi) in Lees creek, Australian Capital Territory. Mar Freshwat Res 1989;40:11–23.
Traill LW, Lim MLM, Sodhi NS, Bradshaw CJA. Mechanisms driving change: altered species interactions and ecosystems function through global warming. J Animal Ecol 2010;79:937–947.
Treton C, Cauvet E, Charcosset J-Y. Competitive interaction between two aquatic hyphomycete species and increase in leaf litter breakdown. Microb Ecol 2004;48:439–446.
Tsui CKM, Hyde KD, Hodgkiss IJ. Colonization patterns of wood-inhabiting fungi on baits in Hong Kong rivers, with reference to the effects of organic pollution. Antonie van Leeuwenhoek 2001;79:33–38.
Vitousek PM, Mooney HA, Lubchenco J, Melillo JM. Human domination of Earth's ecosystems. Science 1997;277:494–499.
Vörösmarty CJ, Green P, Salisbury J, Lammers RB. Global water resources: vulnerability from climate change and population growth. Science 2000;289:284–288.
Ward JV, Stanford JA.Thermal responses in the evolutionary ecology of aquatic insects. Ann Rev Entomol 1984;27:97–117.
Webster J, Moran ST, Davey RA. Growth and sporulation of *Tricladium chaetocladium* and *Lunulo-sporula curvula* in relations to temperature. Trans Br Mycol Soc 1976;67:491–495.
Weyer HS, Suberkropp K. Fungal and bacterial production during the breakdown of yellow poplar leaves in 2 streams. J N Am Benthol Soc 1996;15:408–420.
Wilcox HS, Wallace JB, Meyer JL, Benstead JP. Effects of labile carbon addition on a headwater stream food web. Limnol Oceanogr 2005;50:1300–1312.
Wood-Eggenschwiler S, Bärlocher F. Aquatic hyphomycetes in sixteen streams in France, Germany and Switzerland. Trans Br Mycol Soc 1983;81:371–379.
Woodward G, Gessner MO, Giller PS, Gulis V, Hladyz S, Lecerf A, et al. Continental-scale effects of nutrient pollution on stream ecosystem functioning. Science 2012;336:1438–1440.
Zemek J, Marvanová L, Kuniak L, Kadleoiková B. Hydrolytic enzymes in aquatic hyphomycetes. Folia Microbiol 1985;30:363–372.
Zhang F, Wu X, Chen Y, Lin H. Application of silver nanoparticles to cotton fabric as an antibacterial textile finish. Fibers Polym 2009;10:496–501.

Felix Bärlocher and Kandikere R. Sridhar

19 Association of animals and fungi in leaf decomposition

19.1 History

In the 1950s and 1960s, limnologists started to appreciate the paramount importance of vascular plant detritus imported from the surrounding riparian vegetation for stream communities (for reviews of the early literature, see Hynes 1963; Kaushik and Hynes 1971; Webster and Benfield 1986). Much of this insight was based on the analysis of stomach or gut contents of invertebrates and vertebrates (Hynes 1950). The first investigation on how allochthonous materials, exemplified by deciduous leaves, were incorporated into food webs was published by Kaushik and Hynes (1968). They incubated autumn-shed leaves of three species in glass bowls or artificial laboratory streams with various antibiotics. Generally, when fungi were allowed to grow (no antifungal antibiotics present), the protein content of leaves increased. In some cases the increase was absolute, indicating nitrogen immobilization from external sources. No such increase was observed in the presence of antifungal antibiotics. Kaushik and Hynes (1968) concluded that fungal growth on dead leaves adds to their protein content, which presumably enhances the leaves' nutritional value. This hypothesis was supported by the demonstration that several leaf-eating stream invertebrates prefer to feed on partly decomposed, fungal-colonized leaves rather than freshly fallen or sterile autumn-shed leaves (Triska 1970; Kaushik and Hynes 1971; Mackay and Kalff 1973). These invertebrates are classified as "shredders"; Fig. 19.1 shows *Gammarus fossarum*, an important member of this functional group.

Kaushik and Hynes (1971) suggested that "the importance of leaves as food for aquatic organisms probably lies in providing an energy source for microbes"; by preferentially consuming such conditioned leaves, invertebrates introduce a vast energy resource into the food web. Cummins (1974) further emphasized the importance of microorganisms by comparing decaying leaves to nutrionally unsuitable "crackers" containing valuable "peanut butter" in the form of protein and lipid rich fungal mycelia. On the other hand, Hynes (1963) had suggested earlier that much feeding may still be direct, based on what he considered to be high protein concentrations in freshly-shed dead leaves.

Aquatic hyphomycetes (or Ingoldian fungi) are an ecologically defined, polyphyletic group of Fungi, most with ascomycete affinities (Belliveau and Bärlocher 2005; Baschien et al. 2006; Campbell et al. 2006). They produce large sigmoid or multiradiate mitospores and they are most common on deciduous leaves in relatively clean

Figs. 19.1–19.6: Main players of leaf decomposition in streams and their effects on the integrity of decomposing leaves. **Fig. 19.1** *Gammarus fossarum*, a leaf-shredding invertebrate. **Fig. 19.2** A collection of spores of aquatic hyphomycetes dominated by *Clavariopsis aquatica*. The large sigmoid spore belongs to *Anguillospora filiformis*. **Fig. 19.3** Fungal maceration: a sugar maple leaf after 4 weeks of stream exposure in a fine-mesh bag (prevented access of macroinvertebrates). Small areas of the leaf's lamina are missing. **Fig. 19.4** Fungal maceration: sugar maple leaf disks after 4 weeks in a fine-mesh bag. Typical "pin-prick" holes due to fungal enzymatic attack. **Fig. 19.5** Invertebrate shredding: a white birch leaf after 4 weeks of stream exposure in a coarse-mesh bag (allowed access of macroinvertebrates, among them *Gammarus tigrinus*). Larger areas of the leaf's lamina have disappeared. **Fig. 19.6** Invertebrate shredding: sugar maple disks after 4 weeks in a coarse-mesh bag. Large areas of the leaf have been removed by invertebrate shredders.

running waters (Ingold 1975). These spores often accumulate in naturally forming foam. Figure 19.2 shows a collection of foam with trapped spores.

Kaushik and Hynes (1968, 1971) performed their experiments under conditions where aquatic hyphomycetes were unlikely to be present; not surprisingly, they found exclusively common soil or epiphyllic fungi. Triska (1970) exposed leaves in a stream and then incubated them in distilled water, which favours sporulation by aquatic hyphomycetes. An SEM (scanning electron microscopy) study of leaves decaying in a stream also showed that the dominant sporulating structures belonged to aquatic hyphomycete species (few vegetative structures were seen and were therefore believed to be within the leaf tissue itself) (Suberkropp and Klug 1974).

By using different media and incubation temperatures, Bärlocher and Kendrick (1974) and Suberkropp and Klug (1976) attempted to evaluate the relative importance of terrestrial fungi that may have been present on senescent leaves before abscission compared to aquatic or soil hyphomycetes arriving after the leaf's deposition in the stream. Generally, conditions simulating ambient stream conditions favoured the observation or isolation of aquatic hyphomycetes, but members from other groups (soil fungi, Zygomycota, Chytridiales, Oomycota) were observed or isolated as well. The dominant role of aquatic hyphomycetes also became apparent by observed correlations between their spore production, mass loss of leaves and total fungal biomass as measured by ergosterol (Gessner and Chauvet 1994; Maharning and Bärlocher 1996; Bärlocher et al. 2012). Nevertheless, other fungal taxa have been reported consistently by traditional (culturing, direct microscopic observation; Bärlocher et al. 2012) and, more recently, by molecular techniques (Nikolcheva et al. 2003; Nikolcheva and Bärlocher 2004; Das et al. 2008; Seena et al. 2008; Duarte et al. 2010), and can degrade submerged leaves under laboratory conditions (Bärlocher and Kendrick 1974; Godfrey 1983; Graça and Ferreira 1995).

Another early topic of investigation was a comparison of bacterial and fungal participation in leaf degradation and invertebrate feeding. Kaushik and Hynes (1971) used specific antibiotics to establish the dominance of bacteria, and Mackay and Kalff (1973) showed that caddis fly larvae preferred fungal over bacterial conditioned leaves. On decaying leaves, ATP levels and fungal but not bacterial counts were correlated, again suggesting a greater role for fungi (Suberkropp and Klug 1976). More accurate comparisons were based on contrasting bacterial and fungal biomass, growth rates and productions (Suberkropp and Weyers 1996; Baldy et al. 2002; Pascoal and Cássio 2004). Again, fungi were shown to dominate, though bacterial biomass (up to 11% of microbial biomass) and production (up to 32% of microbial production) was sometimes substantial.

In summary, early studies showed that a) allochthonous detritus plays a major role in streams providing energy to microbes and invertebrates; b) microbial activity on this detritus is dominated by fungi, primarily by aquatic hyphomycetes; and, c) fungal colonization of leaves tends to increase their nitrogen (protein) levels and make them more palatable to invertebrates. The details of fungus/invertebrate relationships have stimulated considerable research, which has been reviewed repeatedly (Cummins 1974; Bärlocher and Kendrick 1976, 1981; Anderson and Sedell 1979; Cummins and Klug 1979; Bärlocher 1985, 1992; Mann 1988; Suberkropp 1992; Gessner et al. 1999; Graça 2001; Suberkropp 2003; Canhoto and Graça 2008). As Canhoto and Graça (2008) pointed out, much work was done in the 1970s and 1980s. This was followed by a relatively slow period of roughly 20 years. New techniques, studies conducted outside temperate regions, and interest in the consequences of anthropogenic changes have revived interest in these topics. Very broadly, the research we review can be subdivided into bottom-up aspects (effects of the leaf-fungus complex on invertebrate consumers) and top-down aspects (effects of invertebrate consumers on the leaf-fungus complex). Our goal was to distinguish between what is reasonably well established and where gaps in our knowledge or uncertainties in interpretation exist.

19.2 Effects of the leaf-fungus complex on invertebrate consumers

19.2.1 Nutritional value of mycelium vs. leaf substrate

Is the presence of fungi (or bacteria) on leaves essential for survival and growth of leaf-eating invertebrates? Based on the relatively high protein levels in freshly shed leaves, Kaushik and Hynes (1968, 1971) believed this not to be the case, and Bärlocher and Kendrick (1973a) showed that both unconditioned maple (*Acer saccharum*) and elm (*Ulmus americana*) leaves with minimal microbial biomass (≤0.03% of total mass) were eaten and allowed growth of *Gammarus pseudolimnaeus*. *Asellus aquaticus* however did not survive on sterilized leaves (Rossi and Fano 1979; Rossi and Vitagliano-Tadini 1978), and *Pycnopsyche gentilis* (Trichoptera) lost weight on sterile leaves (Chung and Suberkropp 2009). There were clear differences between leaf species: growth on elm leaves was twice as high as on maple leaves (Bärlocher and Kendrick 1973a). Some leaves such as alder (*Alnus glutinosa*) are relatively soft and have a high protein content. They are therefore readily accepted as food by many invertebrates even without or with minimal conditioning (Nilsson 1974; Anderson and Grafius 1975), though sterilized leaves may lack some essential ingredients required in small quantities (Cargill et al. 1985a). On the other hand, tough beech leaves or conifer needles are initially rejected by most consumers (Bärlocher 1985).

Numerous studies have shown that leaves of most species become more palatable after a period of conditioning (Kostalos and Symour 1976; Bärlocher 1985; Canhoto and Graça 2008). The conditioning process generally follows a unimodal pattern: palatability increases to a maximum (within 2–6 weeks) and then declines again (Arsuffi and Suberkropp 1984; Suberkropp and Arsuffi 1984). Ignorance of this variability and unsuitable incubation conditions (e.g. lack of natural inoculum, insufficient incubation period, inadequate supply of external nutrients) may result in weak or absent conditioning effects.

While unconditioned leaves of many species may allow survival and growth of leaf-eating invertebrates, they contain large quantities of structural polymers such as cellulose and lignin, which are largely indigestible to stream invertebrates (Cummins and Klug 1979; Bärlocher 1985). Mycelia of terrestrial and aquatic hyphomycetes generally contain more lipids and proteins, and can be assimilated more efficiently. The effective nutritional value per unit of microbial cells weight can be 4–10 times higher than that of freshly fallen leaves (Bärlocher and Kendrick 1975a; Bärlocher 1985). Enrichment of plant detritus with fungal cells should therefore raise its nutritional value, and most studies have indeed demonstrated better survival and higher growth of invertebrates on conditioned than on unconditioned leaves (Bärlocher 1985, but see Graça et al. 1993b). The suitability of fungal mycelium as sole food source for various *Gammarus* and Trichoptera species is variable: on some fungal diets, invertebrate growth was higher than on leaves, other fungi proved to be toxic, some gave similar results as conditioned leaves (Bärlocher and Kendrick 1973a; Willoughby and Sutcliffe 1976; Sutcliffe et al. 1981 Cargill et al. 1985a). When given a choice, pure mycelium of some species was preferred over leaf detritus (Bärlocher and Kendrick 1973b; Arsuffi and Suberkropp 1988; De Lange et al. 2005), while other invertebrates preferred conditioned leaves over fungal mycelium (Graça et al. 1993a). There is evidence that in leaves decaying naturally in streams, sections with higher fungal density are consumed preferentially (Bärlocher and Kendrick 1974), which increases the proportion of fungal biomass in the consumed food. On the other hand, feeding selectively and exclusively on fungal hyphae is generally not feasible for any but the smallest invertebrates, though *Asellus aquaticus* feeds by scraping at leaf surfaces and avoids even fine veins (Graça et al. 1993a, 1994). In the mesophyll layers of tough, thick leaves, such as eucalypt or rhododendron, small Chironomidae are common, where they may consume fungal hyphae and/or leaf tissues (Canhoto and Graça 1999, 2008). Leaf-mining chironomids are believed to be important fungal consumers in some tropical streams (Rueda-Delgado et al. 2006; Wantzen and Wagner 2006), and can use leaf litter at least as a complementary food source (Callisto et al. 2007), while larger shredders (shrimps, crabs, insects) ingest leaf tissues with associated microorganisms (Crowl et al. 2006; Ferreira and Graça 2006).

19.2.2 Modifications of leaf substrate

The question remains to what extent the fungal biomass on naturally decaying leaves meets the nutritional requirements of leaf-consuming invertebrates. Lawson et al. (1984) attributed only 11%–27% of larval growth of *Tipula abdominalis* to microbial biomass. Fungal carbon contributed between 0.05% and 57% of the carbon respired by *Lirceus* sp. (Isopoda, Findlay et al. 1986a), and 25% of the carbon respired by

Peltoperla sp. (Plecoptera, Findlay et al. 1986b). On the other hand, in *Pycnopsyche gentilis* (Trichoptera) fungal carbon accounted for 50% (5th instar) or 100% (3rd instar) of growth (Chung and Suberkropp 2009). Clearly, the actual contribution of fungal mycelia to the nutrition of leaf-eating invertebrates is variable; there are currently no comprehensive hypotheses that relate its magnitude to environmental or physiological factors. A further complication is the fact that these estimates are based on leaves whose condition was optimized in the laboratory. In leaves randomly collected from a stream between September and July, the average protein content and digestibility were considerably lower than those achieved in lab-conditioned leaves (Bärlocher 1983). However, there was considerable variability; a mobile consumer could conceivably search out the most suitable leaf patches. In addition, many so-called leaf-shredders are in fact opportunistic feeders. Bärlocher (1983) showed that vascular-plant remains in *Gammarus fossarum* dropped from 85% to between 23% and 46% of the gut content between fall and summer.

Since fungal biomass rarely fulfills all carbon and energy needs the leaf-eating invertebrates, other mechanisms must be involved. Some of the original leaf substance may be used directly or become more accessible due to conditioning. While freshly shed maple or elm leaves contain large amounts of indigestible structural polymers, *G. pseudolimnaeus* nevertheless assimilated approx. 10% of their mass (Bärlocher and Kendrick 1973a); for *G. pulex* on alder leaves the corresponding number is 30% (Nilssen 1974). During conditioning, leaves accumulate fungal mycelia, as discussed above, which has been associated with an increase in nitrogen content. However, not the entire nitrogen gain is associated with microbial cells. A substantial fraction may be fixed in highly refractory complexes formed by leaf tannins and microbial excretions and secretions, e.g. fungal exoenzymes (Suberkropp et al. 1976; Odum et al. 1979). They can be formed artificially by mixing extracted leaf polyphenols and fungal exoenzymes (Bärlocher et al. 1989). These complexes may be of use to invertebrates with strongly alkaline gut pH, very active proteases and surfactants (Martin et al. 1980; Martin and Martin 1984), which are found in larvae of Tipulidae (see below). Conversely, having captured tannins these complexes may prevent inactivation of invertebrate digestive enzymes by inactivating plant tannins.

During conditioning, leaves lose 10–30% of their initial mass, they become softer as their polymers are being attacked by fungal enzymes (Suberkropp and Klug 1980; Bärlocher 1985; Chamier 1985). Degradation of plant polymers (cellulose, hemicelluloses, pectin) is generally based on the interaction of several synergistically acting enzymes. During this process, long chains of molecules are converted into progressively smaller subunits. As mentioned earlier, most invertebrates have weak or no activities toward native, intact chains of polymers (e.g. cellulose) but their gut enzymes are often active toward subunits of these chains (e.g. oligomers or dimers, such as, cellobiose) (Bärlocher 1985; Bärlocher and Porter 1986). Thus, invertebrates may benefit by consuming "pre-digested" leaves, or, by ingesting leaves with fungal enzymes that continue their activity in their guts (acquired enzymes; Bärlocher and Kendrick 1975b). Indeed, *G. pseudolimnaeus* preferred leaves or filter paper exposed to fungal excretions and secretions or hot HCl (which hydrolyzed polymers, but did not add microbial biomass) over untreated leaves or filter paper (Bärlocher and Kendrick 1975b). Gut extracts of *G. fossarum* released larger amounts of reducing substances (sugars) and amino acids from conditioned than from unconditioned leaves, though the sources of the released monomers were not determined (Bärlocher 1982a).

In a number of studies, the gut extracts of several stream invertebrates were shown to have high levels of alkaline proteinases, significant activity toward fungal polysaccharides and negligible activity toward plant polymers (*Tipula abdominalis*: Martin et al. 1980; stonefly nymphs: Martin et al. 1981; blackfly larvae: Martin et al. 1985). This suggested that the emphasis was on efficient protein extraction and digestion, and that carbon requirements were satisfied by fungal cells. However, this neglects the potential contribution of endosymbiotic microorganisms, which may be present in anaerobic fermentation chambers in the hindgut (Lawson and Klug 1989). These internal microorganisms may assist in the breakdown of cellulose and other leaf polymers (Canhoto and Graça 2006), but their activities may be limited by the presence of tannins (Graça and Bärlocher 1998) and oils (Canhoto and Graça 2006).

Fungal enzymes also contribute to substrate softening and maceration. This is largely due to enzymes attacking pectin, the "glue" that binds plant cells together. Some of these enzymes are generally more active at alkaline pH values (Suberkropp and Klug 1980; Chamier 1985); consequently, in acidic streams, leaf softening and maceration are reduced though fungal biomass may remain high (Jenkins and Suberkropp 1995; Dangles and Chauvet 2003). Softening of leaves or filter paper with hot HCl increased their attractiveness to *G. pseudolimnaeus* (Bärlocher and Kendrick 1975b), though the same treatment also partially hydrolyzed cellulose and other polymers. At least in the laboratory, fungal maceration can result in "pin-prick" holes in leaves by releasing FPOM (fine particulate organic matter; Suberkropp and Klug 1980). Similar holes appear in leaves exposed in fine-mesh bags that exclude macroinvertebrates (Figs. 19.3 and 19.4). The released particles are no longer available to shredders and instead serve as food for filter-feeders (Cummins and Klug 1979).

Aquatic hyphomycetes growing on leaves also release huge numbers of conidia: up to 50% of fungal production may be channeled into spore production (Findlay and Arsuffi 1989; Suberkropp 1991; Sridhar and Bärlocher 2000). The freshwater bivalve *Elliptio complanata* (Unionidae) efficiently removed and ingested conidia from the water column, though their contribution to its nutrition was minimal (Bärlocher and Brendelberger 2004). As pointed out by Canhoto and Graça (2008), leaf softening can sometimes be disadvantageous to invertebrates. Small invertebrates may gain access to inner softer tissues and be protected from predators if the tough outer layers remain intact. This applies, for example, to eucalypt or rhododendron leaves. In addition, harder leaves are better suited for case-building by *Trichoptera* larvae, and *Phylloicos* sp.

(Rincón and Martínez 2006) and *Lepidostoma complicatum* (Kochi and Kagaya 2005) preferred tougher over softer leaves for that purpose.

19.2.3 Do invertebrates differ in their feeding strategies?

Bärlocher and Porter (1986) attempted to correlate digestive enzymes and feeding strategies in three leaf-eating detritus-feeders (*Gammarus tigrinus*, Amphipoda; *Hydropsyche betteni*, Trichoptera; *Tipula caloptera*, Diptera). *Tipula caloptera* had the highest proteinase activity in gut extracts, particularly at the pH of 11 (close to the maximum value of 10.5 observed in the hindgut). Only gut extracts of *G. tigrinus* released sugars from unconditioned leaves. All three invertebrates digested starch and laminarin (potential fungal polymers). Enzymes capable of breaking glycosidic linkages in small chains of monomers were present in the gut fluids of all three invertebrates. Fungal polysaccharidases, ingested with maple leaves, remained active in the guts of *G. tigrinus* and *H. betteni* but not in *T. caloptera*. Combined with other studies (Martin et al. 1980; Sinsabaugh et al. 1981; Bärlocher and Porter 1986; Canhoto and Graça 2006), this suggests that Tipulid larvae have specialized in efficient protein extraction. Their carbon requirements may be satisfied by digestion of fungal polysaccharides, or, via fermentation by endosymbiotic bacteria. When ingesting leaf materials they may therefore be less dependent on optimal fungal conditioning, which, in any case, they may be unable to acquire given their limited motility. Gammarids, on the other hand, potentially derive greater benefit from optimal fungal conditioning by gaining access to easily digested lipids and proteins, by fungal pre-digestion of native polysaccharides and by acquiring fungal exoenzymes that continue their activity after ingestion. To benefit, they should be highly mobile and capable of searching out well conditioned leaves (which may be uncommon in natural streams, Bärlocher 1983) or have alternative sources of food; both conditions seem to be met by Gammarids. Further research into potential correlations between digestive physiology, food preferences and motility is required to evaluate such correlations.

Arsuffi and Suberkropp (1988) compared the relative importance of fungal biomass vs. enzymatically changed substrates by offering caddis fly larvae a choice between aspen leaves (*Populus tremuloides*) colonized by one of three aquatic hyphomycete species in the presence or absence of 1% glucose. The addition of glucose inhibited the production of fungal exoenzymes and leaf softening, but increased the leaf mass by between 33% and 97%. Fungi formed a thick mat of mycelium over the surface of the leaves. While *Psychoglypha* sp. larvae ingested more of the leaf material with a fungus grown with glucose, the larvae in fact "fed primarily on the mycelial mat covering the surface of these leaves". Thus, the choice was between fungal mycelium and conditioned leaves (which contained fungal biomass and were modified by fungal enzymes). It seems premature to conclude that, in the field, fungal biomass is more important for invertebrate nutrition than fungally modified leaf substrate.

19.2.4 What factors ultimately determine food choice and feeding selectivity?

The assumption that fungal conditioning improves the acceptability of leaves to invertebrates seems fairly robust, with the following caveats: a) leaves will eventually enter a post-conditioned state where palatability declines; b) some leaves (e.g. alder) are reasonably well accepted even without conditioning; and, c) green leaves are often more palatable than senescent or dead leaves, and may not improve or even deteriorate with conditioning (Anderson and Sedell 1979; Stout et al. 1985; Yeates and Barmuta 1999; but see Kochi and Kagaya 2005). Even without conditioning, invertebrates clearly prefer certain leaf species over others (e.g. ash preferred over maple preferred over oak, Bärlocher and Kendrick 1973b; alder leaves preferred over most other leaves, review by Anderson and Sedell 1979). Invertebrates also differentiate among leaf-fungus combinations, and by choosing a suitable fungus, the preference for a given leaf species can be overturned. For example, oak conditioned by *Anguillospora longissima* was chosen over ash with *Tetracladium marchalianum* (Bärlocher and Kendrick 1973b). No direct relationship was found between mass loss, changes in protein content and palatability, except when conditioning occurred at 0°C. Not surprisingly, under these conditions only fungi causing mass loss changed palatability. The amount of external nutrients also affected palatability. Thus, temperature and nutrients directly affected fungal growth and degradative activities on leaves. This in turn influenced the palatability of the fungus/leaf complex to invertebrates, but there was no straightforward and intuitively obvious connection between the two.

Larvae of *Hesperophylax* and *Psychoglypha* (Trichoptera) were able to discriminate among leaf patches colonized by different fungal species (Arsuffi and Suberkropp 1985). Patches were prepared by fastening colonized disks onto sterile leaves and incubating. The fungi grew into the areas of the sterile leaf, which produced leaves with areas colonized by a pure culture interspersed with sterile areas. This approach is closer to what invertebrates might encounter in a stream – leaves are mosaics of patches that may be sterile or colonized by different fungal species. Patch size, however, was relatively large (diameter = 12 mm); on stream-collected leaves, intermingling of mycelia by different species occurs at a much smaller scale (Shearer and Lane 1983; Chamier et al. 1984). It is unknown if invertebrates are capable of resolving these mycelial mixtures, or whether they ultimately choose between different combinations of fungal species. In recent experiments, fungi were inoculated by adding conidial suspensions (A. L. Gonçalves, University of Coimbra, personal communication). Two invertebrate species showed clear preferences for some fungus/leaf combination when only one fungus was present on a given leaf disk. When mixed conidial suspensions were used, there was no evidence of preferential feeding for any of three fungal mixtures used as inoculum. This suggests that the joint effect of a combination of fungal species on palatability cannot be predicted from the effects by individual species in isolation.

To summarize, the various effects of fungal colonization on leaf properties are well known: softening, predigestion of polymers, presence of fungal exoenzymes, increased nutrient concentrations due to fungal growth, resulting in lower (more beneficial) C:N and C:P ratios (Cross et al. 2005). But which of these changes dominates food choice? No clear, consistent answer has emerged. For example, there was no correlation between palatability and protein increase or mass loss (as indicator of overall enzymatic activity, Bärlocher and Kendrick 1973b). In a number of studies, various caddis fly larvae (*Hesperophylax* sp., *Psychoglypha* sp., *Lepidostoma* sp., *Limnephilus* sp., *Pycnopsyche guttifer, Sericostoma personatum*) and amphipods (*Gammarus tigrinus, G. pulex*) were given a choice between leaves conditioned by pure cultures of individual fungal species (Suberkropp et al. 1983; Suberkropp and Arsuffi 1984; Butler and Suberkropp 1986; Arsuffi and Suberkropp 1989; Friberg and Jacobsen 1994; Rong et al. 1995). In general, a minimum amount of fungal activity (measured as leaf mass loss, fungal growth, fungal enzymatic activities) was required to improve the substrate's palatability. Once this was achieved, there were no significant correlations between various indicators such as fungal biomass, enzyme activities toward various polysaccharides, or nitrogen enrichment (exception: Assmann et al. 2011). *Pycnopsyche gentilis* (Trichoptera) did not differentiate among fungal species (Chung and Suberkropp 2009), nor did *Tipula caloptera* (Rong et al. 1995). Cargill et al. (1985a, b) drew attention to the role of lipids during the last instar of insects. Larvae of several caddis fly species preferred alder leaf disks coated with 18- and 20-carbon fatty acids, either as crude lipids or when provided as neutral lipid fraction extracted from aquatic hyphomycetes. Similarly, in *G. tigrinus* and *Pycnopsyche guttifer* the sequence of preference for certain fungi could be reproduced by applying non-polar extracts of the same fungi to unconditioned leaf disks (Rong et al. 1995). Assmann and von Elert (2009) documented the presence of repellents in methanol extracts. Consumption of disks treated with fungal extracts, regardless of whether they contained attractants or repellents, was considerably lower than consumption of conditioned disks. This again suggests that a threshold of conditioning has to be reached, beyond this level, the effects of individual fungi is decisive and may be influenced by attractants and/or repellants. *Tipula abdominalis* did not prefer any lipids during the last instar, again pointing to its relative independence of fungal conditioning (Cargill 1985a, b).

The degree of food selectivity may correlate negatively with the degree of endosymbiosis (Jones 1984): endosymbionts are postulated to even out resource quality. This trend agrees with observations on the three invertebrates studied by Rong et al. (1995): *G. tigrinus* had the lowest concentrations of bacteria in the digestive system, and *T. abdominalis* had the highest. When comparing six shredders of differing motility, Arsuffi and Suberkropp (1989) concluded that lower selectivity may be associated with lower "prey-finding" abilities.

It seems clear that food choice and selectivity are potentially influenced by many interacting factors, among them phylogenetic differences and constraints, physiological and developmental status of the consumer and the extent of endosymbiosis. We must also be aware of the fact that there are multiple actors in these trophic interactions, with generally conflicting interests.

19.2.5 Stoichiometric considerations

Fungal colonization of dead leaves raises their protein and nitrogen level (Kaushik and Hynes 1968, 1971). In theory, this could be achieved by fungal degradation of leaf carbon and recycling of leaf nitrogen. However, the observed N increase is often absolute, i.e. remaining leaf material contained more N than was present in the original mass. Fungi therefore must have captured additional N from external sources. This was shown explicitly in laboratory experiments by supplying leaves incubating in stream water enriched with $(NH_4)_2SO_4$, KNO_3, KH_2PO_4, or with crushed *Hyalella*. In general, leaves provided with crushed *Hyalella* or with both N and P lost more mass and accumulated more nitrogen, implying that fungal activity and growth were stimulated most by simultaneous addition of both nutrients. The same pattern was found with pure fungal cultures (Bärlocher and Kendrick 1973b), suggesting that in many streams one or both of these nutrients may limit fungal growth and/or activity. This has since been confirmed in numerous laboratory and field studies (Elwood et al. 1981; Newbold et al. 1983; Suberkropp 1995; Suberkropp and Chauvet 1995; Chauvet et al. 1997; Sridhar and Bärlocher 2000; Robinson and Gessner 2000; Bärlocher and Corkum 2003; Gulis and Suberkropp 2003; Ferreira et al. 2006).

Ecological stoichiometry seeks to understand diverse ecological concepts, including trophic interactions, by considering the combined dynamics of key elements such as carbon, nitrogen and phosphorus (Sterner and Elser 2002; Cross et al. 2005; Hladyz et al. 2009). Of the food sources, detritus from terrestrial plants (leaves, needles, twigs) has a much higher C: nutrient ratio than periphyton. Among consumers, bacteria are generally most nutrient-rich, followed by fishes, invertebrates and fungi (Cross et al. 2005). Clearly, both microorganisms and detritus-feeders feed on nutritionally impoverished sources. Microbes potentially counteract this imbalance by acquiring inorganic nutrients from the water column, while invertebrates select microbially-enriched leaf patches. Overall, litter stoichiometry (C:N:P) may strongly affect microbial N and P cycling in litter, and mineralization of these two elements may be tightly coupled to help maintain cellular homeostasis of microbial communities (Mooshammer et al. 2012). However, aquatic hyphomycetes exhibit considerable flexibility in the C:N:P ratios that permit growth and reproduction (Danger and Chauvet 2013; Grimmett et al. 2013).

19.2.6 Stimulation of fungi by invertebrate feeding

Despite ingesting food with low nutrient contents, invertebrate excretions generally contain inorganic nitrogen and phosphorus. These may benefit aquatic hyphomycetes and other microbes colonizing and degrading leaf litter. Thus, the presence of two caddis fly shredders increased ammonium concentration in the water and fungal biomass on (protected) leaves (Villanueva et al. 2012). Similarly, leaves conditioned with water containing tadpoles had reduced C:N ratios and allowed better growth of the isopod *Asellus hilgendorfii* (Iwai and Kagaya 2007).

19.2.7 Anthropogenic changes

Low nutrient levels in streams appear to restrict fungal diversity (Suberkropp and Chauvet 1995). The addition of nutrients via eutrophication can therefore be expected to increase fungal diversity, microbial decompositional activity and growth, which in turn should provide more nourishment to invertebrates. However, there are several complications: nutrient amounts and ratios strongly affect the balance between growth and reproduction of aquatic hyphomycetes (Sridhar and Bärlocher 2000), their elemental composition (Danger and Chauvet 2013) and may shift the balance between fungi and bacteria (Güsewell and Gessner 2009). The exact form in which the nutrients are provided may also be crucial (Bärlocher et al. 2013), and eutrophication is often accompanied by other anthropogenic changes such as increased load of heavy metals and other pollutants, which may impoverish the fungal community and counteract the stimulation by nutrients (Sridhar et al. 2001; Fleituch 2013). The complex interactions between pollution and fungal diversity and activity have been reviewed by Pascoal and Cássio (2008).

An important point to consider is the range of nutrients being considered. Bärlocher (2005) suggested that the fungal response is likely to be unimodal, i.e. a maximum response is expected at some intermediate value. Ferreira et al. (2006) described an asymptotic relationship between whole-stream nitrate addition and decomposition, while Gulis et al. (2006) found a positive linear relationship between decomposition rates and nitrogen and phosphorus. Pérez et al. (2013) found no correlation between nutrient availability and breakdown rates. Moderate levels of eutrophication enhanced leaf-litter quality but reduced shredder densities. In a pan-European field experiment in 100 streams, breakdown rates were slower at both extremes (Woodward et al. 2012).

Positive effects of nutrient addition on fungal growth and diversity may result in higher invertebrate densities. For example, in a stream with N and P addition continued over 2 years, invertebrate biomass increased by 2%–300% (Greenwood et al. 2007). An increase was also found in streams with moderate eutrophication (Gulis et al. 2006), but no such effect was reported by Ferreira et al. (2006). Clearly, the consequences of fungal activities and growth on higher trophic levels are context-specific. For example, eutrophication may be associated with increased levels of heavy metals and other toxins, which may have a greater direct impact on invertebrates than on fungi. Thus, the addition of copper ions did not significantly affect fungal sporulation and biomass, but had a strongly negative effect on invertebrates (Roussel et al. 2008). Similar results were found in uranium contaminated streams in Portugal (Gonçalves et al. 2011). These results are not unexpected, since invertebrates generally capture more heavy metals through direct uptake from water than through their food (Abel and Bärlocher 1988). Fungal diversity can be surprisingly high, even in very highly polluted streams, and species from such streams have evolved sophisticated adaptations to survive (Sridhar et al. 2001; Krauss et al. 2001, 2010; Azevedo et al. 2007; Ferreira et al. 2010a).

If both fungal growth and invertebrate density are stimulated, the net outcome may still be variable. For example, nutrient addition in an alpine stream resulted in initially higher N and P levels of leaves. At later stages of decay, similar values were found regardless of treatment (Robinson and Gessner 2000). Possibly, the more nutritious patches in the fertilized treatment were rapidly exploited, which depressed the average nutrient content of the leaves. In this case, one fungal species, *Flagellospora curvula*, dominated the communities in all treatments (> 90% of all conidia; Robinson and Gessner 2000). A similar transfer of P to invertebrates via the microbial pathway was described by Connolly and Pearson (2013). Red maple leaves accumulated more fungal biomass in a nutrient-enriched stream than in a reference stream, and the fungal communities were significantly different, though in this case only one sample was taken due to the rapid disappearance of the leaves (Chung and Suberkropp 2008).

The response of stream communities to nutrient enrichment is complex and may be compounded if the treatments continue over several years. In a 5-year experiment, N + P addition increased microbial biomass and activity on a per unit leaf mass basis soon after leaf-fall. Both then declined more steeply than in the reference stream (Suberkropp et al. 2010), partly due to increased consumption by shredders (Cross et al. 2007). This resulted in lower standing crops of leaf litter and therefore less substrate for microbial colonization. Overall, fungal biomass on leaves was negatively affected, while overall fungal production increased (though annual production seemed to decrease gradually during the 5 years). Intra-annual variability of fungal biomass on decaying leaves also increased with nutrient addition. Overall, the data suggest that aquatic hyphomycetes may allocate more resources to reproduction (conidia) than to mycelia at higher nutrient levels (see also Sridhar and Bärlocher 2000). Chung and Suberkropp (2008) also observed that under nutrient-rich conditions, species with larger conidia may be favoured.

In a related 2-year study, nutrient enrichment increased concentrations of conidia in the water column by factors of between 4.5 and 6.9 (Gulis and Suberkropp 2004). It also increased the number of species that were detected on each sampling date (the sampled volume were kept constant; thus, species density was measured; Gotelli and Ellison 2013).

Due to rising CO_2 levels in the atmosphere, global temperature is expected to increase and become more variable. Both changes will directly impact fungi and invertebrates and will modulate their interactions. Literature on this topic is large (e.g. Fernandes et al. 2009; Dang et al. 2009; Ferreira et al. 2010b; Woodward et al. 2010; Bärlocher et al. 2013) and a thorough review is beyond the scope of this article. It is likely, however, that an anticipated increase of 2°C–4°C will generally accelerate microbial decomposition, resulting in faster turnover of leaf and other detritus (Bärlocher et al. 2008). At least in a hyporheic region of a stream, this lowered the occurrence and diversity of aquatic hyphomycetes. In a global experiment, Boyero et al. (2011a) measured leaf decomposition in 22 streams across a latitudinal gradient. Their results suggest that higher temperature will not change overall decomposition rates, however, the microbial component will increase at the expense of invertebrates.

19.2.8 Research outside temperate regions

Research on leaf decomposition and fungal/invertebrate interactions has been dominated by studies in temperate regions of Europe and North America. Are these processes fundamentally different in other regions, which include various subtropical/tropical areas? Boulton et al. (2008) caution that wide variation in climate, geomorphology and geological history may mask patterns that might be considered unique to tropical streams. In addition, much research has been concentrated in relatively small sections of the tropics, in particular, research in the African tropics has been neglected.

The observation that biological diversity in the tropics far exceeds that of temperate zones has been called ecology's oldest pattern (Hawkins 2001). This is certainly the case for riparian forests, and leaves decaying in streams in the tropics are more diverse phylogenetically and biochemically (Wantzen et al. 2008). The basic steps in leaf processing are similar: an early, brief leaching phase is followed by microbial colonization and conversion to fine particles due to physical fragmentation, microbial maceration and invertebrate feeding. The combination of these processes leads to typical skeletonized leaves (Figs. 19.5 and 19.6).

There is some evidence that some of the groups that dominate litter decay in temperate regions are less diverse than in the tropics. To begin with, leaf-eating detritus-feeders in Europe and North America are dominated by insect larvae (Trichoptera, Plecoptera) and amphipods (*Gammarus* spp.). There is good evidence that several aquatic insect orders may have evolved in cool running waters (Hynes 1970). Irons et al. (1994) demonstrated that, if measured in degree-days, decay rates were higher in Alaska than in Costa Rica. They concluded that the role of macroinvertebrates increases with latitude, while the microbial contribution decreases. Results from a global experiment, where alder leaves were exposed in 22 streams, suggested that higher temperature shifts the balance from invertebrate to microbial decomposition (Boyero et al. 2011a). Boyero et al. (2011b) examined diversity patterns of leaf-shredders at 129 stream sites from 14 regions on six continents and confirmed greater abundance and diversity at higher latitudes. These trends were confirmed in a study in the Andean mountains of Colombia, where leaves decayed rapidly, fungal biomass was high but shredders were rare (Mathuriau and Chauvet 2002). By contrast, in a Black Forest stream (Germany), invertebrates were estimated to have made a greater contribution to mass loss (51%–64%) than fungi (15%–18%) or bacteria (7%–9%) (Hieber and Gessner 2002). However, local conditions can drastically change these numbers, e.g. eutrophication in temperate streams can greatly depress the contributions of leaf-shredders (Pascoal et al. 2005). Conversely, shredders are generally common in Australian rain forests (Cheshire et al. 2005).

In many tropical streams, the function of leaf-shredders is filled by a variety of other taxa, including omnivorous decapods, snails, semi-terrestrial cockroaches or even fish (Covich 1988; Chandrashekar et al. 1989; Rueda-Delgado et al. 2006; Wantzen and Wagner 2006; Wantzen et al. 2006; Boulton et al. 2008; Yule et al. 2009; Coughlan et al. 2010; Sridhar and Sudheep 2011).

In broad-leaf deciduous forests of temperate regions, litter input is strongly seasonal; in coniferous forests it is more evenly spaced through the entire year. By contrast, the timing in tropical forest is much more variable and strongly depends on local conditions (Wantzen et al. 2008). The path that a leaf takes from tree to stream is not always direct (Covich 1988); leaves may be trapped by aerial rhizomorphs, or accumulate on the forest floor or empty stream bed during dry seasons. Microbial (or even animal) attack may occur before the leaf becomes available to stream consumers, which may affect its nutritional quality.

The beneficial effects of in-stream conditioning appear to be common but not universal in the tropics. Various caddis fly larvae from Australian rain-forest streams discriminated between different leaf species, and generally preferred conditioned leaves (Nolen and Pearson 1993; Bastian et al. 2007). The same pattern was shown by the crayfish *Cherax cairnsensis* (Coughlan et al. 2010). The freshwater shrimp *Atyaephyra desmarestii*, common in eutrophic streams in Europe, North Africa and the Middle East, preferred conditioned leaves but did not differentiate among fungal species (Duarte et al. 2012), while a leaf-eating fish showed clear preference for rubber leaves colonized by certain species of aquatic hyphomycetes and differentiated between preconditioned, fully conditioned and postconditioned leaves (Chandrashekar et al. 1989). Two tropical and two temperate shredders generally showed the same distinct preference for conditioned over unconditioned leaves, and gained more weight when fed on conditioned leaves (Graça et al. 2001). Very little is known how and to what extent fungal conditioning helps digestion by leaf-eating decapods or fish; hepatopancrease extracts of the crayfish *Paranephrops zealandicus* showed activity toward cellulose, pectin, mannan, laminarin and chitin, i.e. storage polymers of algae, fungi and higher plants (Musgrove 1988). However, conditioning detritus still improved the assimilation efficiency, suggesting that microbial pretreatment was beneficial.

In addition to leaf-shredders, aquatic hyphomycetes may also be impoverished in tropical streams. Based on a literature review, Shearer et al. (2007) suggested that the greatest diversity of aquatic fungi occurs in temperate areas, followed by Asian tropical areas. They cautioned that this may be an artifact of uneven sampling efforts, but suggested that the pattern may be real and reflect a higher diversity in ecological niches due to greater seasonality in temperate regions. The least sampled geographic areas include Africa, Australia, China, South America and boreal and tropical regions worldwide. Variability within broad geographic regions is undoubtedly great. For example, only about 90 species of aquatic hyphomycetes have been reported from South America (including records from Argentina, Brazil, Chile, Ecuador, Peru and Venezuela; Schoenlein-Crusius and Grandi 2003). By contrast, Thomas (1992), in the course of a Ph.D. thesis, found 136 species in several streams of New South Wales, Australia, of which 55 could be assigned to described species or genera. About 100 species are known from two regions in India, the Western Ghats and Himalayas (Sridhar 2009). In a leaf decomposition study in Chile, Mathuriau and Chauvet (2002) found 16 species; in five Australian streams, the diversity varied between 12 and 28 (Bärlocher et al. 2011), both within the usual range found in temperate streams.

Jabiol et al. (2013) tested the hypothesis by Shearer et al. (2007) by sampling fungal communities in five streams along a latitudinal gradient from the subarctic to the tropics. They observed the greatest diversity (number of species) in temperate and/or Mediterranean sites. As potential factors for the low diversity in the tropical stream (French Guiana), they listed the low quality of most tropical litter species and the low nutrient concentrations in the water. Jabiol et al. (2013) point out that more studies involving several streams along the gradient are needed to confirm this pattern.

Aquatic hyphomycetes grow primarily on leaves, but may colonize wood, where ascomycetes are more common (Sridhar et al. 2010). This group seems to be more diverse in tropical streams (Ho et al. 2001), but again, additional studies are needed to confirm this pattern.

19.3 Effects of invertebrate consumers on the leaf-fungus complex

Research on leaf decomposition has been dominated by process-oriented questions, and has primarily addressed the impact of fungal colonization on leaf decomposition and invertebrate nutrition. Few studies have investigated the reverse, i.e. how does consumption by invertebrates affect the biomass, production and diversity of aquatic hyphomycetes? Two processes have a potentially direct impact: consumption of conidia in the water column by filter-feeding invertebrates, and consumption of leaves permeated by fungal hyphae.

19.3.1 Invertebrate ingestion of conidia

To our knowledge, there is only one study documenting removal of conidia by an aquatic filter-feeder. In laboratory experiments, the freshwater bivalve *Elliptio complanata* efficiently cleared suspended conidia (Bärlocher and Brendelberger 2004). Based on water volume and flow, and invertebrate density, this filter-feeder may remove substantial proportion of naturally occurring conidia, under some conditions approaching 100%. The conidia of the seven tested species were removed at significantly different rates, potentially causing distinct shifts in the remaining spore assemblage. This reduction and changing composition of the stream inoculum may affect the rate at which new substrates are colonized and conditioned (GC Malay, unpubl. obs). There is some evidence that competition among aquatic hyphomycetes begins at the settlement stage (Treton et al. 2004; Nikolcheva et al. 2005); successful colonization by different species is generally influenced by the relative quantities of inocula (Raviraja et al. 2006). The mechanisms described above were observed in laboratory experiments, and their impact on activities and diversity of fungal communities in streams is currently unknown.

19.3.2 Invertebrate ingestion of the leaf-fungus complex

Fungal biomass on decaying leaves may approach 20% of total mass, and ≥50% of fungal production may be converted to conidia (Gessner et al. 2003; Krauss et al. 2010). While conidia are captured by filter-feeders and presumably digested, their contribution to invertebrate nutrition is probably low (estimated at <1% for *Elliptio complanata*; Bärlocher and Brendelberger 2004), while the impact of fungal mycelia on shredder nutrition can be substantial (see Section 19.2; Effects of the leaf-fungus complex on invertebrate consumers). To study how shredder consumption affects fungal biomass and diversity on leaf substrates, two types of studies have been described. The more easily conducted experiments involve exposing leaves in fine-mesh (generally ≤0.5 mm mesh) and coarse-mesh (generally 8–10 mm mesh) bags (Bärlocher 2008). The coarse mesh admits both microbes and shredders, while the fine mesh excludes shredders. Comparing leaves in the two bags should therefore reveal the impact of shredders. Potential pitfalls include reduced water flow and oxygenation in fine-mesh bags, and exclusion of very large shredders in coarse-mesh bags (Boulton and Boon 1991). Instead of using bags with different mesh sizes, insect shredders have been manipulated in laboratory stream experiments (Ferreira and

Graça 2006), or eliminated (or at least drastically reduced) from entire streams by applying an insecticide (Cuffney et al. 1990).

When shredders are common in a stream, leaves generally disappear more quickly in coarse-mesh bags (Webster and Benfield 1986; Graça 2001). Fungal biomasses in the two treatments generally do not differ (Howe and Suberkropp 1994; Ferreira and Graça 2006; Chun and Suberkropp 2008). Nevertheless, shredders clearly "prey" on fungal biomass. Predators potentially affect the diversity of their prey species; the intermediate disturbance hypothesis suggests that prey diversity should be highest at an intermediate level of predation. Most studies thus far have failed to reveal a consistent effect. Fungal diversity (based on spore identification) was higher in fine-mesh bags when oak leaves were exposed; with larch and spruce needles, results were variable (Bärlocher 1980). No clear patterns were found by Howe and Suberkropp (1994) on tulip poplar leaves (fungal diversity was very low in their stream) or by Ferreira and Graça (2006) on alder leaves in artificial streams. Jabiol et al. (2013) likewise found no effect of mesh-size (and by implication, of shredders) on fungal diversity in five streams along a latitudinal gradient (however, their study was based on a single sampling date per stream).

Higher sporulation rates were again reported from oak in fine-mesh bags (Bärlocher 1980, 1982b), while Howe and Suberkropp (1994) reported the opposite for tulip poplar leaves, and Ferreira and Graça (2006) found no difference. Ferreira and Graça (2006) also compared total numbers of spores released during the experiment and found no difference due to the presence/absence of shredders. However, at the end of their experiment, leaves protected from shredders had retained more biomass and presumably continued to release spores for a longer period of time than leaves exposed to shredders. This suggests a second role of shredders with regard to aquatic hyphomycetes: they generally shorten the residence time of leaves in streams and thereby act as competitors of fungi (Bärlocher 1980). If this is common, and if we assume that shredders do not affect sporulation rates, more aquatic hyphomycete spores will be released from a leaf before its conversion to fine or dissolved organic matter. Larger sample or population sizes almost always result in more species being found (demonstrated in species accumulation and species area curves; also known as "more individuals hypothesis"; Wright 1983; Rosenzweig 1999; Waide et al. 1999; Gönczöl et al. 2001; Evans et al. 2005; Gotelli and Ellison 2013). Not surprisingly, cumulative species richness was greater in leaves from fine-mesh bags (Bärlocher 1980; Chung and Suberkropp 2008). Jabiol et al. (2013) found no difference, however, they stopped their study after litter in coarse-mesh bags had lost 50%–60% of its initial mass. They also rarefied diversity by estimating average species numbers found in random collections of 50 conidia from each sample. Rarefaction equalizes sample bias by adjusting it to its most extreme value (i.e. bias of smallest sample; Rosenzweig 1999; Bärlocher 2005).

The insecticide methoxychlor applied to a small headwater stream caused massive invertebrate drift and reduced abundances of most insect taxa and functional groups

(Cuffney et al. 1990; Lugthart and Wallace 1992). Shredders declined by 87%. In untreated reference streams they consumed 28%–47% of the standing crop of leaves; in the treated stream, this declined to 4%. This must have increased substrate availability to fungi. Not surprisingly, during much of the year the number of conidia in transport in the insecticide-treated stream was higher (Suberkropp and Wallace 1992). The composition of the fungal community did not seem to change much in the absence of shredders; the same phenomenon was observed by Jabiol et al. (2013). This seems to conflict with the often documented ability of invertebrates to selectively feed on some fungi and reject others (see above). Possibly, the different preferences by various invertebrates cancel each other out, or, invertebrates feeding on stream-conditioned leaves cannot choose among individual fungi. Due to the close intermingling of fungal colonies (Shearer and Lane 1983), they may feed on different combinations of species among which they do not discriminate (AL Gonçalves, unpublished observations).

In summary, it appears that the effect of shredders on fungal communities on individual leaves is generally negligible or variable. By accelerating leaf decay, they nevertheless depress cumulative fungal output of conidia. Depending on the severity of this effect, shredders may restrict the abundance or occurrence of rare species. How important are these rare species? Cao et al. (1998) found them of particular value in diagnosing anthropogenic impact – they often play the role of a canary in coal mines. Their inclusion or exclusion often has a very minor impact on community analyses relying on multivariate techniques, as many similarity analyses strongly underweight rare species and overweight abundant ones. In addition, the rarity or abundance of a species does not necessarily remain constant over several years (Bärlocher 2000).

19.4 Conclusions

The basic mechanisms by which aquatic hyphomycetes colonize autumn-shed leaves and condition them for invertebrate consumption are well established (addition of proteins, lipids, modification of structural polymers, mitigating stoichiometric constraints). Estimates of biomass and production have demonstrated the substantial fungal contribution to stream food webs. Less is known about the evolutionary ecology of these fungi and their invertebrate consumers. Clearly, invertebrates differ in their food preferences. What mechanisms are responsible and how they may relate to the invertebrates' physiology and behaviour requires further research before robust generalizations are possible. It is also clear that researchers continue to concentrate on temperate streams and their organisms. While some studies have been conducted in lower latitudes, they do not generally encompass the entire range of subtropical and tropical habitats.

Potential top-down aspects have received less attention. There is little doubt that invertebrate consumption of mycelia and leaves exerts considerable pressure on aquatic hyphomycetes and generally reduces fungal biomass and production when considering an entire stream reach or the complete conversion of a leaf to fine particles and dissolved matter. It remains unclear if and to what extent this affects fungal diversity and community structure, both at the level of individual leaves or an entire stream reach.

At the level of individual taxa, potential fungal counter adaptations to minimize the damage caused by feeding invertebrates include the production of repellents, association with unpalatable species, preferential colonization of veins vs. the softer mesophyll, or surviving passage through the digestive tract (Bärlocher 1980, 1985). Only the last of these strategies has been documented under field conditions: viable conidia or conidium-producing structures have been found in the faeces of several amphipods (Bärlocher 1981; Sridhar et al. 2011).

Like all fields in biology, research on aquatic hyphomycetes will benefit from increased use of molecular techniques (Bärlocher 2000). They have already resulted in more accurate and inclusive inventories of fungal taxa during leaf decomposition (Bärlocher 2010; Duarte et al. 2013). Additional applications include characterization of the metabolic potential and its expression in taxa when growing in pure cultures or in natural communities, research on the biogeography of aquatic hyphomycetes and the potential role of sexual reproduction.

Acknowledgments

Some of the research reported here has been generously supported for many years by the Natural Science and Engineering Research Council of Canada.

References

Abel T, Bärlocher F. Uptake of cadmium by *Gammarus fossarum* (Amphipoda) from food and water. J Appl Ecol 1988;25:223–231.

Anderson NH, Grafius E. Utilization and processing of allochthonous material by stream Trichoptera. Verh Int Verein Limnol 1975;19:3022–3028.

Anderson NH, Sedell JR. Detritus processing by macroinvertebrates in stream ecosystems. Ann Rev Entomol 1979;24:351–357.

Arsuffi TL, Suberkropp K. Leaf processing capabilities of aquatic hyphomycetes: interspecific differences and influence on shredder feeding preferences. Oikos 1984;42:144–154.

Arsuffi TL, Suberkropp K. Selective feeding by stream caddisfly (Trichoptera) detritivores on leaves with fungal-colonized patches. Oikos 1985;45:50–58.

Arsuffi TL, Suberkropp K. Effect of fungal mycelia and enzymatically degraded leaves on feeding and performance of caddisfly (Trichoptera) larvae. J N Am Benthol Soc 1988;7:205–211.

Arsuffi TL, Suberkropp K. Selective feeding by shredders on leaf-colonizing stream fungi: comparison of macroinvertebrate taxa. Oecologia 1989;79:30–37.

Assmann C, von Elert E. The impact of fungal extracts on leaf litter on the food preference of *Gammarus roeselii*. Internat Rev Hydrobiol 2009;4:484–496.

Assmann C, Rinke K, Nechwatal J, van Elert E. Consequences of the colonisation of leaves by fungi and oomycetes for leaf consumption by a gammarid shredder. Freshw Biol 2011;56:839–852.
Azevedo M-M, Carvalho A, Pascoal C, Rodrigues F, Cássio F. Responses of antioxidant defenses to Cu and Zn stress intwo aquatic fungi. Sci Tot Environ 2007;377:233–243.
Baldy V, Chauvet E, Charcosset J-Y, Gessner MO. Microbial dynamics associated with leaves decomposing in the mainstem and floodplain pond of a large river. Aquat Microb Ecol 2002;28:25–36.
Bärlocher F. Leaf-eating invertebrates as competitors of aquatic hyphomycetes. Oecologia 1980;47:303–306.
Bärlocher F. Fungi on the food and in the faeces of *Gammarus pulex*. Trans Br Mycol Soc 1981;76:160–165.
Bärlocher F. The contribution of fungal enzymes to the digestion of leaves by *Gammarus fossarum* Koch (Amphipoda). Oecologia 1982a;42:1–4.
Bärlocher F. Conidium production from leaves and needles in four streams. Can J Bot 1982b;60:1487–1494.
Bärlocher F. Seasonal variation of standing crop and digestibility of CPOM in a Swiss Jura stream. Ecology 1983;64:1266–1272.
Bärlocher F. The role of fungi in the nutrition of stream invertebrates. Bot J Linn Soc 1985;91:83–94.
Bärlocher F. Research on aquatic hyphomycetes: historical background and overview. In: Bärlocher F, ed. The Ecology of Aquatic Hyphomycetes. Berlin: Springer 1992:1–15.
Bärlocher F. Water-borne conidia of aquatic hyphomycetes: seasonal and yearly patterns in Catamaran Brook, New Brunswick, Canada. Can J Bot 2000;78:157–167.
Bärlocher F. Freshwater fungal communities. In: Dighton J, J White J, Oudemans P, eds.The Fungal Community. Boca Raton, LA: CRC Press 2005:39–59.
Bärlocher F. Leaf mass loss estimated by litter bag technique. In: Graça MAS, Bärlocher F, Gessner MO, eds. Methods to Study Litter Decomposition. Dordrecht: Springer 2008:37–42.
Bärlocher F. Molecular approaches promise a deeper and broader understanding ofthe evolutionary ecology of aquatic hyphomycetes. J N Am Benthol Soc 2010;29:1027–1041.
Bärlocher F, Brendelberger H. Clearance of aquatic hyphomycete spores by a benthic suspension feeder. Limnol Oceanogr 2004;49:2292–2296.
Bärlocher F, Corkum M. Nutrient enrichment overwhelms diversity effects in leaf decomposition by stream fungi. Oikos 2003;101:247–252.
Bärlocher F, Kendrick B. Fungi in the diet of *Gammarus pseudolimnaeus*. Oikos 1973a;24:295–300.
Bärlocher F, Kendrick B. Fungi and food preferences of *Gammarus pseudolimnaeus*. Arch Hydrobiol 1973b;72:501–516.
Bärlocher F, Kendrick B. Dynamics of the fungal population on leaves in a stream. J Ecol 1974;62:761–791.
Bärlocher F, Kendrick B. Assimilation efficiency of *Gammarus pseudolimnaeus* feeding on fungal mycelium or on autumn-shed leaves. Oikos 1975a;26:55–59.
Bärlocher B, Kendrick B. Leaf conditioning by microorganisms. Oecologia 1975b;20:359–362.
Bärlocher F, Kendrick B. Hyphomycetes as intermediaries of energy flow in streams. In: Jones EBG, ed. Recent Advances in Aquatic Mycology. London: Paul Elek 1976:435–446.
Bärlocher F, Kendrick B. The role of aquatic hyphomycetes in the trophic structure of streams. In: Wicklow DT, Carroll GC, eds. The Fungal Community. New York: Marcel Dekker 1981:743–760.
Bärlocher F, Kebede YK, Gonçalves AL, Canhoto C. Incubation temperature and substrate quality modulate sporulation by aquatic hyphomycetes. Microb Ecol 2013;66:30–39.
Bärlocher F, Porter CW. Digestive enzymes and feeding strategies of three stream invertebrates. J N Am Benthol Soc 1986;5:58–66.
Bärlocher F, Tibbo PG, Christie SH. Formation of phenol-protein complexes and their use by two stream invertebrates. Hydrobiol 1989;173:243–249.

Bärlocher F, Seena S, Wilson KP, Williams DD. Raised water temperature lowers diversity of hyporheic aquatic hyphomycetes. Freshw Biol 2008;53:368–379.
Bärlocher F, Stewart M, Ryder DS. Analyzing aquatic fungal communities in Australia: impacts of sample incubation and geographic distance of streams. Czech Mycol 2011;63:113–132.
Bärlocher F, Stewart M, Ryder DS. Processing of *Eucalyptus viminalis* leaves in Australian streams – importance of aquatic hyphomycetes and zoosporic fungi. Fundam Appl Limnol 2012;179:305–319.
Baschien C, Marvanová L, Szewzyk U. Phylogeny of selected aquatic hyphomycetes based on morphological and molecular data. Nova Hedwig 2006;83:311–352.
Bastian M, Boyero L, Jackes BR, Pearson RG. Leaf litter diversity and shredder preferences in an Australian tropical rain-forest stream. J Trop Ecol 2007;23:219–229.
Belliveau MJ-R, Bärlocher F. Molecular evidence confirms multiple origins of aquatic hyphomycetes. Mycol Res 2005;109:1407–1417.
Boulton AJ, Boon PI. A review of the methodology used to measure leaf litter decomposition in lotic environments: time to turn over an old leaf? Austr J Mar Freshw Res 1991;42:1–43.
Boulton AJ, Boyero L, Covich AP, Dobson M, Lake S, Pearson R. Are tropical streams ecologically different from temperate streams? In: Dudgeon D, ed. Tropical Stream Ecology. San Diego, CA: Karen Academic Press 2008:257–284.
Boyero L, Pearson RG, Dudgeon D, Ferreira V, Graça MAS, Gessner MO, et al. A global experiment suggests climate warming will not accelerate litter decomposition in streams but might reduce carbon sequestration. Ecol Lett 2011a;14:289–294.
Boyero L, Pearson RG, Dudgeon D, Graça MAS, Gessner MO, Albariño RJ, et al. Global distribution of a key trophic guild contrasts with common latitudinal diversity patterns. Ecology 2011b;92:1839–1848.
Butler SK, Suberkropp K. Aquatic hyphomycetes on oak leaves: comparison of growth, degradation and palatability. Mycologia 1986;78:922–928.
Callisto M, Gonçalves JF Jr, Graça MAS. Leaf litter as a possible food source for chironomids (Diptera) in Brazilian and Portuguese headwater streams. Rev Brasil Zool 2007;24:442–448.
Campbell J, Shearer C, Marvanová L. Evolutionary relationships among aquatic anamorphs and teleomorphs: *Lemonniera*, *Margaritispora*, and *Goniopila*. Mycol Res 2006;110:1025–1033.
Canhoto C, Graça MAS. Leaf barriers to fungal colonization and shredders (*Tipula lateralis*) consumption of decomposing *Eucalyptus globulus*. Microb Ecol 1999;37:1631–1672.
Canhoto C, Graça MAS. Digestive tract and leaf processing capacity of the stream invertebrate *Tipula lateralis*. Can J Zool 2006;84:1087–1095.
Canhoto C, Graça MAS. Interactions between fungi and stream invertebrates: back to the future. Fung Div Res Ser 2008;20:305–325.
Cao Y, Williams DD, Williams NE. How important are rare species in aquatic community ecology and bioassessment? Limnol Oceanogr 1998;43:1403–1409.
Cargill AS, Cummins KW, Hanson BJ, Lowry RR. The role of lipids, fungi, and temperature in the nutrition of a shredder caddis fly, *Clistoronia magnifica*. Freshwat Invertebr Biol 1985a;4:64–78.
Cargill AS, Cummins KW, Hanson BJ, Lowry RR. The role of lipids as feeding stimulants for shredding aquatic insects. Freshw Biol 1985b;15:455–464.
Chamier A-C. Cell-wall-degrading enzymes of aquatic hyphomycetes: a review. Bot J Linn Soc 1985;91:67–81.
Chamier A-C, Dixon PA, Archer SA. The spatial distribution of fungi on decomposing alder leaves in a freshwater stream. Oecologia 1984;64:92–103.
Chandrashekar KR, Sridhar KR, Kaveriappa KM. Palatability of rubber leaves colonized by aquatic hyphomycetes. Arch Hydrobiol 1989;115:361–369.
Chauvet E, Fabre E, Elósegui A, Pozo J. The impact of eucalypt on the leaf-associated aquatic hyphomycetes in Spanish streams. Can J Bot 1997;75:880–887.

Cheshire K, Boyero L, Pearson RG. Food webs in tropical Australian streams: shredders are not scarce. Freshw Biol 2005;50:748–769.
Chung N, Suberkropp K. Influence of shredder feeding and nutrients on fungal activity and community structure in headwater streams. Fund Appl Limnol 2008;173:35–46.
Chung N, Suberkropp K. Contribution of fungal biomass to the growth of the shredder, *Pycnopsyche gentilis* (Trichoptera: Limnephilidae). Freshw Biol 2009;54:2212–2224.
Connolly NM, Pearson RG. Nutrient enrichment of a heterotrophic stream alters leaf litter nutritional quality and shredder physiological condition via the microbial pathway. Hydrobiologia 2013;718:85–92.
Coughlan JF, Pearson RG, Boyero L. Crayfish process leaf litter in tropical streams even when shredding insects are common. Mar Freshw Res 2010;61:541–548.
Covich AP. Geographical and historical comparisons of neotropical streams: biotic diversity and detrital processing in highly variable habitats. J N Am Benthol Soc 1988;7:361–386.
Cross WF, Benstead JP, Frost PC, Thomas SA. Ecological stoichiometry in freshwater benthic systems: recent progress and perspectives. Freshw Biol 2005;50:1895–1912.
Cross WF, Wallace JB, Rosemond AD. Nutrient enrichment reduces constraints on material flows in a detritus-based food web. Ecology 2007;88:2563–2575.
Crowl TA, Welsh V, Heartsill-Scalley T. Effects of different types of conditioning on rates of leaf-litter shredding by *Xiphocaris elongata*, a Neotropical freshwater shrimp. J N Am Benthol Soc 2006;25:198–208.
Cuffney TF, Wallace JB, Lugthart GJ. Experimental evidence quantifying the role of benthic invertebrates in organic matter dynamics of headwater streams. Freshw Biol 1990;23:281–299.
Cummins KW. Structure and function of stream ecosystems. BioScience 1974;24:631–641.
Cummins KW, Klug MJ. Feeding ecology of stream invertebrates. Ann Rev Ecol Syst 1979;10:147–172.
Dang CK, Schindler M, Chauvet E, Gessner MO. Temperature oscillation coupled with fungal community shifts can modulate warming effects on litter decomposition. Ecology 2009;90:122–131.
Danger M, Chauvet E. Elemental composition and degree of homeostasis of fungi: are aquatic hyphomycetes more like metazoans, bacteria or plants? Fung Ecol 2013;6:453–457.
Dangles O, Chauvet E. Effects of stream acidification on fungal biomass in decaying beech leaves and leaf palatability. Wat Res 2003;37:533–538.
Das M, Royer TV, Leff LG. Fungal communities on decaying leaves in streams: a comparison of two leaf species. Mycol Progress 2008;7:267–275.
De Lange HJ, Lürling M, Van Den Borne B, Peeters ETHM. Attraction of the amphipod *Gammarus pulex* to water-borne cues of food. Hydrobiologia 2005;544:19–25.
Duarte S, Pascoal C, Alves A, Correia A, Cássio F. Assessing the dynamic of microbial communities during leaf decomposition in a low-order stream by microscopic and molecular techniques. Microb Res 2010;165:351–362.
Duarte S, Fidalgo ML, Pascoal C, Cássio F. The role of the freshwater shrimp *Atyaephyra desmarestii* in leaf litter breakdown in streams. Hydrobiologia 2012;680:149–157.
Duarte S, Seena S, Bärlocher F, Pascoal C, Cássio F. A decade's perspective on theimpact of DNA sequencing on aquatic hyphomycete research. Fung Biol Rev 2013;27:19–24.
Elwood JW, Newbold JD, Trimble AF, Stark RW. The limiting role of phosphorus in a woodland stream ecosystem: effects of P enrichment on leaf decomposition and primary producers. Ecology 1981;62:35–44.
Evans KL, Greenwood JJD, Gaston KJ. Dissecting the species-energy relationship. Proc R Soc B 2005;272:2155–2163.
Fernandes I, Uzun B, Pascoal C, Cássio F. Responses of aquatic fungal communities on leaf litter to temperature-change events. Internat Rev Hydrobiol 2009;94:410–418.

Ferreira V, Graça MAS. Do invertebrate activity and current velocity affect fungal assemblage structure in leaves? Internat Rev Hydrobiol 2006;91:1–14.
Ferreira V, Gulis V, Graça MAS. Whole-stream nitrate addition affects litter decomposition and associated fungi but not invertebrates. Oecologia 2006;149:718–729.
Ferreira V, Gonçalves L, Pratas J, Canhoto C. Contamination by uranium mine drainages affects fungal growth and interactions between fungal species and strains. Mycologia 2010a;102:1004–1011.
Ferreira V, Gonçalves AL, Godbold DL, Canhoto C. Effect of increased atmospheric CO_2 on the performance of an aquatic detritivore through changes in water temperature and litter quality. Glob Change Biol 2010b;16:3284–3296.
Findlay SEG, Arsuffi TL. Microbial growth and detritus transformation during decomposition of leaf litter in a stream. Freshw Biol 1989;21:261–269.
Findlay S, Meyer JL, Smith PJ. Contribution of fungal biomass to the diet of a freshwater isopod (*Lirceus* sp.). Freshw Biol 1986a;16:377–385.
Findlay S, Meyer JL, Smith PJ. Incorporation of microbial biomass by *Peltoperla* sp. (Plecoptera) and *Tipula* sp. (Diptera). J N Am Benthol Soc 1986b;5:306–310.
Fleituch T. Effects of nutrient enrichment and activity of invertebrate shredders on leaf litter breakdown in low order streams. Internat Rev Hydrobiol 2013;98:1–8.
Friberg N, Jacobsen D. Feeding plasticity of two detritivore-shredders. Freshw Biol 1994;32:133–142.
Gessner MO, Chauvet E. Importance of stream microfungi in controlling breakdown rates of leaf litter. Ecology 1994;75:1807–1817.
Gessner MO, Chauvet E, Dobson M. A perspective on leaf litter breakdown in streams. Oikos 1999;85:377–384.
Gessner MO, Bärlocher F, Chauvet E. Qualitative and quantitative analyses of aquatic hyphomycetes in streams. Fung Div Res Ser 2003;10:127–157.
Godfrey, BES. Growth of two terrestrial microfungi on submerged alder leaves. Trans Br Mycol Soc 1983;81:418–421.
Gonçalves AL, Lírio AV, Pratas J, Canhoto C. Uranium contaminated water does not affect microbial activity but decreases feeding by the shredder *Sericostoma vittatum*. Fundam Appl Limnol 2011;179:17–25.
Gönczöl J, Révay A, Csontos P. Effect of sample size on the detection of species and conidial numbers of aquatic hyphomycetes collected by membrane filtration. Arch Hydrobiol 2001;150:677–691.
Gotelli NJ, Ellison AM. A Primer of Ecological Statistics. 2nd edition. Sinauer, Massachusetts: Sunderland 2013:614.
Graça MAS. The role of invertebrates on leaf litter decomposition in streams – a review. Internat Rev Hydrobiol 2001;86:383–393.
Graça MAS, Bärlocher F. Proteolytic gut enzymes in *Tipula caloptera* – interactions with phenolics. Aquat Insects 1998;21:11–18.
Graça MAS, Ferreira RCF. The ability of selected aquatic hyphomycetes and terrestrial fungi to decompose leaves in freshwater. Sydowia 1995;47:167–179.
Graça MAS, Maltby L, Calow P. Importance of fungi in the diet of *Gammarus pulex* and *Asellus aquaticus* I: Feeding strategies. Oecologia 1993a;93:139–144.
Graça MAS, Maltby L, Calow P. Importance of fungi in the diet of *Gammarus pulex* and *Asellus aquaticus* II. Effects on growth, reproduction and physiology. Oecologia 1993b;96:304–309.
Graça MAS, Maltby L, Calow P. Comparative ecology of *Gammarus pulex* (L.) and *Asellus aquaticus* (L.) II. Fungal preferences. Hydrobiologia 1994;281:163–170.
Graça MAS, Cressa C, Gessner MO, Feio MJ, Callies KA, Barrios C. Food quality, feeding preferences, survival and growth of shredders from temperate and tropical streams. Freshw Biol 2001;46:947–957.

Greenwood JL, Rosemond AD, Wallace JB, Cross WF, Weyer HS. Nutrients stimulate leaf breakdown rates and detritivorebiomass: bottom-up effects via heterotrophic pathways. Oecologia 2007;151:637–649.

Grimmett IJ, Shipp KN, MacNeil A, Bärlocher F. Does the growth rate hypothesis apply to aquatic hyphomycetes? Fung Ecol 2013;6:493–500.

Gulis V, Suberkropp K. Leaf litter decomposition and microbial activity in nutrient-enrichedand unaltered reaches of a headwater stream. Freshw Biol 2003;48:123–134.

Gulis V, Suberkropp K. Effects of whole-stream nutrient enrichment on the concentration and abundance of aquatic hyphomycete conidia in transport. Mycologia 2004;96:57–65.

Gulis V, Ferreira V, Graça MAS. Stimulation of leaf litter decomposition and associated fungi and invertebrates by moderate eutrophication: implications for stream assessment. Freshw Biol 2006;51:1655–1669.

Güsewell S, Gessner MO. N:P ratios influence litter decomposition and colonization by fungi and bacteria in microcosms. Funct Ecol 2009;23:211–219.

Hawkins BA. Ecology's oldest pattern? Endeavour 2001;25:133–134.

Hieber M, Gessner MO. Contribution of stream detritivores, fungi, and bacteria to leaf breakdown based on biomass estimates. Ecology 2002;83:1026–1038.

Hladyz S, Gessner MO, Giller PS, Pozo J, Woodward G. Resource quality and stoichiometric constraints on stream ecosystem functioning. Freshw Biol 2009;54:957–970.

Ho WH, Hyde KD, Hodgkiss J, Yanna A. Fungal communities on submerged wood from streams in Brunei, Hong Kong, and Malaysia. Mycol Res 2001;105:1492–1501.

Howe MJ, Suberkropp K. Effect of isopod (*Lirceus* sp.) feeding on aquatic hyphomycetes colonizing leaves in a stream. Arch Hydrobiol 1994;130:93–103.

Hynes, HBN. The food of fresh-water sticklebacks (*Gasterosteus aculeataus* and *Pygosteus pungitius*. J Anim Ecol 1950;19:36–58.

Hynes HBN. Imported organic matter and secondary productivity in streams. Proc. XVI Int Congr Zool Wash 1963;4:324–329.

Hynes HBN. The ecology of running waters. Toronto: University of Toronto Press 1970:555.

Ingold CT. Guide to aquatic hyphomycetes. Freshwater Biology Assoiation, Ambleside, Scientific Publication No. 30 1975:96.

Irons JG III, Oswood MW, Stout RJ, Pringle CM. Latitudinal patterns in leaf litter breakdown: is temperature really important? Freshw Biol 1994;32:401–411.

Iwai N, Kagaya T. Positive indirect effect of tadpoles on a detritivore through nutrient regeneration. Oecologia 2007;152:685–694.

Jabiol J, Bruder A, Gessner MO, Makkonen M, McKie BG, Peeters ETHM, et al. Broad-scale diversity patterns of leaf-associated aquatic hyphomycetes. Fung Ecol 2013;6:439–448.

Jenkins CC, Suberkropp K. The influence of water chemistry on the enzymatic degradation of leaves in a stream. Freshw Biol 1995;33:245–253.

Jones CG. Microorganisms as mediators of plant resource exploitation by insect herbivores. In: Price PW, Slobodchikoff CN, Gaud WS, eds. A New Ecology: Novel Approaches to Interactive Systems. New York, NY: Wiley & Sons 1984:53–99.

Kaushik NK, Hynes HBN. Experimental study on the role of autumn-shed leaves in aquatic environments. J Ecol 1968;56:229–243.

Kaushik NK, Hynes HBN. The fate of the dead leaves that fall into streams. Arch Hydrobiol 1971;68:465–515.

Kochi K, Kagaya T. Green leaves enhance the growth and developmentof a stream macroinvertebrate shredder when senescentleaves are available. Freshw Biol 2005;50:656–667.

Kostalos M, Seymour RL. Role of microbial enriched detritus in the nutrition of *Gammarus minus* (Amphipoda). Oikos 1976;27:512–516.

Krauss G, Bärlocher F, Schreck P, Wennrich R, Glässer W, Krauss G-J. Aquatic hyphomycetes occur in hyperpolluted waters in Central Germany. Nova Hedwig 2001;72:419–428.

Krauss G-J, Solé M, Krauss G, Schlosser D, Wesenberg D, Bärlocher F. Fungi in freshwaters – ecology, physiology and biochemical potential. FEMS Microbiol Rev 2010;35:620–651.

Lawson DL, Klug MJ. Microbial fermentation in the hindguts of two stream invertebrates. J N Am Benthol Soc 1989;8:85–91.

Lawson DL, Klug MJ, Merritt RW. The influence of the physical, chemical and microbiological characteristics of decomposing leaves on the growth of the detritivore *Tipula abdominalis* (Diptera: Tipulidae). Can J Zool 1984;62:2339–2343.

Lugthart GJ, Wallace JB. Effects of disturbance on benthic functional structure and production in mountain streams. J N Benthol Soc 1992;11:138–164.

Mackay RJ, Kalff J. Ecology of two related species of caddis fly larvae in the organic substrates of a woodland stream. Ecology 1973;54:499–511.

Maharning AR, Bärlocher F. Growth and reproduction in aquatic hyphomycetes. Mycologia 1996;88:80–88.

Mann KH. Production and use of detritus in various freshwater, estuarine, and coastal marine ecosystems. Limnol Oceanogr 1988;33:910–930.

Martin MM, Martin JS. Surfactants: their role in preventing the precipitation of proteins by tannins in insect guts. Oecologia 1984;61:342–345.

Martin MM, Martin JS, Kukor JJ, Merritt RW. The digestion of protein and carbohydrate by the stream detritivore, *Tipula abdominalis* (Diptera, Tipulidae). Oecologia 1980;46:360–364.

Martin MM, Martin JS, Kukor JJ, Merritt RW. The digestive enzymes of detritus-feeding stonefly nymphs (Plecoptera; Pteronarcyidae). Can J Zool 1981;59:1947–1951.

Martin MM, Kukor JJ, Martin JS, Merritt RW. The digestive enzymes of larvae of the black fly, *Prosimulium fuscum* (Diptera, Simuliidae). Comp Biochem Physiol 1985;82B:37–39.

Mathuriau C, Chauvet E. Breakdown of leaf litter in a neotropical stream. J N Am Benthol Soc 2002;21:384–396.

Mooshammer M, Wanek W, Schnecker J, Wild B, Leitner S, Hofhansl F, et al. Stoichiometric controls of nitrogen and phosphorus cycling in decomposing beech leaf litter. Ecology 2012;93:770–782.

Musgrove RJ. Digestive ability of the freshwater crayfish *Paranephrops zealandicus* (White) (Parastacidae) and the role of microbial enzymes. Freshw Biol 1988;20:305–314.

Newbold JD, Elwood JW, Schulze MS, Stark RW, Barmeier JC. Continuous ammonium enrichment of a woodland stream: uptake kinetics, leaf decomposition, and nitrification. Freshw Biol 1983;13:193–204.

Nikolcheva LG, Bärlocher F. Taxon-specific fungal primers reveal unexpectedly high diversity during leaf decomposition in a stream. Mycol Prog 2004;3:41–49.

Nikolcheva LG, Cockshutt AM, Bärlocher F. Determining diversity of freshwater fungi on decaying leaves: comparison of traditional and molecular approaches. Appl Environ Microbiol 2003;69:2548–2554.

Nikolcheva LG, Bourque T, Bärlocher F. Fungal diversity during initial stages of leaf decomposition in a stream. Mycol Res 2005;109:246–253.

Nilsson LM. Energy budget of a laboratory population of *Gammarus pulex* (Amphipoda). Oikos 1974;25:35–42.

Nolen JA, Pearson RG. Factors affecting litter processing by *Anisocentropus kirramus* (Trichoptera: Calamoceratidae) from an Australian tropical rainforest stream. Freshw Biol 1993;29:469–479.

Odum WE, Kirk PW, Zieman JC. Non-protein nitrogen compounds associated with particles ofvascular plant detritus. Oikos 1979;32:363–367.

Pascoal C, Cássio F. Contribution of fungi and bacteria to leaf litter decomposition in a polluted river. Appl Environ Microbiol 2004;70:5266–5273.

Pascoal C, Cássio F. Linking fungal diversity to the functioning of freshwater ecosystems. Fung Div Res Ser 2008;20:1–19.
Pascoal C, Cássio F, Marcotegui A, Sanz B, Gomes P. Role of fungi, bacteria, and invertebrates in leaf litter breakdown in a polluted river. J N Am Benthol Soc 2005;24:784–797.
Pérez J, Basaguren A, Descals E, Larrañaga, Pozo J. Leaf-litter processing in headwater streams of northern Iberian Peninsula: moderate levels of eutrophication do not explain breakdown rates. Hydrobiologia 2013;50:988–997.
Raviraja NS, Nikolcheva LG, Bärlocher F. Fungal growth and leaf decomposition are affected by amount and type of inoculum and by external nutrients. Sydowia 2006;58:91–104.
Rincón J, Martínez I. Food quality and feeding preferences of *Phylloicos* sp. (Trichoptera: Calamoceratidae). J N Am Benthol Soc 2006;25:209–215.
Robinson CT, Gessner MO. Nutrient addition accelerates leaf breakdown in an alpine springbrook. Oecologia 2000;122:258–263.
Rong Q, Sridhar KR, Bärlocher F. Food selection of three leaf-shredding invertebrates. Hydrobiologia 1995;316:173–181.
Roussel H, Chauvet E, Bonzom J-M. Alteration of leaf decomposition in copper-contaminated freshwater mesocosms. Env Toxic Chem 2008;27:637–644.
Rosenzweig ML. Species diversity. In: McGlade J, ed. Advanced Ecological Theory. Oxford: Blackwell 1999:249–281.
Rossi L, Fano EA. Role of fungi in the trophic niche of two congeneric detritivorous isopods, *Aselllus aquaticus* and *A. coxalis* (Isopoda). Oikos 1979;32:380–385.
Rossi L, Vitagliano-Tadini G. Role of adult feces in the nutrition of larvae of *Asellus aquaticus* (Isopoda). Oikos 1978;30:109–113.
Ruedo-Delgado G, Wantzen KM, Tolosa MB. Leaf-litter decomposition in an Amazonian floodplain stream: effects of seasonal hydrological changes. J N Am Benthol Soc 2006;25:231–247.
Seena S, Wynberg N, Bärlocher F. Fungal diversity during leaf decomposition in a stream assessed through clone libraries. Fungal Divers 2008;30:1–14.
Schoenlein-Crusius IH, Grandi RAP. The diversity of aquatic hyphomycetes in South America. Braz J Microbiol 2003;34:183–193.
Shearer CA, Lane L. Comparison of three techniques for the study of aquatic hyphomycete communities. Mycologia 1983;75:498–508.
Shearer CA, Descals E, Kohlmeyer B, Kohlmeyer J, Marvanová L, Padgett D, et al. Fungal biodiversity in aquatic habitats. Biodivers Conserv 2007;16:49–67.
Sinsabaugh RL, Benfield EF, Linkens AE. Cellulase activity associated with the decomposition of leaf litter in a woodland stream. Oikos 1981;36:184–190.
Sridhar KR. Freshwater hyphomycetes. In: Sridhar KR, ed. Frontiers in Fungal Ecology, Diversity and Metabolites. New Delhi: Intern Publ House 2009:1–16.
Sridhar KR, Bärlocher F. Initial colonization, nutrient supply, and fungal activity on leaves decaying in streams. Appl Environ Microbiol 2000;66:1112–1119.
Sridhar KR, Sudheep NM. Do the tropical freshwater fishes feed on aquatic fungi? Front Agric China 2011;5:77–86.
Sridhar KR, Karamchand KS, Hyde KD. Wood-inhabiting filamentous fungi in 12 high altitude streams of the Western Ghats by damp incubation and bubble chamber incubation. Mycoscience 2011;51:104–115.
Sridhar KR, Beaton M, Bärlocher F. Fungal propagules and DNA in feces of two detritus-feeding amphipods. Microb Ecol 2011;61:31–40.
Sridhar KR, Krauss G, Bärlocher F, Raviraja NS, Wennrich R, Baumbach R, et al. Decomposition of alder leaves in two heavy metal-polluted streams in central Germany. Aquat Microb Ecol 2001;26:73–80.

Sterner RW, Elser JJ (2002) Ecological stoichiometry: the biology of elements from molecules to the biosphere. Princeton: Princeton Univ Press 2002:439.

Stout RJ, Taft WH, Merritt RW. Patterns of macroinvertebrate colonization on fresh and senescent alder leaves in two Michigan streams. Freshw Biol 1985;15:573–580.

Suberkropp K. Relationships between growth and sporulation of aquatic hyphomycetes on decomposing leaf litter. Mycol Res 1991;95:843–850.

Suberkropp K. Interactions with invertebrates. In: Bärlocher F, ed. The Ecology of Aquatic Hyphomycetes. Berlin: Springer 1992:118–134.

Suberkropp K. The influence of nutrients on fungal growth, productivity, and sporulation during leaf breakdown in streams. Can J Bot 1995;73:S1361–S1369.

Suberkropp K. Methods for examining interactions between freshwater fungi and macroinvertebrates. Fung Div Res Ser 2003;10:159–171.

Suberkropp K, Arsuffi TL. Degradation, growth and changes in palatability of leaves colonized by six aquatic hyphomycetes. Mycologia 1984;76:398–407.

Suberkropp K, Chauvet E. Regulation of leaf breakdown by fungi in streams: influences of water chemistry. Ecology 1995;76:1433–1445.

Suberkropp KF, Klug MJ. Decomposition of deciduous leaf litter in a woodland stream. I. A scanning electron microscopic study. Microb Ecol 1974;1:96–103.

Suberkropp K, Klug MJ. Fungi and bacteria associated with leaves during processing in a woodland stream. Ecology 1976;57:707–719.

Suberkropp K, Klug MJ. The maceration of deciduous leaf litter by aquatic hyphomycetes. Can J Bot 1980;58:1025–1031.

Suberkropp K, Wallace JB. Aquatic hyphomycetes in insecticide-treated and untreated streams. J N Am Benthol Soc 1992;11:165–171.

Suberkropp K, Weyers H. Application of fungal and bacterial production methodologies to decomposing leaves in streams. Appl Environ Microbiol 1996;62:1610–1615.

Suberkropp K, Godshalk GL, Klug MJ. Changes in the chemical composition of leaves during processing in a woodland stream. Ecology 1976;57:720–727.

Suberkropp K, Arsuffi TL, Anderson JP. Comparison of degradative ability, enzymatic activity, and palatability of aquatic hyphomycetes grown on leaf litter. Appl Environ Microbiol 1983;46:237–244.

Suberkropp K, Gulis V, Rosemond AD, Benstead JP. Ecosystem and physiological scales of microbial responses to nutrients in a detritus-based stream: Results of a 5-year continuous enrichment. Limnol Oceanogr 2010;55:149–160.

Sutcliffe DW, Carrick DR, Willoughby LG. Effects of diet, body size, age and temperature on growth rates in the amphipod *Gammarus pulex*. Freshw Biol 1981;11:183–214.

Thomas K. The ecology of aquatic hyphomycetes in an Australian upland stream. Ph.D. thesis, Canberra, Australia: Australian National University 1992:215.

Treton C, Chauvet E, Charcosset JY. Competitive interaction between two aquatic hyphomycete species and increase in leaf litter breakdown. Microb Ecol 2004;48:439–446.

Triska FJ. Seasonal distribution of aquatic hyphomycetes in relation to the disappearance of leaf litter from a woodland stream. Ph.D. Thesis, University of Pittsburg 1970:189.

Villanueva VD, Albariño R, Canhoto C. Positive effect of shredders on microbial biomass and decomposition in stream microcosms. Freshw Biol 2012;57:2504–2513.

Waide RB, Willig MR, Steiner CF, Mittelbach G, Glugh L, Dodson SI, et al. The relationship between productivity and species richness. Ann Rev Ecol Syst 1999;30:257–300.

Wantzen KM, Wagner R. Detritus processing by invertebrate shredders: a neotropical-temperate comparison. J N Am Benthol Soc 2006;25:216–232.

Wantzen KM, Ramírez A, Winemiller KO. New vistas in Neotropical stream ecology – preface. J N Am Benthol Soc 2006;25:61–65.

Wantzen KM, Yule CM, Mathooko JM, Pringle CM. Organic matter processing in tropical streams. In: Dudgeon D, ed. Tropical Stream Ecology. San Diego: Academic Press 2008:43–64.
Webster JR, Benfield EF. Vascular plant breakdown in freshwater ecosystems. Ann Rev Ecol Syst 1986;17:567–594.
Willoughby LG, Sutcliffe DW. Experiments on feeding and growth of the amphipod *Gammarus pulex* (L.) related to its distribution in the River Duddon. Freshw Biol 1976;6:577–586.
Woodward G, Perkins DM, Brown LE. Climate change and freshwater ecosystems: impacts across multiple levels of organization. Phil Trans R Soc B 2010;365:2093–2106.
Woodward G, Gessner MO, Giller PS, Gulis V, Hladyz S, Lecerf S, et al. Continental-scale effects of nutrient pollution on stream ecosystem functioning. Science 2012;336:1438–1440.
Wright DH. Species-energy theory: an extension of species-area theory. Oikos 1983;41:496–506.
Yeates LV, Barmuta A. The effects of willow and eucalypt leaves on feeding preference and growth of some Australian aquatic macroinvertebrates. Aust J Ecol 1999;24:593–598.
Yule CM, Leong MY, Liew C, Ratnarajah L, Schmidt K, Wong HM. Shredders in Malaysia: abundance and richness are higher in cool upland tropical streams. J N Am Benthol Soc 2009;28:404–415.

Diego Libkind, Gabriel Russo and María Rosa van Broock

20 Yeasts from extreme aquatic environments: hyperacidic freshwaters

20.1 Introduction

Yeasts are unicellular fungi included in the phyla Ascomycota and Basidiomycota (See Chapter 4). They are heterotrophic microorganisms typically involved in the degradation of organic compounds, which are made available mainly by the metabolic activity of bacteria and other fungi (Spencer and Spencer 1997). Yeasts are ubiquitous, having been isolated from numerous environments including soil, animals, phylloplane, and other plant-associated habitats (Hagler and Ahearn 1987; Lachance and Starmer 1998; Rosa and Ganter 2006).

Moreover, their unicellular type of growth and saprophytic nature make yeasts particularly suited for colonizing and exploiting liquid substrates, or habitats with high moisture contents (Lachance and Starmer 1998); thus, they are common inhabitants of marine and freshwater environments (Hagler and Mendonça-Hagler 1981). Some yeasts, known as extremophiles, can even colonize extreme aquatic habitats such as the deep sea (Nagahama et al. 2001; Gadanho and Sampaio 2005), glacial melt-waters (Vishniac 2005; de Garcia et al. 2007), hypersaline lakes (Butinar et al. 2005) and ultra-oligotrophic mountain lakes exposed to increased UV radiation (Libkind et al. 2009). Yeasts have also been found to inhabit both natural and anthropic hyperacidic (pH <3) aquatic environments such as São Domingos mines (Gadanho and Sampaio 2006; Gadanho et al. 2006), river Tinto (López-Archilla et al. 2001; Gadanho et al. 2006), Smoking Hills tundra ponds (Nakatsu and Hutchinson 1988), hot spring running waters (Nguyen et al. 2001), leachate of uranium mineral heaps (de Siloniz et al. 2002), acid drainage from copper mines (Ehrlich 1963), and sediment from a metal mine (Holland et al. 2011). Yeasts living in these environments are often polyextremophiles, i.e. they tolerate many different extreme conditions (i.e. low pH, high temperatures, osmotic pressures, high concentration of heavy metals resulting from rock weathering caused by the acid waters and chemolithotrophic microbial activity).

Although, as shown above, occurrence of yeasts in acidic aquatic environments has been well documented, systematic approaches to study cultivable yeast diversity in such habitats are scarce, and mainly restricted to isolation and identification of strains from ponds at the São Domingos mines in Portugal (Gadanho et al. 2006) and the river Tinto in Spain (López-Archilla et al. 2004; Gadanho et al. 2006), two acidic systems originated by past mining activities. As well, Russo et al. (2008) studied yeast diversity and ecology of the acidic River Agrio-Lake Caviahue environment.

However, these kinds of studies are scarce and limited to a few environments. Very little is known for naturally occurring acidic lakes and rivers, in particular those of volcanic origin.

The study of microbial diversity in extreme environments, such as acid and/or heavy metal contaminated water, is important from both scientific and biotechnological approaches. These investigations contribute to expand our knowledge on the planet's microbial diversity, in particular from environments which are considered to have been predominant in the early stages of the planet development (Johnson 1998); the study and comprehension of organisms isolated from these locations could help to elucidate the development of archaic life forms. Explorations of microbial biodiversity of extreme environments may also reveal new biocatalysts, such as extremozymes and/or extremolytes, or inspire novel bioremediation strategies (Sharp and Munster 1986; Herbert 1992; Gupta et al. 2000; Watanabe et al. 2002; Lentzen and Schwarz 2006; Buzzini et al. 2012; Gabani and Singh 2013). Furthermore, yeasts isolated from these unfriendly habitats may serve as models for physiological and genetic research, aiding the elucidation of mechanisms involved in eukaryotic cell tolerance to low pH and heavy metals.

The present chapter focuses on the occurrence and diversity of yeasts in naturally originating acidic aquatic systems, with special attention to the River Agrio and the Lake Caviahue (Neuquén, Argentina). A comparison with similar studies in anthropogenically acidified water bodies is also carried out, as well as discussion on the physiological aspects and ecological implications of the yeasts from these extreme environments.

20.2 The River Agrio-Lake Caviahue acidic aquatic system

The River Agrio-Lake Caviahue system (RAC) is a complex environment composed by the upper River Agrio (URA), the Lake Caviahue (LC) and the lower River Agrio (LRA). Lake Caviahue has a horseshoe-like shape and collects water from the URA at the Southeast shore. The lake has a second tributary, the River Dulce (RD), which pours its water on the Northeast shore of the lake. This river has a near neutral pH and its physicochemical characteristics are completely different from the rest of the system. The lake drains into the lower River Agrio (LRA), where pH gradually raises to neutrality ($pH \approx 7$) due the inflow of many tributaries into the river several kilometers away from the outflow of the lake.

The RAC gradually increases its pH from ~1.5 in the URA source to near neutrality a few kilometers after the lake in the LRA (Pedrozo et al. 2001). These water bodies were sampled at seven sites, each with different pH values. Sampling site locations, pH values and site name abbreviations are depicted in Fig. 20.1.

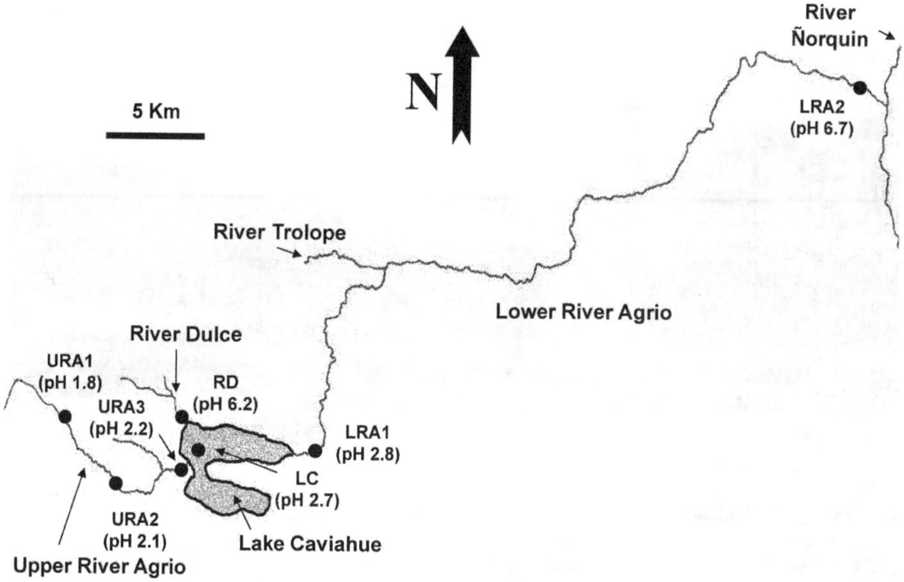

Fig. 20.1: Location and pH values of the sampling sites studied in River Agrio-Lake Caviahue aquatic system.

20.2.1 Yeast occurrence

All the samples, even those with the lowest pH (pH = 1.8), contained viable yeasts in variable numbers (Fig. 20.2). In the URA, yeast densities, as well as the water pH, increased with distance from the source (68 to 713 cells/L). In the RD, a non-acidic river, yeast count was 107 cells/L. The number of yeast cells dropped to 62 cells/L in the LC. In the LRA yeast counts raised again up to values of 825 cells/L. The cell number reduction observed in LC has also been reported for bacteria (Wendt-Potthoff and Koschorreck 2002). Possibly, cells entering the lake precipitate thus remaining mainly in the lake sediments, where yeast quantitative studies have not yet been carried out. Boguslawska-Was and Dabrowski (2001) observed higher yeast counts in sediments than in surface waters of a non-acidic lake, a fact that reinforces this hypothesis. The low number of yeasts observed could also be related to the nutrient concentrations in the URA and the LC. Data obtained simultaneously with the present yeast study indicated that most macro-nutrients (phosphorous, nitrogen) and micro-nutrient concentrations (sulfur, zinc, nickel, manganese) were approximately 10-fold higher in the URA than in the LC (Baffico 2006). Reduced availability of nutrients could affect yeast propagation and survival in the lake, being a probable explanation for the low density of yeasts observed. A dilution effect due to the volume of water of the lake must not be discounted.

Fig. 20.2: Photograph of the Gigant's Cascade, River Agrio (Cascada del Gigante), Neuquén, Patagonia, Argentina. Petri dishes with growing yeast colonies on filters from the upper river Agrio samples (URA3) using conventional culture medium with pH adjusted to 3 (left) or the same medium prepared with URA3 water (right).

A comparison of yeast cell count data from RAC with that of other acidic and non-acidic aquatic environments was performed (Tab. 20.1). Despite the stressing conditions of the acidic aquatic environments, yeast densities at RAC are comparable to or even sometimes higher than other near-neutral or neutral water bodies from Patagonia. It is worth noting that the highest yeast cell values were observed by Gadanho et al. (2006) for the São Domingos mines (over 14,000 cells/L). However, in that study yeast counts varied significantly in different times of the year.

20.2.2 Yeast diversity

Our results indicated that the RAC yeast flora is dominated by basidiomycetous species. These fungi comprised 84 percent of the total isolates. The predominance of this group is not surprising, as in general, basidiomycetous yeasts are more nutritionally versatile, and tolerate harsher environmental conditions than ascomycetous yeasts (Sampaio 2004). A low proportion of ascomycetous yeasts have been also

Tab. 20.1: Yeast occurrence in acidic and non acidic aquatic systems.

Aquatic environment	Characteristics	Origin	pH	Yeast density (cells/L)*
Caviahue-Agrio system				
Upper river Agrio	Volcanic	Natural	1.8–2.2	68–713 (324)
Lake Caviahue	Volcanic	Natural	2.7	65
Patagonian lakes and rivers				
Ultra-oligotrophic lakes	Glacial	Natural	6.5–7.5	2–150 (35)
Mountain rivers	Glacial	Natural	6.8–7.2	130–162 (146)
Iberian Pyrite Belt				
Sao Domingos mines	Mine ponds	Anthropic	1.8–2.9	0–14.2 (2.1)
River Tinto	Acid rock drainage contaminated	Anthropic	2.4–2.5	17–827 (270)

* All data was obtained with the same culture medium. Cell density values indicate ranges and numbers in brackets indicate average values.

found in anthropogenically originated acidic environments such as the river Tinto (López-Archilla et al. 2004).

Yeast identification was performed employing a polyphasic approach as described in Russo et al. (2008), which included phenotypic (morphology, sexuality and assimilation tests) and genotypic characteristics (PCR fingerprinting and rDNA sequencing). Eight genera and 34 yeast species were found in water samples from different sites at RAC. The list of isolated yeast species, and their distribution along the sampling sites is presented here (Tab. 20.2). Members of the heterogeneous genera *Cryptococcus* and *Rhodotorula* were, by far, the most abundant yeasts, collectively comprising over 90 percent of the isolates. Less frequent genera were *Candida* (20 isolates), *Rhodosporidium* (17), *Cystofilobasidium* (2), *Sporobolomyces* (1), and an unclassified yeast related to Microbotryomycetous genus *Bauerago* (2).

Among the 34 species found, 11 could not be assigned to any known yeast species, given that showed enough genetic differences (more than 1% divergence in LSU rRNA) with the closest known species to be considered undescribed taxa. Novel species were named by their genus followed by a consecutive number when more than one species for a given genus were found (Tab. 20.2).

The most abundant yeast species in RAC was *Cryptococcus agrionensis* (Russo et al. 2009), a novel species phylogenetically related to *Cryptococcus ibericus* and the Acid Rock Drainage (ARD) ecoclade described for a group of yeasts species found in S. Domingos mines (Gadanho and Sampaio 2009). The implications of this ecoclade will be discussed further in this chapter. *Cryptococcus agrionensis* also had the widest distribution, being present at all six sampling sites of the aquatic system (Tab. 20.2), though most of the isolates were obtained in the URA (pH = 1.8–2.2).

The second most abundant and widely distributed species was *Rhodotorula mucilaginosa*. With the exception of the most acidic site (URA1; pH = 1.8), this yeast was

Tab. 20.2: Yeast diversity at the Caviahue-Agrio aquatic system.

Yeast species/Sampling sites	URA1 (pH 1.8)	URA2 (pH 2.1)	URA3 (pH 2.2)	LC (pH 2.7)	LRA1 (pH 2.8)	LRA2 (pH 6.7)	RD (pH 6.2)
*Cryptococcus agrionensis**	7	23	61	3	3	3	
Cryptococcus albidisimilis							
Cryptococcus antarcticus	14	25	14	51			
Cryptococcus cylindricus							
Cryptococcus laurentii							
Cryptococcus saitoi							
Cryptococcus sp. 2*							
Cryptococcus sp. 3*							
Cryptococcus sp. 4*			34	66			
Cryptococcus victoriae				10		90	
Rhodotorula colostri							
Rhodotorula mucilaginosa		23	63	3	4	7	
Rhodotorula sp. 1*		3		3	81	13	
Rhodotorula sp. 2*							
Rhodotorula sp. 3*							
Rhodosporidium sp.*							
Rhodosporidium babjevae							
Rhodosporidium toruloides	7	86	7				
Holtermanniella festucosa						50	50
Cystofilobasidium capitatum							
Cystofilobasidium macerans							
Candida austromarina							
Bauerago sp.*							
Sporobolomyces aff. *roseus*							
Total number of species	4	9	5	7	5	11	2
Number of novel species	1	6	2	3	3	3	0

Numbers indicate percentage of strains of the respective species isolated in the corresponding sampling site. Light-grey bars indicate species only recovered from non-acidic sampling sites.* Novel species.

present in all RAC sampling sites (Tab. 20.2). The isolation of *Rh. mucilaginosa* from the acid waters of RAC was not surprising, it has previously been isolated from acidic aquatic environments (López-Archilla et al. 2004; Gadanho et al. 2006) as well as other extreme environments (de Siloniz et al. 2002; Lahav et al. 2002; Libkind et al. 2004). This species is considered ubiquitous and displays a large phenotypic plasticity (Fell and Statzell-Tallman 1998). Libkind et al. (2003) found that *Rh. mucilaginosa* was the most frequent pigmented (carotenoid producing) species in aquatic environments from northern Patagonia, which was also the case of RAC.

Rhodotorula sp. 1, a species closely related to *Rhodosporidium sphaerocarpum*, was also found in high numbers, and had a relatively wide distribution, being found in four out of six studied sites (Tab. 20.2). However, *Rhodotorula* sp. 1 was mostly present in the acidic water samples of the LRA, with only a few isolates in URA and LC. *Rhodotorula* sp. 1 was firstly considered an autochthonous yeast of acidic environments, since it was isolated in high numbers from both São Domingos and RAC environments, showing in the latter a wide distribution (it was present in four out of six studied sites; Tab. 20.2). However, recently Huang et al. (2011) isolated a single strain of the same species from phylloplane in Taiwan, and described it as *Rhodotorula taiwanensis*. Nevertheless, this species remains of particular interest since its close relationship with acidic aquatic environments, and further studies should be carried out in order to elucidate its metabolic features that make it adapted to acidic environments.

Next, in order of isolation frequency, was *Rhodosporidium toruloides*, a species apparently exclusive of the URA, and as will be explained later in this chapter, with a marked preference for growth in culture media amended with river Agrio water.

Other frequent species were *Cryptococcus antarcticus* and *Cr. victoriae*, both with a similar number of isolates, though the former was present in the URA, and the latter was present in the LRA. A few strains of both species were also found in LC (Tab. 20.2).

Seven strains of *Cryptococcus* sp. 2, which is distantly related to *Cr. skinneri*, were recovered from URA2 (pH = 2.1). These strains exhibited a quite remarkable property: pure cultures did not grow on regular Malt extract, Yeast extract, soya Peptone (MYP) culture media (pH = 5), but did so on acidified MYP (pH = 3). A yeast displaying the same preference for acidified media was isolated by Gadanho et al. (2006) from São Domingos mines. Interestingly rDNA sequence analysis revealed that both River Agrio and São Domingos "acidophilic" isolates belong to the same species. To our knowledge, the requirement of acidified culture media for growth has not been described for any other yeast species in the past.

Among ascomycetous yeasts *Candida austromarina* was the most frequently represented one (10 isolates). Even though one isolate was recovered from the most acidic sampling site (pH 1.8), the remaining isolates were obtained from the last sampling site of the river (pH 2.2). This sampling site is the most disturbed one since it is close to the village area and touristic paths. Ascomycetous yeasts in aquatic environments are often related to pollution (Almeida 2005). It is possible that the presence of the species reported here is the result of anthropogenic rather than natural origin. The rest of the species found were less frequent, with one to three isolates obtained for each (Tab. 20.2).

It is interesting to note that clear differences in yeast diversity were found between the acidic and near-neutral sampling sites of the RAC. While 12 species were isolated only at pH < 3 from the former, seven species were recovered in the same conditions from the latter (Tab. 20.2, bars with light grey shade). Only four species were common for both the acidic and near-neutral sampling sites of the RAC: *Cryptococcus agrionensis*, *Rhodotorula* sp. 1, *Rh. mucilaginosa* and *Cr. victoriae*. It is also worth noting, that while six undescribed species (out of 12) were detected in the acidic part of the river, only one novel species (out of 7 exclusive species) was found in the near-neutral sampling site. With the exception of the last species, all yeast taxa found exclusively in the near-neutral part of the RAC, are common to other non-acidic lakes and rivers from northwestern Patagonia (Tabs. 20.2, 20.3).

In summary, a specific yeast community, adapted to the RAC acidic conditions was found. As the river's pH gradually raises, and environmental conditions become less extreme, highly acid-tolerant yeast species are replaced by less extremophilic ones. Finally, as water pH reaches near-neutral values, the yeast flora changes remarkably, becoming similar to that of other non-acidic rivers and lakes of the Patagonian region (see Tabs. 20.2, 20.3).

Tab. 20.3: Systematic classification of yeasts isolated from the Caviahue-Agrio aquatic system and comparison with other aquatic environments.

Species found in the Caviahue-Agrio aquatic system	São Domingos and R. Tinto	Glacial lakes and rivers from Patagonia
Ascomycetes		
Candida austromarina	–	+
Candida fluviatilis	+	–
Candida pseudolambica	+	–
Lecythophora sp.	+	–
Basidiomycetes		
Pucciniomycotina		
Rhodotorula colostri	–	+
Rhodosporidium babjevae	–	+
Rhodotorula mucilaginosa	+	+
Rhodosporidium toruloides	+	–
Rhodosporidium sp.	–	+
Rhodotorula sp. 1 (*taiwanensis*)	+	–
Rhodotorula sp. 2	–	–
Rhodotorula sp. 3	–	–
Bauerago sp.	–	–
Sporobolomyces aff. *roseus*	–	+

(*continued*)

Tab. 20.3: *(continued)*

Species found in the Caviahue-Agrio aquatic system	São Domingos and R. Tinto	Glacial lakes and rivers from Patagonia
Agaricomycotina		
Cryptococcus agrionensis	–	–
Cryptococcus albidisimilis	+	–
Cryptococcus antarcticus	–	+
Cryptococcus cylindricus	–	+
Cryptococcus laurentii	–	+
Cryptococcus saitoi	–	+
Cryptococcus victoriae	–	+
Cryptococcus sp. 2	+	–
Cryptococcus sp. 3	–	–
Cryptococcus sp. 4	–	–
Holtermanniella festucosa	–	+
Cystofilobasidium macerans	–	+
Cystofilobasidium capitatum	–	+

20.3 Comparative yeast diversity study between RAC and the acidic environments of the Iberian Pyrite Belt (IPB)

Yeasts do not occur randomly throughout the biosphere. On the contrary, yeasts form communities of species defined by its habitat which is the sum of all physical, chemical and biotic factors required for their successful existence (Lachance and Starmer 1998). For yeast diversity studies, it is important to distinguish among species found in a habitat, those that are essential components of the community, from those that are transient or present fortuitously. The former are termed autochthonous and the latter allochthonous. The comparison of the yeast diversity occurring in physico-chemically similar habitats may shed light on to the autochthonous or allochthonous nature of the species found. The comparison of yeast communities from different acidic environments is also of interest for the detection of highly adapted species, important for physiological and genetic studies aiming at the elucidation of mechanisms that allow the colonization of such extreme environments.

To precisely assess yeast diversity in a given environment, an accurate identification of the isolates must be obtained. For yeast identification, this task can only be accomplished by using molecular techniques (PCR based fingerprinting methods, plus DNA sequencing analysis). Unfortunately, yeast diversity studies in acid aquatic environments employing this approach are scarce, mainly reduced to the report of Gadanho et al. (2006), who surveyed the river Tinto (Spain), and the São Domingos mines (Portugal) and that of Russo et al. (2008) carried out in the River Agrio-Lake Caviahue acidic system. The former two environments are located at the Iberian Pyrite Belt (IPB), and as previously discussed; both are a consequence of human activities. The last one is a naturally acidified aquatic environment created by the activity of the Copahue volcano.

The comparison of yeast diversity at RAC and IPB acidic environments shows that at least eight yeast species (*Candida fluviatilis*, *C. pseudolambica*, *Cryptococcus albidisimilis*, *Cryptococcus* sp. 2, *Lecythophora* sp., *Rhodotorula mucilaginosa*, *Rhodotorula* sp. 1 and *Rhodosporidium toruloides*) are present in both extreme habitats.

Occurrence of *Rh. mucilaginosa* in both acid aquatic systems is undoubtedly related to its ubiquitous nature and to its adaptability to extreme environments, which were addressed earlier in this chapter. *Cryptococcus albidisimilis*, a less ubiquitous species, had a low frequency of occurrence both in RAC and in IPB. This is also the case for *Candida fluviatilis* and *C. pseudolambica*.

Gadanho et al. (2006) classified São Domingos mines species in terms of the environmental stress level to which they were associated. Thus, all species isolated from IPB aquatic system were sorted into three groups of high, intermediate or low environmental stress. According to this hierarchical organization, *Cr. albidisimilis* was found in microhabitats of low environmental stress. The strain isolated in the RAC could be also considered of low environmental stress, using a similar criterion. These results together with the fact that it has been previously isolated from other non-acidic habitats (Vishniac and Kurtzman 1992), suggest that *Cr. albidisimilis* might be an allochthonous member of acidic aquatic systems.

Another common species to both acidic aquatic environments was *Rh. toruloides*. Gadanho et al. (2006) associated this species to intermediate/high environmental stress. These authors reported that isolates could only be recovered when culture media amended with water from the sampling site were employed. As mentioned in the previous section, this situation was also encountered for the RAC-isolated strains. The hierarchic organization of the community structure from RAC showed *Rh. toruloides* as one of the most adapted yeasts in this environment (Russo et al. 2008). López-Archilla et al. (2004) and Gadanho et al. (2006) suggested that culture media amended with water from the studied environment would be more suitable for the isolation of autochthonous yeasts. These results suggest that these *Rh. toruloides* strains are well adapted to the extreme environments of RAC and IPB, and support the assumption that this species is an autochthonous member of the yeast community in these acidic aquatic environments.

The three remaining yeasts species common to both RAC and IPB acidic environments are undescribed ones, as indicated by the phylogenetic analysis of rDNA sequences (Fig. 20.3). The first common novel species is *Rhodotorula* sp. 1 (species 2 according to Gadanho et al. (2006) or *R. taiwanensis* according to Huang et al. (2011) which, as previously described, showed a high frequency of isolation from RAC and also a wide distribution. However, more than 80% of the isolates of this species were recovered at pH 2.8 (Tab. 20.2), and not at the lower pH values of the URA. Interestingly, Gadanho et al. (2006) also suggested this species to be more associated to lower environmental stress levels, in comparison with other species. Nevertheless, the authors considered it as a probable autochthonous species of the IPB acidic

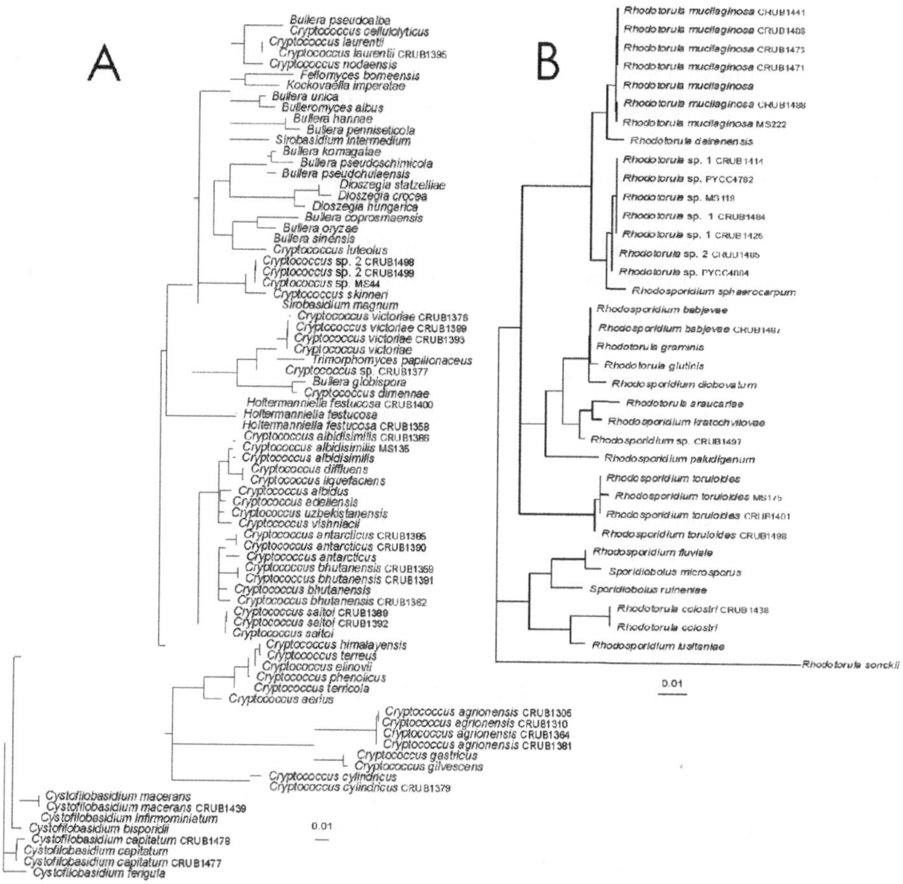

Fig. 20.3: Phylogenetic tree of isolated basidiomycetous yeasts constructed with 26S rDNA partial sequences of the D1/D2 domain. Phylogenetic analysis was performed using PAUP*4.0 version b8 software (Sinauer Associates, Inc. Publishers. Sunderland, MA, USA), the Neighbor-joining method and the Kimura two-parameter algorithm. The scale indicates the number of substitutions accumulated every 100 nucleotides. (**A**) Agaricomycotina yeasts, (**B**) Pucciniomycotina yeasts. Strains with CRUB (Centro Regional Universitario Bariloche culture collection) and MSD code correspond to RAC and São Domingos mines isolates, respectively. Species names without a code correspond to the sequence of the type strain (sequences retrieved from GenBank), except for *S. magnum* and *T. papilionaceus*.

aquatic system, which could be also true for the RAC. Two other conspecific strains (PYCC 4782 and PYCC 5597) from non-aquatic environments are known (Gadanho and Sampaio 2002). The relationship between these strains and those inhabiting extreme acidic waters is unknown.

The second novel species common to both acidic aquatic systems RAC and IPB, and undoubtedly the most interesting one is *Cryptococcus* sp. 2 (designated as *Cryptococcus* sp. 5 according to Gadanho et al. 2006). Based on their findings in the São

Domingos mines, Gadanho et al. (2006) classified this species at the highest level of their environmental stress hierarchic organization. Similar results were obtained for the RAC strains. This, together with the fact that the only known habitats for this species are acidic aquatic environments (anthropogenic or natural), support the idea that *Cryptococcus* sp. 2 is highly adapted to acidic conditions, and may be considered an acidophilic yeast.

The third novel species common to both acidic aquatic environment is *Lecythophora* sp. According to Gadanho et al. (2006) this species is adapted to high environmental stress. One remarkable characteristic of strains isolated both from the IPB and the RAC is their high tolerance to metals, in particular to copper (Cu^{2+}). The isolates from RAC were obtained only when a selective process using culture media amended with metal ions was employed. The high tolerance to metals displayed by *Lecythophora* sp. isolates makes it a good candidate to be considered as autochthonous yeast of acidic aquatic environments.

In summary, several common species to RAC and IPB acidic aquatic environments, either allochthonous or autochthonous acid-dwelling ones were isolated, and they are of special interest for future physiological and genetic studies.

20.4 Acidic rock drainage (ARD) yeasts ecoclade

Gadanho and Sampaio (2009) noticed that three novel yeasts species, namely *Cryptococcus ibericus*, *Cr. metallitolerans* and *Cr. aciditolerans*, although being phylogenetically distantly related, conformed a clade clearly separated from the *Cryptococcus gastricus* clade (Order Filobasidiales), as mentioned previously. The authors named this particular arrangement as the acid rock drainage ecoclade (ARD Ecoclade). An ecoclade is defined as a group of phylogenetically related species that inhabit physicochemically similar environments and show evident physiological adaptations to such environments. A fourth species, *Cryptococcus agrionensis*, was added later to the ARD Ecoclade (Russo et al. 2009) which is phylogenetically related to the three above mentioned species and was isolated from RAC (Fig. 20.4). The new member of the ARD ecoclade, although sharing genetic and physiological features with the previous members, comes from a geographically distant acidic environment (South America vs. Europe), providing further support of the autochthonous nature of this group of species to acidic environments.

Among the basidiomycetous yeasts an additional ecoclade composed exclusively of psychrophilic yeasts can also be defined, which includes the species *Rhodotorula glacialis*, *Rh. psychrophila* and *Rh. psychrophenolica* isolated from alpine glacial environments (Margesin et al. 2007) and *Rh. himalayensis* isolated from mountainous areas of Himalaya (Shivaji et al. 2008).

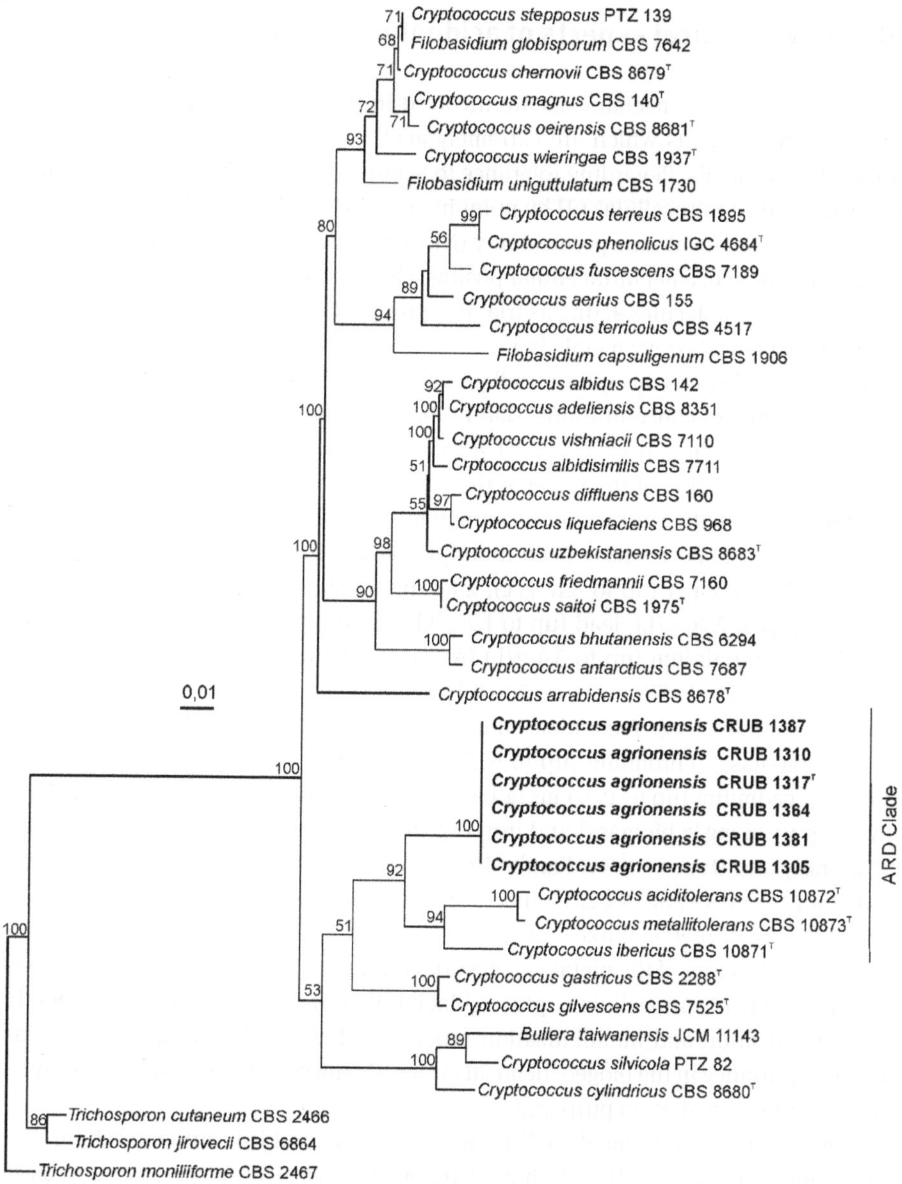

Fig. 20.4: Phylogenetic tree of the order Filobasidiales (Agaricomycotina) including the *Cryptococcus* ARD Ecoclade inferred using the Neighbor-Joining method based on an alignment of nuclear DNA sequences from the D1/D2 region of the 26S rRNA gene. Consensus tree was rooted with *Trichosporon moniliiforme*. Branch lengths are in the units of the number of changes over the whole sequence. Numbers on the branches are bootstrap values (1,000 replicates).

20.5 Physiological aspects of acidophilic yeasts

Several physiological mechanisms are involved in the ability of microorganisms to thrive in environments which are extremely acidic and/or contain high concentrations of heavy metals. Regarding tolerance to acidity, yeasts are known to maintain a relatively neutral intracellular pH by pumping protons out of the cell, and by establishing low proton membrane permeability (Nicolay et al. 1987). This is partly achieved by the generation of a net intracellular positive charge as a result of the protonation of various types of molecules acting as triatable groups (Cavicchioli and Torsten 2000).

Toxicity of most heavy metals is a matter of concentration, because most of them actually play indispensable roles in cell growth and metabolic functions, but become inhibitory at higher concentrations (Gadd 1992; Villegas et al. 2004). Yeast tolerance mechanisms to high concentration of heavy metals include bioaccumulation and precipitation/chelation of the metal at the cell wall (biosorption), thus preventing it from reaching sensitive intracellular components (Raspor and Zupan 2005).

Several yeast species have been found to tolerate high concentrations of heavy metals such as copper (up to 3.17 g/L), zinc (up to 2.93 g/L), nickel (up to 3.1 g/L), cadmium (up to 2.36 g/L), lead (up to 1.2 g/L), iron (up to 5.03 g/L), mercury (up to 2.4 g/L) and aluminium (up to 5.4 g/L) (Abe et al. 2000; Salinas et al. 2000; Aksu and Donmez 2001; de Siloniz et al. 2002; Podgorskii et al. 2004; Rehman and Anjun 2010; Wang et al. 2013). Yeast bioaccumulation has been reported for copper (up to 20 mg/g biomass), chromium (up to 63.73 mg/g biomass), nickel (up to 46.8 mg/g biomass), cadmium (up to 15.63 mg/g biomass), uranium (up to 138 mg/g biomass), silver (up to 26 mg/g biomass) and iron (up to 0.5 mg/g biomass) (Strandberg et al. 1981; Brady and Duncan 1994; Kambe-Honjoh et al. 1997; Nakajima 2001; de Siloniz et al. 2002; Gómes et al. 2002; Li and Yuan 2006; Bankar et al. 2009; Villegas et al. 2009, 2011). Yeast metal uptake typically occurs in two main steps: the first involves a non-specific binding of the metal to cell surfaces; the second involves a slower metabolism-dependent intracellular uptake (Blackwell et al. 1995). Resistance to metals (e.g. solubilization and immobilization processes) developed by fungi is of fundamental importance in the biogeochemical cycles (Wainwright 1988) and is certainly of interest for bioremediation purposes.

RAC yeast isolates, incubated in water (without added nutrients) from the most acidic site (URA1, pH 1.8) showed different survival responses. Several yeast species were able to survive up to 2 months (maximum experimentation time) in the river water incubated at 20°C, whereas others did not survive for even 2 days (Russo et al. 2008). Species such as *Cryptococcus* sp. 2 and *Cr. agrionensis* were the most tolerant, followed by *Rhodotorula mucilaginosa* and *Rhodotorula* sp. 1. With the exception of *Cryptococcus* sp. 2, these species were isolated in high frequency and showed the broadest distribution in the RAC. Species displaying the lowest survival in the river water samples were originally isolated in the near-neutral part of the river (LRA2, pH 6.7), and include *Cystofilobasidium capitatum*, *Cys. macerans* and *Holtermanniella festucosa*.

Tab. 20.4: pH relationships of selected Caviahue-Agrio aquatic system yeasts.

Species	pH of origin	pH growth range	Optimum pH
Cryptococcus agrionensis	1.8–6.7	3.5–6.5	4.5
Rhodotorula mucilaginosa	2.1–6.7	2.5–6.5	5.5
Holtermanniella festucosa	6.2–6.7	3.5–6.5	4.5–5.5
Cryptococcus sp. 2	2.1	2.5–4.5	3.5

These species are clearly not adapted to survive in the acidic conditions of the River Agrio and are most certainly allochthonous. The rest of the species showed intermediate survival times. As a common trait, species isolated from the upper part of the river were more tolerant than those from the lower part, in agreement with the different environmental conditions present at both sites.

The pH growth ranges and optima of certain species clearly reflected their origin of isolation and degree of adaptation to the acidic conditions. For example, species such as *Cryptococcus agrionensis* and *Rhodotorula mucilaginosa*, which were present in almost all sampling sites (i.e. along the whole pH range), showed the broadest pH growth ranges (2.5–6.5) (Tab. 20.4). In contrast, *Holtermanniella festucosa*, a species found only in the near-neutral sites (RD and LRA2) showed a pH growth range from 3.5 to up to 6.5 (Tab. 20.4).

The putative acidophilic yeast *Cryptoccocus* sp. 2 was able to grow in a very narrow pH range (2.5–4.5) with slight growth even at pH 1.5. The optimum pH growth for this species was observed at pH 3. These data confirmed previous observations suggesting a preference for acidified culture media.

20.6 Possible ecological roles of yeasts in acidic aquatic environments

It is very difficult to define the precise role and importance of each organism as part of a dynamic population in extremely acidic environments. They are typically considered, like other fungi, to be exclusively involved in the decomposition of organic matter, and cycling of nutrients. In fact, as obligate heterotrophs, yeasts rely, to a lesser or greater extent, on carbon originating as leakage or lysis products from chemolithotrophic acidophiles (Johnson 1998). However, several field observations and laboratory experiments may shed light on to probable additional ecological role of yeasts in these acidic habitats.

Among the eukaryotes, fungi (including yeasts) seem to play an important role in these environments because, together with other microorganisms, they form biofilms on the surface of rocks (Baffico et al. 2004). The biofilms develop where metal and mineral precipitation occurs, and provides a substrate for other microbial populations (González-Toril et al. 2003). This is of importance for the entire ecosystem, because reducing the concentration of solubilized metals can allow less tolerant species to survive and develop.

Yeasts are also known to interact with other microorganisms in extreme acidic environments. An apparent mutualism between several strains of the alga *Euglena mutabilis* and a yeast of the genus *Cryptococcus* isolated from acidic and heavy metal polluted habitats has been described (Nakatsu and Hutchinson 1988). Growth was strikingly enhanced when the two organisms were cultivated together, compared with growth when cultured separately, for all metals tested and over a wide range of pH values. Furthermore, the initial difficulty for isolating both organisms and culturing them separately and their regular co-occurrence in acid waters from very different locations provide circumstantial support for a functional relationship. This algal-yeast association may be a major factor allowing colonization of these extreme acid habitats.

A less essential interaction has been observed between the bacterium *Thiobacillus ferrooxidans* and the yeast *Rhodotorula mucilaginosa*. It has been shown that ferrous iron oxidation carried out by the bacteria was stimulated by the presence of the yeast (Fournier et al. 1998).

Finally, yeasts may also serve as feed for multicellular organisms. Ehrlich (1963) reported that amoebae thriving in acid drainage from a copper mine fed on yeasts as well as on bacteria. The fact that several zooplanktonic organisms can feed on yeasts is well known (Perez et al. 2006). In the case of lake Caviahue, rotifers were the only zooplanktonic organisms found (Pedrozo et al. 2001).

20.7 Final remarks

The current knowledge of yeast diversity and ecology in acidic aquatic environments is scarce and limited to a few anthropogenically originated environments. The aim of the present chapter was to provide information on yeast occurrence in these extreme environments with especial focus on the naturally acidic River Agrio-Lake Caviahue (RAC, Neuquén, Argentina). The extremely acidic waters of the RAC were shown to harbor viable yeast cells in variable numbers. An interesting diversity of yeasts was revealed, and several previously undescribed species found. The particular distribution of yeasts along the pH gradient of the aquatic systems showed an apparent correlation with the physiological data collected for each species, allowing, in several cases, the discrimination of autochthonous and allochthonous species of the studied extreme habitat. Among the former, different levels of adaptation were observed. For example, a species with apparent acidophilic features such as *Cryptococcus* sp. 2, which seems well adapted to the extreme environmental conditions, was found in the upper River Agrio; while others seem to be adapted to relatively less extreme waters, and were mostly isolated from Lake Caviahue and lower River Agrio. High plasticity to adaptation to the different environmental conditions of the RAC was displayed by

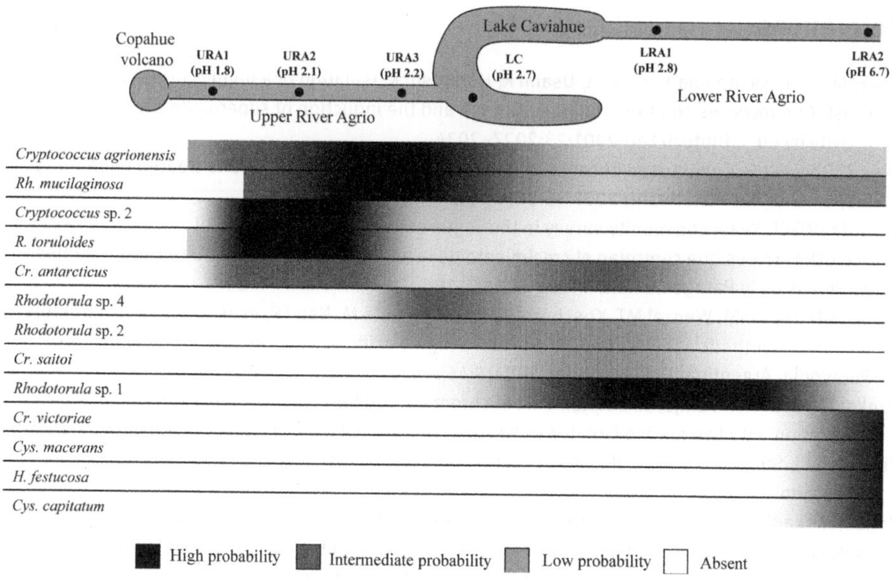

Fig. 20.5: Schematic illustration of the probable distribution of yeasts in the River Agrio-Lake Caviahue aquatic system.

widely distributed species such as *Rhodotorula mucilaginosa* and *Cryptococcus agrionensis*, which showed wide pH growth ranges. Thus, based on data obtained, an organization of the yeasts species structure along the RAC, which predicts species distribution in this peculiar environment, is hypothesized (Fig. 20.5).

Natural acidic systems are usually much older than mining-impacted waters, and it is possible that a well-adapted micro-flora has evolved in such systems (Koschorreck et al. 2003). Acidic water bodies of volcanic origin, with a chemical composition similar to that of mining waters, are promising habitats for the search for acid-adapted microorganisms. The bioprospecting of these naturally selected organisms represents an important strategy to obtain agents for bioremediation processes and other fundamental and applied purposes.

Acknowledgments

This work was funded by the Universidad Nacional del Comahue (Project B171) and CONICET (PIP424) grants to M. van Broock. G. Russo was supported by a CONICET PhD fellowship. We would like to thank authorities of Parques Nacionales (Argentina), for providing permission for water sampling in the National Park. We are most grateful to Jose Paulo Sampaio and Mario Gadanho for their help and advice. Celia Tognetti and Silvia Brizzio are thanked for critical review of the original manuscript.

References

Abe F, Miura T, Nagahama T, Inoue A, Usami R, Horikoshi K. Isolation of a highly copper-tolerant yeast, *Cryptococcus* sp., from the Japan Trench and the induction of superoxide dismutase activity by Cu^{2+}. Biotech Lett 2001;23:2027–2034.
Aksu Z, Dönmez G. Comparison of copper (II) biosorptive properties of live and treated *Candida* sp. Int J Environ Res Publ Health 2001;36:367–381.
Almeida JMGCF. Yeast community survey in the Tagus estuary. FEMS Microb Ecol 2005;53:295–303.
Baffico GD. Factores que controlan el crecimiento de la actividad perifítica en distintos ambientes acuáticos de la Patagonia. PhD Thesis 2006:1–104.
Baffico GD, Díaz MM, Wenzel MT, Koschorreck M, Schimmele M, Neu TR, et al. Community structure and photosynthetic activity of epilithon from a highly acidic (pH < 2) mountain stream in Patagonia, Argentina. Extremophiles 2004;*8*:463–473.
Bankar AV, Kumar AR, Zinjarde SS. Removal of chromium (VI) ions from aqueous solution by adsorption onto two marine isolates of *Yarrowia lipolytica*. J Hazard Mater 2009;170:487–494.
Blackwell KJ, Singleton I, Tobin JM. Metal caption uptake by yeast: a review. Appl Microbiol Biotechnol 1995;43:579–584.
Boguslawska-Was E, Dabrowski W. The seasonal variability of yeasts and yeast-like organisms in water and bottom sediment of the Szczecin Lagoon. Int J Hyg Environ Health 2001;203:451–458.
Brady D, Duncan J R. Bioaccumulation of metal cations by *Saccharomyces cerevisiae*. Appl Microbiol Biotechnol 1994;41:149–154.
Butinar L, Santos S, Spencer-Martins I, Oren A, Gunde-Cimerman N. Yeast diversity in hypersaline habitats. FEMS Microbiol Lett 2005;244(2):229–234.
Buzzini P, Branda E, Goretti M, Turchetti B. Psychrophilic yeasts from worldwide glacial habitats: diversity, adaptation strategies and biotechnological potential. FEMS Microbiol Ecol 2012;82:217–241.
Cavicchioli R, Torsten T. Extremophiles. In: Lederberg J, ed. Encyclopaedia of Microbiology Vol 2, 2nd edn. San Diego, CA: Academic 2000:317–337.
de Garcia V, Brizzio S, Libkind D, Buzzini P, van Broock M. Biodiversity of cold-adapted yeasts from glacial meltwater rivers in Patagonia, Argentina. FEMS Microbiol Ecol 2007;59:331–341.
de Siloniz MI, Payo EM, Callejo MA, Marquina D, Peinado JM. Environmental adaptation factors of two yeasts isolated from the leachate of a uranium Mineral heap. FEMS Microbiol Lett 2002;210(2):233–237.
Dopson M, Baker-Austin C, Koppineedi PR, Bond PL. Growth in sulfidic mineral environments: metal resistance mechanisms in acidophilic micro-organisms. Mycrobiology 2003;149:1959–1970.
Ehrlich HL. Microorganisms in acid drainage from a copper mine. J Bacteriol 1963;86:350–352.
Fell JW, Statzell-Tallman A. *Cryptococcus*, Vuillemin. In: Kurtzman CP, Fell FW, eds. The Yeasts, A Taxonomic Study. 4th Ed. Amsterdam, NL: Elsevier Science Publishers 1998:742–767.
Fournier D, Lemieux R, Couillard D. Essential interactions between *Thiobacillus ferrooxidans* and heterotrophic microorganisms during a wastewater sludge bioleaching process. Environ Pollut 1998;101:303–309.
Gabani P, Singh OV. Radiation-resistant extremophiles and their potential in biotechnology and therapeutics. Appl Microbiol Biotechnol 2013;97:993–1004.
Gadanho M, Sampaio JP. Polyphasic taxonomy of the basidiomycetous yeast genus *Rhodotorula*: *Rh. glutinis* sensu stricto and *Rh. dairenensis comb. nov.* FEMS Yeast Res 2002;2:47–58.
Gadanho M, Sampaio JP. Occurrence and diversity of yeasts in the Mid-Atlantic Ridge hydrothermal fields near the Azores archipelago. Microb Ecol 2005;50:408–417.

Gadanho M, Sampaio JP. Microeukaryotic diversity in the extreme environments of the Iberian Pyrite Belt: a comparison between universal and fungi-specific primer sets, temperature gradient gel electrophoresis and cloning. FEMS Microbiol Ecol 2006;57:139–148.

Gadanho M, Sampaio JP. *Cryptococcus ibericus* sp. nov., *Cryptococcus aciditolerans* sp. nov. and *Cryptococcus metallitolerans* sp. nov., a new ecoclade of anamorphic basidiomycetous yeast species from an extreme environment associated with acid rock drainage in São Domingos pyrite mine, Portugal. Int J Syst Evol Microbiol 2009;59:2375–2379.

Gadanho M, Libkind D, Sampaio J P. Yeast diversity in the extreme acidic environments of the Iberian Pyrite Belt. Microb Ecol 2006;52(3):552–563.

Gadd G M. Metals and microorganisms: a problem of definition. FEMS Microbiol Lett 1992;100:197–204.

Gómes NCM, Rosa CA, Pimentel PF, Mendoza-Hagler LCS. Uptake of free and complexed silver ions by different strains of *Rhodotorula mucilaginosa*. Braz J Microbiol 2002;33:62–66.

González-Toril E, Llobet-Brossa E, Casamayor EO, Amann R, Amils R. Microbial ecology of an extreme acidic environment, the Tinto River. Appl Environ Microbiol 2003;69:4853–4865.

Gupta R, Ahuja P, Khan S, Saxena RK, Mohapatra H. Microbial biosorbents: meeting challenges of heavy metal pollution in aqueous solutions. Curr Sci 2000;78:967–973.

Hagler AN, Ahearn DG. Ecology of aquatic yeasts. In: Rose AH, Harrison JS, eds. The Yeasts, Yeasts and the Environment. Vol. 2. London: Academic Press 1987:181–205.

Hagler AN, Mendonça-Hagler LC. Yeasts from marine and estuarine waters with different levels of pollution in the state of Rio de Janeiro, Brazil. Appl Environ Microbiol 1981;41:173–178.

Herbert RA. A perspective on the biotechnological potential of extremophiles. Trends Biotechnol 1992;10:395–402.

Holland SL, Dyer PS, Bond CJ, James SA, Roberts IN, Avery SV. *Candida argentea* sp. nov., a copper and silver resistant yeast Species. Fungal Biol 2011;115:909–918.

Huang C-H, Lee F-L, Tien C-J, Hsieh P-W. *Rhodotorula taiwanensis* sp. nov., a novel yeast species from a plant in Taiwan. Antonie van Leeuwenhoek 2011;27:297–302.

Johnson BD. Biodiversity and ecology of acidophilic microorganisms. FEMS Microbiol Ecol 1998;27:307–317.

Kambe-Honjoh H, Sugawara A, Yoda K, Kitamoto K, Yamasaki M. Isolation and characterization of nickel-accumulating yeasts. Appl Microbiol Biotechnol 1997;48:373–378.

Koschorreck M, Wendt-Potthoff K, Geller W. Microbial sulfate reduction at low pH in sediment of an acid lake in Argentina. Environ Sci Technol 2002;37:1159–1162.

Lachance M-A, Starmer WT. Ecology and yeasts. In: Kurtzman CP, Fell FW, eds. The Yeasts, A Taxonomic Study. 4th Ed. Amsterdam, NL: Elsevier Science Publishers 1998:21–30.

Lahav R, Fareleira P, Nejidat A, Abeliovich A. The identification and characterization of osmotolerant yeast isolates from chemical wastewater evaporation ponds. Microb Ecol 2002;43(3):388–396.

Lentzen G, Schwarz T. Extremolytes: Natural compounds from extremophiles for versatile applications. Appl Microbiol Biotechnol 2006;72:623–634.

Li Z, Yuan H. Characterization of cadmium removal by *Rhodotorula* sp. Y11. Appl Microbiol Biotechnol 2006;73(2):458–463.

Libkind D, Brizzio S, Ruffini A, Gadanho M, van Broock MR, Sampaio JP. Molecular characterization of carotenogenic yeasts from aquatic environments in Patagonia, Argentina. Antonie van Leeuwenhoek 2003;84:313–322.

Libkind D, Brizzio S, van Broock M. *Rhodotorula mucilaginosa*, a carotenoid producing yeast strain from Patagonian high-altitude lake. Folia Microbiol 2004;49:19–25.

López-Archilla AI, Marín I, Amils R. Microbial community composition and ecology of an acidic aquatic environment: the Tinto River, Spain. Microb Ecol 2001;41:20–35.

López-Archilla AI, González AE, Terrón MC, Amils R. Ecological study of the fungal populations of the acidic Tinto River in southwestern Spain. Can J Microbiol 2004;50(11):923–934.

Margesin R, Fonteyne P-A, Schinner F, Sampaio JP. *Rhodotorula psychrophila* sp. nov., *Rhodotorula psychrophenolica* sp. nov. and *Rhodotorula glacialis* sp. nov., novel psychrophilic basidiomycetous yeast species isolated from alpine environments. Int J Sys Evol Microbiol 2007;57:2179–2184.

Nagahama T, Hamamoto M, Nakase T, Takami H, Horikoshi K. Distribution and identification of red yeasts in deep-sea environments around the northwest Pacific Ocean. Antonie van Leeuwenhoek 2001;80:101–110.

Nakajima A. Effect of selenium on uranium biosorption. J Radioanal Nuc Chem 2001;246:347–350.

Nakatsu C, Hutchinson TC. Extreme metal and acid tolerance of *Euglena mutabilis* and an associated yeast from Smoking Hills, Northwest territories, and their apparent mutualism. Microb Ecol 1988;16:213–231.

Nguyen VAT, Senoo K, Mishima T, Hisamatsu M. Multiple tolerance of *Rhodotorula glutinis* R-1 to acid, aluminum ion and manganese ion, and its usual ability of neutralizing acidic medium. J Biosci Bioeng 2001;92:366–371.

Nicolay K, Veenhuis M, Douma AC, Harder W. A ^{31}P NMR study of the internal pH of yeast peroxisomes. Arch Mikrobiol 1987;147:37–41.

Pedrozo F, Kelly L, Diaz M, Temporetti P, Baffico G, Kringel R, et al. First results on the water chemistry, algae and trophic status of an Andean acidic lake system of volcanic origin in Patagonia (Lake Caviahue). Hydrobiol 2001;452:129–137.

Pérez P, Libkind D, Diéguez MC, Summerer M, Sonntag B, Sommaruga R, et al. Mycosporines from freshwater yeasts: a trophic *cul-de-sac*?. Photochem Photobiol Sci 2006;5:25–30.

Phaff HJ, Starmer WT. Yeasts associated with plants, insects and soil. In: Rose AH, Harrison JS, eds. The Yeasts, Biology of Yeasts. Vol. 1, London: Academic Press 1987:123–180.

Podgorskii VS, Kasatkina TP, Lozovaia OG. Yeasts—biosorbents of heavy metals. Mikrobiol Z 2004;66:91–103.

Raspor P, Zupan J. Yeasts in extreme environments. In: Rosa CA, Ganter P, eds. Yeast Handbook. Biodiversity and Ecophysiology of Yeasts. Berlin: Springer-Verlag 2006:371–417.

Rehman A, Anjun MS. Cadmium uptake by yeast, *Candida tropicalis*, isolated from industrial effluents and its potential use in wastewater clean-up operation. Water Air Soil Pollut 2010;205:149–159.

Rosa CA, Ganter P. Biodiversity and ecophysiology of yeasts. In: Rosa CA, Ganter P, eds. The Yeast Handbook Series. Heidelberg: Springer-Verlag 2006:1–579.

Russo G, Libkind D, Sampaio JP, van Broock MR. Yeast diversity in the acidic Rio Agrio - Lake Caviahue volcanic environment (Patagonia, Argentina). FEMS Microbiol Ecol 2008;65:415–424.

Russo G, Libkind D, Ulloa JR, de Garcia V, Sampaio JP, van Broock MR. *Cryptococcus agrionensis* sp. nov., a basidiomycetous yeast of the acidic rock drainage ecoclade, isolated from River Agrio– Lake Caviahue (Argentina), an acidic aquatic environment of volcanic origin. Int J Syst Evol Microbiol 2009;60:996–1000.

Salinas E, Orellano M, Rezza I, Martinez L, Marchesvky E, Tossetti M. Removal of cadmium and lead from dilute aqueous solutions by *Rhodotorula rubra*. Bioresour Technol 2000;72:107–112.

Sampaio JP. Diversity, phylogeny and classification of basidiomycetous yeasts. In: Agerer R, Blanz P, Piepenbring M, eds. Frontiers in Basidiomycote Mycology. Eching: IHW-Verlag 2004:49–80.

Sharp RJ, Munster MJ. Biotechnological implications for microorganisms from extreme environments. In: Herbert RA, Codd GA, eds. Microbes in Extreme Environments. London: Academic Press 1986:215–295.

Shivaji S, Bhadra B, Rao RS, Pradhan S. *Rhodotorula himalayensis* sp. nov., a novel psychrophilic yeast isolated from Roopkund Lake of the Himalayan mountain ranges, India. Extremophiles 2008;12:375–381.

Spencer JFT, Spencer DM. Ecology: where yeasts live. In: Spencer JFT, Spencer DM, eds. Yeasts in Natural and Artificial Habitats. 1st edn. Berlin: Springer-Verlag 1997:33–58.

Strandberg GW, Shumate II SE, Parrott JR. Microbial cells as biosorbents for heavy metals: accumulation of uranium by *Saccharomyces cerevisiae* and *Pseudomonas aeruginosa*. Appl Environ Microbiol 1981;41:237–245.

Villegas LB, Amoroso MJ, de Figueroa LIC. Selection of tolerant heavy metal yeasts from different polluted sites. In: Spencer JFT, Ragout de Spencer AL, eds. Environmental Biology: Methods and Protocols. 1st edn. Newark, NJ: Springer Protocols 2004:249–256.

Villegas LB, Amoroso MJ, de Figueroa LIC, Siñeriz F. Cu(II) removal by *Rhodotorula mucilaginosa* RCL-11 in sequential batch cultures. Water Sci Technol 2009;60:1225–1232.

Villegas LB, Amoroso MJ, Figueroa LIC. Interaction of copper or chromium with yeasts: potential application on polluted environmental clean up. In: Mason AC, ed. Bioremediation, Engineering and Environmental Management. New York, NY: Nova Science Publishers 2011:177–206.

Vishniac HS. Yeast biodiversity in the Antarctic. In: Rosa CA, Ganter P, eds. Yeast Handbook. Biodiversity and Ecophysiology of Yeasts. Berlin: Springer-Verlag 2005:419–440.

Vishniac HS, Kurtzman CP. *Cryptococcus antarticus* sp. nov. and *Cryptococcus albidisimilis* sp. nov., Basidioblastomycetes from Antarctic soils. Int J Sys Bacteriol 1992;42:547–553.

Wainwright M. Metabolic diversity of fungi in relation to growth and mineral cycling in soil: a review. Mycol Res 1088;90:159–170.

Wang C, Zaho X-Q, Aizawa T, Sunairi M, Shen R-F. High aluminum tolerance of *Rhodotorula* sp. RS1 is associated with thickening of the cell wall rather than chelation of aluminum ions. Pedosphere 2013;23:29–38.

Watanabe K, Futamata H, Harayama S. Understanding the diversity in catabolic potential of microorganisms for the development of bioremediation strategies. Antonie van Leeuwenhoek 2002;81:655–663.

Wendt-Potthoff K, Koschorreck M. Functional groups and activities of bacteria in a highly acidic volcanic mountain stream and lake in Patagonia, Argentina. Microb Ecol 2002;43:92–106.

Nattawut Boonyuen, Somsak Sivichai and E. B. Gareth Jones

21 Decomposition of wood in tropical habitats

The dynamics of leaf litter decomposition has been extremely well studied in temperate freshwater habitats (Triska and Sedell 1976; Bärlocher 1992; Suberkrop 1995; Gessner et al. 1997; Seena et al. 2008), but little is known about tropical ecosystems (Wong et al. 1998; Shearer et al. 2007). The decomposition of woody tissue has received scant attention, despite the fact that numerous studies have documented the colonization of wood by freshwater fungi (Jones and Oliver 1964; Lamore and Goos 1978; Sanders and Anderson 1979; Shearer and Bodman 1983; Shearer 1992; Sridhar et al. 1992; Hyde and Goh 1999; Sivichai et al. 2000, 2002; Tsui et al. 2000, 2001; Kane et al. 2002; Cai et al. 2006). Biodiversity studies have shown a species rich occurrence of freshwater fungi (See Chapters 2–6, 15); however, few have examined their role in the degradation of wood under natural conditions (Triska and Cromack 1980; Shearer and Bodman 1983; Shearer 1992; Gessner and Van Ryckegem 2002). Eaton and Jones (1971) documented the decay of wood baits in water cooling towers and reported losses of 59.7% and 52.9% for *Fagus sylvatica* (beech) and *Pinus sylvestris* (Scots pine), respectively at Connah's Quay, and 56.2% and 56.6%, for the same timbers at Ince Power Station, after exposure for 108 weeks. However, weight losses of pine (22.3%) and beech (15.6%) test blocks in the River Severn were much lower, over 92 weeks' submergence (Kane et al. 2002). These low weight losses, in comparison to the cooling tower data, can be accounted for by the differences in water temperature at the two locations.

The ability of freshwater fungi to degrade lignocellulose *in vitro*, has been examined providing data on weight losses and enzymatic activity (Eaton and Jones 1971; Jones 1972, 1981; Yuen et al. 1999). Jones (1981) reported that six lignicolous hyphomycetes could cause weight loss of beech and birch wood test blocks (0.5%–11.9%), but two did not form soft rot cavities in the secondary cell wall (*Articulospora tetracladia* and *Heliscus lugdunensis*). Higher weight losses were observed for balsa wood (6.2%–24%). Most studies have focused on the ability of freshwater fungi to cause soft rot cavities of small timber test blocks (Leightley and Eaton 1977; Zare-Maivan and Shearer 1988a, b; Yuen et al. 2000). Many studies have examined the enzymatic activities of freshwater fungi and reported amylase, β-glucosidase, β-xylosidase, endoglucanase, endoxylanase, lipase, pectinase, and protease activity (Chamier 1985; Hyde and Jeewon 2003). Their ability to degrade lignin remains to be proven as the assay methods involves extracted lignin's (Shearer 1992).

Although *in vitro* degradation of wood has been reported, comparatively little is known about *in vivo* activity. Ballie et al. (1998) have suggested it takes 25 years for a log to decay in a stream, but most submerged wood in streams and rivers are of smaller dimensions, e.g. twigs. How long does it take for them to be decomposed?

Our objectives in this chapter are to:
1. review the diversity of fungi reported from exposed wood;
2. follow the colonization of 15 timbers exposed at two locations in Thailand; and,
3. determine the rate of decay of selected timbers in two contrasting ecosystems, a rapidly flowing river, and a peat swamp in Thailand.

21.1 Review of fungal diversity on wood in freshwater streams

Filamentous freshwater fungi are often substrate dependent with asexual morphs common on senescent leaves (Ingold 1952, 1974; Goh 1997; Wong et al. 1998; Jones et al. 1999; Sridhar and Bärlocher 2000), with ascomycetes more prevalent on woody tissues (Webster 1959; Eaton and Jones 1971; Hyde and Goh 1998, 1999; Sivichai et al. 1998; Wong et al. 1998a, b; Raja et al. 2012), while basidiomycetes are infrequent on submerged or trapped wood in freshwater habitats (Nawawi and Webster 1982; Marvanová and Suberkropp 1990; Sivichai and Jones 2004). One of the first reports of freshwater fungi on wood was by Jones and Oliver (1964) who demonstrated that "Ingoldian fungi" are early colonizers of submerged wood. Later, Jones and Eaton (1969), and Jones and Stewart (1972), reported many species of freshwater hyphomycetes and ascomycetes on wood in water cooling towers, while Lamore and Goos (1978) documented the wood-inhabiting fungi in a stream in Rhode Island. These reports were all from temperate waters. Subsequently, filamentous freshwater fungi have been documented for a variety of tropical locations: China (Hu et al. 2010), Hong Kong (Ho et al. 2000, 2004), Thailand (Hu et al. 2010; Boonyuen et al. 2011, 2012), Peru (Raja et al. 2012) and USA (Raja et al. 2010). Most studies have focused on the description of new taxa and distribution on drift and trapped wood (Tsui et al. 2003), the ultrastructure of ascospores and their appendages (Wong et al. 1998a, 1999; Ho et al. 2004), their ability to decompose wood (Zare-Mairan and Shearer 1988b; Yuen et al. 2000) and to their role in freshwater ecosystems (Wong et al. 1998).

Few have followed the sequential colonization of tropical fungi on wood by the exposure of test blocks in streams and rivers (Sivichai et al. 2000, 2002; Tsui et al. 2000; Ho et al. 2002). Such studies have been undertaken in temperate water ecosystems: when the effect of different timber species was documented (Eaton and Jones 1971) and bait size, on fungal colonization (Sanders and Anderson 1979). Sivichai et al. (2000) followed the colonization *Dipterocarpus alatus* and *Xylia dolabriformis* test blocks over 12 months in a stream at Khao Yai National Park, Thailand, with monthly retrieval. Eighty-nine fungi were recorded: 53 on *D. alatus* and 62 on *X. dolabriformis* on the test blocks. Fungi present on over 60% of the test blocks included: *Helicomyces roseus*, *Trematosphaeria* sp. 2 on *D. alatus*, and *Savoryella aquatica*, *H. roseus*, *Helicosporium griseum* and *Ellsembia opaca* on *X. dolabriformis*. In a similar study at Tad Ta Phu stream in the Khao Yai National Park, 73 species were documented, 48 on *D. alatus* and 47 on *X. dolabriformis* (Sivichai et al. 2002). *Helicomyces roseus* was present at over 60% on both timbers, and *Aniptodera aquadulcis* (as *Halosarpheia aquadulcis*) on *Xylia*. Fungi could be grouped as early, intermediate and late colonizers which represents their sequential sporulation on the wood. These studies have shown that the two timbers exposed supported a wide range of taxa, and a high percentage colonization of the test blocks (over 60%). The population was dominated by asexual morphs and the timbers supported a divergent fungal community with little overlap, with *H. roseus* the only dominant species on both timbers. The data confirms that the fungi colonizing wood in temperate and tropical locations are different (Eaton and Jones 1971; Kane et al. 2002).

There was also a significant difference in the fungal community on test blocks of *Machilus velutina* and *Pinus massoniana* exposed at a stream in Tai Po Kau, Hong Kong (Ho et al. 2002). They exposed the test blocks for up to 21 months and noted three distinct types of fungal communities: pioneer, early and later successional groups, with 175 species. Pioneer and early colonizers were similar, but differed later in the succession: *Nectria* cf *byssicola*, *Sporochisma nigroseptatum* and *S. unisetatum* were early colonizers on both woods. On *Machilus velutina*, *Savoryella lignicola* was a late colonizer, while *Dictyosporium digitatum*, *Lophiostoma bipolaris* (as *Massarina bipolaris*) and *Lindgomyces ingoldianus* (as *Massarina Ingoldiana*) appeared later on *Pinus massoniana*.

These studies have shown that wood-inhabiting fungi are diverse with asexual morphs dominant, that they differ from timber to timber, geographical location, and cause little decay of the wood over the period of incubation (up to 1 year).

21.2 Colonization of 15 timbers exposed at two locations in Thailand

21.2.1 Materials and methods

Sterilized test blocks 15 × 2.5 × 2.5 cm^3 of *Azadirachta indica* var. *siamensis*, *Erythrophlleum teysmannii*, *Melaluca cajuputi*, *Shorea obtusa*, *S. roxbordhii*, *S. siamensis*, *Wrightia tomentosa*, *Xylia xylocarpa*, and *Zollingeria dongnaiensis*, were submerged and exposed in two selected streams at the Sirindhorn peat swamp forest (6.00295N, 101.9571889E) and Bala Hala Wildlife Sanctuary (5.92833N, 101.88333E) in Narathiwat Province, Thailand on March 12, 2001. Topographically, the Sirindhorn peat swamp forest is low-lying flatland and water logged year long. Rain-fed water drains out of the forest floor very slowly. Due to the tannic decomposition and a pH 4.5–6.0, the water is brownish and acidic. The water at a selected stream from Bala Hala Wildlife Sanctuary was fast moving, especially in the rainy season and slow moving water in the dry season. Due to the wet, a seasonal climate, the site supports wet evergreen forest at elevations below 600 m above sea level, with hill and lower montane evergreen forest at higher elevations. Twelve sets of test blocks, five blocks per set for each timber species, without wood preservative, were threaded on a nylon rope and autoclaved three times at 1.5 kgf/cm^2, 121°C, for 15 min. Each string of test blocks were attached to bank-side trees and weighed down with stones to keep them under the water.

Test blocks were recovered after 1 month, 6 months, 1 year, 2 years and 3 years on April 12, 2001 to March 12, 2004, and each set of blocks transported to the laboratory in a clean polystyrene foam box. Test blocks were washed with stream water to get rid of silt and mud from the surface. Each set of test blocks was separated and placed in pre-sterilized plastic boxes with moist tissue papers layered on the bottom. Incubation was in a cabinet with cool white fluorescent light (20°C).

To examine, isolate and identify sporulating fungi, test blocks were examined under stereo and compound microscopes after removal from the streams at 1 week, and at 1, 2 and 3 months for the presence of spore-producing structures. Fungal voucher slides were mounted in lactophenol-cotton blue, and sealed with polyvinyl alcohol.

21.2.2 Results

Fungi identified on test blocks recovered at 1 month, 6 months, 1 year, 2 years and 3 years are listed in Tab. 21.1. One hundred and thirty species were observed (both test sites) including 89 asexual morphs, 38 Ascomycota and 3 Basidiomycota. Species diversity was high at the sites, with 86 and 71 species at Bala Hala and the peat swamp forest, respectively. While the number of fungi recovered at the peat swamp was high, most were represented with occurrence on a single timber species. The most widely occurring fungi at Bala Hala were those on nine timbers (*Candelabrum brocchiatum*, *Dictyochaeta* sp.1, *Ekuvispora bipolaris*, and *Trematosphaeria* sp.1), eight timbers (*Anthostomella aquatica*, *Dinemasporium* sp.1,

Tab. 21.1: List of freshwater fungi and fungal records occurring in nine different types of timbers in two contrasting ecosystems.

Fungal identification	Collecting sites	
	Bala Hala Wildlife sanctuary	**Peat swamp forest**
Acrodictys globolosa	2* (*This fungus found on two timber species from nine timber species)	–* (*No fungi recorded on the nine timber species)
Acrogenospora sp.1	3	1
Acrogenospora sp.2	1	1
Acrogenospora sphaerocephala	3	2
Annulatascus sp.1	1	1
Annulatascus sp.2	1	1
Annulatascus triseptatus	1	–
Annulatascus velatisporus	1	7
Anthostomella aquatica	8	–
Aquaticola sp.1	1	1
Ascotaiwania sp.1	–	1
Basidiomycete sp.1	–	1

(*continued*)

Tab. 21.1: (*continued*)

Fungal identification	**Collecting sites**	
	Bala Hala Wildlife sanctuary	**Peat swamp forest**
Berkleasmium sp.1	1	–
Bionectria sp.1	3	–
Bombadia sp.1	–	1
Camposporium sp.1	1	–
Canalisporium sp.1	7	–
Canalisporium caribense	1	–
Canalisporium elegans	2	–
Cancellidium applanatum	1	9
Candelabrum brocchiatum	9	4
Chaetopsina fulva	6	–
Chaetopsina sp.1	1	–
Chaetopsina sp.2	5	1
Coelomycete sp.1	4	1
Copinus sp.1	1	–
Copinus sp.2	2	–
Cryptophiale kakombensis	1	–
Cryptophialoidea unilateralis	1	6
Dactylaria sp.1	2	1
Dactylaria sp.2	2	–
Dactylaria sp.3	2	2
Dictyochaeta gyrosetula	–	3
Dictyochaeta sp.1	9	–
Dictyochaeta sp.2	1	–
Dictyochaeta sp.3	–	3
Dictyosporium digitatum	1	–
Dictyosporium elegans	–	1
Dictyosporium giganticum	1	–
Dictyosporium heptasporum	1	–
Dictyosporium nigroapice	1	–
Dinemasporium sp.1	8	1
Ellisembia brachypus	–	1
Ellisembia opaca	4	–
Eluviespora bipolaris	9	–
Geniculosporium sporodochiale	8	–
Gonytrichum sp.1	1	1
Halosarpheia aquadulcis	2	–
Halosarpheia sp.1	1	–
Helicomyces roseus	5	9
Helicosporium sp.1	–	1
Helicosporium griseum	2	–
Helicosporium vegetum	8	–
Hypocrea sp.1	–	1
Jahnula appendiculata	–	8

(*continued*)

Tab. 21.1: (*continued*)

Fungal identification	Collecting sites	
	Bala Hala Wildlife sanctuary	Peat swamp forest
Jahnula sp.1	–	1
Massarina ingoldiana	6	–
Massarina thalassioidea	8	–
Menisporopsis pirozynskii	2	–
Menisporopsis profusa	3	–
Monodictys sp.1	–	5
Monodictys sp.2	1	2
Monodictys sp.3	–	2
Monodictys sp.4	–	2
Monotosporella sp.1	1	–
Monotosporella sp.2	–	1
Nawawia filiformis	–	5
Nectria chaetopsinae	4	–
Nectria sp.1	3	1
Nectria sp.2	4	–
Nectria sp.3	1	–
Ophioceras dolichostomum	4	–
Penicilium-like sp.1	–	1
Phaeoisaria clematidis	5	1
Savoryella aquatica	3	–
Savoryella sp.1	2	–
Savoryella sp.2	1	–
Savoryella verrucosa	1	–
Sporidesmium sp.1	1	–
Sporidesmium sp.2	1	–
Sporoschisma uniseptatum	3	–
Stilbella holubovae	6	–
Thozetella nivea	5	3
Thozetella sp.1	–	1
Thozetella sp.2	1	–
Trematosphaeria sp.1	9	1
Trichoderma sp.1	3	4
Trichoderma sp.2	4	1
Verticillium sp.1	–	2
Verticillium sp.2	–	2
Verticillium sp.3	2	–
Verticillium sp.4	–	1
Xylomyces chlamydosporis	–	8
Total species on each timber and test site	86	71
Total ascomycetes	38	
Total basidiomycetes	3	
Total asexual morphs	89	
Total species in this study	130	
% Similarity between sites	$(27/130) \times 100 = 20.76$	

Geniculosporium sporodochiale, *Helicosporium vegetum*, and *Massarina thalassioidea*) and seven timbers (*Canalisporium* sp.1). At the peat swamp forest, *Cancellidium applanatum* and *Helicomyces roseus* occurred on nine timbers, while *Jahnula appendiculata* and *Xylomyces chlamydosporus* were present on eight timbers. There was little overlap in the fungal communities at the two sites (20.76% similarity), and this may well reflect the differing environmental conditions. The stream in Bala Hala has fast flowing water especially in the rainy season, while at the peat swamp water flow is marginal.

Figure 21.1 documents the progressive colonization of the test blocks at Bala Hala, with the greatest number of species during examination on the first month (35) and the sixth month (41). Additional species colonized the timbers over the 3 year period indicating the continued recruitment of fungi. At the peat swamp site, more fungi were identified at the 6-month stage (30) with continued recruitment of fungi over the subsequent months (Fig. 21.2). Tables 21.2 and 21.3 detail the number of fungi on each timber at the two sites over the 3 year exposure periods. At Bala Hala, only 17 fungi were recorded on *Melaluca cajuputi*, with 29–36 species on most timbers (Tab. 21.2), while at the peat swamp fewer species were reported on all timbers with lowest on *Xylia xylocarpa* (Tab. 21.3).

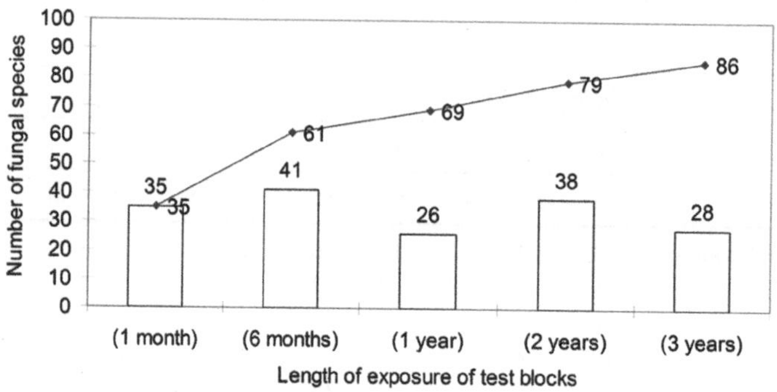

Fig. 21.1: Cumulative increases in species at each period at Bala Hala site.

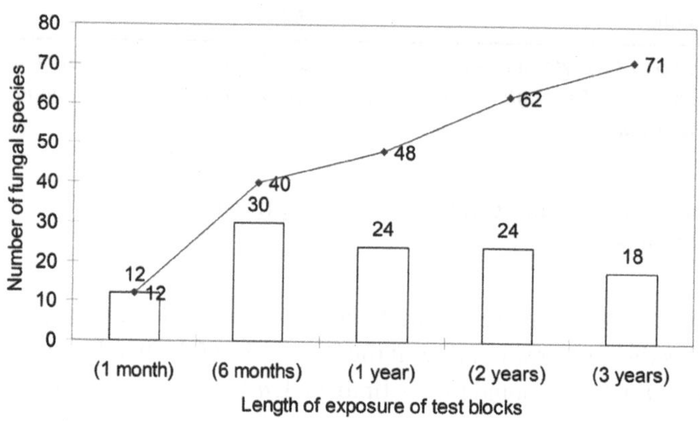

Fig. 21.2: Cumulative increase in species at each period at the peat swamp site.

Tab. 21.2: Number of fungi reported from the nine timbers exposed for 3 years at the Bala Hala stream.

Bala-Hala Wildlife sanctuary	**Timber species**								
	1	2	3	4	5	6	7	8	9
	AI	*ET*	*SO*	*SR*	*SS*	*WT*	*XX*	*ZD*	*MC*
Total species of fungi on each timber	35	20	29	34	26	33	36	26	17
Ascomycetes on each timber	10	5	11	12	11	10	14	8	8
Basidiomycetes on each timber	1	2	0	0	0	0	0	0	0
Asexual morphs on each timber	24	13	18	22	15	23	22	18	9
Total species on all timbers					86				
Total number of Ascomycetes					28				
Total number of Basidiomycetes					2				
Total number of asexual morphs					56				

Remark: *AI* = *Azadirachta indica*; *ET* = *Erythrophlleum teysmannii*; *SO* = *Shorea obtuse*; *SR* = *Shorea roxbordhii*; *SS* = *Shorea siamensis*; *WT* = *Wrightia tomentosa*; *XX* = *Xylia xylocarpa*; *ZD* = *Zollingeria dongnaiensis*; *MC* = *Melaluca cajuputi*.

Tab. 21.3: Number of fungi reported from the nine timbers exposed for 3 years at the peat swamp forest.

Peat swamp forest	**Timber species**								
	1	2	3	4	5	6	7	8	9
	AI	*ET*	*SO*	*SR*	*SS*	*WT*	*XX*	*ZD*	*MC*
Total species of fungi on each timber	22	22	22	19	16	12	11	19	13
Ascomycetes on each timber	4	5	3	3	2	4	2	2	4
Basidiomycetes on each timber	0	0	1	0	0	0	0	0	0
Asexual morphs on each timber	18	17	18	16	14	8	9	17	9
Total species on all timbers					71				
Total number of Ascomycetes					17				
Total number of Basidiomycetes					1				
Total number of asexual morphs					53				

Remark: *AI* = *Azadirachta indica*; *ET* = *Erythrophlleum teysmannii*; *SO* = *Shorea obtuse*; *SR* = *Shorea roxbordhii*; *SS* = *Shorea siamensis*; *WT* = *Wrightia tomentosa*; *XX* = *Xylia xylocarpa*; *ZD* = *Zollingeria dongnaiensis*; *MC* = *Melaluca cajuputi*.

21.2.3 Rate of decay of selected timbers at two contrasting freshwater ecosystems in Thailand

Weight losses for all timbers were greater at the Bala Hala stream than at the peat swamp site which reflects the greater water flow at the former site (Tab. 21.4, Fig. 21.3). High weight losses of over 80% were noted for the timbers *Azadirachta indica* var. *siamensis*, *Zollingeria dongnaiensis* and *Melaluca cajuputi* at the Bala Hala stream, with an average of 70% losses for the nine timber species. At the peat swamp, the highest

Tab. 21.4: Weight loss (%) of nine timbers submerged at two sites in Thailand.

Timber species	Bala Hala wildlife sanctuary	Peat swamp forest
Azadirachta indica var. *siamensis*	81.1	Not available
Erythrophlleum teysmannii	47.4	21.9
Shorea obtusa	69.4	22.6
Shorea roxbordhii	71.4	32.3
Shorea siamensis	78.1	40.8
Wrightia tomentosa	52.4	35.7
Xylia xylocarpa	71.2	10.0
Zollingeria dongnaiensis	80.9	26.5
Melaluca cajuputi	89.0	20.2

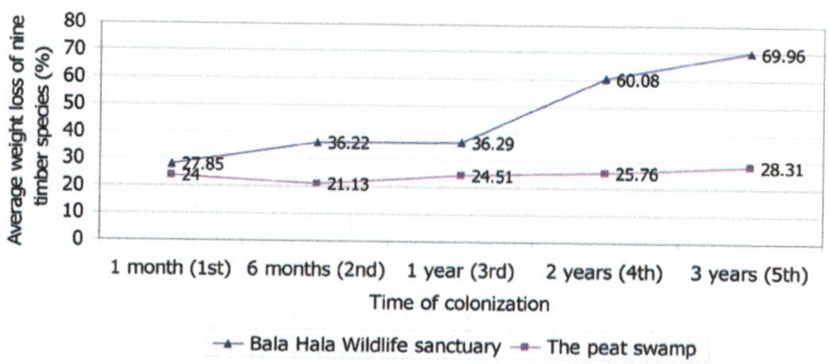

Fig. 21.3: Average weight loss of test blocks for all timbers over 3 years at the Bala Hala Wildlife sanctuary and the peat swamp.

weight losses of 40.8% and 35.7% were reported for *Shorea siamensis* and *Wrightia tomentosa*, respectively. Table 21.4 presents data on the average weight loss percentage for the nine timbers over the 3-year period, with those at Bala Hala three times greater than at the peat swamp site.

21.2.4 Discussion

Two aspects are worthy of discussion, firstly the fungal community reported from submerged timbers, and secondly the decay of timber at the two environmentally contrasting sites.

21.2.4.1 Fungal community

A comparison of fungi colonizing submerged wood shows fewer species in temperate waters ranging from 19 to 77 species, while tropical waters the number was 121–178

Tab. 21.5: A comparison of fungal diversity in different surveys.

Study	Areas	Substrata	Number of species
Sanders and Anderson (1979)	Temperate	Oak	19
Shearer and Bodman (1983)	Temperate	Various	33
Lamore and Goos (1978)	Temperate	Various	59
Willoughby and Archer (1973)	Temperate	Various	40
Gönczöl and Rėvay (1993)	Temperate	Three timbers	77
Kane (1980)	Temperate	Two timbers	61
Ho (1998)	Tropical	Various	178
Sivichai (1999)	Tropical		121
– at Km 29.2		– two timbers	
– Tad Ta Phu		– two timbers	
This study	Tropical		130
– at Bala Hala		– nine timbers	
– Peat Swamp		– nine timbers	

(Tab. 21.5). The taxa reported in the tropics differ from those reported in temperate waters: common temperate water species include: *Camposporium pellucidum, Dictyochaeta parva, Pseudohalonectria lignicola* and *Trichocladium lignicola*, while those in tropical waters are *Anthostomella aquatica, Cancellidium applanatum, C. brocchiatum, Dinemasporium* sp., *Geniculosporium sporodochiale, Helicomyces roseus, Helicosporium griseum, Massarina thalassioidea* and *Trematosphaeria* sp. (Kane et al. 2002; this study). Sivichai et al. (2002) compared the dominant fungi colonizing a range of timber species submerged at five locations worldwide (2 tropical, 1 subtropical and 2 temperate) with no species found occurring at all sites/timbers. In Sivichai et al. (2002), *H. roseus, Biflagellospora gracilis* and *Dictychaeta* sp.1 were the only species common in all timbers at both sites. Similarly, there were few common species on two timbers exposed at two locations in the UK (Eaton and Jones 1971; Kane et al. 2002).

A number of conclusions can be drawn with respect to the colonization of exposed timbers test blocks: 1) fewer fungi occur on submerged wood compared to naturally occurring timber, e.g. 58 and 155, respectively, in a study by Ho et al. (2002); 2) test blocks are sequentially colonized by fungi, and dependent on the exposure period; and 3) there is little overlap between fungi occurring on different timbers, e.g. only 20.76% similarity between the two sites in the current study (Tab. 21.1).

21.2.4.2 Decay of wood in freshwater habitats

Wood as twigs, branches and logs constitute a significant portion of allochthonous material entering freshwater ecosystems (Triska and Cromack 1980; Shearer 1992). Little data is available as to the rate of transformation of wood in such habitats, especially in tropical locations. It is generally assumed that decay of wood in freshwater is slow and has little impact on stream energy budgets (Sridhar et al. 1992; Gessner and Ryckegerm 2002). This is based on the fact that freshwater fungi are slow to cause decay of wood under laboratory conditions, e.g. *Anguillospora crassa* and *Lunulospora curvula* causing weight loss of beech test block of 7.6% and 10.8% after 13 weeks incubation, respectively (Jones 1981). Freshwater fungi are soft rot organisms and lack the lignolytic enzymes of terrestrial fungi that cause white and brown rot of wood (Leightly and Eaton 1977; Zare-Maivan and Shearer 1988a; Yuen et al. 2000). Furthermore, few freshwater animals burrow deeply into wood such as the teredinids (Mollusca) do in marine habitats (Turner and Johnson 1971). Gessner and Ryckegerm (2002) noted that information is scarce on the biomass of fungi associated with submerged wood, and stress that fungi are an important component of the microbial decomposer assemblage. This study provides data on the decay of selected timbers in a tropical location and enables comparison with other studies, primarily from temperate locations.

A comparison of percentage decay of test blocks submerged in various freshwater habitats is listed in Tab. 21.6. At water cooling towers operating with elevated temperatures, beech test blocks at 108 weeks lost 38.6% to 59.7% weight loss, while wood blocks exposed in river water lost only 15.6% to 22.3%. Our test samples exposed at the Bala Hala river site lost an average of 70.0%, ranging from 47.4 to 89% over a 156-week period. Noticeably lower weight loss was incurred in test samples exposed at the peat swamp site. Various factors affect weight lost as can be seen from the data presented in Tab. 21.6, with temperature the dominant parameter. The almost stagnant

Tab. 21.6: Published data on weight loss of wood.

Location	Timber species	Weight loss (%)	Temperature °C	Exposure period (weeks)	Reference
Connah's Quay: Cooling tower	Beech	59.7	25–33	108	Eaton and Jones (1971)
	Scots pine	52.9	108		
Ince cooling tower	Beech	56.2	25–33	108	
	Scots pine	56.6		108	
Agecroft cooling tower	Beech	46.0	22–30	108	
Bold cooling tower	Beech	38.6	21–32	108	
River Severn	Beech	15.6	Circa 10–20	92	Kane et al. (2002)
	Scots pine	22.3	92		
Bala Hala site	Average for nine timber species	71.21	20–30	104	Current study
Peat swamp site		23.33	20–30	104	

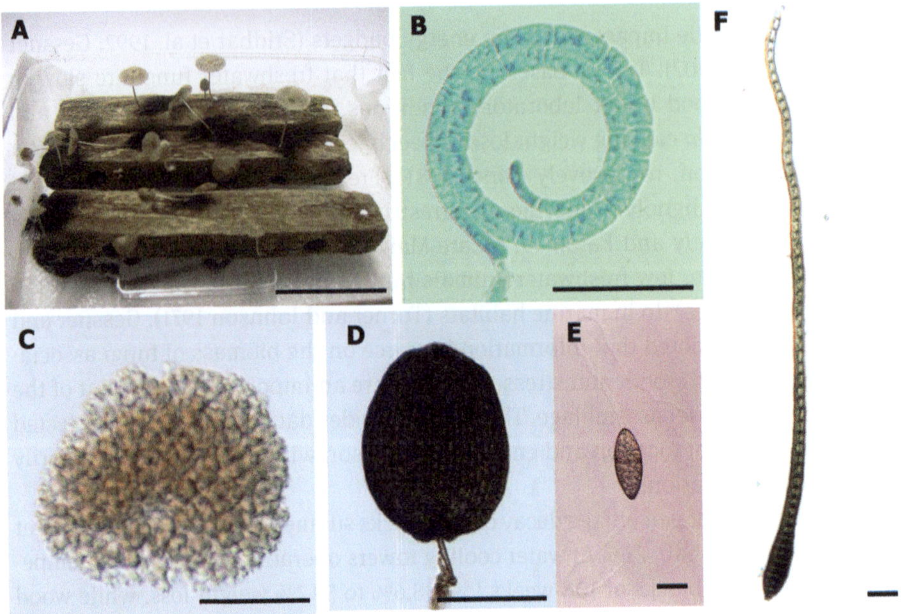

Fig. 21.4: Some freshwater fungi found in this study. A. *Coprinus* sp.1, B. *Helicomyces roseus*, C. *Candelabrum brocchiatum*, D. *Cancellidium applanatum*, E. *Jahnula appendiculata*, and F. *Ellisembia opaca*. Measure bars for A = 5 cm. B, E = 20 µm. C = 40 µm. D = 80 µm = 40 µm. F = 10 µm.

water conditions at the peat swamp with its acidic pH, and slow water flow through the location, markedly affects weight loss of the samples.

The data presented in this study confirms that wood in freshwater habitats is decayed to a significant extent, with little variation between the nine timbers submerged, and support a rich diversity of microfungi such as *Coprinus* sp.1, *Candelabrum brocchiatum*, *Helicomyces roseus*, *Cancellidium applanatum*, *Jahnula appendiculata*, and *Ellisembia opaca* (Fig. 21.4). As Sanders and Anderson (1979) showed, size of the exposed wood determines not only species diversity but also the rate at which decay takes place.

Acknowledgments

Prof. Gareth Jones is supported by the Distinguished Scientist Fellowship Program (DSFP), King Saud University, Saudi Arabia. Nattawut Boonyuen thanks Prof. Morakot Tanticharoen and Dr. Kanyawim Kirtikara for their support in our study. Associate Prof. Ka-Lai Pang is thanked for his suggestions and comments on this manuscript. Mr. Sunthorn Taodam for support with the field work at Bala Hala Wildlife Sanctuary and Sirindhorn Peat Swamp. Dr. Satinee Suetrong is thanked for re-seizing Fig. 20.4 and research funding from the TRF/BIOTEC special Program for Biodiversity Research and Training grants BRT (BRT R_145006, BRT R_252057, BRT R_647001 and BRT R_251009) with co-financing from the National Center for Genetic Engineering and Biotechnology (BIOTEC).

References

Ballie B, Collier K, Halliday J. Woody debris in pumice-bed-streams – how long does it last and its use by aquatic insects. LIRO Rep 1998;23:1–5.

Bärlocher F. The Ecology of Aquatic Hyphomycetes (ed.). Berlin: Springer-Verlag 1992.

Boonyuen N, Chuassharonnachai C, Suetrong S, Sri-indrasuthdhi V, Sivichai S, Jones EBG, et al. Savoryellales (Hypocreomycetideae, Sordariomycetes): a novel lineage of aquatic ascomycetes inferred from multiple-gene phylogenies of the genera *Ascotaiwania*, *Ascothailandia* and *Savoryella*. Mycologia 2011;103:1351–1350.

Boonyuen N, Sri-indrasuthdhi V, Suetrong S, Sivichai S. Jones EBG. *Annulatascus aquatorba* sp. nov., a lignicolous freshwater ascomycete from Sirindhorn Peat Swamp Forest, Narathiwat, Thailand. Mycologia 2012;104:746–757.

Cai L, Hyde KD, Tsui CKM. Genera of Freshwater Fungi. Hong Kong: Fungal Diversity Press 2006.

Chamier AC. Cell-wall-degrading enzymes of aquatic hyphomycetes – a review. Bot J Linn Soc 1985;91:67–81.

Eaton RA, Jones EBG. The biodeterioration of timber in watercooling towers. 1. Fungal ecology and the decay of wood at Connah'sQuay and Ince. Mat Organis 1971;6:51–80.

Gessner MO, Suberkropp K, Chauvet E. Decomposition of plant Litter by fungi in marine and freshwater ecosystems. In: Wicklow DT, Söderström B, eds. The Mycota, Volume 4. Berlin: Springer-Verlag 1997:303–322.

Gessner MO, Van Ryckegem G. Fungal decomposers in freshwater ecosystems. In: Bitton G, ed. Encyclopedia of Environmental Microbiology. New York, NY: Wiley 2002:1–20.

Goh TK. Tropical freshwater hyphomycetes. In: Hyde KD, ed. Biodiversity of Tropical Microfungi. Hong Kong: Hong Kong University Press 1997:189–228.

Gönczöl J, Rèvay. A Further studies on fungal colonization of twigs in the Morgo-stream, Hungary. Nova Hedwig 1993;56:531–542.

Ho WH. Biodiversity, ecological and ultrastructural observations of fungi on wood submerged in tropical streams. Ph.D. Thesis. Department of Ecology and Biodiversity, University of Hong Kong, 1998.

Ho WH, Hodgkiss IJ, Hyde KD. *Cheiromyces lignicola*, a new cheirosporous anamorphic species from Hong Kong. Mycologia 2000;92:582–588.

Ho WH, Yanna, Hyde KD, Hodgkiss IJ. Seasonality ad sequential occurrence of fungi on wood submerged in Tai Po Kau Fortest stream, Hong Kong. Fungal Divers 2002;10:5–10.

Ho WH, Hyde KD, Hodgkiss IJ. *Cataractispora receptaculorum* a new freshwater ascomycete from Hong Kong. Mycologia 2004;96:411–417.

Hu D-M, Cai L, Hyde KD. *Sirorthecium triseriale*, a new chirosporous anamorphic species from China. Cryptog Myco 2010;28:311–314.

Hyde KD, Goh TK. Fungi on wood submerged in the Tiriere Sr Marie-Louis, The Seychelles. South Africa J Bot 1998;64:330–336.

Hyde KD, Goh TK. Fungi on submerged wood in the River Coln, the Cotswold, UK. Mycol Res 1999;103:1561–1574.

Hyde KD, Jeewon R. Physiological studies and molecular diversity of freshwater lignicolous fungi. Fungal Diversity Research series 2003;10:173–193.
Ingold CT. *Actinospora megalospora* n. sp., an aquatic hyphomycete. Trans Br Mycol Soc 1952;35:66–70.
Ingold CT. *Tricladium chaetocladium* sp. nov. Trans Br Mycol Soc 1074;63:624–626.
Jones EBG. The decay of timber in aquatic environments. London: British Wood Preserving Association, Annual Convention. Cambridge, UK 1972:1–18.
Jones EBG. Observations on the ecology of lignicolous aquatic hyphomycetes. In: Wicklow DT, Carroll GC, eds. The Fungal Community. New York, NY: Marck-Dekker 1981:191–212.
Jones EBG Eaton RA. *Savoryella lignicola* gen. *et* sp. nov. from water cooling towers. Trans Br Mycol Soc 1969;52:161 174.
Jones EBG, Stewart R. *Tricladium varium*, an aquatic hyphomycete on wood in water-cooling towers. Trans Br Mycol Soc 1972;59:163–167.
Jones EBG, Oliver AC. Occurrence of aquatic hyphomycetes on wood submerged in fresh and brackish water. Trans Br Mycol Soc 1964;47:45–48.
Jones EBG, Wong S-W, Sivichai S, Au DWT, Hywel-Jones NL. Lignicolous freshwater Ascomycota from Thailand: *Micropeltopsis quinquecladiopsis* sp. nov. Mycol Res 1999;103:729–735.
Kane D. The effect of sewage effluent on the growth of microorganisms in the marine environment Ph.D. Thesis. Portsmouth Polytechnic 1980.
Kane DF, Tam WY, Jones EBG. Fungi colonising and sporulating on submerged wood in the River Severn, UK. In*:* Hyde KD, Jones EBG, eds. Fungal Succession. Fungal Divers 2002;10:45–55.
Lamore BJ, Goos RD. Wood-inhabiting fungi of a freshwater stream in Rhode Island. Mycologia 1978;70:1025–1034.
Leightley LE, Eaton RA. Mechanism of decay of timber by aquatic micro-organisms. Br Wood Preserv Assoc, Ann convention. Cambridge, UK 1977:1–26.
Marvanová L, Suberkropp S. *Camptobasidium hydrophilum* and its anamorphs *Crucella subtilis*: a new heterobasidiomycete from streams. Mycologia 1990;82:208–217.
Nawawi A, Webster J. *Sistotrema hamatum* sp. nov., the teleomorph of *Ingoldiella hamata*. Trans Br Mycol Soc 1982;78:287–291.
Raja HA, Violi HA, Shearer CA. Freshwater ascomycetes: *Alascospora evergladensis*, a new genus and species from the Florida Everglades. Mycologia 2010;102:33–38.
Raja HA, Miller AN, Shearer CA. Freshwater ascomycetes: Natipusllaceaem a new family of tropical fungi, including *Natipusilla bellaspor*a sp. nov. from the Peruvian Amazon. Mycologia 2012;104:569–573.
Sanders PF, Anderson JM. Colonization of wood blocks by aquatic hyphomycetes. Trans Br Mycol Soc 1979;73:103–107.
Seena S, Wynberg N, Bärlocher F. Fungal diversity during leaf decomposition in a stream assessed through clone libraries. Fungal Divers 2008;30:1–14.
Shearer CA. The role of woody debris. In: Bärlocher F, ed. The Ecology of Aquatic Hyphomycetes. Berlin: Springer-Verlag 1992:77–98.
Shearer CA, von Borman S. Pattern of occurrence of ascomycetes associated with decomposing twigs in a midwestern stream. Mycologia 1983;75:518–530.
Shearer CA, Descals E, Kohlmeyer B, Kohlmeyer J, Marvanová L, Padgett D, et al. Fungal biodiversity in aquatic habitats. Biodivers Conserv 2007;16:49–67.
Sivichai S, Jones EBG. *Stauriella* gen. nov., proposed for a new lignicolous basidiomycetous anamorph from freshwater in Thailand. Sydowia 2004;56:131–136.
Sivichai S, Hywel-Jones NL, Jones EBG. Lignicolous freshwater Ascomycota from Thailand: 1. *Ascotaiwania sawada* and its anamorph state *Monotosporella*. Mycosience 1998;39:307–312.

Sivichai S, Jones EBG. Hywel-Jones NL. Fungal colonization of wood in a freshwater stream ar Khao Yao National Park, Thailand. Fungal Divers 2000;5:71–88.

Sivichai S, Jones EBG, Hywel-Jones NL. Fungal colonisation of wood in a freshwater stream at Tad Ta Phu, Khao Yai National Par, Thailand. In: Hyde KD, Jones EBG, eds. Fungal Succession. Fungal Divers. Hong Kong: Fungal Diversity Press 2002;10:113–129.

Sridhar KR, Bärlocher F. Initial colonization, nutrient supply, and fungal activity on leaves decaying in streams. Appl Environ Microbiol 2000;66:1114–1119.

Sridhar KR, Chandrashekar KR, Kaveriappa KM. Research on the Indian subcontinent. In: Bärlocher F, ed. The Ecology of Aquatic Hyphomycetes. Berlin: Springer-Verlag 1992:182–225.

Suberkropp K. The influence of nutrients on fungal growth, productivity, and sporulation during leaf breakdown in streams. Can J Bot 1995;73:S1361-S1369.

Triska FK, Cromack K. The role of wood debris in forests and streams. In: Forest: Fresh Perspectives from Ecosystem Analysis. Proceedings 40th Biology Colloquium, 1979, Waring RH, ed. Corvallis, OR: Oregon University Press 1980:171–190.

Trisksa FJ, Sedell JR. Decomposition of four species of leaf litter in response to nitrate manipulation. Ecology 1976;57:783–792.

Tsui CKM, Hyde KD, Hodgkiss IJ. Biodiversity of fungi on submerged wood in Hong Kong streams. Aquatic Microbial Ecol 2000;21:289–298.

Tsui CKM, Goh TK, Hyde KD, Hodgkiss IJ. New species or records of *Cacumisporium*, *Helicosporium*, *Monotosporella* and *Bahusutrabeeja* on submerged wood in Hong Kong streams. Mycologia 2001;93:389–397.

Tsui CKM, Hodgkiss IJ, Hyde KD. Three new species of *Aquaticola* from the tropics. Nova Hedwig 2003;77:161–168.

Turner R, Johnson AC. Biology of marine wood boring Molluscs. In: Jones EBG, Eltringham SK, eds. Marine Borers, Fungi and Fouling Organisms of Wood. Paris: OECD 1971:1–367.

Webster J. *Nectria lugdunensis* sp. nov., the perfect stage of *Heliscus lugdunensis*. Trans Br Mycol Soc 1959;42:322–327.

Willoughby LG, Archer JF. The fungal spora of a freshwater stream and its colonization pattern on wood. Freshw Biol 1973;3:219–239.

Wong MKM, Goh TK, Hodgkiss IJ, Hyde KD, Rangoo VM, Tsui CKM, et al. Role of fungi in freshwater ecosystems. Biod Conser 1998;7:1187–120.

Wong SW, Hyde KD, Jones EBG. Annulatascaceae, a new ascomycete family from the tropics. System Ascomyc 1998a;16:17–25.

Wong SW, Hyde KD, Jones EBG. *Halosarpheia heteroguttulata* sp. nov. from submerged wood in streams. Can J Bot 1998b;76:1857–1862.

Wong SW, Hyde KD, Jones EBG. Ultrastructural studies on freshwater ascomycetes. *Fluminicola bipolaris* gen. *et* sp. nov. Fungal Divers 1999;2:189–197.

Yuen TK, Hyde KD, Hodgiss IJ. Soft rot decay in tropical freshwater fungi. Mat Organis 1999;33:155–161.

Zare-Maiavn K, Shearer CA. Extracellular enzymes activity production and cell wall degradation by freshwater lignicolous fungi. Mycologia 1988a;80:365–375.

Zare-Maiavn K, Shearer CA. Wood decay activity and cellulose production by freshwater lignicolous fungi. Intern Biodeterior 1988b;24:459–474.

Ka-Lai Pang, Kevin D. Hyde and E. B. Gareth Jones

22 Epilogue

22.1 Introduction

Freshwater fungi is a general term for those which have part of their lifecycle being found in a freshwater habitat including ice sheets, ice caps, glaciers, icebergs, bogs, ponds, lakes, rivers and streams, and underground as groundwater in aquifers and underground streams. Water cooling towers also are a habitat for freshwater fungi (Jones and Eaton 1969). Terrestrial fungi, although able to grow in freshwater liquid media in the laboratories, are not regarded freshwater fungi. Terrestrial and freshwater habitats exist side-by-side and have intimate relationships in terms of energy and nutrient supply through meteorological, geological and biological means (Likens and Bormann 1974). Some fungal propagules can be washed off from terrestrial habitats but play no role in freshwater environments. Repeated isolations from freshwater habitats which have not been isolated from the terrestrial habitat may show the ecological nature of a fungus (Park 1972; Wong et al. 1998). Taxon-specific fluorescent *in situ* hybridization (FISH) was used to detect active fungal mycelia/spores on leaves in an alpine stream, and this technique can help to determine if a fungus is indigenously freshwater as only active mycelia/spores fluoresce (Baschien et al. 2008).

Leading mycologists have written on various aspects of freshwater mycology in this book. In this concluding chapter, we briefly summarize the current knowledge of freshwater mycology, highlight knowledge gaps in our understanding of freshwater fungi and suggest areas for future research.

22.2 Freshwater fungi

Freshwater fungi are diverse taxonomically and they include ascomycetes, basidiomycetes, zygomycetes, chytridiomycetes and microsporidia, which cover all major phyla of the Kingdom Fungi. Some lineages of freshwater fungi have obviously adapted well to the freshwater environment and radiated into speciose groups such as the Jahnulales. In the Jahnulales, *Manglicola* has even evolved into the marine environment (Suetrong et al. 2011). While a few genera are exclusive freshwater (e.g. *Annulatascus*, *Jahnula*), others, especially many ascomycetes (e.g. *Anthostomella*, *Diaporthe*, *Massarina*, *Pleospora*) belong to terrestrial genera of diverse taxonomic assemblages. These many individual lines of evolution into freshwater environments may suggest that it is comparatively easier to occupy with similar physical and chemical conditions to the terrestrial milieu, than the marine environment (Jones and Pang 2012). Salinity is a key factor in determining fungal diversity in the sea (Jones 2000). Water temperature and pH, however, are factors governing fungal diversity in freshwater. There is seasonality in the occurrence of freshwater fungi, with taxa only appearing in particular seasons. *Flagellospora penicillioides* and *Lunulospora curvula* were only present in the colder months in temperate climates (Webster et al. 1976; Suberkropp 1984) and is probably related to their temperature requirements for growth and sporulation, as well as the ready supply of shed leaves (Bärlocher 1992). Concerning water pH, high diversity of aquatic hyphomycetes was found in waters between pH 5 and 7 (Bärlocher 1987). Acidic water generally does not affect fungal activity but the leaching of heavy metals caused by low pH does (Abel and Bärlocher 1984).

Freshwater Ascomycota and Basidiomycota are mainly saprobic, forming part of the decomposer community. Many freshwater ascomycetes are capable of producing cellulases, hemicellulases and ligninases to degrade wood, some forming soft-rot cavities in balsa wood and other timbers (Jones 1972; Simonis et al. 2008). Aquatic hyphomycetes also produce similar enzymes in leaf litter degradation (Suberkropp and Klug 1980), although lignin mineralization is limited (Fisher et al. 1983). Others form symbiotic relationships with other organisms in the case of lichens. Many freshwater fungi are also parasites of freshwater animals and cause diseases of other organisms. A typical example is the chytrid *Batrachochytrium dendrobatidis* which causes a catastrophic global decline in numbers of amphibians (Olson et al. 2013; See Chapter 11). Another example is Microsporidia which are parasites of arthropod and non-arthropod hosts (Vávra and Lukeš 2013). These ecological roles highlight the importance of fungi in freshwater environments.

22.3 Freshwater fungus-like organisms

Zoosporic organisms are also prevalent in freshwater environments, possibly required for the dissemination of zoosporic propagules (Gleason and Lilje 2009). Oomycota is the major group with many species in the genera *Pythium* and *Saprolegnia* causing diseases of freshwater animals. Others may serve as decomposers which may also improve the palatability of plant material and possibly increase their nutritional value by enriching them with polyunsaturated fatty acids for shredder animals (Leaño 2001). Zoospores can also be food for protozoa, which subsequently are food for organisms at higher trophic levels, a process called trophic upgrading (Neuhauser et al. 2012).

22.4 Knowledge gaps and future work in freshwater mycology

Study of freshwater fungi has mainly been restricted to the description of new taxa from various habitats such as rivers and lakes and the total number of freshwater fungi stands roughly at 3,069–4,145 (See Chapter 1). We can summarize our knowledge on freshwater fungi as follows: (1) Ascomycota is dominant (548–650 species)

(Jones and Pang 2012) in freshwater environments with few Basidiomycota (114 species) (See Chapter 4); true freshwater lichens are few (~250 species) (See Chapter 15); zygomycetes are basically absent; (2) more freshwater fungi have been reported from temperate zones, followed by selected locations in Asia; (3) freshwater fungi evolved into the freshwater environments through multiple evolutionary lines; (4) freshwater fungi have evolved specialized morphology for an aquatic existence, i.e. tetra-radiate conidia, ascospore sheaths but are often the result of convergent evolution; (5) freshwater filamentous fungi on wood have been extensively studied while their occurrence on other substrata is comparatively less known, and, (6) zoosporic fungi and fungal-like organisms are abundant in freshwater and cause diseases of animals.

Although literally all studies on freshwater fungi have suggested the dominance of the Ascomycota on various substrates while other fungal groups including zygomycetes are few, there is a bias in the study of freshwater fungi on a particular substrate such as wood, which favors the growth of ascomycetes with their ability to produce wood-degrading enzymes (Yuen et al. 2000). Moreover, isolation of zygomycetes may need special methods or techniques; dilution plating of sediment/water for zygomycetes (Shearer et al. 2007).

Culture-dependent techniques favor the isolation of a particular group of freshwater fungi, thus giving only a glimpse of the whole fungal diversity. Culture-independent techniques, such as denaturing (temperature) gradient gel electrophoresis, terminal restriction fragment length polymorphism, clone library construction of rRNA genes and the recent next-generation sequencing, are useful in uncovering the unculturable fungi (Pang and Mitchell 2005). These techniques also allow isolation of fungi after knowing what organisms are present in the samples.

A more diverse fungal community results from a number of different environments using culture-independent techniques. In a freshwater environment, a novel lineage of the Cryptomycota was recently detected in an Iowa aquifer in the USA using 454 pyrosequencing after PCR amplification by specific 18S rDNA primers and this community differs from those of the freshwater and marine environments (Livermore and Mattes 2013). *Rozella* is the only genus of the Cryptomycota and primarily known through environmental sequences (Jones et al. 2011). In another study which examined the diversity of fungi in two lakes (Pavin, Aydat) in France using cloning/sequencing of the ITS rDNA and pyrosequencing of the variable regions of the 18S rDNA, the Chytridiomycota was found to be the dominant group over the Ascomycota and the Basidiomycota (Monchy et al. 2011) and it supports the notion that there is a high diversity of chytridiomycetes in freshwater environments (See Chapter 1). Specificity of the DNA primers is crucial in this kind of study during PCR and multiple primer sets comprised of phylum-specific primers are recommended as there is not a pair of DNA primers that is specific to all fungal groups (Pang and Mitchell 2005). Also, with the current capacity to generate a large amount of sequences in a relatively short period of time, methods to analyze ITS datasets have lagged behind and it is hoped that new bioinformatic tools will solve this problem (Taylor and Houston 2011).

Deep aquifer/groundwater is a unique habitat for freshwater fungi, but so far our knowledge on them is limited. Common soil fungi, such as *Penicillium* spp. and *Rhodotorula* spp., are known from shallow aquifers and groundwaters (Lategan et al. 2012). Other obvious locations for freshwater fungi are the polar/subpolar areas with the several possible habitats, such as ice sheets, ice caps, glaciers and icebergs (See Chapters 4, 19). Lichenicolous fungi, macrofungi and soil mycobiota have been studied in terrestrial and freshwater ecosystems of the South Shetland Islands (Köhler and Olech 2011). Recent studies suggest that freshwater fungi are also present: *Lecophagus antarcticus*, a predatory hyphomycete isolated from cyanobacterial mats in a shallow lake, Signy Island (McInnes 2003); yeasts present in meltwater of various lakes in Antarctica although the numbers were low (Vaz et al. 2011, See Chapters 4, 19). Zoosporic organisms seem to be common in lakes of the Antarctica, with Ellis-Evans (1985) isolating eight chytridiomycetes, three oomycetes and one hyphochytridiomycete using baiting techniques from water samples collected in lakes of Signy Island. Whether these organisms can cause infection of the phytoplankton or aquatic plants in the lakes will require further study. Nakai et al. (2012) detected the presence of the Chytridiomycota in moss pillars at Lake Hotoke-Ike, east Antarctica by a culture-independent study using the 18S rRNA gene clone library, where they also reported a novel lineage of fungi. Ascomycetes, basidiomycetes and chytridiomycetes are also present in cyanobacterial mats in both the Arctic and Antarctic freshwater systems, again through the 18S rDNA clone library construction (Jungblut et al. 2012). All these studies suggest the prevalence of fungi in freshwater habitats in the polar areas and Nakai et al. (2012) suggested that fungi might play a decomposer or a parasitic role in these environments. The recent development of metaproteomics might further help to understand the specific roles of fungi in freshwater environments through identification and quantification of expressed proteins (von Bergen et al. 2013).

The study of freshwater fungi has been concentrated in North America, Europe and selected countries in the tropics, i.e. China (including Hong Kong), Thailand and Malaysia after extensive collecting efforts by mycologists working in these places (Shearer et al. 2007). The conclusion from these studies is that there is a higher diversity of freshwater fungi in temperate areas than tropical/subtropical locations (Jones and Pang 2012). The rest of the world (Africa, South America, Australasia, some Asian countries) is largely unsampled for freshwater fungi and future exploration will definitely result in many more new taxa and enhance our understanding on the geographical distribution and substrate preference of freshwater fungi.

Many substrates have been extensively sampled for freshwater fungi, i.e. submerged wood and leaves, water column, foam and animals (Wong et al. 1998) while other substrates including grasses, algae, planktons and macrophytes in freshwater environments have rarely been examined for fungi, either epiphytically or endophytically.

Luo et al. (2004) examined the diversity of fungi on submerged *Phragmites australis*, bamboo and *Pennisetum purpureum* in a polluted lake in China and found between 21 and 32 species. No obvious substrate specificity was observed and some fungal species were more common on a particular host (Luo et al. 2004). *Phragmites australis* has a high diversity of freshwater fungi in a study at Mai Po Marshes, Hong Kong (Wong and Hyde 2001), and 600 taxa have already been documented from *Phragmites australis* litter (Gessner and van Ryckegem 2003). With the worldwide distribution of this grass, it is possible that more species will be found on this plant.

During the last decade, our knowledge of the phylogeny of freshwater fungi has advanced significantly with the use of the rRNA gene cluster. Many new phylogenetic lineages have been found for freshwater ascomycetes, however, only a handful of freshwater fungi were sequenced giving only a partial picture on their evolution (Shearer et al. 2009, Boonyuen et al. 2011, Suetrong et al. 2011, Ferrer et al. 2012, Raja et al. 2012). Taxa sequenced mainly belong to the Pleosporales and the Jahnulales, while many key orders/families have not been sequenced. As mentioned in Chapter 3, accuracy of the phylogeny of freshwater ascomycetes relies heavily on the sequencing of the ex-type cultures but many of them are not available. Recollection of these species, especially many freshwater Sordariomycetes, is urgently required. On the other hand, previous studies on the phylogeny of freshwater fungi were mainly based on the ribosomal RNA gene regions which resulted in unstable placement for many taxa and low branch support values. Protein genes such as elongation factor (TEF1) and RNA polymerase (RPB1, RPB2) are mostly lacking in these studies and should be sequenced.

Information on morphological variability of cosmopolitan species is lacking which leads to confusion on species delimitation. An example is *Massarina ingoldiana* which has been collected in various parts of the world, such as Australia, Brunei, Hong Kong, United States (Shearer and Hyde 1997, Tsui et al. 2000). Hirayama et al. (2010) have revealed that this species is unrelated to the genus *Massarina* and a composite species which belongs to two phylogenetically unrelated groups. Two new genera, *Tingoldiago* and *Lindgomyces* (in the new family Lindgomycetaceae), were consequently introduced to accommodate the two groups and it was suggested that the elaborate ascospore sheath resulted through convergent evolution. Other morphological characters (i.e. tetra-radiate conidia) have evolved for adaptation to life in aquatic environments (Hirayama et al. 2010). Many morphological characters of freshwater fungi have not been examined closely, including peridial wall structure (through sectioning) but they might be useful in the delineation of taxa (Pang et al. 2012).

Only a few freshwater asexual morphs have been sequenced and thus their higher-ranked classification is not known (See Chapter 5). Many of them have not been sequenced and even if sequences are available, it is always difficult to link to a named sexual morph due to the low sequence similarity with the closest sequence match. Cultural study is always required to prove teleomorph/anamorph relationship and it is a tricky task to induce either stage (Zhao et al. 2007). Even if a named species is linked, it is difficult to assign the correct name to use to get in line with the current code of nomenclature in fungi. Many freshwater fungi belong to well-known terrestrial genera and their taxonomical issues are still yet to be resolved and the adoption of many names needs discussion and approval from the fungal committee of nomenclature. Therefore, in many cases in this book, we have listed names for both the sexual and asexual stages of freshwater fungi when known.

22.5 Conclusions

Several areas have been suggested above for further research. The current obstacles are, as in other disciplines in mycology, the lack of funding and trained aquatic mycologists. It is obvious that the number of research groups and published articles has been in decline over the last decade. The above two factors are intimately related to the lack of research money, resulting in fewer students being trained. The ultimate question is: how can we attract money for freshwater mycology? Freshwater fungi and fungal-like organisms that potentially cause diseases especially on economic organisms and humans can attract funding. Research projects on basic research including the study of biodiversity, phylogeny, ecology and taxonomy of freshwater fungi can hardly get approved in most countries. Applied research topics such as fungi as bioindicator organisms in freshwater environments and fungi of aquifers and underground water may be easier to secure funds. Applied-oriented basic research has been the key to get research funds and projects related to bioactive metabolites and enzymes produced by freshwater fungi from diversity studies are more likely to get supported by government agencies and industry. However, freshwater fungi are not strong antimicrobial producers and only a few anti-microbial structures have been identified (Jiao et al. 2006).

References

Abel TH, Bärlocher F. Effects of cadmium on aquatic hyphomycetes. Appl Environ Microb 1984;48:245–251.

Bärlocher F. Aquatic hyphomycete spora in 10 streams of New Brunswick and Nova Scotia. Can J Bot 1987;65:76–79.

Bärlocher F. Community organization. In: Bärlocher F, ed. The Ecology of Aquatic Hyphomycetes. Berlin: Springer-Verlag 1992:38–76.

Baschien C, Manz W, Neu TR, Marvanová L, Szewzyk U. *In situ* detection of freshwater fungi in an alpine stream by new taxon-specific fluorescence *in situ* hybridization probes. Appl Environ Microbiol 2008;74:6427–6436.

Boonyuen N, Chuaseeharonnachai C, Suetrong S, Sri-indrasutdhi V, Sivichai S, Jones EBG, et al. Savoryellales (Hypocreomycetidae, Sordariomycetes): a novel lineage of aquatic ascomycetes inferred from multiple-gene phylogenies of the genera *Ascotaiwania*, *Ascothailandia*, and *Savoryella*. Mycologia 2011;103:1351–1371.

Ellis-Evans JC. Fungi from maritime Antarctic freshwater environments. Br Antarct Surv Bull 1985;68:37–45.

Ferrer A, Miller AN, Sarmiento C, Shearer CA. Three new genera representing novel lineages of Sordariomycetidae (Sordariomycetes, Ascomycota) from tropical freshwater habitats in Costa Rica. Mycologia 2012;104:865–879.

Fisher P, Davey R, Webster J. Degradation of lignin by aquatic and aero-aquatic hyphomycetes. Trans Br Mycol Soc 1983;80:166–168.

Gessner MO, van Ryckegem G. Water fungi as decomposers in freshwater ecosystems. In: Bitton G, ed. Encyclopaedia of Environmental Microbiology. New York, NY: Wiley 2003:1–38.

Gleason FH, Lilje O. Structure and function of fungal zoospores: ecological implications. Fungal Ecol 2009;2:53–59.

Hirayama K, Tanaka K, Raja HA, Miller AN, Shearer CA. A molecular phylogenetic assessment of *Massarina ingoldiana sensu lato*. Mycologia 2010;102:729–746.

Jiao P, Gloer JB, Campbell J, Shearer CA. Altenuene derivatives from an unidentified freshwater fungus in the family Tubeufiaceae. J Nat Prod 2006;69:612–615.

Jones EBG, Eaton RA. *Savoryella lignicola* gen. *et* sp. nov. from water cooling towers. Trans Br Mycol Soc 1969;52:161–174.

Jones EBG, Pang KL. Tropical aquatic fungi. Biodivers Conserv 2012;21:2403–2423.

Jones EBG. Marine fungi: some factors influencing biodiversity. Fungal Divers 2000;4:53–73.

Jones EBG. The decay of timber in aquatic environments. Br Wood Preserv Assoc Convent, Cambridge, UK 1972:1–18.

Jones MDM, Forn I, Gadelha C, Egan MJ, Bass D, Massana R, et al. Discovery of novel intermediate forms redefines the fungal tree of life. Nature 2011;474:200–203.

Jungblut AD, Vincent WF, Lovejoy C. Eukaryotes in Arctic and Antarctic cyanobacterial mats 2012;82:416–428.

Köhler P, Olech M. Polish botanical and mycological studies of the Antarctic terrestrial and fresh water ecosystems in 1977–2009: An overview. Polish Polar Res 2011;32:157–174.

Lategan MJ, Torpy FR, Newby S, Stephenson S, Hose GC. Fungal diversity of shallow aquifers in Southeastern Australia. Geomicrobiol J 2012;29:352–361.

Leaño EM. Straminipilous organisms from fallen mangrove leaves from Panay Island, Philippines. Fungal Divers 2001;6:75–81.

Likens GE, Bormann FH. Acid rain: a serious regional environmental problem. Science 1974;184:1176–1179.

Livermore JA, Mattes TE. Phylogenetic detection of novel Cryptomycota in an Iowa (United States) aquifer and from previously collected marine and freshwater targeted high-throughput sequencing sets. Environ Microbiol 2013;15:2333–2341.

Luo J, Yin JF, Cai L, Zhang KQ, Hyde KD. Freshwater fungi in Lake Dianchi, a heavily polluted lake in Yunnan, China. Fungal Divers 2004;16:93–112.

McInnes SJ. A predatory fungus (Hyphomycetes: Lecophagus) attacking Rotifera and Tardigrada in maritime Antarctic lakes. Polar Biol 2003;26:79–82.

Monchy S, Sanciu G, Jobard M, Rasconi S, Gerphagnon M, Chabé M, et al. Exploring and quantifying fungal diversity in freshwater lake ecosystems using rDNA cloning/sequencing and SSU tag pyrosequencing. Environ Microbiol 2011;13:1433–1453.

Nakai R, Abe T, Baba T, Imura S, Kagoshima H, Kanda H, et al. Eukaryotic phylotypes in aquatic moss pillars inhabiting a freshwater lake in East Antarctica, based on 18S rRNA gene analysis. Polar Biol 2012;35:1495–1504.

Neuhauser S, Glockling SL, Leaño EM, Lilje O, Marano AV, Gleason FH. An introduction to fungus-like microorganisms. In: Jones EBG, Pang KL, eds. Marine Fungi and Fungal-like Organisms. Berlin: De Gruyter 2012:137–151.

Olson DH, Aanensen DM, Ronnenberg KL, Powell CI, Walker SF, Bielby J, et al. Mapping the global emergence of *Batrachochytrium dendrobatidis*, the amphibian chytrid fungus. PLOS One 2013;8:e56802.

Pang KL, Chiang WL, Jheng JS. Morphological Evaluation of peridial wall, ascus and ascospore characteristics in the delineation of genera with unfurling ascospore appendages (Halosphaeriaceae). In: Raghukumar C, ed. Biology of Marine Fungi. Berlin: Springer-Verlag 2012:159–171.

Pang KL, Mitchell JI. Molecular approaches for assessing fungal diversity in marine substrata. Bot Mar 2005;48:332–347.

Park D. On the ecology of heterotrophic microorganisms in freshwater. Trans Br Mycol Soc 1972;58:281–290.

Raja HA, Miller AN, Shearer CA. Freshwater ascomycetes: Natipusillaceae, a new family of tropical fungi, including *Natipusilla bellaspora* sp. nov. from the Peruvian Amazon. Mycologia 2012;104:569–573.

Shearer CA, Descals E, Kohlmeyer B, Kohlmeyer J, Marvanová L, Padgett D, et al. Fungal biodiversity in aquatic habitats. Biodivers Conserv 2007;16:49–67.

Shearer CA, Hyde KD. *Massarina ingoldiana,* a new ascomycete from freshwater habitats. Mycologia 1997;89:114–119.

Shearer CA, Raja HA, Miller AN, Nelson P, Tanaka K, Hirayama K, et al. The molecular phylogeny of freshwater Dothideomycetes. Stud Mycol 2009;64:145–153.

Simonis JL, Raja HA, Shearer CA. Extracellular enzymes and soft rot decay: Are ascomycetes important degraders in freshwater? Fungal Divers 2008;31:135–146.

Suberkropp K, Klug MJ. The maceration of deciduous leaf litter by aquatic hyphomycetes. Can J Bot 1980;58:1025–1031.

Suberkropp K. Effect of temperature on seasonal occurrence of aquatic hyphomycetes. Trans Br Mycol Soc 1984;82:53–62.

Suetrong S, Boonyuen N, Pang KL, Ueapattanakit J, Klaysuban A, Sri-indrasutdhi V, et al. A taxonomic revision and phylogenetic reconstruction of the Jahnulales (Dothideomycetes), and the new family Manglicolaceae. Fungal Divers 2011;51:163–188.

Taylor DL, Houston S. A bioinformatics pipeline for sequence-based analyses of fungal biodiversity. In: Xu JR, Bluhm BH, eds. Fungal Genomics: Methods and Protocols, Methods in Molecular Biology, vol. 722. New York, NY: Humana Press 2011:141–155.

Tsui CKM, Hyde KD, Hodgkiss IJ. Biodiversity of fungi on submerged wood in Hong Kong streams. Aquat Microb Ecol 2000;21:289–298.

Vávra J, Lukeš J. Microsporidia and 'the art of living together'. Adv Parasitol 2013;82:253–319.

Vaz ABM, Rosa LH, Vieira MLA, de Garcia V, Brandão LR, Teixeira LCRS, et al. The diversity, extracellular enzymatic activities and photoprotective compounds of yeasts isolated in Antarctica. Brazilian J Microbiol 2011;42:937–947.

Von Bergen M, Jehmlich N, Taubert M, Vogt C, Bastida F, Herbst FA, et al. Insights from quantitative metaproteomics and protein-stable isotope probing into microbial ecology. ISME J 2013;7:1877–1885.

Webster J, Moran ST, Davey RA. Growth and sporulation of *Tricladium chaetocladium* and *Lunulospora curvula* in relation to temperature. Trans Br Mycol Soc 1976;67:491–495.

Wong MKM, Goh TK, Hodgkiss IJ, Hyde KD, Ranghoo VM, Tsui CKM, et al. Role of fungi in freshwater ecosystems. Biodivers Conserv 1998;7:1187–1206.

Wong MKM, Hyde KD. Diversity of fungi on six species of Gramineae and one species of Cyperaceae in Hong Kong. Mycol Res 2001;105:1485–1491.

Yuen TK, Hyde KD, Hodgkiss IJ. Soft rot decay in tropical freshwater fungi. Mater Organis 2000:33:155–161.

Zhao GZ, Liu XZ, Wu WP. Helicosporous hyphomycetes from China. Fungal Divers 2007;26:313–524.

Index

acid-tolerant yeast 450
acidic aquatic systems 444, 452, 453
acidic environments 447, 449, 451, 452, 454, 457, 458, 461
acidic lakes and rivers 444, 450
Acidic rock drainage 454
acidification 394, 395, 406, 407
acidophiles 457
acidophilic 449, 454, 456, 457, 458, 460, 461
actinomyxidia 162
adaptation 65
Aero-aquatic 109, 126
aeroaquatic 25, 39
agriculturally important disease 161
Akinetes 297
Albuginaceae 184
Albuginales 180, 181, 184, 185, 192, 194, 199
algae 333, 336, 341, 342, 343, 350, 351
algal blooms 284
algal-yeast association 458
alkaline proteinases 419
allochthonous 389, 413, 415, 432, 451, 452, 454, 457, 458, 474
alpha- and beta-tubulins 163
alternative respiratory chain 168, 172
alveolates 4, 279
amitochondriate 159
amoebae 205, 206, 279, 282, 293, 295, 299, 301, 304, 307, 311, 313, 314, 318
Amoebidiales 359, 363, 366
amphibian 245, 246, 247, 249, 250, 251, 252, 253, 254, 255, 256, 257, 258, 259, 260, 261, 262
amphibians 203, 206
amphibious lichens 336, 339, 343, 348
amphipods 422, 426, 432, 439
anabolic metabolism 169
anamorphic fungi 109, 128
Annulatascaceae 48, 56, 57, 58, 59, 60, 62, 63, 66, 67, 68, 71
Antarctica 484, 487, 488
anthropogenic activities 392
Anura 245
Aphelida 279, 298
Aphelidea 279, 280, 298
apical ascal ring 57
apicomplexans 177, 178
appendages 27, 35, 38, 42
aquaculture 203, 210, 211, 217, 219, 234, 235, 236, 242
Aquasporidia 163
aquatic lichens 333, 339, 340, 341, 342, 343, 344, 346, 347, 352, 354, 355
aquifers 481, 484, 486, 487
Archezoa 163
arteritis 264, 266, 270, 277, 278
arthropod 157, 160, 163, 203, 205, 359, 360, 365, 368
Ascomycete 413
ascomycetes 373, 379, 380, 383, 387, 389, 466, 477, 478, 479, 481, 482, 483, 485, 486, 488
ascomycetous yeasts 446, 449
Ascomycota 1, 5, 9, 19
Asellariales 359, 360, 364
Asexual fungi 109, 380, 383
asexual morphs 379, 380, 466, 467, 468
asexual states 25, 28
astaxanthin 294
ATP 164, 168, 169, 176, 415
autochthonous 449, 451, 452, 454, 458
autochthonous yeast 449, 452, 454
autotrophs 279

Bacillariophyta 280
bacteria 391, 398, 407, 408, 409, 410, 412, 415, 416, 420, 422, 423, 424, 427, 435, 437, 438, 439, 440, 443, 445, 458, 463
ballistospory 73, 88, 91
Barcode of Life Database 12
basidiomycetes 380, 383, 389, 466, 481, 484
basidiomycetous yeasts 446, 454, 462
Basidiomycota 5, 6, 9, 15, 19, 73, 83, 92, 93, 95, 98, 99, 103, 104
Batrachochytrium dendrobatidis 245, 257, 258, 260, 261
bicosoecids 178
biflagellate zoospores 177
Bigyra 178
binucleate cells 73, 95, 96
bioaccumulation 456
Bioactive compounds 10
biodiversity 371, 386

biogeochemistry 396
Biogeography 344
bioinformatic tools 484
biological control 305, 314, 315, 316, 322, 325
bioprospecting 459
bioremediation 444, 456, 459, 463
biosorption 456, 462
biotrophs 157, 163, 164, 167, 168, 169, 170, 205
biotrophy 180, 181, 185, 188, 194
bitunicate 25, 36
Blastocladiales 294
Blastocladiomycota 5, 6, 7, 133, 150, 279, 293, 298, 305, 306, 313, 317, 318, 326
brackish 342

cancer-like strategies 169
Caudata 245
cell death 207, 208
cellobiose 418
cellulose 417, 418, 419, 427
cercozoa 279
chelation 456, 463
chemotactic 224, 228, 237
chemotaxis 206, 228, 229, 237, 240, 264, 267, 306, 313
chitin-rich inner wall 166
Chlorophyta 280, 293
chloroplasts 178, 199
Chromalveolata 177, 178, 197
chromalveolates 178, 196
Chromista 177, 195, 199
Chrysophyta 280
chytrid 245, 246, 247, 252, 253, 255, 257, 258, 260, 261, 262
Chytridiomycetes 133, 134, 138, 148, 151, 152, 481, 483, 484
chytridiomycosis 15, 245, 246, 247, 252, 253, 257, 258, 260, 261, 262
Chytridiomycota 5, 6, 7, 9, 15, 21, 133, 134, 138, 140, 147, 149, 150, 151, 152, 153, 245, 257, 279, 282, 283, 292, 293, 298, 299, 302, 303, 304, 305, 306, 307, 318, 323, 326
chytrids 1, 4, 6
ciliates 177, 178, 282, 295
clamp-connections 73
coelomycetes 109, 112, 113, 118, 127
coenocytic hyphae 264
coevolution 298
colonizers 389, 411
commercial fresh fruit juices 162
community structure 391, 393, 395, 402, 404
conidia 389, 393, 398, 399, 406, 409, 412
Conservation 352, 355, 358
consumers 393, 416, 417, 423, 427, 428, 429, 431
convergent evolution 60, 336
cooling towers 481, 487
COX2 187
crabs 417
crayfish 427, 438
Crustacea 360, 363, 365, 366
crustose growth forms 341
cryptic species 7
cryptomonads 177, 178
Cryptomycota 279, 280, 281, 282, 299, 301, 303, 306, 322, 323, 326, 327, 483, 487
Culicidae 367, 368
cultivable yeast diversity 443
Culture-dependent techniques 483
Culture-independent techniques 483
Cyanobacteria 280, 284, 304, 284, 293, 294, 295, 297, 299, 300, 301, 302, 303, 304, 333, 340, 342, 343, 351
cysts 206, 207, 210

D1/D2 regions 86
DDT 398, 409
decomposer community 482
decomposers 145
decomposition 389, 391, 392, 393, 394, 395, 396, 402, 403, 404, 405, 406, 407, 408, 410, 411, 465, 467, 478
deforestation 371
Degradation 418, 440, 465, 479
dematiaceous hyphomycetes 25
dermocystidiosis 203, 211
desiccation 337, 339, 341, 346
detritivores 391, 409
Devonian Rhynie Chert 294, 304
dimorphic 83, 84, 85
dimorphic life cycles 84
dinoflagellates 177, 178
diploid stage 168
diplokaryon 166
Diplopoda 360, 365
discomycetes 47, 64
disease 203, 211, 215, 305, 312, 313, 314, 324, 325, 326, 329
divergent and reductive evolution 166

diversity 424, 426, 427, 428, 429, 430, 432, 433, 434, 435, 437, 438, 439
Dothideomycetes 25, 26, 28, 31, 32, 35, 37, 38, 39, 40, 41, 42, 43, 44, 114, 120, 126, 129, 130, 488
downy mildews 181, 184, 185, 187, 188, 189, 193, 194, 196, 197, 198, 199
Drinking water treatment plants 161

Eccrinales 359, 360, 363, 365, 368
Ecological stoichiometry 423, 435, 440
economic losses 217
emergent aquatic plants 25
emerging infectious diseases 252
Encystment 225
endobiotic 136, 138, 139
endoglucanase 465
endoparasite 140
endophytes 9, 14, 16, 17, 19, 20, 21, 22, 26, 40, 66, 81, 99, 101, 104, 106, 107
endospores 205, 206
endosymbiosis 178, 197, 422, 423
endosymbiotic 419
engineered nanoparticle 396
enzymatic activities 465
Enzyme Linked Immuno-Sorbent Assay 270
epiphytic lichens 348
Epitypification 66
epizootic ulcerative syndrome 217, 237, 241
ergosterol 415
eucarpic 138, 140, 142
Euglenophyta 280
euphotic zones 280, 282, 295
eutrophication 391, 393, 407, 408, 424, 427, 437, 439
evolution 28, 40, 41, 45
Exobasidiomycetes 83, 92
exoenzymes 418, 420, 422
extracellular enzymes 389, 392, 396
extremolytes 444
extremophiles 443, 460, 461
extremozymes 444

facultative parasites 305
fallen leaves 417
fatty acids 422
fecal-oral routes 160
filter feeders 389
filter-feeders 419, 429
fine particulate organic matter 419
fingerprinting 447, 451
fish 157, 160, 161, 163, 167, 170, 172, 176, 203, 205, 206, 207, 208, 209, 210, 211, 212, 213, 214, 215, 481
fissitunicate 25, 36, 39
flagellum 279, 282
fluorescent *in situ* hybridization 481
foliose freshwater lichens 343
food webs 394, 404
fossil 294, 295, 364
freshwater ascomycetes 47, 57, 61, 67, 68, 70, 71
freshwater asexual fungi 12
Freshwater lichens 7, 333, 336, 337, 338, 339, 340, 342, 343, 344, 345, 346, 347, 348, 349, 350, 351, 352, 353, 354, 356
freshwater Sordariomycetes 47, 48, 57, 66
Freshwater yeast 1
fruticose growth forms 346, 348
fumagillin 160
fungal activity 391, 392, 393, 394, 403, 410
fungal biomass 415, 417, 418, 419, 420, 422, 424, 425, 427, 429, 430, 432, 435, 436
fungal colonization 402, 403, 406, 466, 477
fungal communities 425, 428, 429, 431, 433, 434, 435
fungal community 467, 473
fungal decomposers 391, 392
fungal diseases 1
fungal diversity 372, 381, 382, 383, 386, 387
fungal nature of microsporidia 165
Fungal respiration rates 402
fungicide 398

galaxies 179
gelatinous sheaths 26
gene-sparing ancestor 165
genotypic 447
geographical distribution 26
germination 225, 226, 227, 231, 233, 238, 239, 243
glaciers 481, 484
global warming 299
Glomeromycota 7
groundwater 481, 484
gut fungi 359, 365, 367
Gymnophiona 245

haemagglutination 270
Halophytophthora 179, 180, 181, 184, 185, 187, 188, 190, 191, 193, 197, 198
Halosphaeriaceae 62, 63, 70

haplosporidia 162
haptophytes 177, 178, 195
Harpellales 359, 360, 361, 362, 363, 365, 366, 367, 368, 369
heavy metals 391, 393, 394, 407, 408, 443, 444, 456, 462, 463
hemicelluloses 418
herbaceous 26
heterokont 279
heterotrophic microorganisms 443, 460
highly divergent sequences 159, 163
(HMG)-type transcription factor 165
holobasidia 83, 95
holocarpic 136, 138
homing response 224
hosts 217, 223, 224, 225, 228, 229, 234, 235
human pathogens 180
humans 157, 160, 161, 167, 170, 171, 172
hyphochytrids 178
hyphomycetes 6, 9, 12, 13, 14, 17, 18, 19, 20, 21, 22, 85, 87, 91, 94, 97, 98, 99, 100, 101, 103, 105, 106, 107, 108, 109, 110, 111, 112, 114, 117, 118, 119, 123, 125, 126, 127, 128, 129, 131, 389, 390, 391, 392, 393, 394, 395, 398, 404, 405, 406, 407, 408, 409, 410, 411, 412, 413, 414, 415, 417, 419, 422, 423, 424, 425, 426, 427, 428, 429, 430, 431, 432, 433, 434, 435, 436, 437, 438, 439, 440, 465, 466, 477, 478

ice caps 481, 484
ice sheets 481, 484
icebergs 481, 484
immunochromatographic 270, 275
immunodiffusion 270, 271, 274
immunofluorescence 271
immunoperoxidase staining 271
infection 207, 208, 210, 217, 219, 232, 233, 235, 236, 239, 241, 242, 245, 246, 247, 249, 251, 253, 254, 257, 259, 260, 261, 282, 283, 284, 292, 294, 297, 298, 299, 301, 302, 304
Ingoldian fungi 109, 110, 118, 127, 466
inoperculate 139, 142, 145
insect larvae 359, 360, 366, 368
insect pollinators 162
Insecta 360, 363
insecticide 430, 440
insects 417, 422, 434, 435
International Code of Botanical Nomenclature 164, 175
inundation tolerance 337, 340, 348
invertebrate detritivores 389, 397
invertebrates 25, 203, 205, 305, 306, 307, 309, 311, 312, 315, 322, 323, 324, 326, 328, 413, 414, 415, 416, 417, 418, 419, 420, 421, 422, 423, 424, 425, 426, 428, 431, 432, 433, 434, 435, 436, 437, 438, 439, 440, 441
itraconazole 271
ITS 181, 184, 185, 186, 187, 188, 189, 190, 192, 193, 195, 197

keratitis 264, 266, 268, 274
Kickxellomycotina 359, 368
Kill- the-winner 284
kinetosome 143, 145
Kingdom Fungi 157, 481

labyrinthulids 178
Labyrinthulomycota 279
laccases 398, 409
lagenidiosis 180
lateral gene transfer 178
leaf decay 431
leaf decomposition 371
leaf degradation 415
lentic habitats 26
Leotiomycetes 119, 120, 125, 126
leucine aminopeptidase 396
lichen communities 337, 339, 341, 342, 345, 348, 349, 350, 353
lichen symbiosis 340, 342, 353
lichenicolous fungi 334
life cycle 159, 164, 166, 167, 172
lignicolous 26, 42, 43
lignin 417
lignocellulose 465
lignolytic enzymes 475
lipids 417, 420, 422, 431, 434
litter decomposition 402, 403
litter mass loss 389, 402
lotic habitats 26
low proton membrane permeability 456
LSU 179, 181, 184, 185, 186, 187, 188, 189, 190, 191, 192, 193, 194, 198

macrophytes 25
marine and freshwater crustacea 160
Marinosporidia 163, 175
mass loss 415, 421, 422, 427, 433
mastigonemes 279

Melbourne code 63
merogonic stage 166
meront 159, 167
Mesomycetozoa 359
Mesomycetozoea 203, 205, 211, 212, 214, 215, 245, 247, 257, 258, 260, 305, 311, 312, 313, 314, 317, 318, 323, 326, 328
metabolic functions 456
metabolic pathways 167, 169
metal hydroxides 394
metaproteomics 484, 488
metazoans 282
micro-flora 459
microbial activity 397, 403, 408, 409
microbial respiration 393, 395, 411
microbody-lipid complex 283
Microbotryomycetes 83, 85, 86, 93, 94, 107
Microsporidia 7, 15, 19, 157, 159, 160, 161, 162, 163, 164, 165, 166, 168, 171, 172, 173, 174, 175, 176, 280, 481
microsporidian 171, 172, 173, 174, 175, 176
microsporidian genomes 167, 168, 171
microsporidian hexokinases 169
mining 392, 394, 411
mitochondria 177, 198
mitosome 159, 162, 163, 168, 169, 171
mitosporic fungi 109
MLC cisterna 148
Monoblepharidomycetes 133, 134, 148
Monoblepharomycota 279
monocentric 138, 139, 140, 141, 142, 144, 145, 146, 147
monophyletic 33, 35, 36, 57, 58, 59, 60, 61, 65, 66, 70
Mortality 209, 212, 219, 234, 241
mosquito larvae 367
mutualists 203, 365
mycobionts 93
mycoparasite 93, 97
mycoparasitism 94, 103
mycorrhizal association 93
mycotic diseases 217
myxosporidia 161, 162

nanomaterials 395, 396, 408
nanoparticle 391, 395, 396, 397, 404, 406, 409, 410, 411, 412
nanotechnology 395
necrosis 223
necrotrophy 180
Nematocera 365
nematode-trapping fungi 10, 22
Neocallimastigomycota 133, 279
nitrogen immobilization 413
nitrogen 391, 392, 407, 408, 413, 415, 418, 422, 423, 424, 438
nitrous oxides 394
nucleomorph 178, 197
nutrients 389, 391, 392, 393, 395, 404, 405, 407, 408, 411, 412

obligate parasite 133, 143, 147, 179, 180, 196, 305
Ochrophyta 178
oligotrophic 391
Olpidiaceae 294
Olpidium 279, 303
One name one fungus 13
Oomycetes 10, 13, 133, 134, 142, 177, 178, 179, 180, 185, 195, 196, 197, 198, 217, 219, 221, 223, 224, 225, 229, 231, 232, 233, 235, 241, 242, 243, 263
Oomycota 177, 196, 197, 198, 245, 250, 251, 253, 254, 257, 279, 305, 306, 307, 313, 318, 323, 325, 328
oospore 177, 179, 180, 185, 186, 187, 188, 190, 191, 193, 305, 306, 307, 310, 311, 313, 314, 315, 316, 319, 320, 321, 323, 327, 328
operculate 139, 140, 141, 142, 144, 145, 146, 147
Opisthokont 203, 214, 279
Opisthosporidia 279, 280
Orbiliomycetes 120
organic carbon 389, 391, 406
organic matter 391, 393, 396, 404, 408, 411
organic sinks 371
oxygen deficiency 391, 393

palm debris 25
palm fungi 372, 386
Paramicrosporidium 280
parasite evolution 297
parasites 203, 205, 206, 210, 211, 212, 214, 215, 279, 280, 283, 284, 292, 293, 294, 295, 297, 298, 299, 300, 301, 303, 304, 305, 306, 307, 309, 311, 313, 314, 317, 318, 321, 322, 323, 324, 325, 326, 327, 329, 330
parasitism 187, 188, 189, 197, 284, 294, 295, 297, 298, 299, 301, 302, 303, 304

particulate matter 305
pathogen 133, 140, 143, 147, 151, 152, 153, 245, 246, 247, 248, 250, 251, 254, 255, 256, 257, 258, 259, 260, 261
pathogenic 203, 211, 213, 214
pathogens 180, 181, 184, 185, 186, 188, 189, 195, 196, 197, 198, 203, 205, 209, 210, 211, 212, 213, 215, 217, 220, 221, 225, 227, 233, 234, 235, 237, 238, 242, 243, 366, 367
peat swamp 54, 61, 62, 63, 466, 467, 468, 471, 472, 473, 475
Peat swamp forests 371
pectin 418, 419, 427
periphyton 423
perkinsids 177, 178
Perkinsozoa 279
Perkinsozoea 245, 251, 257
Peronosporaceae 181, 184, 187, 188, 189, 196, 199
Peronosporales 177, 179, 180, 181, 184, 187, 188, 190, 192, 193, 194, 197, 198, 199, 217, 225, 233
Peronosporomycetes 177, 196, 197, 198, 199
Peronosporomycetidae 179, 192, 196
pH 246, 251, 255, 256, 259, 481
phenotypic 447, 449
phosphatase 396
phosphorus 391, 392, 394, 406, 411
photobionts 336, 341, 342, 348
photosynthesis 341, 354
photosynthetic organisms 279
Phycomycetes 1, 22, 279, 303
phylloplane 443, 449
phylogeny 135, 150, 153
Phytophthora 177, 180, 181, 184, 185, 187, 188, 189, 190, 192, 193, 194, 195, 196, 197, 198, 199
phytoplankton 279, 280, 283, 284, 292, 295, 296, 297, 298, 299, 300, 301, 302, 303, 304, 305, 329
plant litter 25, 389, 392, 393, 394, 395, 403, 411
plastids 178, 195, 197
polar tube 159, 166, 167, 171, 174, 176
pollutants 392, 393, 394, 397, 398, 404, 405, 411
polycentric 138, 139, 141, 145, 146
polycyclic aromatic hydrocarbons 398
polyextremophiles 443
polyphenols 418
polyphyletic 33, 34, 42, 44, 57, 58, 65, 66, 69, 118, 120, 125
polyunsaturated fatty acids 482
population dynamics 298, 300
posteriorly uniflagellate zoospores 134
Precambrian 294
predatory hyphomycete 484
primary producers 391, 394
prokaryotes 391
proteases 229, 232, 233, 255
proteins 417, 420, 431, 438
proteromonads 178
protozoa 482
pseudopods 279, 299
psychrophilic yeasts 454
public health issue 162
pyrosequencing 7, 12, 95, 371, 483, 487
Pythiales 177, 179, 180, 181, 184, 185, 187, 188, 190, 192, 193, 194
Pythiogetonaceae 184, 186, 193
pythiosis 263, 264, 265, 266, 267, 268, 269, 270, 271, 272, 273, 274, 275, 276, 277, 278
pythiosis insidiosi 263, 276, 277
Pythium 180, 181, 184, 185, 186, 187, 188, 189, 190, 192, 193, 194, 197, 198, 199
Pythium insidiosum 263, 264, 274, 275, 276, 277, 278

rainforests 371
red algal endosymbiont 178
Red Queen Hypothesis 284
Rhipidiales 179, 192, 193
Rhizaria 178
riparian lichens 348
riparian zone 389
river 391, 398, 406, 411, 412, 465, 466, 475
RNA helicase 165
rosette agent 203, 212
Rozella 164, 279, 280, 282, 296, 298, 301
rozellids 164

Salinity 247
Salisapilia 179, 181, 184, 185, 187, 190, 191, 193, 194, 197
Salisapiliaceae 185, 187, 197
Saprolegnia 179, 180, 196, 217, 219, 221, 223, 224, 225, 227, 229, 231, 233, 235, 236, 237, 238, 239, 240, 241, 242, 243

Saprolegniaceae 221, 223, 238, 239, 240, 241, 242
Saprolegniales 179, 180, 217, 219, 221, 223, 225, 228, 238, 240
Saprolegniomycetidae 179, 192, 196, 198
saprolegniosis 217, 219, 220, 224, 234, 237, 240
saprotrophic 73, 81, 104
saprotrophs 283, 389
saxicolous communities 349
septal pore 83, 89, 93, 94, 95, 96, 98
sequential colonization 466
sexual cycle 165, 168, 169, 170
sexual cycle of microsporidia 165
sheath 26, 34, 35, 36, 38, 39
shredders 395, 398, 405, 406, 408, 409, 411, 413, 414, 417, 418, 419, 422, 424, 425, 427, 428, 429, 430, 431, 432, 434, 435, 436, 440
shrimps 417
soft rot cavities 465
Sordariales 48, 57, 58, 59, 61, 68, 69
Sordariomycetes 120, 123, 126, 129, 130, 485, 486, 487
spear-like ejectile organelle 166
species richness 391, 395, 402
Spirogyra 284
Spizellomycetales 138, 140, 141, 148, 151, 153, 294
spores of microsporidia 166
sporozoa 162
SSU 179, 187, 197
stauroconidia 87
stramenopiles 4, 217, 280
Straminipila 6, 177, 311, 325
stream communities 413, 425
streams 389, 391, 392, 393, 394, 395, 396, 397, 398, 399, 403, 404, 405, 406, 407, 408, 409, 410, 411, 412
stromatolites 294
sub-aquatic lichens 341, 346, 347
submerged wood 381
sugar transporter 165
sulphate 394
symbiotic relationships 364

tannins 418, 419, 438
teleomorph 83, 86, 89, 93, 101, 105, 108
teleomorph/anamorph connections 112
teleomorph/anamorph relationship 485
teliospores 83, 85
temperate 26
temperature 246, 247, 249, 251, 252, 255, 256, 257, 258, 259, 260, 262
terbinafine 271
Terresporidia 163, 170
terrestrial palm 372, 381, 383
terrestrial 28, 34, 40
test blocks 465, 466, 467, 468, 471, 473, 474, 475
thalassemic patients 264, 273
transcriptome 235
Treatment with albendazole 160
tree pathogens 189
trehalase activity 168
Tremellomycetes 83, 84, 92
trichomycetes 6, 7, 359, 360, 364, 368, 369
trichospores 360, 361, 363, 367
trophic levels 393, 395, 396, 404
trophic links 297
trophic upgrading 482
tropical streams 417, 426, 427, 428, 434, 435, 436, 441
tropics 26, 41
tubular cristae 177
tumor-suppressor gene 169

uniflagellate zoospores 280, 282, 292
unikont 279
unpolluted streams 395
Ustilaginomycetes 83, 87, 95

vertebrates 203, 205

Warburg effect 169, 172, 176
wastewater treatment plants 161
water moulds 217, 221, 223, 235, 242
Water quality 351
water temperature 391, 399, 402, 403, 404, 405, 406, 407, 410, 411, 481
water-logged 371
weight losses 465, 472
Western blot 270, 273
white rusts 181, 184, 185, 188, 197
wood-degrading enzymes 483
woody debris 25

Xanthophyta 280
xenobiotic 391, 397, 398

yeast community 450, 452
Yeast identification 447
yeasts 73, 81, 82, 83, 86, 94, 101, 102, 103, 104, 106, 107, 108, 443, 451, 458, 460, 461, 462

Zonation 345
zooplankton 282, 283, 295, 299, 305, 306
zoosporangia 187, 191
zoospore discharge 186, 190, 191, 193
zoosporic fungi 305, 328
zoosporic organisms 15, 482, 484
zoosporic parasites 305, 306, 307, 314, 322, 323, 326
zoosporic true fungi 279, 280, 282, 294, 295, 298, 299, 304
zygomycetes 165, 279, 481, 483

Printed and bound by CPI Group (UK) Ltd, Croydon, CR0 4YY
21/09/2026
14966557-0001